**THE LIQUID CRYSTALS BOOK SERIES**

# CROSS-LINKED LIQUID CRYSTALLINE SYSTEMS

**FROM RIGID POLYMER NETWORKS TO ELASTOMERS**

# THE LIQUID CRYSTALS BOOK SERIES

*Edited by*

## Virgil Percec

Department of Chemistry
University of Pennsylvania
Philadelphia, PA

The Liquid Crystals book series publishes authoritative accounts of all aspects of the field, ranging from the basic fundamentals to the forefront of research; from the physics of liquid crystals to their chemical and biological properties; and from their self-assembling structures to their applications in devices. The series will provide readers new to liquid crystals with a firm grounding in the subject, while experienced scientists and liquid crystallographers will find that the series is an indispensable resource.

## PUBLISHED TITLES

Introduction to Liquid Crystals: Chemistry and Physics
*By Peter J. Collings and Michael Hird*

The Static and Dynamic Continuum Theory of Liquid Crystals:
A Mathematical Introduction
*By Iain W. Stewart*

Crystals That Flow: Classic Papers from the History of Liquid Crystals
*Compiled with translation and commentary by Timothy J. Sluckin, David A. Dunmur, and Horst Stegemeyer*

Nematic and Cholesteric Liquid Crystals: Concepts and Physical Properties
Illustrated by Experiments
*By Patrick Oswald and Pawel Pieranski*

Alignment Technologies and Applications of Liquid Crystal Devices
*By Kohki Takatoh, Masaki Hasegawa, Mitsuhiro Koden, Nobuyuki Itoh, Ray Hasegawa, and Masanori Sakamoto*

Adsorption Phenomena and Anchoring Energy in Nematic Liquid Crystals
*By Giovanni Barbero and Luiz Roberto Evangelista*

Chemistry of Discotic Liquid Crystals: From Monomers to Polymers
*By Sandeep Kumar*

Cross-Linked Liquid Crystalline Systems: From Rigid Polymer Networks to Elastomers
*Edited By Dirk J. Broer, Gregory P. Crawford, and Slobodan Žumer*

**THE LIQUID CRYSTALS BOOK SERIES**

# CROSS-LINKED LIQUID CRYSTALLINE SYSTEMS

## FROM RIGID POLYMER NETWORKS TO ELASTOMERS

Edited by

Dirk J. Broer

Gregory P. Crawford

Slobodan Žumer

**CRC Press**
Taylor & Francis Group
Boca Raton London New York

CRC Press is an imprint of the
Taylor & Francis Group, an **informa** business

CRC Press
Taylor & Francis Group
6000 Broken Sound Parkway NW, Suite 300
Boca Raton, FL 33487-2742

First issued in paperback 2019

ISBN-13: 978-1-4200-4622-9 (hbk)
ISBN-13: 978-0-367-38310-7 (pbk)

**Visit the Taylor & Francis Web site at**
**http://www.taylorandfrancis.com**

**and the CRC Press Web site at**
**http://www.crcpress.com**

# Contents

Contents

# Preface

Liquid crystals are a special class of materials in nature, hence their name is a deliberate oxymoron to describe a material system that exhibits hybrid properties—physical properties manifested by both liquids and crystals. Because of their application success in flat-panel displays and application potential in many other areas, they have attracted significant interest from both the applied and basic research communities. Like liquid crystals, polymers are unusual materials in many respects. Polymers have also enjoyed a great deal of research attention because of their vast applications and uses and complex fundamental properties. We have now learned that the combination of liquid crystal and polymer properties produces a broad array of new effects that are not simply manifestations of either native liquid crystals or polymers alone. One embodiment includes densely cross-linked networks created from reactive mesogen materials: the liquid crystalline order can be manipulated by external constraints—such as surfaces, electric or magnetic fields, or shear forces—to create a temporary, and otherwise unstable, configuration that can indefinitely be captured through photo-polymerization. These highly ordered and optically transparent films have found their way into the commercial market on nearly all desktop liquid crystal display screens to compensate for the viewing angle or to improve on their contrast. On the other end of the spectrum, elastomers are weakly cross-linked systems that can give rise to spontaneous shape changes of several hundred percent, and by the application of stress or stain, they can exhibit some unusual mechanical–optical effects. Liquid crystal elastomers are being considered for a number of applications, including such breakthrough possibilities as the artificial muscle.

The worldwide need for flat-panel displays drove much of the fundamental research in the field of liquid crystals, making that application of paramount interest to the liquid crystal community. That field is maturing, and much of the applied research is now looking for alternative uses of liquid crystal materials. With the greatly expanding interest in liquid crystalline polymers and elastomers, researchers are now looking into application areas for these unique materials, which range from passive optical elements on displays to far-reaching concepts such as artificial muscles, tunable lasers, actuators, membranes, and spectrometers. This new application space is broad and far reaching. There is currently no comprehensive account that brings together the fundamental materials aspects of liquid crystalline polymer and elastomer systems, their interconnections and interdisciplinary scope, and their broad application potential.

Liquid crystal displays were discovered in the 1960s, and today we are enjoying the benefits of that fundamental discovery and its translation into products. Densely cross-linked liquid crystal systems and liquid crystal elastomers were proposed in the 1980s, two decades after the liquid crystal display and a century after the discovery of the liquid crystal phase itself. In retrospect, liquid crystal display technology drove much of the fundamental research in liquid crystals. The systems' ability to enforce three-dimensional structure in the molecular order and capture it created a number of compelling application possibilities because it provided necessary control of the molecular order. The methods to produce such structures are limited in number, and no process enabled such control until the discovery of photopolymerizable liquid crystal systems. As these systems became more available, fundamental studies ensued and many other applications followed, beyond the optical compensator, such as broadband polarizers, actuators, reorientable microstructures, and so forth.

Unlike the densely cross-linked liquid crystal polymer, the molecular structure of a liquid crystal elastomer is similar to that of a traditional rubber: it consists of long chains of molecules that can easily slip past one other and so enable the material to be expanded with very little force. Attached to the elastomer chains, like the branches of a tree, are smaller rod-like molecules that are usually found in liquid crystals. These liquid crystal moieties allow the material to interact with light and can align the long chains and give unexpected mechanical properties, such as the ability to change color when they are stretched and the ability to dramatically change their shape when they are heated, or, for certain renditions of compounds, when light impinges upon them. Liquid crystal elastomers have a variety of potential uses. For example, they could provide the basis for a laser which needs only a small amount of power to operate and can change its emission just by being stretched. The natural twisting of their internal structure means liquid crystal elastomers could act as a new system for identifying the difference between right-handed and left-handed forms of drug compounds. On the fundamental side, liquid crystal elastomers experience certain shape changes with little or no energy cost. This has been dubbed "soft elasticity" and places these materials in a unique class—those between liquid and solid in an elastic classification of matter.

As a scientist or engineer, you will find upon reflection that there seems to be no shortage of fundamental science, unusual properties, and far-reaching applications in densely crossed-linked liquid crystal systems and liquid crystal elastomers. After some 25 years of research, the science and application possibilities seem endless. These systems provide a rich new avenue for both fundamental and applied research and continue to fascinate scientists and engineers who think broadly about their complexity, physical phenomena, synthetic features, and uses. In this volume, for the first time, we bring together those systems and their variations. We have attracted some of the most creative and influential experts in the field to contribute on their innovative science and application. The field, much like traditional liquid crystals, is one of an interdisciplinary nature with a broad spectrum, from the very fundamental questions of nature to a myriad of practical uses. We hope that our efforts in compiling this text will advance the understanding of basic science behind these systems, accelerate some of the proposed applications to the marketplace, and inspire generations of scientists to think broadly about these exciting and useful materials.

We thank Mr. M. Vavpotic for conversion and improvement of figures that were not prepared according to the T&F prescription and for preparation of the Index.

**Dirk J. Broer**
**Gregory P. Crawford**
**Slobodan Žumer**

# Editors

**Dirk J. Broer** is a polymer chemist (PhD—University of Groningen, the Netherlands) and is specialized in polymer structuring and self-organization. He joined Philips Research in Eindhoven, the Netherlands in 1973 where he worked on a wide variety of topics. Some representative examples are vapor-phase polymerization, optical data storage, telecommunication fibers, and liquid crystal networks. Between 1990 and 1991 he worked at the DuPont Experimental Station in Delaware, United States on nonlinear optical materials and vapor-phase deposition of $\pi$-conjugated polymers. In 1991 he began developing optical films for LCD enhancement at Philips Research and in 2000 he started his work on new manufacturing technologies of LCDs for large area displays and electronic wallpaper. From 2003 to 2010, Dirk J. Broer was appointed as senior research fellow and vice president at the Philips Research Laboratories specializing on biomedical devices and applications of polymeric materials.

In 1996 Dirk J. Broer became a part-time professor at the Eindhoven University covering research topics as liquid crystal orientation, polymer waveguides, solar energy, organic semiconductors, nanolithography, soft lithography, and polymer actuators. In 2010 he became a fulltime professor in Eindhoven University specializing in functional organic materials for clean technologies as energy harvesting, water treatment, and healthcare applications. Special emphasis is given to responsive polymers that morph under the action of heat, light, electrical, and magnetic fields or driven by contacts with agents and/or changes in their ambient conditions and to membrane technology with monodisperse nano-porosity.

He has coauthored more than 200 publications in peer-reviewed journals and holds more than 80 U.S. Patents.

**Gregory P. Crawford** became dean of the College of Science and professor of physics at the University of Notre Dame in 2008. He was previously professor of engineering and physics (1996–2008) and dean of engineering (2006–2008) at Brown University. He joined Brown University in 1996 after working at the Xerox Palo Alto Research Center (PARC) as a member of the research staff. His research focuses on soft matter materials, photonic materials, medical devices, displays, nano-science, magnetic resonance, and photonic devices. The creative deployment of his basic research has resulted in two biotechnology start-up companies, Myomics, Inc. and Corum Medical, Inc., and his inventions have been licensed to companies. During the 2003–2004 academic year, Dean Crawford was on sabbatical at the Technical University of Eindhoven in the Netherlands, working on the underlying physics of novel phenomena of patterned liquid crystals and polymers with potential use in displays and security applications. In the summer of 1999, he was a visiting scientist professor at the Philips Research Laboratory (Natlab) in Eindhoven, working on the physics and applications of emissive and lasing materials. Dean Crawford holds 18 U.S. patents and has authored and coauthored more than 300 research and education publications including review articles and book chapters, and is the editor of a number of books: *Liquid Crystals in Complex Geometries formed by Polymer and Porous Networks* (Taylor & Francis, 1996); *Flexible Flat Panel Displays*; *Liquid Crystals; Frontiers in Biomedical Applications* (World Scientific, 2007); and *Cross-Linked Liquid Crystalline Systems: From Rigid Polymer Networks to Elastomers* (this contribution). He also served as editor of a special edition on nanotechnology in displays and medical displays for the *Journal for the Society for Information Display*. He is a fellow of the Society for Information Display. Dean Crawford's teaching interests include optics, photonics, waves, soft matter, and high technology entrepreneurship at both the undergraduate and graduate levels.

**Slobodan Žumer** received his BSc and PhD (1973) from the University of Ljubljana. During 1975–1976 he held a postdoctoral position at the Université libre de Bruxelles and later during 1984–1986 he held a visiting position at the Liquid Crystal Institute of KSU. In 1987 he became professor of physics at the University of Ljubljana and in 1986 the Scientific Counselor at the Jozef Stefan Institute. His teaching interests include statistical physics, thermodynamics, optics, and soft matter physics at both the undergraduate and graduate level. He served as a head of the physics department (1997–1999) and as a dean of the Faculty of Mathematics and Physics (2005–2007) at the University of Ljubljana. In 2008 he started his four-year term as a president of the International Liquid Crystal Society. His research interests are theory, modeling, and simulations of soft matter with particular stress on liquid crystals, polymer-dispersed liquid crystals, liquid crystalline elastomers, and colloidal dispersions. He is the head of the Research Group for Physics of Soft and Partially Ordered Matter at the University of Ljubljana and the head of the Research Group Physics of Soft Matter, Surfaces, and Nanostructures at Jozef Stefan Institute. Professor Žumer with coworkers has published more than 200 research publications including review articles and book chapters that were cited more than 5000 times. He is a coauthor of three U.S. Patents, and coeditor of four books (*Liquid Crystals in Complex Geometries: Formed by Polymer and Porous Network*, Taylor & Francis, 1996; *Defects in Liquid Crystals: Computer Simulations, Theory and Experiments*, Kluwer, 2001; *Computer Simulations of Liquid Crystals and Polymers*, Kluwer, 2005; *Novel NMR and EPR Techniques*, Springer, 2006). During the period 2001–2007 he was a member of the editorial board of the *European Physical Journal E* and currently a member of the editorial board of *Liquid Crystals*. He is a fellow of the IOP and a member of APS.

# Contributors

**Ichiro Amimori**
Frontier Core-Technology
  Laboratories
Fuji Photo Film Co Ltd
Minamiashgara, Kanagawa, Japan

**Cees W.M. Bastiaansen**
Department of Chemistry and
  Chemical Engineering
Technische Universiteit Eindhoven
Eindhoven, The Netherlands

**Gerben Boer**
ARCOPTIX S.A.
Neuchâtel, Switzerland

**Helmut R. Brand**
Theoretische Physik III
Universität Bayreuth
Bayreuth, Germany

**Dirk J. Broer**
Department of Chemistry and
  Chemical Technology
Eindhoven University of Technology
Eindhoven, The Netherlands

**Darran R. Cairns**
Department of Mechanical and
  Aerospace Engineering
West Virginia University
Morgantown, West Virginia

**Carmen Otilia Catanescu**
Department of Mechanical and
  Aerospace Engineering
West Virginia University
Morgantown, West Virginia

**Liang-Chy Chien**
Department of Mechanical and
  Aerospace Engineering
West Virginia University
Morgantown, West Virginia

**Gregory P. Crawford**
College of Science
University of Notre Dame
Notre Dame, Indiana

**Heino Finkelmann**
Institut für Makromolekulare
  Chemie
Universität Freiburg
Freiburg, Germany

**Tomiki Ikeda**
Chemical Resources Laboratory
Tokyo Institute of Technology
Yokohama, Japan

**Patrick Keller**
Institut Curie
CNRS UMR, Laboratoire
  Physico-Chimie Curie
Paris, France

**Jae-Hoon Kim**
Department of Electronics and
  Computer Engineering
Hanyang University
Seoul, Korea

**Theo Kreouzis**
Queen Mary
University of London
London, United Kingdom

**Zdravko Kutnjak**
Laboratory for Calorimetry
Jožef Stefan Institute
Jamova, Ljubljana, Slovenia

**Sin-Doo Lee**
School of Electrical Engineering
Seoul National University
Seoul, Korea

**Lanfang Li**
Department of Mechanical and
    Aerospace Engineering
West Virginia University
Morgantown, West Virginia

**Min-Hui Li**
Institut Curie
CNRS UMR, Laboratoire
    Physico-Chimie Curie
Paris, France

**Johan Lub**
Department of Biomolecular
    Engineering
Philips Research Laboratories
Eindhoven, The Netherlands

**Tom C. Lubensky**
Department of Physics and
    Astronomy
University of Pennsylvania
Philadelphia, Pennsylvania

**Jun-ichi Mamiya**
Chemical Resources Laboratory
Tokyo Institute of Technology
Yokohama, Japan

**Philippe Martinoty**
Laboratoire de Dynamique des Fluides
    Complexes
Université Louis Pasteur
Blaise Pascal, Strasbourg, France

**Iain McCulloch**
Chemistry Department
Imperial College London
London, United Kingdom

**Tokuju Oikawa**
Flat Panel Display Material Research
    Laboratories
Fujifilm Corporation
Kanagawa, Japan

**L. Oriol**
Departamento de Química Orgánica,
Facultad de Ciencias—ICMA
Universidad de Zaragoza—CSIC
Zaragoza, Spain

**M. Piñol**
Departamento de Química Orgánica
Facultad de Ciencias—ICMA
Universidad de Zaragoza—CSIC
Zaragoza, Spain

**Harald Pleiner**
Theory Group
Max Planck Institute for Polymer
    Research
Mainz, Germany

**Toralf Scharf**
Swiss Federal Institute of Technology
    (EPFL)
Institute of Applied Optics (IOA)
Lausanne, Switzerland

**José Luis Serrano**
Departamento de Química Orgánica
Facultad de Ciencias—ICMA
Universidad de Zaragoza—CSIC
Zaragoza, Spain

**Maxim N. Shkunov**
Advanced Technology Institute
University of Surrey
Guildford, United Kingdom

**G. Skačej**
Fakulteta za Matematiko in Fiziko
Univerza v Ljubljani
Jadranska, Ljubljana, Slovenia

**Matthew E. Sousa**
3M Company
Optical Systems Division
St. Paul, Minnesota

**Olaf Stenull**
Department of Physics and Astronomy
University of Pennsylvania
Philadelphia, Pennsylvania

**Peter Strohriegl**
Makromolekulare Chemie I
Universität Bayreuth
Bayreuth, Germany

**Toshikazu Takigawa**
Department of Materials Chemistry
Kyoto University
Kyoto, Japan

**Eugene M. Terentjev**
Cavendish Laboratory
University of Cambridge
Cambridge, United Kingdom

**Kenji Urayama**
Department of Materials Chemistry
Kyoto University
Kyoto, Japan

**Casper L. van Oosten**
Department of Chemistry and
    Chemical Engineering
Technische Universiteit
    Eindhoven
Eindhoven, The Netherlands

**Mark Warner**
Cavendish Laboratory
University of Cambridge
Cambridge, United Kingdom

**Yanlei Yu**
Department of Materials
    Science
Fudan University
Shanghai, China

**Boštjan Zalar**
Faculty of Mathematics and
    Physics
Jožef Stefan Institute
Jamova, Ljubljana, Slovenia

**Claudio Zannoni**
Dipartimento di Chimica Fisica ed
    Inorganica and INSTM
Università di Bologna
Bologna, Italy

**Slobodan Žumer**
Department of Physics
University of Ljubljana
Jadranska, Ljubljana, Slovenia

# Part I

## Densely Cross-Linked Systems

# 1 Densely Cross-Linked Liquid Crystal Networks by Controlled Photopolymerization of Ordered Liquid Crystal Monomers

## *Properties and Applications*

*Johan Lub and Dirk J. Broer*

## CONTENTS

## 1.1  INTRODUCTION

Polymers with a controlled molecular organization in all three dimensions are of interest because of their unusual, but very accurately adjustable and addressable optical, electrical, and mechanical properties. An established method to produce 3-D ordered polymers is the photoinitiated polymerization of liquid crystal (LC) monomers [1–4]. The variety in possible LC phases of low-molar-mass reactive mesogens provides diversity in the choice of the type of the molecular order, all being accessible to be fixed by the polymerization process [5–8]. Known techniques to establish monolithic molecular order in LCs, such as rubbed surfaces, surfactant-treated surfaces, external electric, or magnetic fields or flow, can be applied or even can be combined with each other to create films of even more complex molecular architectures [9]. The molecular structure of LC monomers can be tailored, for example, to optimize on the mechanical and optical properties of the films. Furthermore, blends of monomers can be made to adjust the properties in the monomeric state, such as the LC transition temperatures and the flow viscosity, and in the polymeric state, such as the elastic modulus, the glass transition ($T_g$) temperature, and the refractive indices.

When polyfunctional (more than one polymerizable group) LC monomers are polymerized in the bulk, the so-called LC networks are produced which, especially at higher cross-link densities, exhibit a stable molecular organization up to the degradation temperature of the polymers. In that sense they distinguish themselves from a class of LC networks that in literature normally is denoted as LC elastomers, often siloxane based, by the absence of phase transitions and the much higher moduli at room temperature. The history of densely cross-linked LC networks goes back to the end of the 1960s where different authors suggested polymerizing and cross-linking LCs in their mesophase aiming highly ordered polymers [10,11]. Thermally initiated bulk polymerization (thermosetting) of LC diacrylates was reported to yield three-dimensionally cross-linked polymer networks with a strong optical anisotropy and the molecular order of the frozen-in monomer phase was retained to the decomposition temperature [12–15]. However, the use of high temperatures to initiate the polymerization often conflicts with the temperatures at which the LC phases of many of the reactive LCs appear. During heating to, and processing in, the LC phase the polymerization starts prematurely before the desired monolithic orientation is established and defects become permanently frozen-in. It is for this reason that photoinitiation is highly preferable for the bulk polymerization and network formation. In the presence of small amounts of polymerization inhibitors, typically of the order of 100 ppm, the monomers can be processed at elevated temperatures until the desired LC phase and long-range order has been obtained. As soon as the desired molecular order has been established in the monomer it can be rapidly fixed by exiting a dissolved photoinitiator, which normally has a concentration between 0.05 and 5 wt% depending on the application and sample thickness, with actinic light of an appropriate wavelength, mostly around 360 nm.

The first reports on the bulk photopolymerization of reactive LCs were on monoacrylates forming linear LC side-chain polymers [16–20]. Real fixation of the molecular order in this case often does not occur and the formed polymer still exhibits various mesophases and can be heated to an isotropic state, although the transition temperatures differ from those of the initial monomer. Consequently, phase transitions might also occur during the polymerization process with a change of order, type, and degree, as a result. In addition phase separation might take place if the LC state of the polymer does not mix with that of the initial monomer, which results in defects in the aimed monolithic order. It is for this particular reason that the bulk photopolymerization of polyfunctional LC monomers (photosetting LCs) became so important. For instance, the photoinitiated free-radical polymerization of monolithically ordered nematic LC diacrylates produces a stable polymer with the same texture as observed for the monomer and almost the same degree of molecular order [1–3]. But also the use of LC diepoxides [21,22], LC divinylethers [23–26], and LC dioxetanes [27] polymerized by photocationic mechanisms essentially lead to the same results.

The optical properties of thin films of monolithically ordered LC networks are quite similar to those of the low-molar-mass LCs. They are transparent for light in the visible wavelength region, exhibit a high birefringence up to 0.25 at 589 nm [3], are able to show selective reflection in the case of a cholesteric order [6,8], and exhibit a half-wave optical retardation in the case of a twisted nematic molecular arrangement [5,28]. These properties make the materials ultimately suitable for the creation of various optical devices as polarizing beam splitters, organic Wollaston prisms, retardation films, color and infrared filters, and so on. All the properties are temperature insensitive and stable against environmental (light, humidity, temperature) aging. The mechanical properties are, apart from their anisotropic nature, of the same class as those of the isotropic acrylate and epoxide networks meaning that the modulus and strength are of the same order and depend strongly on the molecular parameters like cross-link density and the ratio between stiff and flexible units [29]. An interesting mechanical feature is that due to a unidirectional or planar molecular order the thermal expansion can be brought back to essentially zero into one or two directions. This implies that built-up of thermal stresses are avoided when the networks are applied as coating or encapsulate on an inorganic substrate [30].

This chapter will discuss structure–property relationships of the LC monomers and their polymers. By means of some of their applications and the desired properties we will demonstrate further how the properties can be tuned further by the combination of molecular design, alignment technology, polymerization conditions, and eventual additional photochemistry.

## 1.2 NEMATIC NETWORKS

### 1.2.1 STRUCTURE PROPERTY RELATIONSHIP IN NEMATIC (DI) ACRYLATES

Figure 1.1 shows the structure of diacrylate **1** that we prepared more than 20 years ago. It is, as most LC monomers appeared to be, crystalline at room temperature. Upon heating it melts at 108°C and becomes nematic up to 155°C [1]. During supercooling, a monotropic smectic-C phase is observed below 88°C before it crystallizes again. The synthetic steps to make this molecule are outlined in Figure 1.2. The intermediate ethyl ester **4** is normally not isolated. In comparison with a similar molecule with the same aromatic core but with two hexyloxy end groups [31], the presence of the acrylate moiety hardly changed the crystalline melting temperature but destabilized

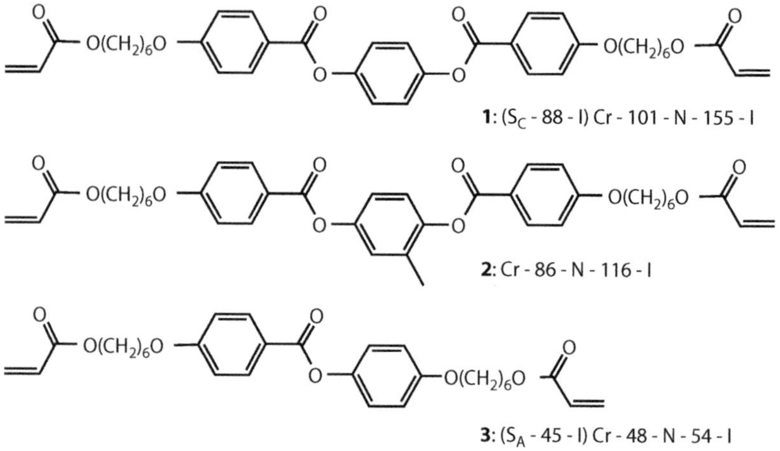

**FIGURE 1.1** Structure and phase transitions of 1,4-phenylene bis(4-(6-acryloyloxyl-hexyloxy)benzoate) **1**, 2-methyl-1,4-phenylene bis(4-(6-acryloyloxylhexyloxy)benzoate) **2**, and 4-(6-acryloyloxylhexyloxy)phenyl 4-(6-acryloyloxylhexyloxy)benzoate **3**. (Cr = crystalline, N = nematic phase, $S_C$ = smectic-C phase, I = isotropic phase.)

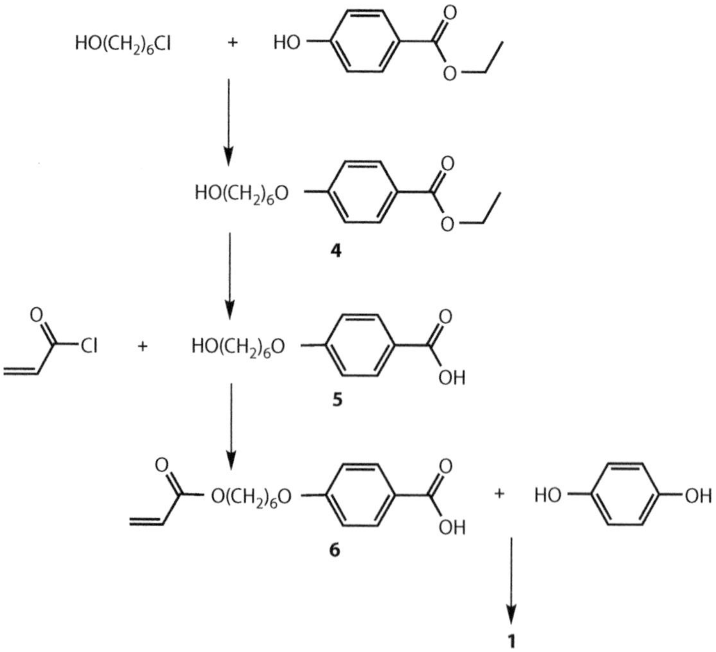

**FIGURE 1.2** Synthesis of 1,4-phenylene bis(4-(6-acryloyloxylhexyloxy)benzoate) **1** (see Figure 1.1).

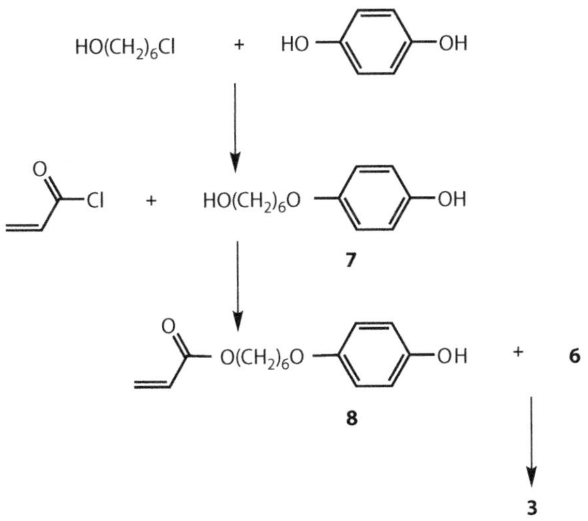

**FIGURE 1.3** Synthesis of 4-(6-acryloyloxylhexyloxy)phenyl 4-(6-acryloyloxylhexyloxy) benzoate **3** (see Figure 1.1).

both the nematic and the smectic phase by reducing these phase transitions with 50°C and 32°C, respectively. The relatively high melting point of **1** allows only processing at elevated temperatures. In order to lower the processing temperature, as well as to modulate the properties of the polymer films of interest for their applications, several chemical adaptations were carried out to the structure of this compound. These adaptations include changes to the central mesogenic group, to the spacer, to the polymerizable group, and of course to combinations of these.

Upon substitution of one of the hydrogen atoms of the central ring of **1** by a methyl group, large changes in thermal transitions were observed. Synthesis of these molecules proceeds very similar; during one of the synthetic steps in Figure 1.2 hydroquinone was replaced by the commercial available methyl hydroquinone [32]. This compound **2** exhibits a melting point 22°C lower and an isotropic transition 39°C lower than that of **1**. Moreover, the influence of such a methyl group on the ordering results in complete disappearance of the monotropic transition to the smectic-C phase observed in **1**. Monomer **2** crystallizes extremely slow compared to **1**, which makes it possible to process it in the supercooled nematic phase at moderate temperatures. Mixtures of this compound with other compounds (e.g., other reactive LCs) can be more crystallization resistant and are therefore, very well suited for processing at room temperature (e.g., after spincoating from a solution) to form highly aligned films that can subsequently be photopolymerized. Also, other liquid crystalline diacrylates derived from several hydroquinones have been described [33].

Other mesogenic groups have been studied as central group for nematic liquid crystalline diacrylates. Compound **3** (Figure 1.1) shows a smaller mesogenic group. It exhibits only a very small nematic phase. Its synthesis is described in Figure 1.3. Monoalkylated intermediate **7** is easily separated from the di-alkylated product [34]. Conditions were found to form the acrylate group in **8** without the reaction of the

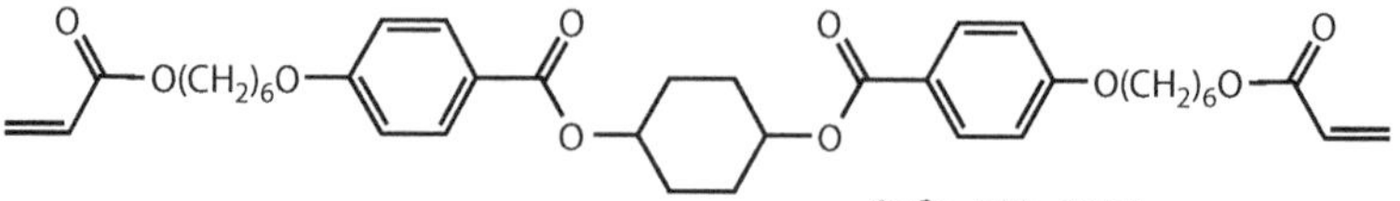

**9:** Cr - 123 - 155 I

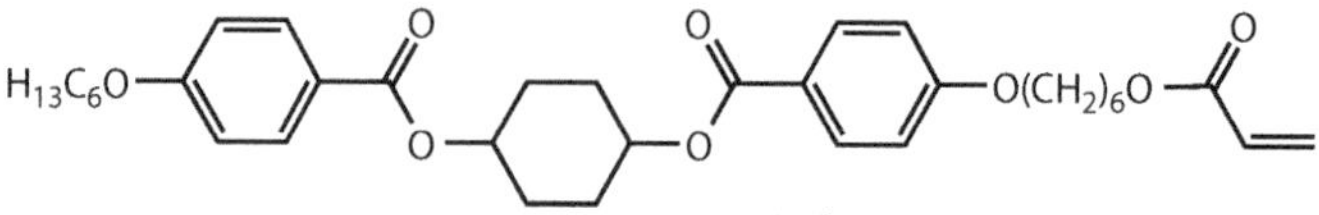

**10:** Cr - 99 - N - 119 I

**FIGURE 1.4**   Structure and phase transitions of (*trans*)-1,4-di(4-(6-acryloyloxyhexyloxy) benzoyloxy)cyclohexane **9** and (*trans*)-4-(4-hexyloxybenzoyloxy)cyclohexyl 4-(6-acryloyloxyhexyloxy)benzoate **10**. (Cr = crystalline, N = nematic phase, I = isotropic phase.)

phenolic group with acryloyl chloride. The nematic to smectic-A transition can be suppressed by decreasing the spacer length, however, at the expense of the isotropic transition, leaving compounds with low applicability. Monoacrylates derived from this mesogenic group, exhibit broader liquid crystalline phases and higher transition temperatures to the isotropic phase. Depending on the type of monoacrylates, some of these molecules are well suited to make mixtures with **2** to obtain room temperature processing mixtures [34].

In order to alter the optical properties and to improve on the light stability under high intensity light loads, the cyclohexane analog of **1** was prepared. This compound **9** (see Figure 1.4) [35] and also compounds with other spacer length are unfortunately not liquid crystalline. Blending it with monomer **10** that exhibits a nematic phase and is also derived from *trans*-cyclohexane diol (Figure 1.4) required large amounts of the latter to keep blends that can be processed in their nematic phase. Nevertheless, the use of 20% of **9** results in a stable nematic monomer blend, which upon irradiation with high intensity near UV light after polymerization indeed shows much slower yellowing than a network formed from **1** alone. The synthesis of **9** is similar to that of **1**: hydroquinone is replaced by *trans*-cyclohexane diol [35]. Monoacrylate **10** is made according to Figure 1.5. The mono-esterified product is separated easily from diacrylate **9** making use of the fact that **9** is completely insoluble in methanol whilst **11** is soluble.

Other rod-like monomers, although not liquid crystalline by themselves are often of interest and can be used in LC formulations. For instance, monoacrylate **12** (Figure 1.6) derived from cyanobiphenyl is not liquid crystalline but can be added to an LC formulation in a considerable amount to enhance the dielectric anisotropy and to reduce viscosity for a better alignment in an electrical field. Moreover, the large polarizability of the cyano group along the long axis of the molecule increases the birefringence of most blends [36]. An interesting combination is formed by blending this monomer with monomer **13** (Figure 1.6). Blends of these two materials do not crystallize at all at room temperature. Coating this blend on a glass plate covered with rubbed polyamide forms a nematic film with a splayed structure (see Figure 1.24). The tilt angle of the film can be tuned by the composition of the mixture and the

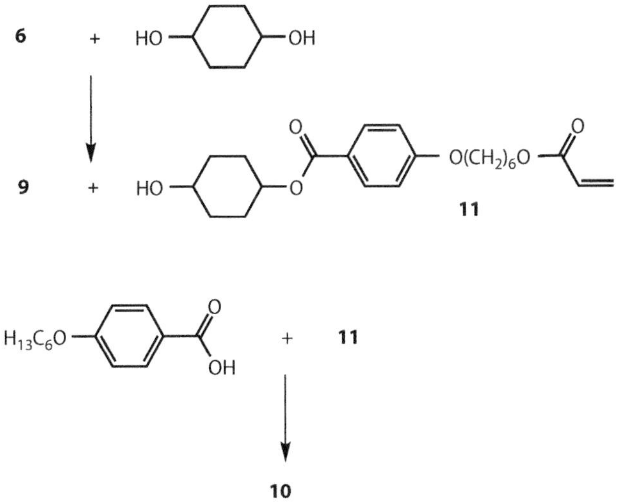

**FIGURE 1.5** Synthesis of **9** and **11** (see Figure 1.4).

film thickness and is frozen in by photopolymerization. Such films find their use for viewing angle improvement of LC displays [37,38].

To optimize further on the properties of liquid crystalline diacrylates the effect of the spacer length on the liquid crystalline transition has been studied. The change of the hexamethylene group in molecules **1**, **2**, and **3** was performed using shorter or longer oligomethylene groups. In some cases these spacers contain chiral groups, inducing a cholesteric phase; this will be discussed later. The replacement of hexamethylene groups by an alkyl group results in the formation of monoacrylates, which find application in mixtures with liquid crystalline diacrylates as discussed above. The hexamethylene spacer in molecules **1**, **2**, and **3** is derived from the commercially available 6-chlorohexanol. It is relatively easy to make derivatives with

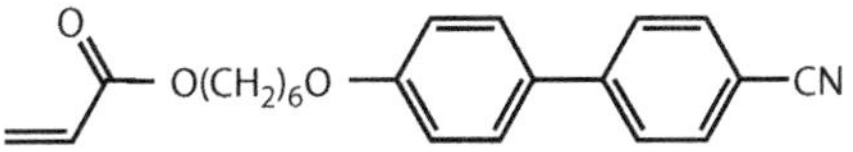

**12:** (N - 44 - I) Cr - 76 - I

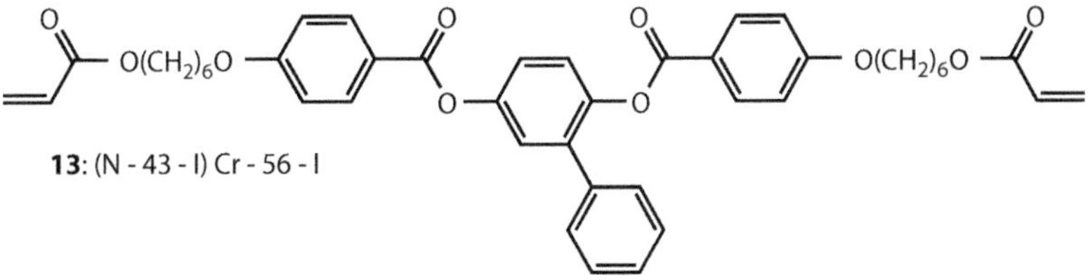

**13:** (N - 43 - I) Cr - 56 - I

**FIGURE 1.6** Structure and phase transitions of 6-(4′-cyanobiphenyl-4-yloxy)hexyl acrylate **12** and biphenyl-2,5-diyl bis(4-(6-(acryloyloxy)hexyloxy)benzoate) **13**. (Cr = crystalline, N = nematic phase, I = isotropic phase.)

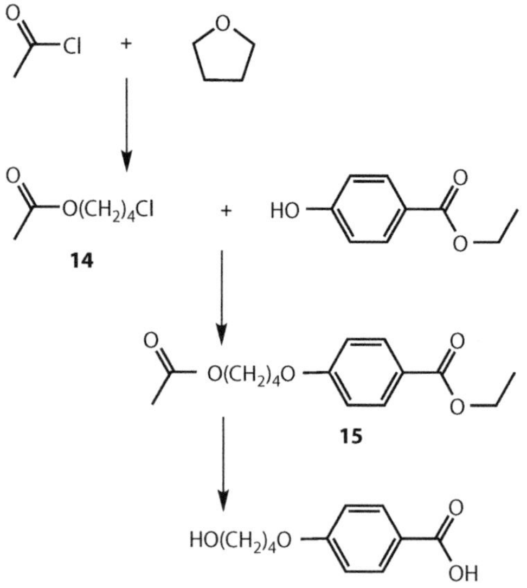

**FIGURE 1.7**   Synthesis of 4-(4-hydroxybutoxy)benzoic acid **16**.

undecylmethylene spacers or trimethylene spacers, making use of the commercially available 11-bromoundecanol and 3-chloropropanol, respectively. 5-Chloropentanol and 4-chlorobutanol are also easy accessible. However, formation of liquid crystalline diacrylates derived from the tetramethylene or pentamethylene spacers is more laborious because a simple alkylation reaction of the alcohols with phenolic compounds mentioned earlier, is not possible due to cyclization of these alcohols under the alkylation conditions. However, this can be performed using the corresponding acetates as outlined for the tetramethylene spacer in Figure 1.7. Intermediate **16** can be treated as described for **5** in Figure 1.2 to obtain all kind of acrylates with a tetramethylene spacer. Thus, analogs of **1, 2**, and **3** containing tetramethylene or pentamethylene spacers were prepared but more synthetic steps are needed compared to the formation of the hexamethylene derivatives to circumvent this cyclization problem. The properties of these tetramethylene or pentamethylene derivatives does not deviate much from the trimethylene or hexamethylene derivatives. Therefore, their use is not justified from a synthetic perspective [1,3]. Compounds containing two mesogenic groups similar to that of compound **2** with various spacers have been prepared [39]. Some of these molecules do not crystallize and are processable at room temperature [40].

   In Figures 1.8 and 1.9 the trimethylene and undecylene analogs of compounds **1** and **2** are compared, respectively. Especially compound **19** turned out to be a very interesting compound. It has a broad nematic phase and, when mixed with other liquid crystalline diacrylates such as **2**, forms stable blends that can be processed at room temperature. The undecylene derivative **18**, exhibits stable smectic phases. This is a known effect when the end groups of LCs are enlarged. However, when the symmetry of the molecule is broken by the introduction of the methyl group

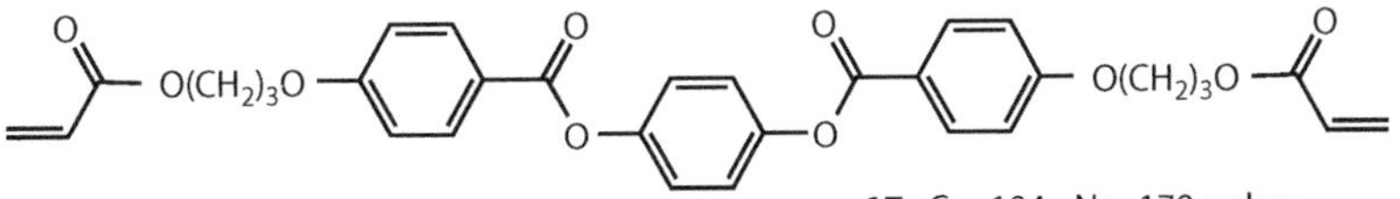

**17:** Cr - 104 - N > 170 polym.

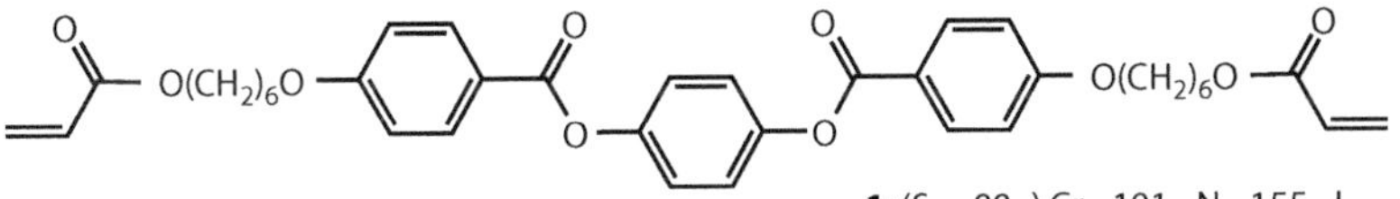

**1:** ($S_C$ - 88 -) Cr - 101 - N - 155 - I

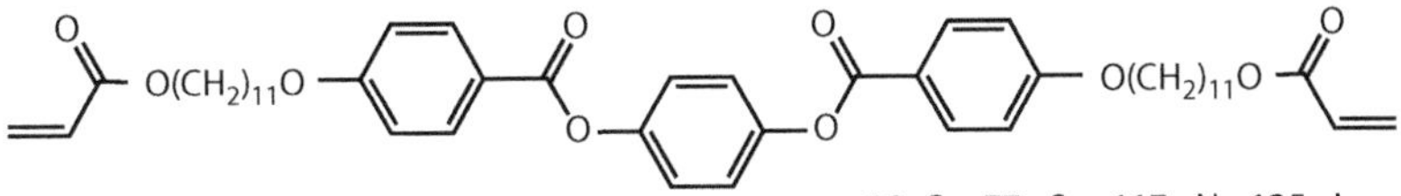

**18:** Cr - 77 - $S_C$ - 117 - N - 135 - I

**FIGURE 1.8** Structure and phase transitions of 1,4-phenylene bis(4-(3-(acryloyloxy) propoxy)benzoate) **17**, 1,4-phenylene bis(4-(6-acryloyloxylhexyloxy)benzoate) **1**, and 1,4-phenylene bis(4-(11-(acryloyloxy)undecyloxy)benzoate) **18**. (Cr = crystalline, N = nematic phase, $S_C$ = smectic-C phase, I = isotropic phase.)

in the central ring the smectic formation is suppressed as is shown by compound **20**. An unwanted side effect of the increased spacer length is that the monomers are susceptible to crystallization. This means that the monomer **20** is less applicable for film formation. Still it has found application in some optical components and mechanical applications [41].

In general, the acrylate-based monomers are general-purpose materials that can be rapidly cured in the presence of a photoinitiator that absorb light of wavelengths

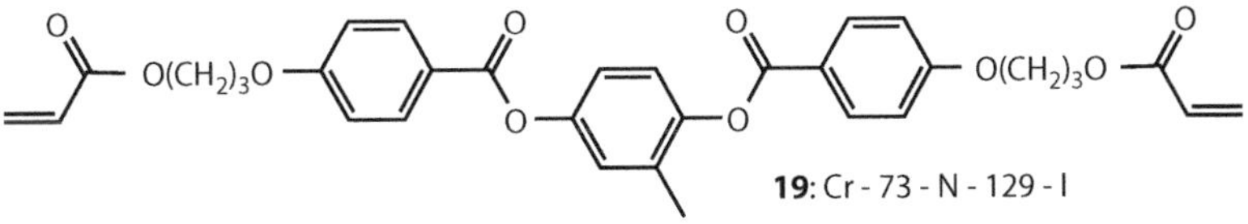

**19:** Cr - 73 - N - 129 - I

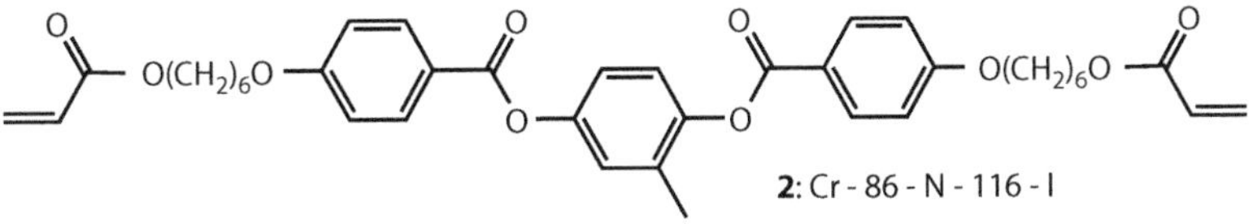

**2:** Cr - 86 - N - 116 - I

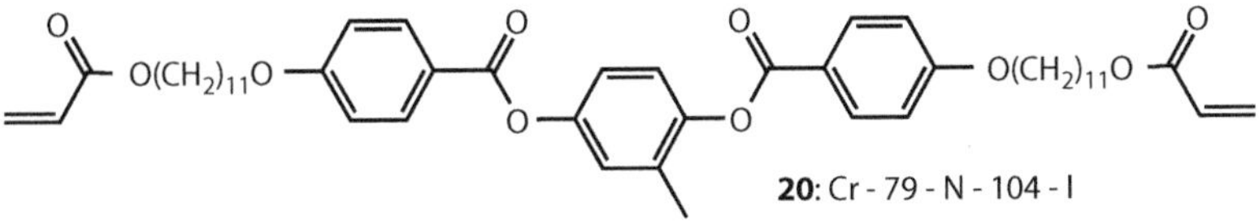

**20:** Cr - 79 - N - 104 - I

**FIGURE 1.9** Structure and phase transitions of 2-methyl-1,4-phenylene bis(4-(3-(acryloyloxy)propoxy)benzoate) **19**, 2-methyl-1,4-phenylene bis(4-(6-acryloyloxylhexyloxy)benzoate) **2**, and 2-methyl-1,4-phenylene bis(4-(11-(acryloyloxy)undecyloxy)benzoate) **20**. (Cr = crystalline, N = nematic phase, I = isotropic phase.)

just below the visible spectrum. This means that current lamp types, such as, mercury pressure lamps or fluorescent lamps provided with near UV phosphors can be utilized. In the absence of UV light, the materials are relatively stable and can be stored in the presence of a minor amount of inhibitor for years without any problems of polymerization or degradation. However, the choice of the acrylate group for polymerization also exhibits a few disadvantages. The free-radical polymerization is extremely sensitive to the presence of oxygen. Especially, when cured in contact with air and in a thin film, oxygen might inhibit polymerization when no precautions are taken. Moreover, acrylates show considerable shrinkage during polymerization.

To circumvent the use of inert gases during polymerization, a cationic polymerization process can be applied. Thereto polymerizable LCs have been developed containing the vinylether [25], epoxide [21,22], or oxetane [27] groups as cationic polymerizable group. An interesting property of LCs derived from vinylethers and epoxides is the relatively broad liquid crystalline phases compared to liquid crystalline diacrylates. This is clearly seen by comparison of the data in Figure 1.10 of some vinylethers with the data of the corresponding acrylates in Figure 1.1. However, the relative instability of vinylethers to moisture in combination with cationic initiators still needs care choosing the process atmosphere and epoxides polymerize relatively slowly giving rise to production problems and negative effects on the optical performance of the birefringent films as sometimes the phase separation is not kinetically suppressed [22].

Liquid crystalline oxetanes exhibit lower isotropic transition temperatures than their corresponding acrylates. Examples of these molecules are shown in Figure 1.11. The lower isotropic transition temperatures are due to the more bulky structure of the oxetane group compared to other polymerizable groups. Comparison of the properties of compounds **25** and **26** reveal that derivatives of 2-methyl-oxetane exhibit

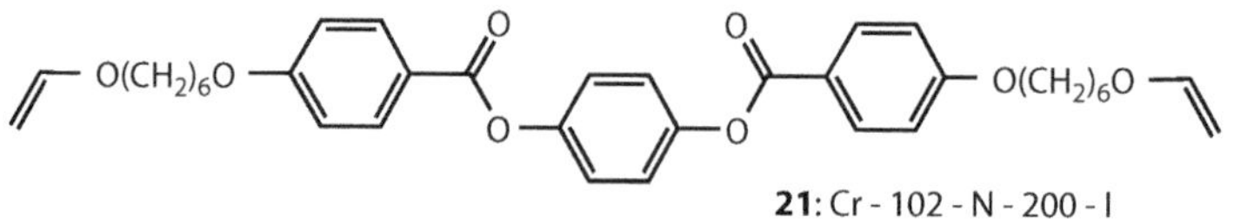

**21**: Cr - 102 - N - 200 - I

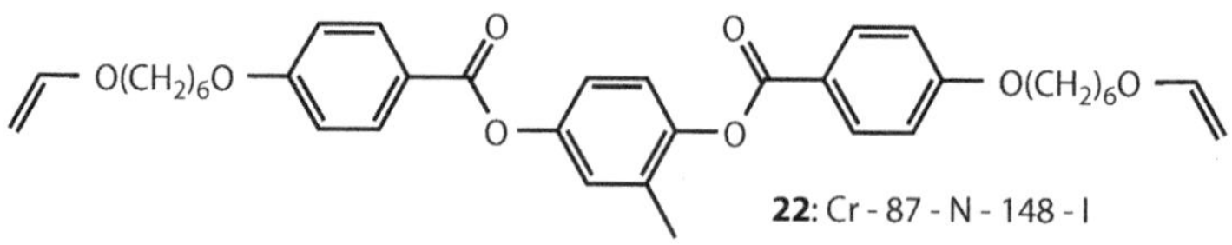

**22**: Cr - 87 - N - 148 - I

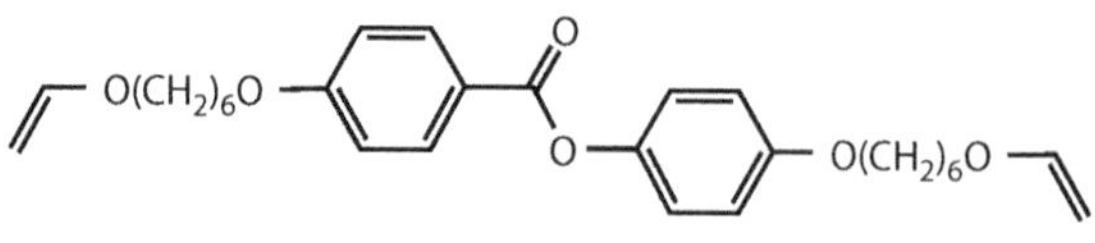

**23**: Cr - 56 - S$_A$ - 72 - N - 82 - I

**FIGURE 1.10** Structure and phase transitions of 1,4-phenylene bis(4-(6-(vinyloxy) hexyloxy)benzoate) **21**, 2-methyl-1,4-phenylene bis(4-(6-(vinyloxy)hexyloxy)benzoate) **22**, and 4-(6-(vinyloxy)hexyloxy)phenyl 4-(6-(vinyloxy)hexyloxy)benzoate **23**. (Cr = crystalline, S$_A$ = smectic-A phase, N = nematic phase, I = isotropic phase.)

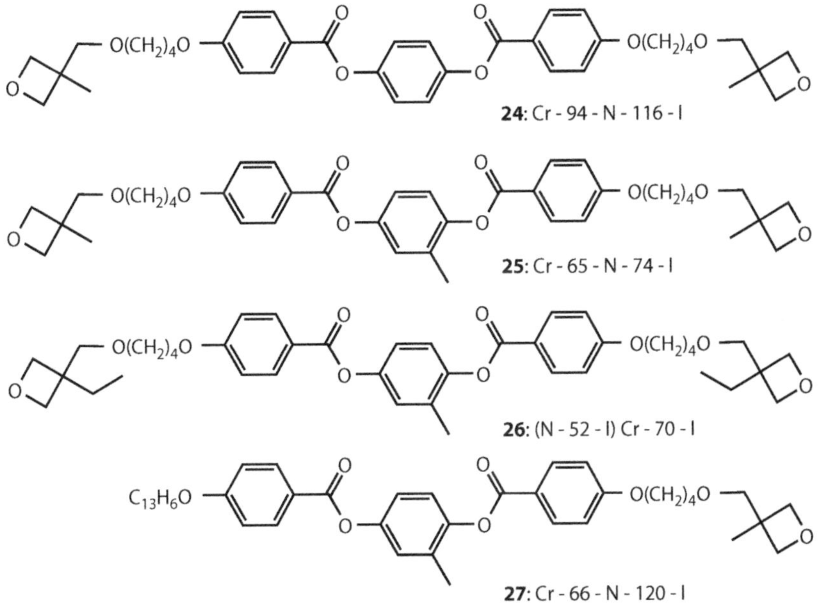

**FIGURE 1.11** Structure and phase transitions of 1,4-phenylene bis(4-(4-((3-methyloxetan -3-yl)methoxy)butoxy)benzoate) **24**, 2-methyl-1,4-phenylene bis(4-(4-((3-methyloxetan-3-yl) methoxy)butoxy)benzoate) **25**, 2-methyl-1,4-phenylene bis(4-(4-((3-ethyloxetan-3-yl)meth-oxy)butoxy)benzoate) **26**, and 4-(4-(hexyloxy)benzoyloxy)-2-methylphenyl 4-(4-((3-methylo-xetan-3-yl)methoxy)butoxy)benzoate **27**. (Cr = crystalline, N = nematic phase, I = isotropic phase.)

broader nematic phases than derivatives of 2-ethyl-oxetane, respectively [27]. Mix-tures of compounds **25** and **27** are applicable by spincoating from a solution without crystallization, mainly due to the methyl group on the aromatic ring [42]. These molecules polymerize fast in a normal atmosphere. As an example the synthetic path for the formation of **25** is given in Figure 1.12. The monomer blend of the di- and mono-oxetane can be processed at room temperature into a highly ordered thin film with large birefringence. Both the processability in air as well as the crystallization suppression of the mixture makes oxetanes an interesting alternative to acrylates for a variety of applications [42].

## 1.2.2 LIQUID CRYSTAL NETWORKS

Photoinitiated polymerization of the polyfunctional monomers or monomer blends in their LC state leads in general to LC networks of which the texture is identical to that of the monomer. Thus, when the monomer is in its monolithically aligned state, for instance by alignment at a rubbed polyimide interface, the polymer network obtained is also monolithically ordered and basically defect-free. It is therefore, a single-step polymerization that directly leads to an aligned film unlike the cross-linked elastomers where the formed polymers first is aligned for example, by stretching, and

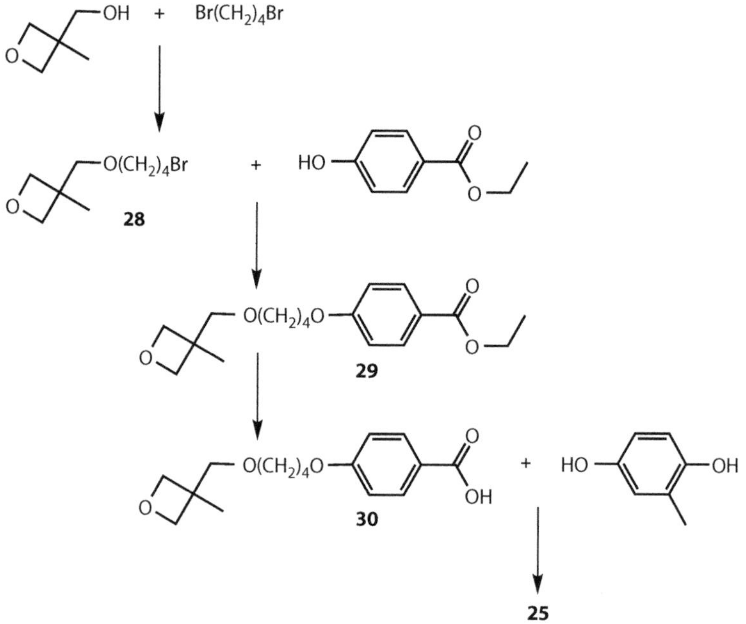

**FIGURE 1.12**  Synthesis of 2-methyl-1,4-phenylene bis(4-(4-((3-methyloxetan-3-yl)meth-oxy) butoxy)benzoate) **25**.

subsequently cross-linked in the stretched state. The benefits of photopolymerization of oriented monomers are obvious. The LC network film can directly be formed at or between substrates that are part of the device that is aimed for.

The polymerization proceeds to high conversion, especially when carried out at elevated temperatures. However, one should keep in mind that when the curing is performed at room temperature vitrification occurs during the polymerization process. The glass transition $(T_g)$ is passed, thus limiting the monomer mobility and obstructing further reaction possibilities of unreacted groups. Therefore, in such case it is recommended to give a post-cure at a temperature above $T_g$, for instance a post-cure temperature of 100–120°C is sufficient for most LC networks. Figure 1.13 shows the polymerization rate of **1** and **2** as a function of temperature. The polymerization rate in general is at its maximum at a temperature of 100°C and the type of phase, in this case either the smectic-C or the nematic phase, does not play an important role. As with isotropic bulky monomers the polymerization kinetics at low temperatures are controlled by viscosity effects and diffusion limitations and at high temperatures by the thermodynamic equilibrium when the ceiling temperature is approached [3]. At temperatures around 220–240°C polymer and monomer of (meth)acrylates are in a thermodynamic equilibrium and polymerization rate goes to zero. Higher-order smectic phases were found to have a more pronounced effect on the rate of polymerization [43].

During polymerization, despite the fact that texture is preserved, the material undergoes some essential changes. This is best illustrated by refractive index measurements

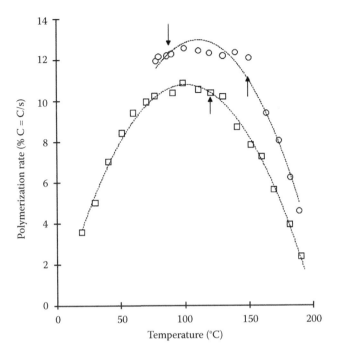

**FIGURE 1.13** Maximum polymerization rate during isothermal polymerization at various temperatures of diacrylates **1** (○) and **2** (□) as measured by Photo-DSC (Philips TL05 lamp, 360 nm, 0.5 mW cm$^{-2}$). The upwards arrows indicate the transition temperature from the nematic to the isotropic phase; the down arrow indicates the transition from the smectic-C phase to the nematic phase.

of **1** before and after polymerization shown in Figure 1.14. In the monomeric state the LC diacrylate has a large birefringence, defined by the difference between the extraordinary ($n_e$) and ordinary ($n_o$) refractive index. The monomeric birefringence is highly temperature dependent as is the case with conventional low molar mass LCs and with LC elastomers. During polymerization the $n_o$ increases. This is caused by polymerization shrinkage as a result of the conversion of Van der Waals distances into covalent bond distances, which increases the density and thereby the refractive index. Also $n_e$ increases during polymerization, in this case even more than $n_o$. Here the increase in density also plays a role, but on top of that there is an effect of the order parameter which in this particular case, when polymerized at 120°C, increases somewhat during polymerization. The change of the order parameter during polymerization depends on a number of factors. Figure 1.15 shows the change in the order parameter for compounds **1** and **18** when they are polymerized at various temperatures. It shows that at a low initial state of order the order parameter tends to increase as rotational molecular motions freeze during network formation whereas, at an initially more highly ordered state the order parameter remains constant. In addition it is found that sometimes at high initial order a decrease in order can be measured when steric factors are playing a role, such as when the monomers are provided with a

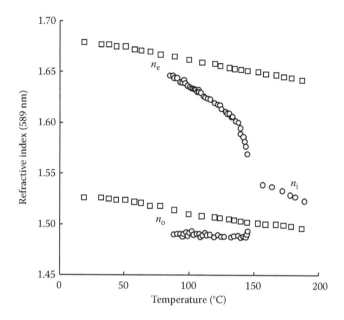

**FIGURE 1.14**  Ordinary, extraordinary and isotropic refractive indices of compound **1** before (○) and after (□) polymerization at 120°C.

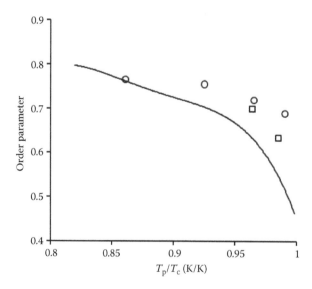

**FIGURE 1.15**  Order parameter as measured by polarization UV–VIS as a function of the reduced temperature given by the polymerization temperature divided by the nematic to isotropic transition. The measured monomers all satisfy one general master curve black line. The value of the polymer networks formed from **1** (○) and **18** (□), is measured at room temperature but in the graph it is shown at its polymerization temperature.

methyl substituent in the central ring [5]. As a conclusion one might state that the orientation and type of order in general is preserved, although the degree of order might be affected by the packing of molecules when they are brought closer together during the formation of the polymer chains. The polymerization shrinkage of these types of monomers is typically between 3 and 8 vol% depending on the molecular weight of the units that connect the two acrylate groups and the polymerization temperature. At an initially highly ordered monomeric state the order parameter therefore shows a small tendency to decrease during polymerization because of steric factors. Di-epoxides and di-oxetanes normally show lower shrinkage values, which is typical for the ring-opening polymerization reactions [21,27]. Often, somewhat higher-order parameters are observed, also caused by the larger number of atoms in a repeating unit of the polymer chain giving more space for the mesogenic rods to pack in a more favorable configuration.

An important difference between the polymeric and the monomeric state is that refractive indices of the polymer networks decrease only slightly and monotonically when heated to higher temperatures (*cf.* Figure 1.14). The presence of the network in this case does not allow large reorganization of the mesogenic moieties. The minor decrease of the refractive indices can be mostly explained by a decrease in density because of thermal expansion and there is only a small contribution to a small reversible decrease of the order parameter when the temperature comes above the glass transition of the LC network. Table 1.1 presents the glass transition temperature of monomers with different spacer length similar to **1** (R = H and n = 6) and **2** (R = CH$_3$, n = 6) and shows the relative value of the order parameter at 150°C as

---

**TABLE 1.1**

**Some Examples of the Glass Transition of LC Networks, Measured by Dynamic Mechanical Measurements and by Volume Expansion, and the Relative Change of the Order Parameter when Heated from 20°C to 150°C Derived from Birefringence Measurements**

LC Network

| R | n | Tg (tan δ) (°C) | Tg Range (Expansion) (°C) | $S^{150}/S^{20}$ |
|---|---|---|---|---|
| H | 4 | 118 | 10–130 | 0.98 |
| H | 5 | | 5–110 | 0.96 |
| H | 6 | 83 | 0–90 | 0.95 |
| H | 8 | 71 | −10–50 | 0.93 |
| H | 10 | 55 | −20–50 | 0.92 |
| CH$_3$ | 4 | | 5–110 | 0.95 |
| CH$_3$ | 5 | | | 0.94 |
| CH$_3$ | 6 | 83 | 5–100 | 0.92 |

obtained from optical measurements. Only marginal changes of the order parameters are measured and they appeared to be reversible after a first heating cycle. During the first cycle small nonreversible changes are observed because of relaxation effects attributed to a reduction of a frozen-in free volume and molecular rearrangements as induced by steric factors. During the rapid polymerization the glass transition is passed whereby the volume, decreasing due to polymerization shrinkage, does not reach equilibrium. LC networks with a high cross-link density, that is, the ones with the smaller molecular lengths of the alkylene spacer between the polymerized acrylate groups and the mesogenic central unit, show only a minor reduction of the degree of orientation. Both by increasing the length of the alkylene spacer between the aromatic central unit and the substitution of the lateral methyl group the reversible decrease of the order parameter becomes somewhat larger when the samples are heated to elevated temperatures. But still decrease in order parameter does not exceed 10% of the value at room temperature.

The small reversible decrease of the order parameter during heating causes an odd linear-expansion behavior. Although the volume expansion of the LC networks compares with that of isotropic polymer networks, that is, a constant and relatively low thermal expansion of the order of $10^{-4}\,\mathrm{K}^{-1}$ in the glassy state and a jump to a higher value around the glass transition temperature and higher. The linear thermal expansion measured parallel and perpendicular to the orientation direction of polymerized **1** is shown in Figure 1.16. Below the glass transition the thermal expansion in both directions is low, the one measured parallel to the director even close to zero. When the glass transition is passed the thermal expansion becomes negative parallel to

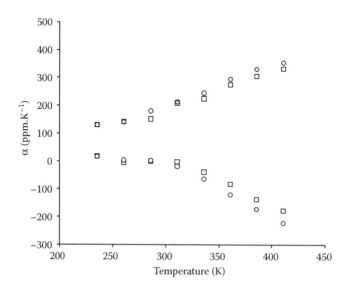

**FIGURE 1.16**  Order linear thermal expansion measured parallel (e) and perpendicular (o) to the director of **1** after polymerization. ($T_\mathrm{p}/T_\mathrm{c} = 0.96$ (□) and $T_\mathrm{p}/T_\mathrm{c} = 0.86$ (○).) $T_\mathrm{p}$ is the temperature of polymerization, $T_\mathrm{c}$ the nematic to isotropic transition temperature of the monomer.

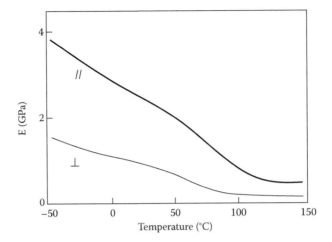

**FIGURE 1.17**   Elastic modulus of **1** after polymerization measured parallel and perpendicular to the average molecular orientation.

the director whereas, perpendicular to that a somewhat excessive high expansion is measured which does not reach a constant value as is the case with isotropic materials. This all is related to the minor reversible decrease in the order parameter at increasing temperatures (*cf.* Table 1.1) that causes the reduction of the average end-to-end distance of the rod-like units projected on the orientation axis. A higher degree of initial order, established by polymerization at a lower reduced temperature enhances this effect (Figure 1.16). Also, a longer length of the alkylene spacer leads to more dramatic effects in this respect. Application of this effect for soft actuator devices is discussed further in this book [44]. Director-controlled low thermal expansion is also useful for adhesives and encapsulation where thermal stresses between the organic and an often inorganic substrate need to be avoided.

Unlike main-chain LC polymers, the modulus and strength of LC networks does not reach the extreme values as sometimes measured for oriented systems. Both the fact that the polymer main chains achieve a more random conformation and the presence of the less ordered flexible alkylene units, the mechanical properties are of the same order as classical highly cross-linked polymer networks such as epoxides and polyacrylates, that is, of the order of a few GPa's. Of course the modulus below the glass transition depends largely on the cross-link density and amount to a few hundreds of MPa's in case of a cross-linked mesogenic diacrylate. Figure 1.17 shows that the modulus is anisotropic and is parallel to the average orientation a factor of 2 to 3 higher than perpendicular to that.

### 1.2.3   BIREFRINGENCE AND POLARIZATION OPTICS

The highly birefringent film, obtained after photopolymerization of monolithically aligned molecules as illustrated in Figure 1.14, finds its application in optical components. An example of such a component is the polarizing beam splitter shown in

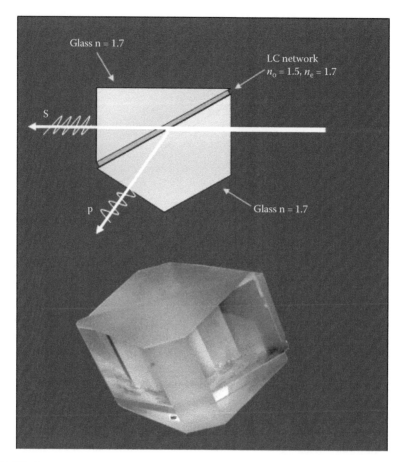

**FIGURE 1.18** Schematic diagram of a polarizing beam splitter based on two glass bodies adhered together with an LC network and the splitting of unpolarized light into two orthogonally polarized components. The photograph shows the actual device. The dimension of the LC network layer measured along its longest side is 6 cm.

Figure 1.18. The monomers provided with a small amount of a photoinitiator are aligned in their nematic state between two glass bodies that contain a thin rubbed alignment layer on their surfaces. The two glass bodies are "glued together" by the polymerization of the aligned monomers resulting in a highly birefringent adhesive layer. The extraordinary refractive index of the film equals the high (isotropic) refractive index of the lead containing glass bodies. Consequently, the ordinary refractive index of the film is much smaller than that of the glass.

Figure 1.18 shows the optical path of a beam of unpolarized light entering the device. The linearly polarized p-component of the light beam (parallel to the surface normal) experiences a decrease in the refractive index at the glass-polymer interface. This leads to complete reflection of the p-component of the light beam resulting in a polarized beam leaving the lower part of the device. The s-component of the light (perpendicular to the film surface but here parallel to molecular director) does not experience any change in refractive index. For this reason the light travels through the film and leaves the device at the upper glass body. Thus, this device separates (splits) the original unpolarized light beam into two beams of polarized light. One of the beams can be rotated by a $\lambda/2$ plate, for example, conveniently made by a film of a similar LC network. This results in a beam of one polarization direction which can find application in high brightness LC display systems such as front projectors or rear projecting television sets. The main advantage of this polarizing system is that no excessive heat is generated by the polarizer because no absorption of light takes place.

Similar principles have been elaborated for creating polarizing backlight for LC displays [45]. Small polarization separation elements are then integrated in the surface of a waveguide. The side lit backlight guides the light which on its path couples out light of the desired polarization while the unwanted orthogonally polarized light is kept in the waveguide by total internal reflection. Because of multiple reflections and a small birefringence in the plastic waveguides the state of polarization of the guided light changes and becomes ultimately also coupled out with the right state of polarization. As a result a very efficient backlight system is obtained that does not hamper the drawback of light loss in the polarizer mounted on the liquid crystal display (LCD) cell. Also applications in the field of optical recording are known [46].

### 1.2.4 TUNING THE LOCAL BIREFRINGENCE BY LITHOGRAPHIC PHOTOCHEMICAL MEANS

An additional benefit of the use of photoinitiated polymerization of the reactive LC monomers is that the polymerization can be restricted to a confined location by locally blocking the light using a photomask. This lithographic approach can generate complex designs where birefringent patterns can be alternated with isotropic area either by the removal of the unreacted monomer or by the polymerization of the blocked area in their isotropic state at elevated temperatures. The formation of a patterned retarder is schematically shown in Figure 1.19. This retarder finds its application in transflective displays where the reflective field of a pixel is given an additional optical retardation of a quarter wave to make the electro-optical response similar to that of the transmissive field [47,48]. It was found that the two step polymerization easily leads to the formation of a complex surface profile at the location of the isotropic/aligned transition, originating from polymerization-induced diffusion of the still unreacted monomers to the polymerizing areas and surface tension effects. This makes the use of planarization layers necessary for this application involving an additional process step.

An alternative to the thermal patterning process is the so-called photochemical patterning process. In order to perform this patterning process, part of the nematic (di-)acrylate(s) is replaced by a photoactive compound that is derived from cinnamic acid. This compound can be photoisomerized from the original E structure to the Z structure. The Z structure breaks the rod-like symmetry and decreases the isotropic transition temperature when photoisomerized in a liquid crystalline mixture. Thus, upon irradiation of the mixture through a mask, the irradiated parts will become optically isotropic and the nonirradiated parts remain birefringent. In order to avoid photopolymerization during the isomerization process, the first exposure, using the

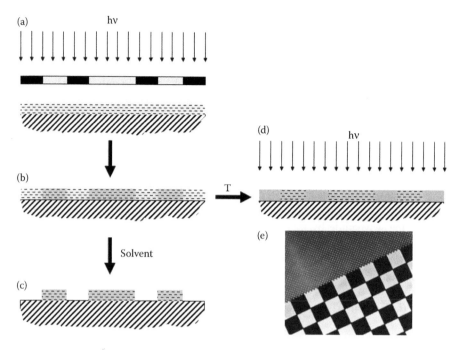

**FIGURE 1.19** The formation of patterned retarders. The reactive mesogen is UV exposed in its aligned nematic state (a) forming cross-linked areas (dark gray) alternated with nonreacted mesogens (light gray) (b). The nonreacted mesogens can be dissolved (c) or UV exposed at a temperature above its nematic to isotropic transition (d). The photograph (e) shows a sample with 1 and 10 mm patterns of aligned and isotropic polymer networks made following method (f) as observed by a polarizing microscope. The bright areas are birefringent; the black areas are isotropic.

photomask, is performed in air; oxygen being the polymerization inhibitor. After isomerization, the atmosphere is changed to nitrogen and upon irradiation a fast polymerization occurs that fixes the optical properties of the pixels. An advantage of this process over the thermal patterning process is that no critical temperature steps are involved and that the surface remains smoother than in the case of the thermal patterning process. However, the critical steps are now encountered in using different atmospheres. Furthermore, the photochemical reaction by itself is not clean [49]. Often, by-products are formed causing yellowing of the polymer film, often not being accepted in optical applications. This is mainly due to the fact that cinnamate esters derived from phenolic groups suffer from unwanted photochemical side reactions. The cinnamic esters derived from cyclohexanol **31** (see Figure 1.20) undergoes a clean photoisomerization reaction to its isomer **32** (also shown in Figure 1.20) and is therefore, more suitable for the patterning process [49]. The synthesis of this compound is shown in Figure 1.21. Due to the almost equal reactivity of hydroxyl groups on the cyclohexane ring and at the hexyl spacer, the acrylate group at the right side has to be introduced in the last step of the synthesis. Cinnamic acid derivative **34** is

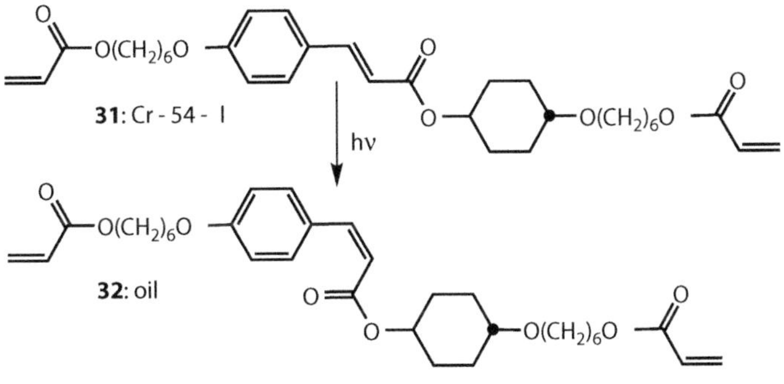

**FIGURE 1.20** Structure and phase transitions of (*trans*)-(E)-4-(6-(acryloyloxy)hexyloxy) cyclohexyl 3-(4-(6-(acryloyloxy)hexyloxy)cinnamate **31**, and (*trans*)-(Z)-4-(6-(acryloyloxy) hexyloxy)cyclohexyl 3-(4-(6-(acryloyloxy)hexyloxy)phenyl)acrylate) **32**, obtained by irradiation of **31**. (Cr = crystalline, I = isotropic phase.)

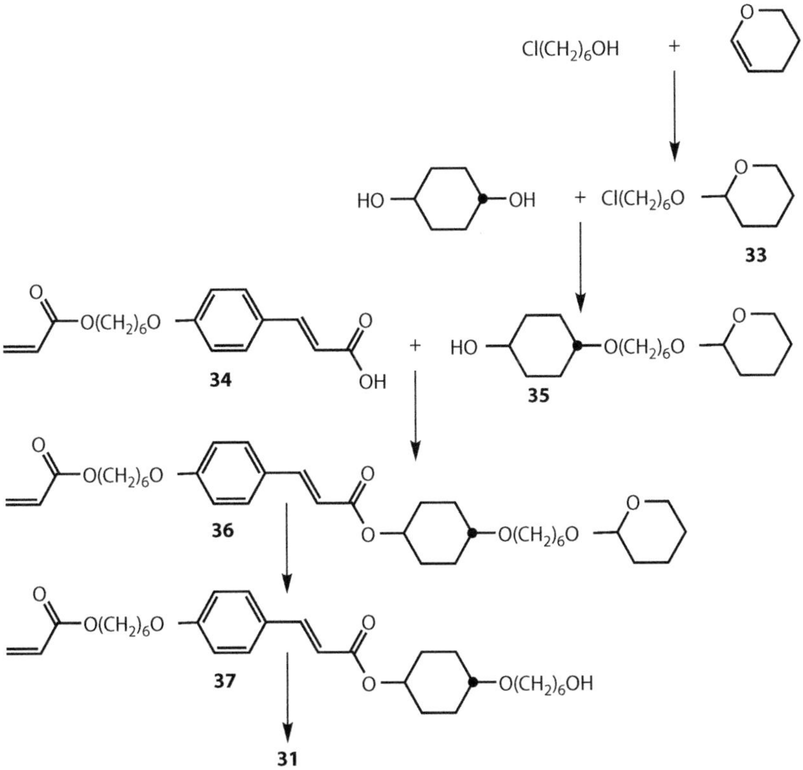

**FIGURE 1.21** Synthesis of (*trans*)-(E)-4-(6-(acryloyloxy)hexyloxy)cyclohexyl 3-(4-(6-(acr-yloyloxy)hexyloxy)cinnamate **31** (see Figure 1.20).

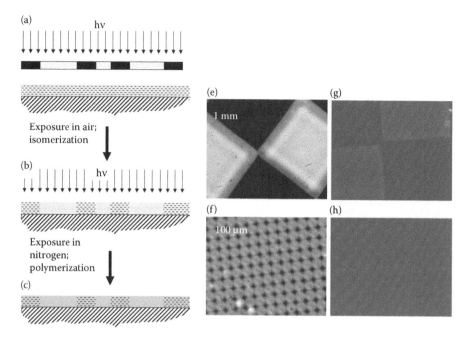

**FIGURE 1.22** The formation of patterned retarders by photoisomerization. The aligned reactive mesogen containing the photoisomerizable compound is UV exposed in the presence of oxygen (a) forming isotropic areas alternated with still aligned mesogens (dashed) (b). Both the isotropic monomer and the aligned monomer are polymerized by a single exposure step in nitrogen yielding a patterned retarder (c). The polarization microscopy pictures shows the patterns with the alignment under 45° (e, f) and parallel (g, h) with the crossed polarizers. The bright areas are birefringent; the black areas are isotropic.

made in a similar manner as described for benzoic acid derivative **6** shown in Figure 1.2, using the ethyl ester of 4-hydroxycinnamic acid instead of the ethyl ester of 4-hydroxy benzoic acid as starting compound. The formation of a patterned retarder by this photoisomerization process is outlined in Figure 1.22.

### 1.2.5 TUNING POLARIZATION OPTICS BY CONTROLLING THE TRANSVERSAL DIRECTOR PATTERNS

Besides modulating the director in the lateral directions of the film by lithographic means and controlled photochemistry we also have the possibility to adjust and/or to modulate the director over the cross-section of the thin films. Figure 1.23 shows some examples. The twisted nematic structure can be obtained by polymerization of the nematic reactive mesogens between two orthogonally rubbed polyimide orientation layers, eventually supported by the addition of a small amount of chiral dopant. When optimized on thickness and birefringence these films form wavelength independent half-wave retarders that are used in displays and optical data storage [50]. In a similar way, the twist can be given at angles other than 90° such as 180° or 240° which prove

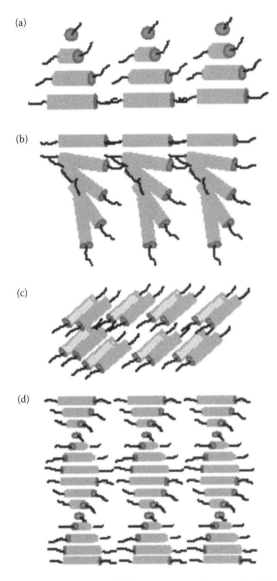

**FIGURE 1.23** Examples of transversal director patterns that can be locked into an LC network. From top to bottom a twisted director structure (a), a splayed director (b), a tilted director (c), and a chiral nematic (cholesteric) (d).

to be useful to compensate for color aberrations in the so-called supertwisted nematic (STN) displays [28].

When polymerized between two substrates, one with a planar orientation layer and the other with a homeotropic orientation layer, the splayed LC network is obtained. Similar structures are obtained when a single substrate is used and one of the components in the LC monomer mixture has an amphiphilic nature, for example, compound

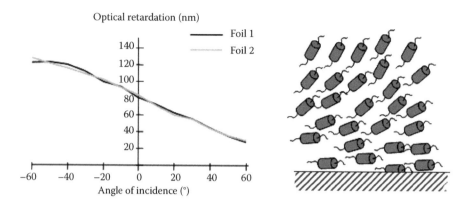

**FIGURE 1.24** The angular dependence of the optical retardation of two splayed films formed from LCs **12** and **13** (see Figure 1.6), stabilized by photopolymerization. The right insert shows schematically the splay direction of these films. For optical compensation of an addressed TN display the two films are placed orthogonal to each other and a set of these two layers is placed at each side of the TN cell.

**12** (see Figure 1.6). In that case the LC tends to orient perpendicular toward the interface with air. The parameters that control the splay profile are the tilt at the substrate interface, the anchoring strengths, the elastic constants for splay and bend deformation, and the layer thickness. By optimizing these parameters a full 90° splay rotation can be made and also any other desired intermediate rotation between 0° and 90° can be achieved. This property proved to very valuable to make the so-called tilted retarders for twisted nematic (TN) displays to improve their viewing angle [37,38]. Figure 1.24 shows an example of a splayed film that was developed as viewing angle film. To compensate the addressed state of a TN display, which can be considered as a stack of two splayed layers, basically four splayed LC network films are needed, stacked two by two.

A tilted network can be achieved by polymerizing the film between two substrates with a high pretilt orientation layer. The optical behavior with respect to angular dependent optical retardation resembles that of the splayed film. The chiral nematic film will be discussed later.

## 1.3 CHOLESTERIC NETWORKS

### 1.3.1 CHIRAL REACTIVE MESOGENS

For a wide range of optical applications cholesteric polymeric materials are of interest because of their ability to reflect circularly polarized light of a given wavelength. The helical structure of cholesteric materials is induced by the presence of chiral components in nematic LCs. The reflection wavelength of such a cholesteric (or chiral-nematic) mixture is defined by:

$$\lambda = \bar{n}.p = \frac{\bar{n}}{\sum_{i} HTP_i.x_i}, \quad \Delta\lambda = \Delta n.p = \frac{\lambda.\Delta n}{\bar{n}} O$$

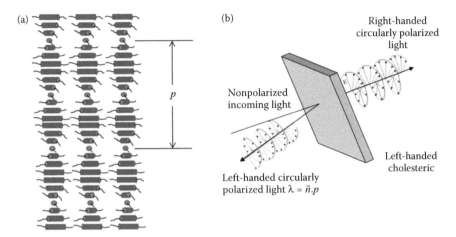

**FIGURE 1.25** Representation of the cholesteric phase with a pitch $p$ (a) and a cholesteric film as a circular polarizer (b). Note that the molecules in the actual films are less well organized where the angular distribution around a common local director equals that of a normal nematic phase.

where $\bar{n}$ is the mean refractive index of the cholesteric mixture, $p$ is the pitch of the helix, $x_i$ is the fraction of chiral component $i$ present in the cholesteric material, and $HTP_i$ is the helical twisting power of chiral component $i$, defined as the reciprocal of the pitch of the helix for $x_i = 1$. The reflected light is circularly polarized and the handedness of the polarization of the reflected light is the same as the handedness of the helix of the cholesteric material. The bandwidth $\Delta\lambda$ of the reflection band scales linearly with the wavelength and the birefringence $\Delta n$ of the LC polymer and is usually between 40 and 80 nm in the visible wavelength range. Light with the opposite handedness of circular polarization is transmitted through the material [51]. These optical effects, outlined in Figure 1.25, make cholesteric films very useful for screening light on wavelength and polarization. Therefore, embedding these properties in a stable polymer film opens the way toward applications in, for instance, LC displays using them as color filter and/or polarization filter.

In order to obtain the cholesteric phase, nematic molecules have to be mixed with chiral molecules that have mesogenic properties but not necessarily exhibit a liquid crystalline phase. If the reactive nematic molecule **1** is substituted with two methyl groups on the spacers, thus introducing two chiral groups, with high enantiomeric excess, cholesteric diacrylate **38** (Figure 1.26) is obtained which exhibits a relatively broad liquid crystalline phase for processing in a reasonable temperature range. The synthesis is similar to that of nematic diacrylate **1**. However, the chiral spacer is formed from 3-methyl-6-hydroxy-bromohexane (**42**) instead of 6-hydroxy-chlorohexane (*cf.* Figure 1.2). The former bromide is not commercially available and was prepared starting from citronellol as outlined in Figure 1.27 [8,52]. The chiral benzoic acid derivative **44** can be derivatized in the same manner as nonchiral derivate **5** (Figure 1.2), to obtain diacrylate **38**. The helical twisting power (HTP) of **38** is approximately $7\,\mu\text{m}^{-1}$ corresponding to a pitch of around 180 nm, as demonstrated

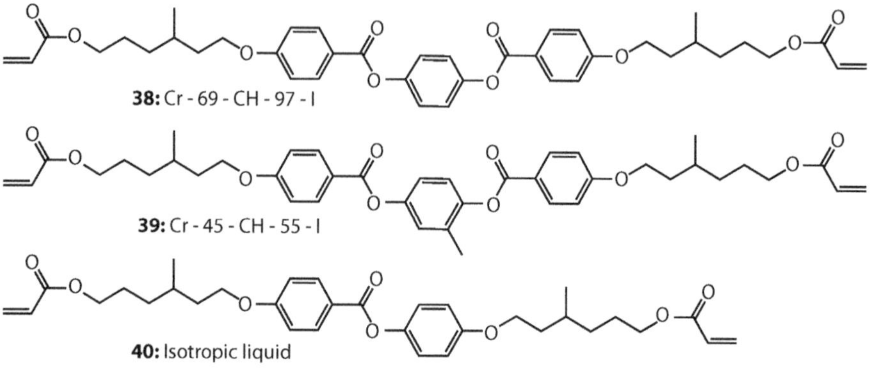

FIGURE 1.26 Structure and phase transitions of 1,4-phenylene bis(4-(6-(acryloy-loxy)-3-methylhexyloxy)benzoate) **38**, 2-methyl-1,4-phenylene bis(4-(6-(acryloyloxy)-3-methylhexyloxy)benzoate) **39**, and 4-(6-(acryloyloxy)-3-methylhexyloxy)phenyl 4-(6-(acryl-oyloxy)-3-methylhexyloxy)benzoate **40**. (Cr = crystalline, CH = cholesteric phase, I = isotropic phase.)

by scanning electron microscopy (SEM) of the fracture surface of a polymerized sample [8]. Several cholesteric diacrylates derived from the same chiral spacer with various mesogenic groups are presented in Figure 1.26. They all exhibit approximately the same HTP. Compound **39** has the advantage of a lower melting point than **38**, which makes processing at lower temperature possible. Compound **40** has in its pure state no liquid crystalline phase and can therefore only be used in conjunction with nematic monomers with relatively high clearing points in order to obtain mixtures with sufficiently broad liquid crystalline processing window. Furthermore, it is an oily substance that makes purification more difficult than the other two compounds that have a crystalline nature at room temperature. The synthesis of **40** closely resem-bles that of **3** [53]. Also, in this case, 6-hydroxy-chlorohexane used in Figure 1.3 was replaced by 3-methyl-6-hydroxy-bromohexane (**42**) of which the synthesis is outlined in Figure 1.27.

Comparison with the nonchiral molecules of Figure 1.1 reveals that the chiral monomers have lower melting points and much lower isotropic transition tempera-tures. The methyl side groups in the spacer clearly destabilize the liquid crystalline phase. Mixtures of compounds **38** or **39** with nematic diacrylates such as **2** form color reflecting layers of which the color depends on the composition. Figure 1.28 shows the reflection color as a function of the composition of a mixture of **38** and **2**. The measurements were performed after photopolymerization. In this way, films with stable reflection properties were obtained. Before polymerization the reflection wave-lengths were temperature dependent. Due to this thermochromic effect, the reflection wavelength after polymerization not only depends on the composition, but also on the polymerization temperature [8]. The thermochromic effect in combination with the formation of a polymer film can be used to make color filters as will be outlined below.

An HTP of about $6\,\mu m^{-1}$ of compounds such as shown in Figure 1.26 is rel-atively low, which means that large quantities of chiral monomers (50% or more)

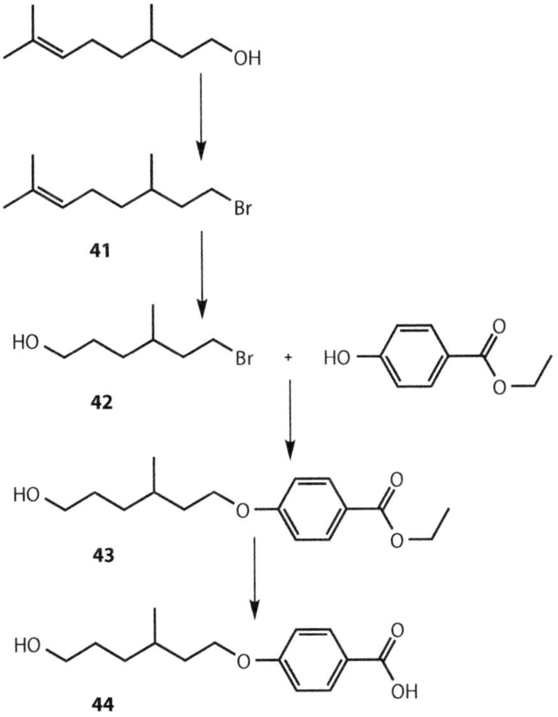

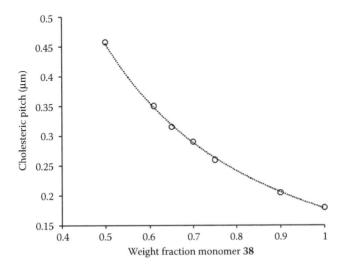

**FIGURE 1.27** Synthesis of 4-(6-hydroxy-3-methylhexyloxy)benzoic acid (**44**), an important intermediate for the synthesis of **38**, **39**, and **40** (see Figure 1.26) starting from citronellol.

**FIGURE 1.28** Dependence of the pitch and reflection color of a cholesteric mixture of **38** and **2** after photopolymerization at 60°C as a function of the fraction of **38**.

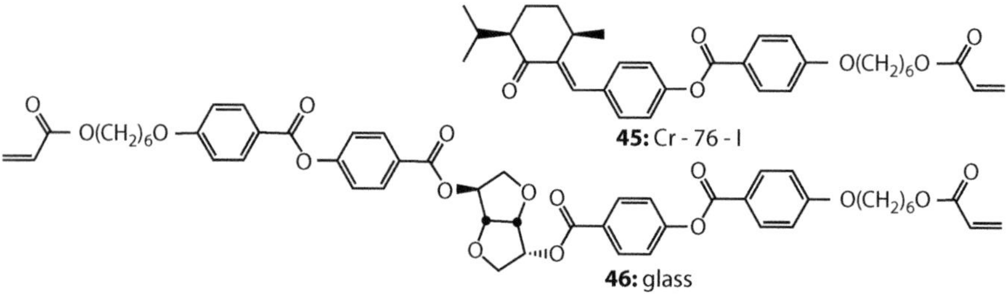

**FIGURE 1.29** Structure and phase transitions of menthone derivative **45** and isosorbide derivative **46**. (Cr = crystalline, I = isotropic phase.)

have to be mixed with nematic molecules to obtain a pitch capable of reflecting visible wavelength (see Figure 1.28). Several molecules derived from other chiral moieties have been developed that have much higher HTP and can therefore be mixed in smaller amounts with nematic molecules to obtain relatively short pitches. Among these molecules are monoacrylates derived from menthone **45** with an HTP of $20\,\mu\mathrm{m}^{-1}$ [54] and diacrylates derived from isosorbide as chiral moiety such as **46** with a HTP of $55\,\mu\mathrm{m}^{-1}$ [55]. Both molecules are outlined in Figure 1.29. The syntheses of **45** and **46** are outlined in Figures 1.30 and 1.31, respectively.

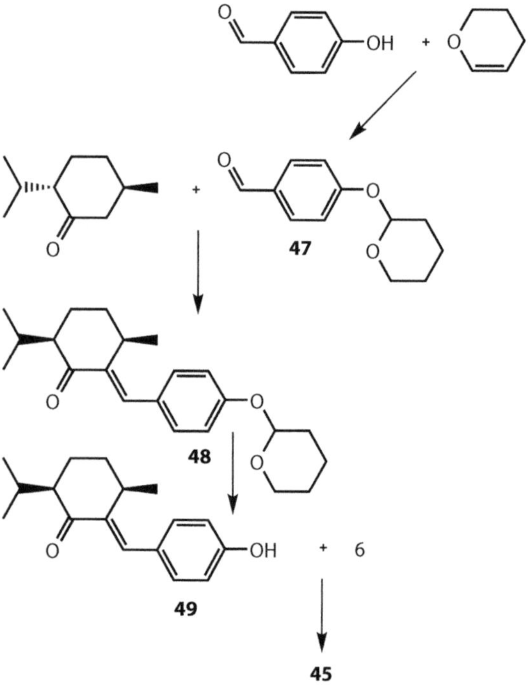

**FIGURE 1.30** Synthesis of 4-((E)-((3R,6R)-3-isopropyl-6-methyl-2-oxocyclohexylidene) methyl) phenyl 4-(6-(acryloyloxy)hexyloxy)benzoate **45** (see Figure 1.29).

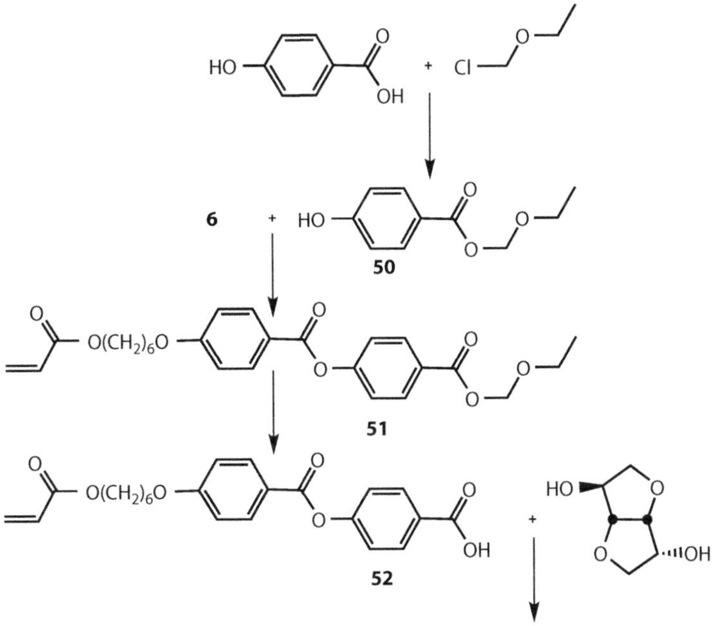

**FIGURE 1.31** Synthesis of (3R,3aR,6S,6aR)-hexahydrofuro[3,2-b]furan-3,6-diyl bis(4-(4-(6-(acryloyloxy)hexyloxy)benzoyloxy)benzoate) **46** (see Figure 1.29).

Owing to the relatively high HTP of **45** and **46** compared to that of **38**, mixtures of **2** with only 15% and 6% of these compounds, respectively, will form a blue reflecting film. The chiral moieties of these two molecules are derived from natural occurring chiral molecules namely menthone and isosorbide. For this reason only one helical direction is accessible when applied in cholesteric films because the enantiomers of these molecules are not naturally occurring and are difficult to prepare. The compounds presented in Figure 1.26 that are derived from citronellol (see Figure 1.27) can be prepared from both enantiomers that are commercially available and thus both helical directions are possible in the cholesteric layer. Also enantiomeric compounds derived from 1,2-phenylethanediol can be prepared easily due to availability of both enantiomeric starting compounds. An advantage of these compounds over the ones derived from citronellol is their approximately 3 times higher HTP [56]. The helical direction may play a role in the optical properties of complex films containing cholesteric layers.

### 1.3.2 Broadband Circular Polarizers

Nearly all LC displays visualize their images by making use of polarized light. Therefore, one of the principal components of a LC display is the polarizer. In a transmissive LC display light is emitted by a backlight, often consisting of a planar waveguide side-lit by a cold cathode lamp or a light-emitting diode, toward the viewer. A sheet polarizer polarizes transmitted light by absorbing the unwanted polarization. This means a loss of the backlight intensity of at least 50%.

Cholesteric films are capable to generate polarized light by transmitting one polarization direction and reflecting the other. The reflected light can be recycled in the backlight as it can be converted into the desired polarization such that a much more efficient polarization device can be made. Disadvantages of the cholesteric films are that they generate circularly polarized light instead of linearly polarized light and they are only effective for a limited bandwidth and not for the whole visible spectrum. The first problem is solved easily by converting the circularly polarized light into linearly polarized light with the aid of a quarter wave plate, in a preferred design also consisting of a uniaxially aligned LC network with appropriate thickness and birefringence. To solve the second problem, the bandwidth of the reflection band of the cholesteric films, which is normally about 60 nm, should be increased by at least three times. In theory this can be realized by an increase in birefringence of the LCs with the same factor. However, such materials are difficult to make and may have the disadvantages such as low stability and absorption bands in the wavelength region of interest.

A better solution to this problem is the production of a cholesteric film in which the pitch and thus reflection wavelength has a gradient such that the blue part of the spectrum is polarized at one side of the film and the red part of the spectrum at the opposite side. Consequently, green becomes polarized halfway. To produce such a film kinetics of the photopolymerization reaction can be used [57,58]. A film is made from a mixture of cholesteric diacrylate **38** and nematic monoacrylate **53** shown in Figure 1.32. This monoacrylate is made by the esterification of phenolic compound **67** (see Figure 1.42 in Section 1.4) with 4-octyloxybenzoic acid. The composition is chosen so as to obtain a green reflection band. If such a mixture is photopolymerized, a green light reflecting

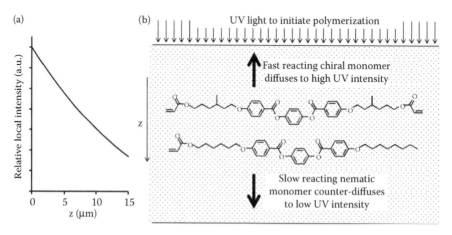

**FIGURE 1.32** Intensity gradient for UV light in a cholesteric film (a) and diffusion direction of 1,4-phenylene bis(4-(6-(acryloyloxy)-3-methylhexyloxy)benzoate) (**38**, upper structure) and 4-(4-(6-(acryloyloxy)hexyloxy)benzoyloxy)phenyl 4-(octyloxy)benzoate (**53**, lower structure) during polymerization (b).

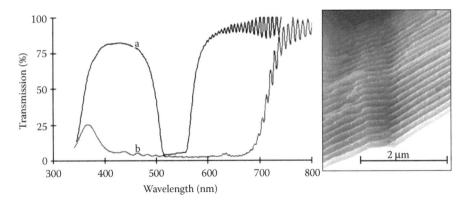

**FIGURE 1.33** Reflection of right-handed circularly polarized light measured as transmission loss of a 1:1 mixture of compounds **38** and **53** before (a) and after (b) polymerization using an UV intensity gradient. The SEM picture shows a cross-section of this polymer film obtained by freeze-fracturing.

layer is obtained. By adding a UV absorbing dye, an intensity gradient of the UV light in the transverse direction is obtained shown on the left side of Figure 1.32. Due to the UV intensity gradient, the polymerization at the top proceeds faster than at the bottom of the layer. The cholesteric component **38** is a diacrylate, and therefore, has twice as high a probability for reaction as nematic monoacrylate **53** that needs to be incorporated in the polymer. If the overall polymerization rate is tuned to the diffusion kinetics (relatively low UV intensity), depletion of the chiral diacrylate near the top of the layer generates a concentration gradient of this diacrylate in the transverse of the film. This in turn starts diffusion of this compound toward the top of the layer. The ultimate result after complete photopolymerization is that the top of the layer contains more chiral material and thus has a shorter reflection wavelength than at the bottom of the layer, which is formed of relatively more of the nonchiral compounds. Although, the film is made partly of monoacrylate **53**, there are still enough cross-links to obtain a similar stability as shown in Figure 1.14 for the polymerized nematic di-acrylate **1**. SEM results show a pitch gradient in the film [57] (see Figure 1.33). The effect of the pitch gradient on the optical properties is apparent from Figure 1.33 in which the green reflection band before polymerization is shown together with the broadband reflection after polymerization. It is clear that such a "molecular architecture" in the film is impossible to make with nonpolymerizable LCs or liquid crystalline polymers. The photopolymerization reaction is not only responsible for the formation of a stable film, but it also plays a role in the formation of this architecture. By selection of the right liquid crystalline structures it is possible to form sheets in a continuous coating process, which can be cut and incorporated in LC displays.

The method of increasing the brightness using the recycling principle is schematically shown in Figure 1.34. Light with the right polarization direction is transmitted while the other polarization direction is reflected to the lamp system. It is then depolarized and can, after again being reflected in the direction towards the viewer, in turn form light of the right polarization. Theoretically in this way a 100% gain in light

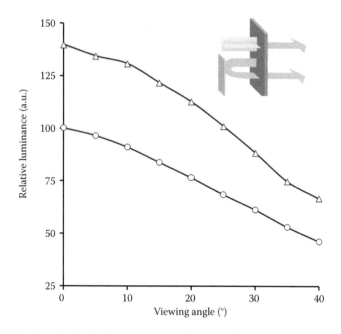

**FIGURE 1.34**  The relative luminance measured at a LC display with (△) and without (○) the presence of a broadband cholesteric filter/quarter wave combination added in between the backlight and the first dichroic polarizer. The insert shows schematically the recycling principle of conversion of unpolarized light into polarized light.

intensity can be obtained. The abovementioned sheets were placed between the active switching part of a display and the backlight system, which was optimized for the use of these sheets. The brightness of such a display is typically 1.6 times as high as with a conventional polarizer [57]. Figure 1.34 shows the results as a function of viewing angle. The luminance is integrated over the three colors.

### 1.3.3  CHOLESTERIC COLOR FILTERS

In addition to their polarizing properties their ability to reflect light within a well-adjusted wavelength region is an important property for the use of cholesterics in color filters, for example, in LC displays. Thereto, the reflection colors must be patterned, preferably in a single layer or, when used in transmission, in a double layer. For the manufacturing of conventional color filters, based on combinations of absorbing dyes or pigments in a polymer matrix, the three colors red-green-blue (RGB) are applied successively, for example, by lithography, in a multistep process. This process can be simplified by the use of cholesteric materials. But there are more benefits by exchanging conventional absorbing color filter with nonabsorbing reflective cholesterics. For instance they can be directly used in low-energy-consuming reflective LC displays. These displays do not require a backlight and are of relevance for mobile devices such as PDA's or mobile phones. The design of these reflective LCDs can be simplified by using the unique combination of polarization and color selection properties of cholesterics [59]. For application in transmissive LCDs, cholesteric color filters have an extra advantage. The reflective nature of the color filter offers the opportunity to recycle the two unwanted primary colors from the backlight. In conventional transmissive LC displays' unwanted light (i.e., at least 66%) is absorbed by the color filter. This means that when cholesteric color filters are used in combination with the cholesteric polarizer described in Section 1.3.2, theoretically a six times higher light intensity can be obtained compared with the use of absorbing components [60].

Cholesteric materials offer the possibility to obtain an RGB array by changing the pitch ($p$) and making use of the thermochromic effect. This process is very similar to the thermal patterning process presented in Section 1.2 (Figure 1.19). Figure 1.35 shows the effect of temperature on the reflection wavelength of a film before polymerization, prepared from a 1:2 mixture of chiral diacrylate **39** (see Figure 1.26) and nonchiral diacrylate **1** (see Figure 1.1). Upon photopolymerization using a mask at 20°C ($T_1$) the red pixels are cross-linked and therefore, stabilized. After increasing the temperature to 55°C ($T_2$) and subsequent photopolymerization through a mask, the green pixels are formed and stabilized. After polymerization at 90°C ($T_3$) the remaining pixels are also stabilized, exhibiting the red color. This process has rather critical parameters because small changes in the temperature change the reflection color. Therefore, we developed another, more reliable, process based on photochemical alteration of the pitch. This process is similar to the photochemical process forming subpixelated retarders also outlined in Section 1.2 (Figure 1.22). It offers the possibility to obtain an RGB array in a single step, by photochemically changing the relative fractions of the chiral compounds, locally in the layer to change the pitch ($p$) [59]. In

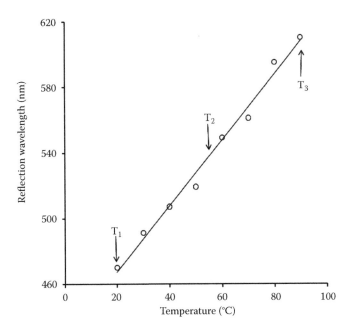

**FIGURE 1.35** Dependence of the reflection color of a 1:2 mixture of **39** and **1** as a function of temperature.

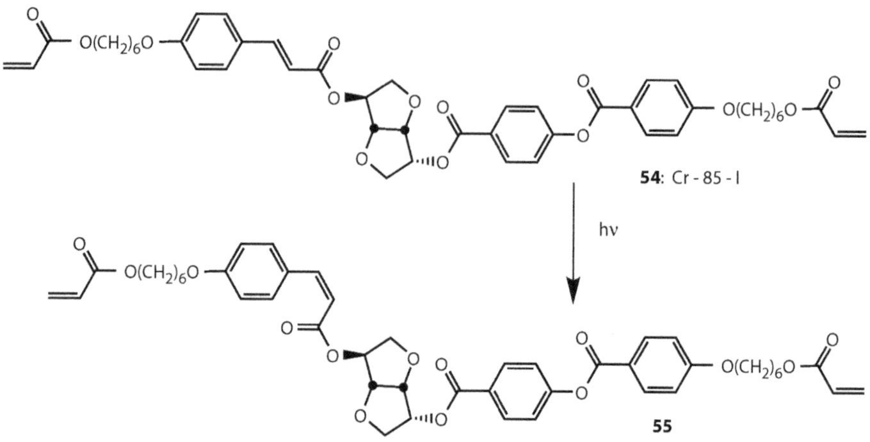

FIGURE 1.36 Structure and phase transitions of isosorbide derivative 54 and its Z-isomer 55 formed upon photoisomerization of 54. (Cr = crystalline, I = isotropic phase.)

this process the isomerizable cinnamic acid derivative 54 (Figure 1.36) is used [61]. It is derived from isosorbide instead of cyclohexane diol in the case of the nonchiral compound 31 (Figure 1.20) used in Section 1.2. The isosorbide moiety induces a high HTP before photoisomerization similar to compound 46, while the cinnamate moiety is the photoactive part. The synthesis of 54 is outlined in Figure 1.37. This figure shows that the monoprotected isosorbide derivative 56 is able to esterify isosorbide on one side with the nonisomerizable part 52 (see Figure 1.31) and with the isomerizable part 34 (Figure 1.21) on the other. Upon photoisomerization compound 55, which is the Z-isomer of 54, is obtained that exhibits a much lower HTP [62]. Thus, upon irradiation of a mixture of 54 (E-isomer) and a nematic diacrylate such as 2, of which the pitch causes a blue reflection, the reflection color shifts to the red part of the spectrum. By exposing the same sample through a gray-scale mask, modulating the intensity, the three RGB colors are obtained in a single exposure step. Polymerization of the acrylate group is postponed by the presence of oxygen that inhibits radical reactions. This process is shown in Figure 1.38. This figure also shows that the color change is dependent on the total UV doses. The colored pixels are fixed by a rapid photopolymerization reaction, by UV exposure in the absence of oxygen.

Apart from isosorbide- and cinnamic-derived molecules, compound 45 (see Figure 1.29) can be used for this purpose. Upon irradiation of this compound a Z-isomer is found that exhibits a much lower HTP [54]. Also cinnamic acid derivatives derived from phenyl ethanediol instead of isosorbide are applicable in this process. The advantage of the cinnamate derivatives over the menthone derivative is the higher thermal stability of the color filters in the case of the cinnamate compounds [56,61,62]. Cholesteric color filters made in this manner reflect colors of high color purity, related to their steep and well-defined reflection bands, and are promising materials for new reflective and transmissive LCDs [60].

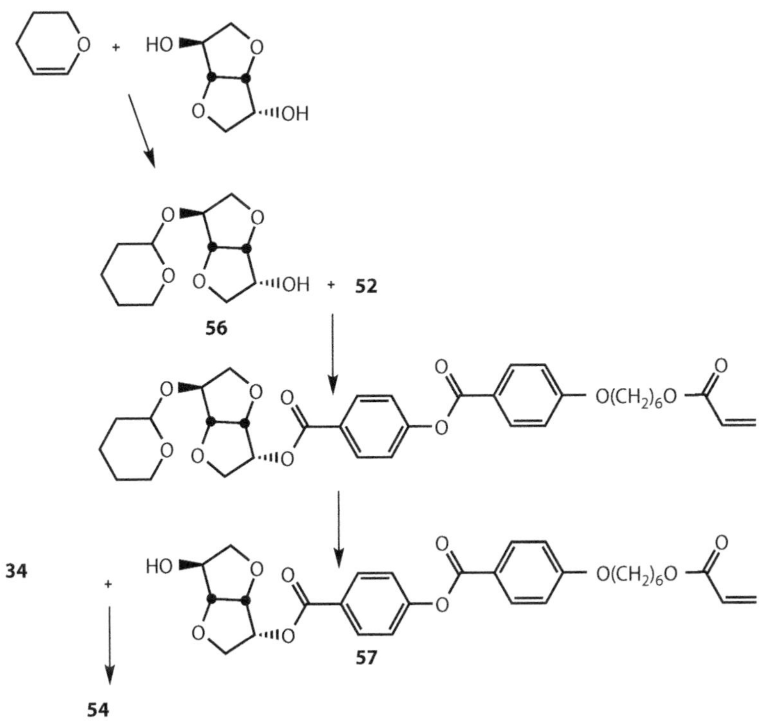

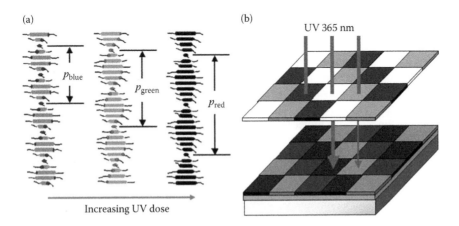

**FIGURE 1.37** Synthesis of (3R,3aR,6S,6aR)-6-((E)-3-(4-(6-(acryloyloxy)hexyloxy)phenyl) acryloyloxy)hexahydrofuro[3,2-b]furan-3-yl  4-(4-(6-(acryloyloxy)hexyloxy)  benzoyloxy) benzoate **54** (see Figure 1.36).

**FIGURE 1.38** Increase of the pitch as a function of the irradiation dose of a cholesteric mixture made with photoisomerizable compound **54** (a). Irradiation through a gray scale mask of a blue reflecting film containing compound **54** (b).

## 1.4   GUEST–HOST LC NETWORKS

The performance of LCDs is strongly influenced by the quality of the polarizer. The most elegant way to define the polarization performance is the dichroic ratio (DR) in absorbance. The DR is a materials property and therefore, it is independent of the thickness of the polarizer unlike the contrast ratio, which increases with the thickness at the expense of the transmittance. The DR in absorbance can be determined through polarized absorption spectroscopy and is defined as

$$DR = \frac{A_{//}}{A_{\perp}}$$

where $A_{//}$ and $A_{\perp}$ are defined as the absorbance parallel and absorbance perpendicular to the average orientation axis of the dye molecules, respectively.

For mobile applications of LC displays, polarizers with DRs of approximately 35 are standard in current products. For the high-end applications such as TFT-monitors and LC-TV, the requirements for the polarization performance are more demanding. Polarizers with DRs exceeding 40–50 are currently used in these applications.

The most widely used polarizers for LC display applications are based on uniaxially stretched poly(vinylalcohol) impregnated with iodine or doped with dichroic dyes. The water sensitive poly(vinylalcohol) film is protected at both sides from moisture and relaxation by laminating a birefringence-free triacetyl cellulose. The polarizer is fixed to the display glass by an adhesive layer. The necessary use of protective layers and adhesives in these polarizers add unnecessary thickness to the LC display. Numerous advantages are foreseen when these traditional sheet polarizers are replaced by ultrathin coatable polarizers located inside the LCD cell that is formed by the two glass plates enclosing the LC, color filter, electrodes, and the thin film transistors (in-cell). Apart from a significant reduction in display thickness and weight, the positioning of the polarizers inside the cell is beneficial to the robustness of the display as it avoids the scratching of the sensitive cellulose-triacetate film on the exterior. As is illustrated in Figure 1.39, where we added an azo dichroic dye to a model LC host with rich phase behavior, exhibiting a nematic and several smectic phases, the order in the nematic phase is too low to provide sufficiently high contrast ratios. Even the smectic-A phase does not perform well enough for this purpose. And only the highly ordered smectic-B phase provides DRs well above 50. On the basis of these results we developed high-contrast thin coatable polarizers based on smectic-B guest-host reactive LCs. Upon alignment of the host liquid crystalline di-acrylates, the guest dye molecules, due to their designed elongated structure, co-align along the director resulting in large absorption of the dye molecules, for light with the electrical-field vector parallel to the high-polarization axes of the dye molecules. Upon photopolymerization, the dichroic properties of the film are stabilized [63]. In order to induce higher ordered phases, the structure of the mesogenic group was further optimized. Compounds **58, 59**, and **60** shown in Figure 1.40 exhibit smectic phases. Furthermore, they also exhibit a nematic phase at higher temperatures, which is important to establish defect-free long range alignment, which is more difficult in the highly viscous smectic phase. The alignment is maintained during cooling of the smectic phase although the molecular arrangement and packing changes.

The synthesis of **58** is outlined in Figure 1.41 [64]. Key intermediate compound is **62**. This compound is used to couple the phenolic derivative **8** (see Figure 1.3) and the benzoic acid derivative **6** (see Figure 1.2) in the right manner. Alternatively, coupling of the two-ring acid **52** (see Figure 1.31) and **8** will lead to compound **58**.

The synthesis of **59** is outlined in Figure 1.42 [35]. The acid derived from the cyclohexane ring was obtained after hydrogenation of aromatic precursor **4** the synthesis of which is described in Figure 1.2. The phenolic compound **67** is made in a similar way as compound **64** (see Figure 1.41). Compound **60** was made in a similar manner as compound **3**. The hydroquinone used for the formation of **3** (see Figure 1.3) was replaced by 4,4′-dihydroxybibenzyl [65].

Polymer films made from compounds **58**, **59**, and **60** modified with dichroic dye **61** exhibit a DR of 10, 30, and 32 respectively. Consequently, the smectic-A phase here of **58** does not result in a high DR. In the case of the smectic-B compounds **59** and **60** the DR had increased considerably, but is still somewhat lower than those of our model compound shown in Figure 1.39 and in fact too low for high-end applications. It was observed that before polymerization the films prepared with **59** and **60** exhibit considerably higher DRs while upon polymerization a decrease of these values up to 50% was observed. One could speculate that the high cross-link density prohibits the highly ordered packing of the mesogenic cores. However, polymerization of a monoacrylate of similar chemical structure, that is, with similar mesogenic core and spacer between core and acrylate group, reveals that the order parameter also decreases substantially. Thus, it is an effect of the polymer formation and its steric influence on the packing of the mesogens rather than the formation of a high cross-link density.

A way of introducing highly ordered smectic phases, such as, the smectic-B phase, after polymerization is by increasing the length of the spacer. Thereto, the hexamethylene spacers of the compounds **1**, **58**, **59**, and **60** were replaced by undecamethylene

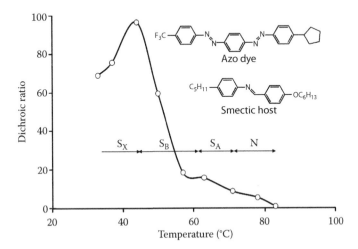

**FIGURE 1.39** Dichroic ratio of an azo dye guest compound (1 wt%) in a model LC host showing rich phase behavior. The guest–host system is uniaxially aligned in a glass cell and the absorbance parallel and perpendicular to a high-contrast polarizer are measured as a function of temperature. Large DRs are only obtained in the higher-ordered phases.

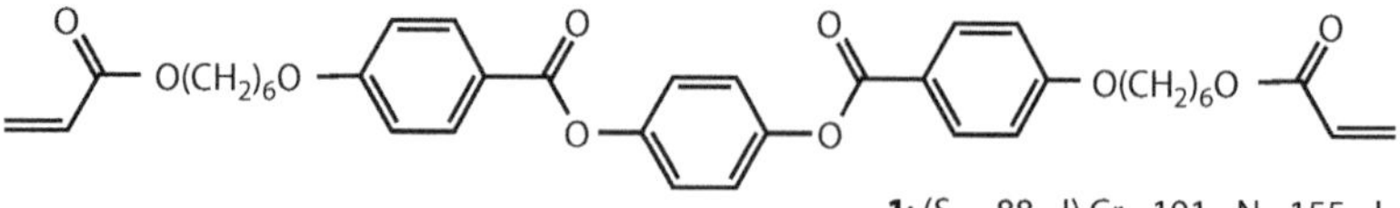

**1**: ($S_C$ - 88 - I) Cr - 101 - N - 155 - I

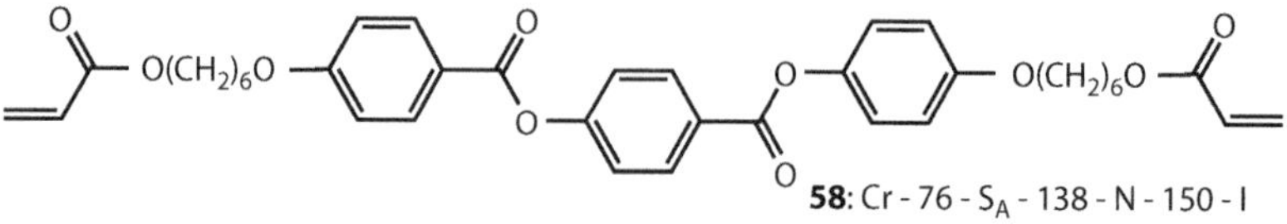

**58**: Cr - 76 - $S_A$ - 138 - N - 150 - I

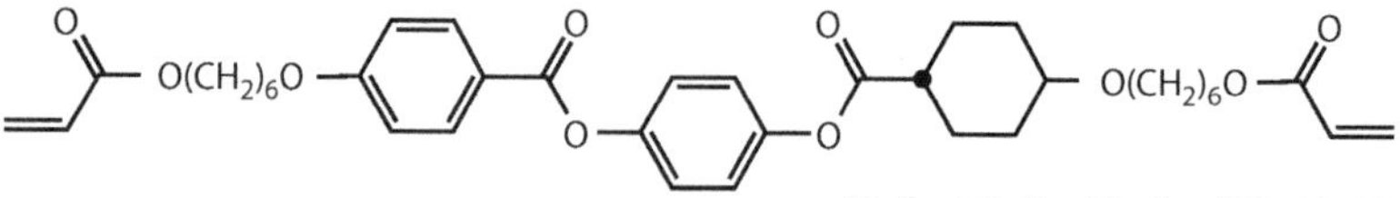

**59**: Cr - 54 - $S_B$ - 78 - $S_A$ - 102 - N - 121 - I

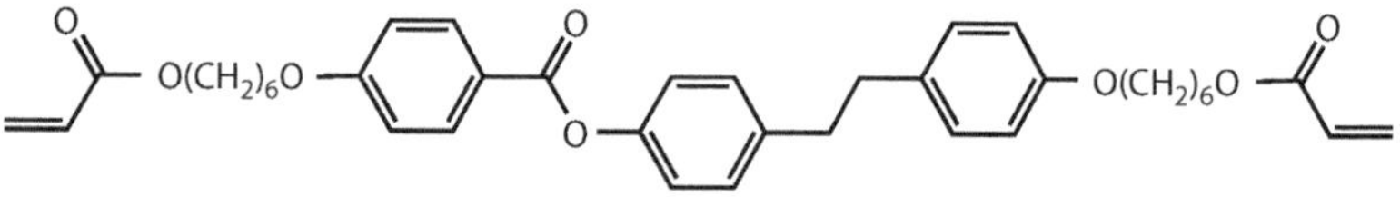

**60**: Cr - 72 - $S_B$ - 88 - $S_A$ - 105 - N - 107 - I

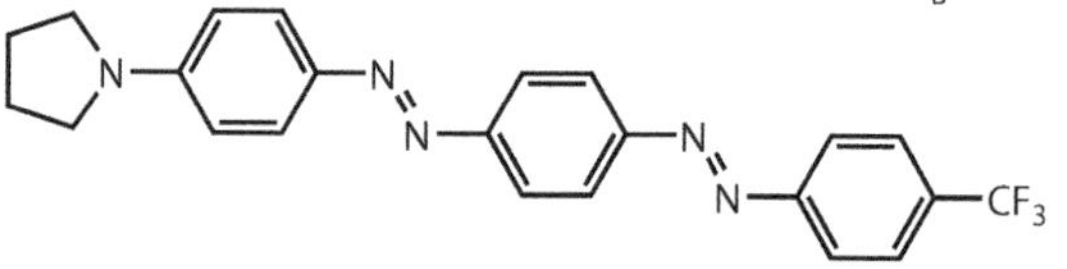

**61**

**FIGURE 1.40** Structure and phase transitions of 1,4-phenylene bis(4-(6-(acryloyloxy)hex-yloxy) benzoate) **1**, 4-((4-(6-(acryloyloxy)hexyloxy)phenoxy)carbonyl)phenyl 4-(6-(acryloy-loxy)hexyloxy) benzoate **58**, *trans*-4-(4-(6-(acryloyloxy)hexyloxy)cyclohexanecarbonyloxy) phenyl 4-(6-(acryloyloxy)hexyloxy) benzoate **59** and 4-(4-(6-(acryloyloxy)hexyloxy) phenethyl)phenyl 4-(6-(acryloyloxy)hexyloxy)benzoate **60**, and of dichroic dye 1-(4-((E)-(4-((E)-(4-(trifluoromethyl)phenyl)diazenyl)phenyl)diazenyl)phenyl)pyrrolidine **61**. (Cr = crystalline, $S_A$ = smectic-A, $S_B$ = smectic-B, $S_C$ = smectic-C, I = isotropic phase.)

spacers. In order to make these compounds, chlorohexanol was replaced by bro-moundecanol in various synthetic steps outlined in Figures 1.2, 1.37, and 1.38. In this way, a series of diacrylates **18, 70, 71**, and **72** is obtained as is presented in Figure 1.43.

Compound **18** exhibits a smectic-C phase and higher-order tilted phases that are not useful for the opted application. Compound **70** exhibits a smectic-B phase next to a smectic-A and a nematic phase and is potentially an interesting compound. The same is for **71**. Films made from these compounds exhibit a DR over 50 after polymer-ization, which means that the compound is very suitable for polarizer applications. Compounds **59** and **71** can be polymerized at 30°C in their supercooled state, leading to high-order parameter and related high DRs. Because the polymerization at this temperature does not reach full conversion, a post cure is needed at elevated tem-peratures. This can be done without affecting this high DR. Compound **60** exhibits a broad smectic-B phase. However, it lacks the nematic phase that is needed for easy

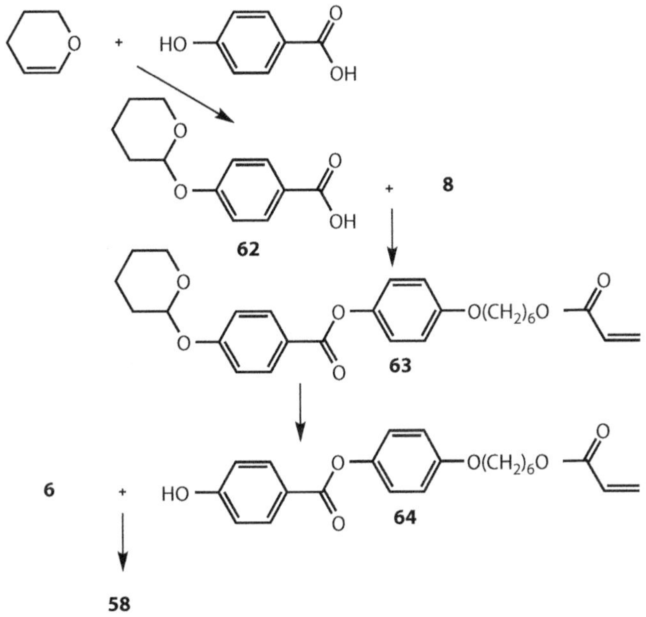

**FIGURE 1.41** Synthesis of 4-((4-(6-(acryloyloxy)hexyloxy)phenoxy)carbonyl)phenyl 4-(6-(acryloyloxy)hexyloxy)benzoate **58** (see Figure 1.40).

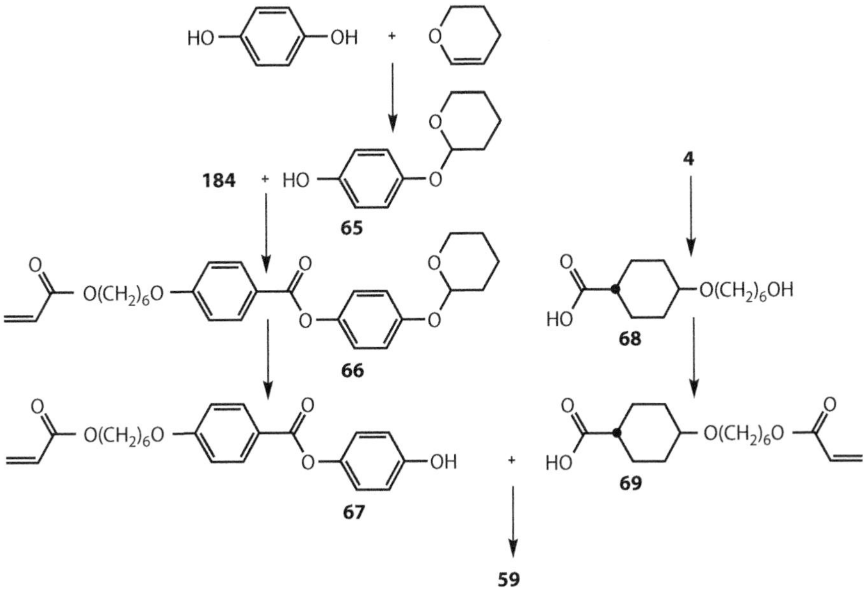

**FIGURE 1.42** Synthesis of *trans*-4-(4-(6-(acryloyloxy)hexyloxy)cyclohexanecarbonyloxy) phenyl 4-(6-(acryloyloxy)hexyloxy)benzoate **59** (see Figure 1.40).

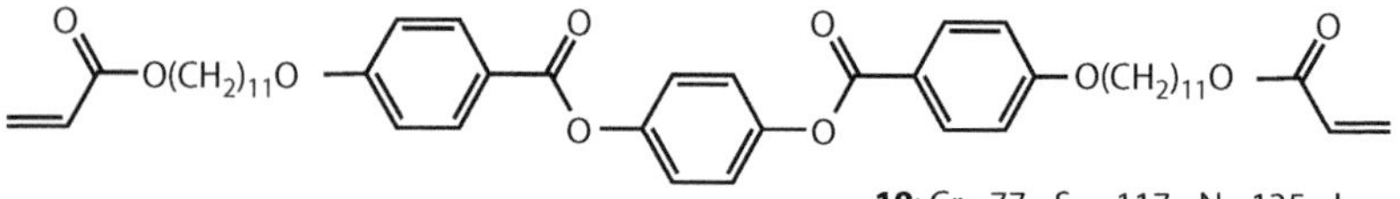

**18**: Cr - 77 - $S_C$ - 117 - N - 135 - I

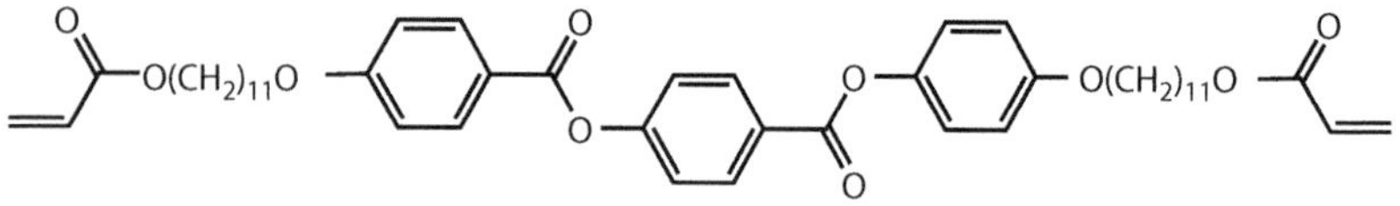

**70**: Cr - 64 - $S_B$ - 81 - $S_A$ - 131 - N - 134 - I

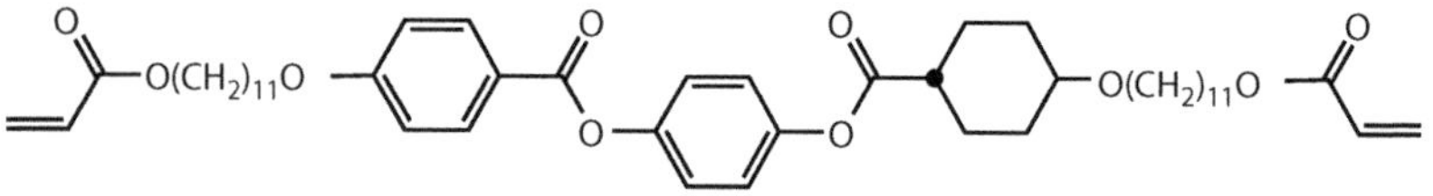

**71**: Cr - 46 - $S_B$ - 98 - $S_A$ - 112 - N - 114 - I

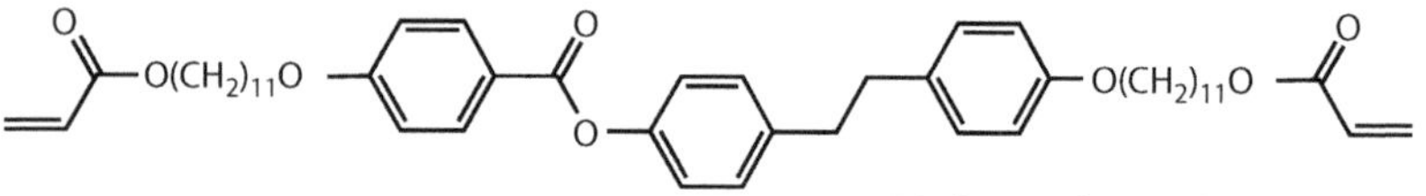

**72**: Cr - 63 - $S_B$ - 99 - $S_A$ - 103 - I

**FIGURE 1.43** Structure and phase transitions of 1,4-phenylene bis(4-(11-(acryloyl-loxy)undecyloxy) benzoate) **18**, 4-((4-(11-(acryloyloxy) undecyloxy)phenoxy)carbonyl) phenyl 4-(11-(acryloyloxy)undecyloxy)benzoate **70**, trans-4-(4-(11-(acryloyloxy)undecyloxy) cyclohexanecarbonyloxy)phenyl 4-(11-(acryloyloxy)undecyloxy)benzoate **71**, and 4-(4-(11-(acryloyloxy) undecyloxy)phenethyl)phenyl 4-(11-(acryloyloxy)undecyloxy) benzoate **72**. (Cr = crystalline, $S_A$ = smectic-A, $S_B$ = smectic-B, $S_C$ = smectic-C, I = isotropic phase.)

alignment and is therefore difficult to use. After mixing it with monoacrylates of similar structure, the nematic phase is retained that is needed for defect-free uniaxial planar alignment prior to polymerization. Also, these mixtures give polymerized films DRs higher than 50.

Films made from compound **71** show DRs > 70 before polymerization. An example is shown in Figure 1.44 where this compound is copolymerized in a 1:1 mixture with its monoacrylate equivalent in its smectic-B state. The use of longer spacers has a positive effect on the order parameter when compared to compounds that contain a hexamethylene spacer. Furthermore, upon polymerization the decrease in DR of the undecyl-derived compounds is much less dramatic. Figure 1.44 shows a 16% reduction of the DR. For reference, the reduction was 26% for the derivatives with the hexamethylene spacer. The longer spacer apparently decouples the order of the central aromatic core form the steric effects of the polymeric chains formed after cross-linking. We anticipate that the interaction of the aromatic cores with the dye molecules determine the order of the latter, which explains the improved DR. Thus, this thin film polarizer technology based on liquid crystalline (di-)acrylates exhibiting the smectic-B phase is highly promising and may prove to be an attractive alternative for traditional sheet polarizers in LCD applications.

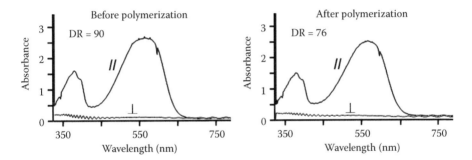

**FIGURE 1.44** Polarized UV/VIS absorption spectra parallel (normal curve) and, perpendicular (dotted curve) to molecular orientation of an LC-cell filled with a 1:1 mixture of C11 spaced LC mono- and di-acrylate mixed with 2% of the dichroic dye **61** measured before and after polymerization at 30°C.

## 1.5 CONCLUSIONS

It has been demonstrated that LC networks provide highly anisotropic polymer materials that are temperature stable and are therefore of interest for use in optical and mechanical applications. Their properties can be tuned to a large extent by optimizing their chemical structures. Localized photopolymerization using masks or dyes create patterns with control over the molecular order in all the three dimensions, which are useful as optical-retarder films in LC displays. The order parameter can be brought to high values, for example, needed to make polarization filters, by introducing a smectic-B phase next to a smectic-A, and a nematic phase. Chirality brought into the monomeric LCs introduces cholesteric phases which, when frozen-in in the polymer network, are applied as nonabsorbing reflective polarizer and color filters. Careful tuning of the chemical structure makes it possible to optimize on processing properties as well as on functional properties such as band width. By the introduction of photosensitive centers in the monomer the pitch can be lithographically adjusted such that patterned filters are made in a single-layer film. It is this special combination of access to the chemistry and molecular design of the LC monomers and knowledge of the structure–property relationships that could lead this to a large variety in applications and products.

## ACKNOWLEDGMENTS

We wish to express our sincere thanks to many of our colleagues at Philips Research who contributed to the results in this chapter. Our thanks to, among others, to Rifat Hikmet, Jannet van der Veen, Bauke Zwerver, Titie Mol, Peter van de Witte, Bianca van der Zande, Emiel Peeters, Wim Nijssen, René Wegh, Joost Vogels. In addition we are indebted to our many Masters, PhD, and postdoctoral students who spent some of their study time at Philips Research and performed experiments leading to the data presented in this chapter.

## REFERENCES

1. Broer, D.J., Boven, J., Mol, G.N., and Challa, G. 1989. *In situ* photopolymerization of oriented liquid-crystalline acrylates, 3. Oriented polymer networks from a mesogenic diacrylate, *Makromol. Chem.*, 190, 2255.
2. Broer, D.J., Hikmet, R.A.M., and Challa, G. 1989. *In situ* photopolymerization of oriented liquid-crystalline acrylates, 4. Influence of a lateral methyl substituent on monomer and oriented network properties of a mesogenic diacrylate, *Makromol. Chem.*, 190, 3201.
3. Broer, D.J., Mol, G.N., and Challa, G. 1991. *Insitu* photopolymerization of oriented liquid-crystalline acrylates, 5. Influence of the alkylene spacer on the properties of the mesogenic monomers and the formation and properties of oriented polymer networks, *Makromol. Chem.*, 192, 59.
4. Broer, D.J. 1993. Photoinitiated polymerization and crosslinking of liquid crystalline systems in *Radiation Curing in Polymer Science and Technology—Vol III, Polymerization Mechanisms*, (J.P. Fouassier and J.F. Rabek, Eds.), Chapter 12, pp. 383, Elsevier Science Publishers Ltd, London and New York.
5. Broer, D.J. and Heynderickx, I. 1990. Three-dimensionally-ordered polymer networks with a helicoidal structure, *Macromolecules*, 23, 2474.
6. Hikmet, R.A.M. and Zwerver, B.H. 1991. Cholesteric networks containing free molecules, *Mol. Cryst. Liq. Cryst.*, 200, 1997.
7. Kitzerow, H.S., Schmid, H., Ranft, A., Heppke, G., Hikmet R.A.M., and Lub, J. 1993. Observation of blue phases in chiral networks, *Liq. Cryst.*, 14(3), 911–916.
8. Lub, J., Broer, D.J., Hikmet, R.A.M., and Nierop, K.G.J. 1995. Synthesis and photopolymerization of cholesteric liquid crystalline diacrylates, *Liq. Cryst.*, 18, 319.
9. Broer, D.J. 1995. Creation of supramolecular thin film architectures with liquid-crystalline networks, *Mol. Cryst. Liq. Cryst.* 261, 513.
10. Wendorff, J.H. 1967. *Liquid Crystalline Order in Polymers*, (A. Blumstein, Ed.), Academic Press, New York.
11. DeGennes, P.G. 1969. Possibilites offertes par la reticulation de polymeres en presence d'un cristal liquide, *Phys. Lett.*, 28A, 725.
12. Strzelecki, L. and Liebert, L. 1973. Sur la synthèse de quelques nouveaux monomères mésomorphes. Polymérisation de la *p*-acryloyloxybenzylidène *p*-carboxyaniline, *Bull. Soc. Chim. Fr.*, 2, 597 and 605.
13. Bouligand, Y., Cladis, P., Liebert, L., and Strzelecki, L. 1974. Study of sections of polymerized liquid crystals, *Mol. Cryst. Liq. Cryst.*, 25, 233.
14. Clough, S.B., Blumstein, A., and Hsu, E.C., 1976, Structure and thermal expansion of some polymers with mesomorphic ordering, *Macromolecules*, 9, 123.
15. Arslanov, V.V. and Nikolajeva, V.I. 1984. Fixation of the structure of a liquid-crystalline monomer based on aromethine by polymerisation in the mesophase, *Vysokomol. Soedin., Ser. B*, 26, 208.
16. Shannon, P.J. 1984. Photopolymerization in cholesteric mesophases, *Macromolecules*, 17, 1873.
17. Broer, D.J., Finkelmann, H., and Kondo, K. 1988. *In situ* photopolymerization of oriented liquid-crystalline acrylates, 1. Preservation of molecular order during photopolymerization, *Macromol. Chem.*, 189, 185.
18. Broer, D.J., Mol, G.N., and Challa, G. 1989. *In situ* photopolymerization of oriented liquid-crystalline acrylates, 2. Kinetic aspects of photopolymerization in the mesophase, *Makromol. Chem.*, 190, 19.

19. Hoyle, C.E., Chawla, C.P., and Griffin, A.C. 1988. Photopolymerization of a liquid crystalline monomer, *Mol. Cryst. Liq. Cryst.*, 157, 639.
20. Hoyle, C.E., Chawla, C.P., and Griffin, A.C. 1989. Photoinitiated polymerization of a liquid crystalline monomer: Order and mobility effects, *Polymer*, 30, 1909.
21. Broer, D.J., Lub, J., and Mol, G.N. 1993. Synthesis and photopolymerization of a liquid-crystalline diepoxide, *Macromolecules*, 26, 1244.
22. Jahromi, S., Lub, J., and Mol, G.N. 1994. Synthesis and photoinitiated polymerization of liquid crystalline diepoxides, *Polymer*, 35, 621.
23. Johnson, H., Anderson, H., Sundell, P.E., Gudde, U.W., and Hult, A. 1991. Photoinitiated cationic bulk-polymerization of mesogenic vinyl ethers, *Polym. Bull.*, 25, 641.
24. Andersson, H., Gedde, U.W., and Hult, A. 1992. Preparation of ordered, crosslinked and thermally stable liquid crystalline poly(vinyl ether) films, *Polymer*, 33, 4014.
25. Hikmet, R.A.M., Lub, J., and Higgins, J.A. 1993. Anisotropic networks obtained by *in situ* cationic polymerization of liquid-crystalline divinyl ethers, *Polymer*, 34, 1736.
26. Lub, J., Omenat, A., Ruiz-Melo, A., and Artal, M.C. 1997. Synthesis and photopolymerization of cholesteric liquid crystalline vinyl ethers, *Mol. Cryst. Liq. Cryst.*, 307, 111.
27. Lub, J., Recaj, V., Puig, L., Forcén, P., and Luengo, C. 2004. Synthesis, properties and photopolymerization of liquid crystalline dioxetanes, *Liq. Cryst.*, 31, 1627.
28. Heynderickx, I. and Broer, D.J. 1991. The use of cholesterically ordered polymer networks in practical applications, *Mol. Cryst. Liq. Cryst.*, 203, 113.
29. Hikmet, R.A.M. and Broer, D.J. 1991. Dynamic mechanical properties of anisotropic networks formed by liquid crystalline acrylates, *Polymer*, 32, 1627.
30. Broer, D.J. and Mol, G.N. 1991. Anisotropic thermal expansion of densely crosslinked oriented polymer networks, *Polym. Eng. Sci.*, 31, 625.
31. Dewar, M.J.S. and Schroeder, J.P. 1965. *p*-Alkoxy- and *p*-Carbalkoxybenzoates of diphenols. A new series of liquid crystalline compounds, *J. Org. Chem.*, 30, 2296.
32. Lub, J., Van der Veen, J.H., Broer, D.J., and Hikmet, R.A.M. 2002. poly[1,4-di-(4-(6-acryloyloxyhexyloxy)benzoyloxytoluene]., *Macrom. Synth.*, (Ober, C.K and Mathias, L.J., Eds.), 12, 101.
33. Hikmet, R.A.M. and Lub, J. 1996. Anisotropic networks and gels obtained by photopolymerization in the liquid crystalline state: Synthesis and applications, *Prog. Pol. Sci.*, 21, 1165.
34. Van der Zande, B.M.I., Lub, J., Leewis, C.M., Steenbakkers, J., and Broer, D.J. 2005. Mass transport phenomena during lithographic polymerization of nematic monomers monitored with interferometry, *J. Appl. Phys.*, 97, 123519-1–123519-8.
35. Lub, J., Van der Veen, J.H., and ten Hoeve, W. 1996. The synthesis of liquid-crystalline diacrylates derived from cyclohexane units, *Recueil des Travaux Chimiques des Pays-Bas*, 115, 321–328.
36. Hikmet, R.A.M., Lub, J., and Maassen van der Brink, P. 1992. Structure and mobility within anisotropic networks obtained by photopolymerization of liquid crystal molecules, *Macromolecules*, 25(16), 4194.
37. Van de Witte, P., Stallinga, S., and Van Haaren, J.A.M.M. 2000. Viewing angle compensators for liquid crystal displays based on layers with a positive birefringence, *Jpn. J. Appl. Phys., Part 1*, 39, 101.
38. Van de Witte, P., Van Haaren, J., Tuijtelaars, J., Stallinga, S., and Lub, J. 1999. Preparation of retarders with a tilted optic axis, *Jpn. J. Appl. Phys., Part 1*, 38, 748.
39. Kűrschner, K. and Strohriegl, P. 2000. Polymerizable liquid crystalline twin molecules: Synthesis and thermotropic properties, *Liq. Cryst.* 27, 1595.

40. Kürschner, K., Strohriegl, P., Van De Witte, P., and Lub, J. 2000. Glass forming nematic win molecules synthesis and polymerization kinetics, *Mol. Cryst. Liq. Cryst.*, 352, 301.
41. Harris, K.D., Cuypers, R., Scheibe, P., Mol, G.N., Lub, J., Bastiaansen, C.W.M., and Broer, D.J. 2005. Molecular orientation control for thermal and UV-driven polymer MEMS actuators, *Proceedings of SPIE—The International Society for Optical Engineering*, 5836(Smart Sensors, Actuators, and MEMS II), 493.
42. Van der Zande, B.M.I., Roosendaal, S.J., Doornkamp, C., Steenbakkers, J., and Lub, J. 2006. Synthesis, properties, and photopolymerization of liquid-crystalline oxetanes: Application in transflective liquid-crystal displays, *Adv. Funct. Mat.*, 16(6), 791.
43. Guymon, C.A. and Bowman, C.N. 1997. Mechanisms of accelerated polymerization rate behavior in polyemer/smectic liquid crystal composites, *ACS Polymer Preprints*, 38, 292.
44. Van Oosten, C.L., Bastiaansen, C,W.M., and Broer, D.J., Polymer MEMS, In *Cross-Linked Liquid Crystalline Systems: From Rigid Polymer Networks to Elastomers*, Chapter 9, 247–281, Taylor & Francis, New York.
45. Blom, S.M.P., Huck, H.P.M., Cornelissen, H.J., and Greiner, H. 2002. Towards a polarized light-emitting backlight: Micro-structured anisotropic layers, *J. Soc. Inf. Display*, 10(3), 209.
46. Stapert, H.R., del Valle, S., Verstegen, E.J.K., Van der Zande, B.M.I., Lub, J., and Stallinga, S. 2003. Photoreplicated anisotropic liquid-crystalline lenses for aberration control and dual-layer readout of optical discs, *Adv. Funct. Mat.*, 13(9), 732.
47. Roosendaal, S.J., Van der Zande, B.M.I., Nieuwkerk, A.C., Renders, C.A., Osenga, J.T.M., Doornkamp, C., Peeters, E., Bruinink J., and van Haaren, J.A.M.M. 2003. Technologies towards patterned optical foils, *SID Symposium Digest Tech Papers*, 78, (Baltimore, USA, meeting 20–22 May 2003).
48. Van der Zande, B.M.I., Doornkamp, C., Roosendaal, S. J., Steenbakkers, J., Hoog, A. Op't; Osenga, J.T.M., Van Glabbeek, J.J., et al. 2005, Technologies towards patterned optical foils applied in transflective LCd's, *J. Soc. Inf. Display*, 13(8), 627.
49. Van Der Zande, B.M.I., Lub, J., Verhoef, H.J., Nijssen, W.P.M., and Lakehal, S.A. 2006. Patterned retarders prepared by photoisomerization and photopolymerization of liquid crystalline films, *Liquid Crystals*, 33(6), 723–737.
50. Van de Witte, P., Neuteboom, E.E., Brehmer, M., and Lub, J. 1999. Modification of the twist angle in chiral nematic polymer films by photoisomerization of the chiral dopant., *J. Appl. Phys.*, 85(11), 7517.
51. Kuball, H.G. and Höfer, T. 2001. *Chirality in Liquid Crystals* (Kitzerow, H.S. and Bahr, C., Eds.), Springer-Verlag, New York, p. 74.
52. Lub, J., Broer, D.J., Van der Veen, J.H., and Gracia Lostao, A.I. 2002. poly[1,4-di-(4-(6-acryloyloxyhexyloxy)benzoyloxytoluene-*co*-(*S*, *S*)-(-)-1,4- poly[1,4-di-(4-(6-acryloyloxy-3-methylhexyloxy)benzoyloxybenzene]., *Macrom. Synth.* (Ober, C.K., and Mathias, L.J., Eds.), 12, 109.
53. Lub, J., Ferrer, A., Larossa, C., and Malo, B. 2003. Synthesis and properties of chiral stilbene diacrylates, *Liquid Crystals*, 30(10), 1207.
54. Van de Witte, P., Galan, J.C., and Lub, J. 1998. Modification of the pitch of chiral nematic liquid crystals by means of photoisomerization of chiral dopants, *Liquid Crystals*, 24, 819.
55. Meyer, F., Ishida, H., and Schuhmacher P. 1998. Polymerisierbare chirale Verbindungen und deren Verwendung, *DE Patent*, 19, 843, 724.
56. Lub, J., Ezquerro, M.P., and Malo, B. 2006. Photo-isomerizable derivatives of phenylethanediol and cinnamic acid: Useful compounds for single-layer R, G, and B cholesteric color filters, *Mol. Cryst. Liq. Cryst.*, 457, 161.

57. Broer, D.J., Lub, J., and Mol, G.N. 1995. Wide-band reflective polarizers from cholesteric polymer networks with a pitch gradient, *Nature*, 378, 467.
58. Broer, D.J., Mol, G.N., Van Haaren, J.A.M.M., and Lub, J. 1999. Photo-induced diffusion in polymerizing chiral-nematic media, *Adv. Mat.*, 11(7), 573.
59. Lub, J., Van de Witte, P., Doornkamp, C., Vogels, J.P.A., and Wegh, R.T. 2003. Synthesis and properties of photoisomerizable derivatives of isosorbide and their use in cholesteric filters, *Adv. Mat.*, 15, 1420.
60. Lub, J., Broer, D.J., Wegh, R.T, Peeters, E., and van der Zande, B.M.I. 2005. Formation of optical films by photo-polymerisation of liquid crystalline acrylates and applications of these films in liquid crystal display technology, *Mol. Cryst. Liq. Cryst.*, 429, 77.
61. Lub, J., Nijssen, W.P.M., Wegh, R.T., Vogels, J.P.A., and Ferrer, A. 2005. Synthesis and properties of photoisomerizable derivatives of isosorbide and their use in cholesteric filters, *Adv. Funct. Mat.*, 15, 1961.
62. Lub, J., Nijssen, W.P.M., Wegh, R.T., De Francisco, I., Ezquerro, M.P., and Malo, B. 2005. Photoisomerizable chiral compounds derived from isosorbide and cinnamic acid, *Liquid Crystals*, 32, 1031.
63. Peeters, E., Lub, J., Steenbakkers, J.A.M., and Broer, D.J. 2006. High-contrast thin-film polarizers by photo-crosslinking of smectic guest–host systems, *Adv. Mat.*, 18(18), 2412.
64. Hikmet, R.A.M., Lub, J., and Tol, A.J.W. 1995. Effect of the orientation of the ester bonds on the properties of three isomeric liquid crystal diacrylates before and after polymerization, *Macromolecules*, 28(9), 3313.
65. Pinol, R., Lub, J, Garcia, M.P., Peeters, E., Serrano, J.-L., Broer, D.J., and Sierra, T. 2008. Synthesis, properties, and polymerization of new liquid crystalline monomers for highly ordered guest-host systems, *Chemistry of Materials*, 20(19), 6076.

# 2 Spatially Ordered Polymers Self-Assembled in Ordered Liquid Crystal Templates

*Liang-Chy Chien, Carmen Otilia Catanescu, and Lanfang Li*

## CONTENTS

## 2.1 INTRODUCTION

Among the numerous new anisotropic materials studied in the past two decades, liquid crystals (LCs) and polymer composites constitute the most important segment because it is a unique class of materials that have led to the development of devices, such as privacy windows, spatial light modulators, three-dimensional displays, and large-area flat-panel displays (Doane, 1990; Broer, 1990; West, 2003). Depending on the composition of the LC and the polymer, in general, the composites with polymer as the majority of the two components are commonly described as polymer-dispersed liquid crystals (PDLCs). It has been well established that polymer morphology largely results from phase separation during polymerization. The phase separation is largely influenced by the types of matrices and LCs, concentration of the components, and the rate of temperature ramping, solvent evaporation, and gelation of the matrix material (Drzaic, 1995). The polymers used in forming PDLCs are mostly isotropic, and universally, the reported polymer morphologies are phase-separated LC droplets encapsulated by a polymer matrix or interconnected networks. Because of high prepolymer content in the mixture, normally greater than 15% by the weight of dissolved LC, the phase separation of the polymer and the LC in PDLC systems in general, follows the spin-nodal decomposition mechanism.

Polymer (or polymer-network) stabilized liquid crystals (PSLCs) are distinctive in their morphological nature from those of PDLCs. The polymers or networks formed in such systems are mostly obtained from bifunctional mesogenic monomers. These polymers are generally dispersed in a small amount (typically between 1 and 10 wt%) in conventional LCs to improve the electro-optic properties of the existing LC matrix. The merit of such a composite system is more apparent when the polymer concentration lies at the low end of PSLC concentration spectrum, where the issue of refractive indices matching between the polymer and the LC becomes negligible. Increasing the polymer concentration in PSLC causes an increase in the response times and a deteriorated optical property by inducing-light scattering. The low-content polymer suspended in the LC host was initially described as a gel (Hikmet, 1990). Those polymers (or networks) that are linked to both the substrates are described as PSLCs (Yang et al., 1992; Chien, 1995; Crawford and Zumer, 1996), in which the monomers are dissolved in a low-molar-mass LC, aligned in a suitable orientational state either by surface alignment induced by an electric field or photopolymerized at the ordered state. The polymer-modified twisted nematic devices have shown significantly lower threshold voltage. Moreover, these polymer networks have eliminated the striping texture at high twist states of supertwisted birefringent-effect devices or supertwisted-nematic devices (Bos, 1993). Polymer-stabilized cholesteric textures (PSCT) whose planar state reflects across the entire visible spectrum thus, producing a black-on-white or color image have also been developed (Yang, 1991; Yang et al., 1992; Agaski et al., 1998). Most of the PSLC devices suggest that the performance of these devices relates to the morphology of the polymer network. Recently, the morphologies of PSLCs and their evolution during photopolymerization were reported (Rajaram, 1995; Kang et al., 2001, 2002; Catanescu and Chien, 2005; Kang and Chien, 2007).

The objective of this work is to demonstrate the effect of photopolymerization on the morphology of PSLCs, and its consequence on the formation of functional materials. We found that well-organized polymer walls were formed throughout the whole sample (in the shape of a thin film). It is understood that polymer morphology largely results from phase separation during polymerization, and that many parameters, such as types of matrices and LCs, and the concentrations of the components, influence the phase-separation process. We also found that the morphology of the ordered polymer structures provides a means of "imaging" important and potentially novel aspects of the pattern-forming LC states. Thus, to understand why we obtained those well-organized polymer walls the relationship between polymer morphology and photopolymerization conditions were investigated by carefully controlling parameters such as UV intensity, concentration of components, curing temperature, and the mesophase of the LC host.

## 2.2 LC INFLUENCED SELF-ASSEMBLY IN MESOGENIC POLYMER AND LC COMPOSITES

The system under investigation is a guest-reactive monomer in LC hosts. Nematic LC molecules have an anisotropy shape that are mostly elongated rod-like structures. The molecular packing through van der Waals and dipolar interactions between molecules make the LC bulky that have a partial long-range order. It can flow, that means the system has no positional lattice as those molecules in a crystal structure, but it has an orientational order. The average orientation direction of molecules is referred to as the LC director. From a hydrodynamic point of view, the nematic LC has an anisotropic flow property in that the viscosities are different in different flow directions with respect to the LC director. At the microscopic scale, the diffusing constant is different in different directions. In an electro-optical cell with a homogeneous alignment there is equilibrium in the molecular orientation—the homogeneous director configuration. When subject to an inhomogeneous surface anchoring field, elasticity of nematic LC assures that the LC local director does not vary more than the extrapolation length, $l_e$, which is the ratio of the LC elastic constant and strength of the surface anchoring field. The typical value of $l_e$ is about 1 $\mu$m. Any deviation of the director configuration from a uniform alignment will always result in an increase in free energy and the system will have the tendency to return to an undistorted state. Similarly as in a crystal, defects exist and are stable in a LC. At the locations where director configuration has no definition, or, is singular, a defect will appear at those areas. To remove defects, LC molecules need to be aligned so that the anisotropy is not diminished by the random orientation of molecules.

Having this background information let us return to the composite system; it is important to understand how LCs will behave with dopant molecules. In this case, the dopants or inclusions are reactive-monomer molecules. During polymerization, the monomer molecules grow into oligomers, and at the late stage of polymerization the oligomers grow into high-molar mass molecules (polymer) within the host LC. If a multifunctional mesogenic monomer is used in the composite, a cross-linked polymer (three-dimensional polymer network) will form in the host LC. In the case of isotropic inclusions, they would just cause a distortion in the LC and would prefer to stay either at an already distorted region, or even at the defect area in order to minimize the energy of the system. When the monomer is a mesogenic monomer, which has a similar molecular structure as that of the LC, it thus resembles the LC and will integrate as a part of the LC host with their molecular long axis aligned in the same direction as the LC director prior to polymerization. Thus, the mesogenic monomer molecules have the potential to form an ordered structure.

To demonstrate the formation of an ordered polymer network in the host LC, one may perform photopolymerization of the mesogenic monomer with multiple polymerizable groups such as acrylates, epoxides, oxetanes, or vinyl ethers in LCs. Without minimum disruption of the LC alignment well-organized polymer network or walls can be obtained throughout the cell by the self-assembly of the polymer within the anisotropic host—the LC. The polymerization conditions, which were parameterized as UV intensity, photoreactivity (characterized by photoinitiator concentration), curing temperature, monomer concentration, and mesophase of the LC host, were systematically varied. Different LC hosts have also exhibited a difference in polymer network microstructures. The phase-separated polymer morphology arises from the spinodal decomposition and the anisotropic elastic property of the LC host, induced diffusion of LC and monomer molecules during polymerization.

### 2.2.1 SAMPLES PREPARATION

Samples were prepared according to the following procedures. Mesogenic diacrylate monomer RM257 obtained from Merck was completely dissolved in a nematic LC, selected from 8CB, BDH 13739, TL213, E31, BL006, or MLC6815, by ultrasonification in dichloromethane followed by the evaporation of the solvent under reduced pressure for 2 days. Photoinitiator IRG 651 from Ciba Additive, Inc., which works with long wavelength UV irradiation, was mixed at the same time. These solutions were capillary filled into the cells, which were assembled from two pieces of ITO-coated glass substrates; their inner surfaces were prior coated with polyimide PI 2555 (Nissan Chemical Ltd.) as the alignment layer. The cells were assembled (inside a clean room facility) in an antiparallel manner with respect to the rubbing direction of the alignment layers and separated with fiber spacers to obtain a uniform cell gap of 15 μm. These cells had a thickness variation within 0.2 μm and gave a planar alignment. The cells were filled with the mixtures and put in a UV lamp chamber, as shown in Figure 2.1, which was equipped with a hot stage at the bottom to maintain a constant temperature (with a temperature variation within 0.5°C). Long wavelength UV irradiation (30 min, peak at 365 nm, 0.78 mW cm$^{-2}$) was used to initiate the polymerization. UV intensity was adjusted by placing UV filters between the UV light source and the cells and measured by photometer (detector sensitive to 365 nm UV).

### 2.2.2 POLYMER MORPHOLOGY

It is found that for composites comprising of mesogenic monomer RM257 and low-molecular-weight nematic LCs, a monomer concentration of 5% is sufficient to form well-structured polymer walls using LC hosts to work as templates. Two complementary approaches were used in studying the polymer morphology in these composites.

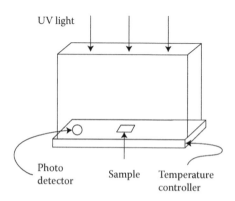

UV light

Photo detector          Sample          Temperature controller

**FIGURE 2.1**  Apparatus for sample preparation.

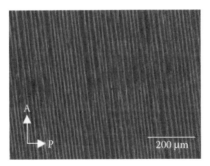

**FIGURE 2.2**  POM picture of polymer structure.

Right after polymerization, the cells were observed by a polarizing optical microscopy (POM) under crossed polarizers. The polymer structure looked darker than the LC under POM if the rubbing direction of every cell is 45° with respect to the polarizers. This observation indicates that the resulting polymer structure is birefringent and possesses less birefringence than the surrounding LC. When the LC director is at a small angle deviated from crossed polarizers, dark stripes with almost equal distance between two strips throughout the cell are observed, as shown in Figure 2.2.

The polymer morphology is studied by using a scanning electron microscope (SEM), which allows the observation of polymer microstructure at higher resolution with a cross-sectional view. For this purpose, the LC should be removed from the cells to allow the observation of the bare-polymer microstructure. The cells were immersed in a mixture of 20% dichlomethane and 80% hexane as solvent for 3 days to allow complete removal of the LCs. The solvent was refreshed each day to ensure the complete removal of the LCs knowing that the LCs may be encapsulated by the polymer network. The solvent was slowly evaporated at a reduced pressure from the cells to prevent the possible damage of the polymer network. Subsequently, the cells were carefully opened up using a razor blade to protect as much as possible the polymer structure. For a cross-sectional view, the cells were first scribed and then broken along the scribing line, while in the liquid nitrogen, to obtain a clear cut of the cross section. The samples then were sputtered with a thin layer of gold to facilitate SEM observation. The SEM analysis of the cells showed that well-organized polymer walls with an almost equal distance were formed, as shown in Figure 2.3.

Polymer walls were formed along rubbing direction, which is also the direction of the LC's director orientation. The mechanism leading to this kind of morphology is understood that during polymerization, as the molecular weight of the oligomers increases, the solubility of the prepolymer in LC decreases dramatically. The attribute of monomer having a nematic phase tends to make monomer molecules to arrange themselves along the LC director. So, in the early stages of polymerization the fibers will grow along the LC director, with active sites mostly at the two ends of the polymer fiber. The polymers start to phase separate or precipitate from the host LC when their molecular weights exceed a critical value. In this situation LC, the major component and actually elastic, will exert anisotropic force upon these relatively low-molecular-weight polymers, and will macroscopically facilitate the monomer

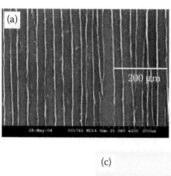

**FIGURE 2.3** SEM observation of polymer morphology: (a) BDH system, (b) 8CB system, and (c) cross section of BDH system.

molecules to diffuse and aggregate at an anisotropic diffusing rate at the polymer active chain ends. For a molecule with an elongated shape moving in a media also made up of elongated molecules, it is easier to diffuse logitudinally with respect to molecular orientation, thus polymer walls are formed along the LC director.

Based on these experiments, it is obvious that different LC hosts give different polymer morphologies because of their distinct physical characteristics, such as order and viscosity. The only thing in common for all polymer structures obtained in the conditions previously specified is that they form ordered fibers or polymer walls along the LC director orientation. The difference in these systems is the length of the polymer fibers (or walls). BDH 13739 and MLC6815 gave the longest polymer strips, almost throughout the cell, while other LCs gave much shorter fibers. The reason for the discrepancy in fibril length is not yet well understood, but it might possibly be related to the elastic properties and the defects of the host LC that resulted from alignment. The polymerization process was carefully controlled to explore the relationship between the polymer structure and polymerization condition. We were interested to study the influence of different parameters on the morphology of polymer walls such as the UV intensity (which is supposed to influence the rate of polymerization), the concentration of monomer (which will determine the final polymer density), and photoinitiator concentration (which is an index of the reactivity of the reactants at early, middle, and final stages of polymerization process, respectively). As one can see from Figure 2.4, the distance between the two polymer walls decreases dramatically with polymer concentration at very low monomer concentration (lower than 2%), and become approximately linear at concentrations higher than 3%. Actually polymer walls can even be seen by bare eyes at as low as 1% monomer concentration.

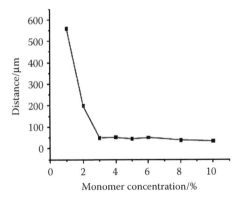

**FIGURE 2.4**  Wall distance versus polymer concentration.

In reference to Figure 2.5, the distance between the polymer walls decreases with the increase in the photoinitiator concentration. Higher photoinitiator concentration will result in more active sites, and thus facilitate the aggregation of polymers because the monomer molecules do not need to diffuse to longer distances. The relationship between UV irradiation and polymer wall distance was also observed. UV irradiation is related to the reactivity of early stage polymers. As the reaction goes on, the polymerization turns from reaction-limited process to a diffusion-limited one, in later stages of network formation, which does not depend much on UV intensity. Thus, higher UV intensity will create more reactive sites, provided there are enough photoinitiator molecules, and therefore will result in shorter distances between fibers. In Figure 2.6 a phenomelogical trend of decreasing polymer wall distance as a function of increasing UV intensity is observed. Figure 2.7 shows the effect of temperature on the polymer structure. The morphology of the polymer network, as a result of phase separation, diffusion, and aggregation of low-molecular-weight polymer, is a thermally sensitive process. Data obtained from our observation in the temperature interval indicates that the distance between fibers increases with temperature. This

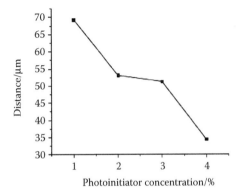

**FIGURE 2.5**  Wall distance versus photoinitiator.

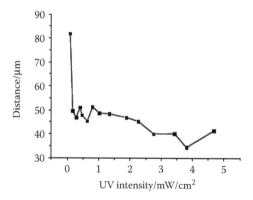

**FIGURE 2.6**   Wall distance versus UV intensity.

is a result of temperature-dependent diffusion. Higher temperatures will endow the oligomer–prepolymer active sites with higher mobility, so that they can diffuse to a longer distance.

From the experimental results we found that the polymer wall distance decreases with high polymer concentration and the increase in photoinitiator concentration and UV intensity, and the distance between polymer walls increases as the temperature increases. The relationship between the variables and the polymer wall distance has been investigated and the results are summarized in Table 2.1.

### 2.2.3   ELECTRO-OPTICAL PROPERTIES

One of the many important applications of polymer stabilized LCs is in electro-optical devices. Frederick's transition voltages of the cells, which are typically the working voltages for PSLC devices and also an indicator of how strong a field the polymer structure exerted on LC molecules, are measured. In this study, exemplary LCs 8CB and BDH were used for electro-optical studies, and the experiments were

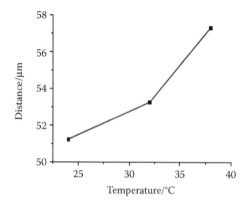

**FIGURE 2.7**   Wall distance versus temperature.

**TABLE 2.1**

**The Distance of Polymer Walls Formed in Different LCs**

| LC Host | $\Delta n$ | $\Delta\varepsilon$ | Observations | Distance between the RM257/TEGDMA Polymer Walls (μm) | |
|---|---|---|---|---|---|
| | | | | 25 μm Cell | 10 μm Cell |
| BDH 13739 | 0.1410 | — | Polymer network almost equally distributed on both substrates | 80–140 | 15–30 |
| TL213 | 0.239 | 5.7 | Polymer network mostly on top of the substrate; good film forming properties | 35–110 | 15–20 |
| E31 | 0.18 | 14.8 | Polymer network mostly on top of the substrate | 18–40 | 8–12 |
| BL006 | 0.2860 | 17.3 | Polymer network mostly on top of the substrate; good film forming properties | 15–40 | 5–12 |
| MLC 6815 | 0.0517 | — | Polymer network almost equally distributed on both substrates; good film forming properties | 50–110 | 25–40 |

performed in the nematic phase. In the case of BDH the experiments were carried out at room temperature, while for 8CB the electro-optical properties were measured in the nematic phase by heating the cells to 38°C. Figure 2.8a shows the relationship between the threshold voltage of Frederick's transition and polymer concentration of three cells containing 8CB at constant 3% photoinitiator with respect to polymer concentration and constant UV illumination (0.78 mW cm$^{-2}$). The field strength that is needed to rearrange 8CB LC molecules in a polymer network increases exponentially with the increase in polymer concentration. But at the concentration range from 1% to 6% there is approximately a linear relationship between Fredersicks' threshold voltage and polymer concentration for the 8CB system. In the cells with BDH LC host (Figure 2.8b) the field strength required to rearrange LC molecules in a polymer network is almost linearly proportional to the polymer concentration.

The relationship between photoinitiator concentration and Frederick's transition voltage is also explored and shown in Figure 2.9. For BDH, one can see that there appears to be a maximum voltage (Figure 2.9b), which coincides with the minimum polymer wall distance. This can be understood as the polymer network anchoring effect for LCs. A denser (in space) polymer network will have a stronger equivalent field exerted on the LC alignment.

To further explore the relationship between curing temperature and threshold voltage of Frederick's transition, the curing temperatures of the diacrylate was chosen to cover the three phases such as smectic A, nematic, and isotropic phases. This study is expected to provide insight information into how the LC phase will influence the polymer network. In the case of BDH, the chosen temperatures are always in a nematic phase. With that, one can obtain the information of the nematic order influence on the polymer network formation with minimum deviation. From the experimental results we can see that the voltage increases as the temperature goes up very slowly for 8CB

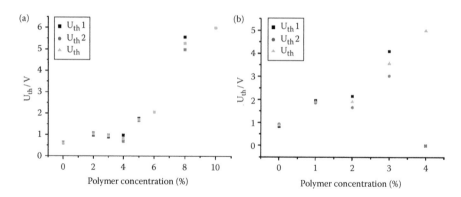

**FIGURE 2.8** Threshold voltage for Frederisck's transition versus polymer concentration of (a) 8CB and (b) BDH system with 3% photoinitiator by the weight of polymer and under UV irradiation for 30 min (365 nm, 0.78 mW cm$^{-2}$).

(Figure 2.10a). The observed fibers are not well shaped, relatively short and ordered in a poorly interconnected network. This indicates that the decreasing of LC order gives a more random polymer network and thus a higher operation voltage. In the case of the BDH system (Figure 2.10b), which remains in nematic phase in the whole temperature interval, the voltage decreases as the temperature increases. For the BDH system, the LC host was divided by the polymer fibers into domains. The bigger these domains are, the less voltage is needed, and a higher curing temperature gives rise to larger LC domains.

The relationship between UV intensity during curing process and Fredersick's transition voltage is also studied. The rate of phase separation increases as the UV intensity increases, which give rise to the shorter and denser polymer fibrils. Figure 2.11 shows the threshold voltage versus UV intensity of 8CB and BDH systems. The threshold voltages were found to increase with the increase in UV intensity regardless of the

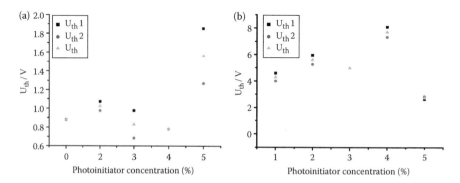

**FIGURE 2.9** Threshold voltage of Frederisck's transition versus photoinitiator concentration for (a) 8CB and (b) BDH system with 4% polymer and under 365 nm, 0.78 mW cm$^{-2}$ UV irradiation for 30 min.

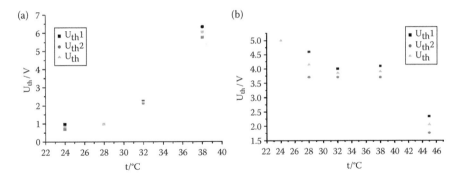

**FIGURE 2.10** Threshold voltage of Frederisck's transition versus curing temperature of (a) 8CB and (b) BDH system with 4% polymer, 3% photoinitiator with respect to polymer, and under 365 nm, 0.78 mW cm$^{-2}$ UV irradiation for 30 min.

LC host. From the morphology study, the trend is approximately the opposite to that of the trend seen in the polymer wall distance versus UV intensity for BDH system.

In summary, the threshold voltage is found to increase with increased polymer and photoinitiator concentration, respectively. The threshold voltage increases very slowly for 8CB-polymer system as the curing temperature rises, while for BDH system the threshold voltage decreases as the curing temperature increases. The morphological studies reveal that increase in polymer and photoinitiator concentration as well as an increase in UV intensity results in smaller distance between polymer walls. The formation of a fiber-like polymer network and the shorter distance between polymer walls results in a lower threshold voltage. However, this is not conclusive for interconnected polymer structure (as we have seen in 8CB).

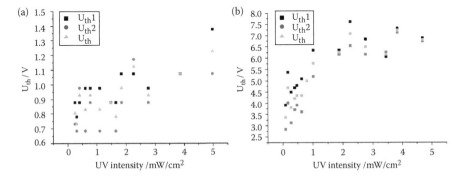

**FIGURE 2.11** Threshold voltage of Frederick's transition versus UV irradiation intensity of (a) 8CB and (b) BDH system with 4% polymer, 3% photoinitiator with respect to polymer and under 365 nm, 0.78 mW cm$^{-2}$ UV irradiation for 30 min at 38°C for 8CB and 24°C for BDH, respectively.

## 2.3    PATTERNING OF CONDUCTIVE POLYMERS USING LC TEMPLATES

Since the discovery of conducting polymers almost three decades ago, many research groups devoted their attention to the field of conductive polymers because of the promise of achieving a new generation of polymers that have a unique combination of properties not available in any other known materials. Conjugated polymers are typically semiconductors. They exhibit the electrical or optical properties of metals or semiconductors and retain the attractive mechanical properties and processing advantages of polymers. Disorder, however, limits the carrier mobility and, in the metallic state, limits the electrical conductivity. Therefore, research directed toward conjugated polymers with improved structural order and hence higher mobility is a focus of current activity in the field. A novel approach for the synthesis of semiconductive polymers is the template synthesis of this type of polymers using LC materials as hosts for alignment purposes (Catanescu and Chien, 2005). LCs are high anisotropic materials with a state of aggregation that is intermediate between the crystalline solid and the amorphous liquid and are unique in possessing the unique properties of high density (promoting compact supramolecular assemblies) and fluidity (enabling self-assembly of molecular structures). Because the molecular ordering, and hence, supramolecular arrangements of these conductive polymers in a well-ordered fashion can be achieved in LC hosts, the patterned high performance polymers with metallic properties could lead to new applications of conductive polymers. Recently, there were research efforts devoted toward synthesis of conductive polymers confined within a mesophase. For example, Shirakawa et al. studied the liquid crystalline templating of the polyacetylenes in a nematic solvent (Shirakawa et al., 1988, 2001).

The aim of this study is to incorporate the conductive polymer of PEDOT in spatially ordered polymer networks using well-ordered LCs as the templates by subsequent *in situ* polymerization of the conductive monomer (EDOT) in the ordered matrix.

### 2.3.1    SAMPLES PREPARATION

In all cases, the LC networks were prepared from a mixture of a LC, diacrylate monomer, and Irgacure 651 photoinitiator in a ratio of 94.75:5:0.25 by weight percentages. The LC host was selected from nematic BDH, E44, 8CB (4-cyano-4'-octyl biphenyl)—a smectic A LC at room temperature, and DSCG—lyotropic nematic LC (disodium cromoglycate, 16% in water). The reactive diacrylate monomers RM257 (1,4-bis[3-(acryloyloxy)propyloxy]-2-methyl benzene), triethylene glycol dimethacrylate (TEGDMA), and tetraethylene glycol diacrylate (TEGDA) are commercially available from Merck and Aldrich, respectively. The mixtures were capillary filled into glass cells ($2 \times 2$ in.), whose inner surfaces were coated with DuPont polyimide PI 2555 as the alignment layers. The cells were assembled (inside a clean room facility) in an antiparallel manner with respect to the rubbing direction of the alignment layers and separated using fiber spacers to obtain a cell gap of 22 or $25 \,\mu$m. The diacrylate monomers in the smectic mixtures were photopolymerized (40 min under 365 nm of ultraviolet light and at an intensity of $0.4 \,\text{mW cm}^{-2}$) in the absence of an electric field. In the case of the cholesteric system, a square-wave voltage was applied across the cell ($2 \times 2$ in., antiparallel rubbed, $25\,\mu$m thickness) and an electric field of $0.27\,V\,\mu m^{-1}$ was kept constant during the photopolymerization of the diacrylate mixture. After the photopolymerization, the cells were immersed in a mixture of solvents (hexane/dichloromethane $= 70:30$ w/w) for several days in order to remove the LC host. The anisotropic and morphological properties of the LC network were studied by POM (Nikon OPTIPHOT2-POL) and SEM (Hitachi S-2600N). The polymer samples for SEM investigation had a vacuum deposited thin layer of gold for enhancing the contrast of images. A smectic A LC (8CB from Merck, 4-cyano-4'octyl-1,1'-biphenyl) and a cholesteric LC (by adding R1011 (0.2 wt%) as a chiral dopant to the nematic LC host BDH 13739, which resulted in a pitch of $10\,\mu$m) were used as the host for forming the porous-ordered polymer network. A solution of EDOT, catalyst (iron (III) p-toluenesulfonate (FepTS)), and dopant [dodecylben-zenesulfonic acid (DBSA) or polystyrene sulfonic acid (PSSA)] in methanol (15 wt% EDOT solution) was filled into the cell and left for 7 days to polymerize. The cell turned gradually to dark blue and after a week was opened and analyzed under the polarized microscope.

### 2.3.2 MORPHOLOGY OF TEMPLATED CONDUCTIVE POLYMERS

This LC templated synthesis method produces a very well-aligned conductive polymer as seen in the SEM images of the polymer. The measured conductivity was found to be uniform throughout the whole film, which implies that the preexisting LC polymer network acts as an ordered template for pattering the resulting conductive polymer at a submicron scale, but also allows it to form a continuous film (more like an interpenetrating polymer network) at a macroscopic scale because of its porous structure.

Figure 2.12 shows the morphologies of the templated conductive polymers prepared using LCs such as BDH, E44, 8CB, and DSCG. As shown in Figure 2.12, LCs exert anisotropic force upon these relatively low concentration prepolymers, and macroscopically facilitates the monomer molecules to diffuse and aggregate at the polymer-active chain ends.

As expected, the liquid-crystalline polymer network formed by photopolymerization of two diacrylates (RM257 and TEGDMA) within the smectic phase (Figure 2.13b and e–g) showed higher molecular orientation compared with the same polymer network obtained in a nematic phase (Catanescu, 2005). The POM images (Figure 2.13c and d) show the PEDOT polymer composite (obtained within the smectic phase) before and after removing the catalyst by rinsing with methanol. SEM images taken for the same samples reveal that the PEDOT not only covers the LC polymer network, but also it was formed inside the polymer network (Figure 2.13h).

In the case of the cholesteric system, large uniform domains were formed by applying a square-wave voltage across the cell after the cell was kept at 30°C on a hot stage for half hour (the fingerprint texture can be seen in Figure 2.14a). The fingerprint texture grew very slowly and after the process was completed within the whole cell, a UV lamp with peak at 365 nm, and intensity of $\sim 0.4\,mW\,cm^{-2}$ was used to polymerize the diacrylate system. The polymer network formed in these

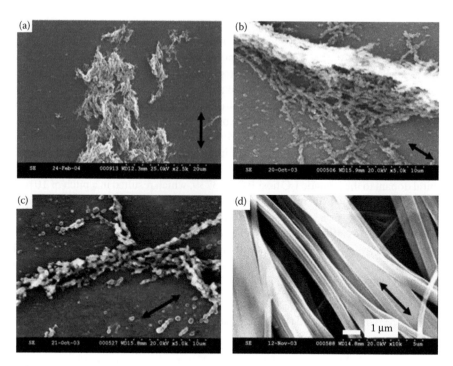

**FIGURE 2.12** The morphologies of templated conductive polymers on (a) BDH, (b) E44, (c) 8CB, and (d) DSCG. The double-headed arrows represent the uniaxially rubbed directions of polyimide alignment layer.

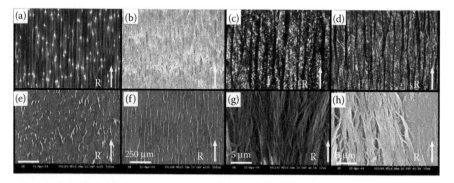

**FIGURE 2.13** POM and SEM images of the RM257/TEGDMA and PEDOT systems, respectively, polymerized within the smectic phase. (a) RM257/TEGDMA dissolved in 8CB, before polymerization (100 × magnification); (b) RM257/TEGDMA after polymerization in the smectic phase (100 × magnification); (c) PEDOT cell before rinsing (the catalyst can be clearly seen) (200 × magnification); (d) PEDOT cell after rinsing with water and methanol (200 × magnification); (e) SEM pictures of the RM257/TEGDMA and PEDOT systems–without PEDOT, (f) SEM pictures of the RM257/TEGDMA and PEDOT systems–with PEDOT; (g) Enhanced image of (e) (RM257/TEGDMA polymer network only); (h) PEDOT cell after rinsing.

conditions followed exactly the fingerprint texture of the cholesteric phase, as seen in Figure 2.14b.

While attempting to remove the LC host by immersing the whole cell in a mixture of solvents, and evaporating the solvent, the POM analysis revealed that the evaporation of the solvent and the subsequent opening of the cell caused a disruption of the polymer network, changing its original morphology. Therefore, the polymer structure of this sample recorded with SEM differs from a helical conformation that should have been normally seen. The polymer network is quite soft and the opening of the cell in addition to the evaporation of the solvent mechanically collapsed the whole structure having helical conformation into a linear one (Figure 2.14c–e).

Profilometer scans of the PEDOT/RM257/TEGDMA composite system obtained in two different conditions (cholesteric- or smectic-induced polymerization) gave additional information on the morphology of the composite system. As one may see from Figure 2.15 and Table 2.2, there are some differences between the two systems. The smectic LC host induced the formation of a polymer network having an average thickness of 1.8–1.9 mm, with the maximum height of the polymer walls ranging between 5.6 mm (on the bottom substrate) and 8.9 mm (on the top substrate), which is less than half of the thickness of the cell. This may suggest that the polymer network was formed primarily near and on the glass substrate and much less in the middle of the cell.

Because the cells were placed on an aluminum foil, the photopolymerization of the reactive monomer took place not only near the top substrate (the closest to the UV incident light), but also on the bottom substrate. In the case of the cholesteric type, because of the applied field that induced the formation of the fingerprint texture within the whole thickness of the cell, the reactive monomer followed the cholesteric helix and therefore, the resulting polymer network filled the entire cell; this supposition is confirmed by the profilometer scans, which shows that the height of the polymer walls is comparable with the thickness of the cell.

The conductivity measurements (that gave a value of $\sim$3S cm$^{-2}$) did not show any macroscopic anisotropy in the conductivity (similar results as in the case of the nematic samples). We assume that even though the PEDOT was aligned at a submicron scale by the preexisting polymer network, it formed a continuous film in both directions. It is also possible that the four-point probe that was aligned in a linear fashion (Pauw, 1958) could not very well detect the differences between the sheet resistance along and across the rubbing direction, while a square alignment of the probes (Montgomery, 1971) should be able to detect these differences. However, a correct characterization supposes the use of a probe with a distance between the tips of order of tenths of microns (Kanegawa et al., 2003). Our measurements used a four-point probe with the distance of 5 mm between the probes, probably too big to detect the anisotropy of the PEDOT polymer composite.

In conclusion, we demonstrated the preparation of spatially oriented conductive polymers in ordered-porous polymer networks. PEDOT formed either in a cholesteric- or smectic-templated polymer network did not exhibit a detectable macroscopic anisotropic conductivity, probably because the PEDOT was formed on large open surfaces and/or because of the system detection used for the measurement of the

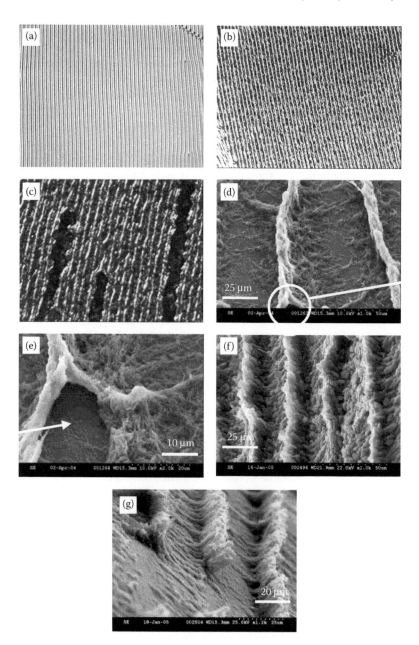

**FIGURE 2.14** POM and SEM images of the RM257/TEGDMA and PEDOT systems, respectively, the former one polymerized within the cholesteric phase. (a) fingerprint texture of the cholesteric phase before polymerization (100 × magnification); (b) fingerprint texture of the cholesteric phase after polymerization (100 × magnification); (c) the ordered network collapsed and coalesced during evaporation of the solvent (200 × magnification); (d) and (e) SEM images of the porous RM257/TEGDMA polymer network; (f) PEDOT polymer composite before rinsing with methanol; (g) PEDOT polymer composite after rinsing with methanol (40° tilt image).

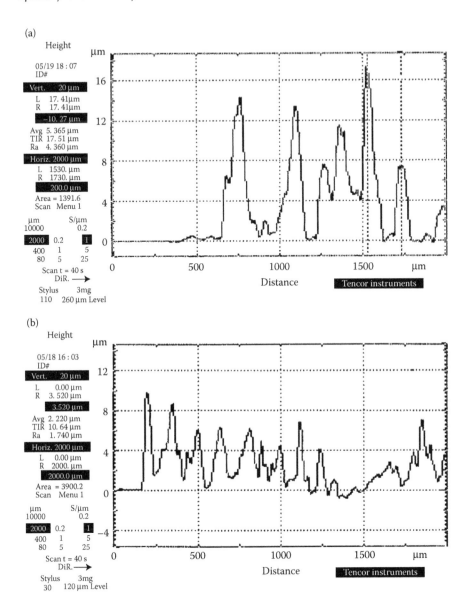

**FIGURE 2.15** The profilometer scans of the PEDOT/RM257/TEGDMA composite system obtained in (a) cholesteric or (b) smectic templated polymer network.

conductivity was not able to detect the anisotropy in conductivity. Future investigation in the improvement of the conductivity measurements along with a change in the elastomer composition and dopant to enhance the conductivity of the resulting composites may lead to a wide range of applications of such materials.

### TABLE 2.2
### Profilometer Data Analysis of the PEDOT/RM257/TEGDMA Composite System

|             | Top of the Substrate | | | Bottom of the Substrate | | |
|-------------|--------|-----|-----|--------|-----|------|
|             | **L**  | **d** | **h** | **L** | **d** | **h** |
| Cholesteric | 50–200 | —   | 25  | 60–200 | —   | 17.5 |
| Smectic     | 50–130 | 1.9 | 8.9 | 50–150 | 1.8 | 5.6  |

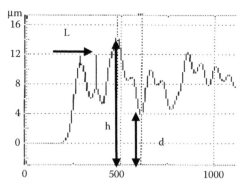

*Note:* L, average distance between polymer walls (in μm); d, average thickness of the bulk polymer network (in μm); h, maximum height of the polymer walls (in μm).

## ACKNOWLEDGMENTS

The LC-templated conductive polymers were supported in part by AFOSR/STTR. The authors would like to gratefully acknowledge Liou Qiu and the Doug Bryant of Clean Room Facility of Liquid Crystal Institute for their valuable help with the SEM studies and cells fabrication, respectively.

## REFERENCES

Akagi, K., Piao, G., Kaneko, S. Sakamaki, K. Shirakawa, H., and Kyotani, M. 1998. Helical polyacetylene synthesized with a chiral nematic reaction field, *Science*, **282**, 1783–1686.

Bos, P., Rahman, J., and Doane, J. 1993. A low threshold voltage polymer network TN device, SID Digest of Technical Papers, 24, 877.

Broer, D. J. and Heynderickx, I. 1990. Three-dimensionally-ordered polymer networks with a helicoidal structure, *Macromolecules*, **23**, 2474–2477.

Catanescu, C. O. and Chien, L. C. 2005. Patterning a poly(3,4-ethylenedioxythiophene) thin film using a liquid crystalline network, *Adv. Mater.*, **17**, 305–308.

Chien, L.-C. 1995. *Recent Advances in Liquid Crystal Polymers*, American Chemical Society, Anaheim, CA; Washington DC.

Crawford, G. and Zumer, S. Eds. 1996. *Liquid Crystals in Complex Geometries*, Taylor & Francis, London, Chapters 3, 5, 9, 10, 13, and 14.

Doane, J. W. 1990. *Liquid Crystals: Applications and Uses*, Vol. 1, B. Bahadur Ed., World Scientific, Singapore, Chapter 14, pp. 361–396.

Drzaic, P. S. 1995. *Liquid Crystal Dispersions*, World Scientific, Singapore.

Hikmet, R. A. M. 1990. Electrically induced light scattering from anisotropic gels, *J. Appl. Phys.* **68**, 4406–4412.

Kanegawa, T., Hobaia, R., Matsuda, I., Tanikawa, T., Notari, A., and Masegawa, S. 2003. Anisotropy in conductance of a quasi-one-dimensional metallic surface state measured by a square micro-four-point probe method, *Phys. Rev. Lett.* **91**, 036805.

Kang, S.-W. and Chien, L.-C. 2007. Field-induced and polymer-stabilized two-dimensional cholesteric liquid crystal gratings, *Appl. Phys. Lett.* **90**, 21110/1–221110/3.

Kang, S.-W., Sprunt, S., and Chien, L.-C. 2001. Ordered polymer microstructures obtained using pattern forming states of a cholesteric liquid crystal as templates, *Adv. Mater.* **13**, 1179–1983.

Kang, S.-W., Sprunt, S., and Chien, L.-C. 2002. Photoinduced localization of orientationally ordered polymer networks at the surface of a liquid crystal host, *Macromolecules.* **35**, 9372–9376.

Montgomery, H. C. 1971. Method for measuring electrical resistivity of anisotropic materials, *J. Appl. Phys.* **42**, 2971.

Pauw, L. J. van der, 1958. A method of measuring specific resistivity and Hall effect of discs of arbitrary shape. *Philips Res. Rep.*, **13**, 1.

Rajaram, C.V., Hudson, S.D., and Chien, L.-C., 1995. Morphology of polymer stabilized liquid crystals. *Chem. Mater.* **7**, 2300–2308.

Shirakawa, H., Akagi, K., Katayama, S., Araya, K., Mukoh, A., and Narahara, T. 1988. Synthesis, characterization, and properties of aligned polyacetylene films. *J. Macromol. Sci. Chem.* **A25**, 643–654.

Shirakawa, H., Otaka, T., Piao, G., Akagi, K., and Kyotani, M. 2001. Synthesis of vertically aligned polyacetylene thin films in homeotropic liquid crystal solvents, *Synth. Met.* **117**, 1.

West, J. L. 2003. Stressed liquid crystals for electrically controlled fast shift of phase retardation, *SID 03 Digest*, 1469–1471.

Yang, D.-K., Chien, L.-C., and Doane, J. W. 1992. Cholesteric liquid crystal/polymer gel dispersion for haze-free light shutters, *Appl. Phys. Lett.* **60**, 3102–3104.

# 3 Responsive Reactive Mesogen Microstructures

*Darran R. Cairns, Matthew E. Sousa,*
*and Gregory P. Crawford*

## CONTENTS

## 3.1 INTRODUCTION

Reactive liquid crystal monomers allow the anisotropic properties of liquid crystal materials to be captured permanently by photopolymerization. This enables the fabrication of a myriad of anisotropic mesostructures with exotic features and properties. In particular, colloidal dispersions of birefringent polymer spheres can be fabricated by suspension and emulsion polymerization; template synthesis can be used to produce self-assembled fibrous mesostructures; dispersed phases within solid films can be used to produce birefringent ellipsoidal inclusions in a polymer matrix for scattering polarizers; and islands of oriented polymer can be embedded in an amorphous matrix to produce thermally responsive surfaces.

The thermal, optical, electrical, and magnetic anisotropies of liquid crystal mesogens allows for a wide range of fabrication techniques to produce anisotropic mesostructures. The anisotropy of these mesostructures then allows the solid structure to be responsive to these same stimuli in the new solid form. This chapter will present a number of these exciting materials and a variety of fabrication techniques.

## 3.2   LIQUID CRYSTAL POLYMER SPHERES

At the turn of the twentieth century the optical textures of liquid crystal droplets were some of the first phenomena to be studied in these intriguing materials. Otto Lehmann's observations (Lehmann, 1904) are the precursors for current research in cross-linked liquid crystal polymer colloids. At the dawn of the twentieth century, polarized light was used as a probe to observe the birefringent texture of liquid crystal droplets; now, at the beginning of the twenty-first century, polarized light is being used to manipulate liquid crystal droplets and solid–liquid crystal spheres in optical tweezers setups (Juodkazis et al., 1999; Wood et al., 2004; Vennes et al., 2006; Diaz-Leyva et al., 2004; Fernandez-Nieves et al., 2005).

### 3.2.1   FABRICATION OF SPHERES

Cross-linked liquid crystal droplets have been fabricated using a number of techniques, including a liquid crystal diacrylate (RM257, EM Industries) (Cairns et al., 2001). The mixture of RM257 was prepared with 2 wt.% photoinitiator (Darocur 1173), heated to 90°C (RM257 exhibits a stable nematic phase for $T > 70$°C and $T < 126$°C) and mixed for 30 s to disperse the monomer and photoinitiator. It was then dispersed in hot 90°C glycerol, agitated and exposed to UV light for 8 min while the temperature was maintained at 90°C. The resulting spheres formed within the mixture had a diameter of approximately 10 μm and exhibited the optical texture characteristic of a bipolar configuration (see Figure 3.1). A rich variety of optical textures are possible depending upon the droplet size, molecular anchoring at the droplet interface, the elastic properties of the liquid crystal, and the direction and magnitude of any external electric or magnetic fields (Erdmann et al., 1990; Crawford et al., 1992). A good summary of the possible optical textures of liquid crystal droplets has been reported by Ondris-Crawford et al. (1991), with the most common textures being the bipolar, radial, and axial (see Figure 3.2).

Weitz and coworkers have produced monodispersed liquid crystal colloidal spheres with both radial and bipolar configurations by extrusion in a surfactant containing fluid, where the choice of surfactant determined the droplet configuration. The developed extrusion technique is particularly powerful for producing monodispersed colloids (Umbanhowar et al., 2000). Polymer droplets are extruded into a surfactant containing a coflowing continuous phase. As droplets form at the tip of the capillary, viscous drag from the coflowing phase detaches the droplet. The size of the droplet is determined by the competition between drag forces and interfacial tension. This technique allows stable emulsions to be formed because surfactants are able to fully coat individual particles before they break off. Polydispersity below 3% and micrometer scale droplets have been obtained. Droplets were produced using RMM14 (Merck) mixed with 10% photoinitiator (Darocur 159, Ciba) in aqueous solution. The temperature required to emulsify RMM14 in water is approximately 40°C and the pressure required to extrude droplets through capillaries is high (2 MPa). For ease of

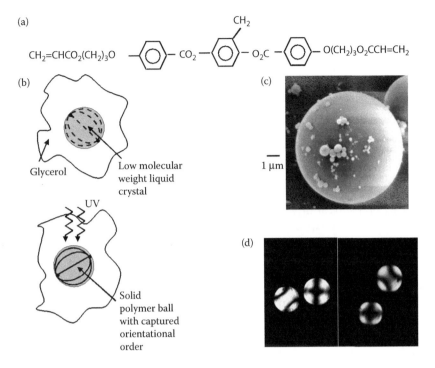

**FIGURE 3.1** Fabrication of liquid crystal polymer spheres. Liquid crystal monomer RM257 Merck (a), fabrication process (b), scanning electron micrograph of a sphere (c), and optical texture of two bipolar spheres (d).

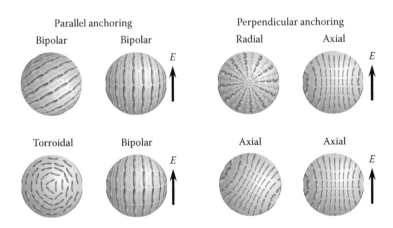

**FIGURE 3.2** Various director configurations of liquid crystal molecules confined to spherical cavities.

processing, the investigators therefore used a solvent, such as chloroform, to reduce the viscosity. The use of a solvent also allows for additional control of the particle diameters. Dilutions of 0.1% by volume were studied extensively by the Weitz group. To obtain colloidal solid spheres with well-defined optical textures it is necessary to evaporate the chloroform for up to 12 h, heat the dispersion to 60°C and then expose to UV light. Using 1 wt.% of polyvinyl alcohol (PVA) as a surfactant yielded bipolar droplets, whereas, using 16 mM sodium dodecyl sulfate (SDS) yielded radial droplets. The authors have reported on radial droplets of diameter $30 \pm 1 \, \mu m$ and bipolar droplets of diameter $16.0 \pm 0.6 \, \mu m$.

Zentel's group, at the University of Mainz, have created liquid crystal polymer colloids using dispersion polymerization (Vennes et al., 2005). In dispersion polymerization the liquid crystal monomer, a steric stabilizer, and a radical initiator are dissolved in a solvent in which the LC polymer is not soluble. Zentel et al. have had success using mixtures of ethanol and 2-methoxyethanol as solvent, hydroxypropyl cellulose (HPC) as a steric inhibitor, and dibenzoyl peroxide (DBPO) as a thermally decomposing initiator. The process consists of mixing these components and then heating them to decompose the DBPO, thereby releasing the free radicals. These radicals initiate the polymerization of the liquid crystal monomers and also abstract hydrogen from the HPC, thus leading to a graft polymerization of the liquid crystal monomer onto the HPC. The resulting graft polymers are sterically stabilized LC polymer spheres. The Zentel group also have synthesized a range of liquid crystal monomers, mostly acrylates and methacrylates, and synthesized solid polymer spheres. Smectic phases have also been focused on to produce spheres that also incorporate straight smectic layers, and onion-like rings.

### 3.2.2 APPLICATIONS OF SPHERES

Liquid crystal polymer spheres are already proving to be useful in fluidic applications. Diaz-Levya et al. have used bipolar liquid crystal droplets of $4 \, \mu m$ diameter to probe the microrheology of complex fluids (Diaz-Levya et al., 2004). The birefringent texture of the individual spheres can be used to image both translation and rotation in a flowing fluid and can thus enable the determination of both translational and rotational diffusion coefficients.

Asymmetries in the spheres can also be used to manipulate them; for instance the reorientation of bipolar spheres using electric fields (Cairns et al., 2001). Using an in-plane electrode configuration spheres can be rotated continuously by sequencing the applied electric field to each electrode (see Figure 3.3). For a number of applications it is the time taken from rest to a final switched orientation that is more important than the steady state frequency. The switching time for such a case is calculated using a simple mechanical argument. The modeled time to switch is given by:

$$t(\phi_2) - t(\phi_1) = \sqrt{\frac{2\rho R^2}{5 \, |\Delta\varepsilon| \, \varepsilon_0 E^2}} \left[ \ln \left[ \tan \left( \frac{\phi_2}{2} \right) \right] - \ln \left[ \tan \left( \frac{\phi_1}{2} \right) \right] \right], \quad (3.1)$$

where $\rho$ is the density, $R$ is the radius, $|\Delta\varepsilon|$ is the dielectric anisotropy of a perfectly aligned nematic, $\varepsilon_o$ is the permittivity of free space, $E$ is the field strength, $\theta_2$ is

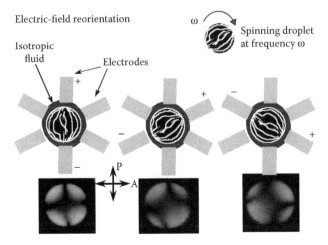

**FIGURE 3.3** Rotation of bipolar liquid crystal droplet by patterned electrodes. Optical textures of rotating droplets are also shown.

the final angle, and $\theta_1$ is the initial angle. A graph showing typical switching times for an applied field strength of $1\,\text{V}\,\mu\text{m}^{-1}$ is shown in Figure 3.4. One potentially exciting application of such spheres is an all-optical switch. One possibility is to write a reflection grating in the sphere and use it as an electrically switchable Bragg reflector (see Figure 3.5). Such spheres have been successfully fabricated using a holographic technique to produce a grating within a single component liquid crystal polymer film.

The formation of gratings in reactive liquid crystals has been studied by examining such films (Kossyrev and Crawford, 2001). Samples were fabricated by preparing an

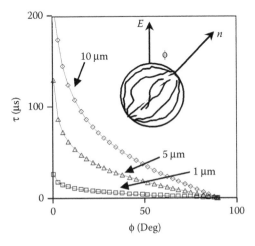

**FIGURE 3.4** Modeled switching time as a function of initial angle for a liquid crystal polymer sphere with a bipolar texture. The field strength is $1\,\text{V}\,\mu\text{m}^{-1}$.

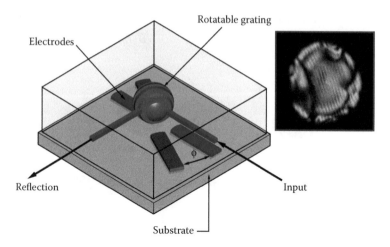

**FIGURE 3.5** Electro-optical switch concept. Reflection grating captured in a liquid crystal polymer sphere (inset).

optical cell with a $5\,\mu m$ cell gap and a rubbed polyimide layer on each of the glass surfaces. The cell was then filled with a reactive liquid crystal (LC242, BASF) and exposed to interfering laser beams from an $Ar^+$ ion laser operating at 351 nm and with a fringe spacing of $0.87\,\mu m$ (see Figure 3.6). The diffraction efficiency of the gratings was then measured using a He–Ne laser at the Bragg angle. We have developed a one-dimensional diffusion model for grating formation in a single component system and assumed three spatiotemporal concentrations: monomer, polymer, and void. The void concentration takes into account the shrinkage of the polymerized material and thus

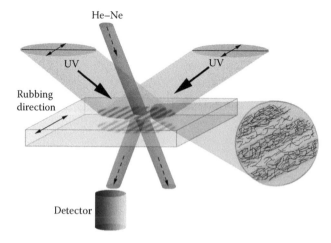

**FIGURE 3.6** Reactive mesogen between glass plates is eradiated by an UV interference pattern (351 nm) to produce a grating. *In situ* diffraction efficiency of the grating is measured by using a 612 nm He–Ne unpolarized laser situated at Bragg angle.

has a higher density than the monomer. We can then assume that the concentrations can be related by

$$\phi_m (x, t) + \phi_p (x, t) + \phi_v (x, t) = \phi_m (x, 0) = 1, \tag{3.2}$$

where the subscripts $m, p$, and $v$ represent monomer, polymer, and voids, respectively. If the polymerization of the monomer is assumed to lead to shrinkage such that from one unit of monomer we can create $1/(1 + z)$ polymer units and $z/(1 + z)$ void units, then the rate of change of concentration, using standard kinetic equations, is predicted to be

$$\frac{d\varphi_m (x, t)}{dt} = -F (x, t) \, \varphi_m (x, t) + \frac{d}{dx} \left( D (x, t) \, \frac{d\varphi_m (x, t)}{dx} \right)$$

$$\frac{d\phi_p (x, t)}{dt} = \frac{1}{1 + z} F (x, t) \, \phi_m (x, t) \tag{3.3}$$

$$\frac{d\phi_v (x, t)}{dt} = \frac{z}{1 + z} F (x, t) \, \phi_m (x, t) - \frac{d}{dx} \left( D (x, t) \, \frac{d\phi_m (x, t)}{dt} \right)$$

Further, the extent of polymerization increases, it is assumed the diffusion constant, $D(x, t)$, and the reaction rate, $F(x, t)$, will both decrease. Thus,

$$D (x, t) = D_0 e^{-\alpha \varphi_p (x, t)}, \tag{3.4}$$

$$F (x, t) = I (x) F_0 e^{-a\phi_p (x, t)}, \tag{3.5}$$

where $I(x)$ is the intensity of the recording light. The refractive index is determined as a function of position and time through:

$$n (x, t) = \langle n_m \rangle \, \phi_m (x, t) + \left\langle n_p \right\rangle \phi_p (x, t) + \langle n_v \rangle \, \phi_v (x, t), \tag{3.6}$$

and the diffraction efficiency for an infinite sinusoidal grating is modeled by

$$\eta (t) = \sin^2 \left( \frac{\pi \Delta n (t) \, d}{\lambda \cos \theta} \right), \tag{3.7}$$

where $\Delta n(t)$ is a refractive index modulation, $d$ is film thickness, $\lambda$ is the wavelength of diffracting light, and $\theta$ the Bragg angle. Experimental data and the fitted model are shown in Figure 3.7, which notes the sample orientation and the temperature of grating during formation. The final grating capability is only due to density modulations resulting from shrinkage in the polymer. This modulation is directly examined using SEM, as shown in the inset in Figure 3.7. The ability to produce diffraction gratings in a single component liquid crystal system is a powerful technique capitalized on to produce diffracting spheres. The combination of simple processing with fast switching times and excellent spatial resolution could open a myriad of exciting applications in the future.

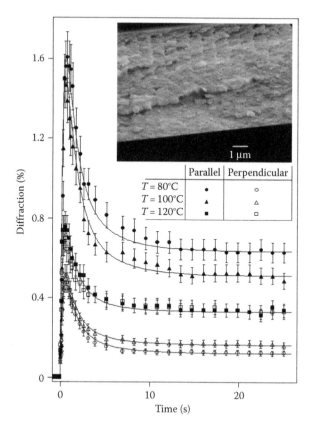

**FIGURE 3.7**  Experimental and modeled diffraction efficiencies versus time of grating formation for different orientation of the sample and at three temperatures show a good agreement. The inset shows a SEM photograph of our sample. The sample for SEM was produced at 100°C with orientation of molecules corresponding to rubbing direction 2.

## 3.3  TEMPLATE SYNTHESIS OF LIQUID CRYSTAL MICROSTRUCTURES

Template synthesis has been used broadly to produce rod like structures using a wide variety of materials. This technique has been used previously to make responsive structures using magnetic particles and solid metal rods. Magnetically activated rod arrays have been produced by Evans and coworkers (Evans et al., 2007). The nano and micro rods were fabricated from poly(dimethylsiloxane) (PDMS) and iron oxide particles. One end of each rod was attached to a substrate and the opposite end was free to move, resulting in various movements of the rod arrays, including back-and-forth and rotational. These magnet containing polymer rods have been stimulated using magnetic fields and been shown to switch with varying magnetic field gradients.

Liquid crystal polymer rods were fabricated with the same template synthesis technique. Just as with liquid crystals confined to spherical geometries there is a

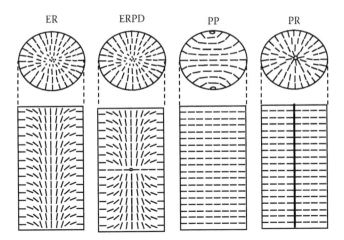

**FIGURE 3.8** Some examples of liquid crystal director configurations confined to cylindrical cavities. Escaped radial (ER), escaped radial with point defects (ERPD), planar polar (PP), and planar radial (PR).

range of molecular configurations possible in cylindrical cavities, some examples of which are shown in Figure 3.8. The configuration can be altered by the use of surface treatments and applied fields, as well as by the diameter of the confining cylinder. A detailed pictorial representation of the steps of fabrication is shown in Figure 3.9. An Anopore membrane (Whatman, Anodisc 13), with pores 200 nm in diameter and 60 μm in thickness, was used as a confining media. The membrane was first filled by capillary action with reactive mesogen liquid crystal, RM257 (Merck), in the nematic state. The director of the liquid crystal aligns parallel to the pore surface in an untreated Anopore membrane (Finotello et al., 1996). Therefore, when confined in a cylindrical cavity the liquid crystal can adopt a director field similar to a biplanar or parallel axial director. The reactive mesogen was then cured in a UV curing chamber while the temperature of the liquid crystal was kept between 71°C and 74°C, which represents the lower boundary of the nematic range of the liquid crystal. This temperature is chosen to maintain the molecular orientation of the liquid crystal parallel to the surface of the template. After curing, the Anopore membrane was etched away in a 0.5 M NaOH solution. The rods are then separated using a sonicator (Branson 1510 R, 42 kHz, 15 min). Finally, the reactive mesogen rods were dispersed in silicone oil (Brookfield, 4.7 cP).

The diameter of the nanorods is almost identical to the diameter of the Anopore membrane—in this case 200 nm. The length of the liquid crystal nanorods varies between a few nanometers and 60 μm depending on whether there was a complete filling of the Anopore membrane and if sonication induced a fracture of the rods into smaller fragments. Electro-optical glass cells were fabricated from indium tin oxide (ITO) coated glass and 60 μm polymethacrylate spacers. The entire system was then completely sealed after the rod dispersion was introduced.

An applied DC voltage tends to align the liquid crystal rods parallel to the DC electric field for a biplanar director field. The parallel axial alignment causes the

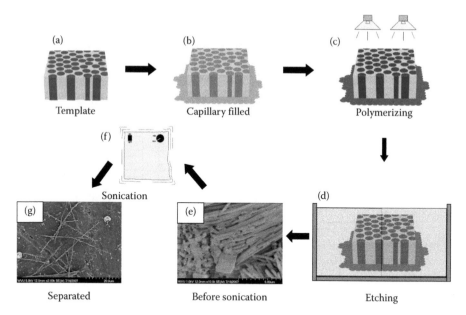

**FIGURE 3.9**  Fabrication process of liquid crystal nanorods. (a) Anopore membrane used as template. (b) Template is capillary filled with RM 257 in nematic state. (c) Polymerization of RM 257 with UV light. (d) Etching away anopore membrane with 0.5 M NaOH. (e) Nanorods clustered before sonication. (f) Sonication of nanorods. (g) Nanorods separated.

rods to switch horizontally with respect to the ITO surface plane. The rods were observed using an optical microscope (Leitz LaborLux 12 Me, Mag. 20×). The reorientation time was measured using a frame grabber (Guppy, Allied Visions Technology), capturing optical images every 0.055s. The apparent length of the rods during switching and relaxation was measured using Matlab image analysis. The angles of the rods with respect to the plane of the glass plate were then calculated from the apparent length measured. The switching procedure initiates when the liquid crystal nanorod lies parallel to the ITO-coated glass plate and finishes when the nanorod is perpendicular to the ITO plane. In addition, the relaxation process when the electric field is removed and the polymer rod returns to its initial position was characterized.

The switching and relaxation times were measured for the rods, as shown in Figure 3.10e, at increasing field strengths. The angle versus time graph, Figure 3.10a, shows the switching and relaxation curves. Figure 3.10b depicts the typical angular variance of the rods during the cycle of switching, fully on, and relaxation with an application of 10 V. The switching time is driven by the electric field, which is the positive slope seen at the beginning of the graph in Figure 3.10a. The responsive nature of the rods during switching is shown to reduce with a decrease of electric field strength. The fastest switching time recorded, 0.11 s, is achieved with field strength of 60 V (1 V $\mu m^{-1}$). The nanorods were still active at a relatively low voltage of 5 V

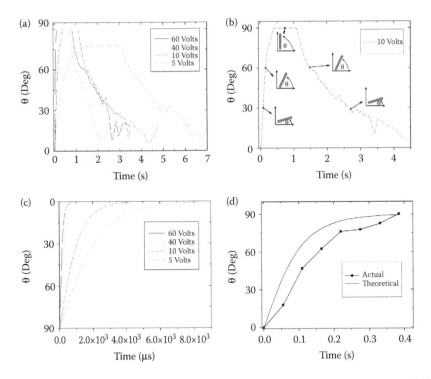

**FIGURE 3.10** Switching and relaxation times. (a) Angle (deg) versus time (s) for various field strengths. (b) Demonstration of rod angle for 10 volt curve. (c) Theoretical switching time. (d) Comparison of theoretical (with different dielectric anisotropy value) and actual switching time (field strength 10 V).

$(0.08~\text{V}~\mu\text{m}^{-1})$; however, this is the threshold voltage. At this low voltage the field strength is not strong enough to affect a complete rotation of the rod; the rod only rotates approximately $75°$.

Also measured was the relaxation time, when the electric field was turned off; this is the portion of Figure 3.10a with a negative slope. Relaxation time is not affected by field strength and is relatively constant for the different previously applied field strengths. Finally, the portion of the graph between the end of the switching curve and the beginning of the relaxation curve is when the rod is held in position by the electric field and is positioned vertically with respect to the ITO plane.

Following the same procedure described above a model was used to calculate the theoretical switching times. Electric torque:

$$T = |\Delta\varepsilon|\,\varepsilon_o E^2 V \cos\theta \sin\theta \tag{3.8}$$

and torque as defined by mechanics:

$$\omega\frac{d\omega}{d\theta} = \frac{T}{I} \tag{3.9}$$

were combined and integrated twice, resulting in an equation relating time and angle:

$$\Delta t = \sqrt{\frac{\rho L^2}{3 \, |\Delta \varepsilon| \, \varepsilon_o E^2}} \left[ \ln \left( \tan \left( \frac{\theta_2}{2} \right) \right) - \ln \left( \tan \left( \frac{\theta_1}{2} \right) \right) \right] \qquad (3.10)$$

where moment of inertia for a slender rod taken about the end of the rod is

$$I = \frac{1}{3} M L^2 \qquad (3.11)$$

and $\rho$ is the density ($1000 \, \mathrm{kg \, m^{-3}}$), $L$ is the length ($30 \, \mu\mathrm{m}$), $|\Delta \varepsilon|$ is the dielectric anisotropy of a perfectly aligned nematic (2), $\varepsilon_o$ is the permittivity of free space, $E$ is the field strength, $\theta_2$ is the final angle ($90°$), and $\theta_1$ is the initial angle.

These equations model the time for the rod to align parallel with the applied field from various initial angles. This model assumes perfect alignment of the liquid crystal, no viscosity and uniform electric field strength. The switching times predicted, Figure 3.10c are considerably faster as compared to the measured values; for example at 60 and 5 V the rod is switching at approximately 700 and 8500 μs, respectively. There is a considerable discrepancy between the theoretical and experimental values. Improperly aligned domains in the rods, which would influence the uniformity of the field director, may cause this. The model, Equation 3.10, calculates switching times assuming a dielectric anisotropy value, $\Delta \varepsilon$, of $-2$ which is for perfect alignment of the director field. We can calculate the effective dielectric anisotropy value by using the actual switching time data and Equation 3.10. This leads to a $|\Delta \varepsilon|$ value of $2.43 \times 10^{-4}$ for a field strength of 10 V; comparing the corrected $|\Delta \varepsilon|$ value to the perfect alignment value of 2 demonstrates that perfect alignment has not been achieved. Now, using this corrected anisotropy value the theoretical switching times can be plotted again. Figure 3.10d compares the actual data for a field strength of 10 V and the corrected theoretical switching times.

In addition to alignment parallel to an applied DC field, the rods responded to AC fields with random movements. Some rods showed simple rotational or vibrational movement, while others were more complex—shown to "swim" and move randomly in and out of the viewing area. A rotating rod was observed at decreasing AC field strengths (Figure 3.11a). It is shown in Figure 3.11b that the voltage and average degree of rotation per second follows a linear relationship, with higher voltage inducing faster rotation. This response to AC fields opens up many new exciting areas in fluidics including micromixing, micro-rheometers, and electro- and magneto-rheological fluids.

## 3.4   SELF-ASSEMBLED MESOSTRUCTURES

As an extension to the template synthesis techniques discussed earlier in this chapter templated dilute solutions of reactive liquid crystal monomers in solvents are discussed (Cairns et al., 2000). This technique has allowed for the fabrication of liquid crystal polymer fibrils, which self assemble into meso-scale structures when the template is removed. A rich variety of structures have been achieved by varying monomer concentration and surface alignment within the template.

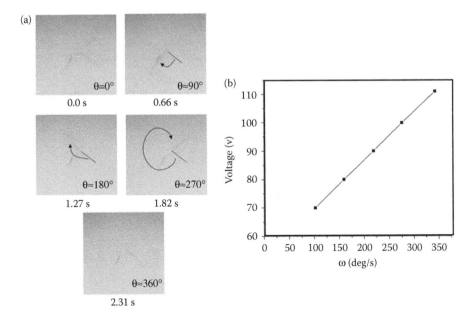

**FIGURE 3.11**   Rotating rod in an applied AC voltage. Optical images of a rod dispersed in a silicone oil with an applied AC voltage of 80 V (a). Angular frequency of rotation of a rod as a function of applied AC voltage (b).

A mixture of the reactive liquid crystal monomer RM257 was prepared with 1 wt.% photoinitiator (Darocur 1173, Ciba) and then diluted with chloroform. The dilute solution was then used to fill an alumina Anopore membrane (Whatman, Anodisc 13). Membranes were used both as received and after surface treatment with lecithin. Low molecular weight liquid crystals are well known to align parallel to an untreated Anopore membrane surface and perpendicular to lecithin-coated surfaces. In all cases, after filling with the dilute solution, the chloroform was evaporated and the membrane was heated so that the liquid crystal monomer was in the nematic phase. The monomer was then polymerized with UV light and the membrane was removed by etching in a 0.4 M solution of NaOH.

For untreated template surfaces, rod and tube-like structures were produced. A 20% solution of RM257 produced hollow tubes (see Figure 3.12c), while when the concentration was reduced to 12% fibrils were produced that self-assembled to form fibrous tubes (Figure 3.12d). For reference, solid structures from undiluted monomer are shown in Figure 3.12a, solid spheres templated in glycerol in Figure 3.12b, and solid rods in an Anopore template.

When the pores were treated with lecithin even more exotic structures were formed. For a solution of 25% RM257 the fibrils self-assembled into toroids (Figures 3.13a and 3.13b). With a 12% solution, yarn balls were created (Figures 3.13c and 3.13d). These yarn balls could potentially be used as novel release devices (Kossyrev and Crawford, 2000).

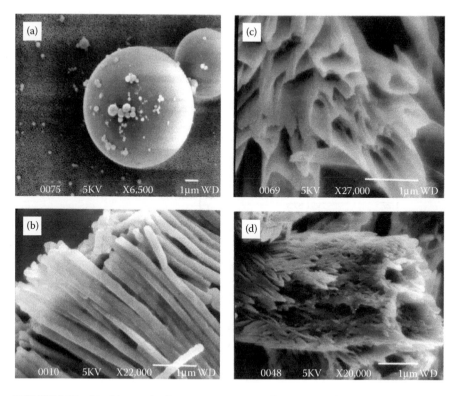

**FIGURE 3.12**  Liquid crystal polymer mesostructures from untreated templates. Spheres (a), rods (b), hollow tubes (c), and fibrous self-assembled tubes (d).

## 3.5  DISPERSIONS IN SOLID FILMS

Additional functionality can be obtained by dispersing liquid crystal polymer microstructures in film matrices. Interesting films have been fabricated with dispersed phases by both photolithography techniques (Sousa et al., 2006) and controlled phase separation (Amimori et al., 2003; Eakin et al., 2003). Both techniques are readily scaled and lend themselves to low cost manufacturing processes.

### 3.5.1  PATTERNED THERMALLY RESPONSIVE REACTIVE MESOGENS

Densely cross-linked and oriented polymer networks, created from the *in situ* photopolymerization of ordered reactive mesogen materials, exhibit highly anisotropic thermal expansion coefficients. The thermal expansion coefficient of uniaxially ordered reactive mesogen films in the direction of the molecular organization is slightly positive below the glass transition of the film and becomes negative as soon as the transition sets in. By introducing a periodic chiral structure oriented orthogonally to the film's surface, a low thermal expansion coefficient is established in two directions in the plane of the film. The thermal expansion coefficient perpendicular to

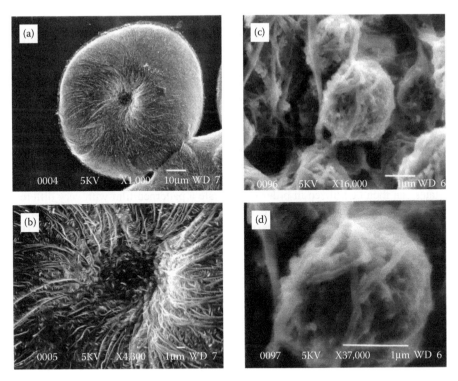

**FIGURE 3.13** Self-assembled liquid crystal polymer mesostructures from lecithin-treated templates. Toriods (a and b) and yarn balls (c and d).

the plane (parallel to the helical pitch axis) is much larger than the in-plane directions in this case (Broer and Mol, 1991).

The use of reactive mesogen materials allows for accurate control of the molecular order in multiple dimensions—molecular order and orientation can be controlled using a variety of methods that successfully manipulate low molar mass liquid crystals. The various liquid crystalline states can each be fixed by the polymerization process (Broer et al., 1995; Broer 1996; Kitzerow et al., 1993). In reactive mesogen materials, the ordering of the molecules in the polymerized liquid crystal network is closely reminiscent of the molecules in their monomeric state.

The basic thermal expansion behavior of planar aligned difunctional liquid crystalline acrylate reactive mesogens has been previously reported—because of their liquid crystallinity, there is an anisotropy in their thermal expansion behavior (Broer and Mol, 1991; Mol et al., 2005). The thermal expansion of the reactive mesogen is substantially lower along the long molecular axis than normal due to the fact that most of the covalent bonds that make up the polymer network are parallel to the long axis of the molecules. The thermal expansion of reactive mesogens normal to the long molecular axis is positive and progressively increases with increasing temperature. In the direction of the long molecular axis, the thermal expansion is close to zero and becomes slightly negative as the temperature increases (Broer and Mol, 1991).

In order to fabricate the films described here, a nonchiral reactive monomer was doped with a chiral monomer in order to induce a helical twist in the molecular organization. The long molecular axes rotate in planes normal to the surface of the sample substrates. The thermal expansion in this plane is approximately zero; however, the thermal expansion normal to the plane remains strongly positive, as shown in Figure 3.14a (Broer and Mol, 1991). If the system is polymerized at high temperatures (greater than the nematic–isotropic transition temperature), then the thermal expansion of the polymer is some average of the thermal expansion parallel to and normal to the long molecular axis. This is due to the disordered nature of the isotropic phase, as shown in Figure 3.14b, which is fixed by photopolymerization.

The different thermal expansion coefficients of the chiral nematic and isotropic states can then be utilized in order to fabricate a patterned film with very different thermal expansion properties. Increasing temperature can be applied to the resulting film, causing the cholesteric regions of the film to rise above the isotropic regions. This is due to the thermal expansion being much larger along the helix in the cholesteric regions than the thermal expansion in the isotropic regions of the film. Upon removing the film from the heat source, the elevated regions revert to their original heights, resulting in a reversible modulation of the surface topology.

### 3.5.1.1   Sample Fabrication

The films were fabricated using a mixture of the reactive mesogen RM 257 (Merck KgaA,) and the chiral material LC 756 (BASF). For photopolymerization, 2 wt.% Irgacure 651 (Ciba Specialty Chemicals) was added as a photoinitiator. The constituents were dissolved in dichloromethane and subsequently mixed. Following the evaporation of the dichloromethane, a droplet of the material in the chiral nematic was placed in the center of a glass substrate. In order to achieve a uniform planar molecular alignment, the glass substrates were spin coated with AL3046 polyimide (JSR Corporation), baked, and uniaxially rubbed with a polyester cloth. The top substrate was oriented such that the rubbing direction of the polyimide was antiparallel to the bottom substrate. Once the droplet of material was allowed to completely fill

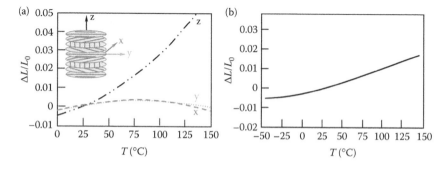

**FIGURE 3.14**   Linear thermal expansion measured parallel and perpendicular to the helical axis of a chiral reactive mesogen film (a) (Broer and Mol, 1991). (b) Linear thermal expansion for an isotropic reactive mesogen film. The values were obtained by averaging the values of thermal expansion parallel and perpendicular to the helical axis.

the cell, the glass plates were sheared parallel to the rubbing direction in order to better align the molecules. During the first UV exposure, the sample was held at 90°C, and a photomask was used to define the desired cholesteric pattern. Subsequently, the sample was exposed to a blanket UV exposure. Prior to the second exposure, the mask was removed, and the sample was heated to 150°C, in order to ensure the remaining unpolymerized regions reached the isotropic phase. The sample was then exposed to the same blanket UV light source for 10 min to completely polymerize the remaining material. Once the sample was fully polymerized, the top substrate was removed using a razor blade, leaving the polymer film intact on a single substrate.

### 3.5.1.2  Sample Characterization

The samples fabricated were characterized using a Veeco white light interference microscope. Upon removing the top glass substrate, some surface relief persisted, as shown in Figure 3.15. This phenomenon has been observed in the temperature process used to create patterned retarders for displays (van der Zande et al., 2005). One possible cause of the formation of these surface relief structures is the diffusion of monomer from the isotropic region to the cholesteric region at elevated temperatures prior to or during polymerization of the isotropic region. A second possible cause of this phenomenon can be attributed to polymerization shrinkage.

Figure 3.15 shows how the surface of the film changes as the temperature of the substrate is increased. At room temperature prior to heating, there was a 20 nm difference in height between the chiral nematic and isotropic regions. Upon heating the film to 155°C, the maximum height difference between the two regions is close to 600 nm. It is clearly observed that the cholesteric regions expand to much greater heights than the isotropic regions. After allowing the sample to return to room temperature, the film returns to its original configuration. Subsequent heating cycles produced similar results.

### 3.5.2  Liquid Crystal Nano-Droplet Films

Polymer dispersed liquid crystal (PDLC) films have attracted much attention in recent years because of their ability to create low-cost electro-optical devices that can be switched from a scattering state to a transmissive state by the application of a voltage. The fabrication of PDLCs is elegant and can be accomplished by controlled phase separation. One particularly attractive technique is to dissolve a liquid crystal monomer in a polymer emulsion and then cast a film. As the film dries, a polymer film forms and liquid crystal droplets phase separate due to the reduced solubility in the polymer compared to the emulsion (Drzaic, 1986). This phase separation technique can be used to form solid liquid crystal polymer nano-droplet films by using a reactive monomer as the liquid crystal and photopolymerizing the droplets after the film is formed.

This technique has been used to produce a robust scattering polarizer. A 20 wt.% aqueous solution of PVA (molecular weight 25,000, degree of hydrolysis 88%, Kuraray Co. Ltd.) was mixed with reactive mesogen (RM257, Merck) and an emulsion was formed by sonication. The films were then cast onto a PET substrate using a Meyer Bar and processed as shown in Figures 3.16 and 3.17. The nano-droplets were

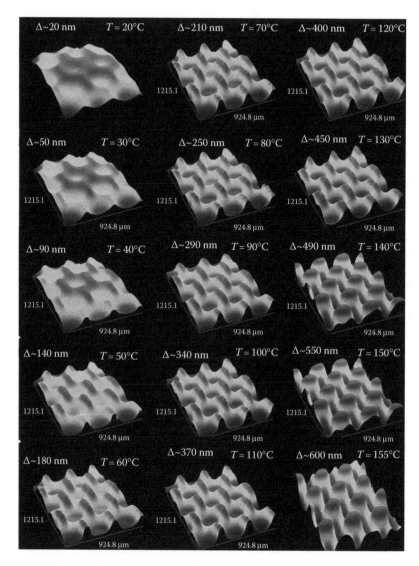

**FIGURE 3.15** Interference microscope images for a reactive mesogen film containing 300 μm isotropic circles in a sea of cholesteric material.

made to be anisotropic by stretching the film uniaxially to 100% strain at a temperature of 130 °C and then curing it by photopolymerization (Eakin et al., 2003). The anisotropic shape of the droplets dispersed in the birefringent PVA matrix produces a scattering polarizer. PDLC polarizers have been discussed previously (Aphonin et al., 1993), but have not been widely adopted because of the strong dependence of the liquid crystal's optical properties on temperature. By using photopolymerization the nano-droplets have no such temperature dependence. An example of such a scattering polarizer is shown in Figure 3.18 for both polarization states.

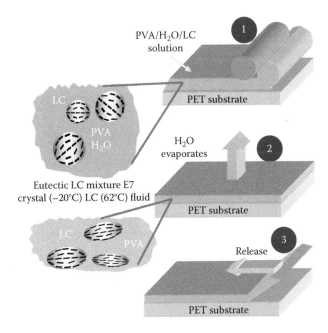

**FIGURE 3.16** Fabrication process for scattering polarizers with aspected liquid crystal polymer inclusion.

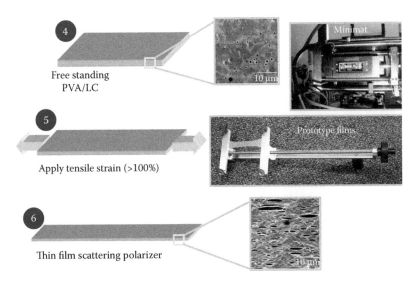

**FIGURE 3.17** Fabrication process for scattering polarizers with aspected liquid crystal polymer inclusion.

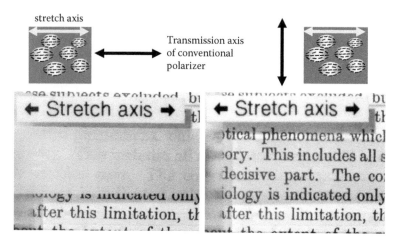

**FIGURE 3.18** Scattering polarizer viewed through a polarizer parallel to the stretch direction (left) and perpendicular to the stretch direction (right).

In addition to producing novel films, this phase separation technique can also be used as a fabrication process for anisotropic liquid crystal polymer microstructures. The anisotropic droplets modify the molecular ordering which is then captured by photopolymerization (Amimori et al., 2005). Uniaxial tension can be used to produce needles with varying aspect ratios. PVA can be stretched to greater than 400% strain to produce needles as long as 10 μm and with aspect ratios greater than 10 (see Figure 3.19). Additionally, disks can be produced by templating with biaxially stressed films.

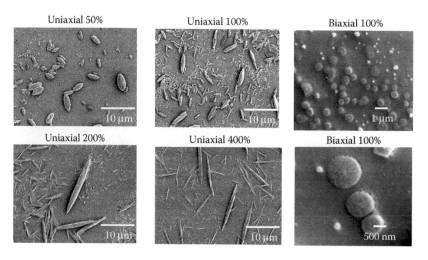

**FIGURE 3.19** Aspected liquid crystal polymer microstructures fabricated by dissolving PVA matrix after polymerization under various film deformations.

## 3.6 CONCLUSIONS

There are a rich variety of microstructures that can be fabricated using reactive liquid crystal mesogens. Template synthesis, surface alignment, self-assembly, and photolithography can all play a role in the fabrication process and it is the interplay of these techniques that makes this field so diverse in potential applications. Despite the diversity of structures already reported there are still many more new combinations to be discovered and this is truly an exciting field.

## REFERENCES

Amimori, I., Priezjev, N. V., Pelcovits R. A., and Crawford, G. P. 2003. Optomechanical properties of stretched polymer dispersed liquid crystal films for scattering polarizer applications, *Journal of Applied Physics* **93**, 3248.

Amimori, I., Eakin, J. N., Qi, J., Skaèej, G., Žumer, S., and Crawford, G. P. 2005. Surface-induced orientational order in stretched nanoscale-sized polymer dispersed liquid-crystal droplets, *Physical Review E* **71**, 031702.

Aphonin, O. A., Panina, Y. V., Pravdin, A. B., and Yakovlev, D. A. 1993. Optical-properties of stretched polymer-dispersed liquid-crystal films, *Liquid Crystals*, **15**, 395.

Broer, D. J. 1996. Liquid crystalline networks formed by photoinitiated chain cross-linking, in *Liquid Crystals in Complex Geometries*, Eds. Crawford, G. P., Žumer, S. Taylor & Francis, London.

Broer, D. J., and Mol, G. N. 1991. Anisotropic thermal expansion of densely cross-linked oriented polymer networks, *Polymer Engineering and Science*, **31**, 625.

Broer, D. J., Mol, G. N., and Lub, J. 1995. Wide-band reflective polarizers from cholesteric polymer networks with a pitch gradient, *Nature*, **378**, 467.

Cairns, D. R., Eichenlaub, N. S., and Crawford, G. P. 2000. Ordered polymer microstructures synthesized from dispersions of liquid crystal mesogens, *Molecular Crystals and Liquid Crystals*, **352**, 709.

Cairns, D. R., Sibulkin, M., and Crawford, G. P. 2001. Switching dynamics of suspended mesogenic polymer microspheres, *Applied Physics Letters*, **78**, 2643.

Crawford, G. P., Steele, L. M., Ondris-Crawford, R., Iannacchione, G. S., Yeager, C. J., Doane, J. W., and Finotello, D. 1992. Characterization of the cylindrical cavities of anopore and nuclepore membranes, *Journal of Chemical Physics*, **96**, 7788.

Diaz-Leyva, P., Perez, E., and Arauz-Lara, J. L. 2004. Dynamic light scattering by optically anisotropic colloidal particles in polyacrylamide gels, *Journal of Chemical Physics*, **121**, 9103.

Drzaic, P. S. 1986. Polymer dispersed nematic liquid crystal for large area displays and light valves, *Journal Applied Physics*, **60**, 2142.

Eakin, J. N., Amimori, I., and Crawford, G. P. 2003. Ordering in highly anisotropic liquid crystal nano-droplets: Scattering polarizer applications, *Society for Information Display Digest of Technical Papers* **XXXIV**, 672.

Erdmann, J. H., Žumer, S., and Doane J. W. 1990. Configuration transition in a nematic liquid crystal confined to a small spherical cavity, *Physical Review Letters*, **64**, 1907.

Evans, B. A., Shields, A. R., Carroll, R. L., Washburn, S., Falvo, M. R., and Superfine, R. 2007. Magnetically actuated nanorod arrays as biomimetic cilia, *Nano Letters*, **7**, 1428.

Fernandez-Nieves, A., Cristobal, G., Garces-Chavez, V., Spalding, G. C., Dholakia, K., and Weitz, D. A. 2005. Optically anisotropic colloids of controllable shape, *Advanced Materials*, **17**, 680.

Finotello, D., Iannacchione, G. S., and Qian, S. 1996. Phase transitions in restricted geometries, in *Liquid Crystals in Complex Geometries*, eds. Crawford, G. P., Žumer, S., Taylor & Francis, London.

Juodkazis, S., Shikata, M., Takahashi, T., Matsuo S., and Misawa, H. 1999. Fast optical switching by a laser-manipulated microdroplet of liquid crystal, *Applied Physics Letters*, **74**, 3627.

Kitzerow, H. S., Schmid, H., Ranft, A., Heppke, G., Himket, R. A. M., and Lub, J. 1993. Observations of blue phases in chiral networks, *Liquid Crystals*, **14**, 911.

Kossyrev, P. A., and Crawford, G. P. 2000. Yarn ball polymer microstructures: A structural transition phenomenon induced by an electric field, *Applied Physics Letters*, **77**, 3752.

Kossyrev, P. A., and Crawford, G. P. 2001. Formation dynamics of diffraction gratings in reactive liquid crystals, *Applied Physics Letters*, **79**, 296.

Lehmann, O., *Flüssige Kristalle*, Wilhelm Engelmann, Leipzig, 1904.

Mol, G. N., Harris, K. D., Bastiaansen, C. W. M., and Broer, D. J. 2005. Thermo-mechanical responses of liquid-crystal networks with a splayed molecular organization, *Advanced Functional Materials*, **15**, 1155.

Ondris-Crawford, R., Boyko, E. P., Wagner, B. G., Erdmann, J. H., Žumer, S., and Doane J. W. 1991. Microscope textures of nematic droplets in polymer dispersed liquid crystals, *Journal of Applied Physics*, **69**, 6380.

Sousa, M. E., Broer, D. J., Bastiaansen, C. W. M., Freund, L. B., and Crawford, G. P. 2006. "Isotropic islands" in a cholesteric "sea": Patterned thermal expansion for responsive surface topologies, *Advanced Materials*, **18**, 1842.

Umbanhowar, P. B., Prasad, V. and Weitz, D. A. 2000. Monodisperse emulsion generation via drop break off in a coflowing stream, *Langmuir*, **16**, 347.

van der Zande, B. M. I., Steenbakkers, J., Lub, J., Lewis, C. M. and Broer, D. J. 2005. Mass transport phenomena during lithographic polymerization of nematic monomers monitored with interferometry, *Journal of Applied Physics*, **97**, 123519.

Vennes, M., Zentel, R., Rössle, M., Stepputat, M. and Kolb, U. 2005. Smectic liquid-crystalline colloids by miniemulsion techniques, *Advanced Materials*, **17**, 2123.

Vennes, M., Martin, S., Gisler, T. and Zentel, R. 2006. Anisotropic particles from LC-polymers for optical manipulations, *Macromolecules*, **39**, 8326.

Wood T. A., Gleeson H. F., Dickinson M. R. and Wright A. J. 2004. Mechanisms of optical angular momentum transfer to nematic liquid crystalline droplets, *Applied Physics Letters*, **84**, 4292.

# 4 Viewing Angle Compensation Films for LCD Using Reactive Mesogens

*Ichiro Amimori and Tokuju Oikawa*

## CONTENTS

## 4.1 INTRODUCTION

Liquid-crystal displays (LCDs) are increasingly being used for monitors and TVs due to their features of thin appearance, lightweight, low power consumption, long lifetime, and high resolution. On the other hand, two major problems still exist; one is the narrow-viewing-angle performance and the other is the slow optical response speed. A viewing-angle compensation-film technology plays an important role for improving the viewing-angle performance only by attaching one or more films onto a liquid crystal (LC) cell. Reactive mesogens are commonly used in viewing-angle compensation films because of their wide variety of optical configurations, which can provide films with various kinds of optical anisotropy. In this chapter, major applications of reactive mesogens to viewing-angle compensation films for LCDs, especially a biaxial film composed of deformed cholesteric structure for vertical alignment (VA) mode LCDs (Amimori et al., 2005) and polymerized discotic material (PDM) hybrid aligned layer for twisted nematic (TN) mode LCDs (Mori, 2005).

### 4.1.1 Viewing Angle of LCDs

A polarization control technology using birefringent media is a basic principle of LCDs. Figure 4.1 illustrates an example of LCDs, which is a normally black mode LCD. In Figure 4.1, unpolarized light through a polarizer becomes linear polarization. The polarization passes through a LC cell without any polarization change at off-state, and rotates its polarization plane by 90° at on-state. The normally black LCD can be easily changed into a normally white mode by rotating the analyzer by 90°. Figure 4.2 shows TN mode (Schadt and Helfrich, 1971), in-plane switching (IPS) mode (Ohta et al., 1995), and VA mode (Ohmuro et al., 1997) LCDs in order to understand how the polarization rotates in the LC cell with applied electric fields.

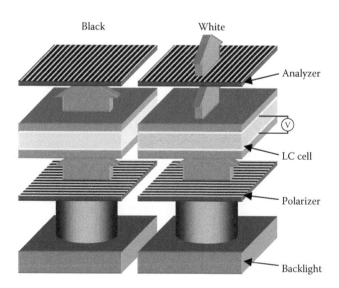

**FIGURE 4.1** Schematic diagram of the structure of LCDs. LCD shown in the figure is a normally black (NB) mode LCD. It can be easily changed into a normally white mode when the second polarizer rotates by 90°.

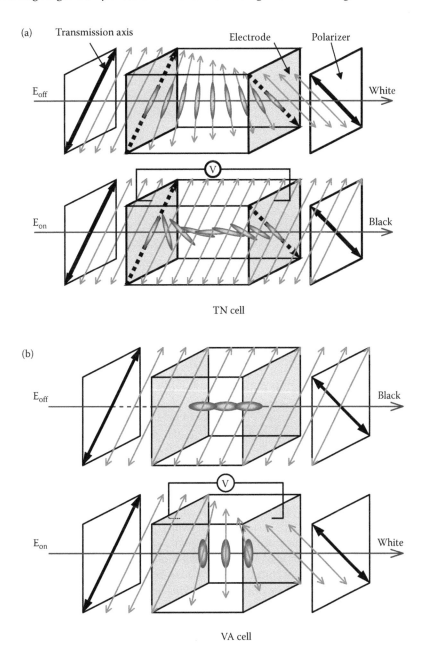

**FIGURE 4.2** Schematic illustrations of an off state and an on state of (a) TN, (b) VA, and (c) IPS cells.

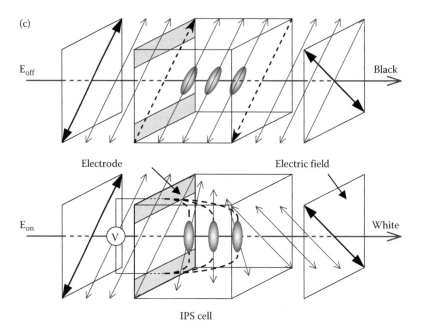

**FIGURE 4.2** Continued

A contrast ratio (*CR*) is one of the most important display properties defined as

$$CR = \frac{\text{highest luminance in the pixels}, L_{\text{max}}}{\text{lowest luminance in the pixels}, L_{\text{min}}}, \tag{4.1}$$

where $L_{\text{max}}$ and $L_{\text{min}}$ correspond to luminance at white and black states, respectively. The *CR* indicates how people recognize display images observed from a viewing angle according to a luminance difference. The viewing-angle performance is expressed by the dependence of *CR* on viewing angle in a polar plot, or an angle range in a horizontal direction at which a specific *CR* value (*CR* = 10 is often used) value is maintained. Sometimes the colors on the display panel may look different from an on-axis image even the *CR* is large enough, and a color viewing angle defined as the angular distribution of color shift from the on-axis direction is also important. In this paper, a viewing angle indicates a contrast viewing angle as long as it is mentioned a color viewing angle.

Even though a display image quality of LCDs at on-axis direction is well optimized, off-axis images are not as good as on-axis images since, undesirable optical phenomena exist in off-axis directions. Therefore, the viewing-angle compensation technology is quite important to improve LCD image qualities. Since a luminance distribution in a white state is not very large compared to that in a black state caused by an off-axis light leakage we designed the viewing-angle compensation such that the off-axis light leakage at black state is minimized. From the viewpoint of LCDs, the viewing angle is designed at an on-state or an off-state in each LCD mode as shown in Figure 4.2 corresponding to its black state.

## 4.1.2 ANISOTROPY OF REACTIVE MESOGENS AND THEIR FILMS

In general, anisotropy of birefringent media is defined with respect to the relationship of refractive indices in three principal axes as shown in Table 4.1 (Born and Wolf, 1999; Yeh and Gu, 1999). Three refractive indices are identical in isotropic media and the polarization propagating in an isotropic medium does not change its polarization state since no retardation exists in any directions.

Anisotropic media have different refractive indices in three principal axes. In case one refractive index is different from two other refractive indices, the medium is called uniaxial. The direction in whose refractive index (called extraordinary index) is different from two other refractive indices (called ordinary index) is called optic axis, and the refractive index in the optic axis is an extraordinary index. If the extraordinary index is greater than the ordinary index, the anisotropy is negative. On the other hand, if the extraordinary index is smaller than the ordinary index, the anisotropy is negative. When uniaxial media are shaped like plates or films, two types of uniaxial properties are defined by an optic direction. $a$-Plates have an optic axis in a plane of films, and $c$-plates have an optic axis in normal to the plane of films.

Anisotropic media whose three refractive indices are different from each other are said to be optically biaxial. Biaxial media can have various retardation properties since three refractive indices can be freely changed. For uniaxial and biaxial films, on-axis

---

**TABLE 4.1**

**Definition of Isotropic and Anisotropic Media with Respect to Three Principal Axes of a Film**

| | Isotropic | Anisotropic Uniaxial $a$-plate | Anisotropic Uniaxial $c$-plate | Biaxial |
|---|---|---|---|---|
| Positive | $n_z$ $n_y$ $n_x$ | $n_x > n_y = n_z$ | $n_x = n_y < n_z$ | $n_x\ n_y\ n_z$ |
| Negative |  $n_x = n_y = n_z$ | $n_x < n_y = n_z$ | $n_x = n_y > n_z$ | |

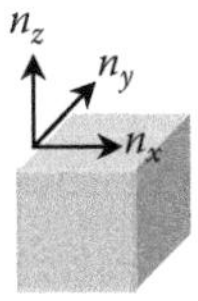

*Note:* $n_x$ and $n_y$ are principal refractive indices in a plane of the film, and $n_z$ is a refractive index in the direction normal to the film.

retardation ($Re$) and out-of-plane retardation ($Rth$) defined as follows are often used.

$$Re = (n_x - n_y)\, d,$$

$$Rth = \left\{ \frac{n_x + n_y}{2} - n_z \right\} d, \tag{4.2}$$

where $d$ is a thickness of an anisotropic layer.

Figure 4.3 shows examples of calamitic and discotic reactive mesogens (Broer, 1993; Kawata, 2002). A single calamitic reactive mesogen molecule has a positive uniaxial anisotropy and a single discotic reactive mesogen molecule has a negative uniaxial anisotropy. Then homogeneously aligned calamitic and discotic reactive mesogen layers are positive and negative $a$-plates, respectively. Similarly, homeotropically aligned calamitic and discotic reactive mesogen layers are positive and negative $c$-plates, respectively. It is known that a narrow pitch cholesteric alignment structure made of calamitic reactive mesogen can also behave as a negative $c$-plate (Allia et al., 1994). A biaxial reactive mesogen layer is able to create using this cholesteric $c$-plate. It is reported that a deformed cholesteric structure has birefringence in the plane of the film and becomes biaxial (Broer et al., 1999). The deformed cholesteric structure is induced by the exposure of polarized UV light with a rod-like dichroic photoinitiator. Viewing-angle compensation of VA mode LCDs using the biaxial deformed cholesteric structure (Kim et al., 2004; Amimori et al., 2005; Suzuki et al., 2006) is discussed in Section 4.1.2.

Reactive mesogens can generate other anisotropic structures such as hybrid (Mori et al., 2003) and twist (Toyooka and Kobori, 2000) structures. Hybrid and twist structures can be practically applied to TN mode and super twisted nematic (STN) mode LCDs, respectively. Hybrid discotic reactive mesogen layer is discussed in Section 4.1.3.

### 4.1.3  VIEWING-ANGLE COMPENSATION MECHANISM

As mentioned above, a deterioration of viewing angle is primarily caused by an off-axis light leakage at black state. There are two major reasons for the light leakage at black state. The first reason is an undesirable retardation at off-axis. The off-axis retardation changes a polarization state that cannot be absorbed by the analyzer. The second reason is a change in an angle between orthogonal polarizer and analyzer as a function of observation direction. The effective angle between orthogonal polarizer and analyzer becomes larger than 90° as shown in Figure 4.4. A viewing-angle compensation mechanism is based on the compensations for these two reasons; one is retardation compensation and the other is polarizer compensation.

Figure 4.5 explains a principle of retardation compensation. An LC cell has positive $c$-plate anisotropy which does not change an incoming polarization at on-axis. A retardation film, which has a negative $c$-plate anisotropy is placed on the LC cell. It is obvious that the retardation film also does not change the incoming polarization passed through the LC cell. If two orthogonal polarizers are placed on the retardation film and beneath the LC cell, an ideal black state can be obtained. Considering an incoming polarization at off-axis, retardations generate in the LC cell and the

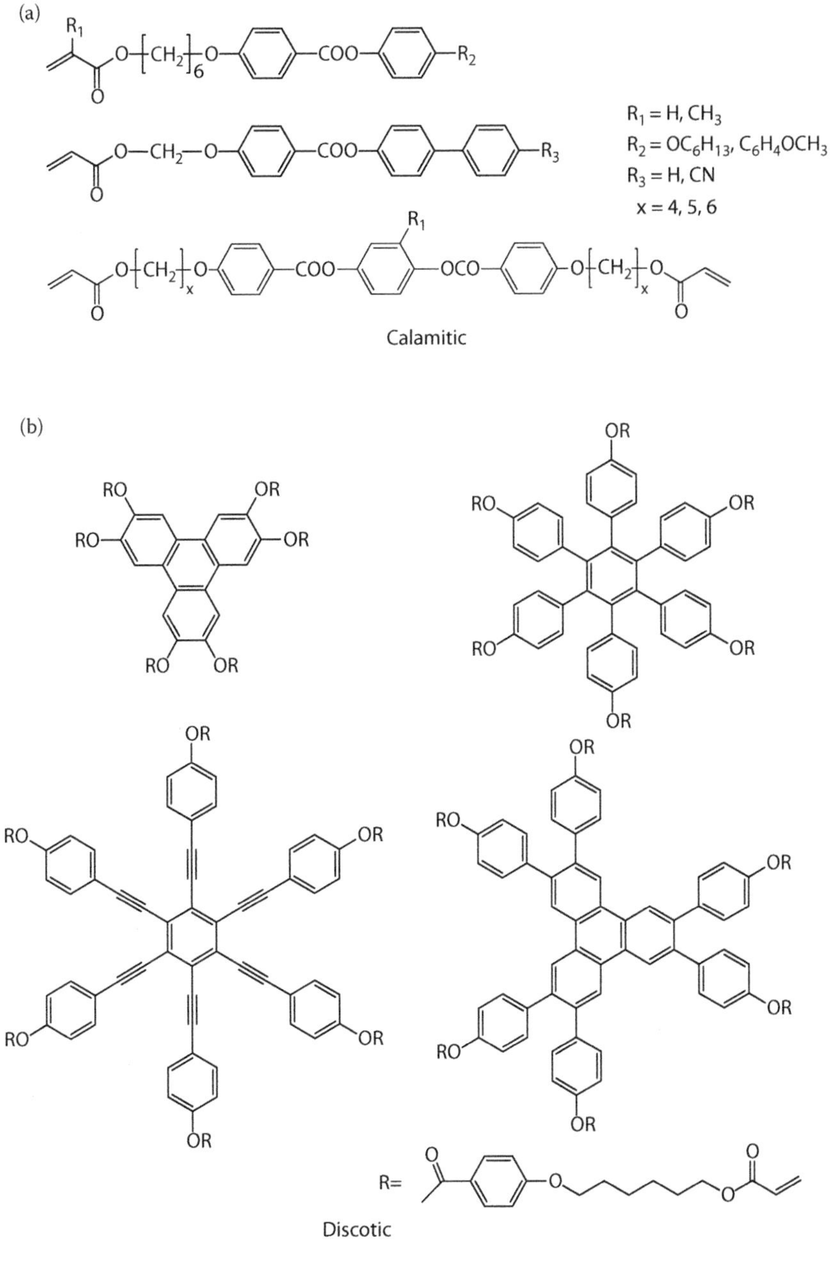

(a)

R₁ = H, CH₃
R₂ = OC₆H₁₃, C₆H₄OCH₃
R₃ = H, CN
x = 4, 5, 6

Calamitic

(b)

R=

Discotic

**FIGURE 4.3** Examples of chemical structures of (a) calamitic reactive mesogens and (b) discotic reactive mesogens.

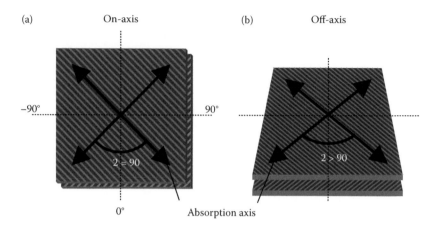

**FIGURE 4.4** Comparison of angles of two orthogonal polarizers observed from (a) on-axis and (b) off-axis. Azimuthal direction of observation is set to be 0° and absorption axes of two polarizers are at ±45°.

retardation film. The retardation in the cell has a slow axis, defined as a larger refractive index axis in a wave front. On the other hand, the retardation in the retardation film has a slow axis orthogonal to the slow axis in the LC cell, resulting in denying the retardation in the LC cell. Thus, the retardation compensation is designed such that the alignment of positive uniaxial LC molecules in a LC cell at black state is compensated by the alignment of negative uniaxial media. If a retardation film consists of liquid crystalline materials, it is clear that discotic mesogens are useful.

Figure 4.6 explains principles of the polarizer compensation using Poincaré sphere (Kliger et al., 1990) observed at an oblique angle. The change of angle between polarizer and analyzer is expressed as a separation of points $P$ (indicating a transmission axis of polarizer) and $A$ (indicating an absorption axis of analyzer) in Figure 4.6, and the polarizer compensation is to change a polarization state from $P$ to $A$ with retardation films. A center of rotation in a Poincaré sphere is determined by $N_z$ factor (Fujimura et al., 1991) defined as

$$N_z = \frac{n_x - n_z}{n_x - n_y}. \tag{4.3}$$

If $N_z$ of a medium is equal to 1.0, the medium is an $a$-plate. If $N_z$ of a medium is infinity ($n_x = n_y$), the medium is a $c$-plate. Other $N_z$ indicates a medium is biaxial.

Figure 4.6a shows a method using a $c$- and an $a$-plate (both positive and negative can be applicable), and Figure 4.6b shows a method using a single biaxial retardation film with $N_z \approx 0.5$ (Chen et al., 1998). These two methods have a large wavelength dependence, which may cause a large color shift. Figure 4.6c shows an improved method using two biaxial retardation films with $N_z \approx 0.28$ and 0.72 (Ishinabe et al., 2002). Utilizing this method, a color shift caused by one biaxial film is compensated by the other biaxial film.

The retardation compensation and the polarizer compensation give us a design rule of a viewing-angle compensation of LCDs, but it is not well enough to optimize

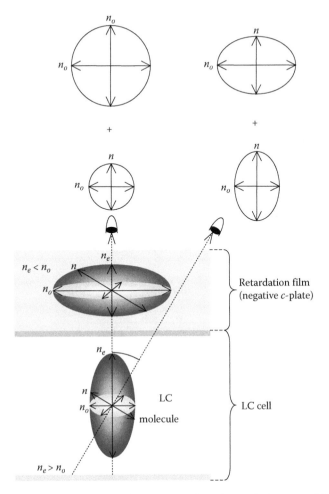

**FIGURE 4.5** A principle of retardation compensation. A retardation caused in a positive uniaxial media at an oblique angle can be compensated by placing negative uniaxial media with the same optic axis direction.

a viewing-angle performance quantitatively. It is possible to simulate a viewing-angle property of LCDs using Jones matrix method (Jones, 1941), Extended Jones matrix method (Lien, 1990) and Berreman's $4 \times 4$ matrix method (Berreman, 1972). Currently, commercial simulation tools are also available.

## 4.2  CALAMITIC COMPENSATION FILMS

### 4.2.1  VIEWING-ANGLE COMPENSATION FOR VA- AND IPS-LCDs

Nowadays, LCDs are utilized in large size TVs over 40 in. due to the recent improvement in viewing-angle technologies. A major LCD mode for large size LCD-TV is VA

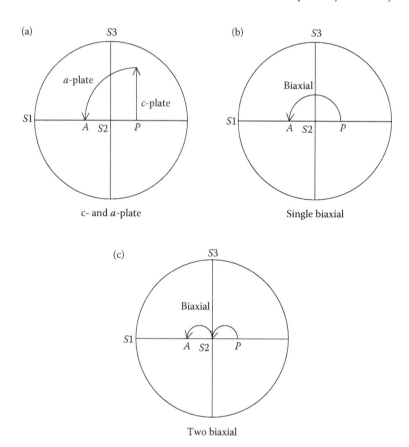

**FIGURE 4.6** Principles of the polarizer compensation using Poincaré sphere (Kliger et al., 1990) observed at an oblique angle: (a) *c*- and *a*-plate, (b) single biaxial (Chen et al., 1998), (c) two biaxial (Ishinabe et al., 2002).

mode and IPS mode known as a wide viewing angle mode LCDs. However, viewing-angle compensation films are still required even though those modes have relatively wide viewing angle compared to a TN mode, and calamitic mesogens are useful for viewing-angle compensation of LCDs, especially VA mode and IPS mode LCDs. In this section, various compensation films using calamitic reactive mesogens and their applications to LCDs are discussed.

## 4.2.2 VIEWING-ANGLE COMPENSATION OF IPS MODE

IPS-LCDs have a good viewing-angle performance and a natural gray scale performance. Viewing-angle compensation configurations for IPS-LCDs (Kajita et al., 2006) using a low retardation TAC film (Nakayama et al., 2005) has been proposed as shown in Figure 4.7. Another method using two positive *a*-plates and a positive *c*-plate made of calamitic reactive mesogens has also been proposed (Skjonnemand et al., 2005).

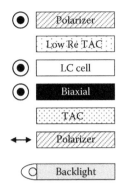

↔ ⊙ : Absorption axis for polarizer/slow axis for retarder

**FIGURE 4.7** Viewing-angle compensation configurations for IPS-LCDs (Kajita et al., 2006) using a low retardation TAC film (Nakayama et al., 2005).

### 4.2.3 VIEWING-ANGLE COMPENSATION OF VA MODE

VA-LCDs also have a good viewing-angle performance and a high on-axis *CR*. Figure 4.8 shows three major viewing-angle compensation configurations for VA-LCDs. The combination of negative *c*- and positive *a*-plate (Ohmuro et al., 1997; Chen et al., 1998) can compensate VA cell. In a *S*1–*S*3 plane of Poincaré sphere observed at an oblique angle, a negative *c*-plate changes from bottom to top and a VA cell as a positive *c*-plate changes it from top to bottom. Then a positive *a*-plate rotates the polarization state with a center at *P* since a slow axis of the positive *a*-plate coincides with a transmission axis of a polarizer. The movement caused by a part of VA cell is compensated by the movement of the negative *c*-plate. The separation of *P* and *A* on the Poincaré sphere is compensated by the rest of the VA cell movement and the whole *a*-plate movement, as explained in Figure 4.6a. The former corresponds to the retardation compensation, and the latter corresponds to the polarization compensation. As mentioned in Section 4.1.1.2, a positive *a*-plate can be made of homogeneously aligned calamitic-reactive mesogens and a negative *c*-plate can be made of homeotropically aligned discotic-reactive mesogens or a narrow pitch cholesteric structure of calamitic-reactive mesogens.

A two-biaxial method and a single-biaxial method utilize biaxial-retardation films with $N_z$ greater than 1.0. The single-biaxial film method is less expensive than the two-biaxial method with similar image performance due to the cost of films. In the next section, the single-biaxial method using a biaxial deformed cholesteric film composed of calamitic-reactive mesogens is discussed.

### 4.2.4 PHOTO-INDUCED BIAXIAL CHOLESTERIC LCs

Recently, VA-LCDs have diversified their usages for various kinds of applications such as TVs, monitors, tablets, car-navigation systems, and so on. VA-LCDs have a wide range of cell-gap for each application and retardation values of viewing-angle

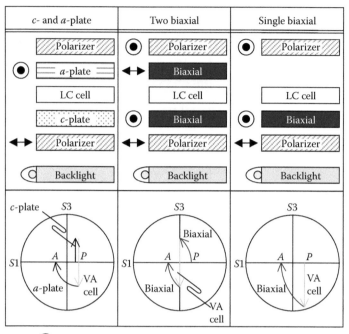

$\longleftrightarrow$ $\textcircled{\bullet}$: Absorption axis for polarizer/Slow axis for retarder

**FIGURE 4.8**   Viewing angle compensation configurations for VA-LCDs (Ohmuro et al., 1997; Chen et al., 1998): (a) *c*- and *a*-plate, (b) two biaxial, (c) single biaxial.

compensation films have to be optimized for each cell-gap. Therefore, an adjustability of on-axis retardation (*Re*) and out-of-plane retardation (*Rth*) of compensation films is required. It has been reported that a deformed cholesteric structure is biaxial (Broer et al., 1999). The deformed cholesteric structure is induced by the exposure of polarized UV light with a rod-like dichroic photoinitiator. These films can be applied for compensation of VA-LCDs (Kim et al., 2004; Amimori et al., 2005; Suzuki et al., 2006). This method has a great potential to control retardations and has an extremely high temperature and moisture stability.

### 4.2.5   OPTICAL MODEL FOR DEFORMED CHOLESTERIC STRUCTURE

According to the theory for form birefringence media (Yeh and Gu, 1999), the refractive indices of a layered stack of different materials shown in Figure 4.9 are described as

$$n_{x,y}^2 = \sum_a n_a^2 f_\alpha$$

$$\frac{1}{n_z^2} = \sum_a \left(\frac{1}{n_\alpha}\right)^2 f_\alpha,$$

(4.4)

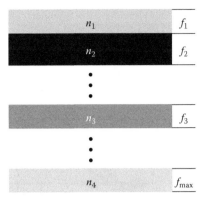

**FIGURE 4.9** Schematic illustration of form birefringent media.

where $\alpha$ denotes a type of materials, $n_\alpha$ is a refractive index of material and $f_\alpha$ is a volume fraction for material $\alpha$. The first equation is for in-plane refractive indices and the second is for a refractive index in the thickness direction.

The on-axis retardation induced by a photo-alignment is explained as a form birefringence of the deformed cholesteric structure. Let us assume an orientation distribution function of the deformed cholesteric structure as

$$f(\phi) = \frac{1 + A\cos(2\phi)}{2\pi}, \tag{4.5}$$

where $A$ is defined as a distortion factor in the range between 0 and 1. Since $A$ is positive, the densities of LC aligning along $\phi = 0$ and $\phi = \pi$ become maximum and the densities perpendicular to them become minimum. Then, $\phi = 0$ corresponds to the exposed polarization direction, according to the mechanism of the photo-alignment in this system.

A refractive index of a cholesteric "mono-" layer, which an incident polarization interacts with, is

$$n(\phi) = \left(\sqrt{\frac{\cos(\phi - \beta)^2}{n_e^2} + \frac{\sin(\phi - \beta)^2}{n_o^2}}\right)^{-1}, \tag{4.6}$$

where $\phi$ is an angle of LC alignment direction in the "mono-" layer and $\beta$ is an angle of polarization plane of observing incident light. Figure 4.10 illustrates the configuration of a deformed cholesteric film with an incident polarization and a cholesteric layer. $n_e$ and $n_o$ are extraordinary refractive index and ordinary refractive index of the LC, respectively. Then, we can obtain the refractive index in the direction $\beta$ from the form birefringence theory extended to the deformed cholesteric structure as

$$n(\beta)^2 = \int_0^{2\pi} n(\phi)^2 f(\phi)\, d\phi. \tag{4.7}$$

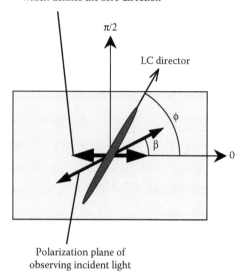

Polarization plane of exposed UV, which defines the zero direction

π/2

LC director

φ

β

0

Polarization plane of observing incident light

**FIGURE 4.10**  Illustration of the configuration of a deformed cholesteric film with an incident polarization and a cholesteric layer. The direction of the LC director varies in the thickness direction.

From Equation 1.6, $n_s$, the refractive index of deformed cholesteric film in the slow axis, is in the case $\beta = 0$,

$$n_s = n(0)$$

$$= \left[ \int_0^{2\pi} n(\phi) f(\phi) d\phi \right]^{1/2}$$

$$= \left[ \int_0^{2\pi} \left( \frac{1 + A\cos(2\phi)}{2\pi} \right) \left( \sqrt{\frac{\cos(\phi - \beta)^2}{n_e^2} + \frac{\sin(\phi - \beta)^2}{n_o^2}} \right)^{-1} d\phi \right]^{1/2} \quad (4.8)$$

$$= \frac{\sqrt{n_e n_o} \{ n_e(1 + A) + n_o(1 - A) \}}{\sqrt{n_e + n_o}}.$$

Similarly, $n_f$, the refractive index of deformed cholesteric film in the fast axis, is in the case $\beta = \pi/2$,

$$n_s = n(\pi/2) = \frac{\sqrt{n_e n_o} \{ n_e(1 - A) + n_o(1 + A) \}}{\sqrt{n_e + n_o}}. \quad (4.9)$$

Therefore, the birefringence becomes,

$$\Delta n = \sqrt{\frac{n_e n_o}{n_e + n_o}} \left\{ \sqrt{n_e (1 + A) + n_o (1 - A)} - \sqrt{n_e (1 - A) + n_o (1 + A)} \right\}.$$

(4.10)

Obviously, the refractive index in the thickness direction always equals to $n_o$. Thus, three refractive indices in principal axes can be obtained.

Another approach based on Berreman's $4 \times 4$ matrix is the following (Amimori and Suzuki, 2007).

$$n_s = \sqrt{n_e^2 \left(\frac{1}{2} + \frac{A}{4}\right) + n_o^2 \left(\frac{1}{2} - \frac{A}{4}\right)}$$

(4.11)

$$n_f = \sqrt{n_e^2 \left(\frac{1}{2} - \frac{A}{4}\right) + n_o^2 \left(\frac{1}{2} + \frac{A}{4}\right)}.$$

The results for $n_s$ and $n_f$ from form birefringence and Berreman's $4 \times 4$ matrix are not completely the same. However, the birefringence from Berreman's $4 \times 4$ is reduced to

$$\Delta n = \sqrt{n_o^2 \left(\frac{1}{2} - \frac{A}{4}\right) + n_e^2 \left(\frac{1}{2} + \frac{A}{4}\right)} - \sqrt{n_e^2 \left(\frac{1}{2} - \frac{A}{4}\right) + n_o^2 \left(\frac{1}{2} + \frac{A}{4}\right)}$$

$$= \frac{A}{2}(n_e - n_o) + \frac{A(A^2 - 4)}{64 n_o^2}(n_e - n_o)^3 - \frac{A(A^2 - 4)}{64 n_o^3}(n_e - n_o)^4 + O((n_e - n_o)^5)$$

$$\cong \frac{A}{2}(n_e - n_o).$$

(4.12)

The results of the birefringence from form birefringence and Berreman's $4 \times 4$ matrix are consistent up to the 4th order of $(n_e - n_o)$.

### 4.2.6  ANALYSIS OF DEFORMED CHOLESTERIC STRUCTURES

Deformed cholesteric structures to investigate their biaxial properties and alignment structures are prepared as follows. A 33 wt.% mixture solution of reactive mesogen monoacrylates, diacrylates, a chiral dopant, surfactants, and a dichroic photoinitiator are dissolved in 2-butanone was prepared. The solution is coated onto a homogeneous alignment layer on TAC film (TD80, *Fuji Photo Film Co., Ltd.*) using Meyer Bar, and then the sample is annealed at 100°C for 2 min and then polarized UV was exposed. X-ray diffraction (XRD) measurements are performed by a grazing incidence in-plane diffractometer (ATX-G, CuK$\alpha$ radiation of 0.154 nm, 50 kV, 300 mA, *Rigaku Corp.*). Soller slits of 0.48° and 0.45° vertical divergences are used for source and detector, respectively. A divergent slit ($1 \times 10$ mm) is used to control the exposure area of incident x-ray, while another divergent slit ($0.1 \times 10$ mm) is used to limit the output signal from a sample.

Before executing a $\phi$ scan, we have done a $2\theta\chi/\phi$ scan to find a specific peak for cholesteric LC alignment. Figure 4.11 shows the XRD patterns of the cholesteric LC

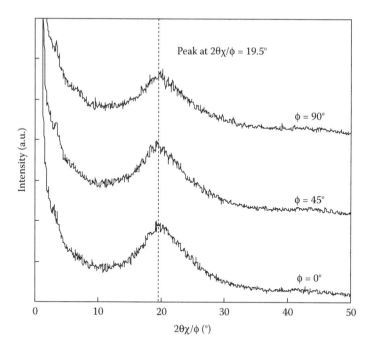

**FIGURE 4.11** XRD patterns of the cholesteric LC film measured by $2\theta\chi/\phi$g can at $\phi = 0°$, $45°$, and $90°$.

film at $\phi = 0°, 45°$, and $90°$. In all patterns, the peaks at $19.5°$ correspond to $0.455$ nm spacing can be considered as a nematic LC orientation along $\phi$.

Figure 4.12 shows the azimuthal XRD intensity distributions at $2\theta\chi/\phi = 19.5°$ of (Figure 4.12a) unpolarized UV exposed sample and (Figure 4.12b) polarized UV exposed sample. The XRD intensity distribution of (Figure 4.12b) has maximum peaks at $0°$ and $180°$ whereas, that of (Figure 4.12a) is flat. The similar XRD results are obtained from deformed cholesteric LC elastomers induced by stretching mechanical

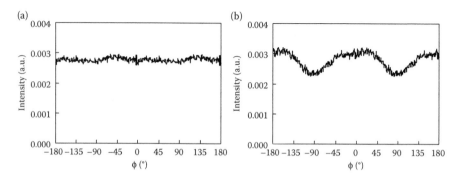

**FIGURE 4.12** Azimuthal XRD intensity distributions at $2\theta\chi/\phi = 19.5°$ of (a) unpolarized UV exposed sample and (b) polarized UV exposed sample.

field and higher intensity indicates that more LC directors align along $\phi$ (Cicuta et al., 2002). Then the deformed cholesteric structure induced by the exposure of polarized UV is observed.

### 4.2.7  RETARDATION ADJUSTABILITY OF DEFORMED CHOLESTERIC FILMS

Figure 4.13 shows the retardation of (Figure 4.13a) unpolarized UV exposed sample and (Figure 4.13b) polarized UV exposed sample as a function of incident angle. The deformed cholesteric film has on-axis retardation, whereas, the unpolarized UV exposed sample does not have. Assuming, $n_e = 1.65$, $n_o = 1.50$ and the thickness is $2.4\,\mu m$, measured data are well described by our optical model for deformed

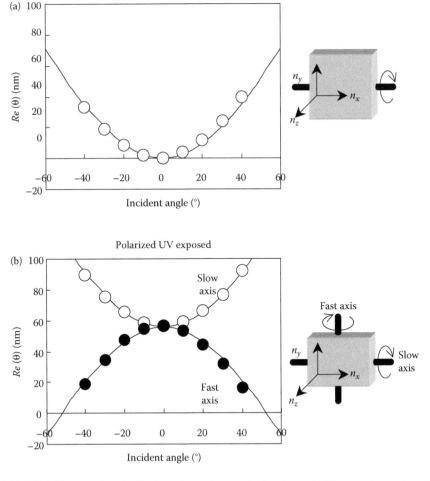

**FIGURE 4.13**  Angular distribution of retardation of (a) polarized UV-exposed sample and (b) unpolarized UV-exposed sample. Fitting curves are obtained from the optical model for deformed cholesteric structure based on form birefringence theory.

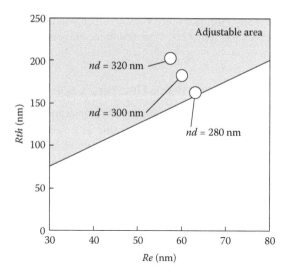

**FIGURE 4.14** Optimum *Re* and *Rth* values for $\Delta nd = 280$, 300, and 320 nm of VA cell. Deformed cholesteric biaxial films can be created in an adjustable area painted in the figure.

cholesteric structure. Deformed factors of the undeformed and deformed cholesteric films are 0.000 and 0.311, respectively.

One of the advantages of biaxial deformed cholesteric films is their retardation-adjustable property. A cell gap ranges from 280 to 320 nm is often used and the optimum *Re* and *Rth* values for various cell gaps are simulated as shown in Figure 4.14. An adjustable area is also shown in the figure. Retardations can be adjusted by changing a composition of mixture, peak irradiance, and energy density of UV and an extinction ratio of polarized UV.

Figure 4.15 shows the retardation properties of three kinds of mixtures. Retardations of cholesteric layers are normalized by on-axis retardations. Retardation properties are well controlled by changing the composition of LC mixtures. The dependence of retardation properties on peak irradiance and energy density is shown in Figure 4.16. Higher peak irradiance and energy density induce larger distortion of molecules. The dependence of retardation properties on extinction ratio is shown in Figure 4.17. Higher extinction ratio induces larger distortion of molecules. These results give us a way to adjust retardations precisely in a fabrication process with a real-time feedback system.

### 4.2.8 VIEWING-ANGLE PERFORMANCE OF VA-LCDs WITH A DEFORMED CHOLESTERIC FILM

The viewing-angle performance of VA-LCD using the photo-induced biaxial cholesteric film is investigated. The VA-LCD used is a commercial 17-inch SXGA VA-LCD monitor (Syncmaster 173P, *Samsung Electronics Co., Ltd.*). Polarizers on both sides of LC cell are replaced as shown in Figure 4.18 since the commercial VA-LCD may have viewing-angle compensation films on both sides. The biaxially

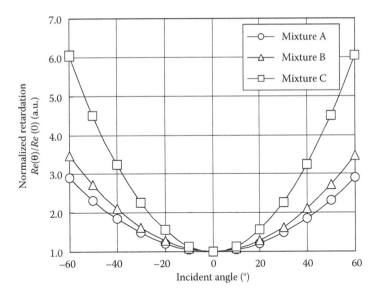

**FIGURE 4.15**  Dependence of normalized retardations $Re(\theta)/Re(0)$ of biaxial deformed cholesteric films on incident angles. Three kinds of mixtures have different biaxial properties.

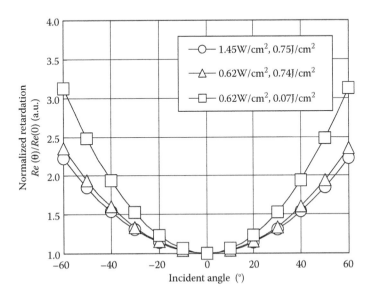

**FIGURE 4.16**  Dependence of normalized retardations normalized retardations $Re(\theta)/Re(0)$ of biaxial deformed cholesteric films with different exposure conditions on incident angles. Higher peak irradiance and energy density induce larger distortion of molecules.

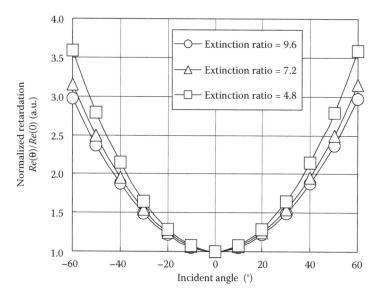

**FIGURE 4.17**  Dependence of normalized retardations $Re(\theta)/Re(0)$ of biaxial deformed cholesteric films with different extinction ratio of a polarized UV on incident angles. Higher extinction ratio induces larger distortion of molecules.

deformed cholesteric film is attached only on the backlight side. Since the deformed cholesteric structure is created by the polarization of UV and the dichroism of the photoinitiator, retardation can be easily controlled by changing an extinction ratio of polarized UV or the chemical structure of dichroic photoinitiator. Figure 4.18a

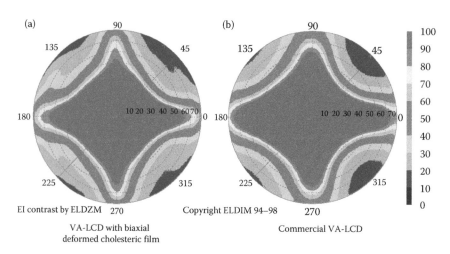

**FIGURE 4.18**  **(See color insert following page 304.)** Comparison of 17-inch SXGA VA-LCD (Syncmaster 173P, *Samsung Electronics Co., Ltd.*) compensated with (a) biaxial deformed cholesteric film to (b) the product as it is. The biaxial deformed cholesteric film is attached only on the backlight side.

and b show the isocontrast maps of the compensated VA-LCD and the product as it is, respectively. The VA-LCD with the deformed cholesteric film shown in Figure 4.18a has a wide viewing angle in all directions and all area keeps *CR* greater than 10. Compared to the commercial product shown in Figure 4.18b, the VA-LCD with the deformed cholesteric film has a similar viewing-angle performance and the on-axis contrast ratio is over 600, which indicates the scattering is almost none and there is no defect of LC alignment in a compensation film sample. Figure 4.11 shows the color shift property of the VA-LCD in black state. The color shift in black state is small and has a good color viewing-angle performance. The adjustability of deformed cholesteric structure has a great potential for the compensation of VA-LCDs.

## 4.3  DISCOTIC COMPENSATION FILMS

### 4.3.1  WIDE VIEW FILM

LCD monitors have almost totally replaced CRT monitors in the PC monitor market. TN-LCDs are popular and widely used in these monitors because of their advantages in high light transmittance, relatively fast response time, and yields of manufacturing processes even though the viewing-angle performance of TN-LCDs is poor compared to that of other LCD modes such as VA- and IPS-LCDs. Many approaches to improve the viewing-angle performance were proposed (Yang, 1991; Kobayashi and Iimura, 1997), and Wide View (WV) film, which consists of discotic reactive mesogen, has been widely used in TN-LCDs. In this section, the WV film as a unique discotic compensation film is discussed (Oikawa et al., 2007).

### 4.3.2  CONCEPT OF VIEWING-ANGLE COMPENSATION USING DISCOTIC REACTIVE MESOGEN

The TN LC layer in the black state has a calamitic hybrid alignment structure in which the direction of the director is continuously changing in the thickness direction. The optical property of this structure cannot be expressed by a single index ellipsoid. The optical compensation film should also have a hybrid alignment structure of disc-like molecules as discussed in Figure 4.5 in terms of retardation compensation. The alignment structure of the disc-like molecules corresponds to that of the black-state TN LC layer and thus three dimensionally compensates it. Figure 4.19 illustrates an idealized and simplified compensation configuration of the on-state TN LC layer compensated with the WV film (Mori et al., 2003). This compensation configuration minimizes the light leakage caused by the birefringence of the TN LC layer, leading to a high contrast ratio in all viewing angles. The WV film is based on this optical compensation concept and has succeeded in giving a remarkably wide-viewing angle since no stretched retardation film, which can only have uniaxial or biaxial index ellipsoid, can compensate the calamitic hybrid alignment in the on-state TN LC layer.

### 4.3.3  STRUCTURE OF WV FILM

Figure 4.20 shows a schematic diagram of WV film. Discotic mesogens are coated on an alignment layer on a TAC substrate in the WV film. A coated discotic mesogen layer

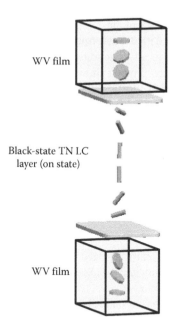

**FIGURE 4.19** The concept of an idealized and simplified compensation configuration of the on-state TN LC layer compensated with the WV film.

is annealed to take on discotic nematic ($N_D$) phase, resulting in a hybrid alignment structure. Then the discotic mesogen layer is cured by UV exposure to fix the hybrid alignment structure composed of PDM layer. According to the measurement by polarized micro-Raman spectroscopy of a WV film, it is suggested that the PDM has more

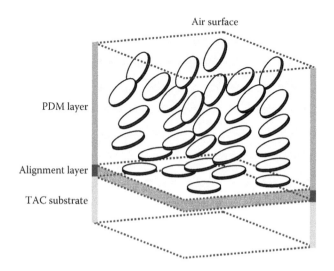

**FIGURE 4.20** Schematic diagram of a cross-sectional view of WV film. A hybrid alignment structure is composed of discotic reactive mesogens are aligned.

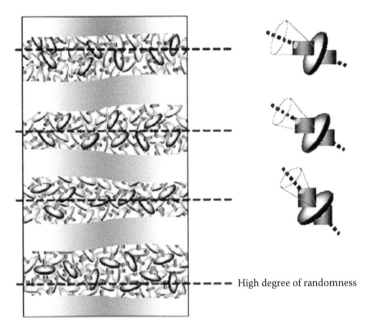

**FIGURE 4.21**  Cross sectional view of molecular alignment in a PDM layer measured by polarized micro-Raman spectroscopy (Takahashi et al., 2004). The PDM layer has both the average director and the degree of order profile in the depth direction.

complicated molecule orientations and that the degree of alignment of PDM molecules changes in terms of the depth. And the degree of alignment of the PDM was lower near the alignment layer and near the air surface, especially near the alignment surface region where there is a high degree of randomness as shown in Figure 4.21 (Takahashi et al., 2004). The WV film combines functions of viewing-angle compensation and surface protection of the polarizing layer composed of iodine-doped uniaxially stretched polyvinyl alcohol (PVA) layer. The TAC substrate of the WV film is directly attached onto the PVA layer in the roll-to-roll process. One of the advantages of WV film is the availability of fast and high yield roll-to-roll process to TAC substrate, alignment layer, PDM layer, and polarizer manufacturing, resulting in low cost.

### 4.3.4  VIEWING-ANGLE PERFORMANCE OF WV FILMS

As is discussed in previous sections, the WV film has a great potential to compensate TN-LCDs. In addition, the WV film also can control its viewing-angle properties for various requirements in display applications. So far, three types of WV films, such as WV-A, WV-SA, and WV-EA have been proposed. Figure 4.22 compares the viewing-angle performance between a conventional TN-LCD without WV film and TN-LCDs with various WV films. It can be seen that WV-A, the first commercialized WV film, resulting in a wider viewing angle, and WV-SA, the second version, results in a much wider viewing angle compared even with WV-A. Typical viewing angles at a contrast ratio of 10:1 for the horizontal and vertical directions are 130° and 100° for WV-A

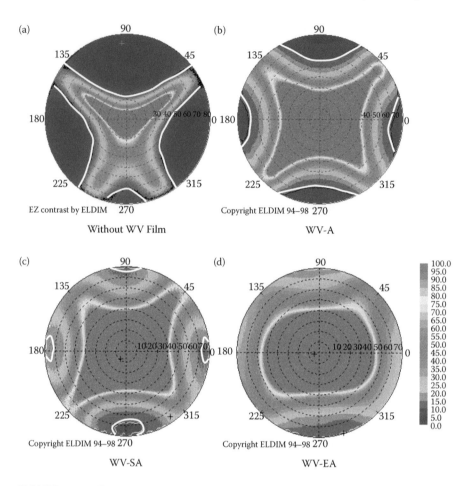

**FIGURE 4.22**  (See color insert following page 304.) Viewing-angle performance of TN-LCDs (a) without WV film, (b) with WV-A, (c) with WV-SA, and (d) with WV-EA.

and 145° and 135° for WV-SA. The difference in performance between WV-A and WV-SA is seen only in the thickness of the PDM layer.

Recent spread of LCDs brings larger and wider screens. Therefore, the viewing-angle characteristics and the color shift have become more important, especially in the horizontal direction. WV-EA is developed for wide-aspect-ratio LCD monitors and TVs. To widen the viewing angle, various parameters including a TAC substrate and a PDM layer with the depth profile of the order parameter are further optimized. In addition, the color shift in the off state of TN-LCDs is improved by optimizing the proportionality of the PDM layer to the TAC film in the total optical compensation of the WV film. As a result, the viewing angle of TN-LCDs with WV-EA exceeds 80° in all azimuthal directions.

Figure 4.23a and b show the horizontal viewing-angle dependence of light leakage in the on-state (black-state) TN-LCDs and the horizontal viewing-angle dependence

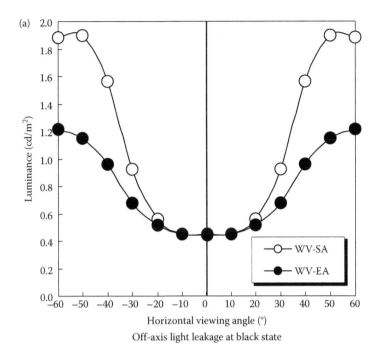

Off-axis light leakage at black state

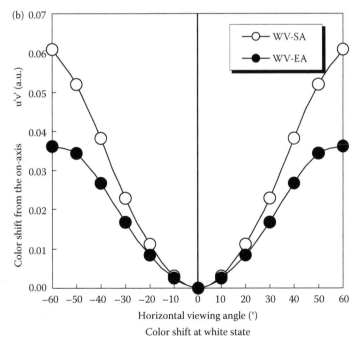

Color shift at white state

**FIGURE 4.23** Comparison of TN-LCD with WV-SA and WV-EA: (a) light leakage in the on-state (black-state) TN-LCDs, (b) color shift in the off-state (white-state) TN-LCDs.

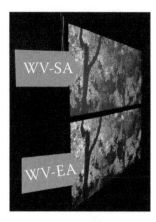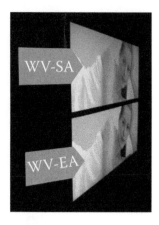

**FIGURE 4.24   (See color insert following page 304.)** Comparison of image quality of TN-LCDs between WV-SA and WV-EA at an oblique angle.

of color shift in the off-state (white-state) TN-LCDs with WV films attached, respectively. The on-state TN-LCD with WV-EA shows less light leakage at horizontal oblique angles than that with WV-SA. Figure 4.24 shows the comparison of the image qualities at oblique angles. The TN-LCD panel with WV-EA shows vivid color, much darker black quality and clear whiteness at horizontal oblique angles than that with WV-SA.

As for the most-recent TN mode, a faster response due to improved LC and a narrower cell gap was achieved without using a special drive circuit. Further progress in WV film will make TN-LCDs compatible with larger-sized LCD-TVs and multi-functional LCD monitors.

## 4.4   SUMMARY

Reactive mesogens have a great potential to the viewing-angle compensation because of their wide variety of alignment configuration as discussed in the chapter. Especially, the alignment of the WV film is sophisticatedly controlled to have a hybrid alignment structure of discotic molecules and leads to the ideal compensation to the complicated alignment structure of the on-state LC layer to minimize off-axis light leakage.

In addition, transflective LCDs, which is used in small size applications such as digital cameras and mobile phones, requires a quarter-wave plate only on the reflective part in order to make them thin, and retardation layers are placed in LC cells (Roosendaal et al., 2003; Imayama et al., 2007). For such an in-cell application, a high thermal stability is required sine annealing processes higher than 200°C exist. Conventional retardation films made of stretched polymers cannot be applied due to thickness, patterning disability, and poor thermal stability. Difunctional or more functional reactive mesogens realize the three-dimensional polymer network resulting in very high thermal stability and are suitable for the in-cell application.

## REFERENCES

Allia, P., Galatola, P., Oldano, C., Rajteri, M., and Trossi, L. 1994. Form birefringence in helical liquid crystals, *Journal de Physique II*, **4**, 333–347.

Amimori, I. and Suzuki, S. 2007. Study on the models for optical properties of photo-induced biaxial cholesteric liquid crystals, *Fujifilm Research & Development*, **52**, 42–46.

Amimori, I., Suzuki, S., Obata, F., and Ruslim, C. 2005. Deformed nanostructure of photo-induced biaxial cholesteric films and their application in VA-mode LCDs, *Journal of the SID*, **13**, 799–804.

Berreman, D.W. 1972. Optics in stratified and anisotropic media: 4×4-Matrix formulation, *Journal of the Optical Society of America*, **62**, 502–510.

Born, M. and Wolf, E. 1999. *Principles of Optics 7th (expanded) edition*. Cambridge: Cambridge University Press, pp. 790–811.

Broer, D.J. 1993. Photoinitiated polymerization and crosslinking of liquid-crystalline systems. In: Fouassier, J.P. and Rabek, J.F. (Eds.) *Radiation Curing in Polymer Science and Technology*. Vol. III. London: Elsevier Applied Science, pp. 383–443.

Broer, D.J., Mol, G.N., van Haaren, J.A.M.M., and Lub, J. 1999. Photo-induced diffusion in polymerizing chiral-nematic media, *Advanced Materials*, **11**, 573–578.

Chen, J., Kim, K. -H., Jyu, J. -J., Souk, J.H., Kelly, J.R., and Bos, P.J. 1998. Optimum Film Compensation Modes for TN and VA LCDs, *Society for Information Display 1998 International Symposium Digest of Technical Papers*, **XXIX**, 315–318.

Cicuta, P., Tajbakhsh, A.R., and Terentjev, E.M. 2002. Evolution of photonic structure on deformation of cholesteric elastomers, *Physical Review E*, **65**, 051704.

Fujimura, Y., Nagatsuka, T., Yoshimi, H., and Shimomura, T. 1991. Optical properties of retardation films for STN-LCDs, *Society for Information Display 1991 International Symposium Digest of Technical Papers*, **XXII**, 739–742.

Imayama, H., Tanno, J., Igeta, K., Morimoto, M., Komura, S., Nagata, T., Itou, O., and Hirota, S. 2007. Novel pixel design for a transflective IPS-LCD with an in-cell retarder, *Society for Information Display 2007 International Symposium Digest of Technical Papers*, **XXXVIII**, 1651–1654.

Ishinabe, T., Miyashita, T., and Uchida, T. 2002. Wide-viewing-angle polarizer with a large wavelength range, *Japanese Journal of Applied Physics*, **41**, 4553–4558.

Jones, R.C. 1941. A new calculus for the treatment of optical systems, *Journal of the Optical Society of America*, **31**, 488–493.

Kajita, D., Hiyama, I., Utsumi, Y., Miyazaki, K., Hasegawa, M., and Ishii, M. 2006. IPS-LCD with High contrast ratio over 80:1 at all viewing angles, *Society for Information Display 2006 International Symposium Digest of Technical Papers*, **XXXVII**, 1162–1165.

Kawata, K. 2002. Orientation control and fixation of discotic liquid crystal, *The Chemical Record*, **2**, 59–80.

Kim, K.-H., Lyu, J.-J., Chung, D.-H., Verrall, M., Slaney, K., Perrett, T., Parri, O., Lee, S. –E., and Lee, H.-K. 2004. Biaxial integrated optical film for VA mode LCD's made from *in-situ* photopolymerised reactive mesogens, *Proceedings of the 24th International Display Research Conference in Conjunction with the 4th International Meeting on Information Display*, **24**, 773–775.

Kliger, D.S., James, W.L., and Randall, C.E. 1990. *Polarized Light in Optics and Spectroscopy*, San Diego, CA: Academic Press, pp. 103–152.

Kobayashi, S. and Iimura, Y. 1997. Multidomain TN-LCD fabricated by photoalignment, *Proceedings of SPIE*, **3015**, 40–51.

Lien, A. 1990. Extended Jones matrix representation for the twisted nematic liquid-crystal display at oblique incidence, *Applied Physics Letters*, **57**, 2767–2769.

Mori, H. 2005. The Wide View (WV) Film for enhancing the field of view of LCDs, *Journal of Display Technology*, **1**, 179–186.

Mori, H., Nagai, M., Nakayama, H., Itoh, Y., Kamada, K., Arakawa, K., and Kawata, K. 2003. Novel optical compensation method based upon a discotic optical compensation film for wide-viewing angle LCDs, *Society for Information Display 2003 International Symposium Digest of Technical Papers*, **XXXIV**, 1058–1061.

Nakayama, H., Fukagawa, N., Nishiura, Y., Nimura, S., Yasuda, T., Ito, T., and Mihayashi, K. 2005. Development of low-retardation TAC FILM, *Proceedings of the 12th International Display Workshop in Conjunction with Asia Display 2005*, 1317–1320.

Ohmuro, K., Kataoka, S., Sasaki, T., and Koike, Y. 1997. Development of super-high-image-quality vertical-alignment-mode LCD, *Society for Information Display 1997 International Symposium Digest of Technical Papers*, **XXVIII**, 845–848.

Ohta, M., Oh-e, M., and Kondo, K. 1995. Development of super-TFT-LCDs with in-plane switching display mode, *Asia Display '95, Proceedings of the 15th International Display Research Conference*, Hamamatsu, Japan, 707–710.

Oikawa, T., Yasuda, S., Takeuchi, K., Sakai, E., and Mori, H. 2007. Novel WV film for wide-viewing angle TN-mode LCDs, *Journal of SID*, **15**, 133–137.

Roosendaal, S.J., van der Zande, B.M.I., Nieuwkerk, A.C., Renders, C.A., Osenga, J.T.M., Doornkamp, C., Peeters, E., Bruinink, J., and van Haaren, J.A.M. M. 2003. Novel high performance transflective LCD with a patterned retarder, *Society for Information Display 2003 International Symposium Digest of Technical Papers*, **XXXIV**, 78–81.

Schadt, M., and Helfrich, W. 1971. Voltage-dependent optical activity of a twisted nematic liquid crystal, *Applied Physics Letters*, **19**, 127–128.

Skjonnemand, K., Parri, O., Harding, R., May, A., and Dunn, C. 2005. Reactive mesogen optical elements in liquid crystal displays, *Proceedings of the International Display Manufacturing Conference 2005*, 97–100.

Suzuki, S., Obata, F., Amimori, I., Katagiri, T., and Mizutani, H. 2006. Retardation-adjustable photo-induced biaxial cholesteric films for VA-LCDs, *Society for Information Display 2006 International Symposium Digest of Technical Papers*, **XXXVII**, 1539–1542.

Toyooka, T., and Kobori, Y. 2000. Application of liquid crystalline polymer film to display device, *Ekisho*, **4**, 159–164.

Takahashi, Y., Watanabe, H., and Kato, T. 2004. Depth-dependent determination of molecular orientation for WV-Film, *Proceedings of the 11th International Display Workshops*, Niigata, Japan, 651–654.

Yang, K.H. 1991. Two-domain twisted nematic and tilted homeotropic liquid crystal displays for active matrix applications, *Proceedings of the 11th International Display Research Conference*, San Diego, CA, 68–71.

Yeh, P., and Gu, C. 1999. *Optics of Liquid Crystal Displays*. New York, NY: John Wiley Interscience, pp. 380–385.

# 5 Interferometric Applications Using Liquid Crystalline Networks

*Toralf Scharf and Gerben Boer*

## CONTENTS

## 5.1  INTRODUCTION

An interferometer is a device that divides a light beam into two (or more) beams, which follow different optical paths before being recombined to cause interference. The most important parameter for interferometers is the optical phase retardation (or shift) or optical path difference (OPD) that results from the fact that different rays travels different optical distances after splitting. In the particular case of a polarization interferometer, the entering light beam is divided by a polarization-division

**119**

device into two beams having orthogonal polarization directions. The first polarization interferometer was made by Jamin in 1868 (Jamin, 1868).

Polarization interferometers can be divided into two categories: The ones with a weak beam-splitting angle, called collinear, and the ones with a high-splitting angle, called noncollinear. In the first category, because of the weak splitting the optical phase shift is mainly created by the birefringence of the medium that is traversed by these beams. The main advantages of these interferometers is that the two beams follow the same geometrical path; they are called the "common-path interferometers". They are usually insensitive to environmental conditions since an external perturbation affects the two paths equally and consequently no unwanted additional phase shift is introduced. In the second category, the polarization-division is mainly performed by a beam splitter that may be wire grids or polarization beam-splitter (PBS) cubes. These systems separate the incoming beam into two orthogonal polarized beams following two distinct geometrical paths known as the Michelson's interferometer. These interferometers offer the possibility to induce large phase shifts but they cannot be considered as common-path interferometers anymore. Several examples of such systems, mainly used in astronomical spectroscopy, can be found in Polavarapu (1998). Polarization interferometers have many applications; most of them are presented in Françon and Mallick (1971). The most popular one is probably the observation of phase objects in microscopy. But there are also other applications such as optical surface and system testing, spectrometry, or optical thickness measurements. The advantage of common-path interferometers is their robustness. Polarization interferometers usually use uniaxial crystals as birefringent material with the disadvantage to increase the price considerably. It is therefore interesting to replace the inorganic crystal by densely cross-linked liquid crystalline networks (LCNs) in some applications.

Presented here are three well-known common-path interferometers with LCN. We start the discussion by introducing the priciples of polarization interferometry. We then will discuss the materials used to fabricate the optical elements. Components were fabricated from different materials. Birefringent elements were fabricted from liquid crystal (LC) mixtures, polymerizable liquid crystal mesogens (PLCM), and cross-linkable side-chain liquid crystal polymers (LCP). In the second part application in optical systems are considered. We present three different systems:

- The first system is a common-path polarization interferometer that is used for optical surface testing. The key element is a double-focusing plano-concave LC lens, which divides the incoming wave front into a testing wave front and a (focused) reference wave front. From the interference pattern the topology of the surface under test can be retrieved.
- The second system we examine is a polarization interferometer for phase-object observation in microscopy. The principle of operation is called the Differential Interference Contrast (DIC) microscopy. DIC microscopy is a widely used method for the observation of transparent specimens with low refractive-index contrast, such as cultured cells in biology for example. We present a design based on two Wollaston prisms made of polymer LC that can be inserted in almost every polarization microscope.

- The third system is a static Fourier transform spectrometer (FTS). It is based on modified Wollaston prisms made from cross-linked LCN liquid. We took an ultra-compact spectrometer head with liquid crystalline polymers and characterized its performance for spectral measurement in the visible region.

## 5.2  OPTICAL INTERFERENCE IN BIREFRINGENT MATERIALS

Polarization interferometers are based on interference of light. Interference is only possible under certain circumstances. There are three parameters that influence the appearance of interference.

- The effect of the spectral composition of light that interferes given by the temporal coherence.
- The effect of the size of the source leading to the spatial coherence.
- The conditions for interference with polarized light given by the state of polarization and the degree of polarization.

An important parameter, which determines the visibility of fringes in an interference experiment, is the spatial coherence (Goodman, 2000). For simulations, extended light sources are usually divided in a multitude of areas where the light within such an area is considered to be coherent. The coherence area is an important parameter that characterizes random light sources and must be considered in relation to other dimensions of the optical system. For example, if the area of spatial coherence is greater than the size of the aperture through which light is transmitted, the light is regarded as spatially coherent. Similarly, if the coherence area is much smaller than the resolution of the optical system, it is regarded as incoherent. In most practical cases it must be regarded as incoherent. Interference occurs when coherent waves are superposed. Those portions of the light field, which produce interference, have to be spatially coherent. A more extensive discussion about spatial coherence that uses the concept of modes can be found in Dändliker (2000). The spatial coherence is particularly important if amplitude splitting leads to change of direction of propagation of the rays. In such a situation, it can be shown (Hariharan, 2003) that the zone of maximal-interference contrast is found at the point where corresponding rays intersect. For thin-film imaging devices such as LC displays it is not of importance because the splitting is marginal and travel distances are small so that interference effects are mainly affected by the temporal coherence and polarization of light. For polarization interferometers of a certain resolution, the phase shift between interfering beams should be large and spatial coherence becomes important and cannot be neglected when system performance is discussed.

With polarized light there are additional conditions to observe the interference effects usually referred to as the rules of Fresnel and Aarago. The following rules apply (Goldstein, 2003):

a. Two waves that are linearly polarized in the same plane can interfere.
b. Two waves that are linearly polarized with perpendicular polarizations cannot interfere.

c. Two waves that are linearly polarized with perpendicular polarizations, if derived form the perpendicular components of unpolarized light and subsequently brought into the same plane, cannot interfere.

d. Two waves that are linearly polarized with perpendicular polarizations, if derived form the same linearly polarized wave and subsequently brought into the same plane, can interfere.

Beside these basic rules there is an additional factor that influences the interference fringe contrast, that is the degree of polarization. For a decreasing degree of polarization the interference fringe contrast decreases too because only the completely polarized part of light leads to useful interference effects. Effects become dramatic if imperfect polarizers are used or depolarization happens (Scharf, 2006). Depolarization effects in LC layers are caused by impurities and local order variations especially in thick layers. For polarization interferometers as discussed here this is an important fact because the contrast of the useful interference effects might become too small for detection.

## 5.3   ORGANIC BIREFRINGENT ELEMENTS TECHNOLOGY

### 5.3.1   Cell Fabrication Technology

The organic materials we used for our experiments are LCs and cross-linkable LCN of two types: polymerazible liquid crystal mesogens (PLCM) and cross-linkable side-chain LCPs. We used conventional technology for cell fabrication. After having an insight in this matter we will discuss details of the material processing. To make thick LC components, we need a cell that confines the liquid crystalline material at all fabrication stages. To produce uniform alignment different methods can be used that are either triggered by external fields, such as magnetic or electric or by the cells surfaces. We used the conventional surface-alignment technique to produce bulk orientation for thicknesses up to 500 μm. Commercial polyimides were deposited and rubbed to implement strong anchoring with a certain alignment direction and pretilt. We produced essentially three types of cells: wedge-shaped cells, planar cells and lenses. Before assembling, the glass substrates were cleaned in a soap solution (Deconnex) in an ultrasonic bath, rinsed with DI water, dried and coated with an alignment layer in a spin-coater. All cells were mounted manually. As shown in Figure 5.1, the wedge-shaped cells were obtained by placing spacers between two glass substrates. Substrate thickness is usually 0.55 mm. The cells were filled with the LC materials by a capillary through inlets and finally sealed with a two-component glue (Araldite) or a UV polymerizable glue from Norland (NOA 81).

The wedge-shaped cells are very useful to study the alignment quality in the functioning of the LC layer thickness. Because they produce interference fringes between crossed polarizes it is easily possible to study stress birefringence and deformations. For an ideal birefringent wedge the fringes are exactly parallel. Figure 5.2 shows a wedge-shaped cell with poor performance. Deformations of the fringes give an indication of the homogeneity of the alignment and stress birefringence.

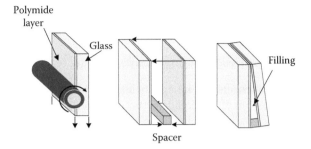

**FIGURE 5.1**  Assembling technology for wedge shaped LC cells. Glass slides coated with polyimide were rubbed and used as substrates in sandwiched cell. On one side a polyimide foil spacer assures a predefined thickness.

The LC lenses were made out of a flat glass substrate and a plano-concave glass lens. Since one cannot fill the lenses by a capillary, they were filled over the edge. The glass cover is smoothly deposed on the lens and the LC excess material is evacuated from the sides. The difficulty results here is to avoid air bubbles being trapped in the center of the assembly. The lens and the cover are finally sealed together with glue. The depth of the plano-convex lenses that can be used is limited because of the rubbing process that gives the alignment, since the rubbing cylinder of a conventional rubbing machine cannot "penetrate" inside the lens. The depth of the concave lens

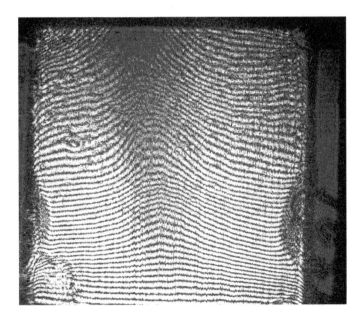

**FIGURE 5.2**  A 25 mm × 25 mm LCP cell seen between crossed polarizers oriented under 45°. The deformation of the fringes indicates stress birefringence that appears when the polymerization is too fast.

cavity is then limited by the pile length of the rubbing cloth. However, deeper lenses can be used when rubbing manually.

### 5.3.2 LIQUID CRYSTALLINE CROSS-LINKABLE MATERIALS

For static optical components, we are mainly interested in a high birefringence and a good optical transmission of the liquid-crystalline material. It is also important that the LC has a nematic phase over a large temperature range, if the final device is used under difficult ambient conditions. Since in static applications the LC molecules are not switched, the elastic properties, viscosity, and electric properties are less relevant. However, the viscosity of the material should not be too high to be able to fill the cells and the elastic properties should not be too weak in order to avoid a high sensitivity to mechanical disturbances.

Densely cross-linked LCNs combine the properties of LCs and polymers. An LCN can be described as a normal flexible polymer with mesogenic groups incorporated into their chains. The mesogenic groups give the liquid-crystalline characteristics to the LCN and the flexible spacers (or chain), link these groups together to give the polymeric properties. LCNs can be aligned with the same methods as used for conventional liquid-crystalline materials. LCN optical components can be prepared by filling a prepared cell with a noncrosslinked LC polymer. The cell is then heated to the nematic phase. After reorientation, there are two methods to freeze the nematic phase: by *in situ* polymerization or by vitrification. All the LCN cells presented in this work were fabricated with the *in situ* polymerization method. In this method, the aligned LCN is irradiated with UV light for cross-linking. This curing process can be controlled by adding adequate quantities of photo-initiator or inhibitor and by adjusting the UV intensity and dose. One of the advantages of this method is that the material and the cell do not have to be heated to high temperatures. Once the LCN is cross-linked, it is very stable and the orientation cannot be disturbed anymore. In the vitrification method, LCNs with a high molar mass that exhibit a glass transition are used. In this method the aligned material is simply cooled down below the glass transition temperature. Of course, for practical use this glass transition must be significantly above ambient temperature. The main problem is the high viscosity of those materials. They must be heated at a very high temperature to align them in a uniform texture. LCNs are mainly used for high-strength plastics because of their chemical stability, poor solubility, high melting point, and especially for their excellent mechanical properties. This disqualifies LCNs for electro-optical applications since the molecules are linked together and consequently cannot be switched by applying any electric or magnetic field, as in the case of LC displays. But the fabrication of thermally-stable birefringent optical (static) components lead to promising applications of those materials for retardation films.

Two different material classes were used for experiments: A mixture of side-chain LCP was provided by Wacker-Chemie GmbH, which has high-weight molecules that are of a silicone backbone (siloxane) with at least five mesogenic side chains containing methacryl groups attached to it through a flexible spacer, similar to the material used in Morito et al. (1999). The material can be cross-linked through methacryl groups by UV illumination. The second material is commercialized LC mesogens from Merck originally developed by Phillips at the University of Eindhoven (Broer et al., 1989a–d). The molecules are constituted of a mesogenic unit with acrylate groups on one or on both ends. The acrylate groups are separated from the central aromatic core by flexible methyl spacers of variable size. These materials are typical members of PLCM.

They main problem of a thick cell fabrication is the deformation during processing and the resulting stress birefringence because of inhomogeneous shrinkage of the material. They appear during the polymerization process, when intermolecular distances decrease because of cross-linking. The thermal shrinkage that occurs during cooling can also induce stress birefringence. In Figure 5.2 we had showed an example of a fringe pattern of a wedge polymerized under nonoptimal conditions. It shows heavily deformed fringes. Such problems can be avoided by carefully choosing the process parameters and especially the temperature gradients when cooling and heating.

Figure 5.3 shows cells fabricated under different cross-linking conditions. The cells are shown between parallel and crossed polarizers. On the left the material contained 1% weight percent of photo-initiator (Irgacure 907, Ciba) and on the right an identical wedge-cell without photo-initiator was prepared. The photo-initiator is activated by UV light and creates monomeric radicals, which initiates the chain polymerization (Atkins, 1994). This means that polymerization velocity increases significantly with the initiator concentration. The small quantity of photo-initiator is added to the liquid crystalline material in order accelerate the polymerization process and to assure long-term stability. In the bottom-left of Figure 5.3 inhomogeneities because of stress birefringence become visible when the sample is observed between crossed polarizers. This is dominant for the cell containing the photo-initiator (left) and forms a bright area at the bottom of the cell. Apparently, the polymerization process was too fast for the cell containing the photo-initiator and the rapid shrinkage of the material caused some distortions of the alignment and stress birefringence. Similar behavior has been observed when illuminating the LCN material with high UV intensities. This also increases the polymerization rate and consequently causes distortions because of shrinkage. Comparable effects appear when the cooling rate is faster than 5 K/min. In this case the stress birefringence is because of thermal shrinkage.

To obtain optimal results, LC wedge cells filled with LCP were treated in the following manner. For high pretilt planar alignment we used rubbed polyimides from Nissan. The cells had to be filled at about 100°C because of the high viscosity of the LC-silicone at room temperature (Morito et al., 1999). The photo-initiator content was set to 0.1%. After filling, the cells must stay in the nematic phase for a certain time for presetting of the texture. Usually it takes 15–60 min at 100°C. The cells were then cooled down to room temperature or to the limit of crystallization temperature, which was close to room temperature. Finally, the cells were polymerized with UVA light at 365 nm at 2 mW/cm$^2$ for about 1 h. Good cell quality can be obtained by slowly cooling the LCP to room temperature and by reducing the polymerization velocity, which means that photo-initiator concentration and UV intensity were as low as possible. The drawback of this method may be a lower cross-linking of the material because of a reduced mobility of the molecules during polymerization.

**FIGURE 5.3** **(See color insert following page 304.)** LC polymer in a wedge cell polymerized under different conditions. The cells have planar alignment. A 25 mm × 25 mm LCP cell is seen between parallel and crossed polarizers on the top and bottom respectively. The quality of the cells is extremely sensitive on cross-linking conditions and amount of photoinitiator.

For PLCM a similar process was used. Alignment, photo-polymerization kinetics, birefringence, temperature dependence, and phase transitions of such materials have already been reported extensively in Broer et al. (1989a–d), Hikmet (1992), Broer et al. (1991), Broer and Moll (1991), and Schultz and Chartoff (1998). We used two raw materials for our fabrication that are called RM-82 and RM-257 (Merck). RM-82 has a nematic phase from 86°C to 116°C (when heated) and RM-257 from 70°C to 125°C. During cooling, the crystallization temperatures occur at lower temperatures and start at about 35°C for RM-257 and about 63°C for RM-82. At lower temperatures, the material rapidly begins to crystallize. Mixtures of mesogens generally show an extended nematic range. We lowered the crystallization temperature by mixing RM-82 and RM-257 in equal proportions. In this way, we were able to decrease the crystallization temperature of the mixture to close to room temperature (25°C). Note that this mixture has not been optimized and it does not necessarily correspond to the eutectic point with the lowest crystallization temperature. After polymerization in the nematic phase the molecules form a highly cross-linked network. Since the mesogenic units are uniaxially oriented along the rubbing direction in the nematic phase, the material shows the same optical properties as conventional LCs. The material is solid at room temperature. We filled the cells by a capillary at 110°C. After 15 min relaxation time, until domain walls disappeared, the cells were cooled down to room temperature at 3°C/min and polymerized by UV illumination ($\sim$1 mW/cm$^2$). As already observed, sometimes for the LCP, the cells show important alignment inhomogeneities and stress birefringence, which has been already mentioned for polymerized films (Broer and Mol, 1991; Hikmet and Zwerver, 1992). Best results were obtained when the polymerization temperature and polymerization rate were kept as low as possible. This can be realized by polymerizing at fairly low UV intensity at lower temperature and by using very low concentration of photo-initiator. For interferometric applications it is necessary to know the effective temperature dependence of the birefringence of the cross-linked material. Because of the very high stability of the LCP and PLCM in the cross-linked state, we could not observe any phase shift by varying the temperature with transmission measurements between polarizers (Boer, 2003). A more accurate measurement method is necessary. Nevertheless, we can state that the effective temperature dependence of the birefringence is lower than $10^{-5}$ K$^{-1}$, which is in accordance with Broer et al. (1989a).

### 5.3.3  ABSORPTION AND SCATTERING LOSSES OF THE POLYMER MATERIALS

The main differences between inorganic crystalline materials and liquid crystalline polymers for interference applications are losses because of imperfections. Such imperfections are intrinsic or because of manufacturing issues such as misalignment, pollution or material inhomogeneities. All these effects lead to absorption and scattering. Nematic LC samples of a certain thickness look generally milky; indicating that light is being scattered. In LCs, scattering is because of thermal fluctuations, causing a random variation of the director in time. We should keep in mind that the thermal fluctuations are not present anymore in polymerized LC materials. It is therefore worthwhile to check the diffusion properties of light for this class of material. To do this we measured the ratio between the scattered light and the total transmitted light with a Perkin-Elmer Lamda900 spectrometer. The scattered light was measured by placing a light trap in the integration sphere that removes the light within a cone of angle 10° around the directly transmitted light beam. We present here some indicative measurements of absorption and scattering losses of thick planar-aligned cells made of PLCM and LCP. We prepared two 300 μm thick cells with 0.55 mm glass substrates and filled them with RM-257 and LCP-silicone from Wacker–Chemie. A small amount ($c_w = 0.1\%$ of the total weight) of photo-initiator (Irgacure 819) was added into the LCP silicone to initiate the polymerization and assure high cross-linking. The results are reported in Figures 5.4 and 5.5. The transmission dip in Figure 5.4 of the LCP-silicone is between 380 and 500 nm because of the presence of photo-initiator in the LCP. The transmission is in general high and will not limit the performance in applications. Polymerizable liquid crystalline materials apparently do not scatter as

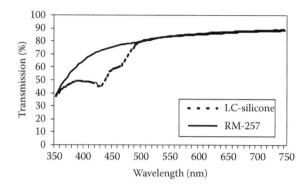

**FIGURE 5.4**   Transmission of a 0.3 mm thick cell filled with different LCN materials.

much as conventional LC's as can be seen in the measurements presented in Figure 5.5. To compare, E7 shows scattering losses of 45% at 400 nm for the same measurement geometry (Boer, 2003). The LC-silicone that has a birefringence of $\Delta n = 0.2$ that is comparable to the LC mixture E7 (Merck) shows only 20% losses for the same wavelength. Scattering is mainly because of imperfections. Figure 5.6 shows some typical defects that appear. In Figure 5.6a, air bubbles are shown that enter the cell during the filling process. These bubbles strongly influence their neighborhood because the trapped oxygen inhibits the polymerization process and produce important deformations around them. Figure 5.6b shows a defect because of crystallization. Actually, when the LC reaches a certain temperature it starts slowly to crystallize. In our case, crystallization started after about 30 min at room temperature when cooling down from nematic phase. Such defects can be avoided by curing above the crystallization temperature. The last type of defect presented on the right, Figure 5.6 shows pollution that might already be present in the original materials or that is created afterwards by chemical reactions between the different components of the mixture. It can be partially removed by dissolving and filtering the material by standard methods. Finally, by choosing an adequate amount of photo-initiator (0.1% of Irgacure 907), and by

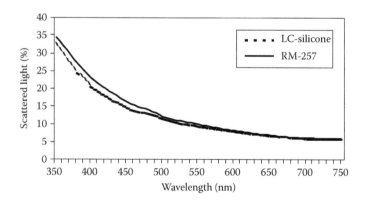

**FIGURE 5.5**   Scattering properties of a 0.3 mm thick cell filled with different LCN materials.

(a)

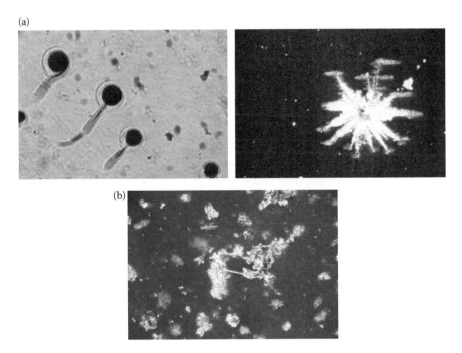

(b)

**FIGURE 5.6** Typical defects found because of processing errors in polymerizable LC materials. Besides bubbles (a) and local crystallizations (b) inhomogeneities form for unknown reasons that scatter the light and alter the optical performance of thick cells. (Thickness 0.3 mm.)

curing with adapted UV intensity ($\sim 1 \, mW/cm^2$) at low temperatures (22°C) and careful processing (avoiding air bubbles, crystallization, and pollution), we realized cells with a good optical quality as shown already in Figure 5.3. The cells show a high contrast between polarizers and good homogeneity.

## 5.4  COMMON PATH INTERFEROMETER

### 5.4.1  PRINCIPLE OF OPERATION

A double-focus interferometer is a polarization interferometer that uses a birefringent lens as beam splitter (Françon and Mallick, 1971). One beam, the reference beam, is focused on a small spot of the surface under test and the second beam, the testing beam, is expanded over the whole aperture. The main advantage of such common-path interferometers is its insensitivity to mechanical vibrations and thermal drift. One of the first double-focus interferometer for testing optical components had been built up by Dyson (Dyson, 1957). Since that time, other similar interferometers have been built and tested (Down et al., 1985; Kikuta et al., 1992; Iwata and Nishikawa, 1989; Kikuta et al., 1994). One of the principal problems of these systems is the use of expensive birefringent lenses made of inorganic crystal. Calcite is the preferred material here because of its high birefringence. To avoid the use of expensive components lenses incorporating LC mixtures can be used. Such lenses very demonstrated first in (Sato,

1979). Nose et al. built an interferometer using such a lens made of a LC mixture (Nose et al., 1997, 2001). In the present work, we have realized an interferometer with cross-linkable LCNs instead of conventional LCs.

The working principle of the interferometer is schematically depicted in Figure 5.7. Monochromatic light produced by a HeNe laser (633 nm) is polarized at 45° by a sheet polarizer, collimated and expanded to a diameter of more than 1 cm. This expansion defines together with the diameter of the birefringent lens the maximum size of the active surface. After beam expansion a polarized quasi-plane wave is obtained. This wave passes through a calibrated LC phase shifter and a beam splitter before attaining the birefringent plano-convex lens. The optical axis of the lens is oriented perpendicular to the propagation direction and at an angle of 45° with respect to the polarization direction. In our case the direction of the extraordinary refractive index is at 0°. Because the entrance polarization is at 45° two polarization component have to be considered separately. The polarization component perpendicular to the optical axis of the lens, which corresponds to the test wave front T, will experience the ordinary index. In our setup the ordinary refractive index of the LC is equal to the refractive index of the surrounding glass because we designed an index matching situation. So, the propagation of this polarization will not be affected until it reaches the test surface. The diameter of the beam stays the same. The polarization component parallel to the optical axis of the birefringent lens, which corresponds to the reference wave front R, will experience the extraordinary index. Because the extraordinary refractive index is larger than the ordinary in our case, it will be focused on a small spot on the test surface. The interferometer has a certain working distance and the sample has to be adequately positioned, so that the beam does not suffer from any appreciable wave-front distortion. When reflected from the test surface, the two wave fronts travels back through the birefringent lens and the beam splitter. Ideally, the reference beam should become planar again and the test wave front carries the information about the topology

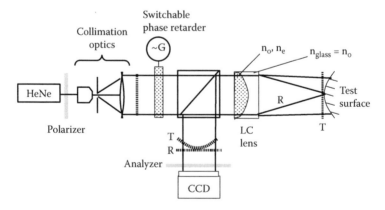

**FIGURE 5.7** Schematics of a common-path interferometer with a birefringent lens. The lens splits the beam into two parts: One is focalized on the surface and serves as a reference R. The second part T is modified by the surface under test. The two beams are brought to interference on the CCD camera by using a beam splitter and an analyzer.

of the test surface. The two wave fronts pass finally through a second polarizer oriented at ±45° to permit them to interfere. The interference pattern is recorded with a 12-bit CCD camera (Dalsa CA-D1). The surface profile is reconstructed by using a phase-shifting algorithm. To obtain easy phase retrieval with numerical algorithms, phase shifts between the two polarization components can be introduced by the switchable LC phase retarder. Because we are working in reflection, the measured wave front deformation W corresponds to twice the deformation of the surface profile. When the tested, the surface varies more than one wavelength (corresponding to $2\pi$ phase shift), W shows discontinuities caused by the periodicity of the trigonometric functions needed for the reconstruction. In order to retrieve the real surface profile a phase unwrapping method was applied (Boer, 2003).

The heart of the system is the birefringent lens. That lens can be made with polymer LC material or conventional LC mixtures. In our experiment, the birefringent lenses were made manually as described in Chapter 3. For best performance and most compact design, it is recommended to match one of the liquid crystalline refractive index and the surrounding media. We used BK7 glass lenses with the refractive index $n_{BK7} = 1.5167$ as the substrate lens. The extraordinary refractive index of the LC materials is close to that and used for matching. One has to assure that the birefringence is large in order to get a reasonably short focal length. Otherwise the working distance becomes very long and the system is not well suited for integration. A large number of LC mixtures are available and the index matching can be done with certain accuracy. Of course, the LC should also have a wide nematic range around room temperature and show good alignment even for thick cells, up to half a millimeter. Table 5.1 proposes a collection of three nematic LCs with a birefringence larger than 0.13 for which the ordinary refractive index is relatively close to the one of BK7.

The focal length $f$ of an immersed lens system with the extraordinary polarization component is given by

$$\frac{1}{f} = (n_e - n_{BK7})\frac{1}{R_{lens}} \tag{5.1}$$

where $R_{lens}$ is the radius of curvature of the concave substrate lens with and $n_{BK7}$ is its refractive index. Using the values for ZLI-1132 and $R_{Lens} = 24.81$ mm into Equation 5.1, one obtains a focal length $f = 200$ mm. In this case, the LC lens has a maximum thickness of 0.62 mm in the center and a diameter of 12 mm. To obtain a shorter focal length, LC mixtures with larger birefringence should be used. For example,

---

**TABLE 5.1**
**Refractive Indices of Different LC Materials Compared to the Refractive Index of BK7 Glass**

| Material | $n_e$ | $n_o$ | Mismatch |
|---|---|---|---|
| ZLI 1132 | 1.6280 | 1.4910 | 0.026 |
| ZLI 1738 | 1.7051 | 1.5168 | 0.001 |
| ZLI 1289 | 1.707 | 1.517 | 0.003 |

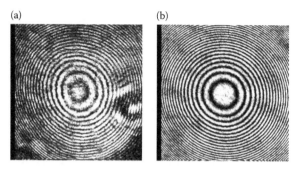

**FIGURE 5.8** Interference fringes produces by a birefringent lens placed in between crossed polarizers at an orientation of 45° with respect to the optical axis. (a) a lens made of LCP-Silicone (Wacker) and (b) a lens made of a LC mixture BL006 (Merck) are shown.

with an extraordinary refractive index of $n_e = 1.8$ one would get a focal lengths of only $f = 100$ mm. A smaller radius of curvature can also decrease the focal length, but the lens would become too thick to align the LC easily.

Several lens prototypes were fabricated by using LC mixtures and cross-linkable LCNs. A good quality test is provided by inspecting the interference fringes between crossed polarizers in the polarization microscope. Figure 5.8 shows the corresponding interference fringes. The captured area is $5 \times 5$ mm. The lens shown in Figure 5.8a is made with a commercial LC mixture BL007 and the one in Figure 5.8b with a LCP from Wacker. The polarizers are set at 45° with respect to the orientation of the optical axis of the lens. On the left, noisy and deformed fringes clearly show that the LCP lens has a poor quality for the present application. The poor quality is mainly because of stress birefringence. The right picture shows a regular and clear interference pattern, which reveals good optical quality. In the present prototype we present results obtained with a LC lens made of ZLI-1132. The small mismatch between the ordinary refractive index of ZLI-1132 and the glass BK7 makes the T wave front slightly divergent when illuminated with collimated light. It is therefore recommended to use an additional converging lens between the test surface and the birefringent lens, as shown in Figure 5.9. In the present setup, a lens with a focal distance of $f = 1.33$ m was used to compensate the divergence of the beam. The lens allows shortening the working distance.

Birefringent lenses in general show strong aberrations (Lesso et al., 2000). But there is a fundamental difference between lenses fabricated from crystals and this one using LC materials. A lens fabricated from crystals always has its optical axis in the same direction because it is machined out of a uniform block of material. Curved surfaces cut the optical axis at different angles, which affects the local refraction properties. For a LC lens, and given by the LC alignment technology, the optical axis always has the same angle with the surface. This angle is equal to the pretilt angle and follows the form of the limiting substrates. In LC cells this fact causes a spatial distribution of the optical axes within the cell and alters the imaging properties. In general a birefringent bifocal lens shows strong astigmatism, that is, the extraordinary wave front that has traversed the lens is curved differently along the x-axis and the y-axis.

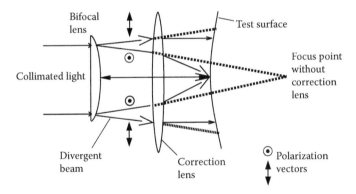

**FIGURE 5.9**  Setup with an additional correction lens. The correction lens has two purposes: It assures that the light stays collimated after leaving the interferometer and additionally it shortens the working distance.

As already been studied by Nose et al. (2001), this asymmetric effect is because of aforementioned alignment of the molecules. Figure 5.10 shows the alignment of the LC molecules that follow the curvature of the lens surface. If one studies the molecule orientation along the cuts through the center of the cells, the molecules are all oriented parallel to $y$-direction. In $x$-direction however, the orientation of the molecules changes making an angle $\theta$ with the normal to the incoming light at the lens rim and $0°$ at the center. The extraordinary refractive index changes with this angle $\theta$ in a well-defined manner (Yeh and Gu, 1999). So, there is phase-shift difference for one and the same polarization in points A and B. This will produce an asymmetric phase-shift pattern over the lens area. Note that for a lens made of inorganic crystal where the optical axis is oriented homogeneously over the whole lens, we will not observe this asymmetry. One can determine the astigmatism of the LC lens experimentally if a planar test surface, a mirror, is positioned perpendicularly to the propagation direction in the interferometer setup. If the optical system has no aberrations and if the mirror is in the focal point of the reference beam, a uniform interference pattern should be obtained. Instead of this, one obtains interference patterns as shown in Figure 5.11. In

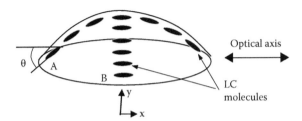

**FIGURE 5.10**  The alignment of the LC molecules in a cell with curved surfaces. At the surface the angle between the molecules and the surface is given by the pretilt angle and fixed. Along x the curvature causes a deformation profile that alters the optical properties compared to an inorganic crystal lens.

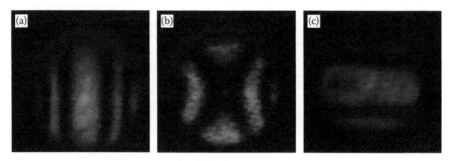

**FIGURE 5.11** Interference patterns obtained with a flat mirror positioned at 0 mm (a), at 1 mm (b), and at 2 mm (c).

Figure 5.11a, the interference pattern corresponds to the situation where the extraordinary wave (or reference beam) is perfectly focused along the $x$-direction but not along the $y$-direction. In Figure 5.11b and c the test surface is positioned 1 and 2 mm away from the position corresponding to the interference pattern, respectively. The interference pattern in Figure 5.11c corresponds to the inverse situation when the reference beam is focused along $y$-direction. In Figure 5.11b the mirror is placed in between the two focal lines and the intensity varies in both directions. From these measurements, we can deduce that the focal lengths corresponding to the $x$- and the $y$-axis are separated by approximately 2 mm. That is distance between interference patterns in Figure 5.11a and b which is in agreement with a theoretical value of 1.5 mm for such a situation (Boer, 2003).

### 5.4.2 INTERFEROMETER PERFORMANCE

We measured different reflective test surfaces with the setup in Figure 5.7. Figure 5.12 shows the surface profile obtained for a flat mirror over a surface of about 16 mm$^2$ resolved with $250 \times 300$ pixels. The measured curvature in x direction is a result of the already discussed astigmatism of the LC double-focus lens that was not corrected. Since we cannot distinguish between the phase shift because of the surface and the phase shift because of the lens, the additional retardation caused by the asymmetry of the lens on the reference beam is directly visible in the measurement. This measurement inaccuracy can be subtracted and a calibration can be done. From the local irregularities, which lead to fluctuations of less than 50 nm, we can conclude that the measurement precision for the present interferometer is at least $\lambda/10$ (Boer, 2003). Actually, this precision is relatively modest compared to other high-end common-path interferometers that are mostly dedicated to special surfaces and which have precisions up to $\lambda/1000$. The present version of the interferometer also shows some other limitations: it cannot be miniaturized below a certain dimension because of the long working distance. This distance is determined by the curvature of the concave glass lens and the extraordinary refractive index of the LC material used to form the bifocal lens. Because the LC lens should not exceed a thickness greater than 0.7 mm, the curvature defines the size of the active area of the interferometer.

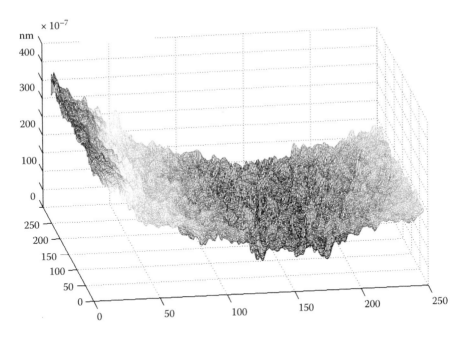

**FIGURE 5.12** Surface profile measurement for a metal mirror. The active area is about $4 \times 4$ mm wide.

The measurement presented is very noisy because of the poor quality of the polarizers and LC lens. The fact that for such an interferometer the whole surface of the lens has to have an excellent optical properties to obtain high resolution is not in line with the difficulties of element fabrication made from cross-linked LCNs. The stress birefringence and inhomogeneities worsen the interference pattern considerably and do not allow, at least in the present stage, to use such elements in surface-testing common-path interferometers. For other applications were the demands are less stringent birefringent lenses from LC materials can be very useful (Stapert et al., 2003; Hain et al., 2001). The astigmatism produced by the asymmetric alignment of the LC molecules alters the performance of the interferometer considerably and special measures have to be taken to correct for these.

Another important and more general limitation of all type of interferometric measurements is that the tested surfaces should not have large gradients of depth variations. If not , the measured interference pattern will show frequencies that the CCD camera cannot resolve and aliasing effects will occur.

## 5.5 INTERFERENCE MICROSCOPY

### 5.5.1 Principle of Operation

The system that we present here is a polarization interferometer for phase object observation in microscopy, often called DIC microscopy, which is a widely used method for observation of colorless transparent objects such as cultured biological cells for example. These objects show no clear contrast when observed with a conventional microscope. A DIC microscope is relatively expensive equipment because it consists of precise birefringent prisms mostly made of inorganic crystals, such as quartz or calcite. We present here a DIC microscope that is made of prisms containing cross-linkable LCN.

In DIC microscopy, two lateral sheared images are created. One image is related to the ordinary polarization component and the other image to the extraordinary polarization of the light. There are two types of shearing interferometers: The total doubling interferometer, where the shear is greater than the linear dimension of the phase object, and the differential interferometer, where the shear is much smaller than the object dimension. The first interferometer type is not very practical since for large phase objects the lateral shear that can be obtained with the interferometer may not be sufficient for total doubling. We will consider a differential interferometer with a lateral shear that is roughly equal to the resolution limit of the microscope.

A widely used DIC microscope is the Smith interferometer with two modified Wollaston prisms for polarization splitting. Figure 5.13 illustrates the principle. The light passes through a polarizer oriented at 45° with respect to the optical axis of the Wollaston prisms in the system. It is then sent through the first modified Wollaston prism (or Normarski prism) N1 to produce a lateral shear between the two polarization

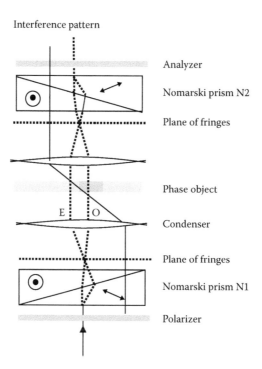

**FIGURE 5.13**  Principle of operation of a DIC microscopy configuration also known as Smith interferometer.

components. This is indicated by E and O and the dashed lines. The prism N1 is placed so that the front focal plane of the condenser coincides with the plane of the fringes of the prism. The two rays traverse the phase object at slightly different points. The light is then collected by the objective and the rays are recombined with a second modified Wollaston (or Nomarski) prism N2. The prism N2 is placed so that the rear focal plane of the microscope objective coincides with the plane of fringes. Finally, the light passes trough the second polarizer to allow the two polarization components to interfere. Nomarski prisms or modified Wollaston prisms are used to assure that the plane of maximum fringe contrast is outside of the prisms (Françon and Mallick, 1971). The Nomarski prisms N1 and N2 can be replaced by conventional Wollaston prisms. The main difference between modified Wollaston prisms and standard one is a tilt of the optical axis in the propagation direction (Montarou and Gaylord, 1999). Since the back focal length of the objective is generally very small, the Wollaston prism, that has its plane of apparent splitting inside the system, must be placed very close to the objective. This is for practical reasons not always possible and modified Wollaston prisms should be used. One has to be aware that off-axis rays (dotted line) will have different phase shifts than the on-axis rays because they pass through N1 at different positions. N2 compensates the phase shift produced by N1 when the following condition is fulfilled:

$$\alpha_1 f_1 = \alpha_2 f_2, \tag{5.2}$$

where $\alpha_1$ and $\alpha_2$ are the splitting angles of the prisms N1 and N2, $f_1$ is front focal distance of the condenser and $f_2$ the rear focal distance of the objective. If condition in Equation 5.2 is fulfilled, the prisms are said to be conjugated. A similar setup can also be used for reflection DIC microscopy. In this case, the setup corresponds roughly to the top of Figure 5.13. The same Nomarski prism is used for splitting and recombination of the incident and reflected beam. Note that the DIC system is extremely sensitive to phase shifts and has to be used with strain-free objectives in order to avoid any additional distortion of the phase shift because of residual birefringence.

For our purposes, we needed a design that can be implemented with LC technology and can easily be installed in a conventional polarization microscope. The main problem of the Smith interferometer is the need of a tilted optical axis in the first wedge of the Nomarski prisms. Such tilt angles are realized in LC technology with the pretilt angle that is given by the alignment at the substrate surfaces. Only small pretilt angles up to 7° were obtained with conventional rubbing and LC polymer technology. In addition, the prisms had to be placed very precisely and the back focal plane of the objective is not always accessible. To avoid these difficulties, Françon and Yamamoto (1962) had designed a system that was based on two conventional Wollaston prisms, which does not need to be positioned at a specific place. The schematic setup of the system is shown in Figure 5.14. As for the Smith interferometer, the two Wollaston prisms must be conjugated in order to compensate the phase shift of off-axis rays. The conjugation condition Equation 5.2 can now be expressed as

$$v_{\mathrm{W1}} M_{\mathrm{cond}} = v_{\mathrm{W2}} M_{\mathrm{obj}} \tag{5.3}$$

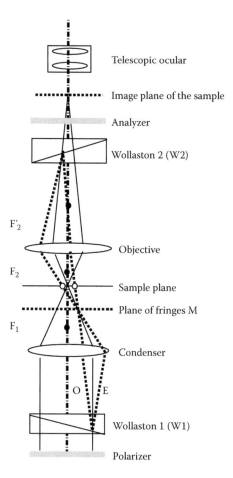

Telescopic ocular

Image plane of the sample

Analyzer

Wollaston 2 (W2)

$F'_2$

Objective

$F_2$

Sample plane

Plane of fringes M

$F_1$

Condenser

O    E

Wollaston 1 (W1)

Polarizer

**FIGURE 5.14** Schematic setup of the Françon–Yamamoto interferometer. The dashed rays show the paths followed by the ordinary (O) and extraordinary (E) rays. The continuous lines represent the light paths without considering the polarization splitting.

where $\nu_{W1}$, $\nu_{W2}$ are the frequencies of the fringes produced by the Wollaston prisms on its own, $M_{cond}$ and $M_{obj}$ are the magnification of the fringes realized by the condenser and the objective, respectively. These magnification terms depend on the position of the Wollaston prisms. If this condition is fulfilled and the prisms are oriented so that their birefringence is compensated, one obtains a uniform illuminated field. The dashed lines O and E represent the ordinary and extraordinary rays created by the splitting for an off-axis incoming ray at the Wollaston prism W1. They intersect in the plane of fringe localization. They then pass through the sample at two slightly separated points indicated by the white dots. After passing the objective, the two rays are set parallel again by the Wollaston prism W2. Finally, the analyzer combines the two rays to interfere. The main difference of this system compared with the Smith interferometer is that the fringes are localized near the front focal plane of the objective and not in the back focal plane. This configuration has several advantages:

- W1 and W2 are not required to be placed at specific positions. However, the objective and the condenser have to be adjusted so that the fringes belonging to the two systems are both imaged in M.
- The optical axis of W1 and W2 do not need a particular tilt angle, which is particularly advantageous when using LC technology.
- Since the birefringence induced by W2 is compensated by W1 (if W1 and W2 are conjugated) the whole free aperture can be used (as for the Smith interferometer).

### 5.5.2  PERFORMANCE OF THE DIC MICROSCOPE

We realized the Françon–Yamamoto configuration with the LC technology, as described above. We fabricated a pair of Wollaston prisms with LCP from Wacker having thickness of 125 and 36 μm, respectively. The prisms were used in a Leica DMR XP polarization microscope. We adjusted the positions of W1 (125 μm), W2 (36 μm), the objective (20×, NA = 0.4), and the condenser to fulfill the condition in Equation 5.3. Experimentally, this condition is satisfied when a uniform illumination is observed in the object plane without object. We could not obtain a perfect compensation over the whole aperture because of slight phase deformations present in the LCP Wollaston prisms. Phase objects were fabricated by immersing transparent microstructures in index matching oil. These transparent samples cannot be observed with a microscope, which is not equipped with DIC system. Figure 5.15a shows measurements of a replicated grating with a period of about 10 μm that is immersed in index-matching fluid. The replication was done in a UV adhesive optical glue (NOA 65) and we used index matching oil with $n = 1.518$. Figure 5.15b shows that by turning the same grating by 90°, the observed pattern disappears. This can be understood if we consider that only optical path gradients in the plane of splitting can be observed with this interferometer.

**FIGURE 5.15**  In (a) a 10 μm grating is immersed in index matching fluid and observed with the Françon–Yamamato interferometer. (b) The same grating turned by 90°, the grating is now parallel to the splitting plane of the Wollaston prisms and not visible anymore.

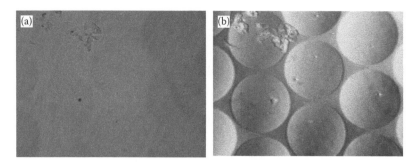

**FIGURE 5.16** In (a) an array of micro-lenses is immersed in an index-matching fluid and observed under the polarization microscope. The array can hardly be observed. With DIC microscopy in (b) the immersed lenses are well seen.

Figure 5.16a and b show an array of micro-lenses with a diameter of 0.145 mm replicated in optical glue (NOA 65) and immersed in the index-matching fluid. The picture renders a three-dimensional impression of the structures and also reveals the presence of small defects. Figure 5.16a shows the sample when the analyzer is removed, which is equivalent to "switching off" the interferometer.

To summarize, we have demonstrated that it is possible to replace conventional inorganic birefringent elements by LC elements to perform high-quality DIC measurements. Moreover, in principle, this application does not need very thick elements compared to other applications like the double-focus interferometer describe in Section 5.4. The Françon–Yamamoto configuration is well suited for the LC technology, because it does not need birefringent elements with highly tilted optical axis and has the advantage that it can be installed in any polarization microscope having adjustable condenser and sample position.

## 5.6  STATIC POLARIZATION SPECTROMETER

### 5.6.1  INTRODUCTION

In this section, we present the realization of FTS with LC polymer technology. It is based on a polarization interferometer that has a Wollaston prism made of cross-linkable liquid crystalline material as a key component. We show the consequences of realizations with conventional LC's and cross-linked LC networks for the most suitable design. The performances change because of the use of liquid crystalline polymer components will be discussed in detail, especially the temperature dependence will be addressed.

In comparison to other types of spectrometers the FTSs have the well-known throughput and multiplex advantages. This leads to a better signal-to-noise ratio and a high throughput even at high resolutions. We developed a monolithic low-resolution static FTS that is based on a polarization interferometer using a Wollaston prism with moderate resolution. Since the fabrication of these devices uses potentially inexpensive LC technology and needs no particularly precision adjusting or assembling procedures, it may become an interesting candidate for applications in the visible

(VIS) or near-infrared (NIR) region. Applications such as colorimetry in the graphic industry, quality inspection, or environment monitoring might be envisaged.

### 5.6.2 Principle of Operation

All FTSs are based on the relation between the temporal coherence of the source and the spectrum of the incident light. The present FTS is a static common-path polarization interferometer. A principle setup with imaging lenses of such a FTS is schematically shown in Figure 5.17. Divergent incident light is collimated with collimation optics. The light is linear polarized at 45° with a dichroic sheet polarizer. When entering the Wollaston prism, the light is split into two linear polarization components along the principle polarization direction of that prism, in this case along x and y direction. Because of the birefringence and the perpendicular (x- and y-) orientation of the optical axes of the wedges that form the Wollaston prism, the system produces a spatially varying phase shift in y direction. The two polarization components are recombined with the help of an analyzer, which is oriented at ±45° to obtain maximal contrast. With the help of the imaging optics the plane of maximum interference contrast, which is virtually localized inside the Wollaston prism, is focused onto a photo detector array (PDA). Finally, a Fourier transformation of the recorded spatial intensity distribution on the PDA gives the spectral intensity. To use the FTS in the visible region for colorimetry, it is recommended to have a resolution $\Delta\lambda$ better than 20 nm. The resolution $\Delta\lambda$ of the FTS is determined by the maximum phase shift or OPD between the interfering beams. One finds the relation

$$\Delta\lambda = \frac{\lambda^2}{\text{OPD}}. \tag{5.4}$$

By introducing $\lambda = 700\,\text{nm}$ into Equation 5.4, a minimal path difference of $\text{OPD}_{min} = 24.5\,\mu\text{m}$ is calculated to obtain $\Delta\lambda = 20\,\text{nm}$ resolution. The OPD in the present configuration is determined by the difference of the ordinary and extraordinary refractive index of the material and the thickness of the Wollaston prism. One

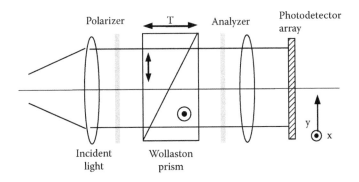

**FIGURE 5.17** Optical principle of the Fourier transform spectrometer based on a Wollaston prism.

**TABLE 5.2**

**List of the Resolutions that can be Obtained for Different LCs and for a Fixed Thickness of the Wollaston of 200 μm**

| Liquid Crystalline Material | Birefringence ($\Delta n$ at 547 nm) | Maximal Phase Shift ($\text{OPD}_{max}$) (μm) | Theoretical Resolution ($\Delta\lambda$) (nm) at 700 nm |
|---|---|---|---|
| BL006 | 0.28 | 56 | 9 |
| E7 | 0.22 | 44 | 11 |
| ZLI-1132 | 0.14 | 28 | 18 |
| Wacker (LCP) | 0.19 | 38 | 13 |
| RM-257 (LCP) | 0.14 | 28 | 18 |

can write

$$\text{OPD}(\lambda) = (n_e(\lambda) - n_o(\lambda))T = \Delta n(\lambda)T, \qquad (5.5)$$

where $T$ is the maximum useful thickness of the wedges used in the Wollaston prisms. Note that the refractive index is depended on the wavelengths. Therefore, for the same thickness of the wedges different OPD's result for different wavelengths. Table 5.2 gives a list of maximal phase shifts $\text{OPD}_{max}$ and corresponding resolutions $\Delta\lambda$ for some LC materials. The thickness $T$ of the Wollaston prism was set to $T = 200\,\mu\text{m}$, which represents a good compromise between resolution and proper alignment of the LC material.

The values in Table 5.2 were calculated for 700 nm, which represents the red end of the visible spectra and gives the case with lowest resolution. Note again that the resolution decreases with the square of the wavelength, this means that in the near infrared region at a wavelength of 1.5 μm choosing $T = 300\,\mu\text{m}$ we would have only a resolution of about 60 nm for the same device.

In practical applications for our interferometer, sources are often spatially extended and incoherent. It is known that for such sources the interference fringes are localized in space at the plane of maximal contrast (Goodman, 2000). For best performance the detector has to be put at this position. If this plane is localized inside the system, additional optics is needed to focus it onto the detector. This makes the system bulky. A better way to overcome this problem is given by properly tilting the optical axis of the first wedge in the Wollaston prism (Françon and Mallick, 1971; Montarou and Gaylord, 1999). With this technique the plane of maximal contrast can be pulled out of the system. This is illustrated in Figure 5.18. In the first variant (Figure 5.18a), the LC molecules (and so the optical axis) are parallel to the surface, the plane of maximal contrast is localized inside the system. The two rays that correspond to perpendicular polarization components intersect virtually inside the Wollaston prism. In the second variant (Figure 5.18b), the molecules are tilted and the plane is situated outside the system. The LC molecules have to have a certain pretilt angle if a compact system is envisaged. This pretilt angle is an optical design parameter and has to be adjusted carefully. Note that the maximum value of the pre-tilt was limited at about 7° if polyimide rubbing is used as the alignment mechanism. It is therefore necessary to find an optical design that is adapted to this constrain.

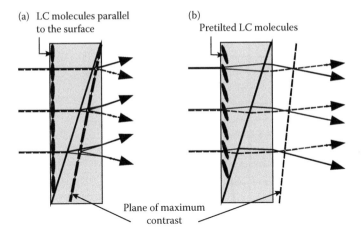

**FIGURE 5.18** Localization of the plane of maximal contrast for the classical Wollaston configuration. (a) The LC molecules are parallel to the surface and the plane is localized inside the system. (b) The LC molecules are tilted and the plane is localized outside the system.

The localization of the plane can be found analytically by geometrical considerations (Montarou and Gaylord, 1999). For a configuration with more complex source geometries, simulations become necessary. For simulations, we used a ray-trace program that is based on Gaussian beam propagation called ASAP from Breault Research Corporation. This program is able to model correctly the splitting of beams in an ordinary and extraordinary ray when entering a uniaxial medium and to track modification of the polarization state. Moreover, each ray is constituted of several elementary rays and is modeled as a Gaussian beam of a certain spatial extension having a defined phase (Arnaud, 1985). A quasi-plane wave can be modeled by a superposition of such beams. So, coherent propagation and polarization effects can be treated, which is necessary for simulation of polarization interferometer properties (Boer et al., 2002; Scharf, 2006). Figure 5.19 shows a ray trace trough a spectrometer illuminated with an extended source having two wavelengths (632 and 800 nm). The pretilt angle in the simulation is 10°. The simulation gives the interferogram on the detector plane and its Fourier transform containing two peaks corresponding to the two wavelengths.

### 5.6.3 EXPERIMENTAL PERFORMANCE OF THE SPECTROMETER

In this section we will discuss the optical performance of a prototype spectrometer. The spectrometer module contains two prisms made of cross-linked material and a LC twist cell. Details on such a configuration are published in Boer et al. (2002) and Boer et al. (2004). The presented results were obtained by using the setup that is schematically described in Figure 5.20. Light from a high-power xenon lamp or a He–Ne laser is coupled into a light guide. A spatially incoherent source is created by illuminating a diffusing plate. Before being analyzed by the spectrometer, the size of the source is limited by an aperture. The birefringent elements were made of silicone LC except the twisted nematic cell that is filled with of ZLI-1132 from Merck. The

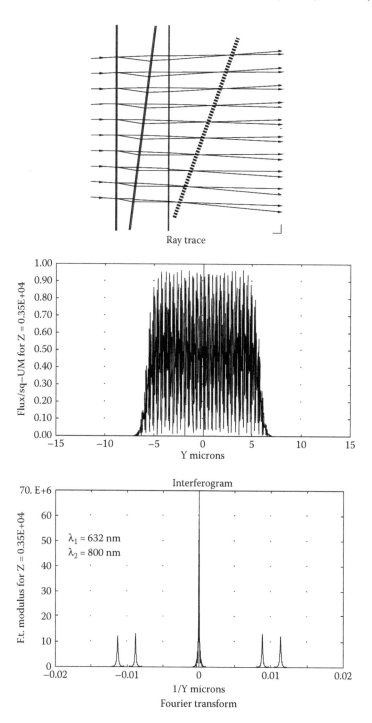

**FIGURE 5.19**   Polarization ray trace through the spectrometer with the resulting interferogram and its Fourier transform.

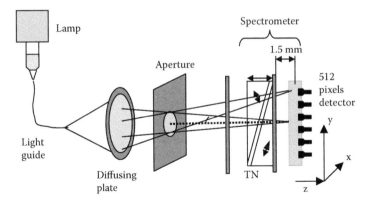

**FIGURE 5.20** Test setup to evaluate the optical performance of the miniaturized FT spectrometer.

whole arrangement is sandwiched by two sheet polarizers oriented at ±45°. A 512 pixels line detector array from Hamamatsu was placed at about 1.5 mm behind the analyzer (exit polarizer) to register the interferogram. This position corresponds to the plane of maximal contrast. We measured the signal-to-noise ratio of the spectrometer with two different light sources: a He–Ne laser (633 nm) and a xenon lamp combined with a GG495 filter from Schott, which blocks all wavelengths below 495 nm of more than 40 dB. The integration time was set to 160 ms. The numerical aperture corresponds to an angle of ±20° or an NA = 0.35. No nonlinearity correction of the detector was performed. The signal-to-noise ratio between 490 and 420 nm was better than 20 dB. Because of the large absorption losses, noise becomes important for wavelengths below 420 nm and stray-light suppression is only 10 dB. The total transmission of the system is influenced by the transmission of the polarizers and the LCP wedges. Ideally, for unpolarized incoming light, the first polarizer absorbs 50% of the light and the second polarizer absorbs again 50% of the remaining light because of the regular interference fringes that are formed for each wavelengths. Supposing that all the other elements are perfectly transparent the total transmission should be 25%. As we have seen above, the birefringent elements absorb about 30% of the light intensity. So, one expects a total transmission of about 20%. We measured the transmission of the complete system (prisms and polarizers) with a Perkin-Elmer spectrometer. Since we used an integration sphere, the forward-scattered light is also included in the measurements. The transmission decreases drastically below 450 nm. This is not only because of absorption of the LC elements but also because of a larger absorption of the polarizers starting below 420 nm. The cut-off wavelength of standard polarizers used in LC display technology is 400 nm (Scharf, 2006). Figure 5.21 shows the measurement of the spectrum of a xenon lamp, which can be considered as a "cold white" source. It is clearly seen that absorption causes the intensity to decrease significantly below 450 nm. In a commercial instrument, used for reflection colorimetry the high absorption losses can be compensated by using adequate illumination, which has higher power in the blue region.

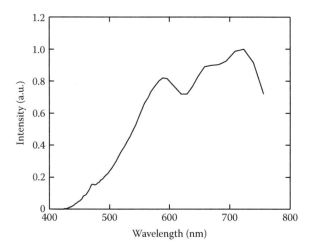

**FIGURE 5.21**   Measurement of a spectrum of a xenon lamp. The throughput is drastically reduced below 450 nm.

The theoretical resolution of the spectrometer is given by Equation 4.4. By introducing the actual parameters $\lambda = 633$ nm, $\Delta n = 0.19$ and $T = 200$ μm we obtain a theoretical resolution of $\Delta\lambda = 10$ nm. We measured the resolution by fully illuminating the spectrometer with a He–Ne laser. The full-widths at half-maximum (FWHM) found in the measurement is 11 nm, which is very close to the expected value.

For applications it is important to have a device that is not sensitive to temperature variations. The birefringence of LC is known to be very much temperature dependent and so a spectrometer made from LCs is also expected to find big effects. In spectrometry, the interesting quantity is the change of the meausred wavelengths as a function of the temperature. Our compact FTS uses the birefringence to produce phase shift. Fringes are produced between crossed polarizers and the period of the fringes gives the wavelengths position in the Fourier transform. The period is proportional to the birefringence. The wavelength's temperature dependence is directly related to the temperature dependence of the birefringence. A model can be developed to estimate the influence of the temperature on the position of the wavelengths (Boer, 2003). Rather large values are predicted by theory, which indicates that a LC mixture cannot be used for interferometric applications. To see the difference we investigated experimentally two prototypes of the FTS: one made with the commercial mixture ZLI-1132 and a second one made with LCP from Wacker. To measure the temperature dependence, the FTS prototype was put into a temperature-controlled test chamber (Vötsch VT4004). Monochromatic light (spectral width of $\Delta\lambda \approx 5$ nm) is provided by a monochromator and injected into a plastic wave guide that is connected to the inlet of the spectrometer. We recorded the measured wavelengths at every 5°C temperature variance. The results are shown in Figure 5.22. The wavelength drift rate at around 25°C for 500 nm can be reduced from 1 nm/K for LC mixtures to an acceptable rate of 0.1 nm/K for cross-linked LCPs.

To summarize, the main characteristics of the spectrometer are listed in Table 5.3.

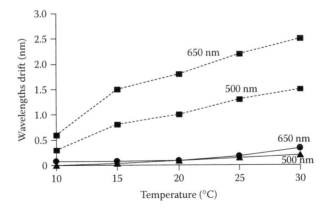

**FIGURE 5.22** Wavelengths drift as function of temperature for two central wavelengths (500 and 650 nm). Dashed lines are measurements performed with a spectrometer prototype with conventional LC and continuous lines show the performance for the prototype made with LCN components.

**TABLE 5.3**
**Overview of the Performances of the Investigated Spectrometer**

| | |
|---|---|
| Size | $25 \times 25 \times 6 \, \text{mm}^3$ (without detector) |
| Resolution | 11 nm (at 630 nm) |
| Field of view | $>35°$ (depending on OPD) |
| Stray light suppression | $<22 \, \text{dB}$ (for $\lambda > 420 \, \text{nm}$) |
| Average transmission (VIS) for unpolarized light | $\sim15\%$ |
| Cut-off wavelength | 410 nm |
| Temperature dependence | 0.1 nm/K |

Several characteristics can still be improved and are not optimized yet. The size of the spectrometer can be reduced to the size of the detector. Signal-to-noise ratio can be enhanced by optimizing the fabrication process of the LC cells. Temperature dependence can be reduced by using higher cross-linked LCP or by performing post-curing processing steps.

## 5.7   CONCLUSION

We have demonstrated a wide range of applications where cross-linkable LCN can be used. In all cases we could show that inorganic crystals can be replaced by organic components. However, because of technological limitations, the performance of the interferometers made with LCN materials is not always comparable with the ones known from crystal optics. The use of LCN also imposes some design restrictions because of the fairly low pretilt angles that can be obtained with conventional alignment methods. LCN technology is interesting for not so high demanding applications where the LC layer thickness stays below 200 μm. We realized successfully a DIC microscope and a miniaturized FTS. The use of LCN in both cases was mainly driven by the fact that it leads to high temperature and shock stability. In the case of the FTS, where no temperature dependence is allowed, we showed that it is necessary to use LCN to obtain a system that can be used. The main limitation here is that the fabrication of wedge-shaped elements are not conforming to planar technology used in LC display industry.

## REFERENCES

Arnaud, J. Representation of Gaussian beams by complex rays, *Appl. Opt.* 24, 1985, 538–545.

Atkins, P.W. *Physical Chemistry*, Fifth Edn, Oxford University Press, Oxford 1994, ISBN 0198557302.

Boer, G. Polarization Interferometers using Liquid Crystal as Birefringent Elements, PhD thesis, University of Neuchâtel, 2003.

Boer, G., Ruffieux, P., Scharf, T., Seitz, P., and Dändliker, R. Compact liquid crystal polymer Fourier transform spectrometer, *Appl. Opt.* 43, 2004, 2201–2208.

Boer, G., Scharf, T., and Dändliker, R. Compact static Fourier transform spectrometer with a large field of view based on liquid-crystal technology, *Appl. Opt.* 41, 2002, 1400–1409.

Broer, D.J., Boven, J., Moll, G.N., and Challa, G. *In situ* photopolymerization of oriented liquid-crystalline acrylates, 3. Oriented polymer networks from mesogenic diacrylate, *Makromol. Chem.* 190, 1989a, 2255–2268.

Broer, D.J., Finkelmann, H., and Kondo, K. *In situ* photopolymerization of oriented liquid-crystalline acrylates, 1. Preservation of molecular order during photopolymerization, *Makromol. Chem.* 190, 1989b, 19.

Broer, D.J., Hikmet, R.A.M., and Challa, G. *In situ* photopolymerization of oriented liquid-crystalline acrylates, 4. Influences of a lateral methyl substituent on monomer and oriented network properties of a mesogenic diacrylate, *Makromol. Chem.* 190, 1989c, 3201.

Broer, D.J., Mol, G.N., and Challa, G. *In situ* photopolymerization of oriented liquid-crystalline acrylates, 2. Kinetic aspects of photopolymerization in the mesophase, *Makromol. Chem.* 190, 1989b, 19.

Broer, D.J. and Mol, G.N. Anisotropic thermal expansion of densely cross-linked oriented polymer networks, *Pol. Eng. Sci.* 31, 1991, 625–631.

Broer, D.J., Moll, G.N., and Challa, G. *In situ* photopolymerization of oriented liquid-crystalline acrylates, 5. Influence of the alkylene spacer on the properties of the mesogenic monomers and the formation and properties of oriented polymer networks, *Makromol. Chem.* 192, 1991, 59.

Dändliker, R. The concept of modes in optics and photonics, *SPIE* 3831, 2000, 193–198.

Down, M.J., McGiven, M.H., and Ferguson, H.J. Optical systems for measuring the profiles of super-smooth surfaces, *Prec. Eng.* 7, 1985, 211–215.

Dyson, J. Common-path interferometer for testing purposes, *J. Opt. Soc. Am.* 47, 1957, 386–390.

Françon, M. and Yamamoto, T. Un nouveau et très simple dispositif interférentiel applicable au microscope (A new, very simple interference device to be applied to microscopy), *Opt. Acta* 9, 1962, 395–408.

Françon, M. and Mallick, S. *Polarization Interferometers*, Wiley-Interscience, New York, 1971, ISBN 0471274704.

Goldstein, D. *Polarized Light*, Marcel Dekker, New York, 2003, ISBN 0-8247-4053-x.

Goodman, J.W. *Statistical Optics*, John Wiley & Sons, New York, 2000, ISBN 0-471-39916-7.

Hain, M., Glockner, R., Bhattacharya, S., and Tschudi, T. Fast switching liquid crystal lenses for a dual focus digital versatile disc pickup, *Opt. Commun.* 188, 2001, 291–299.

Hariharan, P. *Optical Interferometry*, 2nd ed., Elsevier/Academic Press, Amsterdam 2003, ISBN 0-12-311630-9.

Hikmet, R.A.M. and Zwerver, B.H. Anisotropic shrinkage behaviour of liquid-crystalline diacrylates, *Polymer* 33, 1992, 89.

Iwata, K. and Nishikawa, N. Profile measurement with a phase shifting common-path polarization interferometer, *Proc. SPIE*, 1162, 1989, 389–384.

Jamin, M.J. Sur un réfracteur pour la lumière polarisée, *Comptes Rendus Acad. Sc. Paris*, 67, 1868, 814.

Kikuta, H. and Iwata, K. Phase-shifting common-path interferometers using double focus lenses for surface profiling, *Proc. SPIE* 1720, 1992, 133–143.

Kikuta, H., Tanabe, T., and Iwata, K. Double focus interferometer and the influence of their sample setting errors, *Opt. Rev.* 1, 1994, 266–269.

Lesso, P., Duncam, A.J., Sibbett, W., and Padgett, M.J. Aberrations introduced by a lens made from a birefringent material, *Appl. Opt.* 39, 2000, 592–598.

Montarou, C.C. and Gaylord, T.K. Analysis and design of modified Wollaston prisms, *Appl. Opt.* 33, 1999, 6604–6613.

Morito, Y., Stockley, J.E., Johnson, K.M., Hanelt, E., and Sandmeyer, F. Active liquid crystal devices Incorporating liquid crystal polymer thin film waveplates, *Jpn. J. Appl. Phys.* 38, 1999, 95–100.

Nose, T., Masuada, S., and Sato, S. Application of a liquid crystal lens to a double focusing common path interferometer, *Proc. SPIE* 3143, 1997, 165–175.

Nose, T. Honma, M., and Sato, S. Influence of the spherical substrate in the liquid crystal lens on the optical properties and molecular orientation state, *Proc. SPIE* 4418, 2001, 120–128.

Polavarapu, P.L. *Principles and Applications of Polarization-Division Interferometry*, Wiley, New York, 1998. ISBN 0-471-97420-X.

Sato, S. Liquid-crystal lens-cells with variable focal length, *Jpn. J. Appl. Phys.* 18, 1979, 1679–1684.

Scharf, T. *Polarized Light in Liquid Crystals and Polymers*, John Wiley and Sons, New York, 2006, ISBN 0471740640.

Schultz, J.W. and Chartoff, R.P. Photopolymerization of nematic liquid crystal monomers for structural applications, *Polymer* 39, 1998, 319–324.

Stapert, H.R. del Valle, S., Verstegen, E.J.K., van der Zande, B.M.I., Lub, J., and Stallinga, S. Photoreplicated anisotropic liquid-crystalline lenses for aberration control and dual-layer readout of optical discs, *Adv. Func. Mat.* 13, 2003, 732–738.

Yeh, P. and Gu, C. *Optics of Liquid Crystal Displays*, John Wiley and Sons, New York, 1999, ISBN 0-471-18201-X.

# 6 Anisotropic Emitting Cross-Linked Polymers Based on Liquid Crystals

*L. Oriol, M. Piñol, and José Luis Serrano*

## CONTENTS

## 6.1 INTRODUCTION

In the last few years there has been increasing interest in the field of polarized-light-emitting layers because these systems can provide the linearly polarized emission that is needed for liquid crystal displays (LCDs), thus simplifying their manufacture and reducing costs. There are a number of existing flat-panel display technologies in direct competition and, of these LCDs have gained a strong position for most applications because their low cost, lower power consumption, low operating voltage, and low weight constitute advantages that are not easily surpassed [1]. However, any position in the market cannot be sustained without continuous improvement in performance and cost, assisted by developments in liquid crystal materials, science, and technology [2]. Color LCD commercial devices are based on the manipulation of polarized light that is generated by transmitting light from an independent light source (either internal or external) through a combination of sheet polarizers and color filters. Such a combination of elements reduces the brightness and overall energy efficiency of the device because a substantial fraction of the incident light is converted into thermal energy.

One approach to overcome these limitations was proposed by Weder et al. [3,4]. The incorporation within a conventional LCD configuration of photoluminescent polarizers, which both polarize light and generate bright colors, enables the design of devices with improved brightness, contrast efficiency, and viewing angle.

In the last few decades, rapid improvements have been made and a great deal of commercial effort has been focused on organic light-emitting diodes (OLEDs), which have become the major competitor to LCDs in the flat-panel display market [5,6]. These systems are intrinsically emissive displays based on electroluminescent materials that have superior brightness and viewing angle dependence than LCDs for some applications. Since the first report of polarized electroluminescence by Dyreklev et al. [7] a number of approaches have been reported to achieve polarized OLED devices. Such systems can provide a source of bright linearly polarized light for use as backlight in LCDs. This would remove the need for polarizers and color filters, and also improve the power consumption of the device. Thus, the dominant LCD technology and its main competitor OLEDs could combine to mutual benefit to create hybrid LCD-OLEDs for which the performance can be optimized by replacing conventional light sources and polarizers.

## 6.2   POLARIZED LUMINESCENCE FROM ORIENTED MATERIALS

Photoluminescence (PL) and electroluminescence (EL) are related physical phenomena as they involve the emission of light under photonic or electric excitation, respectively. Indeed, PL and EL emission spectra of a particular organic material or luminophore are almost always identical since they originate from the same $n-\pi^*$ or $\pi-\pi^*$ transitions. Linearly polarized light emission, both PL and EL, can be achieved using organic materials that are optically anisotropic, provided that they can be oriented in a specific direction on a macroscopic scale [8]. Anisotropic rodlike luminophores usually exhibit direction-dependent absorption and emission of light if the optical dipole transition moment is oriented along the long molecular axis of the molecule [9]. The emission will be linearly polarized if the luminophores are uniformly disposed in a preferential uniaxial arrangement with their transition moments parallel to each other. Under these circumstances, the polarized emission is controlled by the macroscopic order, with maximum emission parallel to the molecular orientation ($\parallel$) and minimum emission perpendicular to it ($\perp$). The anisotropy of the emission can be given by the polarization ratio of photoluminescence ($PL_\parallel/PL_\perp$) or electroluminescence ($EL_\parallel/EL_\perp$).

In principle, orientation can be induced by a variety of methods, which were discussed by Grell and Bradley in 1999 [8]. Langmuir–Blodgett (LB) deposition has enabled the construction of oriented films but two significant drawbacks are associated with this approach—the anisotropy is inadequate for practical applications and the technique is arduous for large-scale production. For instance, $EL_\parallel/EL_\perp$ values in the range 3–4 have been reported for LB films of poly($p$-phenylene) derivatives [10]. For low-molar mass materials, epitaxial evaporation into specific substrates has been used to fabricate aligned films with polarized emission [11]. From a practical point of view, a useful approach that is mainly used for polymeric films is mechanical alignment by the application of uniaxial stress or by rubbing. A significant example was reported by Weder et al., who fabricated polarized luminescent layers by solution casting and subsequent tensile drawing of 2% poly($p$-phenyleneethynylene) derivative **1** in ultrahigh molecular weight poly(ethylene) (UHMW-PE); $PL_{||}/PL_{\perp}$ ratios of 60:1 were obtained [12]. Nevertheless, the method can be problematic for thin films because mechanical damage is possible.

The direct rubbing of emitting films obtained by evaporation was employed by Hamaguchi and Yoshino with poly($p$-phenylenevinylene) (PPV) and poly(thiophene) derivatives, and PL polarization ratios of about 5 were achieved [13,14]. However, an effective rubbing-alignment process requires optimum fluidity in the polymeric film so that polymeric chains can reorient. In cases where the layer was excessively soft or tough, the conjugated polymers did not fully orient. As an alternative, Jandke et al. reported better results on mechanically rubbing and orienting a partially converted PPV **2**. Highly oriented films were prepared with a high PL ratio of 18 and an EL ratio of 12 obtained for a 40 nm thick film [15].

In relation to ordered materials, liquid crystalline materials (LCs) have been recognized as a promising alternative in terms of implementation and process scale-up. Based on their molecular self-assembly capability, uniaxial arrangements on a macroscopic scale can easily be achieved through LC phase formation.

## 6.3    LIQUID CRYSTALS FOR POLARIZED EMISSION

From the point of view of structural order, the self-organizing capability of LCs offers excellent control over molecular orientation through the use of appropriate processing techniques. Linear polarized emission has been demonstrated using a variety of thermotropic LCs, which include low-molar mass compounds, oligomers, side-chain polymers, polymer blends, main-chain polymers, and cross-linked polymers.

In general, there are several feasible approaches to prepare a film that actively emits polarized light on the basis of liquid crystalline behavior. One possibility is the preparation of intrinsically luminescent liquid crystals, which are emitters in the pure state [16]. The other alternative is based on the incorporation of an anisotropic luminophore into a liquid crystal matrix, in which it can be aligned as a result of a guest–host effect [17,18]. On using this second approach, the structural anisotropy of the luminophore, either LC or non-LC, also improves the compatibility with the liquid crystalline matrix and the orientational ability.

The conventional method to achieve macroscopic alignment in LCs is to use alignment layers where macroscopic orientational order is commonly achieved by surface effects. Uniaxial alignment can be obtained by depositing the LCs onto an appropriate alignment layer and subsequently performing a suitable thermal treatment. The alignment can be stabilized by rapidly quenching below the glass transition temperature ($T_g$). Alignment of luminescent LCs has most commonly been achieved on polyimide (PI) [19,20], nylon [21] or PPV [22] layers mechanically rubbed in one direction. As a representative example, highly polarized emission has been obtained for **3** with $PL_{||}/PL_\perp$ ratio of about 10, which in some cases can rise up to 25 for $EL_{||}/EL_\perp$ [20,22]. Despite the fact that this is a rapid and simple technique, mechanical rubbing can cause damage to the surface or generate dust and electrostatic charge [23].

Alternatively, the alignment of LCs is possible through the photo-alignment of polymeric layers, a technique in which polarized UV light can be used to generate surface anisotropy that leads to the uniform alignment of a LC deposited on the surface [24]. A diverse range of related alignment layers are based on this noncontact approach. For instance, the anisotropic photo-degradation of thin PI films with polarized UV light can be used to induce uniform alignment—although this approach can lead to chemical decomposition. Recently, polymers containing photo-addressable azobenzene units, which experience orientational changes through *cis–trans* photo-isomerization, have been employed as alignment layers (see Figure 6.1). For instance, PI containing azobenzene units within the backbone structure (see example a in Figure 6.2) have been employed as alignment layer after photo-induced orientation of the polymer using linearly polarized UV light. PL polarization ratios of 30 were reported for the end-capped poly(fluorene) **4** using this alignment layer [25]. Films of side-chain azobenzene polymers (see example b in Figure 6.2) have also been used for alignment layers. In this case the alignment of the azobenzene layer is not permanent and can be reoriented by subsequent irradiation with light. Poly(fluorene) **5** has been reported to have PL ratio of 12:1 using the mentioned side-chain azobenzene [26].

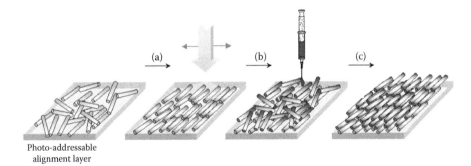

Photo-addressable
alignment layer

**FIGURE 6.1** Schematic representation of macroscopic uniaxial orientation using photo-addressable alignment layers: (a) Illumination with linearly polarized light induces a preferred in-plane orientation either parallel or perpendicular to the electric field vector of the incident light; (b) the luminescent LCs is spin-coated on the surface of the alignment layer; (c) after thermal annealing at the mesophase the overlaying luminescent LCs becomes oriented.

The exposure to UV light of films of polymers containing cinnamate or coumarin cores can also generate homogeneous alignment because of anisotropic photore-actions [27,28]. In particular, coumarin-containing films (see examples c and d in Figure 6.2) undergo angle-dependent [2+2] cycloaddition of the coumarin moieties on irradiation with polarized UV radiation, which induces the surface anisotropy responsible of the alignment properties of the layer [29,30].

All-in-one layers have recently been reported and these were produced from mul-tifunctional side-chain polymers with photo-addressable (e.g., azobenzene, stilbene, or cinnamate) and luminescent units to give anisotropy PL values of 8 [31]. The poly-mer is designed in a way that combines the photo-generation of anisotropy because of photo-addressable unit, the thermal amplification based on its liquid crystallinity

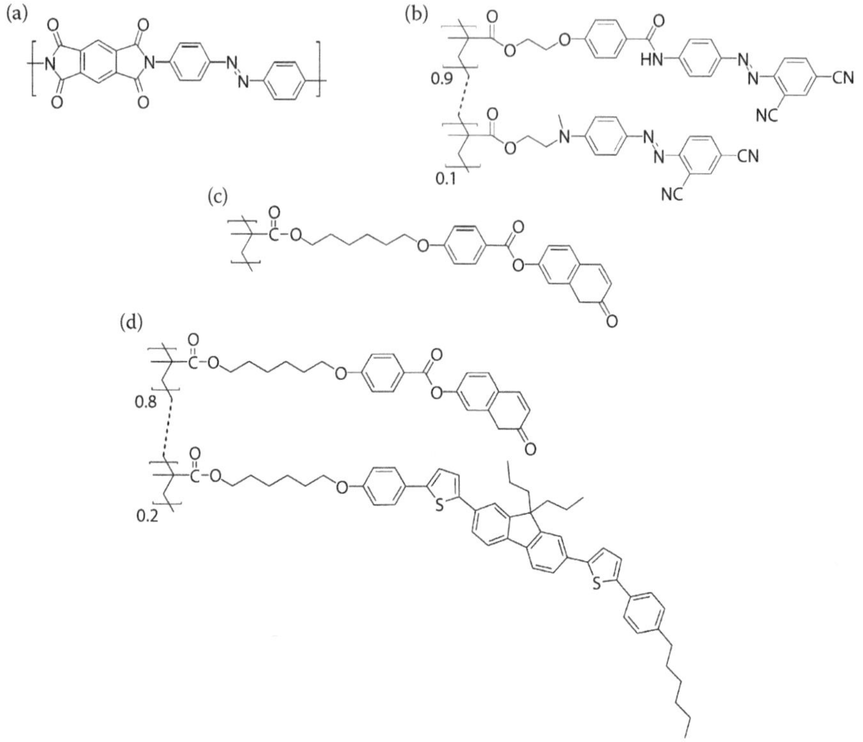

**FIGURE 6.2** Examples of photo-addressable polymers used as alignment layers. (a) and (b) are representative examples of main-chain and side-chain polymers containing azobenzene moieties; (c) and (d) are examples of polymers containing coumarin moieties.

and the anisotropic emission of rod-like luminophores, which become cooperatively aligned (see Figure 6.3). In this approach the use of an alignment layer is redundant, although the appropriate selection of the polymeric components is required to avoid quenching of the emission by the photo-addressable component [32,33].

A different approach has recently been described and is based on the fact that, in some cases, uniaxially oriented fibres spun from solution are comparatively easier than to obtain films with a controlled macroscopic orientation. Therefore, anisotropic films of poly(p-phenylenebenzobisoxazole) have been prepared by fusion of aligned fibres, which retain the uniaxial orientation of the original fibre. Anisotropies of 5:1 for PL were reported [34].

The practical application of light-emitting films requires solid and stable layers of good optical quality at the device operation temperature and, consequently, low-molar mass compounds are not recommended because of their rather high fluidity. The uniaxial molecular arrangement induced in the LC phases needs to be fixed in order to avoid ordering relaxation and also to impart robustness. Stabilization of the LC state has been accomplished through the development of luminescent glassy liquid crystals and liquid crystalline polymers, in which it is possible to quench the

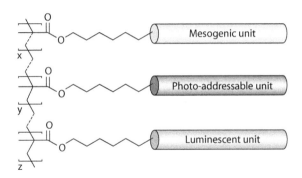

**FIGURE 6.3**   Schematic representation of multifunctional polymers for all-in-one layer where the anisotropic emission is because of light-induced orientation of the photo-addressable unit.

LC state into an ordered glassy state at room temperature. These systems are capable of preserving molecular alignment in the solid state on cooling below the $T_g$ while avoiding crystallization.

From the point of view of the structure, polymers with extended $\pi$ conjugation and stiff backbones are ideal materials because the intrinsic anisotropy of their electronic structure means that, after suitable alignment, polarized emission will occur. Polymers have the advantage of high $T_g$ and good film-forming ability. However, some disadvantages are also associated with polymers when compared to low molecular weight LCs. These drawbacks mainly concern the high viscosity, which may lead to long annealing times at high temperatures to achieve the satisfactory macroscopic alignment that must be frozen before crystallization occurs. Alignment stability can also be achieved with oligomers that form glassy LC films, an approach that combines the ease of processing into well-aligned films of polymers and the monodispersity and solubility of small molecules. For instance, oligo(fluorene)s that form glassy nematic films have been oriented on mechanically rubbed-surfaces and $PL_{||}/PL_{\perp}$ ratios of 20:1 and $EL_{||}/EL_{\perp}$ ratios of 27 have been reported for some of them [35,36]. Though they represent an unquestionable alternative to polymers, the synthesis of monodisperse oligomers involves a larger number of steps and has proven to be more difficult. Therefore, an efficient alternative to stabilize the LC order is the polymerization of reactive liquid crystals in an oriented state, an approach that gives good oriented anisotropic polymeric networks and enables secure control over the macroscopic ordering of the film.

## 6.4   ANISOTROPIC CROSS-LINKED POLYMERS BY *IN SITU* PHOTOPOLYMERIZATION

Reactive LCs are low-molar mass molecules that possess polymerizable groups attached to the rigid liquid crystalline core by a flexible aliphatic spacer. These materials generally melt at low temperatures and, because of their low viscosity, are relatively easy to orient into a macroscopic monodomain through the use of existing alignment techniques. Under controlled polymerization, this orientation can be maintained on forming a polymeric structure. Polymerization can take place either by thermal or light-induced processes. From both alternatives, the *in situ* photopolymerization of reactive liquid crystals is a well-established procedure in display technologies and has been used for the production of side-chain liquid crystal polymers, liquid crystal gels, and slightly or densely cross-linked liquid crystalline polymers—namely, elastomers and networks, respectively [37,38]. Inherent advantages over thermal polymerization are that this method allows the production of any kind of desired pattern in the film, allows the selection of the polymerization temperature and, consequently, the mesophase of the monomer in which the polymer will be produced. During light-induced polymerization, the ordering of the monomers is transferred to the polymer, yielding uniaxially oriented polymeric films or, alternatively, other configurations [39].

In order to produce cross-linked polymeric films by photopolymerization, formulations are required that contain monoreactive, direactive, or multireactive LCs, with the latter acting as cross-linkers (see Figure 6.4). The formulation usually contains a photoinitiator, which leads to polymerization on illumination with the appropriate light source by generating reactive species (either radicals or cations), and a thermal inhibitor that decouples the polymerization process from temperature and allows the manipulation of the monomers in the mesophase before polymerization. Within the polymeric films, cross-linking imposes constraints on the segmental motion of the polymeric chains and imparts a high degree of mechanical, thermal, and chemical stability. It has been reported that low cross-linking densities produce elastomers whose properties are temperature dependent but are restored by cooling down to the polymerization temperature. High levels of cross-linking give rise to

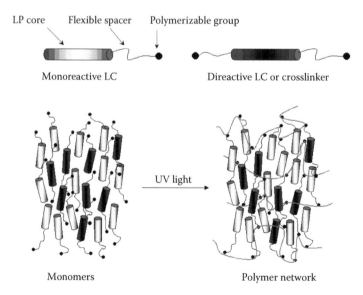

**FIGURE 6.4** Schematic representation of the processing of oriented films by *in situ* photopolymerization.

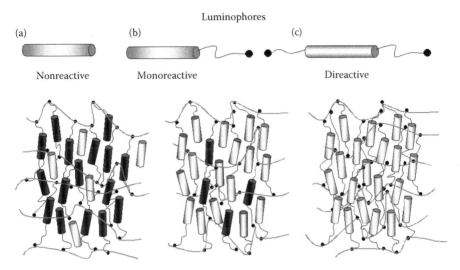

**FIGURE 6.5**   Schematic examples of anisotropic emitting cross-linked networks. (a) Nonreactive luminophore incorporated as guest–host in the polymeric network; (b) monoreactive luminophore incorporated covalently bonded to the polymeric network; (c) polymeric network obtained from a direactive liquid crystalline luminophore.

networks, the properties of which are almost temperature independent because of the very restricted motion of the molecules preventing reorientation.

The *in situ* photopolymerization of reactive liquid crystals in an oriented mesomorphic state has proved to be a versatile and advantageous method for processing oriented polymers with polarized emission. For instance, Hikmet et al. dispersed a 0.1% weight perylene dye into a reactive mesogen that, after annealing on the surface of a rubbed PEDOT layer and UV photopolymerization, was used in a device with an EL polarization ratio of about 2 [40].

Different approaches are possible in order to apply the concept of *in situ* photopolymerization to the production of polarized emitting networks. Variable amounts of an efficient luminophore can be dispersed into a LC matrix that would give the cross-linked polymer upon polymerization. The luminophore can be incorporated into the polymer either as guest–host system or covalently bonded to the polymer backbone if the luminophore carries a suitable polymerizable group. An alternative methodology uses emitting-reactive LCs with two polymerizable groups as single component that give rise to a fully cross-linking network without the requirement of a LC matrix (see Figure 6.5).

## 6.5   CHEMICAL DESIGN OF LUMINESCENT LIQUID CRYSTALS FOR CROSS-LINKED NETWORKS

One major problem encountered in the production of light-emitting polarized films by *in situ* photopolymerization is the selection of a luminophore that combines in a single molecule several requirements. In fact, there is only a narrow range of chemical structures that have been applied to the production of LC-based emissive films, probably because display applications demand a combination of requirements that it is not always easy to fulfil.

Amongst the optical requirements, a narrow emission at the appropriate wavelength—red, green, or blue for a full color display—with a high quantum efficiency is necessary. Besides, the luminescence should be highly stable under illumination and heating (up to moderate temperatures such as 100°C). On the other hand, the films should have good optical quality (to minimize light scattering) and the molecular orientation must be as high as possible (to maximize the polarization ratios) and stable, both in time and with respect to temperature changes. Furthermore, solubility of the material in a particular solvent is normally needed for most processing procedures.

In addition, the structure has to be compatible with the formation of mesophases. This condition is probably one of the most difficult to realize because the structural criteria for efficient emission and liquid crystallinity are often mutually exclusive and a delicate balance of parameters is required. There are a large number of organic molecules of both low- and high-molar mass that emit light when activated by light [41]. However, the design of intrinsically luminescent liquid crystals, where their self-organization capability is combined with emission properties, involves the appropriate selection of the rigid core [16].

In fact, it is not easy to establish a straightforward relationship between structure and emissive properties, and there are luminophores with a wide variety of structures. Nevertheless, there are some structural characteristics that favor the appearance of emission. Most of the compounds are highly conjugated, planar, and rigid structures that incorporate aromatic and/or heteroaromatics rings. Structural rigidity with a degree of molecular planarity and little inter-annular twisting may preclude undesirable emission deactivation processes. Amongst the structural features that influence wavelengths are conjugation and electron-donating/electron-withdrawing substitution.

Several features should be considered in the design of LCs. The aromatic core should have a high length-to-breadth ratio in order to induce liquid crystallinity. The melting points should be low—preferably close to or at room temperature—to allow straightforward processing of the material. For this reason, the presence of long aliphatic chains in terminal and sometimes lateral positions is advantageous in order to reduce the interactions between the aromatic cores and to increase the intermolecular distance. The presence of substituents in lateral positions can achieve the same effect, although they can disrupt the planarity of the π-core. For processing purposes, smectic phases are much more difficult to align macroscopically for polarized emission and, consequently, the nematic phase is usually preferred. High melting points can be tolerated if glassy phases form on cooling. However, polymerization of the end groups of reactive mesogens is much slower in the glassy nematic state than in the nematic state because of the much higher effective viscosity in an organic glass.

With these requirements in mind, a very limited number of luminescent cores with acrylate, methacrylate, or diene polymerizable groups have been investigated to date. The majority of these reactive mesogens emit blue or blue-green light. The synthetic pathways should preferentially be as short as possible to facilitate commercialization of this technology and to guarantee final compounds with a high purity by avoiding the presence of contaminants because of synthetic procedure. Besides, chromophores should not suffer photo-degradation or bleaching of the luminescence during the photopolymerization.

In the following sections we give an overview of several examples concerning the fabrication of cross-linked layers for use as emissive films and these are categorized by the chromophore.

## 6.6   PHENYLETHYNYL AND PHENYLVINYL-BASED LUMINOPHORES

The three-ring bis(*p*-phenylethynyl)benzene system, also known as bis-tolane, is an interesting dichroic and highly efficient blue-emitting luminophore for polarized emission, with reported quantum yields close to unity for some derivatives [42,43]. This core also yields materials that have high optical anisotropy and liquid crystalline properties, where the incorporation of lateral or terminal substituents lowers the melting temperatures while retaining the desired optical anisotropy [44–46]. Indeed, examples of oriented films prepared by drawing blends of bis(*p*-phenylethynyl)benzene derivatives in a poly(ethylene) matrix show a highly polarized emission, with values up to 50:1 on using unpolarized excitation [47]; nevertheless, this approach faces a major limitation concerning compatibility and large-scale phase segregation [43]. In addition, polyacrylate **6** containing a bis(phenylethynyl)benzene core has been used to produce polarized blue light with PL anisotropies of about 10:1 on rubbed PI [48].

The first example found in the literature of a LC network processed by *in situ* photopolymerization concerns the use of the bis(phenylethynyl)benzene chromophore **7** [49]. Oriented films of 10 μm thickness were produced at 130°C by surface alignment on nylon 66 layers from blends of 5% of **7** in the nematic direactive compound **8**. Linearly polarized violet-blue emitted light was observed with polarization ratios of 13:1. The processing method was effective up to 15% of the luminophore with no sign of phase separation or photo-degradation.

Diacrylates containing a bis(phenylethynyl)benzene moiety with the alkyl chains located in the long axis of the core **9** have also been reported [50]. Oriented layers of the monomers were obtained by spin-coating and annealing the monomers on a uniaxially rubbed surface of poly(ethylenedioxythiophene)/poly(styrenesulfonic acid) (PEDOT/PPS). These films showed anisotropic violet-blue emissions with PL anisotropies of about 11. Unfortunately, details of the polarized PL emission of the films after *in situ* photopolymerization were not reported. Nonetheless, an OLED

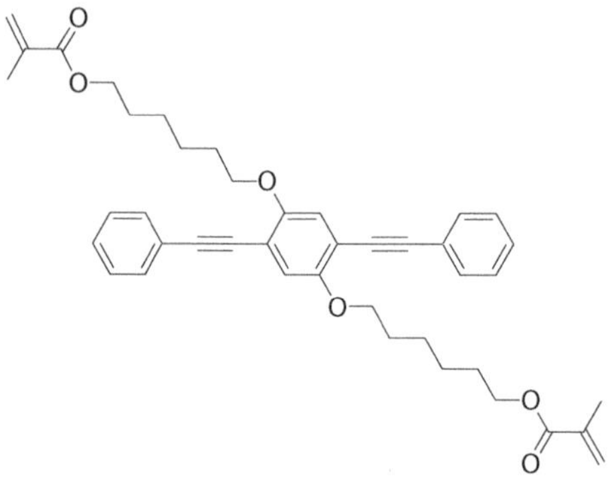

**7**

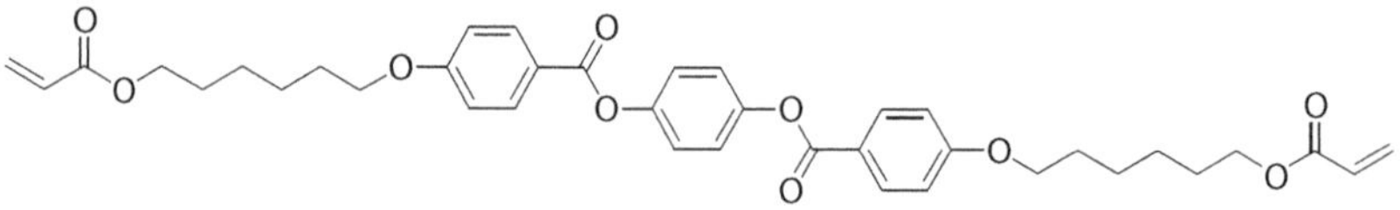

**8**

device was fabricated using a polymeric film that was produced by *in situ* photopoly-merization at room temperature of one of the monomers ($R_1$ = H, $R_2$ = $CH_2CH_3$), with the film acting as the emissive layer. The OLED showed polarized EL that was around one order of magnitude greater when recorded parallel to the rub-bing direction than perpendicular to it, but the luminance of the display was not acceptable for practical applications. The authors related the poor performance with some level of photo-degradation of the monomer upon exposure to UV light during photopolymerization.

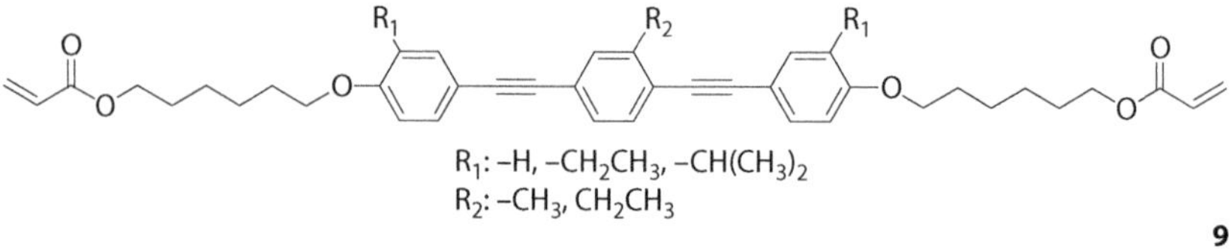

$R_1$: –H, –$CH_2CH_3$, –$CH(CH_3)_2$
$R_2$: –$CH_3$, $CH_2CH_3$

**9**

In terms of chemical and optical properties, compounds that are closely related to those described above contain the bis(*p*-phenylvinyl)benzene core—in particular the *trans*-isomers because the *cis*-isomers usually display only weak luminescence [41]. There is a very large number of literature reports on the production of emissive films from PPV and its derivatives, and these have shown considerable promise as materials for commercial displays because of their good PL efficiencies and tuning of color emission. However, there is some concern about the susceptibility of the vinylene linkages to degradation. There are also a number of reports on polymeric systems containing isolated bis(*p*-phenylvinyl)benzene or well-defined oligomeric *p*-phenylenevinylenes (OPV) fluorescent segments either in the main chain or as pendant groups [51–54]. In addition, several OPVs have been reported to be LCs

[55–58]. However, there are very few reports concerning polarized emission of OPV-based systems. In one example, the LC polymer **10** that contains an OPV mesogenic core was spin-coated and annealed on the surface of uniaxially rubbed PI and this system showed a polarized blue-green emission with polarization ratio about 5.7 (using unpolarized excitation) [59]. The polymer shows a smectic A phase above the $T_g$, which is at 77°C.

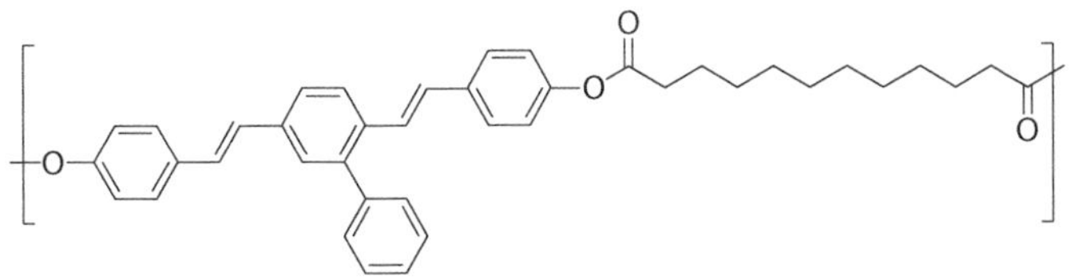

**10**

A conjugated luminescent OPV-based mesogen was investigated for polarized luminescence applications by Turner and coworkers [60,61]. Aligned cross-linked networks were prepared by thermally induced polymerization (in the absence of an initiator) of the spin-coated reactive mesogen **11** on the surface of an alignment layer under nitrogen atmosphere. The process was controlled to give a network with dichroic ratio of $D \approx 3.5$, as determined from the absorption spectra. The films gave rise to blue PL but data on the polarization of the emission were not provided.

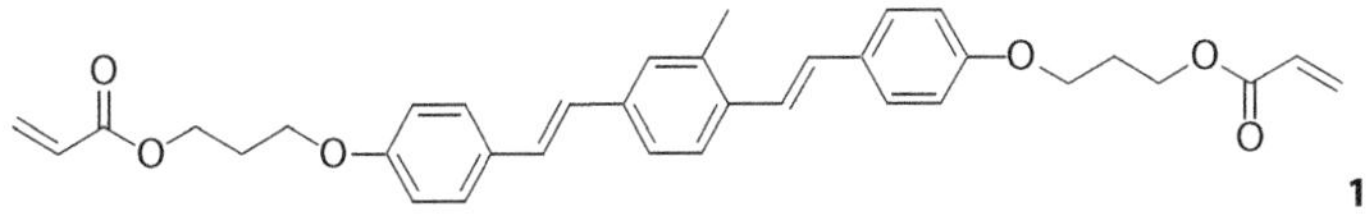

**11**

Gin and coworkers designed a series of fluorescent LCs based on OPVs under different structural criteria. Mesogens of general formula **12** were synthesized and this system contains an OPV core linked to two dihydrocholesterol mesogens. The films were mechanically sheared to produce oriented glassy films with polarized PL ratio of 7.5 under polarized excitation [62]. The same authors also reported the properties of a series of hexacatenar LCs that form columnar mesophases and are highly efficient blue-emitters [63]. Recently, the same group incorporated polymerizable groups into these phasmidics LCs to produce stabilized networks that are not sensitive to temperature [64]. It was found that diene reactive groups are tolerated but acrylate groups disturb the formation of the columnar hexagonal mesophases. UV-photoinitiated radical polymerization leads to photo-degradation but thermally initiated polymerization afforded cross-linked materials with retention of LC order and emission properties for compounds **13** and **14**. It is known that a uniaxial orientation of the columnar structures is readily attainable [65], but the optical properties of oriented cross-linked films of such systems have not been studied to date.

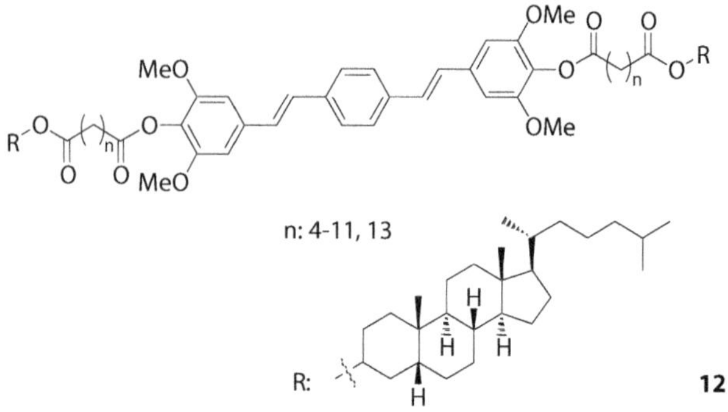

n: 4-11, 13

R:                                                                    12

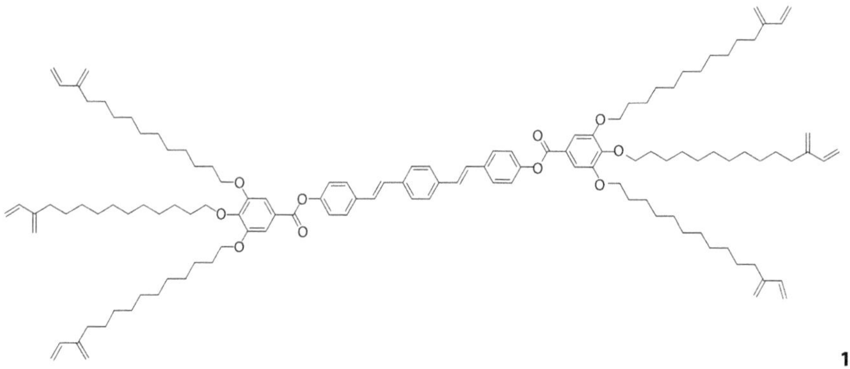

13

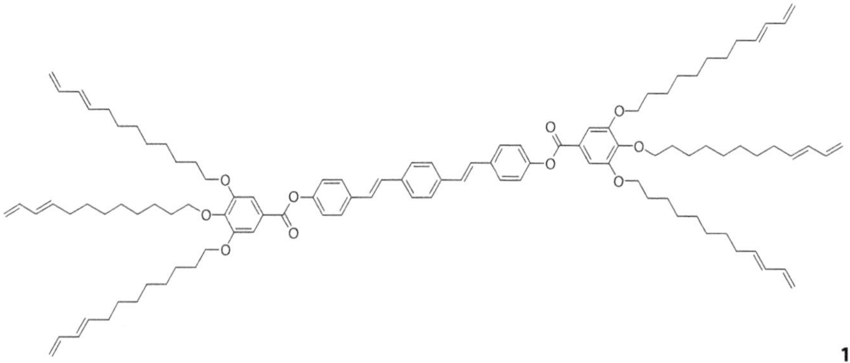

14

## 6.7 TERPHENYL-BASED CHROMOPHORES

Polyphenyls are a major class of efficient linear luminophores in which the absorption and emission properties are determined by the size of the π-system. On changing from benzene to biphenyl and $p$-terphenyl, the emission shift towards longer wavelengths from the violet to the blue region of the visible electromagnetic spectra is accompanied by an increase in the quantum yield of the emission from 0.07, 0.18 to 0.93, respectively [41]. In particular, $p$-terphenyl has found practical uses in scintillation counting. The rod-like structure of the *p*-terphenyl core is also suitable to generate LC phases and provides highly birefringent LCs with good thermal, chemical, and photochemical stability widely employed in display devices [66–68].

Linearly and circularly polarized emissive films have been pursued using vitrifiable LCs and LCs polymers functionalized with the 4-cyanoterphenyl luminescent unit [21]. This chromophore has a featureless absorption band centered at 315 nm and a blue emission band located at 410 nm (in dichloromethane) with the quantum efficiency reported to be 0.60. In particular, films of **15** with thickness of 4 μm, in which a uniaxial arrangement of the chromophores was induced on nylon 66 alignment layers, presented modest polarization ratios of emission of ca. 3.7.

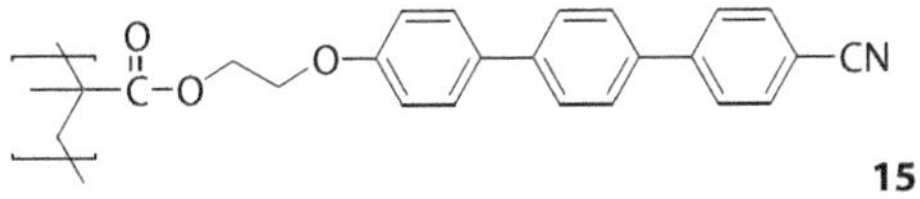

**15**

Polarized PL in cross-linked films obtained by *in situ* photopolymerization of materials containing 4-cyanoterphenyl, either functionalized or not, have been studied by Sánchez et al. [69,70]. These authors studied the influence of the composition and polymerization temperature on the stability and the anisotropy of the emission using reactive and nonreactive luminophores **16**, **17**, and **18**.

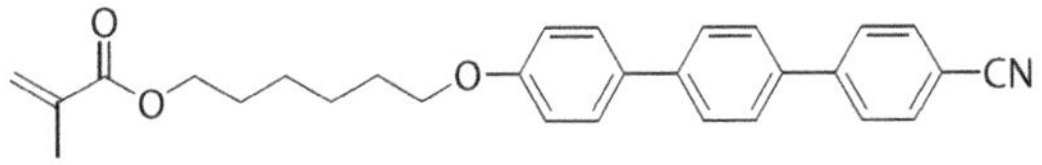

**16**

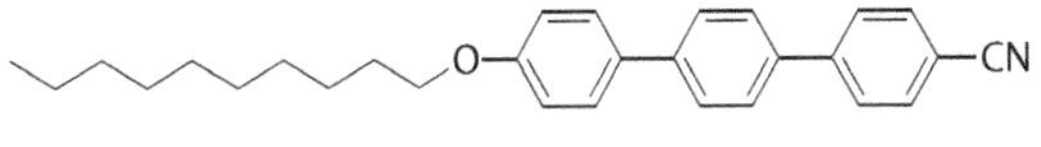

**17**

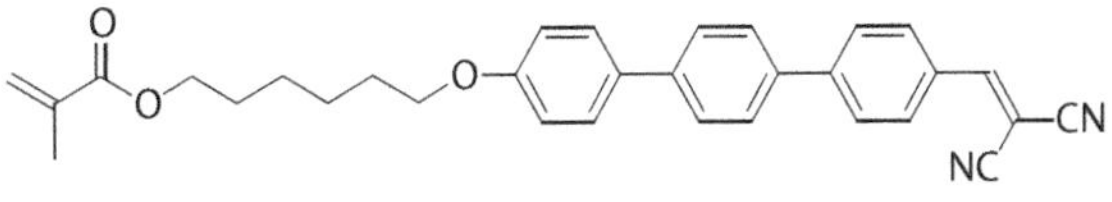

**18**

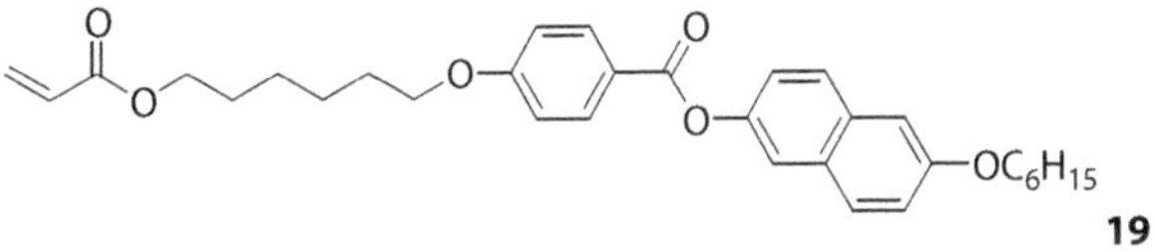

**19**

20

21

The polymerizable formulations combine monoreactive (**19** or **20**) and direactive (**21**) nonemissive LC monomers to control the cross-linking degree, and 1% or 5% of the luminophore. Previous reports on these systems have demonstrated that polymeric films with low cross-linking degrees show a memory effect that it is characteristic of liquid crystalline elastomers but materials with cross-linking densities above 10% (expressed as the molar percentage of cross-linker **21** in the formulation) seem to behave as polymeric networks [71]. Films of 5 μm thickness were prepared by *in situ* photopolymerization of samples in commercial cells for planar alignment at different temperatures. Under these circumstances, a preferential orientation of the matrix and the chromophore was demonstrated by polarized UV–VIS spectroscopy. The emission spectra of blends of **19** and **21** with the luminophore **16** were recorded under polarized excitation light (360 or 370 nm wavelength) in order to preclude effects associated with the absorption of the matrix; furthermore, a low excitation intensity was used in order to prevent optical bleaching of the luminescence.

The emission anisotropy appeared to be independent of the chromophore content and almost independent of the polymerization temperature. The emission anisotropy was, however, dependent on the matrix composition. The results show that the emission anisotropy ratio increases on decreasing the cross-linking degree. PL anisotropy values of 9.2 were determined for oriented films with 10% cross-linker and polymerized at 100°C but only 6.2 for those with 95% cross-linker. This finding was explained in terms of the distortions induced in the films at a molecular level by linking of the individual polymeric chains and, consequently, the deviation of the mean molecular axis with respect to the rubbing direction. Nevertheless, for high cross-linking densities, the PL polarization ratio can be slightly improved by lowering the polymerization temperature, with polarization ratio up to 7.3 at 60°C. Such changes are possible as long as the process takes place in the nematic phase, otherwise the optical quality of the films becomes poor.

The study described above also highlighted the influence that the illumination on the excitation band and thermal treatments up to 200°C have on the intensity and anisotropy of the emission. In this case, thermal quenching and optical bleaching seem to be related to the composition of the polymer matrix but not to the cross-linking density. As such, the lack of stability either with temperature or under strong illumination observed for 10% cross-linked films containing the monoreactive monomer **19**

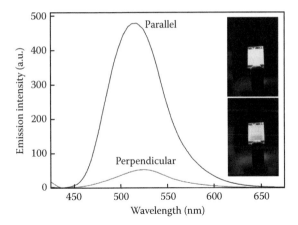

**FIGURE 6.6**   (**See color insert following page 304.**) Polarized emission of a film processed by *in situ* photopolymerization of a LC formulation of **18** (5%), **19** (85%), and **21** (10%) in a commercial glass cell for planar alignment. Emission was registered parallel and perpendicular to the preferred molecular orientation direction using polarized excitation.

was overcome by replacing it with the **20** monomer, which has the same mesogenic core as the cross-linker.

Luminescence is very sensitive to structural changes and the replacement of a cyano substituent by the more strongly electron-withdrawing the dicyanovinyl group in chromophore **18** gives rise to a strong red shift in the emission to 535 nm. However, this is also associated with a decrease in the emission efficiency, and a quantum yield of 0.18 in tetrahydrofurane (THF) was calculated. In any case, a stable green emission was obtained in a 10% cross-linked film with $PL_{||}/PL_{\perp}$ ratio of 8.5 using the nonemissive monoreactive LC **19** (see Figure 6.6) [72]. These results confirm that it is possible to keep the high anisotropy in emission for films with low cross-linking degree and prevent the lack of stability of the emission by appropriate changes in the composition of the polymeric matrix.

The chromophores can be incorporated into the polymer linked to the polymer backbone, as in the examples described above, or as a guest dispersed in the polymeric matrix. Investigations into films with similar compositions in a guest–host configuration using the nonfunctionalized chromophore **17** have proven that polarization of the emitted light is higher, mainly because of a better orientation of the molecules within the film.

## 6.8   FLUORENE-BASED LUMINOPHORES

Both polymers and oligomers based on the 2,7-disubstituted-9,9-dialkylfluorene chromophore have emerged as being amongst the most promising classes of light-emitting organic materials because of their exceptional features [73,74]. A variety of these systems have been tested in different electro-optical devices with a good correspondence found between solution and solid-state emission spectra as well as between PL and EL spectra. Of all the conjugated systems studied to date, fluorene-containing materials have superior thermal and photostability than PPV derivatives [75]. These systems have exceptional chemical and physical versatility, as well as extremely high emission efficiencies and the potential for uniaxial alignment mediated by LC behavior. From the point of view of the chemical structure, when compared to the $\pi$-isoelectronic biphenyl unit, the fluorene ring has a more planar and rigid structure with a lower level of conformational freedom because of the presence of the methylene bridge. This structural feature gives a red shift in the emission and improves the emission efficiency. For the same reason, the fluorene unit is slightly bent and the noncoaxial structure of the 2,7-disubstituted fluorene core has an adverse effect on the LC properties. The presence of alkyl chains at the bridging 9-position of the fluorene units increases the solubility in organic solvents, lowers the melting point and imparts high tendency for glass formation. In addition, the alkyl chains at the bridging methylene give rise to larger intermolecular distances and they hinder the molecular packing required for the development of liquid crystallinity, in short oligomers (in general, from dimers to pentamers) with low length-to-breadth ratios [76,77]. In this sense, branched lateral chains are preferred over linear counterparts in terms of mesophase formation, glass formation, and polarization ratios in PL and EL [19,35,73]. It seems that branched chains at the 9-position of the fluorene core lower the effective molecular diameter and affect the solid morphology giving glass-forming materials with good alignment properties.

On the basis of the information discussed above, nematic 9,9-dialkyl substituted poly(fluorene)s and oligo(fluorene)s have been actively pursued with the aim of polarizing PL and EL. Uniaxial alignment of the conjugated backbones has been accomplished by thermal annealing for several hours of spin-cast films on different alignment layers. The feasibility of polarized emission has been proved on mechanically rubbed PI (with or without dopants), PPV or PEDOT/PPS layers. Several relevant examples are outlined below. Grell et al. reported blue polarized EL from poly(fluorene)s processed by the thermal annealing of films spin-cast onto the surface of a rubbed PI layer doped with a hole transporter with $EL_{||}/EL_{\perp}$ ratios of 15 [19]. In a similar approach, Whitehead et al. used rubbed-PPV layers to orient and report $EL_{||}/EL_{\perp}$ polarization ratios of 25 [22]. LC poly(fluorene)s have also been aligned on photo-addressable polymeric LC azobenzene layers (see b in Figure 6.2) [26]. In these photo-addressable films anisotropy is induced by linearly polarized illumination, with a preferential alignment perpendicular to the electric vector of the incident light. PL polarization ratios of about 12 were determined using polarized excitation and this photo-alignment technique.

Regarding mesomorphic oligo(fluorene)s, derivatives that form glassy nematic films have been oriented both in PI and PEDOT/PSS layers [35,36]. The dodecamer **22** was oriented onto a rubbed PEDOT/PSS layer to give, for a film of 73 nm thickness, $PL_{||}/PL_{\perp}$ ratio of 20 at 424 nm. The EL polarization ratio $EL_{||}/EL_{\perp}$ evaluated at the same wavelength was 27. Emission polarization values are similar to those determined for LC poly(fluorene)s.

Based on interesting results obtained for poly(fluorene)s and oligo(fluorene)s, reactive luminophores containing the 2,7-disubstituted fluorene chromophore have also been investigated by several research groups. A number of papers produced at the University of Hull by Kelly and O'Neill concern systematic investigations on

**22**

the structure/mesomorphic behavior relationships of diverse light-emitting and/or charge-transporting liquid crystals bearing a 2,7-disubstituted-9,9-dialkylfluorene as an essential part of the central chromophore. A large number of structural modifications have been investigated in an attempt to optimize the optical and electronic properties, and to design reactive mesogens with either vitrifiable or LC phases at room temperature for use in OLED devices [78–81]. As a consequence of systematic property/structure investigations, reactive luminescent mesogens were synthesized to produce polymeric networks by UV photopolymerization.

The authors have compared the photopolymerization of a range of luminescent reactive LCs, with the same luminescent fluorene-based core and a variety of polymerizable end groups—methacrylate and dienes [80]. UV radiation of 300 nm at room temperature in the absence of photoinitiator was used to photopolymerize nematic glasses of these compounds formed upon cooling. Table 6.1 collects the structure of some of the investigated monomers and the general structure of the polymer backbones formed by polymerization. The results show that the methacrylate group polymerizes faster giving a relatively flexible polymer backbone—compound **23** requires fluence of 3 J cm$^{-2}$ to form an insoluble network. Photopolymerization of dienes is generally slower—compound **24** requires fluence of 100 J cm$^{-2}$—and provides a particularly rigid backbone. In spite of this, the photopolymerization of **23** is accompanied with a decrease of the PL intensity; however, an increase in the quantum efficiency of PL after cross-linking is noted for **24**. The authors have related these observations with the conformation changes imposed by the polymer backbone formed by polymerization and some level of photo-degradation of the compound **23** during polymerization.

Compounds with photoreactive diene end groups were polymerized by UV irradiation in the glassy nematic state but an optimal photochemical cross-linking requires long exposure times. Therefore, Kelly, O'Neill and coworkers obtained reactive LCs with a nematic mesophase at room temperature by appropriate substitution on the 9,9-dialkylfluorene or, alternatively, by mixing homologous compounds. The results indicated that in some cases photopolymerization in the fluid nematic state requires half the fluence necessary for polymerization in the glassy state [79,82].

The same authors also found that the process of photochemical cross-linking of reactive mesogens having diene end-groups, such as **24**, leads to improved charge-transport and emission properties, thus improving the efficiency of the OLED devices [81,83,84].

On the basis of the described results, polarized PL and EL was generated from uniformly aligned nematic networks obtained by polymerization of some fluorene-based reactive luminophores. These systems were characterized by the incorporation of thiophenes and long alkyl chains at the fluorene 9-position, and the presence of diene

**TABLE 6.1**

**Structure of Luminescent Reactive LCs Based on Fluorene Cores Studied by Kelly and O'Neill**

|    | R–O | General Structure of the Polymer Backbone |
|----|-----|-------------------------------------------|
| 23 | | |
| 24 | | |
| 25 | | |
| 26 | | |

polymerizable groups. Macroscopic orientation was achieved by photo-alignment techniques using coumarin-containing polymer layers (see c and d in Figure 6.2) [29,85,86]. The exposure of such photoactive coumarin-containing layers to UV polarized light induces a surface anisotropy that it is transferred to the overlaying LC deposited by spin-coating and annealed in the mesophase. The LC director aligns parallel to the electric field vector of the incident UV beam and the polarization anisotropy depends on the composition and processing conditions of the alignment layer [29,30,87,88]. Using a coumarin-containing alignment layer (c in Figure 6.2), PL polarization ratios up to 8:1 have been reached for blue emitter monomer **24**, which has a nematic phase that readily vitrifies at room temperature ($T_g = 39°C$) [80]. Alignment was achieved in the mesophase by slowly cooling from the isotropic phase and it was then quenched into the glassy state. After photopolymerization, an anisotropic network of compound **24** was incorporated as the emissive layer in an OLEDs device with $E_{||}/E_{\perp}$ ratio of 11:1 and brightness values of 60 cd m$^{-2}$ [29].

Appropriate structural modifications to extend the rigid core and modification of the side chains and end chains can be performed to tune the LC, electronic, and optical properties. Because the order parameter of nematic materials increases with molecular length, the same authors have synthesized compounds with a related chemical structure but extended π-cores to increase the emission anisotropy [89]. Reactive compounds **27, 28,** and **29** were prepared that form nematic glasses on quenching from the nematic phase. The eight-ring chromophore **27** exhibits relatively low melting and clearing points. This compound has a nematic phase that can be easily supercooled to room temperature ($T_g = 0°C$) [81]. Polarized PL and EL ratios of 13:1 were obtained using an alignment layer based on coumarin-containing side-chain polymers (see d in Figure 6.2) [30,88].

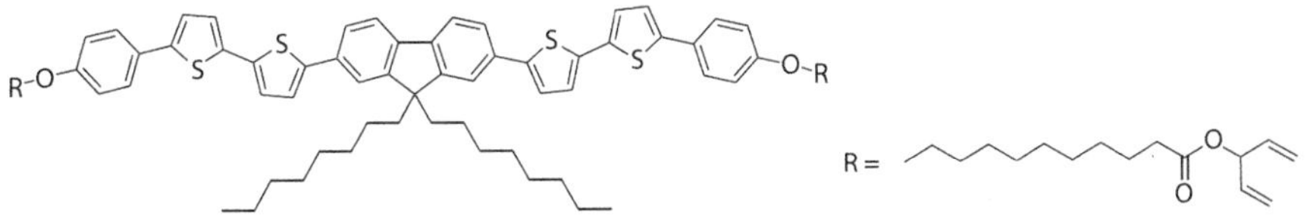

**27**

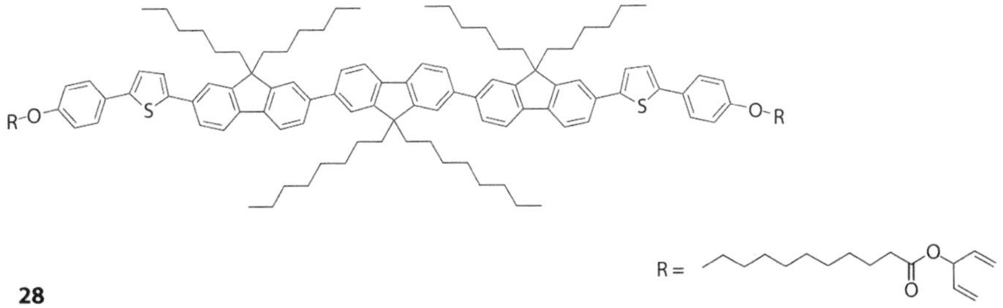

**28**

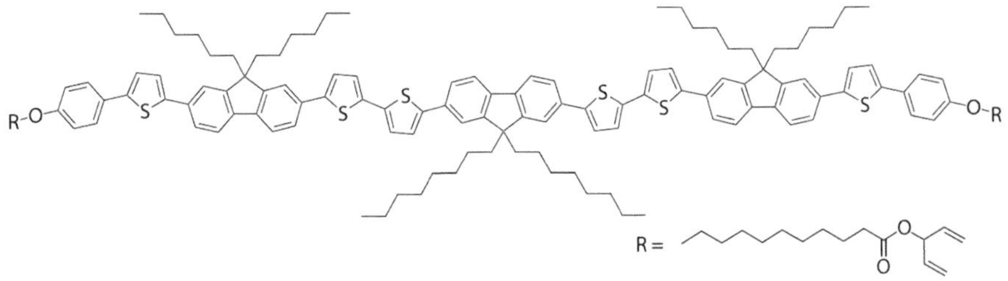

**29**

Data of the emission anisotropy of aligned polymeric networks based on photopolymerizable compounds **28** and **29**, with 10 and 14 aromatic rings respectively, have not been given to date. Nevertheless, a nonpolymerizable homologue of **29**, with

R=octyl, aligned on a rubbed PEDOT/PSS substrate showed a maximum PL polarization ratio of 30 [89]. This is an extremely high value for a nematic material that is equivalent to the highest values obtained for LC polymers and oligo(fluorene)s.

The chemical structure of the aromatic core was also tailored for different colored emission and, for instance, green emission was obtained from chromophore **27**. Red emission was obtained from the reactive fluorene-based compound **30**. However, the compound crystallizes at room temperature rather than retaining a supercooled nematic phase restricting the formation of good optical quality films. To overcome this limitation, blends were prepared of **30** in the green-emitter **27** (1:3 ratio by weight). Because absorption spectrum of **30** overlaps the emission spectrum of **27**, efficient Förster energy transfer is possible and, as a result, only red emission was observed. A maximum $PL_{||}/PL_{\perp}$ value obtained was 7 using a coumarin-based alignment layer (see d in Figure 6.2) [30].

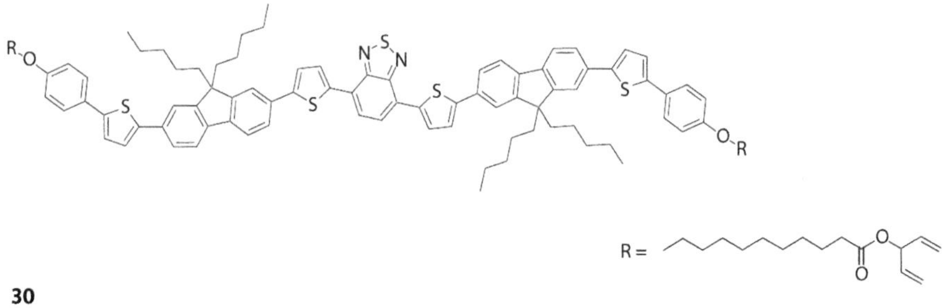

**30**

Strohriegl and coworkers have developed a series of reactive mesogens with fluorene units and two acrylate terminal groups that include monodisperse oligomers as well as oligomeric mixtures obtained by end-capping to control the molecular weight and therefore, the mesophase interval [77,90,91]. The oligomeric mixtures were synthesized as an attempt to overcome synthetic difficulties and low overall yields associated to the synthesis of monodispersed oligomers. Anisotropic films were prepared from **31** and **32** by UV-initiated polymerization of the monomers spin-cast onto the surface of several alignment layers. Polarization ratios (using nonpolarized excitation) of 15:1 were obtained for the blue-emitting anisotropic films in both cases using 30 nm thickness films on rubbed PI. On increasing the thickness to 90 nm, the polarization ratios were found to decrease to 9:1.

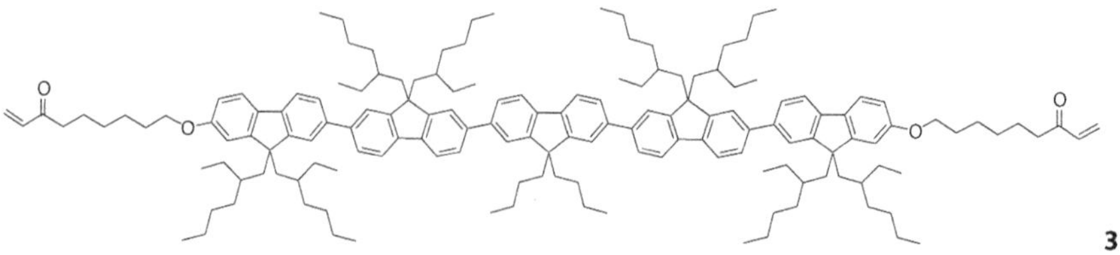

**31**

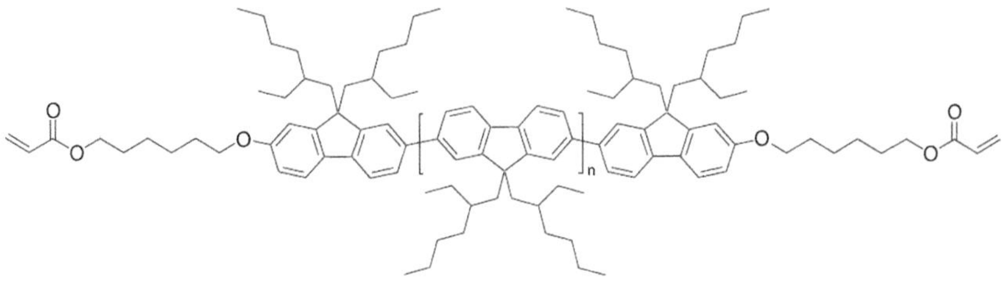

**32**

Serrano and coworkers have reported results on liquid crystalline networks with linearly polarized blue emission from fluorene-based mono- and dimethacrylates. The films were prepared by *in situ* photopolymerization of nonemissive reactive liquid crystalline matrixes containing reactive fluorene emitters. The material was oriented by capillarity forces in commercial cells for planar alignment with unidirectionally rubbed PI coatings. The nonemissive nematic matrix is composed of monomers **20** and **21**, which proved to be suitable components in related studies carried out with *p*-terphenyl emitters (see Section 6.7).

Systematic synthesis of monoreactive fluorene-based emitters with different structural modifications has revealed that the presence of alkyl chains at the 9-position of the fluorene ring does not alter the optical properties either in solution or in the solid state as far as the position of the absorption/emission bands is concern, but does improve the photo-stability of the luminophore. However, the incorporation of such chains is an adverse structural element for the development of liquid crystallinity and also polarization of the emission, which is lower for the same structural core. Therefore, films containing 5% **33** with $n = 0$ along with 10% of a cross-linker **21** show $PL_{||}/PL_{\perp}$ values of 9:1; however, in the case of **33** with $n = 8$ the $PL_{||}/PL_{\perp}$ values obtained were only 4:1. This limitation can be overcome by extending the molecular anisotropy of the emitting core. In spite of their lack of mesogenicity, polarized films containing 5% of **34** or **35** along with 10% of cross-linker **21** were prepared and gave emission polarization ratios $PL_{||}/PL_{\perp}$ of 8:1 [92].

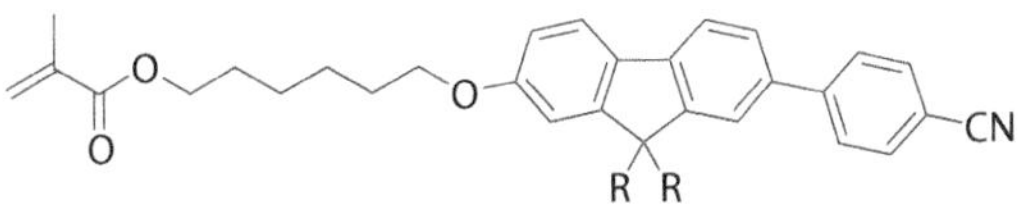

R: $C_nH_{2n+1}$ with n: 0, 4, 8, 12

**33**

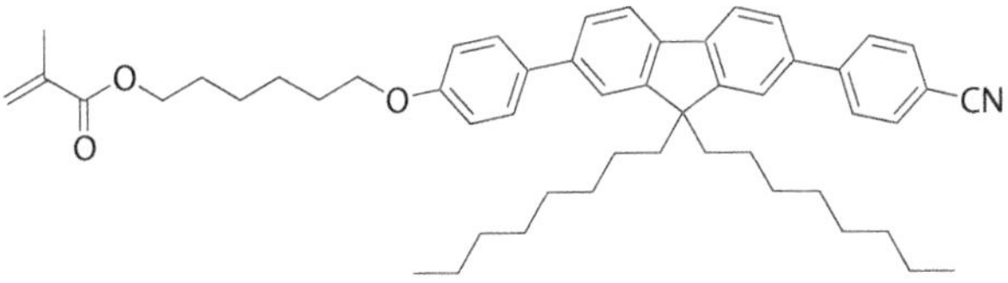

**34**

**35**

The same group has also reported the synthesis and behavior of fluorene-based direactive **36**, **37**, and **38**, which act as emitting cross-linkers in photopolymerizable mixtures with **20**, in a 5:95 molar ratio, respectively. The anisotropy of the emission have been investigated under polarized excitation, with the emission found to be about five-to-six times higher in the direction of the molecular orientation [93,94].

R: C$_2$H$_5$, C$_8$H$_{17}$                                    **36**

**37**

**38**

## 6.9   HETEROCYCLE-BASED LUMINOPHORES

A diverse range of heterocyclic compounds with diene end groups have been reported by Kelly and O'Neill. The aim of preparing these materials was to adjust the optical properties of reactive mesogens for emissive layers in full-color devices, that is, red-green-blue pixellated displays [78,95,96]. A number of heterocycles such as benzothiazole, benzothiadiazole, and pyridine, pyrimidine or tetrazine have been incorporated into the aromatic core to give smectic-reactive mesogens. However, extensive details on the polarizing PL or EL of these heterocyclic compounds are not available because of the fact that smectic polymer networks are less efficient than corresponding OLEDs using nematic LCs based on fluorenes. Polarized emission at 525 nm was reported for **39**, which has smectic A phase between 57°C and 73°C. The material was uniaxially oriented on rubbed PEDOT and $PL_{||}/PL_{\perp}$ ratios of 5:1 were obtained [95].

**39**

Serrano and coworkers reported 2-phenylbenzoxazoles-based luminophores **40** that were incorporated in a guest–host configuration in anisotropic films obtained by *in situ* photopolymerization of the nematic diacrylate **21** [97]. Oriented films were produced in commercial cells from mixtures of 5% (w/w) 2-phenylbenzoxazole and these films actively emit violet-to-blue linearly polarized light. The optical properties of the networks were found to be dependent on the substitution pattern at the 2-phenylbenzoxazole core. Amongst the investigated structural modifications, the authors have concluded that the placement of an electron-donor substituent at the 5-position of the benzoxazole heterocycle and an electron-withdrawing substituent at the *p*-position of the phenyl radical gives blue efficient emission with a suitable photo-stability in solid state. $PL_{||}/PL_{\perp}$ ratios between 3 and 4 were reported under polarized excitation.

**40**

## 6.10   CONCLUDING REMARKS

The high demand of flat-panel displays for consumer equipments in everyday applications and the competitiveness amongst the LCD and new emerging technologies based on OLEDs require for an increasing performance on new materials such as polarized light emitters. In this field, liquid crystalline order has proved to be an adequate mean to orient luminophores in order to achieve polarized light emission. In particular, *in situ* polymerization of reactive LCs to yield anisotropic networks is a low-cost processing method, which can be easily scaled-up. Nevertheless, as it has been shown along this review, this technique has not been thoroughly investigated and some drawbacks still need to be overcome. Thus, the polarization ratios of the emitted light reported for emitting LCs networks processed by *in situ* polymerization are in most cases relatively low for commercial applications. For instance, the use of polarized OLEDs as backlight in LCDs requires polarization ratios in the range 30–40 even if values of 10 and more are acceptable with the use of a clean-up polarizer. For this reason, efforts should be made for obtaining better degrees of molecular order. Optimization of the calamitic structure of chromophores, polymerization processes or alternatively the investigation of or photopolymerization in highly ordered mesophases (smectic phases instead of nematic mesophase) are aspects that can still be investigated. In fact, most of the reported results concern to nematic materials that because of their low viscosity are easily oriented. However, *a priori*, higher degree of molecular orientation can be obtained from smectic mesophases provided that a good oriented monodomain can be achieved. Although this technique may be limited by lower values of polarization of the emitting light than those obtained by means of mechanical alignment reported for polymers, it may be competitive because of the advantages of the method. However, an effort to develop optimum materials emitting in the red, green, or blue region must be done.

Furthermore, new alternatives can arise from the exploration of other mesophases such as chiral calamitic or columnar phases for circularly polarized emission, which have been poorly investigated to date. In particular, columnar phases in which cofacial stacking is avoided by rotation of chromophores around the column axis, may be of interest for producing LC networks of highly luminescent chromophores (e.g., pyrenes).

There are added features related to the production of highly oriented networks with anisotropic emitting properties that have not been undertaken in this review and where the combination of optical anisotropy with a stable network can provide additional advantages. For instance, it has been described that incorporation of in-plane aligned emitting layers into OLEDs can improve the efficiency of the device relative to nonoriented layers. In a different approach, the photopolymerization of reactive LCs allows the production of multilayer structures that, in addition, can be easily patterned by photolithographic processes. In this last application, orientation is not essential but can provide additional benefits.

In conclusion, the combination of anisotropic properties of LCs, emissive and transport properties of luminophores and the easy processing of reactive LCs may report new improvements on interesting devices as well as new properties of interest provided that a complete background of basic knowledge would be undertaken.

## REFERENCES

1. S.M. Kelly, *Flat Panel Displays: Advanced Organic Materials*, Ed. J.A. Connor, The Royal Society of Chemistry, Cambridge, 2000.

2. S. Naemura, Introduction special section on advances in liquid crystal materials and electro-optic effects. *J. Soc. Inf. Display* 2006, **14**, 515.

3. C. Weder, C. Sarwa, A. Montali, C. Bastiaansen, P. Smith, Incorporation of photoluminescent polarizers into liquid crystal displays. *Science* 1998, **279**, 835.

4. A. Montali, C. Bastiaansen, P. Smith, C. Weder, Polarizing energy transfer in photoluminescent materials for display applications. *Nature* 1998, **392**, 261.

5. K. Müllen, U. Scherf (Eds.), *Organic Light Emitting Devices: Synthesis, Properties and Applications*, Wiley-VCH, Weinheim, 2006.

6. M. O'Neill, S.M. Kelly, Liquid crystals for charge transport, luminescence, and photonics. *Adv. Mater.* 2003, **15**, 1135.

7. P. Dyreklev, M. Berggren, O. Inganäs, M.R. Andersson, O. Wennerström, T. Hjertberg, Polarized electroluminescence from an oriented substituted polythiophene in a light-emitting diode. *Adv. Mater.* 1995, **7**, 43.

8. M. Grell, D.D.C. Bradley, Polarized luminescence from oriented molecular materials. *Adv. Mater.* 1999, **11**, 895.

9. W. Schnabel, Absorption of light and subsequent photophysical processes. *Polymers and Light. Fundamentals and Technical Applications*, Wiley-VCH, Weinheim, 2007, p. 22.

10. V. Cimrova, M. Remmers, D. Neher, G. Wegner, Polarized light emission from LEDs prepared by the Langmuir-Blodgett technique. *Adv. Mater.* 1996, **8**, 146.

11. Y. Gotou, I. Kakinoki, M. Noto, M. Era, Efficient polarized blue electroluminescent device using oriented p-sexiphenyl thin film. *Curr. Appl. Phys.* 2005, **5**, 19.

12. C. Weder, C. Sarwa, C. Bastiaansen, P. Smith, Highly polarized luminescence from oriented conjugated polymer/polyethylene blend films. *Adv. Mater.* 1997, **9**, 1035.

13. M. Hamaguchi, K. Yoshino, Polarized electroluminescence from rubbing-aligned poly(2,5-dinonyloxy-1,4-phenylenevinylene) films. *Appl. Phys. Lett.* 1995, **67**, 3381.

14. M. Hamaguchi, K. Yoshino, Polarized fluorescence and electroluminescence from rubbing-aligned conjugated polymers and their composites with fluorescent dyes. *Polym. Adv. Technol.* 1997, **8**, 399.

15. M. Jandke, P. Strohriegl, J. Gmeiner, W. Brütting, M. Schwoerer, Polarized electroluminescence from rubbing aligned poly(*p*-phenylenevinylene). *Adv. Mater.* 1999, **11**, 1518.

16. See R. Giménez, M. Piñol, J.L. Serrano, Luminescent liquid crystals derived from 9,10-bis(phenylethynyl)anthracene. *Chem. Mater.* 2004, **16**, 1377, and references cited therein for examples of intrinsic luminescent liquid crystals.

17. N.S. Sariciftci, U. Lemmer, D. Vacar, A.J. Heeger, R.A.J. Janssen, Polarized photoluminescence of oligothiophenes in nematic liquid crystalline matrices. *Adv. Mater.* 1996, **8**, 651.

18. B. Bahadur, Guest-host effect. in *Handbook of Liquid Crystals*, Vol. 2A. *Low Molecular Weight Liquid Crystals I*; Eds. D. Demus, J. Goodby, G.W. Gray, H.W. Spiess, V. Vill, Wiley-VCH, Weinheim, 1998.

19. M. Grell, W. Knoll, D. Lupo, A. Meisel, T. Miteva, D. Neher, H.G. Nothofer, U. Scherf, A. Yasuda, Blue polarized electroluminescence from a liquid crystalline polyfluorene. *Adv. Mater.* 1999, **11**, 671.

20. M. Grell, D.D.C. Bradley, M. Inbasekaran, E. Woo, A glass-forming conjugated main-chain liquid crystal polymer for polarized electroluminescence applications. *Adv. Mater.* 1997, **9**, 798.

21. B.M. Conger, J.C. Mastrangelo, S.H. Chen, Fluorescence behavior of low molar mass and polymer liquid crystals in ordered solid films. *Macromolecules* 1997, **30**, 4049.

22. K.S. Whitehead, M. Grell, D.D.C. Bradley, M. Jandke, P. Strohriegl, Highly polarized blue electroluminescence from homogeneously aligned films of poly(9,9-dioctylfluorene). *Appl. Phys. Lett.* 2000, **76**, 2946.

23. D.H. Chung, Y. Takanishi, K. Ishikawa, H. Takezoe, B. Park, Visualization of rubbing nonuniformity by double surface treatments of polyimide-coated substrate for liquid crystal alignment. *Jpn. J. Appl. Phys.* 2001, **40**, 1342.

24. K. Ichimura, Photoalignment of liquid-crystal systems. *Chem. Rev.* 2000, **100**, 1847.

25. K. Sakamoto, K. Usami, Y. Uehara, S. Ushioda, Excellent uniaxial alignment of poly(9,9-dioctylfluorenyl-2,7-diyl) induced by photoaligned polyimide films. *Appl. Phys. Lett.* 2005, **87**, 211910.

26. D. Sainova, A. Zen, H.G. Nothofer, U. Asawapirom, U. Scherf, R. Hagen, T. Bieringer, S. Kostromine, D. Neher, Photoaddressable alignment layers for fluorescent polymers in polarized electroluminescence devices. *Adv. Funct. Mater.* 2002, **12**, 49.

27. N. Kawatsuki, H. Ono, H. Takatsuka, T. Yamamoto, O. Sangen, Liquid crystal alignment on photoreactive side-chain liquid-crystalline polymer generated by linearly polarized UV light. *Macromolecules* 1997, **30**, 6680.

28. P.O. Jackson, M. O'Neill, W.L. Duffy, P. Hindmarsh, S.M. Kelly, G.J. Owen, An investigation of the role of cross-linking and photodegradation of side-chain coumarin polymers in the photoalignment of liquid crystals. *Chem. Mater.* 2001, **13**, 694

29. A.E.A Contoret, S.R. Farrar, P.O. Jackson, S.M. Khan, L. May, M. O'Neill, J.E. Nicholls, S.M. Kelly, G.J. Richards, Polarized electroluminescence from an anisotropic nematic network on a non-contact photoalignment layer. *Adv. Mater.* 2000, **12**, 971.

30. M.P. Aldred, P. Vlachos, A.E.A. Contoret, S.R. Farrar, W. Chung-Tsoi, B. Moonsor, K.L. Woon, R. Hudson, S.M. Kelly, M. O'Neill, Linearly polarised organic light-emitting diodes (OLEDS): Synthesis and characterisation of a novel hole-transporting photoalignment copolymer. *J. Mater. Chem.* 2005, **15**, 3208.

31. R. Rosenhauer, J. Stumpe, R. Gimenez, M. Piñol, J.L. Serrano, A.I. Viñuales, All-in-one layer: Anisotropic emission due to light-induced orientation of a multifunctional polymer. *Macromol. Rapid. Commun.* 2007, **28**, 932.

32. R. Rosenhauer, Th. Fisher, J. Stumpe, R. Gimenez, M. Piñol, J.L. Serrano, A.I. Viñuales, D. Broer, Light-induced orientation of liquid crystalline terpolymers containing azobenzene and dye moieties. *Macromolecules* 2005, **38**, 2213.

33. R. Gimenez, M. Piñol, J.L. Serrano, A.I. Viñuales, R. Rosenhauer, J. Stumpe, Photo-induced anisotropic films based on liquid crystalline copolymers containing stilbene units. *Polymer* 2006, **47**, 5707

34. C.C. Wu, P.Y. Tsay, H.Y. Cheng, S.J. Bai, Polarized luminescence and absorption of highly oriented, fully conjugated, heterocyclic aromatic rigid-rod polymer poly-p-phenylenebenzobisoxazole. *J. Appl. Phys.* 2004, **95**, 417.

35. Y. Geng, S.W. Culligan, A. Trajkovska, J.U. Wallace, S.H. Chen, Monodisperse oligofluorenes forming glassy-nematic films for polarized blue emission. *Chem. Mater.* 2003, **15**, 542.

36. S.W. Culligan, Y. Geng, S.H. Chen, K. Klubek, K.M. Vaeth, C.W. Tang, Strongly polarized and efficient blue organic light-emitting diodes using monodispersed glassy nematic oligo(fluorene)s. *Adv. Mater.* 2003, **15**, 1176.

37. D.J. Broer, Photoinitiated polymerization and cross-linking of liquid crystalline systems. in *Radiation Curing in Polymer Science*. Vol. III *Polymerization Mechanisms*, Eds. J.P. Fouassier, J.F. Rabek, Elsevier Science, London, 1993, p. 383.

38. S.M. Kelly, Anisotropic networks, elastomers and gels. *Liq. Cryst.* 1998, **24**, 71.

39. D.J. Broer, Creation of supramolecular thin film architectures with liquid-crystalline networks. *Mol. Cryst. Liq. Cryst.* 1995, **261**, 513.

40. R.A.M. Hikmet, D.B. Braun, A.G.J. Staring, H.F.M. Schoo, J. Lub, An electroluminescent device useful as backlight for liq-crystal displays-comprises anisotropic layer of melted electroluminescent organic cpd between two electrode layers. Int. Patent Apl. WO 97/07654, 1995.

41. B.M. Krasovitsky, B.M. Bolotin, (Eds.) *Organic Luminescent Materials*, VCH, Weinheim, 1988.

42. H. Li, D.R. Powell, R.K. Hayashi, R. West, Poly((2,5-Dialkoxy-*p*-phenylene)ethynylene-*p*-phenyleneethynylene)s and their model compounds. *Macromolecules* 1998, **31**, 52.

43. A.R.A. Palmans, M. Eglin, A. Montali, C. Weder, P. Smith, Tensile orientation behavior of alkoxy-substituted bis(phenylethynyl)benzene derivatives in polyolefin blend films. *Chem. Mater.* 2000, **12**, 472.

44. S.T. Wu, C.S. Hsu, K.F. Shyu, High birefringence and wide nematic range bis-tolane liquid crystals. *Appl. Phys. Lett.* 1999, **74**, 344.

45. C.S. Hsu, K.F. Shyu, Y.Y. Chuang, S.T. Wu, Synthesis of laterally substituted bistolane liquid crystals. *Liq. Cryst.* 2000, **27**, 283.

46. C. Sekine, K. Iwakura, N. Konya, M. Minai, K. Fujisawa, Synthesis and properties of some novel high birefringence phenylacetylene liquid crystal materials with lateral substituents. *Liq. Cryst.* 2001, **28**, 1375.

47. M. Eglin, A. Montali, A.R.A. Palmans, T. Tervoort, P. Smith, C. Weder, Ultra-high performance photoluminescent polarizers based on melt-processed polymer blends. *J. Mater. Chem.* 1999, **9**, 2221.

48. S.W. Chang, A.K. Li, C.W. Liao, C.S. Hsu, Polarized blue emission based on a side chain liquid crystalline polyacrylate containing bis-tolane side groups. *Jpn. J. Appl. Phys.* 2002, **41**, 1374.

49. A.P. Davey, R.G. Howard, W.J. Blau, Polarised photoluminescence from oriented polymer liquid crystal films. *J. Mater. Chem.* 1997, **7**, 417.

50. Y.H. Yao, L.R. Kung, S.W. Chang, H.S. Hsu, Synthesis of UV-curable liquid crystalline diacrylates for the application of polarized electroluminescence. *Liq. Cryst.* 2006, **33**, 33.

51. I.D. Rees, K.L. Robinson, A.B. Holmes, C.R. Towns, R. O'Dell, Recent developments in light-emitting polymers. *MRS Bull.* 2002, **27**, 451.

52. L. Akcelrud, Electroluminescent polymers. *Prog. Polym. Sci.* 2003, **28**, 875.

53. A. Kraft, A.C. Grimsdale, A.B. Holmes, Electroluminescent conjugated polymers-seeing polymers in a new light. *Angew. Chem. Int. Ed.* 1998, **37**, 402.

54. D.Y. Kim, H.N. Cho, C.Y. Kim, Blue light emitting polymers. *Prog. Polym. Sci.* 2000, **25**, 1089.

55. T. Maddux, W. Li, and L. Yu, Stepwise synthesis of substituted oligo(phenylenevinylene) via an orthogonal approach. *J. Am. Chem. Soc.* 1997, **119**, 844.

56. W. Zhu, W. Li, L. Yu, Investigation of the liquid crystalline isotropic phase transition in oligo(phenylenevinylene) with alkyl side chains. *Macromolecules* 1997, **30**, 6274.

57. J.F. Eckert, J.F. Nicoud, D. Guillon, J.F. Nierengarten, Columnar order in thermotropic mesophases of oligophenylenevinylene derivatives. *Tetrahedron Lett.* 2000, **41**, 6411.

58. J.F. Hulvat, M. Sofos, K. Tajima, S.I. Stupp, Self-assembly and luminescence of oligo(*p*-phenylenevinylene) amphiphiles. *J. Am. Chem. Soc.* 2005, **127**, 366.

59. G. Lüssem, R. Festag, A. Greiner, C. Schmidt, C. Unterlechner, W. Heitz, J.H. Wendorff, M. Hopmeier, J. Feldmann, Polarized photoluminescence of liquid-crystalline polymers with isolated arylenevinylene segments in the main-chain. *Adv. Mater.* 1995, **7**, 923.

60. S.A. Bacher, P.G. Bentley, D.D.C. Bradley, L.K. Douglas, P.A. Glarvey, M. Grell, K.S. Whitehead, M.L. Turner, Synthesis and characterisation of a conjugated reactive mesogen. *J. Mater. Chem.* 1999, **9**, 2985.

61. S.A. Bacher, P.G. Bentley, P.A. Glarvey, M. Grell, K.S. Whitehead, D.D.C. Bradley, M.L. Turner, Conjugated reactive mesogens. *Synth. Met.* 2000, **111–112**, 413.

62. A.C. Sentman, D.L. Gin, Fluorescent trimeric liquid crystals: Modular design of emissive mesogens. *Adv. Mater.* 2001, **13**, 1398.

63. B.P. Hoag, D.L. Gin, Fluorescent phasmidic liquid crystals. *Adv. Mater.* 1998, **10**, 1546.

64. B.P. Hoag, D.L. Gin, Polymerizable hexacatenar liquid crystals containing a luminescent oligo(p-phenylenevinylene) core. *Liq. Cryst.* 2004, **31**, 185.

65. T. Yasuda, K. Kishimoto, T. Kato, Columnar liquid crystalline p-conjugated oligothiophenes. *Chem. Commun.* 2006, 3399.

66. M. Hird, K.J. Toyne, G.W. Gray, S.E. Day, D.G. McDonell, The synthesis and high optical birefringence of nematogens incorporating 2,6-disubstituted naphthalenes and terminal cyano-substituents. *Liq. Cryst.* 1993, **15**, 122.

67. M. Goulding, S. Greenfield, O. Parri, D. Coates, Liquid crystals with a thiomethyl end group: Lateral fluoro substituted 4-(trans-4-(n-propyl)cyclohexylethyl-4'-thiomethylbiphenyls and 4-n-alkyl-4''-thiomethylterphenyls. *Mol. Cryst. Liq. Cryst.* 1995, **265**, 27.

68. G.W. Gray, K.A. Harrison, J.A. Nash, Wide-range nematic mixtures incorporating "4"-n-alkyl-4-cyano-para-terphenyls. *Chem. Commun.* 1974, 431.

69. C. Sánchez, B. Villacampa, R. Cases, R. Alcalá, C. Martínez, L. Oriol, M. Piñol, Polarized photoluminescence and order parameters of in-situ photopolymerized liquid crystals films. *J. Appl. Phys.* 2000, **87**, 274.

70. C. Sánchez, R. Cases, R. Alcalá, L. Oriol, M. Piñol, Photoluminescence stability of a cyanoterphenyl chromophore in liquid crystalline polymeric systems. *J. Appl. Phys.* 2000, **88**, 7124.

71. C. Sánchez, B. Villacampa, R. Alcalá, C. Martínez, L. Oriol, M. Piñol, J.L. Serrano, Mesomorphic and orientational study of materials processed by in-situ photopolymerization of reactive liquid crystals. *Chem. Mater.* 1999, **11**, 2804.

72. M. Millaruelo, L. Oriol, M. Pelegrín, M. Piñol, PhD Thesis, University of Zaragoza, 2003.

73. D. Neher, Polyfluorene homopolymers: Conjugated liquid-crystalline polymers for bright blue emission and polarized electroluminescence. *Macromol. Rapid. Commun.* 2001, **22**, 1365.

74. U. Scherf, E.J.W. List, Semiconducting polyfluorenes-towards reliable structure-property relationships. *Adv. Mater.* 2002, **14**, 477.

75. V.N. Bliznyuk, S.A. Carter, J.C. Scott, G. Klärner, R.D. Miller, D.C. Miller, Electrical and photoinduced degradation of polyfluorene based films and light-emitting devices. *Macromolecules* 1999, **32**, 361.

76. Y. Geng, A. Trajkovska, D. Katsis, J.J. Ou, S.W. Culligan, S.H. Chen, Synthesis, characterization, and optical properties of monodisperse chiral oligofluorenes. *J. Am. Chem. Soc.* 2002, **124**, 8337.

77. H. Thiem, M. Jandke, D. Hanft, P. Strohriegl, Synthesis and orientation of fluorene containing reactive mesogens. *Macromol. Chem. Phys.* 2006, **207**, 370.

78. M.P. Aldred, A.J. Eastwood, S.P. Kitney, G.J. Richards, P. Vlachos, S.M. Kelly, M. O'Neill, Synthesis and mesomorphic behaviour of novel light-emitting liquid crystals. *Liq. Cryst.* 2005, **32**, 1251.

79. M.P. Aldred, A.J. Eastwood, S.M. Kelly, P. Vlachos, A.E.A. Contoret, S. Farrar, B. Mansoor, M. O'Neill, W.C. Tsoi, Light-emitting fluorene photoreactive liquid crystals for organic electroluminescence. *Chem. Mater.* 2004, **16**, 4928.

80. A.E.A. Contoret, S.R. Farrar, M. O'Neill, J.E. Nicholls, G.J. Richards, S.M. Kelly, A.W. Hall, The photopolymerization and cross-linking of electroluminescent liquid crystals

containing methacrylate and diene photopolymerizable end groups for multilayer organic light-emitting diodes. *Chem. Mater.* 2002, **14**, 1477.

81. K.L. Woon, M.P. Aldred, P. Vlachos, G.H. Mehl, T. Stirner, S.M. Kelly, M. O'Neill, Electronic charge transport in extended nematic liquid crystals. *Chem. Mater.* 2006, **18**, 2313.

82. A.E.A. Contoret, S.R. Farrar, S.M. Kelly, J.E. Nicholls, M. O'Neill, G.J. Richards, Photopolymerisable nematic liquid crystals for electroluminescent devices. *Synth. Met.* 2001, **121**, 1629.

83. S.R. Farrar, A.E.A. Contoret, M. O'Neill, J.E. Nicholls, G.J. Richards, S.M. Kelly, Nondispersive hole transport of liquid crystalline glasses and a cross-linked network for organic electroluminescence. *Phys. Rev. B* 2002, **66**, 125107.

84. A.E.A. Contoret, S.R. Farrar, S.M. Khan, M. O'Neill, G.J. Richards, M.P. Aldred, S.M. Kelly, Photoluminescence study of crosslinked reactive mesogens for organic light emitting devices. *J. Appl. Phys.* 2003, **93**, 1465.

85. A.E.A. Contoret, S.R. Farrar, P.O. Jackson, M. O'Neill, S.M. Kelly, G.J. Richards, Crosslinked reactive mesogens and photo-chemical alignment for organic polarised EL. *Synth. Met.* 2001, **121**, 1645.

86. A.E.A. Contoret, S.R. Farrar, P.O. Jackson, S.M. Kelly, S. Khan, E. Nicholls, M. O'Neill, G.J. Richards, Electroluminescent nematic polymer networks with polarised emission. *Mol. Cryst. Liq. Cryst.* 2001, **364**, 511.

87. P.O. Jackson, M. O'Neill, W.L. Duffy, P. Hindmarsh, S.M. Kelly, G.J. Owen, An investigation of the role of cross-linking and photodegradation of side-chain coumarin polymers in the photoalignment of liquid crystals. *Chem. Mater.* 2001, **13**, 694.

88. M.P. Aldred, A.E.A. Contoret, S.R. Farrar, S.M. Kelly, D. Mathieson, M. O'Neill, W.C. Tsoi, P. Vlachos, A full-color electroluminescent device and patterned photoalignment using light emitting liquid crystals. *Adv. Mater.* 2005, **17**, 1368.

89. K.L. Woon, A.E. Contoret, S.R. Farrar, A. Liedtke, M. O'Neill, P. Vlachos, M.P. Aldred, S.M. Kelly, Material and device properties of highly birefringent nematic glasses and polymer networks for organic electroluminescence. *J. Soc. Inf. Display* 2006, **16**, 557.

90. P. Strohriegl, D. Hanft, M. Jandke, T. Pfeuffer, Reactive mesogenes: Synthesis and application in optoelectronic devices. *Mat. Res. Soc. Symp. Proc.* 2002, **709**, 31.

91. M. Jandke, D. Hanft, P. Strohriegl, K. Whitehead, M. Grell, D.D.C. Bradley, Polarized electroluminescence from photocrosslinkable nematic fluorene bisacrylates. *Proc. SPIE-Int. Soc. Opt. Eng.* 2001, **4105**, 338.

92. M. Millaruelo, L.S. Chinellato, J.L. Serrano, L. Oriol, M. Piñol, Fluorene-based liquid crystalline networks with linearly polarized blue emission. *J. Polym. Sci.: Part A. Polym. Chem.* 2007, **45**, 4804.

93. M. Millaruelo, L. Oriol, M. Piñol, P.L. Sáez, J.L. Serrano, Polarised luminescent films containing fluorene cross-linkers obtained by *in situ* photopolymerisation. *J. Photochem. Photobiol. A: Chem.* 2003, **155**, 29.

94. M. Millaruelo, L. Oriol, J.L. Serrano, M. Piñol, P.L. Sáez, Emissive anisotropic polymeric materials derived from liquid crystalline fluorenes. *Mol. Cryst. Liq. Cryst.* 2004, **411**, 451.

95. M.P. Aldred, M. Carrasco-Orozco, A.E.A. Contoret, D. Dong, S.R. Farrar, S.M. Kelly, S.P. Kitney, D. Mathieson, M. O'Neill, W.C. Tsoi, P. Vlachos, Organic electroluminescence using polymer networks from smectic liquid crystals. *Liq. Cryst.* 2006, **33**, 459.

96. P. Vlachos, S.M. Kelly, B. Manssor, M. O'Neill, Electron-transporting and photopolymerisable liquid crystals. *Chem. Commun.* 2002, 874.

97. R. Giménez, L. Oriol, M. Piñol, J.L. Serrano, A.I. Viñuales, T. Fisher, J. Stumpe, Synthesis and properties of 2-phenylbenzoxazole-based luminophores for *in situ* photopolymerized liquid-crystal films. *Helv. Chim. Acta* 2006, **89**, 304.

# 7 Photomechanical Effects of Cross-Linked Liquid-Crystalline Polymers

*Jun-ichi Mamiya, Yanlei Yu, and Tomiki Ikeda*

## CONTENTS

## 7.1   INTRODUCTION

Liquid crystals (LCs) represent thermodynamically stable phases situated conceptually between an ordinary isotropic liquid and a crystalline solid. Due to a fact that LCs possess simultaneously the optical anisotropy of crystals and the fluidity of liquid, they show many exclusive properties (optical, electrical, and magnetic) and alignment changes by external fields at surfaces and interfaces [1]. Control of the LC alignment is the most important technology in most devices using LCs as active media. Especially, alignment of nematic LCs can be easily changed and controlled by weak external fields. It is suggested that cooperative motion of LCs may be most advantageous in changing the alignment of LC molecules by external stimuli. If a small portion of LC molecules change their alignment in response to an external stimulus, the other LC molecules also change their alignment. This means that only a small amount of energy is needed to change the alignment of whole LC molecules: such a small amount of energy as to induce an alignment change of only 1 mol % of LC molecules is enough to bring about the alignment change of the whole system [2]. According to this principle, photomanipulation of the LC alignment has also been reported for polymer films containing photochromic molecules such as azobenzenes, in which a small amount of azobenzene molecules change their shape by light and then the cooperative change of LC alignment takes place [2]. Photoalignment of LCs has attracted much attention due to its high light sensitivity and large change in refractive index. The effects of structure and density of azobenzene molecules on the photoalignment have been explored systematically [3].

Since a huge amplification is possible in LC systems, a macroscopic shape change could be induced if one can find a good way, such as cross-linking, to transfer the alignment change in a microscopic level to the whole LC systems. Cross-linked liquid-crystalline polymers (CLCPs) are a new type of materials that has both properties of LCs and elastomers arising from polymer networks [4,5]. Due to the LC properties, mesogens in CLCPs show alignment and this alignment of mesogens is coupled with polymer network structures. This coupling gives rise to characteristic properties of CLCPs. Depending on the mode of alignment of mesogens in CLCPs, they are classified into nematic CLCPs, smectic CLCPs, cholesteric CLCPs, and so on. If one heats the nematic CLCP films toward the nematic–isotropic phase-transition temperature, the nematic order will decrease and when the temperature is reached above the phase-transition temperature, one has a disordered state of mesogens. By this phase transition, the CLCP films show, in general, contraction along the alignment direction of mesogens, and if the temperature is lowered below the phase-transition temperature, the CLCP films revert to their original size (expansion) [6]. This anisotropic deformation of the CLCP films is sometimes very large, which provides the CLCP materials with a promising nature as artificial muscles [7]. By incorporating photochromic moieties into CLCPs, which can induce a reduction in the nematic order and in an extreme case a nematic–isotropic phase transition of LCs, a contraction of CLCP films has been observed upon exposure to UV light to cause a photochemical reaction of the photochromic moiety [8–10].

These kinds of photomobile smart materials that can undergo a shape or volume change in response to light are attracting increasing attention [11]. By using the movement of these materials, one can convert light energy into work directly (photomechanical effects). In this chapter, we describe photomechanical effects observed in many kinds of materials, focusing our attention on a hot topic the light-driven CLCP actuators.

## 7.2 PHOTORESPONSIVE BEHAVIOR OF LCs CONTAINING PHOTOCHROMIC MOLECULES

It is well known that photochromic molecules such as azobenzenes, spiropyrans, and diarylethenes can undergo a reversible photochemical reaction between two forms. When a small amount of photochromic molecules is incorporated into LC molecules and the resulting guest–host mixtures are irradiated to cause photochemical reactions of the photochromic guest molecules, a LC to isotropic phase-transition of the mixtures can be induced isothermally [2]. The *trans* form of the azobenzenes, for instance, has a rod-like shape, which stabilizes the phase structure of the LC phase, while its isomer (*cis* form) is bent and tends to destabilize the phase structure of the mixture. Photochromic reactions are usually reversible, and with *cis*–*trans* back isomerization the sample reverts to the initial LC phase. This means that phase transitions of LC systems can be induced isothermally and reversibly by photochemical reactions of photoresponsive guest molecules [12], which is called photochemical phase transition. The photochemical phase transition is interpreted in terms of a change in the phase-transition temperature of LC systems on accumulation of one isomer of the photochromic guest molecule [13,14].

LC polymers possess properties of both polymers and LCs, and currently are regarded as promising photonic materials because of their advantageous properties. Wendorff et al. reported the first example of a photon-mode photoresponse of LC polymers: a holographic recording in LC polymer films containing azobenzene moieties and mesogenic groups [15,16]. Ikeda et al. reported the first example of the photochemical phase transition in LC polymers, which were doped with low-molecular-weight azobenzene molecules [17–19]. However, it soon became apparent that LC copolymers are superior to the doped systems because phase separation occurred in the doped systems when the concentration of the photochromic molecules was high. A variety of LC copolymers were prepared and examined on their photochemical phase-transition behavior [18–22]. One of the important factors of the photoresponsive LCs is their response to optical stimuli. In this respect, the response time of the photochemical phase transition has been explored by time-resolved measurements [22,23].

Some azobenzene derivatives showing LC properties were developed [24–27]. In these azobenzene LCs, the azobenzene moiety could play both roles as a mesogen and as a photosensitive moiety (Figure 7.1). These azobenzene LCs show a stable LC phase only when the azobenzene moiety is in the *trans* form, and they never show any LC phase when all the azobenzene moieties are in the *cis* form. Studies on the photochemical phase transition of these azobenzene LC polymers have revealed that the nematic–isotropic phase transition is induced in 200 ns under optimized conditions and in a wide temperature range [26].

Recently, Feringa et al. have demonstrated a light-driven molecular motor embedded in a cholesteric LC film that can rotate microscopic-scale objects placed on the film by harvesting light energy [28]. Photochemically and thermally induced topological changes in a chiral dopant can cause a rotational reorganization of the cholesteric LC film. This rotational movement can be applied to rotate objects that are 4 orders of magnitude larger than the chiral dopant.

## 7.3   PHOTOMECHANICAL EFFECTS IN POLYMERS

Polymers are one of the most superior materials in view of high processability, ability to form self-standing films with a variety of thicknesses from nanometer to centimeter scales, lightweight, diversity of chemical structure, precisely controlled synthesis, and many kinds of polymers have been put to practical use in daily life and industry. From this point of view, polymer actuators capable of responding to external stimuli and deforming are most desirable for practical applications. Various chemical and physical stimuli, such as temperature [29], electric field [30,31], and solvent composition [32], have been applied to induce deformation of polymer actuators.

Since 1960s, photoinduced deformation (contraction and expansion) of amorphous polymers has been studied intensively [33–36]. If light can be used as an external stimulus, remote control and rapid performance of deformation can be brought

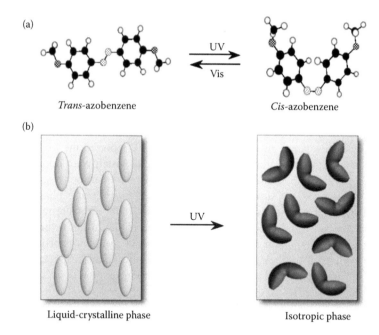

(a)

Trans-azobenzene                                                        Cis-azobenzene

(b)

Liquid-crystalline phase                                                Isotropic phase

**FIGURE 7.1** Photoisomerization of azobenzene (a) and photochemical phase transition of azobenzene LCs (b).

about. Merian first reported that a nylon filament fabric dyed with an azobenzene derivative shrank upon photoirradiation [37]. The change in size of the polymers is ascribed to the photochemical structural change of the azobenzene moiety absorbed on the nylon fibers. However, the observed shrinkage was too small, only about 0.1%. Following this work, much effort was made to find new photomechanical systems with an enhanced efficiency [38].

Eisenbach investigated the photomechanical effect of poly(ethyl acrylate) networks cross-linked with azobenzene moieties (**1**) [39]. The polymer network contracted upon exposure to UV light to cause the *trans–cis* isomerization of the azobenzene cross-links and expanded by irradiation with Vis light to cause *cis–trans* back isomerization (Figure 7.2). This photomechanical effect was ascribed to the conformational change of the azobenzene cross-links by the *trans–cis* isomerization of the azobenzene chromophore. However, the degree of deformation was again so small (0.2%).

Matejka et al. synthesized several types of photochromic polymers based on a copolymer of maleic anhydride with styrene containing azobenzene moieties both in the side chains and in the cross-links of the polymer network [40–42]. The photomechanical effect was enhanced with an increase in the content of photochromic groups. In case of the polymer with 5.4 mol % of the azobenzene moieties, the photoinduced contraction of the polymer amounted to 1%.

Recent development of single-molecular force spectroscopy by atomic force microscopy (AFM) techniques has enabled quite successfully to measure mechanical

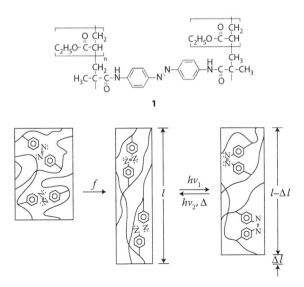

**FIGURE 7.2**  Schematic illustration of the photomechanical effect in the poly(ethyl acrylate) network with azobenzene cross-linkers upon irradiation.

force generated at a molecular level. Ends of a polymer with azobenzene moieties in its main chain were covalently bonded to the AFM tip and a supporting glass substrate [43,44]. The force (pN) and extension (nm) produced in a single polymer were measured in total internal reflection geometry using the slide glass as a wave guide. This excitation way is very useful to avoid thermomechanical effects on the cantilever. Individual polymer chains were lengthened and contracted by switching the azobenzene moieties between their *trans* and *cis* forms by irradiation with UV (365 nm) and Vis (420 nm) light, respectively. The mechanical work generated by *trans–cis* photoisomerization of the azobenzene polymer strand was $W \approx 4.5 \times 10^{-20}$ J. This mechanical work observed at molecular level results from a macroscopic photoexcitation. A maximum efficiency of the photomechanical energy conversion at a molecular level can be estimated as 0.1, if it is assumed that each switching of a single azobenzene unit is initiated by a single photon carrying an energy of $Ex = 5.5 \times 10^{-19}$ J ($\lambda = 365$ nm) [43,44].

A photoinduced expansion of thin films of polymers containing azobenzene chromophores was explored in real-time by single wavelength ellipsometry [45]. An initial expansion of the azobenzene polymer films was found to be irreversible with an extent of relative expansion of 1.5–4% in films of thickness ranging from 25 to 140 nm. A subsequent and reversible expansion was observed with repeated irradiation cycles, achieving a relative extent of expansion of 0.6–1.6%.

A photoinduced deformation of polymer colloidal particles was reported by Wang et al. [46]. They observed that the spherical polymer particles containing azobenzene moieties changed their shapes from a sphere to an ellipsoid upon exposure to interfering linearly polarized laser beams and the elongation of the particles occurred along the polarization direction of the actinic laser beam.

Photomechanical effects in polymers with other photochromic molecules have also been explored in detail. Smets et al. studied the mechanical properties of polymer matrices containing spirobenzopyran cross-links (Figure 7.3) [38]. Irradiation with UV light to cause the spiropyran (closed form)–merocyanine (open form) isomerization led to a contraction of the sample under isothermal conditions, while in the dark, the sample reverted to its initial length [38,47,48]. The photomechanical response of polystyrene and poly(methyl methacrylate) doped with a spirobenzopyran derivative was reported by Blair et al. [49]. They studied the effects of the photoisomerization of the photochromic dopant on the stress of the polymer sample. Upon exposure to UV light to cause the spiropyran–merocyanine isomerization, the initial stress applied to the sample decreased, indicating an expansion of the sample, while in the dark, the length of the sample recovered to its initial state by a subsequent contraction, which was evidenced by an increase in stress. It is interesting to note that the spiropyran

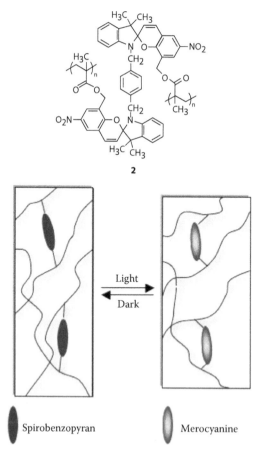

**FIGURE 7.3** Chemical structure of a spirobenzopyran derivative and the photomechanical effect of polymer networks containing spirobenzopyran cross-links.

chromophore exhibited an opposite effect on the polymer substrates by spiropyran–merocyanine isomerization, depending on the way of incorporation: when it is located at the cross-links, it affects the polymer material in such a way that it induces contraction of the polymer while it causes relaxation of the polymer substrate, resulting in a decrease in stress, when doped into polymer matrices.

In this respect, Athanassiou et al. reported an interesting result on the photomechanical phenomena of a polymer by photoisomerization of a spiropyran dopant [50]. They prepared a polymer film of poly(ethyl methacrylate-*co*-methyl acrylate) doped with 5.0 wt % spiropyran derivative. Figure 7.4 shows the deformation of this film (70 μm thick) upon exposure to laser pulses (UV pulses at 308 nm with a pulse width of 30 ns and Vis (green) pulses at 532 nm with 5 ns) at the intensity of 70 mJ/cm$^2$. It is clearly seen that the film bent toward the actinic light source upon exposure to green laser pulses. The film exhibited its maximum bending after 40 pulses, and continuing green pulses led to the recovery of the film after 160 pulses. However, no deformation was observed when the sample was irradiated with green laser pulses without preceding UV pulses. The volume remained unaffected by the initial UV pulses to cause the spiropyran–merocyanine isomerization, whereas the green pulses are fully responsible both for its contraction and recovery. Thus they interpreted that the bending of the film was not induced directly by the spiropyran–merocyanine isomerization, but caused by the formation of aggregates of merocyanine isomers. The photomechanical effects occurring in polymers doped with spiropyrans seem more complicated than those observed in the azobenzene-containing systems.

Lendlein et al. prepared polymers containing cinnamic groups (**4**) [51]. Similar to thermally induced shape-memory polymers, the photoresponsive polymer film was first stretched by an external force. Then exposure to UV light longer than 260 nm led to the fixation of the elongated shape due to a photoinduced [2+2] cycloaddition

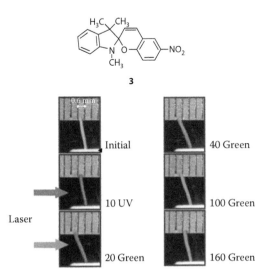

**FIGURE 7.4** Bending cycle of a polymer film doped with spiropyran (**3**). The photographs were taken after irradiation of the film with the number of laser pulses shown.

reaction. Even though the external stress was released, the film could remain in an elongated form for a long time. When the elongated sample was irradiated with UV light shorter than 260 nm at ambient temperature, the cleavage of the cross-links occurred and the film reverted to its original shape. Furthermore, when only the top side of a polymer film in a stretched state was irradiated with UV light longer than 260 nm, a corkscrew spiral shape was obtained due to the formation of two layers, in which the elongation is fixed well for the top layer and the bottom layer keeps its elasticity as shown in Figure 7.5.

As mentioned above, the polymer films containing photochromic molecules are potential materials for application. However, in these amorphous polymer materials, deformation in response to external stimuli takes place in an isotropic way; no preferential direction for deformation. Also the degree of the deformation is in general small. Therefore, it is of great importance to develop novel photomechanical systems that can undergo fast and large deformations. If the materials possess any anisotropy, their deformation in response to external stimuli could be induced in an anisotropic way with a preferential direction of deformation, which could produce much larger deformation than those observed in the amorphous materials.

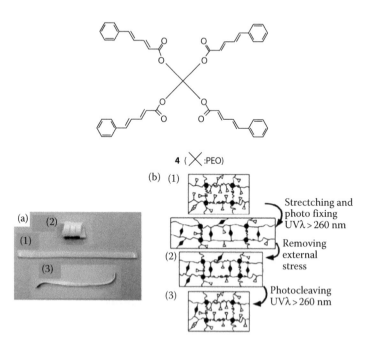

**FIGURE 7.5** Shape-memory effect of photoresponsive polymers. (a) A polymer film doped with **4**: (1) original shape; (2) a corkscrew shape obtained; (3) the original shape recovered by photoirradiation at <260 nm. (b) Mechanism of the shape-memory effect of the grafted polymer network. The steps (1) to (3) demonstrate the fixation of the stretched state of the film and the recovery of the initial state.

## 7.4　DEFORMATION OF CLCPS BY EXTERNAL STIMULI

Anisotropic deformation of CLCPs has been a subject of extensive experimental and theoretical studies [6,52–54]. The possibility of using CLCPs as artificial muscles, by taking advantage of their substantial uniaxial contraction in the direction of the director axis, was theoretically proposed by de Gennes [55]. Kupfer and Finkelmann first demonstrated the anisotropic deformation of monodomain CLCPs induced by thermal LC-isotropic phase transition [4]. A nematic CLCP was found to contract by about 26% due to the change in molecular alignment of mesogens. The origin of the contraction is a subtle decrease in the microscopic order upon phase transition; however, this deformation is closely related to a coupling between the LC order of mesogens and the elastic properties of the polymer network.

Artificial muscle-like materials have been developed in recent years [7,56–62]. A new type of LC coelastomers composed of LC side-chain and main-chain polymers (**5a–c**) as network strands was prepared by Finkelmann et al. [7]. A remarkable increase in the thermoelastic response was observed by increasing the concentration of main-chain segments owing to the direct coupling of the LC main-chain segments to the network anisotropy. Compared to the length of the networks in an isotropic state, an elongation of up to a factor of almost 4 occurred in the nematic state along the direction of the director [7]. Thermoelastic properties of CLCPs with laterally attached side-chain mesogens (**6a, 6b**) were studied by Ratna et al. [58]. The CLCPs were shown to contract by 35–40% through a nematic–isotropic phase transition. The maximum retractive force generated was 270 kPa, which is comparable to that of a skeletal. This increase in the stress during the isostrain measurements is related to the entropy change of the polymer chains and interpreted as a result of the worm-like (prolate) to coil transition in the polymer backbones with laterally coupled mesogenic side chains.

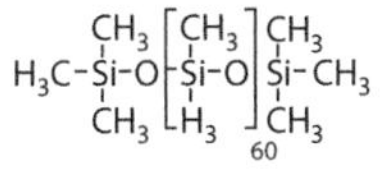

**5a** 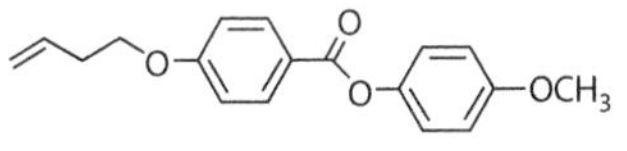　　　　　　　　　　　**5b**

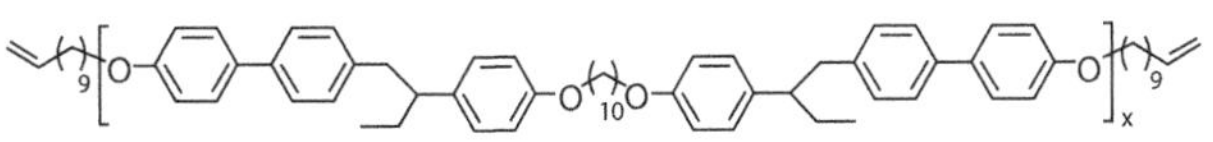

**5c**

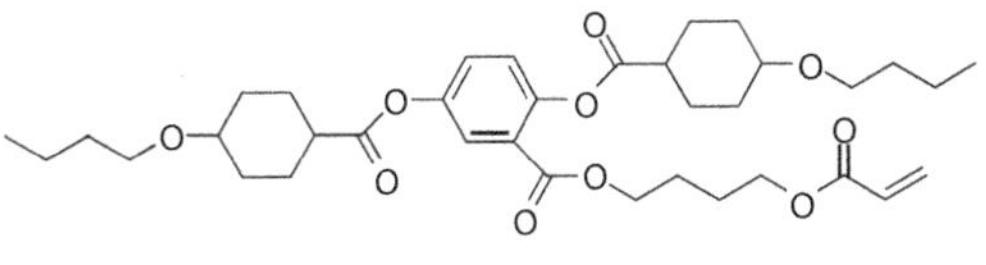

**6a**

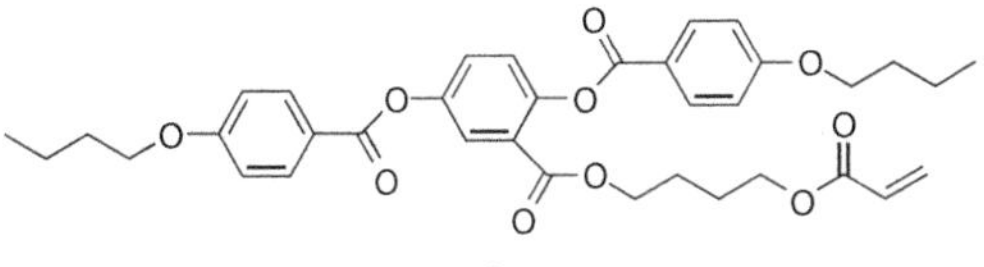

**6b**

Skeletal muscles are made of many bundles of fibers and their anisotropic contraction and elongation are induced along the fiber axis. Naciri et al. described a method of preparing LC fibers from a side-chain LC terpolymer containing two side-chain mesogens and a nonmesogenic group that acts as a reactive site for cross-linking (Figure 7.6) [59]. Fibers were drawn from a melt of the polymer and a cross-linker. The obtained fibers were found to show high LC alignment by polarizing optical microscopy. The thermoelastic response exhibited strain changes of about 35% through the nematic–isotropic phase transition. A retractive force of nearly 300 kPa was also observed in the isotropic phase.

CLCP films with a splayed or twisted molecular alignment display a well-controlled deformation as a function of temperature [60]. The twisted films show a complex macroscopic deformation owing to the formation of saddle-like geometries, whereas the deformation of the splayed films is smooth and well controlled [60].

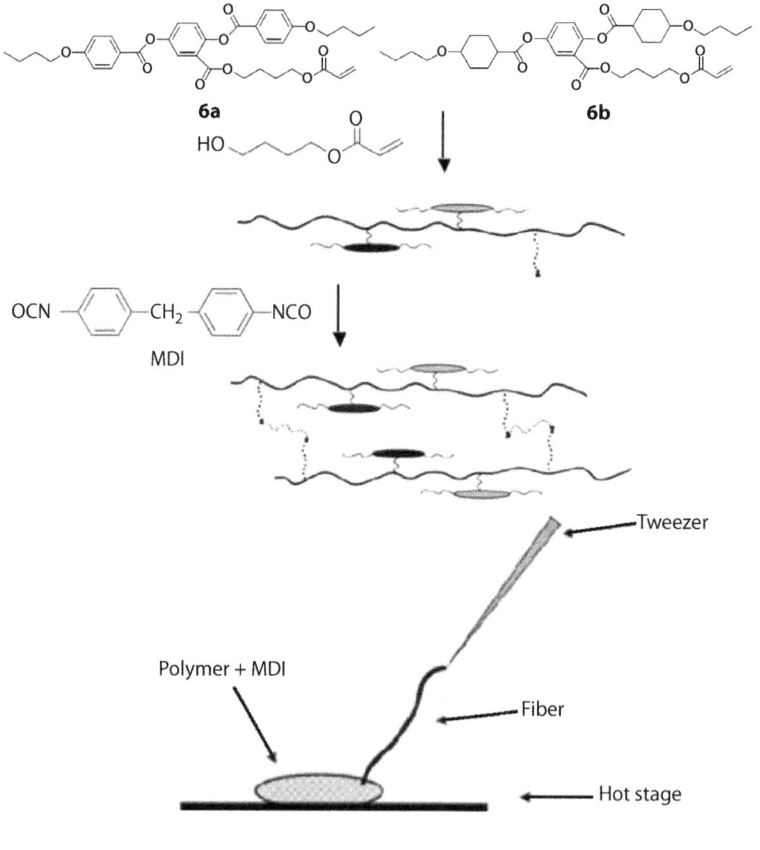

**FIGURE 7.6** Preparation of a CLCP fiber from laterally attached side-chain mesogens. The polymer mixture with the MDI crosslinker is heated at a hot stage. During the crosslinking reaction a fiber is drawn by dipping a tweezer in the reacting melt and subsequently controlled withdrawing.

Keller et al. have successfully prepared an array of micro-sized CLCPs by using a soft lithography technique. The CLCPs contracted and expanded in response to small temperature changes around a well-defined transition temperature [62].

To induce deformation of CLCPs, many kinds of external stimuli have been applied, such as an electric field and humidity [63,64]. Kremer et al. developed ultrathin ferroelectric CLCP films that showed high and fast strains of 4% by electrostriction under a much lower applied electric field (only 1.5 MV/m) than those reported previously [63]. Broer et al. proposed pH- or water-controlled actuators based on LC networks [64]. LC polymer films with three configurations were prepared for controlled bending. The uniaxially aligned film responded equally to water or pH in all regions of the film and only elongated by a small amount when exposed to a uniform stimulus. Differences in the pH or humidity of the upper and lower surfaces were sufficient to induce large bending. The twisted and splayed configurations do not require environmental gradients to produce macroscopic motion. In both cases, the preferred expansion directions on the opposite surfaces of a film were offset by 90°, and with a uniform stimulus, this resulted in expansion gradients over the thickness of the film and bending similar to thermal deformation in a metallic bilayer. Anisotropic and large deformation in response to external stimuli was achieved by using CLCPs with various structures.

Elias et al. prepared CLCP films by using surface alignment and photopatterning techniques and investigated the themomechanical properties of the CLCP films with microscaled dimension [65]. LC polymers can be patterned through a photomask during photopolymerization. This exposure preserves LC ordering in the polymerized areas. The surface profile of the patterned CLCP films was measured as a function of temperature. The initial thickness of the film was 16.6 μm, which increased by 8.4% (1.4 μm) when the sample was heated to 200°C. The change in thickness was reversible. They concluded that photopatterning of LC polymers is a promising method for fabrication of microactuators.

## 7.5   PHOTOMECHANICAL EFFECTS IN CLCPS

As light is a good energy source to be controlled rapidly, precisely, and remotely, photodeformable CLCPs have attracted increasing attention. CLCPs show thermoelastic properties: across the nematic–isotropic phase-transition, they contract along the alignment direction of mesogens and by cooling below the phase-transition temperature they show expansion. By the combination of this property of CLCPs with photochemical phase transition (or photochemically induced reduction of nematic order), it is expected that one can induce deformation of CLCPs by light [8–10]. In fact, Finkelmann et al. have succeeded in inducing contraction by 20% in a nematic CLCP containing polysiloxane main chain and azobenzene chromophores at cross-links (**7**) upon exposure to UV light (Figure 7.7) [8]. From the viewpoint of the photomechanical effects, the subtle variation in the nematic order by *trans–cis* isomerization of the azobenzene moiety causes a significant uniaxial deformation of LCs along the director axis, if the LC molecules are strongly associated by covalent cross-linking to form a three-dimensional polymer network. Various azobenzene derivatives (**8**) were incorporated as photoresponsive moieties into CLCPs. Terentjev et al. studied their deformation upon exposure to UV light, and analyzed in detail these photomechanical effects [9,10]. Keller et al. synthesized side-on nematic monomers (**9**) containing an azobenzene [66]. The photopolymerization was performed on the aligned azobenzene monomers in conventional LC cells. Thin films of these CLCPs showed fast (less than 1 min) photochemical contraction up to 18% by irradiation with UV light and a slow thermal back reaction in the dark (Figure 7.8).

Ikeda et al. have demonstrated three-dimensional movements of CLCP films: photoinduced bending of LC gels [67] and CLCPs [67–73] containing azobenzenes. In comparison with the contraction mode that is a one-dimensional movement, the bending mode, a three-dimensional movement, could be advantageous for artificial hands

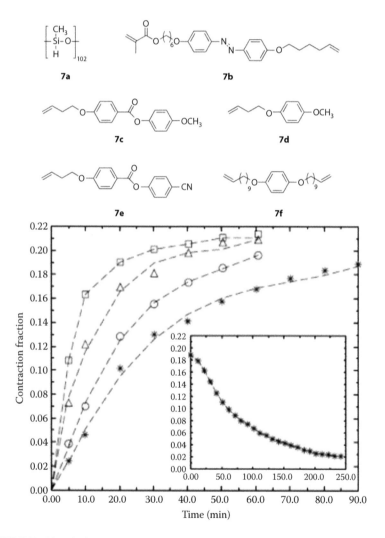

**FIGURE 7.7** Photoinduced contraction of CLCPs prepared from **7a–7f**. Inset: recovery of the contracted CLCP at 298 K after irradiation was switched off.

and microrobots that are capable of particular manipulations. The CLCP films prepared by in situ polymerization of LC monomers and cross-linkers showed the bending and unbending processes induced by irradiation with UV and Vis light, respectively. It was observed that the monodomain CLCP film bent toward the actinic light source along the rubbing direction, and the bent film reverted to the initial flat state after exposure to Vis light. This bending and unbending was reversible just by changing the wavelength of the incident light.

**8a**

**8b**                     **8c**

Exposure to UV light leads to *trans–cis* isomerization of azobenzene moieties and destabilization of the nematic phase (reduction in nematic order), even a nematic–isotropic phase transition of LC systems. However, incident UV light is absorbed only at the film surface, because the extinction coefficient of the azobenzene moieties at around 360 nm is large. Then the reduction in nematic order occurs only in the surface

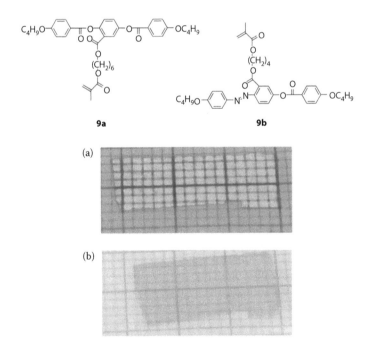

**9a**                     **9b**

(a)

(b)

**FIGURE 7.8** (**See color insert following page 304.**) Photographs of photodeformation of the azobenzene CLCP before UV light irradiation (a) and under UV light irradiation (b).

region facing the incident light, whereas in the bulk of the film the *trans*-azobenzene moieties remain unchanged. As a result, the volume contraction is generated only in the surface layer, causing the bending toward the actinic light source. Furthermore, the azobenzene moieties are preferentially aligned along the rubbing direction of the alignment layers, and the decrease in the alignment order of the azobenzene moieties is thus produced just along this direction, contributing to the anisotropic bending.

The monodomain CLCP films with different cross-linking densities prepared by copolymerization of **10a** and **10b** showed the same bending [68]. However, the maximum bending extents were different among the films with different cross-linking densities. Because the film with a higher cross-linking density holds a higher-order parameter, the reduction in the alignment order of the azobenzene moieties gives rise to a larger volume contraction along the rubbing direction, contributing to a larger bending extent of the film along this direction. CLCP films with a high LC order and a low $T_g$ were prepared from ferroelectric LC monomer and cross-linker [70]. Irradiation with 366-nm light led the films to bend at room temperature toward the actinic light source along the direction with a tilt to the rubbing direction of the alignment layer. The bending process was completed within 500 ms upon irradiation with a laser beam (Figure 7.9).

**10a**

**10b**

Palffy-Muhoray et al. reported a photomechanical response of a CLCP sample (**12**) containing azobenzene dyes. The dye-doped CLCP underwent a large and rapid deformation upon nonuniform illumination by Vis light [74]. When a laser beam from above is shone on the dye-doped CLCP sample floating on water, the CLCP swims away from the laser beam, with a movement resembling that of flatfish (Figure 7.10).

By means of the selective absorption of linearly polarized light in the polydomain CLCP films, Ikeda et al. succeeded in controlling the direction of a photoinduced bending in that a single polydomain CLCP film could be bent repeatedly and precisely along any chosen direction (Figure 7.11) [68]. Upon exposure to linearly polarized light, the selective absorption of light in a specific direction, light polarization, leads to the *trans–cis* isomerization of azobenzene moieties in specific domains where the azobenzene mesogens are aligned along the direction of light polarization. Recently, Terentjev et al. and Dunn compared the photomechanical effect of the monodomain and polydomain systems [75,76]. Terentjev et al. have reported how the intensity and the polarization of light affect photoactuation of the monodomain and

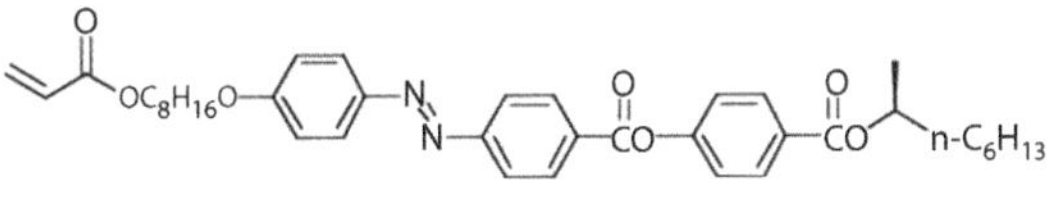

**11a**

K 60 SmCa˙ 95 SmC˙ 104 SmA 113 I

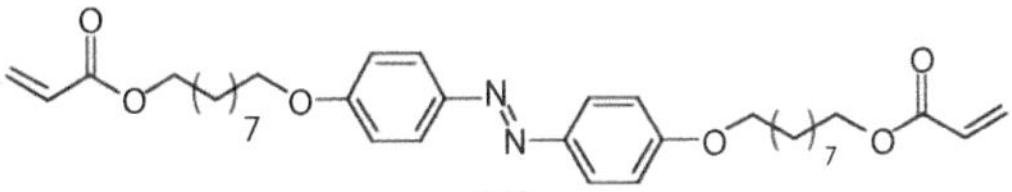

**11b**

K 76 SmA 92 I

K 46 SmC˙ 94 SmA107 I 11a/11b (80:20)

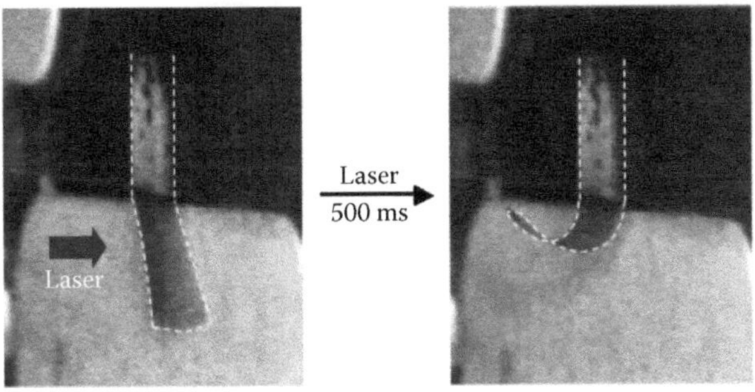

Laser
500 ms

**FIGURE 7.9** Chemical structures, phase-transition temperatures of the LC monomer (**11a**) and the cross-linker (**11b**), and photoinduced bending of the ferroelectric CLCP film upon irradiation with a UV laser at room temperature.

polydomain CLCP films. Dunn also developed a simple analytical model for the deformation of the films with photostrains that arise from photoirradiation. Finite element calculations predicted a much more uniform bending for the polydomain than the monodomain.

An azobenzene CLCP film with an extraordinarily strong and fast mechanical response to a laser beam has been developed [77]. The direction of the photoinduced bending or twisting of CLCP can be controlled by changing the polarization of the laser beam. The photomechanical effect is a result of photoinduced reorientation of azobenzene moieties in the CLCP.

Broer et al. have demonstrated photomechanical phenomena of CLCP films with a densely-cross-linked, twisted configuration of azobenzene moieties [78]. They have shown a large amplitude bending and coiling motion based on the 90° twisted LC alignment upon irradiation with UV light. They also showed that CLCP films with a splayed or twisted molecular alignment exhibited faster bending with a greater amplitude than those of a uniaxial planar alignment with same chemical composition [79].

The alignment of the azobenzene mesogens in CLCP films was explored how it affects the photoinduced bending. Homeotropically aligned films were prepared and irradiated with UV light. It was found that the homeotropic CLCP films showed a

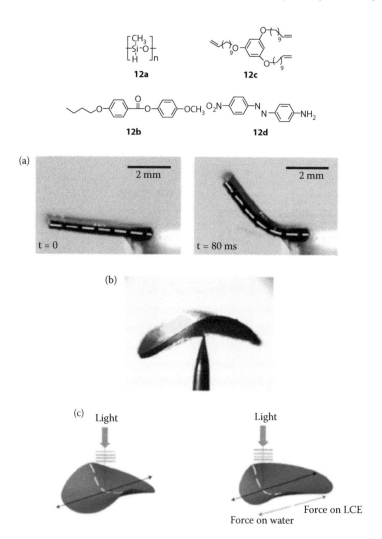

**FIGURE 7.10** **(See color insert following page 304.)** Photomechanical response of the CLCP (a) and shape deformation of the CLCP upon exposure to 514 nm light (b). Schematic illustration is shown in (c) on the mechanism underlying the locomotion of the dye-doped CLCP.

completely different bending; upon exposure to UV light they bent away from the actinic light source (Figure 7.12) [72]. This result indicates that the surface of the homeotropic films expands. The bending direction of the films could be controlled by the alignment direction of mesogens in CLCP films.

Recently, carbon nanotubes have been discovered to exhibit extraordinary photomechanical properties. The carbon nanotube/polymer nanocomposites showed structural expansion in response to light under small prestrains and contraction under

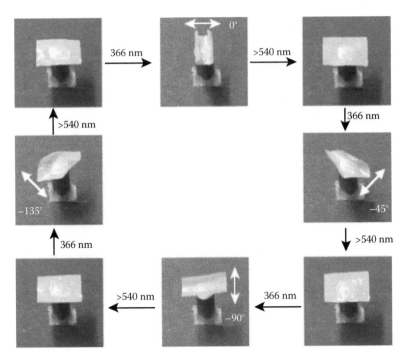

**FIGURE 7.11**   (**See color insert following page 304.**) Precise control of the bending direction of a film by linearly polarized light. Photographic frames of the film in different directions in response to irradiation by linearly polarized light of different angles of polarization (white arrows) at 366 nm, and bending flattened again by Vis light longer than 540 nm.

large prestrains. Photomechanical actuators based on carbon nanotubes could have many applications in various disciplines [80].

## 7.6   SUMMARY

This chapter describes the photomechanical effects observed in cross-linked polymers. Light is a sufficient power source for many applications: it can be localized at a specific point in space and at a specific time, selective, nondamaging, which allows for remote activation and remote delivery of energy to a system. Light-driven actuators can convert light energy directly into mechanical work. In various systems, light-driven actuation has been achieved; however, at the moment the efficiency for light energy conversion is far from satisfactory. CLCPs could be promising materials by improving their photomechanical properties: enhancement of (1) alignment of mesogens, (2) coupling of LC order with polymer networks, and (3) change in order by light. Light-driven actuators based on CLCPs are promising for application in driving micromachines and nanomachines, because their photodeformations require neither batteries nor controlling devices on the materials themselves. The research field of photomobile polymers is still at its beginning, but substantial advance can be expected in the near future.

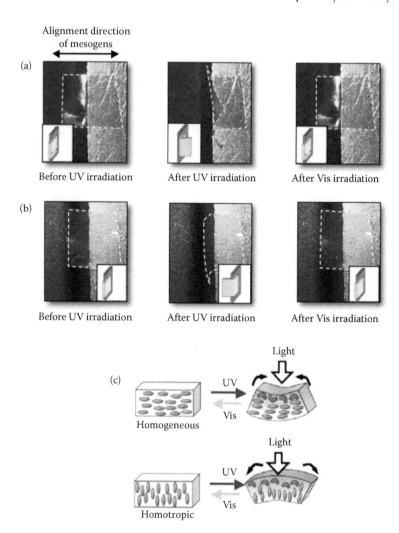

**FIGURE 7.12** (**See color insert following page 304.**) Photographs of the homogeneous film (a) and the homeotropic film (b) that exhibit photoinduced bending and unbending. The white dash lines show the edges of the films and the inset of each photograph is a schematic illustration of the film state. (c) Schematic illustration of the bending mechanism in the homogeneous and the homeotropic films.

## REFERENCES

1. Gray, G. W., In *Handbook of Liquid Crystals, Vol. 1, Fundamentals*, Eds. Demus, D., Goodby, J., Gray, G. W., Spiess, H.-W., Vill, V., Wiley-VCH, Weinheim, 1998.
2. Ikeda, T., Photomodulation of liquid crystal orientations for photonic applications. *J. Mater. Chem.*, *13*, 2037–2057, 2003.
3. (a) Wu, Y., Demachi, Y., Tsutsumi, O., Kanazawa, A., Shiono, T., Ikeda, T., Photoinduced alignment of polymer liquid crystals containing azobenzene moieties in the side chain.

3. Effect of structure of photochromic moieties on alignment behavior. *Macromolecules*, *31*, 4457–4463, 1998; (b) Wu, Y., Zhang, Q., Kanazawa, A., Shiono, T., Ikeda, T., Nagase, Y., Photoinduced alignment of polymer liquid crystals containing azobenzene moieties in the side chain. 5. Effect of the azo contents on alignment behavior and enhanced response. *Macromolecules*, *32*, 3951–3956, 1999.

4. Finkelmann, H., *Physical Properties of Liquid Crystalline Elastomers*, Eds. Demus, D., Wiley-VCH, New York, 1998.

5. Warner, M., Terentjev, E. M., *Liquid Crystal Elastomers,* Oxford University Press, UK, 2003.

6. Kupfer, J., Finkelmann, H., Liquid crystal elastomers: Influence of the orientational distribution of the crosslinks on the phase behaviour and reorientation processes. *Macromol. Chem.*, *195*, 1353–1367, 1994.

7. Wermter, H., Finkelmann, H., Liquid crystalline elastomers as artificial muscles. *e-Polymers*, no. 013, 2001.

8. Finkelmann, H., Nishikawa, E., Pereira, G. G., Warner, M., A new opto-mechanical effect in solids. *Phys. Rev. Lett.*, *87*, 015501, 2001.

9. Hogan, P. M., Tajbakhsh, A. R., Terentjev, E. M., UV manipulation of order and macroscopic shape in nematic elastomers. *Phys. Rev. E*, *65*, 041720, 2002.

10. Cviklinski, J., Tajbakhsh, A. R., Terentjev, E. M., UV isomerisation in nematic elastomers as a route to photo-mechanical transducer. *Eur. Phys. J. E*, *9*, 427–434, 2002.

11. Ikeda, T., Mamiya, J., Yu, Y., Photomechanics of liquid-crystalline elastomers and other polymers. *Angew. Chem. Int. Ed.*, *46*, 506–528, 2007.

12. Tazuke, S., Kurihara, S., Ikeda, T., Photochemically triggered physical amplification of photoresponsiveness. *Chem. Lett.*, 911–914, 1987.

13. Legge, C. H., Mitchell, G. R., Photo-induced phase transitions in azobenzene-doped liquid crystals. *J. Phys. D: Appl. Phys.*, *25*, 492–494, 1992.

14. Sung, J.-H., Hirano, S., Tsutsumi, O., Kanazawa, A., Shiono, T., Ikeda, T., Dynamics of photochemical phase transition of guest/host liquid crystals with an azobenzene derivative as a photoresponsive chromophore. *Chem. Mater.*, *14*, 385–391, 2002.

15. Eich, M., Wendorff, J. H., Reck, B., Ringsdorf, H., Reversible digital and holographic optical storage in polymeric liquid crystals. *Makromol. Chem. Rapid Commun.*, *8*, 59–63, 1987.

16. Eich, M., Wendorff, J. H., Erasable holograms in polymeric. Liquid crystals. *Makromol. Chem. Rapid Commun.*, *8*, 467–471, 1987.

17. Ikeda, T., Horiuchi, S., Karanjit, D. B., Kurihara, S., Tazuke, S., Photochemical image storage in polymer liquid crystals. *Chem. Lett.*, 1679–1682, 1988.

18. Ikeda, T., Horiuchi, S., Karanjit, D. B., Kurihara, S., Tazuke, S., Photochemically induced isothermal phase transition in polymer liquid crystals with mesogenic phenyl benzoate side chains. 1. Calorimetric studies and order parameters. *Macromolecules*, *23*, 36–42, 1990.

19. Ikeda, T., Horiuchi, S., Karanjit, D. B., Kurihara, S., Tazuke, S., Photochemically induced isothermal phase transition in polymer liquid crystals with mesogenic phenyl benzoate side chains. 2. Photochemically induced isothermal phase transition behaviors. *Macromolecules*, *23*, 42–48, 1990.

20. Ikeda, T., Kurihara, S., Karanjit, D. B., Tazuke, S., Photochemically induced isothermal phase transition in polymer liquid crystals with mesogenic cyanobiphenyl side chains. *Macromolecules*, *23*, 3938–3943, 1990.

21. Sasaki, T., Ikeda, T., Ichimura, K., Time-resolved observation of photochemical phase transition in polymer liquid crystals. *Macromolecules*, *25*, 3807–3811, 1992.

22. Tsutsumi, O., Demachi, Y., Kanazawa, A., Shiono, T., Ikeda, T., Nagase, Y., Photochemical phase-transition behavior of polymer liquid crystals induced by photochemical reaction of azobenzenes with strong donor–acceptor pairs. *J. Phys. Chem., B 102*, 2869–2874, 1998.

23. (a) Kurihara, S., Ikeda, T., Sasaki, T., Kim, H.-B., Tazuke, S., Time-resolved observation of isothermal phase transition of liquid crystals triggered by photochemical reaction of dopant. *J. Chem. Soc., Chem. Commun.*, 1751–1752, 1990; (b) Ikeda, T., Sasaki, T., Kim, H.-B., Research Article "Intrinsic" response of polymer liquid crystals in photochemical phase transition. *J. Phys. Chem.*, *95*, 509–511, 1991.

24. Ikeda, T., Tsutsumi, O., Optical switching and image storage by means of azobenzene liquid-crystal films. *Science*, *268*, 1873–1875, 1995.

25. Tsutsumi, O., Shiono, T., Ikeda, T., Galli, G., Photochemical phase transition behavior of nematic liquid crystals with azobenzene moieties as both mesogens and photosensitive chromophores. *J. Phys. Chem. B*, *101*, 1332–1337, 1997.

26. Ikeda, T., Photochemical modulation of refractive index by means of photosensitive liquid crystals. *Mol. Cryst. Liq. Cryst.*, *364*, 187–197, 2001.

27. Tsutsumi, O., Kitsunai, T., Kanazawa, A., Shiono, T., Ikeda, T., Photochemical phase transition behavior of polymer azobenzene liquid crystals with electron-donating and -accepting substituents at the 4,4′-positions. *Macromolecules*, *31*, 355–359, 1998.

28. Eelkema, R., Pollard, M. M., Vicario, J., Katsonis, N., Ramon, B. S., Bastiaansen, C. W. M., Broer, D. J., Feringa, B. L., Molecular Machines: Nanomotor rotates microscale objects. *Nature*, *440*, 163, 2006.

29. Liu, C., Chun, S. B., Mather, P. T., Zheng, L., Haley, E. H., Coughlin, E. B., Chemically crosslinked polycyclooctene: Synthesis, characterization, and shape memory behavior. *Macromolecules*, *35*, 9868–9874, 2002.

30. Otero, T. F., Cortés, M. T., Artificial muscles with tactile sensitivity. *Adv. Mater.*, *15*, 279–282, 2003.

31. Fukushima, T., Asaka, K., Kosaka, A., Aida, T., Fully plastic actuator through layer-by-layer casting with ionic-liquid-based bucky gel. *Angew. Chem. Int. Ed.*, *44*, 2410–2413, 2005.

32. Gao, J., Sansiñena, J.-M., Wang, H.-L., Tunable polyaniline chemical actuators. *Chem. Mater.*, *15*, 2411–2418, 2003.

33. Kumar, G. S., Neckers, D. C., Photochemistry of azobenzene-containing polymers. *Chem. Rev.*, *89*, 1915–1925, 1989.

34. Irie, M., Photoresponsive polymers. *Adv. Polym. Sci.*, *94*, 27–67, 1990.

35. Seki, T., Mono- and multilayers of photoreactive polymers as collective and active supramolecular systems. *Supramolecular Sci.*, *3*, 25–29, 1996.

36. Kinoshita, T., Photoresponsive membrane systems. *J. Photobiol. B: Biol.*, *42*, 12–19, 1998.

37. Merian, E., Steric factors influencing the dyeing of hydrophobic. Fibers. *Textile Res. J.*, *36*, 612–618, 1966.

38. Smets, G., De Blauwe, F., Chemical reactions in solid polymeric systems. Photomechanical phenomena. *Pure Appl. Chem.*, *39*, 225–238, 1974.

39. Eisenbach, C. D., Isomerization of aromatic azo chromophores in poly(ethyl acrylate) networks and photomechanical effect. *Polymer*, *21*, 1175–1179, 1980.

40. Matejka, L., Dusek, K., Iiavsky, M., The thermal effect in the photomechanical conversion of a photochromic polymer. *Polym. Bull.*, *1*, 659–664, 1979.

41. Matejka, L., Ilavsky, M., Dusek, K., Wichterle, O., Photomechanical effects in crosslinked photochromic polymers. *Polymer*, *22*, 1511–1515, 1981.

42. Matejka, L., Dusek, K., Photochromic polymers: Photoinduced conformational changes and effect of polymeric matrix on the isomerization of photochromes. *Makromol. Chem.*, *182*, 3223–3236, 1981.

43. Hugel, T., Holland, N. B., Cattani, A., Moroder, L., Seitz, M., Gaub, H. E., Single-molecule optomechanical cycle. *Science*, *296*, 1103–1106, 2002.

44. Holland, N. B., Hugel, T., Neuert, G., Cattani-Scholz, A., Renner, C., Oesterhelt, D., Moroder, L., Seitz, M., Gaub, H. E., Single molecule force spectroscopy of azobenzene polymers: Switching elasticity of single photochromic macromolecules. *Macromolecules*, *36*, 2015–2023, 2003.

45. Tanchak, O. M., Barrett, C. J., Light-induced reversible volume changes in thin films of azo polymers: the photomechanical effect. *Macromolecules*, *38*, 10566–10570, 2005.

46. Li, Y., He, Y., Tong, X., Wang, X., Photoinduced deformation of amphiphilic azo polymer colloidal spheres. *J. Am. Chem. Soc.*, *127*, 2402–2403, 2005.

47. Smets, G., New developments in photochromic polymers. *J. Polym. Sci.*, *13*, 2223–2231, 1975.

48. Smets, G., Braeken, J., Irie, M., Photomechanical effects in photochromic systems. *Pure Appl. Chem.*, *50*, 845–856, 1978.

49. Blair, H. S., Pogue, H. I., Photomechanical effects in polymers containing 6'-nitro-1,3,3-trimethyl-spiro-(2'H-1'-benzopyran-2,2'-indoline). *Polymer*, *23*, 779–783, 1982.

50. Athanassiou, A., Kalyva, M., Lakiotaki, K., Georgiou, S., Fotakis, C., All-optical reversible actuation of photochromic-polymer microsystems. *Adv. Mater.*, *17*, 988–992, 2005.

51. Lendlein, A., Jiang, H., Jünger, O., Langer, R., Light-induced shape-memory polymers. *Nature*, *434*, 879–882, 2005.

52. Warner, M., Gelling, K. P., Vilgis, T. A., Theory of nematic networks. *J. Chem. Phys.*, *88*, 4008–4013, 1988.

53. Mao, Y., Terentjev, E. M., Warner, M., Cholesteric elastomers: Deformable photonic solids. *Phys. Rev. E*, *64*, 041803, 2001.

54. Stenull, O., Lubensky, T. C., Phase transitions and soft elasticity of smectic elastomers. *Phys. Rev. Lett.*, *94*, 018304, 2005.

55. de Gennes, P. -G., Reflecxions sur un type de polymeres nematiques. *C. R. Acad. Sci. B*, *281*, 101–103, 1975.

56. Li, M.-H., Keller, P., Yang, J., Albouy, P.-A., An artificial muscle with lamellar structure based on a nematic triblock copolymer. *Adv. Mater.*, *16*, 1922–1925, 2004.

57. Clarke, S. M., Hotta, A., Tajbakhsh, A. R., Terentjev, E. M., Effect of crosslinker geometry on equilibrium thermal and mechanical properties of nematic elastomers. *Phys. Rev. E*, *64*, 061702, 2001.

58. Thomsen III, D. L., Keller, P., Naciri, J., Pink, R., Jeon, H., Shenoy, D., Ratna, B. R., Liquid crystal elastomers with mechanical properties of a muscle. *Macromolecules*, *34*, 5868–5875, 2001.

59. Naciri, J., Srinivasan, A., Jeon, H., Nikolov, N., Keller, P., Ratna, B. R., Nematic elastomer fiber actuator. *Macromolecules*, *36*, 8499–8505, 2003.

60. Mol, G. N., Harris, K. D., Bastiaansen, C. W. M., Broer, D. J., Thermo-mechanical responses of liquid-crystal networks with a splayed molecular organization. *Adv. Funct. Mater.*, *15*, 1155–1159, 2005.

61. Saikrasun, S., Bualek-Limcharoen, S., Kohjiya, S., Urayama, K., Anisotropic mechanical properties of thermoplastic elastomers in situ reinforced with thermotropic liquid-crystalline polymer fibers revealed by biaxial deformations. *J. Polym. Sci. Part B: Polym. Phys.*, *43*, 135–144, 2005.

62. Buguin, A., Li, M.-H., Silberzan, P., Ladoux, B., Keller, P., Micro-actuators: When artificial muscles made of nematic liquid crystal elastomers meet soft lithography. *J. Am. Chem. Soc.*, *128*, 1088–1089, 2006.

63. Lehmann, W., Skupin, H., Tolksdorf, C., Gebhard, E., Zentel, R., Krüger, P., Lösche, M., Kremer, F., Giant lateral electrostriction in ferroelectric liquid-crystalline elastomers. *Nature*, *410*, 447–450, 2001.

64. Harris, K. D., Bastlaansen, C. W. M., Lub, J., Broer, D. J., Self-assembled polymer films for controlled agent-driven motion. *Nano Lett.*, *5*, 1857–1860, 2005.

65. Elias, A. L., Harris, K. D., Bastiaansen, C. W. M., Broer, D. J., Brett, M. J., Photopatterned liquid crystalline polymers for microactuators. *J. Mater. Chem.*, *16*, 2903–2912, 2006.

66. Li, M.-H., Keller, P., Li, B., Wang, X., Brunet, M., Light-driven side-on nematic elastomer actuator. *Adv. Mater.*, *15*, 569–572, 2003.

67. Ikeda, T., Nakano, M., Yu, Y., Tsutsumi, O., Kanazawa, A., Anisotropic bending and unbending behavior of azobenzene liquid-crystalline gels by light exposure. *Adv. Mater.*, *15*, 201–205, 2003.

68. Yu, Y., Nakano, M., Shishido, A., Shiono, T., Ikeda, T., Effect of cross-linking density on photoinduced bending behavior of oriented liquid-crystalline network films containing azobenzene. *Chem. Mater.*, *16*, 1637–1643, 2004.

69. Yu, Y., Nakano, M., Ikeda, T., Soft actuators based on liquid-crystalline elastomers. *Pure Appl. Chem.*, *76*, 1435–1445, 2004.

70. Yu, Y., Maeda, T., Mamiya, J., Ikeda, T., Photomechanical effects of ferroelectric liquid-crystalline elastomers containing azobenzene chromophores. *Angew. Chem. Int. Ed.*, *46*, 881–883, 2007.

71. Yu, Y., Nakano, M., Ikeda, T., Directed bending of a polymer film by light. *Nature*, *425*, 145, 2003.

72. Kondo, M., Yu, Y., Ikeda, T., How does the initial alignment of mesogens affect the photoinduced bending behavior of the azobenzene liquid-crystalline elastomers? *Angew. Chem. Int. Ed.*, *45*, 1378–1382, 2006.

73. Mamiya, J., Yoshitake, A., Kondo, M., Yu, Y., Ikeda, T., Is chemical crosslinking necessary for the photoinduced bending of polymer films? *J. Mater. Chem.*, *18*, 63–65, 2008.

74. Camacho-Lopez, M., Finkelmann, H., Palffy-Muhoray, P., Shelley, M., Fast liquid crystal elastomer swims into the dark. *Nat. Mater.*, *3*, 307–310, 2004.

75. Harvey, C. L. M., Terentjev, E. M., Role of polarization and alignment in photoactuation of nematic elastomers. *Eur. Phys. J. E*, *23*, 185–189, 2007.

76. Dunn, M. L., Photomechanics of mono- and polydomain liquid crystal elastomer films. *J. Appl. Phys.*, *102*, 013506, 2007.

77. Tabiryan, N., Serak, S., Dai, X.-M., Bunning, T., Polymer film with optically controlled form and actuation. *Opt. Express*, *13*, 7442–7448, 2005.

78. Harris, K. D., Cuypers, R., Sscheibe, P., van Oosten, C. L., Bastiaansen, C. W. M., Lub, J., Broer, D. J., Large amplitude light-induced motion in high elastic modulus polymer actuators. *J. Mater. Chem.*, *15*, 5043–5048, 2005.

79. van Oosten, C. L., Harris, K. D., Bastiaansen, C. W. M., Broer, D. J., Glassy photomechanical liquid-crystal network actuators for microscale devices. *Eur. Phys. J. E*, *23*, 329–336, 2007.

80. Lu, S., Panchapakesan, B., Photomechanical responses of carbon nanotube/polymer actuators. *Nanotechnology*, *18*, 305502, 2007.

# 8 Photoreactive Processes for Flexible Displays and Optical Devices

*Sin-Doo Lee and Jae-Hoon Kim*

## CONTENTS

## 8.1 INTRODUCTION

In recent years, functional organic materials including liquid crystals (LCs) and organic semiconductors have been extensively studied for applications in displays [1–7], electronic and optical devices [8–12], and biomedical sensors [13,14]. In developing organic-based devices, it is very important to manipulate the molecular order in a thin film structure and to pattern the film for fabricating an array of active regions such as pixels and thin-film transistors that can be addressed in an electrical or optical control scheme.

An efficient way of controlling the molecular order and producing patterns on thin films of organic materials is to use a photoreactive process [15,16] since a simple exposure of ultraviolet (UV) light through a photomask can be employed to selectively change the physical and chemical properties of a photoreactive material needed for a variety of applications in the area in optical signal processing, optical data storage, and photoalignment of the LCs.

In particular, combined with the LCs, the photoreactive materials have been used for producing active electro-optic (EO) layers based on the LC/photopolymer composites [3,4,17–20], the surface microstructures in a one- or two-dimensional array for displays [6,7] and microlens [21], the photoalignment layers for the LCs, and passive optical components such as patterned retarders [22,23]. As the first example, for producing a LC/photopolymer composite, a photoreactive monomer (e.g., acrylate, epoxy, or vinylacrylate) is mixed with an LC to form a homogenous solution in a certain range of the concentration. Such LC/photopolymer composite can be used as an EO medium. As the second example, photoreactive polymers forming surface relief structures can be used for fabricating an array of surface microstructures on a substrate. Another approach to the formation of surface microstructures is to employ an imprinting technique combined with the UV exposure process. The surface microstructure array can be used for wide-viewing LC displays and LC microlens arrays. As the last example, the LC aligning capability of the photoreactive polymers by a simple UV exposure enables to produce multidomain structures of LC displays and patterned structures of optical retarders in a noncontact manner.

In this chapter, the physical principle, the fabrication process, and the EO performances of various optical devices using the LCs combined with the photoreactive materials will be described. The second section is devoted to present the physical principle and the fabrication method of flexible LC displays and microlens arrays, through an anisotropic phase separation process of the LC/photopolymer composites. In the third section, dye-doped LC grating structures having inter-polymer networks are described together with the photoinduced dye adsorption phenomenon on a substrate surface. In the fourth, wide-viewing LC displays with surface relief microstructures and flexible LC displays with imprinted surface microstructures are presented. In the fifth, the photoreactive layers for the LC alignment, and the Fresnel lens, and patterned retarders for transflective LC displays are discussed. Some concluding remarks are made in the remaining section.

## 8.2   PHASE-SEPARATED LCs WITH POLYMER STRUCTURES

The LC/polymer composites exhibit interesting features associated with the phase separation phenomena that are useful for new types of LC displays [1–5] and photonic devices [17,18]. The phase separation can be categorized into three processes as follows [24]: polymerization-induced phase separation (PIPS), thermally-induced phase separation (TIPS), and solvent-induced phase separation (SIPS).

Among the three processes, PIPS has proven to be the most useful in forming durable structures showing good EO properties. In the PIPS, the LC is mixed with low molecular-weight monomers or oligomers, which act as a solvent for the LC. By the application of heat, radiation, or light, the monomers can be polymerized and the LC is expelled from the polymerized volume. In particular, the photoinitiated polymerization is very powerful and widely used for fabricating various EO devices because of the ease of control of polymerization and the simple patterning capability.

Depending on the parameters including the concentration, temperature of the phase separation, the spatial gradient in the rate of polymerization, and the diffusion coefficients of the materials being used, several structures can be produced as shown in Figure 8.1. At low polymer concentrations, a polymer stabilized cholesteric texture (PSCT) shown in Figure 8.1a or cellular structures are obtained [25–28]. At higher polymer concentrations, the results depend on the rate and the nature of the phase separation. For example, at a fast rate, the phase separation occurs in the whole area of the sample in a spatially isotropic way. In this case, LC droplets are formed and dispersed in polymer networks, namely, a polymer-dispersed LC (PDLC) is produced as shown in Figure 8.1b [25]. This structure is quite durable and stable but requires a high operation voltage for EO devices. Moreover, it has the extinction ratio as small as 10:1 because it is operated in a scattering mode.

For an anisotropic phase separation, when the polymerization occurs slowly, the LC droplets become larger and are formed closer to the substrate which is far from the illuminating light source. The simplest case is the formation of a phase separated composite organic film (PSCOF) having multilayer structures of the LC and the polymer because of the one-dimensional anisotropic phase separation [1]. Various shapes of polymer structures such as polymer walls at specific areas will be produced under the intensity modulation of the illuminated light. For example, it is possible to isolate the LC molecules in pixels surrounded by inter-pixel vertical polymer walls and horizontal polymer films on the substrate, namely, the pixel-isolated LC (PILC) structure [3] is produced as shown in Figure 8.1c.

In this section, an anisotropic phase separation in a LC/polymer composite is briefly discussed. A mechanically stable flexible LC display with polymer structures, fabricated using the anisotropic phase separation, is demonstrated. As an example

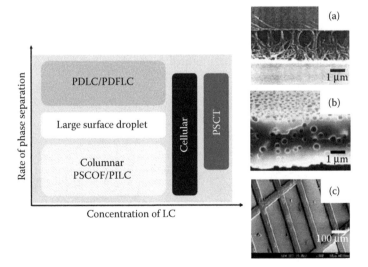

**FIGURE 8.1** Diagram of the phase separation depending on the concentration and the rate of the phase separation, and the SEM images of (a) the PSCT, (b) the PDLC, and (c) the PILC structures.

of two- or three-dimensional anisotropic phase separation, a reconfigurable LC microlens array is then presented.

### 8.2.1 ANISOTROPIC PHASE SEPARATION BY PHOTO REACTION

An anisotropic phase separation method using the PIPS is useful for fabricating the PSCOF of the LC and a polymer [1,2,28]. This PSCOF structure, in general, has adjacent uniform layers of the LC and the polymer parallel to substrates. Figure 8.2 shows the schematic diagram of constructing the PSCOF structure by the photoinitiated phase separation and the scanning electron microscope (SEM) image of a polymer layer in the PSCOF structure formed because of phase separation.

We used a commercially available photocurable prepolymer NOA65 (Norland, USA) and a LC. The prepolymer and the LC were mixed in the weight ratio of 60:40 and introduced into a sandwiched cell by capillary action at a temperature well above the clearing point of the LC. Two glass substrates with transparent electrodes were sandwiched by using the spacers. One of the substrates was spin coated with an alignment layer of a commonly used polymer such polyimide or polyvinyl alcohol (PVA). The phase separation was initiated by exposing the cell to the UV light through the substrate having no alignment layer. The UV light was illuminated at the intensity of 200 W for about 5 min.

In a one-dimensional kinetic model for the phase separation [29], the formation of the PSCOF structure can be understood by the nonuniform polymerization [30] in terms of the diffusions of monomers and LC molecules and the polymerization gradient. In this model, the phase separation depends on the concentrations of three constituents in the mixture; the volume fractions of the LC, the prepolymer, and the immobile polymer. Under the UV illumination, the polymerization rate is higher, close to the illuminated region (top in the Figure 8.2) and a gradient of the prepolymer concentration is produced. This leads to the migration of the prepolymers from the bulk to the illuminated area and causes the migration of the LC molecules toward the opposite direction. The polymers formed during the polymerization process were

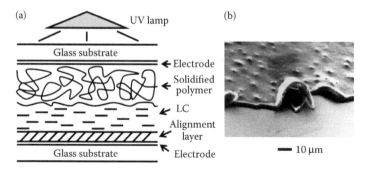

**FIGURE 8.2** (a) Schematic diagram of constructing the PSCOF structure and (b) the SEM image of polymer layer in PSCOF structure. (Reprinted with permission from T. Qian et al., *Phys. Rev.* E 61, 4007, 2000. Copyright 2000 by the American Physical Society.)

treated as immobile and have a local volume fraction that keeps growing until the polymerization is ended.

One interesting point is that the presence of an alignment layer on the substrate in contact with the LC will enhance the formation of uniform layers. The morphology associated with the phase separation was found to be greatly affected by the surface interactions between the LC/prepolymer composite and the surface layer [31]. The phase-separated structures depend not only on the bulk properties of the LC/monomer mixture but also on the surface interactions with the surface layer.

Through the control of the UV illumination and the use of a photomask, various structures of the polymer such as micro-walls or micro-relief arrays can be produced with the LC/prepolymer composites. The resultant polymer structures can be used for constructing various LC-based EO devices such as flexible displays and active microlenses. In the presence of the polymer structures formed by the photoreaction, the mechanical stability of such devices will be much improved.

### 8.2.2 PIXEL-ISOLATED LC MODE FOR FLEXIBLE DISPLAYS

In recent years, LC devices using plastic film substrates have drawn much attention for use in applications such as smart cards, personal digital assistant (PDA), and head mount displays because of their lighter weight, thinner packaging capability, flexibility, and lower manufacturing cost through continuous roll processing than other similar available devices. Different EO modes have been proposed for plastic LCDs including twisted nematic (TN), cholesteric, PDLC, and bistable ferroelectric LC (FLC) modes [32]. However, since plastic substrates do not give solid mechanical supports for the LC alignment, polymer walls and/or networks as supporting structures have been proposed and demonstrated [1,33,34]. These structures were fabricated using an anisotropic phase separation method from the polymer and the LC by applying a patterned electric field or spatially modulated UV light. However, a high electric field is needed to initiate the anisotropic phase separation or residual polymers, deteriorating the optical properties, remain in an unexposed region.

The PILC structure was found to produce a stable LC structure through the anisotropic phase separation [3,4]. In this case, the LC molecules are isolated in pixels surrounded by inter-pixel vertical polymer walls and horizontal polymer films on the substrate. As a result, a good mechanical stability is obtained without diminishing the EO properties of the LC device.

Figure 8.3 shows the fabrication process of a PILC structure. The cell with the LC/monomer mixture was irradiated by the UV light through a photomask with two-dimensional rectangular patterns that block the UV transmission (see Figure 8.3a). With the UV irradiation, the polymer walls were formed in the high intensity region. As shown in Figure 8.3b, the polymer walls were produced along the $z$-direction (vertical) on the substrate. After removing the photomask, the subsequent UV exposure was performed to fully harden the polymer. During this process, the second phase separation of the LC molecules and the polymers was accompanied and the polymer layer was formed in the $x$–$y$ plane on the illuminated substrate (i.e., the upper substrate in Figure 8.3a). Figure 8.3b shows the resultant structure through two anisotropic phase-separation processes by the two-step UV exposure.

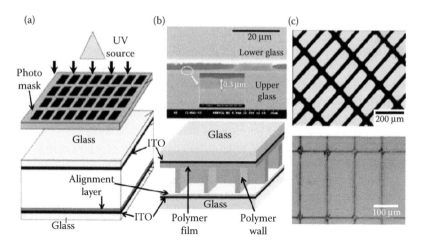

**FIGURE 8.3** Schematic diagram of (a) fabricating PILC structure by anisotropic phase separation, (b) the resultant device configuration upon the UV exposure and cross-sectional SEM image, and (c) the aligned textures of NLC (above) and FLC (below) observed with a polarizing optical microscope. (After J.-W. Jung et al., *Jpn. J. Appl. Phys.* 43, 4269, 2004, and J.-W. Jung et al., *Jpn. J. Appl. Phys.* 44, 8547, 2005.)

The magnified image in Figure 8.3b, obtained with a SEM, shows the vertical polymer walls in inter-pixels and horizontal uniform polymer films on the upper substrate. These polymer structures separate the LC from neighboring pixels and act as patterned spacers. In principle, the PILC mode can be applicable for various LC phases such as the nematic, the ferroelectric, and the cholesteric phases. Figure 8.3c shows microscopic textures of the pixel-isolated nematic LC (NLC) and FLC cells observed with a polarizing optical microscope. Note that the phase-separated LC textures depend on the wetting properties of both the monomer and the LC on the alignment layer.

Two indium tin oxide (ITO)-coated poly(ether sulfone) films of 200 μm thick were used as flexible substrates. It was found that the PILC structures were well formed on these plastic substrates. Figure 8.4 shows the textural changes of a standard LC cell (with no polymer structures) and a PILC cell on plastic substrates in the presence of external point pressure generated with a sharp tip. The LC alignment in the standard cell was severely deformed because of the variation of the cell gap as shown in Figure 8.4a. Such deformation propagates over a quite large area and results in the degradation of the EO properties. In contrast, the PILC cell exhibits no appreciable change since the LC molecular reorientation is restricted in pixels by the vertical polymer walls and the horizontal polymer layer as shown in Figure 8.4b. The dispersed small dots were glass spacers.

Let us now examine the mechanical stability of the PILC cell under bending states. The cell was bent using a pair of linear translation stages and placed between two crossed polarizers. The degree of bending is typically represented by the curvature of the cell ($R$) which is inversely proportional to the degree of bending. The measured transmittance for the standard and PILC cells is shown as a function of the

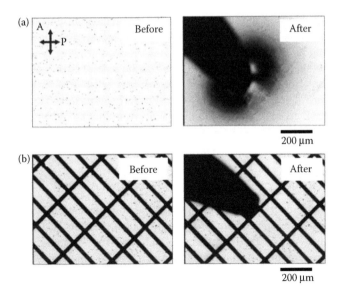

**FIGURE 8.4** Microscopic textures of a standard LC cell and a PILC cell with plastic substrates: (a) before and after point pressure on a standard LC cell and (b) before and after point pressure on a PILC cell. (After J.-W. Jung et al., *Jpn. J. Appl. Phys.* 44, 8547, 2005.)

applied voltage for various bending states in Figure 8.5a and b, respectively. For the standard plastic LC cell, decreasing $R$ reduces the transmittance. For $R \sim 1.1$ cm, the transmittance becomes reduced by 70% compared to that for no bending as shown in Figure 8.5a.

In contrast, under bending deformation, the PILC cell shows nearly the same behavior in a wide range of the applied voltage except for the low-voltage regime as shown in Figure 8.5b. It is clear that LC molecular distortions because of the bending stress is effectively suppressed in the PILC cell. Moreover, the PILC cell

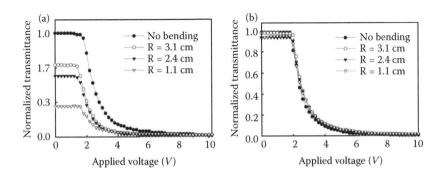

**FIGURE 8.5** Normalized transmittance versus the applied voltage: (a) for the standard LC cell and (b) for the PILC cell in various bending states. The radius of curvature $R$ represents the degree of bending. (After J.-W. Jung et al., *Jpn. J. Appl. Phys.* 44, 8547, 2005.)

has no appreciable shift of the threshold voltage, which is different from a polymer network or a PDLC cell where the threshold voltage increases with the polymer concentration [35].

Figure 8.6 shows a prototype PILC cell of the size of 3 in. under the bending state. The resolution of the prototype flexible display is 124 × 76 with a pixel size of $500 \times 500 \, \mu m^2$. The electrode on the lower substrate was patterned to display specific characters. Red, green, and blue color elements were formed on the upper substrate. The PILC structures formed by a photoinduced anisotropic phase separation would play an important role in fabricating high-performance flexible LC displays.

### 8.2.3 FABRICATION OF RECONFIGURABLE MICROLENS ARRAY

With the advancement in optical computing and communications technology, there is a growing need for real-time reconfigurable optical elements, fast optical switches, beam-steering (e.g., diffractive gratings), and wave-front-shaping (e.g., microlens arrays) devices in high-density data storage, optical interconnects, and beam-modulating and energy-directing applications [21,36–40]. Here, we introduce the fabrication of reconfigurable microlens array with a NLC and a FLC by using the anisotropic phase separation method [17,18,37]. The NLC-based device has an electrically variable focal length in milliseconds while the FLC-based microlens array exhibits the memory effect and modulates the transmitted light within a few microseconds.

The NLC-based microlens configuration is schematically shown in Figure 8.7 [17]. For the incident light with the polarization parallel to the director of the NLC, the beam will be focused because of the refraction at the curved boundary in the absence of applied voltages as shown in Figure 8.7. Under an applied voltage, the director of the NLC reorients perpendicular to the substrate and the incident beam simply passes through the LC cell without refraction if the refractive indexes of two mediums, the NLC and the polymer, are identical.

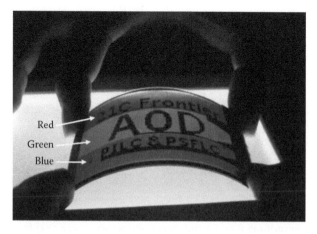

**FIGURE 8.6** A prototype PILC cell of 3 in., fabricated on the PILC structures. (After J.-W. Jung et al., *Jpn. J. Appl. Phys.* 44, 8547, 2005.)

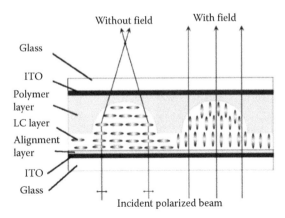

**FIGURE 8.7**    Schematic diagram of a device configuration of NLC microlens and their two operating states. (After H.-S. Ji, J.-H. Kim, and S. Kumar, *Opt. Lett.* 28, 1147, 2003.)

The focal length $f$ of the NLC microlens can be given in terms of the radius of curvature ($R$) of the curved morphology, the effective refractive index of the LC ($n_{lc}$), and that of the polymer ($n_p$) as follows [36].

$$f(V) = \frac{R}{n_{lc}(V) - n_p}. \tag{8.1}$$

The effective refractive index of the NLC is written as [41].

$$n_{lc}(V) = \frac{n_e n_o}{\sqrt{n_e \sin^2\theta(V) + n_o \cos^2\theta(V)}}, \tag{8.2}$$

where $\theta(V)$ is the polar tilt angle of the LC as a function of the applied voltage. This means that the focal length of a NLC-based microlens can be electrically tuned.

The LC materials of NLC E-31 (Merck), and a photocurable prepolymer NOA65 were used for fabricating a NLC microlens. The ordinary and extraordinary refractive indexes of the NLC are 1.533 and 1.792, respectively. The refractive index of the cured NOA65 was given as 1.524. A mixed solution of the NLC and the prepolymer, in the weight ratio of 60:40, was introduced into the cell by capillary action in the isotropic phase of the LC. A rubbed layer of PVA was used for the LC alignment. The thickness of the LC cell was about 5 μm.

Figure 8.8 shows the focusing properties of the NLC microlens for an incident beam a He–Ne laser with the wavelength of 632.8 nm. The polarization of the incident beam is parallel to the LC director. Figure 8.8a is the image of the beam focused by a microlens under no applied voltage. The NLC microlens acts as a convex lens with the focal length of 1.7 mm. The light intensity profile measured at the focal point is presented in Figure 8.8b. At the applied voltage of 3 V, the beam was defocused at a distance of 1.7 mm as shown in Figure 8.8c while refocused at 3.7 mm as shown in Figure 8.8d. The variation of the focal length with the applied voltage is shown in Figure 8.9. The focal length was found to depend quadratically on the voltage above

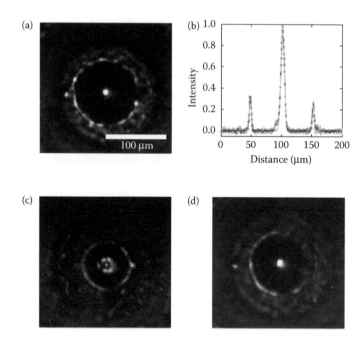

**FIGURE 8.8**  Focusing properties of the NLC microlens for a incident laser beam: (a) focused beam image at 1.7 mm under no applied voltage, (b) light intensity profile at the focal point, (c) defocused beam image at 1.7 mm under 3 V, and (d) refocused beam image at 3.7 mm under 3 V. (After H.-S. Ji, J.-H. Kim, and S. Kumar, *Opt. Lett.* 28, 1147, 2003.)

the threshold of 1.5 V. The field-on and field-off switching times were found to be 30 and 130 ms, respectively. From the measured curvature $R = 381 \pm 20\,\mu\text{m}$, the focal length was calculated as $1.6 \pm 0.1$ mm, which is in good agreement with the measured value of 1.7 mm.

For fast switching LC-based microlens, FLC materials have been widely used [18,37]. In FLCs, the spontaneous polarization is directly coupled with an external electric field and results in fast response. In a surface-stabilized FLC geometry [42], two stable states exist and a memory effect is produced. The bistable switching between the two states is driven by the polarity of an applied electric field. A FLC of Felix 15–100 (Clariant, Germany) and a photocurable prepolymer, NOA65, were used for fabricating a FLC-based microlens array. The FLC alignment layer was Nylon 6 (Sigma Aldrich, USA). The thickness of the FLC cell was $3\,\mu\text{m}$.

Figure 8.10 shows the focusing properties of a FLC microlens array for an incident laser beam. The diameter of each microlens is $335\,\mu\text{m}$. At the applied of voltage 10 V, the focal length was found to be 11 mm, and the focused beam was shown in Figure 8.10a. When the polarity of the electric field was reversed, the focused image became blurred as shown in Figure 8.10b. When the electric field was off from ±10 V, the image in Figure 8.10a and that in Figure 8.10b appeared as shown in Figure 8.10c and Figure 8.10d, respectively. The intensity profiles of the four images are shown in Figure 8.10e. It is clear that the FLC microlens possesses the memory effect with a light change in the light intensity. The slight change is believed to be because of the

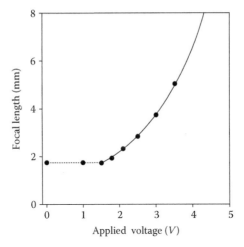

**FIGURE 8.9** Voltage dependence of the microlens' focal length. Focal length increases quadratically for fields higher than the threshold value of 1.5 V. (After H.-S. Ji, J.-H. Kim, and S. Kumar, *Opt. Lett.* 28, 1147, 2003.)

remaining polymer in the FLCs. The switching times for the on-state and the off-state were measured as 150 ms and 88 μs, respectively. These times are about 1000 times faster than the NLC microlens case.

The phase-separated LCs with polymer structures such as polymer walls and microlens, formed through the photoinduced anisotropic phase separation, would be very useful for developing flexible displays and electrically tunable microlens arrays.

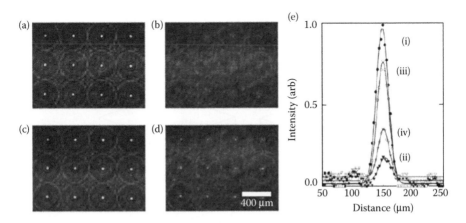

**FIGURE 8.10** Focusing characteristics of the FLC microlens for an incident laser beam: (a) the beam image under 10 V at 11 mm, (b) that under −10 V at 11 mm, (c) the beam image under 0 V switched from +10 V, (d) that under 0 V switched from −10 V, and (e) the intensity profiles for (i) to (iv). (After J.-H. Kim and S. Kumar, *Jpn. J. Appl. Phys.* 43, 7050, 2004.)

## 8.3   DYE-ADSORBED LCs IN PHOTOPOLYMER NETWORKS

Recently, the LCs containing azo-dyes have attracted great attention for the use in optical switching devices [8–12] and optical data storage [43,44] since azo-dye molecules are highly photosensitive and the large birefringence of the LC medium magnifies the optical reaction of azo-dyes. If the azo-dye molecules are exposed to resonant light, they undergo *trans–cis* photoisomerization [45,46] which induces the reorientation of the LC molecules in the bulk.

In the LCs with azo-dyes, the dye molecules can be either coated on the substrate surfaces of the LC cell [8–10, 47, 48] or simply doped in the LC host [43–46] as shown in Figure 8.11. In the case that the azo-dyes are coated on the substrate surface, the photoisomerization process of the dye molecules on the surface predominantly governs the LC alignment which propagates into the bulk. In the case of dye-doped LCs, the dye molecules are in their stable *trans*, rod-like conformation resembling the LC molecules and thus the dye molecules tend to be parallel to the LC director via the guest–host effect [45]. The LC reorientation is dictated by an optical torque generated by the photoisomerization of the azo-dyes in the bulk [46] as well as the anisotropic aligning forces produced from the interaction of the dye molecules with the command surface [19,43,49–55]. The reorientation of the LC molecules by the optical torque in bulk is usually transient while that on the surface is relatively stable.

While the LC/polymer composites such as PDLCs [56,57] or polymer-stabilized LCs [58,59] show the EO switching effect for displays and optical devices, dye-doped LCs where polymer structures are embedded often provide the optical memory resulting from the LC reorientation in the bulk not at the substrate surface [19,60,61]. Thus, when combined with the patterning capability of a photopolymer, a dye-doped LC/photopolymer composite can be used for a variety of photonic applications.

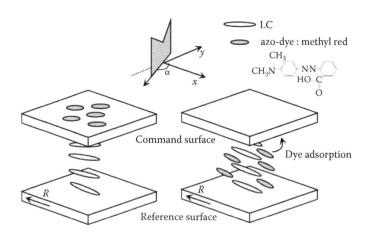

**FIGURE 8.11**   Schematic diagram of the LC cells containing azo-dyes, methyl red where the dye-molecules are (a) coated on the substrate surface or (b) doped in the LC host. (From E. Jang et al., *Mol. Cryst. Liq. Cryst.* 422, 11, 2004. With permission.)

In this section, a brief review of the dye adsorption phenomenon on the substrate surface will be first given. For optical device applications, recent works on the dye adsorption in the dye-doped LCs containing the polymer structures are then presented.

## 8.3.1 DYE ADSORPTION PHENOMENON

The surface effect, associated with the photoinduced anchorage producing the stable LC reorientation, in the dye-doped LCs has been extensively studied on a polymer-coated surface [19,43,49–55] since it can be used for recording high resolution holographic or binary LC gratings with a low energy density of the order of 100 mJ/cm$^2$. The photoinduced anchorage is known to be originated from the azo-dye molecules adsorbed on a command surface under the polarized-light irradiation [19,49,51–55]. It was experimentally observed that a photoinduced easy axis in the dye-doped LC cell is either parallel or perpendicular to the polarization of the pump beam depending on the light intensity and the pumping duration [51,52,55]. This behavior was explained in terms of the competition between two processes, the photoinduced adsorption of the dye molecules and the photoinduced desorption of the adsorbed dye molecules [51–53,55]. In this description, the anisotropic adsorption of the dye molecules (dominant in the strong-intensity regime) produces the easy axis parallel to the polarization of the pump beam while the anisotropic desorption (dominant in the weak-intensity regime) produces the easy axis perpendicular to the polarization of the pump beam. In the weak-intensity regime, a homogeneous layer of the adsorbed dye molecules was observed at the early pumping stage [54]. An easy axis perpendicular to the polarization of the pump beam was then produced. Whereas, at the late pumping stage, the easy axis was found to be parallel to the polarization of the pump beam due to the formation of the photoinduced periodic surface structures of the adsorbed dye molecules [54]. This behavior can be understood by a simple adsorption–desorption mechanism. In the strong-intensity regime, inhomogeneous ribbon-like adsorbents associated with the rapid and random aggregation was found to disturb the LC orientation [54]. The above results imply that the generation of the photoinduced easy axis in the dye-doped LCs depends on different experimental conditions such as the type of the command surface used (fluorinated polyvinylcinnamate [49,51–53,55] or ITO [54], the light intensity [62] and the pumping duration, and the shape of the pump beam.

Among the experimental conditions, the type of the command surface used is most interesting since the photoinduced anchorage in the dye-doped LCs is controlled by the delicate surface interactions of the azo-dyes and the LCs. The photoinduced LC reorientation will depend on how the surface adsorbs or desorbs the dye molecules [19,63]. Figure 8.12 shows the different behavior of the dye-induced LC reorientation depending on the command surface. The photoexcitation and relaxation kinetics of the LC reorientation was monitored using a pump-probe experiment as shown in Figure 8.1a. Two sample cells have a same reference surface of a rubbed PVA layer for one substrate and two different command surfaces of an isotropic PVA layer and a UV-cured photopolymer NOA65 layer for the other substrate. Under the same irradiation condition, the reorientation angle of the LC on the NOA65 surface is much larger than that on the PVA surface. For the polyimide surface (not shown in Figure 8.12), used

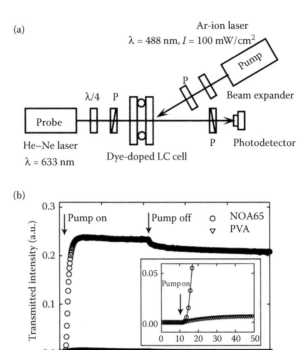

**FIGURE 8.12** (a) Experimental setup for monitoring the photoexcitation and relaxation kinetics of the LC reorientation under a pump beam and (b) the transmitted intensity of the probe beam through the dye-doped LC cell as a function of time. Two sample cells have a same reference surface of a rubbed PVA layer for one substrate and two different command surfaces of an isotropic PVA layer and a UV-cured NOA65 layer for the other substrate. (From E. Jang et al., *Mol. Cryst. Liq. Cryst.* 422, 11, 2004. With permission.)

typically as the LC alignment layer, the reorientation angle was found to lie in the range between them. The surface-commanded LC reorientation in the dye-doped LCs is important for various applications including optically switchable LC gratings [63].

Let us now discuss the LC reorientation in the dye-doped LC systems with the polymer structures as shown in Figure 8.13. In the PDLC with ball-type polymers [60,61], the LC reorientation was found to be stable because of surface interactions between the LC and the dye molecules at the polymer droplets dispersed in the LC medium. However, the PDLC has the intrinsic scattering loss which results from random orientation of the LC droplets as well as the distortions of the LC director around the droplets. This limits the application of dye-doped PDLCs in some cases.

Another type is to use inter-polymer networks produced in a dye-doped LC with a sufficiently low concentration (1–2 wt.%) of a photopolymer material not to disturb the LC orientation [19]. A small amount of the NOA65 was introduced into a dye-doped LC. The dye-doped LC/NOA65 composite cell was exposed to the UV light

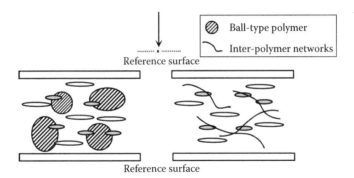

**FIGURE 8.13** Schematic diagrams of dye-doped LC cells including the polymer structures: ball-type polymers (left) and inter-polymer networks (right), providing the interfaces for the dye adsorption.

from a Xe–Hg lamp at 10 mW/cm$^2$ for 50 min to form the inter-polymer networks. An Ar-ion laser with the wavelength $\lambda = 488$ nm at the intensity of 100 mW/cm$^2$ was used as a single pump beam. The pump beam was linearly polarized at an angle of $\alpha$ with respect to the rubbing direction $\boldsymbol{R}$. Figure 8.14 shows microscopic textures of the optically reoriented dye-doped LC with the inter-polymer networks observed with a polarizing optical microscope under crossed polarizers. In the region under no pump beam (outside the circular region), the LC molecules were uniformly aligned along $\boldsymbol{R}$, meaning that a sufficiently low density of the inter-polymer networks has a negligible effect on the LC alignment. The white area inside the circular region in Figure 8.14a represents the photoinduced LC reorientation by the pump beam which was linearly polarized along the direction of 45° to the rubbing direction. The white state became completely dark when the rubbing direction $\boldsymbol{R}$ rotated by an angle of 35° to the analyzer (denoted by "A") as shown in Figure 8.14b. This means that the photoinduced optic-axis of the dye-doped LC with the inter-polymer networks is uniformly rotated by an angle of 35° with respect to the rubbing direction. In other words, under a pump beam, the reorientation of the dye molecules confined within the inter-polymer networks results in a pure optic-axis rotation in a planar LC configuration. The magnitude of the optic-axis rotation depends on the anchoring competition between the dye-induced easy axis perpendicular to the polarization of the pump beam and the rubbing direction $\boldsymbol{R}$ as shown in Figure 8.14c. The optic-axis angle is controllable by simply varying the polarization state of the pump beam.

In contrast to existing dye-doped LCs with the surface-adsorbed dyes [43,49–55], dye-doped LC systems with the inter-polymer networks produce the optic-axis rotation that gives high optical efficiency.

## 8.3.2 BINARY GRATING APPLICATIONS

Binary LC grating devices play an important role in optical systems for information processing such as optical modulators [64] and displays [65]. Among a

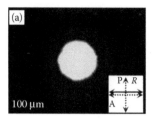
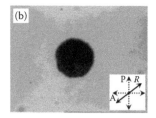

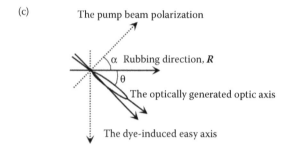

**FIGURE 8.14** Microscopic textures of the optically reoriented dye-doped LC cell with the inter-polymer networks under crossed polarizer after pumping at $\alpha = 45°$ where $\alpha$ is the angle between the polarization of the pump beam and the rubbing direction $R$. The textures were taken such that $R$ makes an angle of (a) 0 and (b) 35 with respect to the polarizer denoted by "A". (c) The magnitude of the optic-axis rotation is governed by the anchoring competition between the dye-induced easy axis perpendicular to the polarization of the pump beam and the rubbing direction $R$.

variety of binary LC gratings, the LC diffraction gratings based on a photochemical effect through *trans–cis* isomerization have been demonstrated for optical switching between two states [10–12]. Such binary LC gratings with spatially periodic LC distributions are easily produced using geometrical patterns of a photosensitive polymer material.

Another type of a binary LC grating can be fabricated using a dye-doped LC cell with the inter-polymer networks [20]. In this type, the periodic patterns of the inter-polymer networks produce the pure optic-axis modulation in the whole LC bulk. The binary grating with the periodic polymer networks, having the optic-axis modulation angle $\delta\theta = \theta_H - \theta_L$, is illustrated in Figure 8.15.

The dye-doped LC/NOA65 composite cell was exposed to the UV light through a patterned photomask to produce spatially periodic inter-polymer networks in the LC bulk. The inter-polymer networks become denser in the exposed region than the unexposed region [58,59]. Figure 8.16 shows microscopic textures of the dye-doped LC binary grating where the optic-axis modulation was produced by a single beam pumped at $\alpha$. For $\alpha = 45°$, as shown in Figure 8.16a and b, "H" region having the high density of the polymer networks (or "L" region having the low density of the inter-polymer networks) was completely dark under crossed polarizer when the rubbing direction $R$ makes an angle of $\theta_H = 35°$ (or $\theta_L = 0°$) with respect to the analyzer

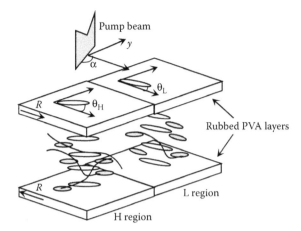

**FIGURE 8.15** A binary LC grating with the optic-axis modulation angle $\delta\theta = \theta_H - \theta_L$ which is induced through the reorientation of the dye molecules confined within spatially periodic inter-polymer networks. The spatially periodic inter-polymer networks are produced by the patterned exposure of UV light. (Reprinted with permission from E. Jang et al., *Appl. Phys. Lett.* 91, 071109, 2007. Copyright 2007, American Institute of Physics.)

denoted by "A." This means that a larger optic-axis rotation is induced in "H" region having denser inter-polymer networks. For $\alpha = 0°$, it was found that $\theta_H = 63°$ and $\theta_L = 9°$. Such well-defined optic-axis rotation in the LC cell with inter-polymer networks results in the rewritable, multistage optical memory in terms of $\alpha$.

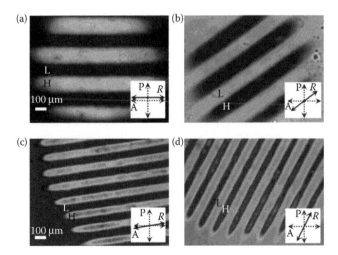

**FIGURE 8.16** Microscopic textures of the optic-axis modulated LC grating under crossed polarizer after pumping at $\alpha$. In (a) and (b), $\theta_H = 35$ and $\theta_L = 0$ when $\alpha = 45$ and the grating period of the patterned inter-polymer networks is 300 $\mu$m. In (c) and (d), $\theta_H = 63°$ and $\theta_L = 9°$ when $\alpha = 0°$ and the grating period of the patterned inter-polymer networks is 150 $\mu$m. (Reprinted with permission from E. Jang et al., *Appl. Phys. Lett.* 91, 071109, 2007. Copyright 2007, American Institute of Physics.)

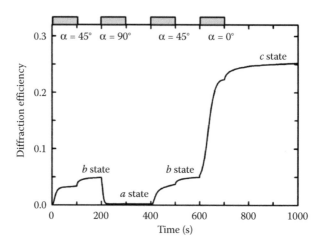

**FIGURE 8.17**   The evolution of the first-order diffraction efficiency in time ($a \rightarrow b \rightarrow a \rightarrow b \rightarrow c$) on sequencially pumping at $\alpha = 45°, 90°, 45°$, and $0°$. The bars above the graph denote $\alpha$ and the pumping duration. (Reprinted with permission from E. Jang et al., *Appl. Phys. Lett.* 91, 071109, 2007. Copyright 2007, American Institute of Physics.)

Figure 8.17 shows the evolution of the first-order diffraction efficiency in time during successive pumping processes with varying $\alpha$. More specifically, on sequentially pumping at $\alpha = 45°, 90°, 45°$, and $0°$ for 100 s in each saturated state, the optic-axis modulated state in the dye-doped LC grating evolves as $a \rightarrow b \rightarrow a \rightarrow b \rightarrow c$. The overshoot behavior, observed just before reaching the saturated state after the pump beam was off, is attributed to a small increase of $\delta\theta$ by the surface relaxation in "L" region because of negligible inter-polymer networks. In the absence of a subsequent pump, the optical state for given $\alpha$ retains long-term memory.

Figure 8.18 shows a schematic diagram of the switching mechanism among various optic-axis modulated states of the LC grating depending on $\alpha$. The optic-axis modulation angle $\delta\theta$ between "H" and "L" regions varies with $\alpha$. Ideally, the reorientation of the LC molecules in "L" region should be fully relaxed toward the rubbing direction ($\theta_L = 0°$ irrespective of $\alpha$) after the pump beam is off as shown in Figure 8.18. However, because of the diffusive nature of the UV light through the photomask, a small amount of the inter-polymer networks was formed in "L" region, resulting in a small deviation of $\theta_L$. This behavior becomes profound for the LC grating with a short period as shown in Figure 8.16a and 8.16c ($\theta_L = 0°$ for the grating period of 300 μm and $\theta_L = 9°$ for the grating period of 150 μm, respectively).

The dye-doped LC system with the inter-polymer networks, producing the pure optic-axis modulation, would be very useful for devising various optical devices that require the optical rewritability in a single pump beam scheme.

## 8.4   FORMATION OF SURFACE MICROSTRUCTURES THROUGH PHOTO REACTION

The microfabrication technologies are needed for microelectronic devices, elements in display, and sensor devices. They are categorized into the photolithography and the

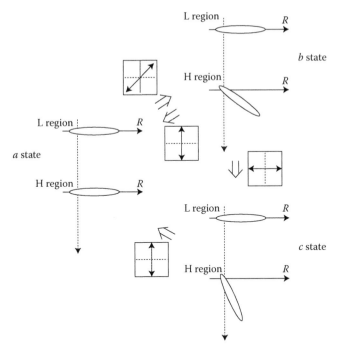

**FIGURE 8.18** Schematic diagram of the switching mechanism among various optic-axis modulated states of the LC grating depending on α. The arrows inside the boxes represent α.

softlithography as shown in Figure 8.19 [66,67]. The photolithography techniques have been extensively used for manufacturing memory devices because of the high integration capability. A shorter wavelength of the light source becomes progressively utilized to produce micropatterns of smaller feature size. Recently, the softlithogrphy techniques have attracted much interest for the fabrication of microstructures because they are applicable for nonplaner surfaces and not limited by the optical diffraction.

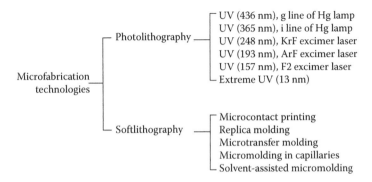

**FIGURE 8.19** Classification of the microfabrication technologies.

The photoreaction process by a UV light exposure is widely used for the lithography because of the simple and easy control of defining micro/nano patterns.

This section is devoted to describe the fabrication processes of two different surface microstructures, one of which is surface relief microstructures formed by selective UV exposure [6], and the other imprinted surface microstructures [7]. The surface relief structures are used for wide viewing LCDs and imprinted surface structures on plastic substrates are useful for flexible LCDs.

### 8.4.1 Surface Relief Microstructures for Wide Viewing LCDs

Although LCDs are currently the most important flat panel displays because of lightness, low-power consumption, and high image quality, they still need to improve the performances such as wide viewing and fast response characteristics. Because of the geometrically anisotropic shapes of the LC molecules, the light propagating through the LC cell experiences asymmetric refractive indices dependent on the incident angle. Thus, for obtaining wideviewing characteristics, a multidomain LC alignment has been developed to compensate the optical asymmetry in each pixel [68]. Since this method of using protrusions in each pixel involves a complex process, a rather simple way of producing multidomains is required. One approach is to use surface relief microstructures fabricated by photoreaction described in this section.

The surface relief microstructures were fabricated by the polymerization of a photoreactive material (NOA65) on an ITO deposited glass substrate as shown in Figure 8.20a. The NOA65 was spin coated on the substrate at 4000 rpm for 100 s (giving the film thickness of 6 μm), and irradiated by the UV light generated from a Xe–Hg lamp with 50 mW/cm$^2$ for 1 s through a chromium photomask. The substrate was subsequently illuminated without the photomask to cure the whole area uniformly with the same UV power for 15 min. The mask has circular apertures of 100 μm in diameter that are arranged in a period of 200 μm. The formation of surface relief microstructures of the photoreactive material can be understood in a diffusion model [69]. As the UV exposure increases, the difference in the photopolymer density between the illuminated area (Region I) and the unilluminated area (Region II) causes the contraction effect to make the polymer move into the illuminated area (Region I) to join the polymerization process. In this case, a microstructure is formed at the center of the illuminated region. The convex shape of the surface relief microstructure is produced when the UV energy is less than 50 mW/cm$^2$, as shown in Figure 8.20b. On further increasing the UV illumination energy (or illumination time), the photopolymerization process takes place so fast that a round rim is formed near the edge of each aperture in the photomask as shown in Figure 8.20b. The SEM image of an array of the surface relief microstructures was shown in Figure 8.20c. The circular patterns in Figure 8.20c correspond to the UV illuminated regions.

The LC cell was assembled using one substrate with surface relief microstructures and the other treated with only homeotropic polyimide as shown in Figure 8.21. The cell thickness was about 5 μm. The LC cell was filled with a NLC, EN 37 (Chisso Petrochemical Co.), which has negative dielectric anisotropy.

The microscopic textures of the LC cell with surface relief microstructures observed under crossed polarizers as a function of the applied voltage are shown

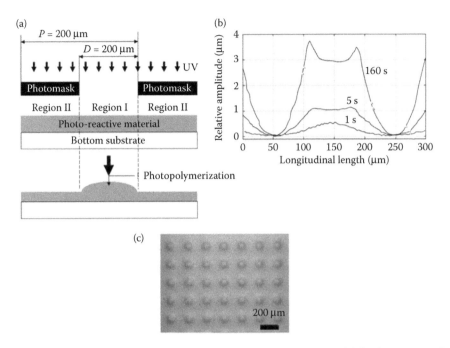

**FIGURE 8.20** The fabrication of surface relief microstructures: (a) the fabrication process of two-dimensional microstructures, (b) the profiles of a single surface relief microstructure for different illumination times, and (c) the SEM image of an array of the microstructures. (After J.-H. Park, J.-H. Lee, and S.-D. Lee, *Mol. Cryst. Liq. Cryst.* 367, 801, 2001. With permission.)

in Figure 8.22. Under no applied voltage, the LC molecules are homeotropically aligned so that a dark state is obtained as shown in Figure 8.22a. When a sufficiently high voltage is applied, the axially symmetric structure appears along the surface relief microstructures as shown in Figure 8.22d. The dark regions under a low applied voltage in Figures 8.22b and 8.22c represent the hills of the surface relief structures since the actual voltage drop in each hill region is larger than that in the valley region.

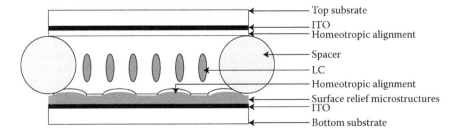

**FIGURE 8.21** The schematic diagram of the LC cell with surface relief microstructures.

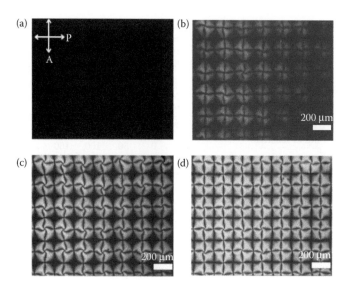

**FIGURE 8.22**  Microscopic textures of the LC cell with surface relief microstructures under crossed polarizers: (a) 0 V, (b) 12.5 V, (c) 12.8 V, and (d) 13.6 V (After J.-H. Park, J.-H. Lee, and S.-D. Lee, *Mol. Cryst. Liq. Cryst.* 367, 801, 2001. With permission.)

Figure 8.23 shows the voltage-dependent normalized transmittance of the LC cell with surface relief microstructures. As shown in Figure 8.23, the optical transmission appears at about 9 V and becomes saturated at about 13 V. The contrast ratio between the dark state at 0 V and the bright state at 13 V was about 100:1. In this LC cell, the threshold of 9 V is somewhat higher than that in a conventional VA cell (typically about 3 V) because of the voltage drop across the surface relief microstructures.

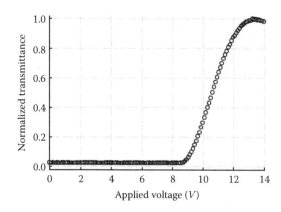

**FIGURE 8.23**  The normalized EO transmission of the LC cell with surface relief microstructures as a function of the applied voltage. (After J.-H. Park, J.-H. Lee, and S.-D. Lee, *Mol. Cryst. Liq. Cryst.* 367, 801, 2001. With permission.)

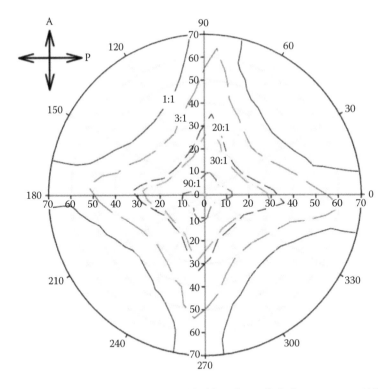

**FIGURE 8.24**  Isocontrast map of the LC cell with surface relief microstructures. (After J.-H. Park, J.-H. Lee, and S.-D. Lee, *Mol. Cryst. Liq. Cryst.* 367, 801, 2001. With permission.)

The iso-contrast map of the LC cell with surface relief microstructures is illustrated in Figure 8.24. As shown in Figure 8.24, the iso-contrast map has fourfold symmetry with respect to the normal incidence as expected from the on-state shown in Figure 8.22d. Moreover, the viewing angles are far extended along the four axes of crossed polarizers as well as the off-axes with respect to two polarizers (*x*- and *y*-axes). Thus, the iso-contrast map in this LC cell with looks like a rhombus.

It is concluded that the use of the surface relief microstructures formed by photoreaction is one of viable ways of improving the viewing characteristics of the LCDs.

## 8.4.2 IMPRINTED MICROSTRUCTURES FOR FLEXIBLE DISPLAYS

The plastic substrate-based flexible LCDs [70–72] are very important for mobile applications because of their unique features such as lighter weight, better portability, and better durability than the glass substrate-based LCDs. Although the LCD technologies have been relatively well established for flexible displays, it is still difficult to obtain the uniform LC alignment using polyimide on a plastic substrate because of the structural change resulting from the baking process over 120°C. Therefore, a new LC alignment method which is compatible with the low-temperature process is required. Here, we demonstrated that imprinted surface microstructures can be used for aligning LC molecules without a thermal process.

For topographically aligning the LC molecules on a plastic substrate without using a polyimide alignment layer, the Berreman approach [73,74] is considered. One-dimensional microgrooves for aligning LC molecules and two-dimensional microstructures for spacers can be simultaneously fabricated by the imprinting technique using a bi-level stamp. The imprinting process of fabricating the microgrooves and spacers onto two plastic substrates is shown in Figure 8.25. Two stamps, I and II, were fabricated by the replica molding technique [69]. The photoreactive polymer, NOA65, was first spin coated onto two plastic substrates at the spinning rate of 5000 rpm for 300 s (giving the film thickness of 5 μm). Using the two stamps, the imprinted microstructures were produced on the NOA65 layers, and they were subsequently irradiated with the UV light for 10 min at room temperature so that the shapes of the imprinted microstructures were well preserved.

In the case of using microgrooves to align LC molecules, the azimuthal anchoring energy can be easily tailored by varying the periods ($d_1$ and $d_2$) and/or the amplitudes ($h_1$ and $h_2$) of the microgrooves shown in Figure 8.25. According to the Berreman description [77,78], the azimuthal anchoring energy is proportional to $h_1/d_1$ (or $h_2/d_2$), meaning that different periods and amplitudes, $h_1 \neq h_2$ and $d_1 \neq d_2$, will generate different azimuthal anchoring energies between the top and bottom substrates. The azimuthal anchoring energy is theoretically evaluated $W_B = 2\pi d^2 K / h^3$, where $d$, $h$, and $K$ represent the period microstructure, the amplitude of each microstructure, and the mean value of the splay and bend elastic constants. Based on the parameters of the amplitude $d = 1\,\mu m$, the period $h = 4\,\mu m$, and the mean elastic constant $K = 15.2 \times 10^{-12}$ N, the theoretical value of the azimuthal anchoring energy is calculated

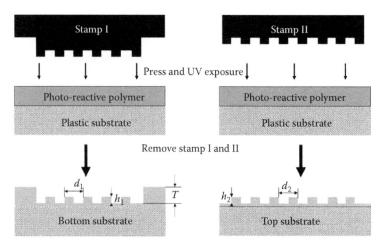

**FIGURE 8.25** Imprinting process using a bilevel stamp I and a stamp II on the top and bottom plastic substrate. The physical dimensions of the period, the amplitude, and the cell gap are $d_1 = d_2 = 4\,\mu m$, $h_1 = h_2 = 1\,\mu m$, and $T = 5\,\mu m$, respectively. (Reprinted with permission from Y.-T. Kim et al., *Appl. Phys. Lett.* 89, 173506, 2006. Copyright 2006, American Institute of Physics.)

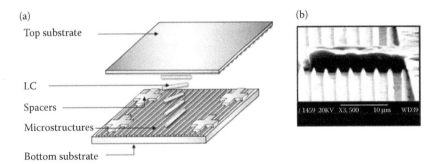

(a)

Top substrate

LC

Spacers

Microstructures

Bottom substrate

(b)

/ 1459  20KV  X3, 500        10 μm        WD39

**FIGURE 8.26** A flexible TN LC cell fabricated by an imprinting method: (a) the schematic diagram of the flexible LC cell with microgrooves and spacers and (b) the SEM image of the bottom substrate duplicated from the bi-level stamper I. (Reprinted with permission from Y.-T. Kim et al., *Appl. Phys. Lett.* 89, 173506, 2006. Copyright 2006, American Institute of Physics.)

as $1.47 \times 10^{-5}$ J/m$^2$. It should be noted that the magnitude of the azimuthal anchoring energy generated by the microstructures is sufficiently strong to uniformly align the LC molecules.

In light of the above idea, the flexible LC cell in the TN geometry was fabricated using two plastic substrates having two-dimensional arrays of spacers and one-dimensional microgrooves that are perpendicular to each other as shown in Figure 8.26a. The images of the microgrooves and the spacer observed with the SEM are shown in Figure 8.26b. The LC material was ZLI 2293 (Merck) whose dielectric anisotropy and birefringence are $\Delta\varepsilon = 10$, $\Delta n = 0.1322$, respectively.

The microscopic textures of the flexible LC cell with imprinted microstructures, observed with a polarizing optical microscope under crossed polarizers, are shown in Figure 8.27. The optic axis polarizer on each substrate coincides with the direction of the microgrooves fabricated on that substrate. Under no applied voltage, a bright state was observed in the initial TN structure as shown in Figure 8.27a. This indicates that the microgrooves with $d = 4\,\mu$m and $h = 1\,\mu$m are capable of uniformly aligning the LC molecules. At a sufficiently high applied voltage (20 V), a dark state was obtained as shown in Figure 8.27b.

The normalized transmittance of the flexible LC cell is shown in Figure 8.28. It was measured using a light source of a He–Ne laser with the wavelength of 632.8 nm

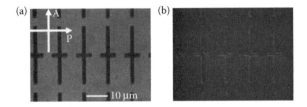

(a)       A
          P

          — 10 μm

(b)

**FIGURE 8.27** Microscopic textures of the flexible LC cell observed under crossed polarizers: (a) a bright state under no applied voltage and (b) a dark state under the applied voltage of 20 V.

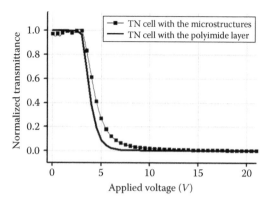

**FIGURE 8.28**  The normalized transmittance as a function of the applied voltage in the flexible LC cell and a conventional TN cell.

at room temperature. In the case of the flexible LC cell with the microstructures, the optical transmission begins to decrease at about 4 V and vanishes beyond 20 V. The driving voltage of the flexible LC cell is somewhat higher than that of a TN cell with a polyimide alignment layer since a voltage drop occurs across the NOA65 layer. The contrast ratio of the flexible LC cell was measured as about 100:1.

The mechanical stability and the reproducibility of the EO properties of a proto- type of the flexible LC panel are shown in Figure 8.29. The prototype LC panel of 1.5 cm × 4 cm, shown in Figure 8.29a, was operated in a direct driving scheme. As shown in Figure 8.29b, a logo of "Molecular Integrated Physics & Devices Lab" was off at 0 V and on at 10 V under the bend environment. The EO properties were found to be well preserved and reproducible in a bent state. Note that defects observed in certain areas of the prototype LC panel result from initially less imprinted microstructures but not from the irreversible deformations of the microstructures in a bent state. It is expected that the imprinting process based on a photoreactive material and a bi-level stamp is applicable for the roll-to-roll fabrication of flexible LCDs.

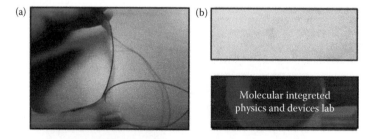

**FIGURE 8.29**  **(See color insert following page 304.)** A prototype of the flexible LC panel of 1.5 cm × 4 cm: (a) a bent state and (b) photographs showing a logo, which disappears at 0 V and appear at 10 V in a direct driving scheme.

## 8.5  PHOTOALIGNMENT TECHNIQUE FOR OPTICAL DEVICES

The alignment of LCs on the confining surfaces is a great importance for both scientific and technical viewpoints [75]. It is generally believed that anisotropic surface interactions in addition to geometrical surface patterns of the substrates dictate the LC alignment.

In this section, several LC alignment techniques are briefly discussed along with the emphasis on the photoalignment technique. The application of the photoalignment technique to the fabrication of optical devices such as the LC Fresnel lens and patterned wave plate are then described. A variety of the LC alignment techniques are classified and summarized in Figure 8.30 [76]. A rubbing process is the most widely used, but it has such disadvantage as generation of electrostatic charges and dust particles from the mechanical contact. As an alternative to the rubbing process, there are several LC alignment techniques including oblique evaporation [77], ion beam exposure [78], surface modification using the atomic force microscopy (AFM) [79], and photoalignment [80–84] as shown in Figure 8.30.

Particularly, as a noncontact method, the photoalignment technique has been extensively studied since it offers a unique way of preparing an alignment layer with complex structures and patterns for the LC molecules.

In the photoalignment process of the LC molecules, a polymer alignment layer is usually irradiated with polarized UV light. For certain materials, this polarized UV light can give rise to the rupture or the combination of bonds of the photoreactive polymers in a specific direction. Examples are the UV light-induced cycloaddition of cinnamate, coumarin derivatives, and the photodegradation of polyimides [80,83,84]. On such UV-exposed alignment layer, the LC molecules can be oriented either parallel or perpendicular to the plane of the UV polarization.

For some applications, this photoalignment technique plays a key role in fabricating a structured birefringent film like an optical waveplate using a polymerizable liquid crystalline (PLC) material. For instance, a birefringent layer with a locally varying optical axis can be produced using the PLC material on the top of a structured alignment layer [22,85].

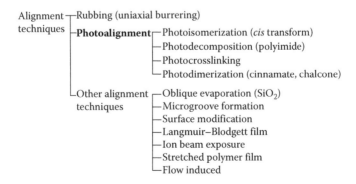

**FIGURE 8.30**  Classification of various LC alignment techniques.

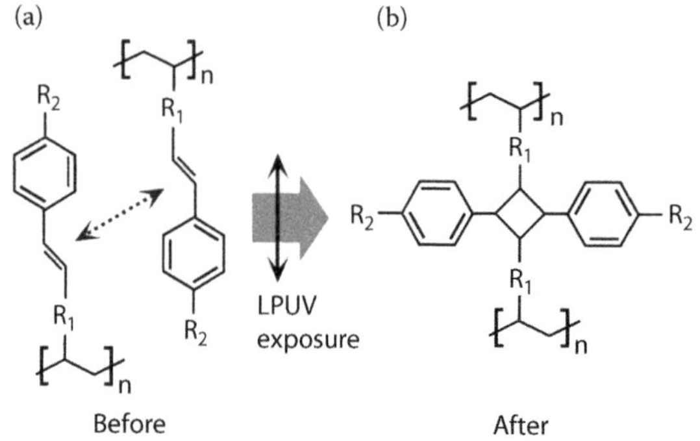

**FIGURE 8.31** The molecular structures of LGC-M1 (a) before and (b) after the LPUV exposure. After the LPUV illumination, dimerization of the side chains, whose direction is perpendicular to the polarization direction of the illuminated LPUV light, produces anisotropic anchoring forces on the substrate.

Let us describe two examples of photodimerizable polymers, LGC-M1 and LGC-M2 (all from LG Cable Ltd. Korea) as shown in Figures 8.31 and 8.32.

The polymethylmethacrylate (PMMA) moiety of LGC-M1 dissolved in cyclopentanon has a capability of aligning the LC molecules homogeneously under the illumination of the linearly polarized UV (LPUV) light and homeotropically on a

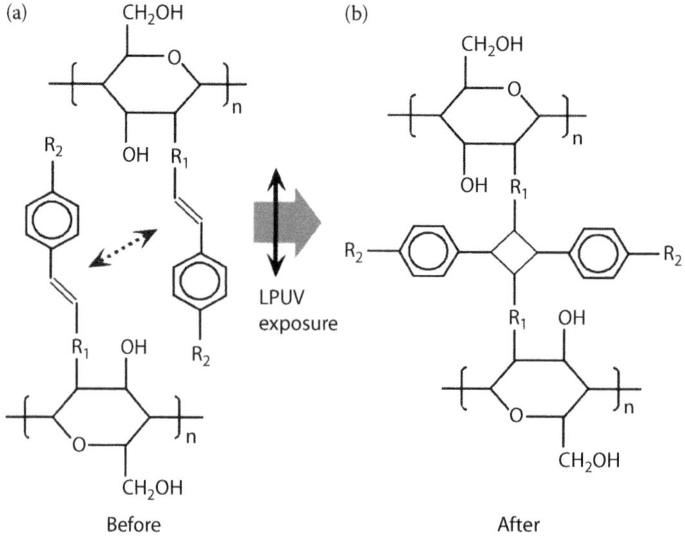

**FIGURE 8.32** The molecular structures of LGC-M2 (a) before and (b) after the LPUV exposure. After the LPUV illumination, dimerization of the side chains, whose direction is perpendicular to the polarization direction of the illuminated LPUV light, produces anisotropic anchoring feature on the substrate.

nontreated substrate with the UV light [64]. It has cinnamoyl containing photosensitive side-chain groups attached to polymetacrylate backbone as shown in Figure 8.31. Upon the LPUV exposure, the photosensitive side chains induce the anisotropic photoreaction along the UV polarization which is perpendicular to the LC molecular director as shown in Figure 8.31. It should be noted that the magnitude of the pretilt angle with respect to the substrate plane depends on both the intensity and the incident angle of the UV light.

Another material, LGC-M2, has cinnamate-containing photosensitive groups attached to the polypyranose backbone as shown in Figure 8.32. It aligns the LC molecules homogeneously under the illumination of the LPUV light and repeatedly alters the direction of the LC alignment depending on the polarization of the UV light [21,86]. The homogeneous alignment can be repeatedly produced by the LPUV illumination where the photosensitive side chains induce the anisotropic photoreaction along the UV polarization direction which is perpendicular to the LC molecular director as shown in Figure 8.32.

In the following, we will demonstrate two optical devices, the LC Fresnel lens [21] and patterned waveplates [84,85], where the two materials of LGC-M1 and LGC-M2 are used for photoalignment layers of the LC molecules.

### 8.5.1　LC Fresnel Lens

The Fresnel lenses [87,88] performing various intensity modulation of an incident light play an important role in many optical applications such as optical interconnection, optical information processing, and three-dimensional display system [89–91]. The static Fresnel lenses have been fabricated by electron-beam writing, thin film deposition [92,93]. Recently, dynamic Fresnel lenses based on the LCs [64,94–96] have attracted great attention because their large electrical and optical anisotropies allow for real-time configurable data processing. However, in general, the uniaxial anisotropic property of the LC molecules results in the polarization-dependent diffraction efficiency. Thus, several fabrication methods including orthogonal cascading of two Fresnel lenses [96] and orthogonal aligning of the LC molecules in neighboring zones [21,95] have been proposed. Here, we demonstrate a polarization-independent LC Fresnel lens in a binary phase type fabricated by photoalignment technique using the LGC-M2 material.

A typical diffractive-type Fresnel lens is in a binary amplitude or phase type. In a binary amplitude type, transparent and opaque zones are alternating in the Fresnel lens, while in a binary phase type, all zones of the Fresnel lens are transparent and induce the phase shift of $\pi$ between neighboring zones. Because of the transparency of the Fresnel lens, the throughput efficiency of a binary phase type is approximately four times better than a binary amplitude type.

In general, the radius of each zone in a binary phase type Fresnel lens (see Figure 8.33) is determined by the optical pass difference between the neighboring zones inducing the phase shift $\pi$ and it is given by [97]

$$R_m^2 = mR_1^2, \tag{8.3}$$

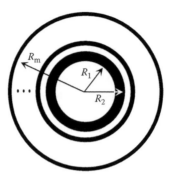

**FIGURE 8.33** Schematic diagram of a binary phase type Fresnel lens. The radius of each zone is determined by the optical pass difference between the neighboring zones inducing $\pi$ phase shift.

where, $R_1$ and $m$ represent the radius of the inner most zone and the number of $m$th zone, respectively. The focal length $f$ of the Fresnel lens is simply related to the innermost radius $R_1$ as $f = R_1^2/\lambda$, where $\lambda$ is the wavelength of the input light. When this binary phase type Fresnel lens is illuminated with a plane monochromatic wave of wavelength $\lambda$, a multitude of diverging and converging spherical waves can be observed behind the lens. In this case, the field distribution $U(x', y', z)$ of outgoing waves at a point behind a Fresnel lens is determined from the Fresnel diffraction equation as follows [81,97]:

$$U(x',y',z) = \iint t(R^2) \exp\left(\frac{i\pi}{\lambda z}[(x-x')^2 + (y-y')^2]\right) dx\, dy. \qquad (8.4)$$

Here, the $t(R^2)$ is the amplitude transmittance function of the Fresnel lens if some unimportant factors are dropped out.

Figure 8.34 shows the basic structure of a polarization-independent LC Fresnel lens. The LC molecules are aligned in an orthogonally aligned hybrid configuration with alternating zones. Note that the orthogonally alternating homogeneous alignment of bottom substrate is easily obtained using the LGC-M2 material. The orthogonality

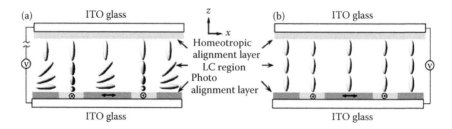

**FIGURE 8.34** Schematic diagram of the polarization-independent LC Fresnel lens in an orthogonally alternating hybrid configuration under (a) no applied voltage and (b) an applied voltage.

between the optic axes in two adjacent zones in an alternating hybrid configuration leads directly to the polarization-independent property of the LC Fresnel lens irrespective of the polarization state of the incident light.

In general, the polarization state of the incident light propagating along the z-axis can be decomposed into two orthogonal polarization components such as

$$\mathbf{E}^{\text{in}} = \begin{bmatrix} \cos \psi \\ \sin \psi \end{bmatrix} \qquad (8.5)$$

where $\psi$ represents the input polarization direction with respect to the x axis. Under no applied voltage as shown in Figures 8.34a, the relative phase shift $\Delta\phi$ of the x-polarized (or the y-polarized) light is different because of an orthogonally alternating hybrid configuration. For the x component, the incident light passing through odd zones experiences the phase retardation of $2\pi n_{\text{eff}} d/\lambda$ where $n_{\text{eff}}$ and $d$ denote the effective refractive index controlled by the applied voltage and the cell thickness, respectively. The phase retardation through even zones remains to be $2\pi n_o d/\lambda$ where $n_o$ is the ordinary refractive index of the LC. As a result, the relative phase shift between the odd and even zones is given by $\Delta\phi = 2\pi(n_{\text{eff}} - n_o)d/\lambda$, which is electrically controllable. Similarly, the phase retardation of the y component of the incident light is determined $2\pi n_o d/\lambda$ and $2\pi n_{\text{eff}} d/\lambda$ in odd and even zones, respectively. However, in a high voltage regime as shown in Figure 8.34b, the LC molecules become reoriented perpendicular to the substrate plane and thus the effective refractive index $n_{\text{eff}}$ approaches $n_o$.

Consequently, the transmittance functions of the x and y components in the Fresnel diffraction equation are defined by [21]

$$t_x\left(R^2\right) = \cos \psi \cdot \sum_{m=0}^{M/2} \left[ \text{rect}\left(\frac{R^2 - 2mR_1^2 - R_1^2/2}{R_1^2}\right) \cdot \exp\left(i\Delta\phi\right) \right.$$

$$\left. + \text{rect}\left(\frac{R^2 - 2mR_1^2 - 3R_1^2/2}{R_1^2}\right) \right],$$

$$t_y\left(R^2\right) = \sin \psi \cdot \sum_{m=0}^{M/2} \left[ \text{rect}\left(\frac{R^2 - 2mR_1^2 - R_1^2/2}{R_1^2}\right) \right.$$

$$\left. + \text{rect}\left(\frac{R^2 - 2mR_1^2 - 3R_1^2/2}{R_1^2}\right) \cdot \exp\left(i\Delta\phi\right) \right],$$

$$(8.6)$$

where $\text{rect}(x)$ is 1 for $|x| \leq 1/2$, and 0 otherwise. Here, $M$ and $\Delta\phi$ represent the number of the zones and the relative phase shift, respectively. The amplitude of the $n$th Fourier component, is then given by [21,97]

$$A_n^{x(\text{or } y)} = \frac{1}{2R_1^2} \int_0^{2R_1^2} t_{x(\text{or } y)}(R^2) \exp\left(-i\pi n\frac{R^2}{R_1^2}\right) d(R^2). \qquad (8.7)$$

The corresponding Fourier components of the $x$ and $y$ states, $A_n^x$ and $A_n^y$, derived from Equations 8.6 and 8.7 are given by

$$
A_n^x = \frac{1}{2} \exp\left(-\frac{i\Delta\phi n}{2}\right)\left[-1 + \exp\left(-i\Delta\varphi n\right)\right] \sin c\left(\frac{n}{2}\right) \cdot \cos\psi
$$

$$
A_n^y = \frac{1}{2} \exp\left(-\frac{i\Delta\phi n}{2}\right)\left[1 - \exp\left(-i\Delta\varphi n\right)\right] \sin c\left(\frac{n}{2}\right) \cdot \sin\psi.
$$

(8.8)

The diffraction efficiency in the focal plane, at $\Delta\phi = \pi$ and $n = -1$, is readily obtained from Equation 8.8 6 as

$$
\eta = \left|A_{-1}^x\right|^2 + \left|A_{-1}^y\right|^2 = \left(\frac{2}{\pi}\right)^2
$$

(8.9)

It is clear that the diffraction efficiency in Equation 8.9 is constant irrespective of the polarization state of the incident light.

The fabrication processes of the orthogonally alternating homogeneous alignment layer are depicted in Figure 8.35. As described previously, the LGC-M2 material produces the homogeneous alignment of the LC molecules along a certain direction upon the latest LPUV exposure. Such aligning capability of LGC-M2 makes it possible to align the LC molecules orthogonally in neighboring zones. As shown in Figure 8.35a, the whole region of the LGC-M2 layer was first illuminated with the LPUV to produce the homogeneous alignment of the LC molecules along the $y$ direction. A subsequent LPUV exposure through a binary amplitude type Fresnel lens photomask was carried out to align the LC molecules perpendicular to the initial alignment direction (the $y$ direction) as shown in Figure 8.35b. The above two-step UV exposure process produces orthogonally alternating neighboring zones on the substrate. The vertical alignment of the LC molecules in the other substrate was produced using the polyimide of JALS-203 (Japan Synthetic Rubber Co.). The LC material used was MLC-6082 of Merck. The dielectric anisotropy $\Delta\varepsilon = 10.0$, the ordinary refractive index $n_o = 1.4935$, and the extraordinary refractive index $n_e = 1.6414$. The cell gap was maintained using glass spacers of 7.5 μm thick. Note that the focal length of

**FIGURE 8.35** Fabrication processes of the orthogonally alternating homogeneous alignment layer using the LGC-M2 material: (a) LPUV exposure on the whole region, (b) orthogonal LPUV exposure through a photomask, and (c) produced orthogonally alternating alignment layer. The arrows in (c) represent two alignment directions of the LC molecules. (Reprinted with permission from D.-W. Kim et al., *Appl. Phys. Lett.* 88, 203505, 2006. Copyright 2006, American Institute of Physics.)

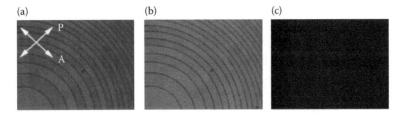

**FIGURE 8.36** Microscopic textures of the polarization-independent LC Fresnel lens with orthogonally alternating hybrid zones under crossed polarizers at (a) 0 V, (b) 1 V, and (c) 10 V. (Reprinted with permission from D.-W. Kim et al., *Appl. Phys. Lett.* 88, 203505, 2006. Copyright 2006, American Institute of Physics.)

the binary phase type Fresnel lens is governed by the physical dimensions of transparent odd zones and opaque even zones in the photomask. The photomask used has a focal length of 30 cm at the wavelength of 543.5 nm and consists of 80 zones in the aperture size of 7.2 mm. The radius $R_1$ of the innermost zone is 403.8 μm.

As shown in Figure 8.36, the microscopic textures of orthogonally alternating hybrid regions in the polarization-independent LC Fresnel lens were observed under crossed polarizers at (a) 0, (b) 1, and (c) 10 V, respectively. The principle axes of the LC layer in even and odd zones were oriented at the angle of ±45° with respect to the polarizer.

The images and corresponding intensity profiles for the dynamic focusing of the incident beam are presented in Figure 8.37a. The images of a laser beam at 1 and 10 V were captured at the $x$- and $y$-polarization state of the incident light with a charge-coupled device (CCD) camera in the focal plane. As the applied voltage increases from 1 to 10 V, the relative phase shift approaches zero and thus the focused laser beam becomes a Gaussian form. The focusing and defocusing times were about 35.4 ms for the voltage from 10 to 1 V, and 10.5 ms from 1 to 10 V, respectively.

The diffraction efficiency was measured at various incident polarizations and applied voltages. As shown in Figure 8.37b, the centro-symmetric polarization independent diffraction efficiency of the polarization-independent LC Fresnel lens is well preserved under any applied voltage. Squares, circles, and triangles represent the diffraction efficiencies at the voltages of 0, 1, and 10 V, respectively. A linearly polarized incident beam was rotated counterclockwise from 0° to 360° with respect to the $x$-axis. The magnitude of the diffraction efficiency is given by the radius of the circle and the polarization direction of the incident beam is denoted by the angle $\psi$. As shown in Figure 8.37b, when the applied voltage is 1 V, the relative phase shift of $\Delta\phi = \pi$ occurs and the maximum diffraction efficiency is about 0.35, somewhat smaller than the theoretical value of 0.41 in Equation 8.9. This discrepancy between experimental and theoretical values comes from the smooth boundaries between alternating zones because of the fluidity and the continuum elasticity of the LC. At 10 V, the LC molecules become oriented perpendicular to the substrate plane and thus the effective refractive index $n_{eff}$ approaches $n_o$. Therefore, the relative phase shift disappears. Thus, the diffraction efficiency of the polarization-independent LC Fresnel lens reduces asymptotically to zero as shown in Figure 8.37b.

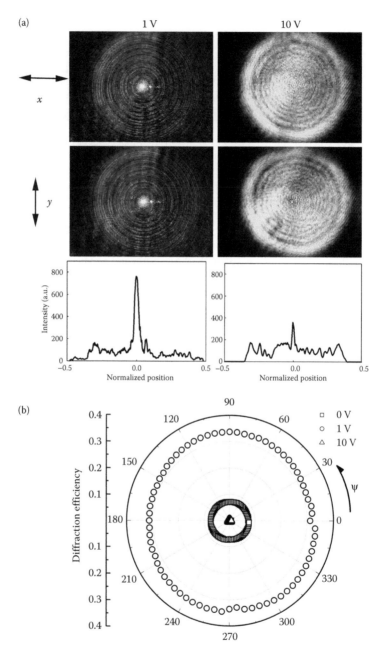

**FIGURE 8.37** (a) The CCD images and corresponding intensity profiles of the $x$- and $y$-polarized light at 1 and 10 V. (b) The polar plots of the centro-symmetrical diffraction efficiencies as a function of the polarization state of the incident light for various applied voltages. (Reprinted with permission from D.-W. Kim et al., *Appl. Phys. Lett.* 88, 203505, 2006. Copyright 2006, American Institute of Physics.)

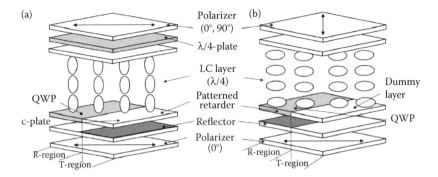

**FIGURE 8.38** Schematic diagram of the transflective LC cells with patterned retarders in (a) a VA LC configuration and (b) a PA LC configuration.

The single-masking photoalignment technique is a powerful tool of fabricating multiordered surface required for various LC devices. The polarization-independent LC Fresnel lens demonstrated here will be applicable for a variety of optical systems.

### 8.5.2 PATTERNED WAVE PLATES FOR TRANSFLECTIVE LCDs

PLC materials have been widely used for interference filters [98], optical retarders [99,100], color filters [100,101], and polarization converters [101,102]. Among the PLC-based devices, thin and patterned optical retarders [102,103] are essential to produce transflective LCDs for mobile applications in the ubiquitous environment [104,105]. Two typical configurations of the transflective LCDs [22,23] are shown in Figure 8.38.

Figures 8.38a and 8.38b show two transflective LC cells in a vertically aligned (VA) configuration [22] and a planar aligned (PA) configuration [23], respectively. Note that each LC cell is realized in a single VA or PA LC configuration. The VA-type transflective LCD is known to have high reflectance in the reflective region (R-region) and fast response in both the R-region and the transmissive region (T-region). The reflectance is about 41% higher than those of existing LCDs [85,106] and the response time is typically about 1.6 ms which is fast enough for video-rate applications [22]. In the PA-type transflective LCD, a single driving scheme can be used since the EO characteristics in the T-region are very similar to those in the R-region [23].

In both cases, the optical path difference between the T-region and the R-region should be compensated by a patterned retarder with two different regions, one of which behaves as a quarter wave plate (QWP), as shown in Figure 8.38. For fabricating such patterned retarder, the use of a photoaligned PLC layer is essential to obtain two different optical regions. Moreover, the photoalignment technique makes it possible to fabricate the patterned retarder on the interior of the LC cell, which eliminates the optical parallax and gives the robustness of the LCD.

In this section, we describe two different types of patterned retarders for the VA-and PA-type transflective LCDs fabricated using two kinds of photopolymer

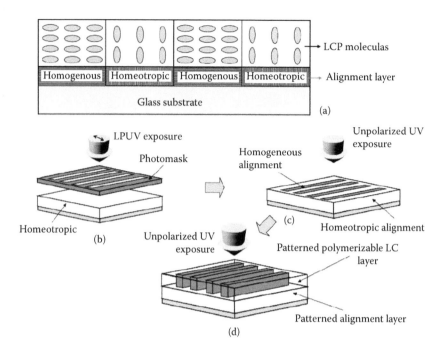

**FIGURE 8.39** Schematic diagrams of (a) the structure and the fabrication process of a patterned retarder for the VA-type transflective LCD by (b) the LPUV exposure through a photomask, (c) the unpolarized UV exposure without a photomask, and (d) the photopolymerization of the PLC molecules onto the patterned alignment layer by the unpolarized UV exposure.

materials, LGC-M1 and LGC-M2. The patterned retarders were produced by aligning and subsequently cross-linking the two photopolymers through successive UV exposure processes.

Figure 8.39 shows schematic diagrams of the structure and the fabrication process of the patterned retarder for the VA-type transflective LCD. The photoreactive polymer layer of LGC-M1 was used for the purpose of patterning and aligning the PLC molecules, LC-242 (BASF) [22].* This photopolymer induces an orientation order of the PLC from the homeotropic to the homogeneous configuration as an alignment layer depending on the UV irradiation. The LGC-M1 material dissolved in cyclopentanone was coated onto the glass substrate and baked at 150°C for 30 min. The LPUV light at the intensity of 20 mW/cm² was selectively irradiated through a photomask for 200 s onto the photopolymer layer as shown in Figure 8.39b. Before coating LC-242 onto LGC-M1, the whole region of the LGC-M1 alignment layer was exposed to the unpolarized UV light without the photomask at 10 mW/cm² for 1 s as shown in Figure 8.39c. The PLC, LC-242, with a photoinitiator was dissolved in chloroform and coated onto the patterned LGC-M1 photoalignment layer. The LC-242 layer was baked at 100°C for 1 min and polymerized by the unpolarized UV irradiation at the

* Data of LGC-M1 provided by LG Cable Ltd., Korea.

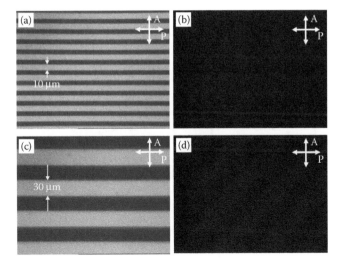

**FIGURE 8.40** Microscopic textures of a patterned retarder for the VA-type transflective LCD having two regions as the QWP and the c-plate, observed under crossed polarizer. The optic axes of the QWPs in (a) and (c) make an angle of 45° and those in (b) and (d) make an angle of 0° to one of the crossed polarizers. The line width of a patterned retarder in (a) is $10\,\mu m$ and that in (c) is $30\,\mu m$.

intensity of $40\,mW/cm^2$ for $100\,s$ as shown in Figure 8.39d. The thickness of the LC-242 was about $1.2\,\mu m$. The line widths of two patterned retarders fabricated in our study were 10 and $30\,\mu m$.

The microscopic textures of the patterned LC-242 layer were observed using a polarizing optical microscope equipped with a CCD camera under the crossed polarizers in Figure 8.40. It is clear that the LC-242 patterned retarder shows periodic homogeneous regions and homeotropic regions. As shown in Figure 8.40a and 8.40b, the homeotropic regions are always in a dark state irrespective of the direction of one of the crossed polarizer. Whereas, the homogeneous regions change the transmission according to the direction of the polarizer. The optical retardation in the homogeneous region and that in the homeotropic region of the patterned retarder were measured using the photoelastic modulation (PEM) technique [107]. Those were found to be about 1.6 and 0.09 respectively, from Figure 8.41. The optical retardation values in the two regions of the patterned retarder correspond to the QWP ($\lambda/4$) and the c-plate (0). This well-defined retarder is an essential optical element of a VA-type transflective LCD.

We now describe another kind of a patterned retarder suitable for the PA-type transflective LCD as shown in Figure 8.38b. The photoreactive polymer layer of LGC-M2 was used for aligning the PLC material, LC-298 (BASF) [23].* Figure 8.42 shows the schematic diagram of the structure and the fabrication process of the patterned retarder with two different homogeneous regions. In order to produce two alternating patterns with the retardation values of $\lambda/4$ and 0 in the two homogeneous regions as

---

* Data of LGC-M2 provided by LG Cable Ltd., Korea.

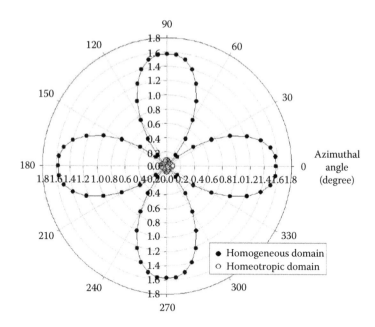

**FIGURE 8.41**  The measured optical retardation values in the homogeneous region and that in the homeotropic region of the patterned retarder.

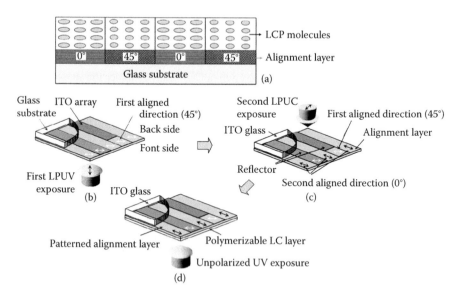

**FIGURE 8.42**  Schematic diagrams of (a) the structure and the fabrication process of a patterned retarder for the PA-type transflective LCD by (b) the first LPUV exposure from the front side, (c) the second LPUV exposure from the rear side through a self-masking process, and (d) the photopolymerization of the PLC molecules onto the patterned alignment layer by the unpolarized UV exposure. (After Y.-W. Lim, J. Kim, and S.-D. Lee, *SID Int. Symp. Digest Techn. Paper* 37, 806, 2006. Reproduced with permission of The Society for Information Display.)

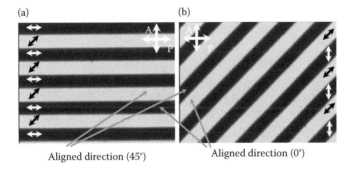

Aligned direction (45°)          Aligned direction (0°)

**FIGURE 8.43** Microscopic textures of a patterned retarder with two different homogeneous, behaving as the QWP with different optic axes, observed with a polarizing optical microscope under crossed polarizers: (a) the easy axis defined by the first LPUV exposure is parallel to one of the crossed polarizers and (b) the easy axis defined by the second LPUV exposure is parallel to one of the crossed polarizers. (After Y.-W. Lim, J. Kim, and S.-D. Lee, *SID Int. Symp. Digest Techn. Paper* 37, 806, 2006. Reproduced with permission of The Society for Information Display.)

shown in Figure 8.42a, two successive LPUV exposure steps are necessary to define periodically two optic axes in the photoalignment layer. The photoreactive polymer, LGC-M2, was coated onto the glass substrates with an array of patterned mirrors and baked at 120 °C for 30 min. The mirror array was used as a photomask. The first LPUV exposure onto the whole substrate from the front side defines one of two easy axes as shown in Figure 8.42b and the second LPUV exposure from the rear side produces the other easy axis which makes an angle of 45° with respect to the first-defined easy axis as shown in Figure 8.42c. Since the array of mirrors (metal reflectors) was served as an amplitude photomask, the second LPUV exposure did not disturb the first-defined easy axis on the photoalignment layer. A solution of the mixture of the PLC material, LC-298, doped with a photoinitiator was spin coated on the patterned LGC-M2 alignment layer at room temperature. The PLC molecules were aligned periodically on the patterned LGC-M2 layer along two easy axes. The unpolarized UV exposure was then carried out to photopolymerize the PLC layer as shown in Figure 8.42d. The thickness of the LC-298 was about 1.2 μm.

Figure 8.43 shows microscopic textures of the patterned retarder with two different homogeneous regions for the PA-type transflective LCD observed with a polarization optical microscope under crossed polarizers. The optic axes in the two regions of the patterned retarder making angle of 0° and 45° to one of crossed polarizers are shown in Figure 8.43a and Figure 8.43b, respectively. Two alternating easy axes of the patterned retarder make angles of 45° and 0° to one of the crossed polarizers. One type of the homogeneous regions, having the easy axis defined by the first LPUV exposure, becomes dark as shown in Figure 8.43a. When the patterned retarder is rotated by an angle of 45° from the polarizer axis as shown in Figure 8.43b, the other type, having the easy axis defined by the second LPUV exposure, becomes dark. Note that the bright and dark stripes of the patterned retarder are alternating at every 45° rotation, since the two easy axes make an angle of 45° to each other. The measured optical

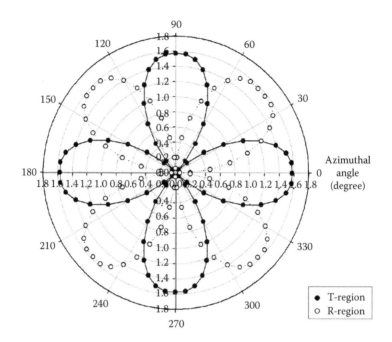

**FIGURE 8.44** The measured optical retardation values in two different homogeneous regions (T- and R-regions) of the patterned retarder.

retardation values in two different homogeneous regions of the patterned retarder were determined as $\pi/2$, which corresponds to $\lambda/4$ for $\lambda = 632.8$ nm, from Figure 8.44. This patterned LC-298 retarder with two alternating optic axes in two homogeneous regions is required for the PA-type transflective LCD as shown in Figure 8.38b.

It should be noted that the photoreactive process used for the polymerization of LC-242 and LC-298 together with the phototreatment of LGC-M1 and LGC-M2 alignment layers plays a key role in fabricating organic optical elements including patterned retarders. Particularly, the patterned retarders described here have several advantages such as the compactness, lightness in weight, and no parallax over an external optical film when fabricated directly onto the interior of the LCD.

## 8.6 CONCLUSION

We have described the photoreaction phenomena and the underlying physics such as the anisotropic phase separation, the photoinduced dye adsorption, and the photoalignment of LC molecules. Based on the photoreactive process, flexible LC displays and various optical devices such as microlens arrays, binary gratings, the Fresnel lens, and retardation plates were demonstrated.

The flexible LC displays with polymer structures, formed through the anisotropic phase separation process, were mechanically stable and showed good EO performances under bending distortions. The formation of surface relief microstructures by the photoreaction provides a simple and powerful way of fabricating wide viewing

LC displays. It was shown that the photoalignment technique allows for complex configurations of various LC-based devices and components such as transflective LC displays, Fresnel lenses, and wave retarders.

In conclusion, the photoreaction process provides a variety of methods of fabricating microstructures such as polymer columns and walls, controlling the structural molecular order at interfaces, and producing an array of patterns of optical elements for practical applications.

## ACKNOWLEDGMENTS

The authors express thanks to Eunje Jang, Yong-Woon Lim, Dong-Woo Kim, Yeun-Tae Kim, Yoonseuk Choi, and Jin-Hyuk Bae for their assistance in manuscript preparation.

This work was supported in part by Samsung Electronics, AMLCD and by the Ministry of Science and Technology of Korea through the 21st century Frontier Research.

## REFERENCES

1. V. Vorflusev, and S. Kumar, Phase-separated composite films for liquid crystal displays. *Science*, **283**, 1903, 1999.
2. F. Matsumoto, T. Nagata, T. Miyabori, H. Tanaka, and S. Tsushima, Color STN-LCD on polymer film substrates. *SID Int. Symp. Digest Techn. Paper*, **24**, 965, 1993.
3. J.-W. Jung, S.-K. Park, S.-B. Kwon, and J.-H. Kim, Pixel-isolated liquid crystal mode for flexible display applications. *Jpn. J. Appl. Phys.*, **43**, 4269, 2004.
4. J.-W. Jung, M. Y. Jin, H.-R. Kim, Y.-J. Lee, and J.-H. Kim, Mechanical stability of pixel-isolated liquid crystal mode with plastic substrates. *Jpn. J. Appl. Phys.*, **44**, 8547, 2005.
5. R. Penterman, S. I. Klink, H. D. Koning, G. Nisato, and D. J. Broer, Single-substrate liquid-crystal displays by photo-enforced stratification. *Nature*, **417**, 55, 2002.
6. J.-H. Park, J.-H. Lee, and S.-D. Lee, Vertically-aligned liquid crystal display with axial symmetry using surface relief gratings on polymer. *Mol. Cryst. Liq. Cryst.*, **367**, 801, 2001.
7. Y.-T. Kim, S. Hwang, J.-H. Hong, and S.-D. Lee, Alignment layerless flexible liquid crystal display fabricated by an imprinting technique at ambient temperature. *Appl. Phys. Lett.*, **89**, 173506, 2006.
8. W. M. Gibbons, P. J. Shannon, S.-T. Sun, and B. J. Swetlin, Surface-mediated alignment of nematic liquid crystals with polarized laser light. *Nature*, **351**, 49, 1991.
9. T. Ikeda and O. Tsutsumi, Optical switching and image storage by means of azobenzene liquid-crystal films. *Science*, **268**, 1873, 1995.
10. B. J. Kim, S.-D. Lee, S. Y. Park, and D. H. Choi, Unusual characteristics of diffraction gratings in a liquid crystal cell. *Adv. Mater*, **14**, 983, 2002.
11. X. Tong, G. Wang, A. Yavrian, T. Galstian, and Y. Zhao, Dual-mode switching of diffraction gratings based on azobenzene-polymer-stabilized liquid crystals. *Adv. Mater*, **17**, 370, 2005.
12. E. Jang, H. Baac, Y.-T. Kim, and S.-D. Lee, Electrically and optically controllable liquid crystal grating with a patterned surface-command layer. *Mol. Cryst. Liq. Cryst,* **453**, 293, 2006.

13. V. K. Gupta, J. J. Skaife, T. B. Dubrovsky, N. L. Abbott, Optical amplification of ligand-receptor binding using liquid crystals. *Science*, **279**, 2077, 1998.

14. J. M. Brake, M. K. Daschner, Y.-Y. Luk, N. L. Abbott, Biomolecular interactions at phospholipid-decorated surfaces of liquid crystals. *Science*, **302**, 2094, 2003.

15. V. V. Krongauz and A. D. Trifunac., *Process in Photoreactive Polymers*, Chapman & Hall, New York, 1995.

16. J.-P. Fouassier, *Photoinitiation, Photopolymerization, and Photocuring: Fundamental and Applications* Hanser Publishers, Munich, 1995.

17. H.-S. Ji, J.-H. Kim, and S. Kumar, Electrically controllable microlens array fabricated by anisotropic phase separation from liquid-crystal and polymer composite materials. *Opt. Lett.,* **28**, 1147, 2003.

18. J.-H. Kim and S. Kumar, Fast switchable and bistable microlens array using ferroelectric liquid crystals. *Jpn. J. Appl. Phys.,* **43**, 7050, 2004.

19. E. Jang, H.-R. Kim, Y.-J. Na, and S.-D. Lee, Stability of optical reorientation of nematic liquid crystals doped with azo-dyes in polymer networks. *Mol. Cryst. Liq. Cryst*, **422**, 11, 2004.

20. E. Jang, H.-R. Kim, Y.-J. Na, and S.-D. Lee, Multistage optical memory of a liquid crystal diffraction grating in a single beam rewriting scheme. *Appl. Phys. Lett.,* **91**, 071109, 2007.

21. D.-W. Kim, C.-J. Yu, H.-R. Kim, S.-J. Kim, and S.-D. Lee, Polarization-insensitive liquid crystal Fresnel lens of dynamic focusing in an orthogonal binary configuration. *Appl. Phys. Lett.,* **88**, 203505, 2006.

22. J. Kim, Y.-W. Lim, and S.-D. Lee, Brightness-enhanced transflective liquid crystal display having single-cell gap in vertically aligned configuration. *Jpn. J. Appl. Phys.,* **45**, 810, 2006.

23. Y.-W. Lim, J. Kim, and S.-D. Lee, Single driving transflective liquid crystal display in a single mode configuration with an inner-patterned retarder. *SID Int. Symp. Digest Techn. Paper,* **37**, 806, 2006.

24. J. W. Doane, N. A. Vaz, B. G. Wu, and S. Zumer, Field controlled light scattering from nematic microdroplets. *Appl. Phys. Lett.,* **48**, 269, 1986.

25. P. S. Drzaic, Ed., *Liquid Crystal Dispersions,* World Scientific, Singapore, 1995.

26. D. K. Yang, L. C. Chien, and J. W. Doane, Cholesteric liquid crystal/polymer gel dispersion bistable at zero field. *SID Int. Symp. Digest Techn. Paper*, **22**, 49, 1991.

27. D. K. Yang, J. L. West, L. C. Chien, and J. W. Doane, Control of reflectivity and bistability in displays using cholesteric liquid crystals. *J. Appl. Phys.,* **76**, 1331, 1994.

28. G. P. Crawford, and S. Zumer, Eds., *Liquid Crystals in Complex Geometries,* Taylor & Francis, London, 1996.

29. T. Qian, J.-H. Kim, S. Kumar, and P. L. Taylor, Phase-separated composite films: Experiment and theory. *Phys. Rev. E* **61**, 4007, 2000.

30. V. V. Krongauz, E. R. Schmelzer, and R. M. Yohannan, Kinetics of anisotropic photopolymerization in polymer matrix. *Polymer*, **32**, 1654, 1991.

31. M. Y. Jin, T.-H. Lee, J.-W. Jung, and J.-H. Kim, Surface effects on photopolymerization induced anisotropic phase separation in liquid crystal and polymer composites. *Appl. Phys. Lett.* **90**, 193510, 2007.

32. G. P. Crawford, Ed., *Flexible Flat Panel Displays*, John Wiley & Sons, Chichester, 2005.

33. Y. Kim, J. Francl, B. Taheri, and J. L. West, A method for the formation of polymer walls in liquid crystal/polymer mixtures. *Appl. Phys. Lett.,* **72**, 2253, 1998.

34. H. Sato, H. Fujikake, Y. Iino, M. Kawakita, and H. Kikuchi, Flexible grayscale ferro-electric liquid crystal device containing polymer walls and networks. *Jpn. J. Appl. Phys.,* **41**, 5302, 2002.

35. P. A. Kossyrev, J. Qi, N. V. Priezjev, R. A. Pelcovits, and G. P. Crawford, Virtual surfaces, director domains, and the Fréedericksz transition in polymer-stabilized nematic liquid crystals. *Appl. Phys. Lett.,* **81**, 2986, 2002.

36. B. E. A. Saleh, and M. C. Teich, *Fundamentals of Photonics,* John Wiley, New York, 1991.

37. J.-H. Kim, and S. Kumar, Fabrication of electrically controllable microlens array using liquid crystals. *J. Lightw. Tech.,* **23**, 628, 2005.

38. Y. Choi, J.-H. Park, J.-H. Kim, and S.-D. Lee, Fabrication of a focal length variable microlens array based on a nematic liquid crystal. *Opt. Mater,* **21**, 643, 2002.

39. T. Nose, and S. Sato, A liquid crystal microlens obtained with a non-uniform electric field. *Liq. Cryst.,* **5**, 1425, 1989.

40. Liang-Chen Lin, Hung-Chang Jau, Tsung-Hsien Lin, and Andy Y.-G. Fuh, Highly efficient and polarization-independent Fresnel lens based on dye-doped liquid crystal. *Opt. Express,* **15**, 2900, 2007.

41. P. Yeh, and C. Gu, *Optics of Liquid Crystal Displays,* John Wiley, New York, 1999.

42. N. A. Clark, and S. T. Lagerwall, Submicrosecond bistable electro-optic switching in liquid crystals. *Appl. Phys. Lett.,* **36**, 899, 1980.

43. A. G. Chen and D. J. Brady, Surface-stabilized holography in an azo-dye-doped liquid crystal. *Opt. Lett.,* **17**, 1231, 1992.

44. F. Simoni, O. Francescangeli, Y. Reznikov, and S. Slussarenko, Dye-doped liquid crystals as high-resolution recording media. *Opt. Lett.,* **22**, 549, 1997.

45. I. Janossy and A. D. Lloyd, Low-power optical reorientation in dyed Nematics. *Mol. Cryst. Liq. Cryst.,* **203**, 77, 1991.

46. I. Janossy, Molecular interpretation of the absorption-induced optical reorientation of nematic liquid crystals. *Phys. Rev. E,* **49**, 2957, 1994.

47. W. M. Gibbons and S.-T. Sun, Optically generated liquid crystal gratings. *Appl. Phys. Lett.,* **65**, 2542, 1994.

48. G. P. Crawford, J. N. Eakin, M. D. Radcliffe, A. Callan-Jones, and R. A. Pelcovits, Liquid-crystal diffraction gratings using polarization holography alignment techniques. *J. Appl. Phys.,* **98**, 123102, 2005.

49. D. Voloshchenko, A. Khyzhnyak, Y. Reznikov, and V. Reshetnyak, Control of an easy-axis on nematic-polymer interface by light action to nematic bulk. *Jpn. J. Appl. Phys.,* **34**, 566, 1995.

50. O. Francescangeli, S. Slussarenko, F. Simoni, D. Andrienko, V. Reshetnyak, and Y. Reznikov, Light-induced surface sliding of the nematic director in liquid crystals. *Phys. Rev. Lett.,* **82**, 1855, 1999.

51. E. Ouskova, D. Fedorenko, Y. Reznikov, S. V. Shiyanovskii, L. Su, J. L. West, O. V. Kuksenok, O. Francescangeli, and F. Simoni, Hidden photoalignment of liquid crystals in the isotropic phase. *Phys. Rev. E,* **63**, 021701, 2001.

52. E. Ouskova, Y. Reznikov, S. V. Shiyanovskii, L. Su, J. L. West, O. V. Kuksenok, O. Francescangeli, and F. Simoni, Photo-orientation of liquid crystals due to light-induced desorption and adsorption of dye molecules on an aligning surface. *Phys. Rev. E,* **64**, 051709, 2001.

53. O. Francescangeli, L. Lucchetti, F. Simoni, V. Stanic, and A. Mazzulla, Light-induced molecular adsorption and reorientation at polyvinylcinnamate-fluorinated/liquid-crystal interface. *Phys. Rev. Lett.,* **71**, 011702, 2005.

54. C.-R. Lee, T.-L. Fu, K.-T. Cheng, T.-S. Mo, and A. Y.-G. Fuh, Surface-assisted photoalignment in dye-doped liquid-crystal films. *Phys. Rev. E,* **69**, 031704, 2004.

55. D. Fedorenko, E. Ouskova, V. Reshetnyak, and Y. Reznikov, Evolution of light-induced anchoring in dye-doped nematics: Experiment and model. *Phys. Rev. E*, **73**, 031701, 2006.

56. H.-S. Kitzerow, H. Molsen, and G. Heppke, Linear electro-optic effects in polymer-dispersed ferroelectric liquid crystals. *Appl. Phys. Lett.*, **60**, 3093, 1992.

57. K. Lee, S.-W. Suh, and S.-D. Lee, Fast linear electro-optical switching properties of polymer-dispersed ferroelectric liquid crystals. *Appl. Phys. Lett.*, **64**, 718, 1994.

58. R. A. M. Hikmet and H. L. P. Poels, An investigation of patterning anisotropic gels for switchable recordings. *Liq. Cryst.*, **27**, 17, 2000.

59. H. Ren and S.-T. Wu, Tunable electronic lens using a gradient polymer network liquid crystal. *Appl. Phys. Lett.*, **82**, 22, 2003.

60. A. Y.-G. Fuh, C.-R. Lee, and Y.-H. Ho, Thermally and electrically switchable gratings based on polymer-ball-type polymer-dispersed liquid-crystal films. *Appl. Opt.*, **41**, 4585, 2002.

61. A. Y.-G. Fuh, C.-R. Lee, and K.-T. Cheng, Fast optical recording of polarization holographic grating based on an azo-dye-doped polymer-ball-type polymer-dispersed liquid crystal film. *Jpn. J. Appl. Phys.*, **42**, 4406, 2003.

62. M. Becchi, I. Janossy, D. S. S. Rao, D. Statman, Anomalous intensity dependence of optical reorientation in azo-dye-doped nematic liquid crystals. *Phys. Rev. E*, **69**, 051707, 2004.

63. S.-Y. Huang, S.-T. Wu, and A. Y.-G. Fuh, Optically switchable twist nematic grating based on a dye-doped liquid crystal film. *Appl. Phys. Lett.*, **88**, 041104, 2006.

64. J.-H. Park, C.-J. Yu, J. Kim, S.-Y. Chung, and S.-D. Lee, Concept of a liquid-crystal polarization beamsplitter based on binary phase gratings. *Appl. Phys. Lett.*, **83**, 1918, 2003.

65. J. Chen, P. J. Bos, H. Vithana, and D. L. Johnson, An electro-optically controlled liquid crystal diffraction grating. *Appl. Phys. Lett.*, **67**, 2588, 1995.

66. S. Okazaki, *J. Vac. Sci. Technol. B*, **9**, 2829, 1991.

67. Y. Xia and G. M. Whiteside, Replica molding with a polysiloxane mold provides this patterned microstructure. *Angew. Chem., Int., Ed.*, **37**, 551, 1998.

68. K. Ohmuro, S. Kataoka, T. Sasaki, and Y. Koike, Super-high-image-quality multi-domain vertical alignment LCD by new rubbing-less technology. *SID. Int. Symp. Digest. Tech. Papers*, **27**, 845, 1997.

69. C. C. Bowley and G. P. Crawford, Diffusion kinetics of formation of holographic polymer-dispersed liquid crystal display materials. *Appl. Phys. Lett.*, 76, 2235, 2000.

70. Y.-T. Kim, J.-H. Jong, T.-Y. Yoon, and S.-D. Lee, Pixel-encapsulated flexible displays with a multifunctional elastomer substrate for self-aligning liquid crystals. *Appl. Phys. Lett.*, **88**, 263501, 2006.

71. I. Shiyanovskaya, A. Khan, S. Green, G. Magyar, and J. W. Doane, Distinguished contributed paper: Single substrate encapsulated cholesteric lcds: coatable, drapable and foldable. *SID. Int. Symp. Digest. Tech. Papers*, **36**, 1556, 2005.

72. D.-W. Kim, C.-J. Yu, Y.-W. Lim, J.-H. Na, and S.-D. Lee, Mechanical stability of a flexible ferroelectric liquid crystal display with a periodic array of columnar spacers. *Appl. Phys. Lett.*, **87**, 051917, 2005.

73. C. J. Newsome, M. O'Neill, R. J. Farley, and G. P. Bryan-Brown. Laser etched gratings on polymer layers for alignment of liquid crystals. *Appl. Phys. Lett.*, **72**, 2078, 1998.

74. D. W. Berreman, Solid surface shape and the alignment of an adjacent nematic liquid crystal. *Phys. Rev. Lett.*, **28**, 1683, 1972.

75. J. Congnard, Alignment of nematic liquid-crystals and their mixtures. *Mol. Cryst. Liq. Cryst.*, **78**, 1, 1982.

76. D.-H. Chung, Studies on nematic liquid crystal alignment by double surface treatment using photosensitive Azo polymer. Ph.D. Thesis, Tokyo Institute of Technology, Tokyo, 2003.

77. J. Janning, Thin film surface orientation for liquid crystals. *Appl. Phys. Lett.*, **21**, 173, 1972.

78. P. Chaudhari et al., Atomic-beam alignment of inorganic materials for liquid-crystal displays. *Nature*, **411**, 56, 2001.

79. J. H. Kim, M. Yoneya, J. Yamamoto, and H. Yokoyama, Surface alignment bistability of nematic liquid crystals by orientationally frustrated surface patterns. *Appl. Phys. Lett.*, **78**, 3055, 2001.

80. K. Ichimura, Y. Suzuki, T. Seki, A. Hosoki, and K. Aoki, Reversible change in alignment mode of nematic liquid crystals regulated photochemically by command surfaces modified with an azobenzene monolayer. *Langmuir*, **4**, 1214, 1988.

81. M. Schadt, K. Schmitt, V. Hozinkov, and V. Chifrinnov, Surface-induced parallel alignment of liquid crystals by linearly polymerized photopolymers. *Jpn. J. Appl. Phys.*, **31**, 2155, 1992.

82. Y. Iimura, T. Saitoh, S. Kobayashi, and T. Hashimoto, Liquid crystal alignment on photopolymer surfaces exposed by linearly polarized uv light. *J. Photopolymer Sci. and Tech.*, **8**, 257, 1995.

83. M. Schadt, Optical patterning of multi-domain liquid-crystal displays with wide viewing angles. *Nature*, **381**, 212, 1996.

84. J.-K. Jang, H.-S. Yu, S. H. Yu, J. K. Song, B. H. Chae, and K.-Y. Han, The synthesis of BTDA-type polyimides and their photo-alignment behavior. *SID Int. Symp. Digest Techn. Paper*, **28**, 703, 1997.

85. J. Kim, D.-W. Kim, C.-J. Yu, and S.-D. Lee, A new transflective geometry of low twisted nematic liquid crystal display having a single cell gap. *Jpn. J. Appl. Phys.*, **43**, L1369, 2004.

86. C.-J. Yu, D.-W. Kim, J. Kim, and S.-D. Lee, Polarization-invariant grating based on a photoaligned liquid crystal in an oppositely twisted binary configuration. *Opt. Lett.*, **30**, 1995, 2005.

87. G. R. Fowels, *Introduction to Modern Optics,* Dover Publisher, New York, 1989.

88. J. W. Goodman, *Introduction to Fourier Optics,* McGraw-Hill, Singapore, 1996.

89. T. Kitaura, S. Ogata, and Y. Mori, Spectrometer employing a micro-fresnel lens. *Opt. Eng.*, **34**, 584, 1995.

90. M. Ferstl, and A.-M. Frisch, Static and dynamic Fresnel zone lenses for optical interconnections. *J. Mod. Opt.*, **43**, 1451, 1996.

91. M. Hain, W. Spiegel, M. Schmiedchen, T. Tschudi, and B. Javidi, 3D integral imaging using diffractive Fresnel lens arrays. *Opt. Exp.*, **13**, 315, 2005.

92. T. Fujita, H. Nishihara, and J. Koyama, Fabrication of micro lenses using electron-beam lithography. *Opt. Lett.*, **6**, 613, 1981.

93. J. Jahns, and S. J. Walker, Two-dimensional array of diffractive microlenses fabricated by thin film deposition. *Appl. Opt.*, **29**, 931, 1990.

94. H. Ren, Y.-H. Fan, and S.-T. Wu, Tunable Fresnel lens using nanoscale polymer-dispersed liquid crystals. *Appl. Phys. Lett.*, **83**, 1515, 2003.

95. J. S. Patel, and K. Rastani, Electrically controlled polarization-independent liquid-crystal Fresnel lens arrays. *Opt. Lett.*, **16**, 532, 1991.

96. G. William, N. J. Powell, A. Purvis, and M. G. Clark, Electrically controllable liquid crystal Fresnel lens. *Proc. SPIE*, **1168**, 352, 1989.

97. K. Rastani, A. Marrakchi, S. F. Habiby, W. M. Hubbard, H. Gilchrist, and R. E. Nahory, Binary phase Fresnel lenses for generation of two-dimensional beam arrays. *Appl. Opt.* **30**, 1347, 1991.

98. M. Schadt, H. Seiberle, A. Schuster, S. M. Kelly, Photoinduced alignment and patterning of hybrid liquid-crystalline polymer-films on single substrates. *Jpn. J. Appl. Phys.*, **34**, L764, 1995.

99. D. J. Broer, G. N. Mol, A. M. M. van Haaren, and J. Lub, Photo-induced diffusion in polymerizing chiral-nematic media. *Adv. Mater.*, **11**, 573, 1999.

100. M. Schadt, H. Seiberle, A. Schuster, and S. M. Kelly, Photo-generation of linearly polymerized liquid-crystal aligning layers comprising novel, integrated optically patterned retarders. *Jpn. J. Appl. Phys.*, **34**, 3240, 1995.

101. P. Van de Witte, M. Brehmer, and J. Lub, LCD components obtained by patterning of chiral nematic polymer layers. *J. Mater. Chem.*, **9**, 2087, 1999.

102. H. Ono, A. Hatayama, A. Emoto, N. Kawatsuki, and E. Uchida, Two-dimensional crossed polarization gratings in photocrosslinkable polymer liquid crystals. *Jpn. J. Appl. Phys.*, **44**, L306, 2005.

103. B. M. I. van der Zande, C. Doornkamp, S. J. Roosendaal, J. Steenbakkers, A. Op't Hoong, J. T. M. Osenga, J. J. Van Glabbeek et al., Technologies towards patterned optical foils applied to transflective LCDs. *J. SID*, **13/8**, 627, 2005.

104. S. T. Wu and C. S. Wu, Mixed-mode twisted nematic liquid crystal cells for reflective displays. *Appl. Phys. Lett.*, **68**, 1455, 1996.

105. S. H. Lee, K-H. Park, J. S. Gwag, T.-H. Yoon, and J. C. Kim, A multimode-type transflective liquid crystal display using the hybrid-aligned nematic and parallel-rubbed vertically aligned modes. *Jpn. J. Appl. Phys.*, **42**, 5127, 2003.

106. K.-H. Park, J.-C. Kim, and T.-H. Yoon, In-plane-switching liquid crystal cell with zigzag electrodes for tansflective displays. *Jpn. J. Appl. Phys.*, **43**, 7536, 2004.

107. J.-H. Lee, C.-J. Yu, and S.-D. Lee, A novel technique for optical imaging of liquid crystal alignment layers. *Mol. Cryst. Liq. Cryst.*, **321**, 317, 1998.

# 9 Polymer MEMS

*Casper L. van Oosten, Cees W.M. Bastiaansen,*
*and Dirk J. Broer*

## CONTENTS

## 9.1   INTRODUCTION

Micro electromechanical systems (MEMS) in the broadest sense are devices that are small, include mechanical elements in the order of micrometers, and convert an input signal into a mechanical movement or vice versa. Their applications range from sensors in automotive airbags, to switching micromirrors for projectors, or optical communication networks to microfluidic applications, such as inkjet heads or lab-on-a-chip devices. Traditionally, MEMS have been dominated by materials and micromachining techniques originating from the semiconductor industry [1]. However, polymers are of increasing interest because they offer a low-cost alternative while increasing the range of possibilities with respect to the potential applications [2–5]. There is a range of processing tools available for producing micrometer sized features, including embossing, lithographic processing and printing. Chemical composition provides a wide control over the properties and polymers are capable of deformations, which are orders of magnitude larger than those of inorganic actuators.

MEMS devices and their processing techniques are basically two-dimensional in nature [1]. In many MEMS applications a motion perpendicular to the plane is desired, for example to influence flow structures in microfluidics or for cantilever beams in microresonator applications. In these cases, out-of-plane bending actuators offer many advantages as they can be manufactured in-plane and small, in-plane strains are sufficient to create large, out-of-plane motion.

Liquid crystal network (LCN) actuators are fully compatible with these require-ments. Low molar mass liquid crystal (LC) acrylates can be aligned using a number of techniques including surface treatments, electrical fields or magnetic fields, all of which are capable of inducing local variations in the molecular director. Subsequently, this order is "frozen in" using photopolymerization. Throughout this chapter, bend-ing will be the most common motion of the actuators, but it will also be shown that the direction of the motion can be manipulated through control over the molecular structure of the material.

Three major classes of LC polymers are shown in Figure 9.1. LCNs are densely cross-linked polymers, where the liquid crystalline units are on both sides linked to the polymer backbone. The LC units in these systems therefore, have limited mobility. Side-chain LC polymers have pendant LC side groups on the polymer backbone and can be converted into LC elastomers (LCEs) by the introduction of a small amount of cross links. In LC-main-chain polymers, the LC units are con-nected head-to-tail by the polymer backbone. This type of LC polymers typically has the highest mobility of the LC units. Most practical systems have a combination of the different types.

One can also distinguish two processing routes to make LC polymers. LCNs are made in a one-step polymerization from the monomers and form a densely cross-linked network [6]. Low molar mass LC acrylates can be aligned using a number of techniques including surface treatments, electrical fields or magnetic fields. Subse-quently, this order is "frozen in" using photopolymerization. The cross-link density is controlled by the ratio of LC-monomers with a single functionality and with a double functionality. The single polymerization-step route offers several benefits. The low molecular weight of the monomers allows alignment of the molecules with relatively weak or local fields. Furthermore, the polymer network can be formed *in situ*, essen-tial for micro-scale systems where assembly is difficult. The high cross-link density of the polymer results in a glassy system and a high room temperature modulus. As a result, the liquid crystalline units in these systems have limited mobility and in the LCNs, a nematic–isotropic transition cannot be achieved.

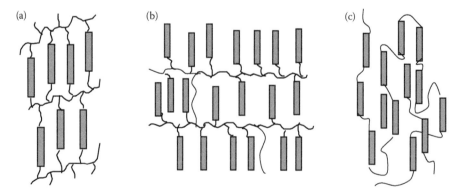

**FIGURE 9.1**  Schematic representations of different types of liquid crystal polymers. (a) LCN, (b) LC side-chain polymer, (c) LC-main chain polymer.

The LCEs are made in a two-step polymerization process using LC-side chain polymers in combination with LC-main chain cross linkers. For the cross linking, nonliquid crystalline cross linkers may be used. These systems are typically aligned using mechanical stretching after the first polymerization step, followed by a second polymerization to fix the system in the aligned state. Often, these systems use a siloxane polymer backbone, which offers high mobility to the liquid crystalline units. Due to the low cross-link density and the flexible backbone, these systems achieve the highest strains, passing through the nematic–isotropic transition. Both LC-side-on elastomer actuators and LCE actuators are treated elsewhere in this book.

To quantify the capacity of LCNs as actuators, general properties of the LCN actuators are compared. Two systems are given as a reference: mammalian muscles and conjugated polymers. Conducting polymers like polypyrrole is a well studied class of polymer actuators suitable for MEMS and various microscale systems have been reported (see, e.g., [7]). The most common set of actuator properties used for comparison are the displacement of the actuator (strain), the force generated (stress and modulus), the speed of displacement (strain rate) and the efficiency at which this is done (work density, power density, and energy efficiency) [8,9]. The area under the stress–strain curve yields the energy per unit volume necessary for a mechanical deformation of the same magnitude, which can thus easily be calculated from stress and strain measurements. In Table 9.1, an estimate of the work density ($W$) is obtained from the room-temperature modulus ($E$) and the maximum strain ($\varepsilon$): $W = 1/2\varepsilon^2 E$. The work-to-volume ratio is an important indicator for MEMS applications, as available space for actuators is typically limited in these applications. As it appears from Table 9.1, there is a trade-off between mobility (strain) and modulus, such that all LC polymer actuators have a similar work density. The deformation kinetics of LC actuators are still largely unexplored, though studies show that these actuators reach their maximum strain fast, within a few tenths of a second for the

## TABLE 9.1
## Comparison of LCN Actuators with Various Inputs, Conjugated Polymer Actuators, and Mammalian Muscles

|  | Typical Input | Maximum Strain | Room Temperature Modulus | Time Constant (s) | Work Density (kJ/m³) |
|---|---|---|---|---|---|
| Thermal LCN actuator | $\Delta T = 175°$ | 20% | 0.1–3 GPa | <1 | 300 |
| Photomechanical LCN actuator | 100 mW/cm² | 2% | 1 GPa | 0.25 | 200 |
| Chemical agent LCN actuator | Water, solvent | 2.5 % | 2 GPa | 5 | 600 |
| Conjugated polymers [8] | 10 V | 12% | 1 GPa | <1 | 1000 |
| Mammalian skeletal muscle [8] | Glucose and oxygen | 40% | 10–16 MPa | <1 | 40 |

photomechanical actuators. On the basis of these studies, it appears that LC actuators are positioned uniquely with respect to alternative actuator systems [10]. More systematic and quantitative data are necessary to obtain a good benchmark.

In this chapter, we will first explain the basic principle of shape deformation in LCNs using the most straight forward case of a thermal actuator. The influence of the chemical composition on the mechanical properties is discussed. Using the thermal actuator as an example, effects of various alignments such as twisted nematic (TN), splay, and cholesteric are shown. LCN actuators can be made sensitive to other inputs than heat. Two input mechanisms are treated here: photoactuation and chemical-agent actuation. The chapter concludes with an example of the miniaturization of actuators for a microfluidic application, where actuators are structured using inkjet printing.

## 9.2 ANISOTROPY IN MECHANICAL PROPERTIES

It is has been known for long that polymers exhibit anisotropic behavior if macroscopic order is generated, that is, showing different properties in the length direction of the polymer chain versus the two-axis perpendicular to the polymer chain [11]. Monolithic (single domain), densely cross-linked polymer networks like the LC acrylate-based systems used throughout this chapter display a large difference between the properties in the length direction and the perpendicular direction of the monomeric unit. In the remainder of this chapter, properties and deformations will be discussed with the molecular director as a reference, using the denotations "parallel" and "perpendicular" with respect to the molecular director.

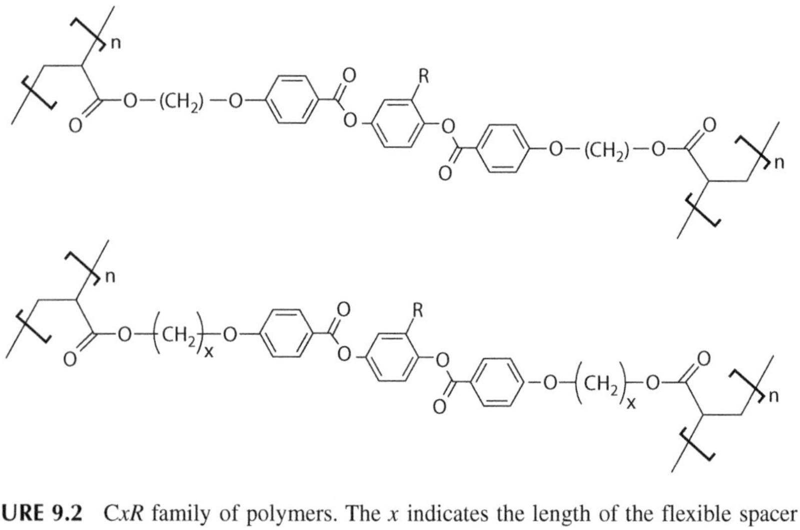

**FIGURE 9.2**  C*x*R family of polymers. The *x* indicates the length of the flexible spacer on both sides of the rigid core, the *R* indicates the sidegroup of the rigid core, either a hydrogen (H) or a methyl group (M).

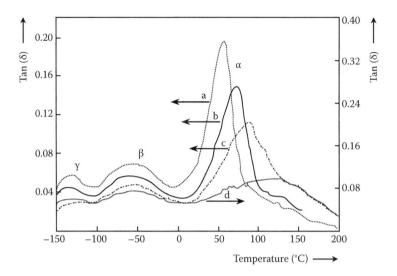

**FIGURE 9.3** Dynamical-mechanical analysis of poly(C*x*H ), with *x* = 10, 8, 6, and 4 for curves a, b, c, and d, respectively. The polymer network shows a glass transition in the range between 50°C for poly(C10H) and 140°C for poly(C4H). β-Relaxations related to the benzene groups are seen around −50°C and a gamma transition is visible at very low temperatures. (After Hikmet, R.A.M. and D.J. Broer, *Polymer*, 32, 1627–1632, 1991.)

The polymer family indicated in Figure 9.2 is by far the best studied glassy liquid crystalline system in relation to MEMS applications [6]. These polymers exhibit a glass transition temperature ($T_g$) above room temperature, typically around 60°C (Figure 9.3). Being in the glassy state, the room temperature modulus is in the order of 1–3 GPa (Figure 9.4). For comparison, nematic elastomers typically have a modulus around 1 MPa. Due to the anisotropy of the system, the compliance of the polymer network is roughly 3 times higher perpendicular than parallel to the director. This is influenced by the length of the flexible spacer: as the longer spacer introduces more mobility into the network, the $T_g$ decreases with an increasing spacer length, resulting in a lower room-temperature modulus (Figure 9.5).

Attaching a methyl group to the middle benzene group has several effects on the properties of the monomer and polymer. In the monomer phase, a methyl side group suppresses the presence of a smectic LC phase of the monomer and thus helps processing. In the polymer, the methyl group does not affect the $T_g$, but it does decrease the order of the network. Upon heating, the sterical hindrance of the methyl group causes the network to show more pronounced expansion behavior than the nonsubstituted analogues.

Like any polymer, the volume of the LCNs increases with heating. If the LCN has monodomain order, however, this deformation will not be uniform. Similar to their elastomeric counterparts, LCNs display a strong anisotropy in their thermal expansion coefficients [12]. Figure 9.6 shows this behavior for the CnR systems. Below the $T_g$, the thermal expansion coefficient in the direction parallel to the director is close to zero but on passing the $T_g$, the thermal expansion becomes negative. Orthogonal to the

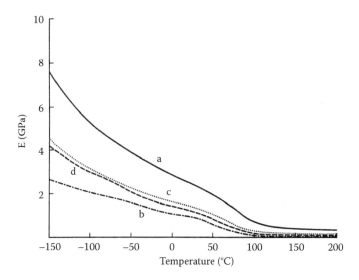

**FIGURE 9.4** Dynamic modulus of poly(C6H) versus temperature. Curves a and b were measured at a uniaxially aligned sample parallel and perpendicular to the molecular director respectively. Curve c was measured at a TN sample, curve d was measured at an isotropic sample of the same material. (After Hikmet, R.A.M. and D.J. Broer, *Polymer*, 32, 1627–1632, 1991.)

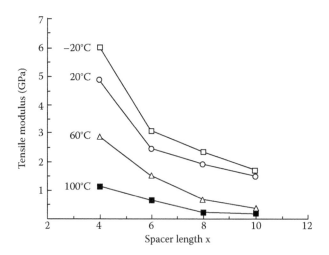

**FIGURE 9.5** Modulus of poly(CxH) at different temperatures, measured parallel to the director, as a function of length of the aliphatic tail, with $x$ denoting the number of carbon atoms in the spacer. (After Hikmet, R.A.M. and D.J. Broer, *Polymer*, 32, 1627–1632, 1991.)

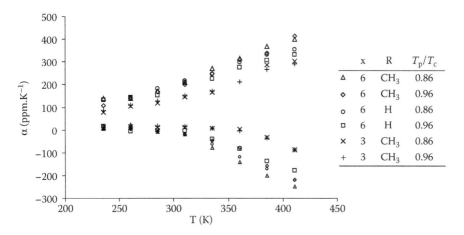

**FIGURE 9.6** Thermal expansion coefficients for uniaxially aligned networks of poly(C*n*R) made at various reduced temperatures measured perpendicular (top) and parallel (bottom) to the director. (After Mol, G.N. et al. *Adv. Funct. Mater.*, 15, 2005.)

director, the thermal expansion rapidly increases above the $T_g$. As can be expected, the systems with a longer aliphatic spacer display a larger temperature response. Furthermore, the systems that were cured close to the nematic–isotropic transition of the monomer show a smaller response than the systems cured further below the clearing temperature. The reduced temperature, the curing temperature divided by the clearing temperature of the monomers, is listed for comparison reasons for the systems in Figure 9.6. A decrease in the curing temperature only slightly affects the order of the polymer network, but leads to a significant increase in the temperature response of the system. For example, for C6M changing the reduced curing temperature from 0.96 to 0.86 increases the order parameter of the network from 0.71 to 0.76, but changes the strain parallel to the director upon heating from −50°C to 150°C from −1.3% to −17%.

An explanation for these phenomena is given in [12] and [13]. Well below the $T_g$, the system expands with temperature due to an increasing molar volume. The preferential expansion direction is perpendicular to the long axis of the molecule, as the expansions are mainly ruled by an increasing intramolecular distance. Around and above the $T_g$, there is a small and reversible loss of the degree of molecular ordering causing additional deformation. A decrease in the molecular order is favorable for entropic reasons but is limited by the polymer network.

The measured changes in birefringence between room temperature and elevated temperatures are small, in the order of a few percent, but geometrical arguments show that these order changes can explain for the observed length changes. Figure 9.7 illustrates this principle: due to an increasing average tilt of the mesogenic unit θ, the projection of the end-to-end length of the monomer onto the director decreases with loss of order. The order parameter is given by $\langle P_2 \rangle = 1/2(3(\cos^2\theta) - 1)$ and the

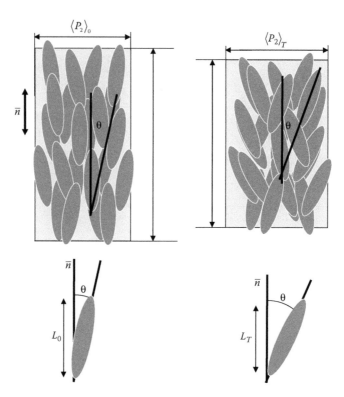

**FIGURE 9.7** Schematic representation of the order network with director n. The length changes from $L_0$ to $L_T$ when the average angle of the mesogenic units with the director increases from $\theta_0$ to $\theta_T$. (Adapted from Hikmet, R.A.M. and D.J. Broer, *Polymer*, 32, 1627–1632, 1991; Broer, D.J. and G.N. Mol, *Polym. Eng. Sci.*, 31, 625–631, 1991.)

length change parallel to the director can be then estimated using

$$\left(\frac{\Delta L}{L_0}\right)^{//}_{\langle P_2 \rangle} = \left(\frac{2\langle P_2 \rangle_T - \langle P_2 \rangle_0}{2\langle P_2 \rangle_0 + 1} + 1\right)^{1/2} - 1.$$

The total length change is then given by the sum of the length change due to molar volume increase and the change due to order loss:

$$\left(\frac{\Delta L}{L_0}\right)^{//} = \left(\frac{\Delta L}{L_0}\right)^{//}_V + \left(\frac{\Delta L}{L_0}\right)^{//}_{\langle P_2 \rangle}.$$

For the a number of LCNs based on the CnH and CnM family, the length estimates obtained in this way are consistent with the experimentally obtained length change (Table 9.2).

**TABLE 9.2**
**Measured Change in Birefringence for Various Polymers**

| Material | $\langle P_2 \rangle$ | $T_g$ (°C) | $\Delta n/(\Delta n)_0$ at 150°C | $(\Delta L/L_0)_{calc}$ at 150°C | $(\Delta L/L_0)_{meas}$ at 150°C |
|---|---|---|---|---|---|
| Poly(C4H) | 0.56 | 118 | 0.98 | −0.005 | −0.003 |
| Poly(C5H) | 0.66 | — | 0.96 | −0.011 | −0.009 |
| Poly(C6H) | 0.68 | 83 | 0.95 | −0.015 | −0.012 |
| Poly(C6M) | 0.58 | 83 | 0.92 | −0.018 | −0.024 |
| Poly(C8H) | 0.68 | 71 | 0.93 | −0.020 | −0.020 |
| Poly(C10H) | 0.68 | 55 | 0.92 | 0.023 | −0.025 |

*Note:* The calculated length change and the measured length change are given [12,13].

## 9.3   MOLECULAR ALIGNMENT CONFIGURATIONS

The linear expansions of a planar uniaxially aligned LCN are relatively small and therefore of limited practical use. During processing in the monomeric phase, molecular alignments like splay, TN, or cholesteric ordering can be generated. After photopolymerization, networks with these alignments will have properties that vary in one or more directions of the system. The techniques available for creating this molecular alignment are mostly borrowed from the display manufacturing industry and range from rubbed polyimide alignment layers to electrical and magnetic fields. One of the first applications that uses the particular temperature expansion behavior of the LCNs is stress-free coatings in integrated circuits [12]. In these applications, it is important to match the thermal-expansion coefficient of the inorganic layer underneath such that the interface between coating and substrate is stress-free over a wide temperature range. In the case of one-dimensional matching of a thermal coefficient the thermal-expansion coefficient of a uniaxial LCN can simply be tuned by choosing the proper length of the flexible spacer. On planar surfaces, such as integrated circuits, the thermal-expansion coefficient has to be matched in two directions. This is possible by rotating the director through the thickness of the film in a helical fashion. The helical order can be induced by adding a trace of a chiral dopant to the reactive mesogens. The degree of director rotation with thickness can be controlled with the concentration of the chiral dopant. Freestanding LCNs with a helicoidal ordering have the same volume thermal expansion as a uniaxially ordered network, but the linear expansions are amplified in the direction parallel to the helicoidal axis. In the other two directions, the thermal expansion coefficients are close to zero.

Figure 9.8 shows a number of alignment configurations and their corresponding deformations upon heating. The TN and cholesteric alignments only differ in the degree of director rotation, but the two alignments show distinct deformation behavior. To visualize this difference, consider a film element the size of a few molecules, small enough such that the twist in the molecules inside the element can be neglected. In TN and cholesteric networks, the element has principle expansion axis that are offset from its neighboring elements just above or below. For films where the helicoidal pitch is small with respect to the film thickness, that is, the director rotation is 180° or more,

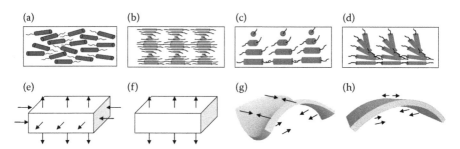

**FIGURE 9.8** Schematic representation of various molecular orderings, planar uniaxial (a), cholesteric (b), twisted nematic (TN) (c), and splay (d), with their corresponding deformations (e–h) upon a decrease of molecular order.

these stresses are compensated by the stresses generated by the elements that are at a rotation of 90° at half the helicoidal pitch above and below. These stresses become unbalanced when the thickness of the film is in the order of the helicoidal pitch. The special case is the TN alignment, where the film thickness is exactly half the helicoidal pitch. In that case, the molecular director makes a gradual rotation in the plane of the film, such that between the top and the bottom the director rotates exactly 90°. The gradient in director orientation results in a gradient in thermal expansions through the thickness of the film. As a result, a freestanding film will deform upon heating. Figure 9.9 shows images of the heat induced deformation of a clamped film. The film is clamped such that at the top of the film the molecular director is parallel to the length axis of the film. Two competing bending axis are present: one in the length direction of the film and one perpendicular to it. Where the film is clamped, one bending axis is suppressed and the film curls up. Here, we will use a two-dimensional analysis of the strains in the film using the coordinate system as depicted in Figure 9.10 to explain this bending behavior. At the top of the film, the molecular director points along the $x$-axis, at the bottom the director points out-of-plane in the $z$-direction, perpendicular to the $x-y$ plane. Upon heating, the top side of the film will thus contract and the bottom will expand in the length direction of the film. This gradient in deformation directions causes the film to curl-up over the $z$-axis as is observed close to the clamp in Figure 9.9. In reality however, the strains are three-dimensional. In fact, following the same reasoning as before, analysis of the strain in the $z$-direction results in the film bending down over the $x$-axis. Two competing bending axis are thus observed, as can be seen from the saddle-shape behavior in Figure 9.9.

As mentioned before, bending thin films are of practical interest in particular for their application as actuators in MEMS. Whereas most bending actuators operate similar to bimetals and need two or more layers to bend, bending LCN actuators are monolithic in nature and therefore, do avoid the adhesion problems caused by the large interlayer shear in bilayer constructions [14,15].

Using LCNs, it is possible to create a gradient in the degree of expansion through a single material. In the TN alignment, this behavior was already visible. The competing bending axes make the behavior unpredictable and therefore, it is not the most suitable molecular configuration. A configuration that does not exhibit the strain gradient in

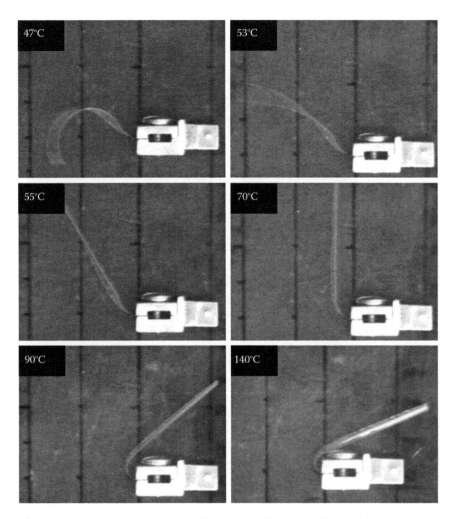

**FIGURE 9.9** Deformation of a 20 μm thick clamped film of poly(C6M) with a TN molecular ordering upon heating. The molecular director at the top of the film is oriented along the x-axis of the film. At 55°C the film starts to display a saddle shape due to the two competing bending axes. (After Hikmet, R.A.M. and D.J. Broer, *Polymer*, 32, 1627–1632, 1991.)

the z-direction is the splay configuration, where the molecular director rotates from in-plane alignment to homeotropic alignment over the thickness of the film. Figure 9.11 shows the bending of a splayed actuator with temperature. Here, the actuator has the molecular director at the top of the film parallel to the x-axis and at the bottom of the film it is parallel to the y-axis. This results in the smooth bending of the film and no clamps are required to enforce bending over a single axis. In the z-direction, the thermal expansion of the film is constant over the thickness of the film. Therefore, no bending is observed over the x-axis of the film.

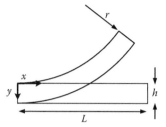

**FIGURE 9.10**  Two-dimensional model of a bending actuator with thickness $h$, length $L$, and bending radius $r$. The top left corner of the flap is chosen as the origin, the $z$-direction is the out-of-plane direction.

In the previous examples, the modulation of the director has only been through the thickness of the film. A large number of techniques are available for varying the director in the plane of the film, such as microrubbing, linear photopolymerizable (LPP) materials, electrical, and magnetic fields (see, e.g., [16–19]). These techniques are compatible with the splayed, twisted, and cholesteric alignments. However, there have been few published studies so far where the molecular director was varied in the plane of the film. Potential application areas are in controllable surface-relief structures and nanoactuators. One example of a controllable surface-relief structure is given by Sousa et al.[19]. Using a mask for the photopolymerization of the monomers partly cures the sample in the ordered cholesteric phase. The sample is then heated into the isotropic phase of the monomer and a flood exposure is used to cure the remaining areas. The result is a LCN polymer that has a high degree of order in the

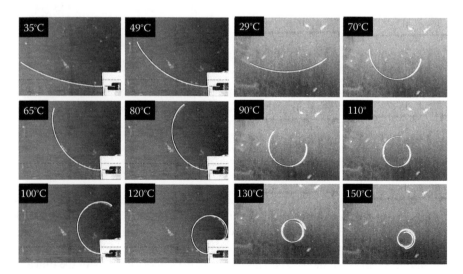

**FIGURE 9.11**  Bending of a clamped and freestanding actuator of poly(C3M) with splayed molecular configuration. The actuator is 40 μm thick and 15 mm long. (After Hikmet, R.A.M. and D.J. Broer, *Polymer*, 32, 1627–1632, 1991.)

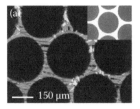
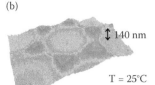

**FIGURE 9.12** Microscopy images (a) and interferometer images (b and c) of cylinders with isotropic ordering surrounded by cholesteric ordered poly(C3M). (a) shows an optical polarizing microscope image, with the inset showing the photomask used in fabrication. Images (b) and (c) are white-light interferometer images taken at room temperature and at 200°C, respectively. (After Sousa, M.E. et al. *Adv. Mat.*, 18 (14), 1842, 2006.)

areas exposed during the first exposure and an isotropic phase around it. Due to their cholesteric organization, the ordered areas will have a large thermal expansion in the direction perpendicular to the plane of the film, while the disordered areas will have the bulk thermal expansion. Thus upon heating of the film, a surface relief appears matching the image of the photomask (Figure 9.12).

The reported change in relief height are in the order of 1%, much smaller than the 5% deformation for an unpatterned freestanding cholesteric film as reported by Broer and Mol [12]. A detailed study by Elias et al. [20] gives more insight in the origin of this difference.

A detailed study by Elias et al. [20] gives more insight in the origin of this difference. In that study, the LCN was polymerized from a mixture of mono- and bifunctional reactive LC's that was optimized to reach large strains. Figure 9.13 shows the design and optical images of the samples in that study. For their material, they find that upon heating from room temperature to 200°C the height relief of the cholesteric versus the isotropic regions increases by 1.6%. A first approximation, using the measured freestanding deformations to predict the relief change would indicate an increase by 20% and thus overestimate the effect. The study shows three main factors for this difference. The surface anchoring of the sample has a major impact on the allowed deformations.

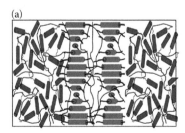
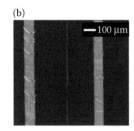
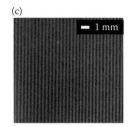

**FIGURE 9.13** Photopatterned LCN structures, schematically (a) cholesteric regions in surrounded by isotropic polymer network. Polarizing microscopy image of 100 μm wide cholesteric lines separated by 400 μm isotropic material (b) and 200 μm wide cholesteric lines separated by isotropic regions of the same width (c). (After Elias, A.L. et al. *J. Mater. Chem.*, 16, 2903, 2006.)

**TABLE 9.3**

**Measured Height Change of Samples Heated from Room Temperature to 200°C**

|  | Freestanding Film | Surface-Anchored Film |
|---|---|---|
| Deformation parallel to director axis | −21% | Homeotropic: 2% |
| Deformation perpendicular to director axis | 19% | Planar uniaxial: 9%; cholesteric: 11% |
| Isotropic | 5.7%[a] | 5% |

[a] Calculated value using the approximation $\varepsilon_{iso} = (\varepsilon_\| + 2\varepsilon_\perp/3)$.

Table 9.3 compares the deformations perpendicular and parallel to the director in freestanding samples and fixed films. Finite element calculations incorporating the anisotropic thermal expansion accurately predict the measured height change of the surface anchored film. The difference in deformation can thus be ascribed to the build-up of stresses in the surface anchored films. As illustrated by the planar uniaxial and cholesteric samples, the molecular alignment determines the compliance of a sample for a certain deformation.

Another factor influencing the deformation is the photopatterning process. This effect was measured by varying the pattern of films. A cholesteric polymer film was created with photolithography and the monomer in the unexposed area of the sample was washed away. The measured deformation is slightly less than the homogeneously exposed cholesteric sample in Table 9.3. Van der Zande et al. [21] have shown that under certain conditions, mass transport can take place upon localized exposure in the photocuring and therefore it is speculated that during photocuring, some of the order is lost due to diffusion of the monomers.

A last factor limiting the actual achieved strain is the embedding of the cholesteric areas in an isotropic "sea." Figure 9.13 shows a schematic and images of photopatterned lines in isotropic surroundings. The data in Table 9.4 predicts an increase in relief between the cholesteric and isotropic areas of 6%, the difference in expansion between isotropic and cholesteric surface anchored films, although only 1.6% is found. Part of this deviation from the expected value is due to the patterning effect as discussed above. The remaining deviation is thought to be due to the extra stress induced by the interface between the isotropic and cholesteric region.

**TABLE 9.4**

**Measured and Expected Height Relief Increase in Patterned Cholesteric Areas upon Heating to 200°C**

|  | Measured | Expected Deformation Based on Uniaxial Samples, Table III |
|---|---|---|
| Cholesteric patterned and etched areas | 8.4% | 11% |
| Cholesteric areas in an isotropic sea | 1.6% | 6% |

## 9.4   LIGHT-INDUCED DEFORMATION

For many MEMS applications, heat is not a practical stimulus because it is difficult to achieve fast reversible actuation in a wet environment and heat loss to the environment can be a problem. Light induced deformation offers an attractive alternative because it is fast and a remote light source can be used to drive the actuator. Azobenzene is a well-known compound to achieve photomechanical effects [22,23]. The azobenzene unit undergoes a *trans–cis* isomerization upon illumination with UV light. Finkelmann and coworkers [23] first showed that it is possible to amplify the conformational change of the azobenzene moiety to the macroscopic domain by incorporating azobenzene units in a nematic LCE. The shape changes are reversible and studies since have shown that it is possible to control the deformation direction by changing the polarization state of the light [24–26]. Similarly, in glassy LCN actuators, including azobenzene moieties shows the same effects but here the molecular alignment of the LC host offers an additional handle on the deformation direction.

It is believed that in systems with relatively low concentration of azo compounds the dimensional change is mainly caused by order reduction of the LCN, similar to the thermal systems. By cooperative interactions the shape change of the azo group affects the average orientation of the neighboring molecular units. At room temperature and under ambient light conditions, the azobenzene unit has a photostationary state that is predominantly in the elongated *trans* form. Upon illumination with light of a specific wavelength, this photostationary state will shift toward a state that is *cis*-dominated. Unlike the *trans*-state, the *cis*-state of the azobenzene moiety does not align with the LC host material and causes a decrease in the molecular order. Depending on the intensity and spectrum of the light source, thermal deformation may contribute significantly to the deformation. Due to the light absorption of the film, the film will heat up and deform like a thermal actuator. Because the host material of the network is the same as that of a thermal actuator, the responses to temperature are very similar.

The wavelength of the peak absorption of the *trans*- and *cis*-state of the azobenzene chromophores depends on the exact molecular structure. For the AxMA system depicted in Figure 9.14, the absorption peaks are around 360 nm for the *trans* state and at 320 and 440 nm for the *cis*-state (Figure 9.15). Typically a pure azobenzene monomer will absorb 90% of the light within the first micrometer.

LCNs incorporating AxMA monomer show strain responses that are in the order of a few percent. Figure 9.16 shows the response of a uniaxially aligned film thermostatted in a water bath, parallel and perpendicular to the director. In this case, the film is cycled with UV and visible light, with a power of $100\,mW\,cm^{-2}$ for the UV and $10\,mW\,cm^{-2}$ for the visible light.

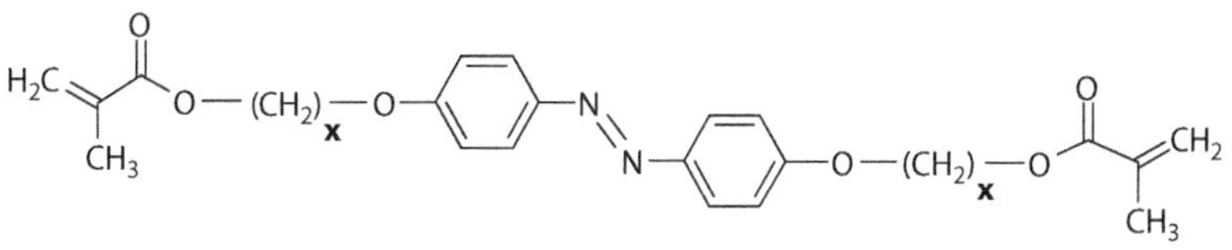

**FIGURE 9.14**   Structure of azobenzene containing monomer AxMA, where x indicates the length of the flexible spacer unit.

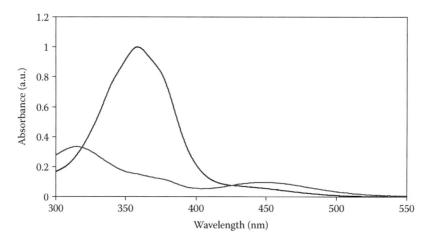

**FIGURE 9.15** Absorption spectra of the monomer A6MA in solution after exposure with visible light (a) corresponding to a preferentially *trans*-state of the monomer, and after exposure with UV light (b) corresponding to a preferentially *cis*-state of the monomer.

After the first cycle, the response is repeatable. The UV response in the first cycle is somewhat larger than the consecutive cycles. This effect is often seen in LCNs and is thought to be free volume effects due to the polymerization shrinkage and small reorganizations of the molecules because of steric effects. It is important to stress that the *cis–trans* isomerization of azobenzene in the LCN is only partly responsible for the experimentally observed deformation. Owing to the light absorption, the film will heat up leading to thermally induced deformations, despite the heat transfer to

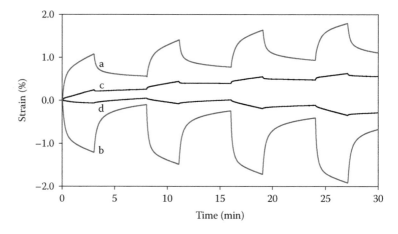

**FIGURE 9.16** Strain response to UV input on a uniaxial planar film made from 8% A3MA perpendicular and parallel to the molecular alignment (a and b) in a LCN host and the response of the pure LC host perpendicular and parallel (c and d) to cycled UV illumination of 100 mW cm$^{-2}$ in water.

the water. To visualize this effect, Figure 9.16 shows the heating of the LCN without azobenzene dye under the same illumination program. In addition to an overall drift due to slight heating of the water bath, the LCN without dye shows a small response to the cycling of the light.

When this temperature effect is subtracted from the measurements, the maximum strains display a linear dependence on the azobenzene concentration for dye loads up to 10% (Figure 9.17). Similarly, the deformation speed scales linearly with the light intensity [26].

Upon illumination, the presence of azo-dye in the film will cause the light to attenuate through the thickness of the film, creating a gradient in the magnitude of the strain through the thickness of the film. When illuminated from the top or bottom, a planar uniaxially aligned film with azo-dye therefore, bends toward the light source. If splayed or TN films are used, the deformation gradient will be much larger and the film can be made to bend to or from the light source as dictated by the molecular alignment in the material. This can be visualized using a splayed system with the molecular director at the top of the film parallel to the length direction of the film and thus homeotropic alignment at the bottom. The contraction at the side of the light source has the same magnitude for both the planar uniaxial as the splayed film. At the bottom of the film the splayed film will expand in the plane of the film, whereas the planar film will contract. This situation is sketched in Figure 9.18. For the same concentration of azobenzene, a splayed film will therefore display sharper bending than a planar film. A similar strain profile can be drawn for the TN case, but for a TN film the three-dimensional strains cause a more complex bending behavior.

In addition to a steeper strain gradient, the splayed system also dictates the deformation direction and the bending axis thus becomes independent from the location of the light source. When the film is in position with the homeotropic side toward

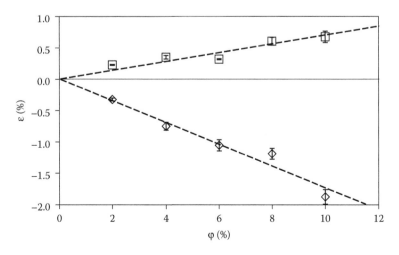

**FIGURE 9.17**  Maximum strain difference between UV-illuminated state and VIS-illuminated state, for increasing A3MA content, perpendicular (□) and parallel (◇).

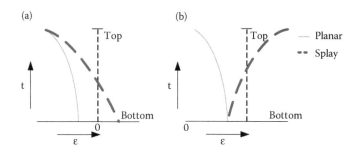

**FIGURE 9.18** Schematic illustration of the photogenerated in-plane strains ε through the thickness of the film for a planar uniaxial and a splayed system with the planar alignment at the top of the film (a) and a splayed system with the homeotropic alignment at the top of the film (b).

the light source, the film will bend away from the source. Figure 9.19 shows the responses of films with varying alignments upon illumination with UV light.

Similar to the thermal actuators, these systems are glassy at room temperature, with a room temperature modulus of 1.3 GPa parallel to the director and 0.6 GPa perpendicular to the director. The responses of the splayed and the TN systems are fast, with the response to the UV-illumination of $100\,\mathrm{mW\,cm^{-2}}$ reaching 70% of its final deformation in about 0.25 s [27]. The backward relaxation under ambient conditions is significantly slower, in the order of 15 s. The speed of this backward relaxation can be increased by illuminating with visible light to address the absorption band of the *cis*-state, or by increasing the temperature. The TN and splayed systems show similar responses in speed as well as in final bending radius. When not restricted by the clamp, the TN system displays a second bending axis similar to what has been reported in the thermal systems.

At a large length to width ratio, a TN film cut at 45° with the alignment direction of the film surfaces will display a helical coiling behavior upon illumination. Figure 9.20 shows the response of such a film upon illumination with UV light. This response illustrates the control over the bending behavior through the molecular structuring of the material, as the film first bends away from the light before it again curls toward the light.

In terms of light input, the splayed system presents a more efficient bending mechanism than the planar system. The planar–uniaxial system loses part of its energy in a length contraction that does not contribute to the bending of the beam. Put in another way, the splayed system will achieve the same stationary bending state upon illumination at a lower concentration of azobenzene dye.

## 9.5 CHEMICAL AGENT-DRIVEN SYSTEMS

In applications such as lab-on-a-chip systems or molecular medicine, reliable stimuli-response materials are of major importance. The delivery of drugs can be achieved by opening of a reservoir upon reaching a certain concentration of disease-specific molecules, for example. In on-chip diagnostics, improved schemes for fluid mixing

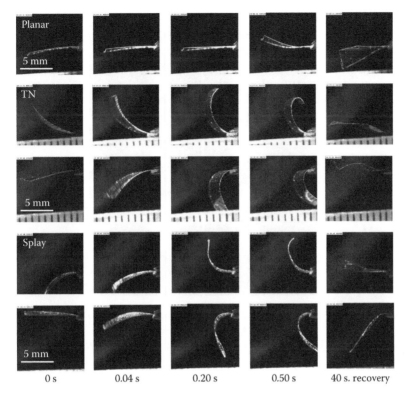

**FIGURE 9.19** Responses in air of films containing 2% A3MA with varying director alignments illuminated from the top. The film dimensions are 10 μm thick, 4 mm wide and 7 mm long. After switching off the UV light, the films were left to recover at ambient conditions. (After Van Oosten, C. L. et al. *Eur. Phys. J. E.*, 23, 329–336, 2007.)

or flow control upon detection, or in response to a molecular trigger are considered beneficial [28–30].

Harris et al. [31] showed that it is possible to incorporate responsiveness to chemical agents into LCNs. Because of the liquid crystalline character of the networks, deformations are anisotropic. The systems are stiff and glassy and can be patterned using microstructuring techniques and are therefore well-suited for use as actuating elements in MEMS devices. As the system only takes up moderate quantities of solvent under equilibrium conditions, the material differentiates itself from a traditional hydrogel. Furthermore, it can be processed in the same manner as the thermal or photomechanical actuators described in previous sections of this chapter, allowing for photopatterning, templating, and soft lithography.

In order to achieve responsiveness to chemical agents, chemical bonds are required that can be reversibly broken upon contact with the agent. Monomers are used that are capable of forming noncovalent bonds such as hydrogen bonds. In their study, Harris et al. used monomers that are not liquid-crystalline by themselves, but are capable to pair under the formation of hydrogen bridges, thereby forming the rod-like structure

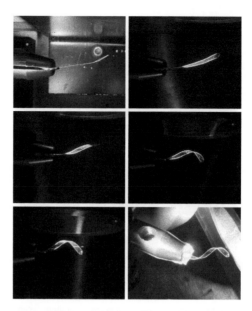

**FIGURE 9.20** UV-induced coiling of a photomechanical actuator. The film has a TN molecular alignment and is composed of 8% A6MA in a C6M host. This deformation is completed in roughly 30 s. The film dimensions are 24 mm×2 mm×17 μm. (After Harris, K. D. et al. *J. Mater. Chem.*, 15, 5043, 2005.)

necessary to induce liquid-crystalline behavior. Figure 9.21 shows the monomers used in the study. The breakable bonds are formed between the carboxylic acid units of the monomers.

Using a photoinitiator, the *n*OBA monomers are polymerized into a network. For stability, permanent cross linker C6M is used in small quantities (12% wt). To decrease the crystallization temperature and widen the temperature range of the nematic phase, monomer *n*OBA is used as a mixture of three homologues with different aliphatic tail lengths ($n = 3, 5$, and 6) in a 1:1:1 weight ratio.

After polymerization, a hydrogen bonded network is obtained that is highly ordered and rigid. The hydrogen bonds are sufficiently strong to withstand immersion in water or most solvents without significant deformation of the network. Talroze et al. [32] performed a mechanical analysis of a system polymerized in the smectic phase. For small strains, the system shows an almost linear stress–strain response. At higher strains, the hydrogen bonds are broken and the layers in the smectic system are

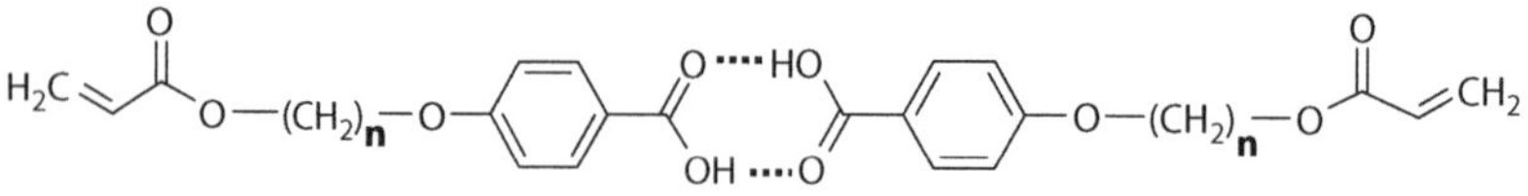

**FIGURE 9.21** Paired monomers *n*OBA with hydrogen bridges. "*n*" indicates the length of the aliphatic chain in terms of the number of carbon atoms. Typically, monomers are used with $n = 3, 5$, and 6 in weight ratio 1:1:1.

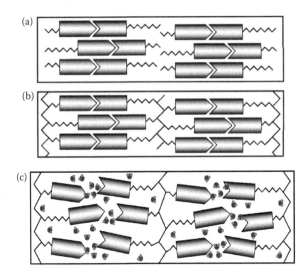

**FIGURE 9.22**   Sketch of the formation of the aligned polymer salt poly($n$OBA/C6M)$^-$K$^+$.
The paired monomers are aligned in the nematic phase (a). Upon photopolymerization a highly
ordered network is created (b). In an alkaline environment, the hydrogen bonds are broken,
reducing the order of the network and causing macroscopic anisotropic swelling (c).

able to slide without any extra stress applied to the system. In sufficiently alkaline
conditions, the hydrogen bonds are broken and the network is converted into a polymer
salt. Upon reduction of the number of hydrogen bonds, the order of the network drops,
causing a macroscopic, pH-induced deformation. The hydrogen bonds can be again
restored by exposure to acids, bringing back the film to its original shape. Figure 9.22
schematically illustrates the formation and the response of the network.

The conversion into a polymer salt "activates" the system: the network becomes
much more hygroscopic and responds rapidly to a number of chemical agents. In
the studies by Harris et al. [31,33], an aqueous solution of pH13 KOH is used to
activate the networks. The networks obtained after this acid–base reaction will be
referred to as poly($n$OBA/C6M)$^-$K$^+$. The acid–base reaction was limited in time,
typically 10–20 s, such that only a fraction of the hydrogen-bonded units are broken
down. Typically after 10 s, the optical retardation is reduced by 30–40%, indicating a
reduction in the order parameter. Mechanically, the modulus of the material changes
upon KOH immersion. Although the modulus at room temperature parallel to the
molecular director decreases slightly from 3.9 to 2.7 GPa after 10–20 s of KOH treatment, the modulus of poly($n$OBA/C6M)$^-$K$^+$ perpendicular to the director increases
from 1.2 GPa to about 1.6–1.9 GPa. Ionic bonding and the resonant character of the
carboxylate ions are thought to contribute to the mechanical strength of the network.
For comparison, the typical room temperature moduli of a pure C6M network are 1.5
and 0.9 GPa for the parallel and perpendicular directions, respectively.

The mechanism of deformation is best illustrated using the actuator response to
water. Figure 9.23 shows the response of a uniaxially aligned film upon full submer-
sion in water, perpendicular and parallel to the director. Water swells the network,

but the swelling in the directions perpendicular to the director is significantly greater than the swelling parallel to the director. In the acid–base reaction in aqueous KOH solution, the carboxylic acid groups are deprotonated, resulting in water in the hygroscopic salt network. In the drying step after KOH immersion, water is only partially removed from the network, leaving a substantial amount of water behind due to the hygroscopic nature of the film. When the film is reimmersed in water, the network is hydrated again and the material swells anisotropically.

Although the networks may be strongly hydrophilic, they are only mildly solvophilic for the less polar solvents. If the films, after air drying, are immersed in a water-miscible solvent for which the network has lower affinity, remaining water is extracted from the film. Because the water is not replaced, the films shrink, with the largest contraction perpendicular to the director. If the film is instead immersed in a water-immiscible solvent such as toluene, the polar components in the film will remain there and little contraction is observed. Figure 9.23 compares the responses of a sample to immersion in water and a number of solvents.

The preferential affinity of the polymer salt to water and acetone, a polar and water miscible solvent, is shown below. In pure acetone, the remaining water in the film is extracted and the film shrinks preferentially in the direction perpendicular to the director. When water is added to the acetone, the film swells again. Figure 9.24 shows this behavior for a film of uniaxially aligned poly($n$OBA/C6M)$^-$K$^+$.

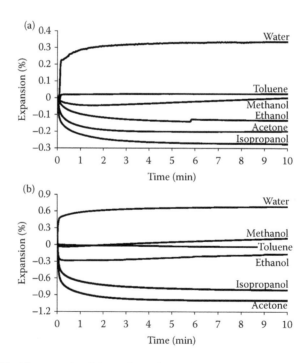

**FIGURE 9.23** Macroscopic deformation of a uniaxial aligned network poly($n$OBA/C6M)$^-$K$^+$ in response to various solvents. The response parallel to the director (a) is typically smaller than the response perpendicular to the director (b).

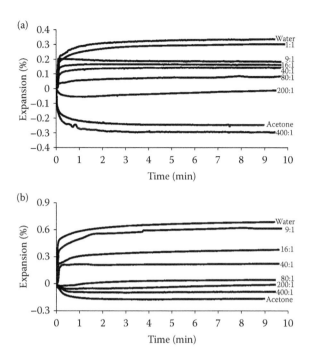

**FIGURE 9.24** Responses of a uniaxially aligned poly($n$OBA/C6M)$^-$K$^+$ film in response to a chaning environment, from a pure acetone environment to a pure water environment, parallel to the molecular director (a) and perpendicular to the molecular director (b). The ratios display the proportion of acetone to water. (After Harris, K.D., C.W.M. Bastiaansen, and D.J. Broer, *Macromol. Rapid Comm.*, 27, 1323–1329, 2006.)

The anisotropic deformations of the agent-responsive networks can be used to construct bending mode actuators. There are two distinct mechanisms with which bending can be achieved. In environments where there is a strong gradient in chemical agent present that swells the film, for example, water vapor, a planar uniaxial film will swell on the side with the high concentration, bending away from the vapor source. Figure 9.25 shows the response to a gradient of water vapor. When the other side of the film is exposed to water vapor, it also bends away from the vapor source as is consistent with anisotropic swelling.

**FIGURE 9.25** Response of a 17 µm thick film of poly($n$OBA/C6M)$^-$K$^+$ to a gradient in water vapor from the vial below. In ambient conditions, the film is almost flat. When the film is brought close to the vapor source, it quickly bends. When the film is removed, the original position is recovered. The water in the vial is at room temperature.

When there is no gradient in the chemical agent present, for example when the system is completely immersed in water, a planar uniaxial film will not bend. In analogy to the thermal and UV-driven systems, bending can be obtained by creating a gradient in the molecular director using TN or splayed alignments. After the photopolymerization of the film on the substrate, the films shrink during the partial dehydrogenation. Due to the anisotropy of the shrinkage, TN and splayed films are therefore curled in their neutralized state. When a chemical agent, in this case water, enters the film, the film swells and unrolls. Figure 9.26 shows the response of a twisted film to changes in the relative humidity of an environment.

Using the splayed or TN alignments, the solvent selective responses as described above can be used to create bending mode actuators. Immersed in water, a twisted sample expands, with the largest expansion perpendicular to the director. Due to the director gradient, the film bends. If the film is then immersed in a different, water-miscible solvent, osmotic pressure differences between the film and the medium cause the extraction of water from the film. The water in the film is not replaced and thus the material shrinks anisotropically, causing the film to bend in the opposite direction. In a solvent that is nonmiscible with water, the water remains in the water soaked film and little response is observed. Figure 9.27 shows this behavior for a number of solvents. For every two images a different sample was used. The film was first completely submersed in water and then transferred into a solvent. The polar solvents, methanol, ethanol, and acetone, show a very pronounced response. The nonpolar solvents, toluene and xylene, show little effect on the bending of the film.

## 9.6  INKJET PRINTING ACTUATORS: TOWARD POLYMER MEMS

We have demonstrated that LC polymers are able to show large and reversible shape deformations in response to a variety of applied stimuli such as heat, light and a change in chemical environment and light. They are therefore, attractive candidate materials for making microactuators: early reports already mention the potential of light-driven actuators for MEMS devices [34] and lab-on-a-chip applications [35]. Numerous MEMS components can already be fabricated from polymers; however, the large-scale manufacturing of polymer microactuators using, for instance, roll-to-roll processing on a single substrate still faces a few challenges. The polymer actuators reported so far are hardly suitable for microstructuring [8,9]: for instance, the multilayer structure of conjugated polymer microactuators complicates the production [3]. A further challenge is that the actuators are often electrically addressed, requiring inorganic (metal) electrodes, thus leading to extreme manufacturing conditions, that

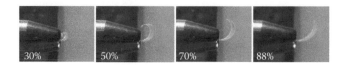

**FIGURE 9.26**   Response of a TN poly($n$OBA/C6M)$^-$K$^+$ film to changes in relative humidity in a homogeneously humid air environment. (After Harris, K.D. et al., *Nano Lett.*, 5, 1857–1860, 2005.)

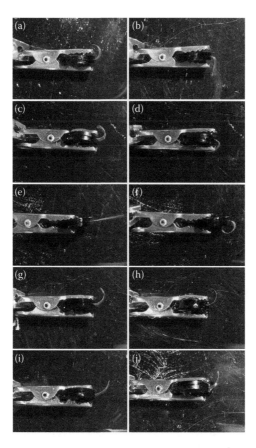

**FIGURE 9.27** Response of a TN actuator of poly($n$OBA/C6M)$^-$K$^+$ films in various solvents compared to their state in water: (a) water, (b) methanol, (c) water, (d) ethanol, (e) water, (f) acetone, (g) water, (h) toluene, (i) water, (j) xylene. For every two images a new sample was used. (After Harris, K.D., C.W.M. Bastiaansen, and D.J. Broer, *Macromol. Rapid Comm.* 27, 1323–1329, 2006.)

is, processing occurs at high temperatures and/or in acidic or basic environments [7]. LCN actuators can overcome these challenges: all-polymer microdevices can be fabricated using inkjet printing technology in combination with self-organizing LCN actuators. The self-assembling properties of the LC allow the creation of large strain gradients, and light-driven actuation is chosen to enable simple and remote addressing. A simple actuator design is adopted, mimicking cilia in nature to act as active pumps or mixers in microfluidic systems [36]. Then, processing by inkjet printing and device integration is demonstrated. By using multiple inks, microactuators with different subunits are created that can be selectively addressed by changing the wavelength of the light [37]. Large deformations of the actuator structure are achieved, and it is shown that the different substructures can be independently addressed using different colors of light.

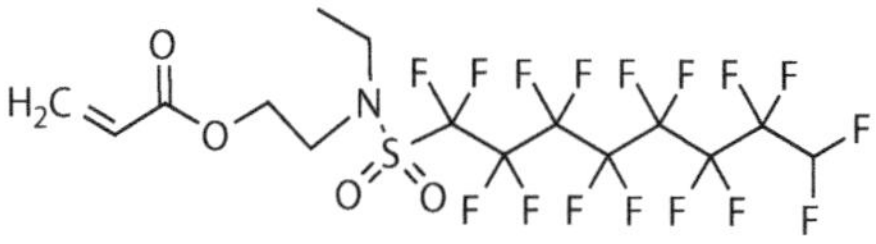

**FIGURE 9.28** The monomers used to form the inkjet printed actuators. The two photoactive monomers are A3MA and DR1A. The host matrix is built from C6M, C6BP, and C6BPN. The surfactant PS16 is used to obtain the desired alignment at the air interface.

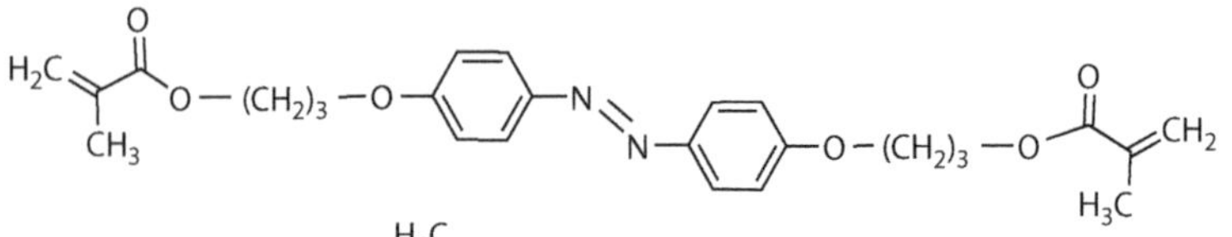

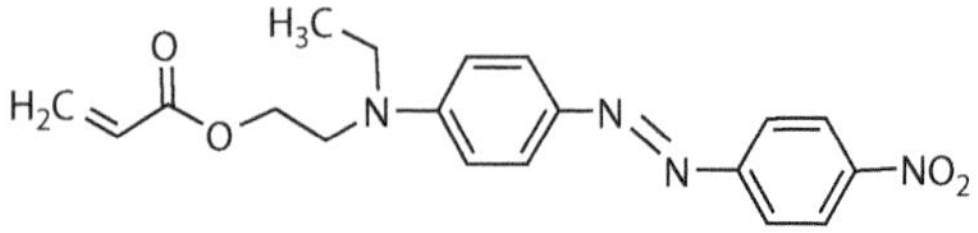

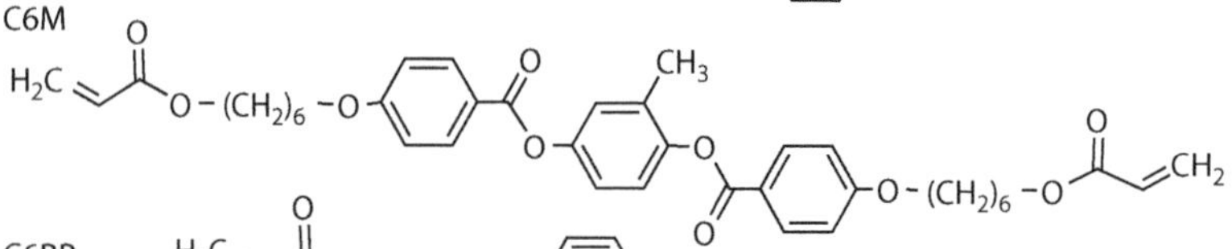

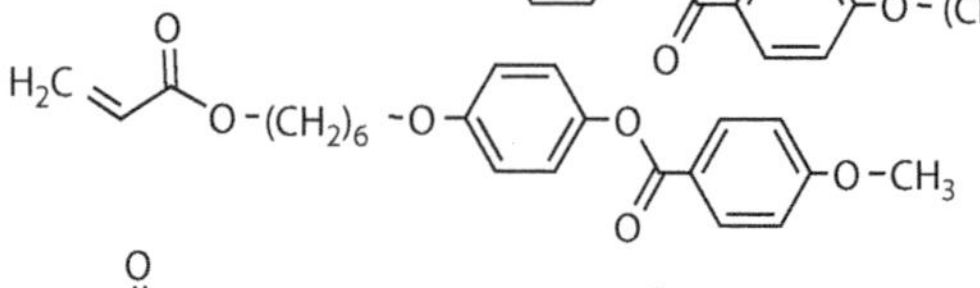

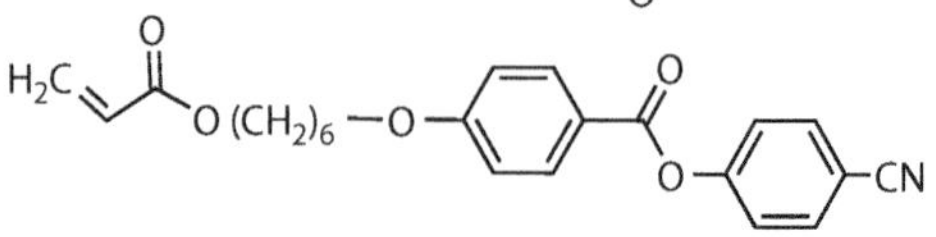

The most common way of obtaining light-driven order change in LC polymers is through the inclusion of molecules incorporating the azobenzene unit. Azobenzene molecules are able to undergo the *trans–cis* isomerization upon irradiation with light. There are a number of azobenzene dyes that can cause shape deformations when included in an LCN or LC rubber [24,25,34,38,39]. Depending on their chemical structure, they have absorption bands for the *trans*-state that vary from UV-light to visible (green) light. Here two dyes are used, A3MA and DR1A (Figure 9.28). A3MA has an absorption maximum at 358 nm for the *trans*-state; whereas the *trans*-absorption band of DR1A is at 490 nm. The optimal dye concentration depends on the molar absorption of the dye and the thickness of the film [40]. An azobenzene concentration of 4 wt% for the A3MA dye and 1 wt% for the DR1A dye was chosen, the concentration being high enough to give a significant response, but low enough so that in a 10 μm thick film the light penetrates deeply enough to affect more than just the top layer.

The host matrix of the actuator is based on a combination of LCs functionalized with acrylates or methacrylates. Both of these compounds can rapidly photopolymerize using a radical initiator while preserving the molecular alignment from the

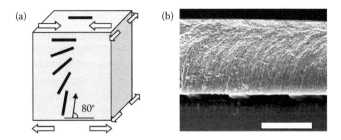

**FIGURE 9.29** The splay-bend molecular organization through the thickness of the film. (a) Schematic representation of a splay-bend molecular orientation through the thickness of the film. The arrows indicate the direction of the material response upon actuation. (b) The molecular alignment is visible in a cross-section of the film that is broken in half. The scale bar indicates 5 μm.

monomers into the polymer [6]. With this one-step manufacturing route, an optimized molecular alignment can be obtained in the monomer phase so that the desired motion of the actuator is obtained [27]. In many actuators, bending is achieved by creating a bi-layer structure, using two materials with a different degree of responsiveness. Here, the self-organizing and anisotropic properties of the LC were exploited to create an internal gradient in a single layer. By changing the director orientation through the thickness of the film in a splay-bend configuration, significantly stronger bending can be achieved than with a uniaxially aligned film of the same chemical composition. A splay-bend configuration is presented in Figure 9.29a: the liquid crystalline units undergo a gradual orientational change through the thickness of the film, from perpendicular to the substrate at the bottom, to parallel at the top. In the polymerized state the splayed molecular alignment dictates both the bending direction and the bend axis of the film, making the direction of the response independent from the direction of the incident light. To facilitate manufacturing the mix of reactive LCs has a balanced ratio of LC elastic constants so that in the monomeric state a splay-bend configuration is energetically permitted using surface alignment techniques on a single substrate. This mixture of monomers serves as a host material for the dyes: the resulting polymer will be referred to as A3MA polymer or DR1A polymer, depending on the dye that is included.

An essential step toward achieving large amplitude actuators is controlling the self assembly of the LCs so that an upward-bending (rather than a downward-bending) splayed actuator is created. To achieve this, the molecular director should be pointing perpendicular to the substrate at the bottom and parallel to the substrate at the top of the film, causing the bottom of the film to expand and the top to contract (Figure 9.29a). Under unforced conditions, the mix of LC monomers will align perpendicular to the air interface, minimizing its surface-free energy. To overcome this, in-plane alignment at the air interface is obtained by adding a small amount of surfactant PS16 to the reactive monomeric mix. The directionality of the alignment at the air interface is introduced by giving the molecular alignment at the bottom surface a slight pre-tilt from the surface normal. A molecular anchoring of 80° with respect to the substrate is obtained by the unidirectional rubbing of a homeotropic-aligning polyimide layer, which reproduces the desired molecular alignment. The alignment becomes visible by breaking the sample and inspecting the fracture surface using SEM (Figure 9.29b). Inkjet printing was chosen as a process for microstructuring the actuators because it allows variations of the material composition in the plane of the substrate in a single processing step, while alternative micropatterning techniques such as lithography require many more steps to achieve the same [41]. Using a commercial inkjet printer, the monomeric LC mix containing one of the two dyes was deposited on the substrate. In our case, the reactive monomeric mix had a crystalline–nematic transition just above room temperature, and the mix was therefore printed using a solvent.

The goal of this procedure is to produce artificial cilia that are driven by light for use as mixers or pumps in microfluidic systems. Artificial cilia are nature-inspired structures and mimic the behavior of cilia that are, for instance, used by paramecia to move through fluids (Figure 9.30a). An individual cilium makes a flapping, asymmetric motion, with a backward stroke different from the forward stroke, causing an effective flow in the surrounding fluid (Figure 9.30b). Artificial cilia can provide effective mixing in microchannels [36]. Using a single dye, asymmetric motion can be introduced into LC azobenzene actuators by varying over time, the light intensity over the actuator surface. Given the small dimensions of the actuator and the large deformations, this is difficult to realize in practice. Instead, using two dyes, varying the composition of the actuator in the plane, is proposed (Figure 9.30c). By selecting the color composition of the light, the position of the actuator can be brought into four positions: in the dark, the actuators are flat. With only UV light, the yellow A3MA dyed part of the actuator bends. At the same time, any DR1A in the *cis*-state absorbs (with peak absorption at 360 nm) and is therefore actively brought to the unbent *trans*-position. When both UV and visible light are used, the flap bends over its total length. Finally, if only visible light is used, only the red DR1A part of the flap bends. Simultaneously, *cis*-A3MA (with peak absorption at 440 nm) absorbs and is therefore actively driven to the flat, *trans*-dominated state. Switching between these four positions produces a cilia-like motion. The weak *cis* absorption in both dyes adds to the thermal-driven recovery to the flat state, speeding up the rate at which the illumination cycle can be run.

To demonstrate the effect on a macroscopic scale, two strips of A3MA polymer and DR1A polymer were cut and glued together at one end. Care was taken to orient the films so that they both had the same molecular alignment and that the strip was cut parallel to in-plane alignment of the molecules. Figure 9.30d provides a schematic representation of the setup and the alignment, with the DR1A polymer mounted at the top and the A3MA polymer mounted at the bottom.

The strip was actuated in air using light of different colors, resulting in the steady state responses shown in Figure 9.30e. When the strip was illuminated with visible light, 455–550 nm, the top part with the DR1A polymer showed a small bend. Then, the strip was illuminated with a combination of UV and visible light, resulting in rapid and strong bending over the whole length, assuming a final state completely bent into the light. With only UV illumination, the top part of the film remained flat and only the bottom part (the A3MA polymer) of the film was bent. With the light switched off, the film completely reverted to its original position. This shows the desired motion of the actuator.

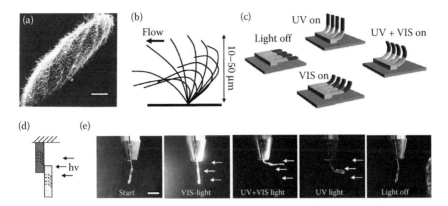

**FIGURE 9.30** Natural and artificial cilia and their motions. (a) Cilia can be found on several micro-organisms, such as paramecia (scale bar 20 μm). (b) A paramecium uses the beating motion of the cilia, characterized by a different forward and backward strokes, for self-propulsion. (c) Artificial, light-driven cilia produce an asymmetric motion controlled by the spectral composition of the light. (d) Schematic representation of the macroscopic setup, showing the orientation of the molecules. (e) Steady-state responses of a 10 μm thick, 3 mm wide and 10 mm long modular LCN actuator to different colors of light (scale bar 5 mm).

To manufacture free-standing miniature actuators, four basic process steps were taken (Figure 9.31). As a release layer, a 1 μm thick polyvinyl alcohol (PVA) layer was used. PVA was chosen because it is soluble in water but not in the solvents used for the processing of the polyimide or the LC monomers. Consequently, the sample was coated with the "soft" homeotropic aligning polyimide and treated to create the desired molecular pretilt angle of 80° to the surface.

The monomeric LC mix was then deposited with the inkjet printer, using separate cartridges for the monomeric mixtures with the two different dyes. The reactive monomers can be cured after printing every separate color, or after printing the full structure, depending on whether a distinct separation of the colors or a strong adhesion

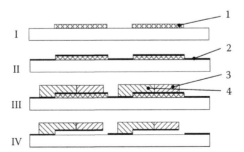

**FIGURE 9.31** Overview of the four basic processing steps to produce the modular cilia. (I) Structured deposition of the PVA release layer 1. (II) Spincoating, curing, and buffing of polyimide alignment layer 2. (III) Inkjet deposition of the monomer mixes containing DR1A 3 and A3MA 4 and curing. (IV) Dissolving the PVA release layer.

between the two parts is desired. For the remainder the first approach was followed for practical reasons: with the available printing setup, only a single mix could be printed at one time. Polymerizing directly after printing helps to preserve the morphology during the exchange of a printer cartridge. The cilia are released from the substrate by dissolving the PVA layer in water. The polyimide layer is thin enough that it breaks, and in some cases it was observed that the polyimide broke upon first actuation of the cilia. The printing process is well controlled and different sizes can be printed (Figure 9.32a). Inspection in the SEM reveals that the structures are indeed partially free standing on the substrate (Figure 9.32b).

A number of microactuators are shown in Figure 9.33a–d. The printed structures have a thickness of around 10 μm at the location of the freestanding flaps, being slightly thinner at the base than at the top. It is believed that the combination of the print geometry and the drop coalescence causes this height difference. At the base of the flaps, where the structures are permanently attached to the substrate, the printed layer has a thickness of 20 μm. In this thicker layer, the monomers do not form a monodomain-oriented structure, as is visible from the crossed polar images (Figure 9.33b). Because this part of the structure does not contribute to the motion, the poor alignment of the base is not expected to harm the performance of the actuators. At smaller line widths, the structures become somewhat irregular due to dewetting of the monomers on the substrate. With the current setup, using a drop spacing of 15 μm, the minimum line width needed to produce a bending flap is 100 μm, but it is expected that this can be decreased with further optimization of the printer, substrate, and monomeric mix to a few tens of micrometers.

The response of A3MA polymer cilia in water to cycled light is shown in Figure 9.33e. All flaps bend in response to illumination with UV light. The forward, UV-induced response is much faster than the thermally driven back reaction. The actuation of the cilia was repeatable without loss of the stroke length. The experiments were carried out in water to minimize initial sticking of the cilia to the substrate and limit the thermal heating from the light source to a maximum increase of 2°C. The difference in responses between the different cilia is mainly attributed to the sticking to the substrate and variations in the thickness of the cilia.

(a)                                                    (b)

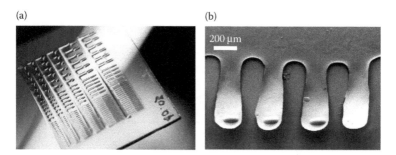

**FIGURE 9.32** (a) Inkjet printed artificial light-driven cilia before their partial release from the substrate. The polymeric, light responsive actuators, with the darker parts being sensitive to green light and the light areas to blue or UV light, are structured in different sizes on a glass substrate partly provided with a sacrificial release layer. (b) SEM image of free-standing LCN actuators.

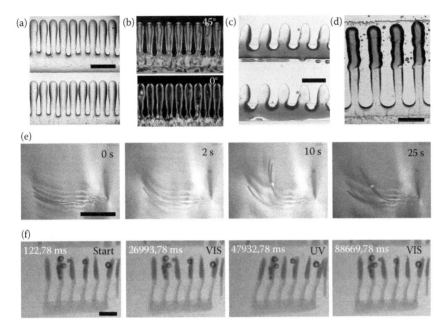

**FIGURE 9.33** Microstructured cilia and their response to light. (a) Shows arrays of inkjet printed A3MA polymer flaps on PVA. (b) Crossed polar microscopy with the sample at 45° and 0° shows that the splayed alignment in the flaps is parallel to the length of the flap. (c) Cilia can be manufactured with a gradient in composition from DR1A to A3MA polymer. (d) Alternatively, cilia are made with two separate parts of DR1A and A3MA polymer by separately polymerizing the two parts. (e) Side view of the actuation of A3MA polymer cilia with UV light (1 W/cm$^2$) in water. (f) Frontal view of actuation of multicolor cilia in water addressed with VIS (4 mW/cm$^2$) and UV (9 mW/cm$^2$) light. All scale bars indicate 0.5 mm.

The response of the dual response material printed cilia is shown in Figure 9.33f. When illuminated with visible light, the DR1A part of the actuators show a small bending. Upon illumination with UV light, the A3MA polymer part bends strongly. Visible actuation drives the A3MA polymer back to its original state and bends the DR1A part again. With the current setup and actuators, movements are still too slow to induce measurable flow. It is expected that 10 times faster movements can be achieved with thinner actuators and higher light intensities. For asymmetric cilia movements, a slight frequency change of a few Hertz will provide effective fluid manipulation in microsystems [42].

To characterize the steady-state light response, uniform flaps with only A3MA dye or only DR1A dye were produced. The incident intensity of the light was varied while a few cilia were monitored using optical microscopy. The images were analyzed by fitting the bend radius of the cilia with a circle. At maximum intensity, the A3MA polymer cilia show a full bend with respect to the surface at an intensity of 2 W/cm$^2$, with an average bending radius of 220 μm. To compare the performance of the inkjet-printed flaps with bulk material, the strain difference is estimated using $\Delta\varepsilon = t/r$, where $t$ is the film thickness, $r$ the bending radius and $\Delta\varepsilon$ the strain difference between

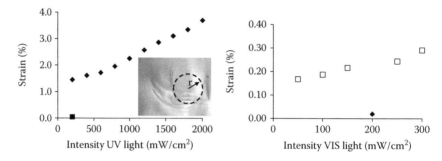

**FIGURE 9.34**   Strain response of printed cilia. Calculated strains from A3MA polymer (◇) and DR1A polymer (□), in response to UV light (250–390 nm), randomly polarized, and to visible light (488 nm), with polarization state parallel to the molecular director. The inset shows how the bending radius was measured by fitting a circle to the bending part of the flap.

the top and the bottom of the film. Figure 9.34 shows this strain as a function of intensity UV and visible light. At an intensity of 100 mW/cm$^2$, the average bend radius for A3MA flaps is 0.7 mm, giving a strain difference of 1.4%. For bulk material, the deformation in a film of 10 μm was measured at the same intensity, and found to be −0.75% and 0.35% for the directions parallel and perpendicular to the director [27]. For a splayed orientation, the predicted in-plane strain difference is thus 1.1%, closely matching the deformation of the microactuator. The slightly better performance of the printed flaps than the bulk film is ascribed to an overestimation of the strain due to thickness variations.

The response of the DR1A polymer to visible light was significantly less than that of the A3MA for UV light of the same intensity. This is due, in part, to the fact that the molar concentration is 3.16 times lower for the DR1A dye than for the A3MA dye in the polymer host. Also it is speculated that the molecular structure, and, in particular, the fact that the DR1A dye acts as a pendant group on the acrylate backbone rather than as a cross linker, results in a less efficient strain transfer to the host material.

Finally, it was verified that the A3MA and DR1A polymers can be addressed independently, by actuating the A3MA polymer flaps with visible light and the DR1A flaps with UV light (Figure 9.34). In both cases, the response of the flaps was minimal, confirming that the response is indeed wavelength specific. The simple addressing of the cilia by changing the color of the light allows a well controlled motion that has the potential to be more effective in mixing than the electrostatic driven counterparts that can only perform an asymmetric motion [36]. By actuation with light, this design overcomes the inherent problems related to the use of electrical fields in wet environments.

## 9.7   CONCLUSION

LCNs provide a platform for a range of actuators for miniaturized systems. Their one-step processing from the monomers to the polymer is fully compatible with current MEMS manufacturing processes and allows actuators to be manufactured *in situ* using photopolymerization. Polymerization in the aligned nematic phase results in a strong anisotropy in the properties of the material, which can be tuned through the control over the molecular alignment of the monomers before polymerization. The anisotropy of the material dictates the direction of the deformation of the actuator. A number of mechanisms can drive the deformation of the material, including a reduction of the molecular order, anisotropic swelling of the material using chemical agents, or amplification of a molecular confirmation change through the LC host. Actuators responding to temperature, UV light, water vapor, pH change, solvent polarity, or water content were made. The high cross-link density of the networks allows for only moderate in-plane strains, but due to the high modulus of the material they have a large work potential. Bending mode actuators are created using alignments such as TN or splay, where the direction of the twist or splay dictates deformation direction. Patterning of aligned submillimeter structures using photolithographic techniques was demonstrated for thermal actuators. Furthermore, in-plane patterning of the alignment of thermal actuators was demonstrated, creating in-plane variations of the response. Inkjet printing of the actuators opens new possibilities, such as color printing in order to vary actuator responsiveness in the plane. Using reactive LCs, the actuators can be made in only a few processing steps, exploiting the self-assembling capacity of the material. The processing allows fabrication of large-area and roll-to-roll active all-polymer devices and opens up possibilities for rapid prototyping of low-cost MEMS. We conclude that LCNs are a promising technique for the production of microactuators.

## ACKNOWLEDGMENT

Some of the work presented here was supported by the Dutch Polymer Institute under project number 532.

## REFERENCES

1. Gad-el-Hak, M. (ed.), *The MEMS Handbook*, CRC Press, Boca Raton, FL, 2002.
2. Quake, S.R. and A. Scherer, From micro- to nanofabrication with soft materials, *Science*, 290, 1536–1540, 2000.
3. Zhang, Q.M., H.F. Li, M. Poh, F. Xia, Z.Y. Cheng, H.S. Xu, and C. Huang, An all-organic composite actuator material with a high dielectric constant, *Nature*, 419, 284–287, 2002.
4. Pelrine, R., R. Kornbluh, Q.B. Pei, and J. Joseph, High-speed electrically actuated elastomers with strain greater than 100%. *Science*, 287, 836–839, 2000.
5. Jager, E.W.H., E. Smela, and O. Inganas, Microfabricating conjugated polymer actuators, *Science*, 290, 1540–1545, 2000.
6. Broer, D.J. in *Polymerisation Mechanisms*, Vol. 3 (Eds. J.P. Fouassier, and J.F. Rabek), Elsevier Applied Science, London, 1993, Ch. 12.
7. Smela, E., O. Inganas, and I. Lundstrom, Controlled folding of micrometer-size structures, *Science*, 268, 1735–1738, 1995.
8. Madden, J.D.W., N.A. Vandesteeg, P.A. Anquetil, P.G.A. Madden, A. Takshi, R.Z. Pytel, S.R. Lafontaine, P.A. Wieringa, and I.W. Hunter, Artificial muscle technology: Physical principles and naval prospects, *IEEE J Oceanic Eng.*, 29, 706–728, 2004.
9. Mirfakhrai, T., J.D.W. Madden, and R.H. Baughman, Polymer artificial muscles, *Mater Today*, 10, 30–38, 2007.

10. White, T.J. et al. A high frequency photodriven polymer oscillator. *Soft Matter*, 4, 1796–1798, 2008.

11. Clough, S.B., A. Blumstein, and E.C. Hsu, Structure and thermal-expansion of some polymers with mesomorphic ordering, *Macromolecules*, 9, 123–127, 1976.

12. Broer, D.J. and G.N. Mol, Anisotropic thermal-expansion of densely cross-linked oriented polymer networks, *Polym. Eng. Sci.*, 31, 625–631, 1991.

13. Hikmet, R.A.M. and D.J. Broer, Dynamic mechanical-properties of anisotropic networks formed by liquid-crystalline acrylates, *Polymer*, 32, 1627–1632, 1991.

14. Broer, D.J. and G.N. Mol, Anisotropic thermal-expansion of densely cross-linked oriented polymer networks, *Polym. Eng. Sci.*, 31, 625–631, 1991; Christophersen, M., B. Shapiro, and E. Smela, Characterization and modeling of PPy bilayer microactuators—Part 1. Curvature, *Sensor Actuat. B–Chem.*, 115, 2, 2006.

15. Timochenko, S., Analysis of bi-metal thermostats, *J. Opt. Soc. Am. Rev. Sci. Inst.*, 11, 233, 1925.

16. Varghese, S., G.P. Crawford, C.W.M. Bastiaansen, D.K.G. de Boer, and D.J. Broer, Microrubbing technique to produce high pretilt multidomain liquid crystal alignment, *Appl. Phys. Lett.*, 85, 230–232, 2004.

17. O'Neill, M. and S.M. Kelly, Photoinduced surface alignment for liquid crystal displays, *J. Phys. D. Appl. Phys.*, 33, R67–R84, 2000.

18. Hikmet, R.A.M., Anisotropic networks and gels formed by photopolymerisation in the ferroelectric state, *J. Mater. Chem.*, 9, 1921–1932, 1999.

19. Sousa, M.E., D.J. Broer, C.W.M. Bastiaansen, L.B. Freund, and G.P. Crawford, Isotropic "islands" in a cholesteric "sea": Patterned thermal expansion for responsive surface topologies, *Adv. Mat.*, 18 (14), 1842, 2006.

20. Elias, A.L., K.D. Harris, C.W.M. Bastiaansen, D.J. Broer, and M.J. Brett, Photopatterned liquid crystalline polymers for microactuators, *J. Mater. Chem.*, 16, 2903, 2006.

21. Van der Zande, B.M.I., J. Steenbakkers, J. Lub, C. M. Leewis, and D.J. Broer, Mass transport phenomena during lithographic polymerization of nematic monomers monitored with interferometry, *J. Appl. Phys.*, 97, 123519, 2005.

22. Holland, N.B., A. Cattani, L. Moroder, M. Seitz, H.E. Gaub, and T. Hugel, Single-molecule optomechanical cycle, *Science*, 296, 1103–1106, 2002.

23. Finkelmann, H., E. Nishikawa, G.G. Pereira, and M. Warner, A new optomechanical effect in solids, *Phys. Rev. Lett.*, 87, 015501, 2001.

24. Tabiryan, N., S. Serak, X.M. Dai, and T. Bunning, Polymer film with optically controlled form and actuation, *Opt. Express*, 13, 19, 2005.

25. Ikeda, T., J. Mamiya, and Y.L. Yu, Photomechanics of liquid-crystalline elastomers and other polymers, *Angew. Chem. Int. Edit.*, 46, 506–528, 2007.

26. Harris, K.D., R. Cuypers, P. Scheibe, C.L. van Oosten, C.W.M. Bastiaansen, J. Lub, and D.J. Broer, Large amplitude light-induced motion in high elastic modulus polymer actuators, *J. Mater. Chem.*, 15, 5043, 2005.

27. Van Oosten, C. L., K. D., Harris, C. W. M. Bastiaansen, and D. J. Broer, Glassy photomechanical liquid crystal network actuators for microscale devices, *Eur. Phys. J. E.*, 23, 329–336, 2007.

28. Thornton, P.D., G. Gail McConnell, and R.V. Ulijn, Enzyme responsive polymer hydrogel beads, *Chem. Commun.*, 5913–5915, 2005.

29. Howse, J.R., P. Topham, C.J. Crook, A.J. Gleeson, W. Bras, R.A.L. Jones, and A.J. Ryan, Reciprocating power generation in a chemically driven synthetic muscle, *Nano Letters*, 6, 73–77, 2006.

30. Miyata, T., N. Asami, and T. Uragami, A reversibly antigen-responsive hydrogel, *Nature*, 399, 766, 1999.

31.  Harris, K.D., C.W.M. Bastiaansen, J. Lub, and D.J. Broer, Self-assembled polymer films for controlled agent-driven motion, *Nano Lett.*, 5, 1857–1860, 2005.

32.  Shandryuk, G.A., S.A. Kuptsov, A.M. Shatalova, N.A. Plate, and R.V. Talroze, Liquid crystal H-bonded polymer networks under mechanical stress, *Macromolecules,* 36, 3417–3423, 2003.

33.  Harris, K.D., C.W.M. Bastiaansen, and D.J. Broer, A glassy bending-mode polymeric actuator which deforms in response to solvent polarity, *Macromol. Rapid Commun.*, 27, 1323–1329, 2006.

34.  Yu, Y.L., M. Nakano, and T. Ikeda, Directed bending of a polymer film by light— Miniaturizing a simple photomechanical system could expand its range of applications, *Nature,* 425, 6954, 2003.

35.  Woltman, S. J., G. D. Jay, and G. P. Crawford, Liquid-crystal materials find a new order in biomedical applications. *Nat. Mater.*, 6, 929–938, 2007.

36.  Den Toonder, J. et al. Artificial cilia for active micro-fluidic mixing. *Lab Chip*, 8, 533–541, 2008.

37.  Van Oosten, C.L., C.W.M. Bastiaansen, and D.J. Broer, Printed artificial cilia from Liquid Crystal Network actuators modularly driven by light, *Nature Mater.,* 8, 677–682, 2009.

38.  Camacho-Lopez, M., H. Finkelmann, P. Palffy-Muhoray, and M. Shelley, Fast liquid-crystal elastomer swims into the dark. *Nat. Mater.*, 3, 307–310, 2004.

39.  White, T.J. et al. A high frequency photodriven polymer oscillator. *Soft Matter*, 4, 1796–1798, 2008.

40.  Warner, M. and Mahadevan, L. Photoinduced deformations of beams, plates, and films. *Phys. Rev. Lett.*, 92, 134302, 2004.

41.  Nie, Z. and Kumacheva, E. Patterning surfaces with functional polymers. *Nat. Mater.* 7, 277–290, 2008.

42.  Khatavkar, V.V., P.D. Anderson, J.M.J. Den Toonder, and H.E.H. Meijer, *Active Micromixer Based on Artificial Cilia, Phys. Fluids.* 19, 083605, 2007.

# 10 Polymerizable Liquid Crystal Networks for Semiconductor Applications

*Maxim N. Shkunov, Iain McCulloch, and Theo Kreouzis*

## CONTENTS

## 10.1   INTRODUCTION

### 10.1.1   π-Conjugated Reactive Mesogens as Organic Semiconductors

Liquid crystalline (LC) self-assembly offers enormous potential for fabrication of electronic, optical, and electro-optical devices by cost-effective solution coating methods. Formation of electro-active LC molecules into desired two-dimensional arrays and three-dimensional networks (Broer et al., 1988, 1989) represents a very attractive opportunity for applications, such as broadband cholesteric polarizers (Broer et al., 1995), color filters (Lub et al., 2003), charge transport layers (Yoshimoto et al., 2002; O'Neill et al., 2003), polarized electroluminescent devices (Whitehead et al., 2000; Contoret et al., 2002), photovoltaics (Carrasco-Orozco et al., 2006), and organic field-effect transistors (FETs) (Katz et al., 1998; Sirringhaus et al., 2000b; Huisman et al., 2003; McCulloch et al., 2003, 2006, 2008; Shkunov et al., 2003, 2004; Mushrush et al., 2003).

For organic electronic applications, where molecular layers play the role of semi-conducting components, and other parts of the functional circuits could also be based on organic molecules (such as conducting polymers for electrodes, and dielectric polymers or small molecules for insulators) there is a window of opportunity to compete with traditional inorganic semiconductor processes based on high-temperature vacuum technologies. Due to the low charge carrier mobility in organic semiconductors (van der Waals bound solids) these applications cannot compete in high-speed applications but are targeting devices where large area, vacuum-free, and low-temperature processing are essential and low charge carrier mobility of less than 1 $cm^2/V$ s can be tolerated.

The intrinsic properties of molecular materials allow new deposition technologies for organic-based circuits such as low-cost printing techniques and large area solution-based coatings to advance novel applications including rollable displays (Granmar et al., 2005), flexible organic light emitting diodes (Yagi et al., 2008), disposable chemical sensors (Dodabalapur, 2006; Wang et al., 2006), elastic robotic skin sensors (Someya et al., 2004), and plastic solar cells (Brabec, 2004).

Extensive research in small molecule organic semiconductors demonstrated that high degree of intermolecular order is vital for efficient charge transport (Dimitrakopoulos et al., 2001; Podzorov et al., 2003).

It has been envisaged that reactive mesogen (RM) (polymerizable liquid crystal) semiconductors will be able to provide high degree of intermolecular order in thin film environment through liquid crystal self-assembly in thermotropic or lyotropic phases and consequently good charge carrier mobility; allows full control of molecular orientation on a substrate through well-studied liquid crystal alignment methods and bring forward the ability to "freeze" molecular assemblies, in virtually all available LC mesophases, using photo or thermal initiation reactions and allow photomasking of the desired deposition area similar to negative photoresist technology to provide insoluble cross-linked areas for multilayer devices.

The processing of RMs involves several steps, and each one might introduce some degree of disorder, impurities, and traps that could have detrimental effects of charge transport properties of the resulting films. As described in Section 10.2, very specific molecular design criteria are necessary to develop efficient semiconductor material including: choice of π-conjugated molecular core, polymerizable groups, and spacer length between them. Additional considerations should be given to molecular shape and side groups to allow particular mesophase behavior with not very high clearing temperatures and also presence of higher-order liquid crystal phases leading to "plastic crystal" formation with shorted intermolecular distances and higher molecular organization. There are still some challenges remaining including RM molecular design to promote optimal sequence of liquid crystal phases for monodomain growth over the entire substrate. Another optimization is the need to minimize the impact of photopolymerization step on electronic transport properties of cross-linked films.

## 10.1.2  MULTILAYER DEVICES AND CROSS-LINKING

A typical fabrication process for an organic semiconductor device involves deposition of a thin film of RMs from solution, followed by a thermal drying step to remove residual solvent. Depending on device architecture, other layers may need to be applied onto the surface of the deposited RM film. These layers could be structured metal contacts, polymer insulators, additional semiconductor coatings, and encapsulation layers. During the process of fabricating the device, the RM film is subjected to heating cycles and exposure to solvents used in either further deposition steps or etching and development. Each event can result in loss of thin film integrity. To retain RM thin film morphology and create a temperature independent orientation and order, the molecules can be cross-linked within the mesophase temperature (Broer, 1995; Lestel et al., 1994). This step "freezes-in" the molecular organization, with the additional benefit that the cross-linked film is also now impervious to solvent.

Cross-linking of the thin film RMs (Broer et al., 1991) can occur either on UV exposure or thermally. In the case of UV cross-linking, the film can be heated into the optimal mesophase, then exposed to irradiation in the presence of a photoinitiator (PI) matched to the exposure wavelength, thus initiating cross-linking. This process allows cross-linking to proceed independently of the film temperature, ensuring that polymerization only occurs at the mesophase temperature, and assuming that there is no molecular reorientation as a result of cross-linking, then the ordering within the mesophase is preserved. Additionally, photopolymerization offers the possibility of lateral patterning, using a mask to define cross-linked and noncross-linked areas, where a development step in an appropriate solvent can remove uncross-linked mesogens, a process analogous to negative photolithography. Most conjugated RMs, however, have significant absorption at the common wavelengths utilized by commercial PIs, namely 248, 365, and 436 nm. This absorption not only reduces the efficiency of initiation, but may lead to photochemical degradation of the chromophore and subsequent loss in performance (McCulloch et al., 2006). Thermal cross-linking on the other hand is not such a selective process. Even if the half-life temperature of the initiator can be chosen to match the mesophase temperature, there is often some cross-linking that occurs on heating into the mesophase, leading to a suboptimal organization.

## 10.2  DESIGN OF SEMICONDUCTING RMs

### 10.2.1  Liquid Crystal Behavior, Morphology, and Charge Transport

The molecular characteristics collectively responsible for the promotion of a LC phase include a central core unit comprising either aromatic or stiff cyclo-aliphatic units connected by rigid links, with either polar or flexible (or both) alkyl or alkoxy terminal groups. This ensures the axial anisotropy required to establish a mesophase. In order to exhibit semiconducting properties, the core unit must have an extended, delocalized π-electron system, preferably conjugated aromatics units. Charge transport occurs in these materials through a hopping mechanism from one molecule to another, in which the charge resides within the delocalized electronic orbitals. The speed of the hop is inversely related to the distance between two semiconducting molecules, hence the prerequisite for close packing, and also for relatively long cores, and especially larger core to tail group ratios, which will improve the intermolecular overlap for hopping. Illustrating this effect, a RM comprised of a quaterthiophene core exhibited two orders of magnitude higher charge-carrier mobility than a similar RM with a shorter phenyl naphthalene core (McCulloch et al., 2003). The morphology of the semiconductor is thus critical to charge transport, with conjugated aromatic cores desirable as they can adopt an essentially coplanar conformation allowing a compressed π-stacked microstructure, promoting extended order as well as ensuring a good intramolecular π-orbital delocalization within the core. Molecular conformation is also responsible for the phase behavior of the mesophases. The composition and shape of the core has a profound effect on crystallization and clearing temperatures, as well as solubility. Long stiff cores with less flexible linkages tend to exhibit higher melting temperature transitions, as well lower solubility, whereas introducing a lateral group on the core decreases the axial ratio of the mesogen, thus suppressing crystallinity, depressing the melt and increases the solubility. This is well illustrated when comparing an oxetane terminated quaterthiophene RM (RM9) with a quinquethiophene (RM10) and its methyl analogue (RM11), as shown in Table 10.1. Extending the conjugated core from four to five linked thiophene units, increases the melt temperature by 14 degrees, but more significantly, increases the clearing temperature by almost 90 degrees. Incorporation of a methyl group in the three position of the central thiophene, dramatically reduces both the melting temperature by 50 degrees, and the clearing temperature by over 130 degrees.

The charge transport properties of calamitic liquid crystals (referred here to as mesogens) are dependent on the order and orientation of the molecules (Funahashi et al., 1997). It has been observed that transport in the more ordered mesophases such as high-order smectics, where there is long-range positional order within lamella phases, is optimal. Preparation of thin films of these materials from solution typically results in a polycrystalline morphology with crystallization occurring either from solution on drying, or from the LC phase on cooling after drying. These crystalline

**TABLE 10.1**

**Phase Behavior of Reactive and Nonreactive Mesogens**

| Notation | Chemical Structure | Phase Transitions | HOMO (−eV) |
|---|---|---|---|
| M1 | 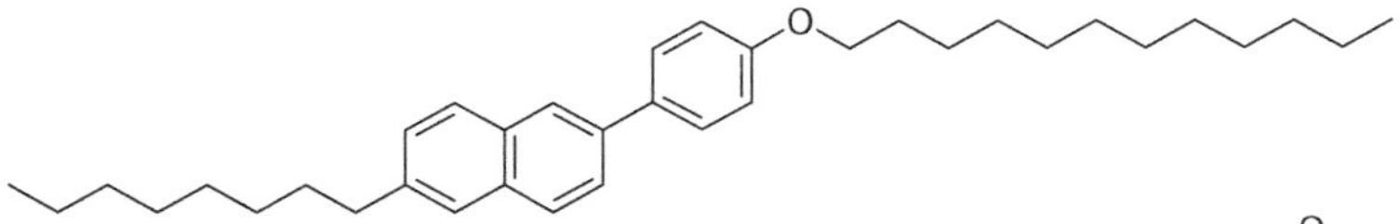 | K-85-SmB-101-SmA-122-Iso Iso-121-SmA-100-SmB-65-SmE-59-K | 5.79 |
| RM1 | 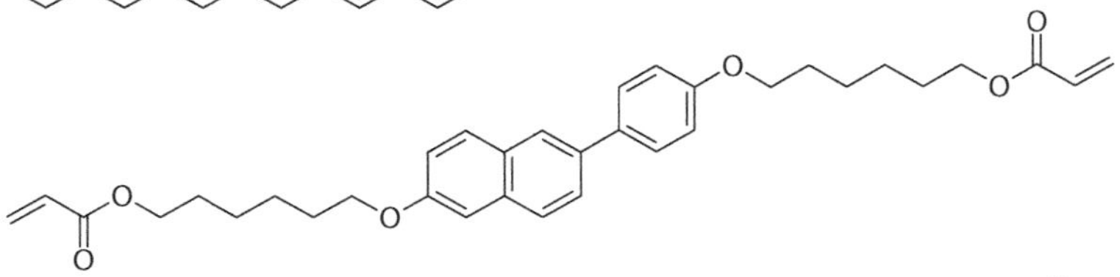 | K-85-SmA-111-Iso Iso-109-SmA-66-SmC-50-K | 5.87 |
| RM2 | 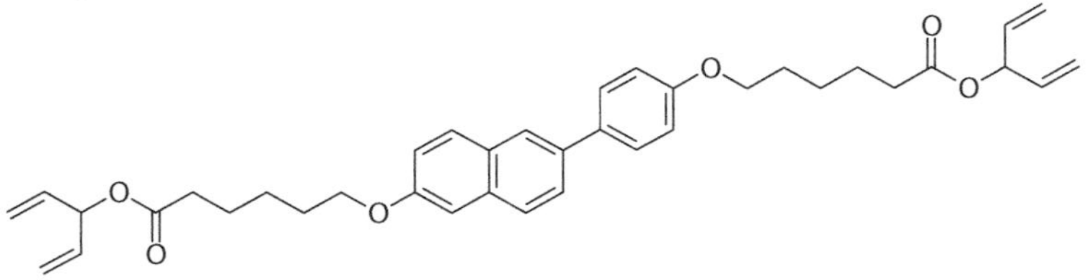 | K-9-Ki-18-SmG-66-SmC-80-Iso Iso-75-SmC-59-SmG-(-10)-K | 5.78 |
| M2 | 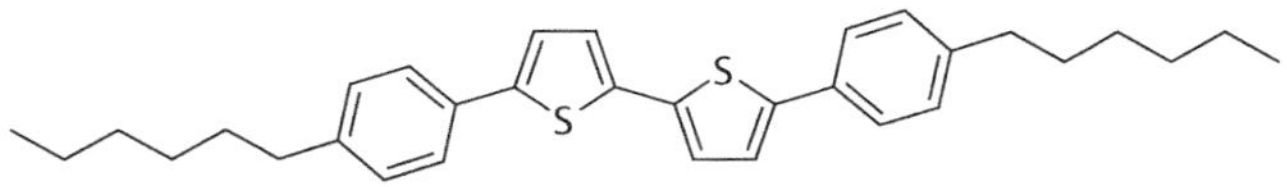 | K-76-SmH-130-SmG-226-SmC-232-SmA-233-Iso Iso-231-SmA-228-SmC-224-SmG-129-SmH-58-K | |

*Continued*

**TABLE 10.1 (Continued)**
**Phase Behavior of Reactive and Nonreactive Mesogens**

| Notation | Chemical Structure | Phase Transitions | HOMO (−eV) |
|---|---|---|---|
| RM3 | | K-104-SmG-187-SmC-202-Iso<br>Iso-201-SmC-184-SmG-80-K | |
| RM4 | | K-75-SmG-175-Iso<br>Iso-169-SmG-65-K | 5.45 eV |
| RM5 | | K-88-SmG-93-SmB-145-Iso<br>Iso-144-SmB-92-SmG-82-K | |
| M3 | | K-66-SmB-182-N-187-Iso<br>Iso-186-N-178-SmB-34-K | |
| M4 | | K-97-SmG-179-Iso<br>Iso-173-SmG-73-K | |
| RM6 | | K-80-SmG-96-Iso | 5.5 |

RM7    K-50-SmG-125-Iso

RM8    K-0-SmH-56-SmG-98-Iso
Iso-96-SmG-50-SmH-(-20)-K

RM9    K-82-SmG-111-Iso
Iso-110-SmG-74-K

RM10    K-96-SmC-198-Iso
Iso-196-SmC-K

RM11    K-46-SmC-62-N-67-Iso
Iso-N-61-SmC-32-K

M5DHTT2FPTT[a]    Iso-250-N-220-SmC-150-SmE-120-K

[a]  The compound was kindly supplied by Prof. M.L. Turner and Dr. F. Jamarillo, School of Chemistry, University of Manchester, M13 9PL, UK.

domains can have lateral dimensions of tens or even hundreds of microns in size. Due to the anisotropy of the crystalline motif, there is typically a strongly preferred direction for charge carrier transport, in which the conjugated cores assemble together in close $\pi$-stacked lamella, allowing charge carriers to hop between molecules as they percolate between the device electrodes. This directional anisotropy can lead to poor device to device uniformity.

### 10.2.2   Polymerization and Choice of Reactive Groups

On polymerization there is a physical movement of the cross-linking groups to accommodate the spatial decrease from van der Waals distances to that of a covalent bond. The cross-linking sites on the molecule must therefore be decoupled from the conjugated core, such that the ordered packing within the $\pi$-stacked lamellae is preserved. This is accomplished by introducing aliphatic, flexible linear spacer chains, connecting the aromatic core with the reactive end groups. The spacer group also extends the linearity of the molecule, helping to promote more ordered mesophases.

The mechanism of polymerization is also important. Both free-radical and cationic polymerization have been demonstrated. Electron-rich polymerizable groups such as vinyl ethers, epoxides, and oxetanes in particular can be polymerized by a cationic initiator. Where photopolymerization is employed, a photoacid generator such as a sulfonium salt can be used as initiator. Cationic polymerization is not sensitive to oxygen, and therefore, polymerization can be carried out in ambient conditions. If the thermal transition into the mesophase occurs at high temperature, oxetane groups, which possess excellent thermal stability, are commonly used. Free-radical polymerization is most efficient when electron poor polymerizable groups are used, such as acrylates and methacrylates, and there is a wide choice of both photo and thermal free-radical initiators. Efficient chain transfer reactions with oxygen hinder the use of this polymerization mechanism in ambient conditions, and therefore, free-radical polymerization is typically carried out under a nitrogen blanket. Inclusion of a polymerizable unit also has a marked effect on the phase behavior of the liquid crystal, with bulky end groups lowering melting and to a lesser extent the clearing, temperatures. This effect is illustrated in Table 10.1 where the oxetane RM9 is compared with the nonreactive version M4. The incorporation of the oxetane group significantly lowers the melting temperature of the mesogen. Other studies have directly compared the effect of range of different polymerizable groups on thermal properties. For example, incorporation of a bulky diene polymerizable group was observed to significantly suppress the mesogen melting point, in comparison to both acrylate and methacrylate analogues with identical core and spacer groups, as well as reduce the clearing point, although less significantly. This can be attributed to the more sterically bulky shape of the diene unit reducing the packing order. Another consequence of the melting point suppression is that high-order smectic mesophases are also exhibited by the diene RMs.

On polymerization, the local environment of the mesogens must accommodate the growing chain, and the volume change associated with the length reduction of neighboring polymerizable units from intermolecular van der Waals distances to that of a covalent bond. Shrinkage, on the one hand may result in closer $\pi$–$\pi$ stacking distances, but in a polydomain film as the domain volumes decrease then the grain boundaries become more pronounced. Therefore, depending on whether the dominant effect is either compressed π stacking, or increased grain-boundary volume, the carrier mobility can either increase or decrease on cross-linking, and in fact evidence of both results has been reported.

Vinyl type free-radical polymerizations are also susceptible to chain transfer side reactions which quench polymerization. The presence of sulfur containing conjugated core can often be a site for chain transfer. Polymerization typically requires an initiator which starts the process. The concentration of initiator required for *in situ* thin film polymerization is often very high (between 1 and 5 wt.%). Molecular fragments of the initiator will remain in the film after polymerization, and may interfere with transport. The ionic PIs commonly used in cationic polymerization are more likely to trap charge, thus lowering mobility, and also bulky fragments may reduce crystallinity. It may be possible to extract these components with a post polymerization wash, which often improves electrical performance.

### 10.2.3   INFLUENCE OF CORE DESIGN ON ELECTRICAL PROPERTIES

Charge carrier transport, injection characteristics, and the ambient stability of a semiconductor device are all influenced by the energy levels of the π-conjugated electron system of the core. In addition to the influence of geometrical overlap factors previously discussed, holes residing in low energy highest occupied molecular orbitals (HOMOs) have greater propensity for trap, filling deeper trap states and therefore, the mobility decreases. Considering simplified charge injection picture, in a p-type transistor device, holes are injected into the HOMO energy level by the source electrode, as shown in the heterojunction energy diagram in Figure 10.1. It is therefore, also important to ensure that there is no significant endothermic energy gap between the electrode and semiconductor (Schottky barrier) to prevent charge injection. The

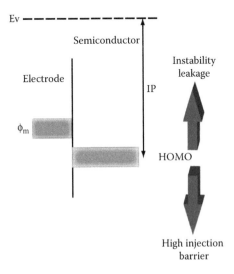

**FIGURE 10.1**   Heterojunction energy diagram for metal–semiconductor interface.

fundamental electrochemical oxidation of the semiconductor in ambient air under saturated humidity has been proposed to occur if the HOMO energy level is less than 4.9 eV from vacuum energy level. Molecular design of the semiconductor requires therefore, to lower the position of the HOMO energy level of the conjugated system in respect to vacuum level (can also be referred to as increasing the ionization potential above 4.9 eV), and yet from injection and trapping perspectives, the HOMO should match well with injecting electrode work function. Therefore, a compromise HOMO energy level is required, and depending on the choice of electrodes, this value is usually between 5.0 and 5.5 eV.

There are several ways in which the design of the core π-electron system can influence its molecular orbital energy levels. Increasing the number of conjugated units or conjugation length along the core manifests as an increase in the HOMO energy level and decrease in the LUMO (lowest unoccupied molecular orbital) energy level (in respect to vacuum level), that is, a band gap reduction. For example, RMs with short phenyl-naphthalene core units exhibit HOMO energy levels of around −5.9 eV, whereas, the more extended conjugation of a quaterthiophene, cores have HOMO energy levels from −4.9 to −5.5 eV. Raising the HOMO energy level through extending conjugation is successful up to a critical "effective" conjugation length, at which point the energy levels remains more or less constant as additional units are added. Conjugation between coupled aromatic units also relies on efficient intramolecular π-orbital overlap, which can be sterically manipulated by lateral functional groups to twist out of a coplanar conformation and thus reduce conjugation pathways. This effect conversely increases the LUMO energy level and decreases the HOMO energy level. Reduction of the electron density of the conjugated system typically reduces both the LUMO and HOMO energy levels. Therefore, for example, substitution of an electron rich thiophene monomer with a more electron poor monomer such as benzene will result in a reduction in the HOMO energy level. For example, the HOMO energy level of a phenyl-napthalene core containing RM1 (Table 10.1) was lower, at about −5.9 eV than the analogous quaterthiophene core RM6 (HOMO at −5.5 eV) as measured by ambient ultraviolet photoelectron spectroscopy. In the same way, introduction of electron withdrawing functional groups as part of the conjugated system will also reduce the HOMO energy level, and conversely, removing electron donating groups from the conjugated system will also reduce the HOMO energy level.

## 10.3   MOLECULAR ALIGNMENT, MORPHOLOGY CONTROL, AND CHARACTERIZATION

### 10.3.1   ALIGNMENT

Calamitic liquid crystals can align (Coates et al., 2000a, 2000b) either orthogonal to the substrate surface, referred to as homeotropic alignment, or parallel to the substrate surface, referred to as planar alignment. In most FET device structures, charge transport occurs within a thin, planar region, close to the semiconductor/insulator interface. Therefore, it is necessary to ensure that the charge transport component of the molecule is aligned within the charge transport layer such that the distances between the π-electron orbitals of adjacent molecules are minimal. This can be achieved by homeotropic alignment in the smectic mesophase, where the molecules are aligned with their long axis normal to the film plane and confined within a lamella, as shown in Figure 10.2, with the domain size persisting over the area of the film. This is the preferred orientation for in-plane transistors. Aligned phases are promoted and their areas extended by the use of self-assembled monolayers such as octyl-trichlorosilane (OTS), hexamethyldisilizane (HMDS), or alignment layers such as rubbed polyimides or friction-transferred polytetrafluoroethylene (PTFE) (Wittmann et al., 1991; van de Craats et al., 2003). Homeotropic alignment layers (OTS and HMDS) are generally hydrophobic in nature, and the low-energy aliphatic tails of the molecule are thermodynamically driven to this interface to minimize the interfacial energy. In addition, alignment layers also promote the formation of large area domains, or monodomains, in which the order and orientation persists throughout the area of the alignment layer. The formation of monodomains is advantageous in optimizing charge carrier mobilities, as grain boundaries can act as charge traps, thus lowering mobility.

Typically, however, for these aromatic mesogens, crystallization from the mesophase occurs above room temperature. This is a kinetically driven event, and nucleation is directionally relatively random, resulting in a polycrystalline film with no long range collective orientation of the crystalline domains. Achieving long range homeotropic alignment when fabricating films from a solution-based fully additive multilayer fabrication process therefore has many challenges.

### 10.3.2 Optical Characterization of Homeotropic Films

As previously discussed, it is optimal for in-plane charge transport to promote homeotropic alignment, shown in Figure 10.2, achieved through pretreatment of the surface on which the mesogen is deposited. Identification of mesogen alignment in thin films can be achieved by a variety of techniques (Coates et al., 2000a, 2000b; Gray et al., 1984). In smectic phases molecules are arranged in layers with molecular long

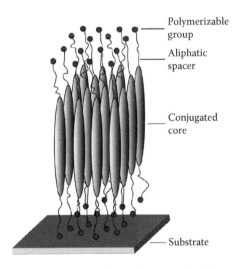

Polymerizable group

Aliphatic spacer

Conjugated core

Substrate

**FIGURE 10.2** Schematic representation of RMs homeotropic alignment on a substrate.

axes perpendicular or tilted in respect to the layer planes. It is difficult to identify the orientation of tilted smectic phases by polarized microscopy since both homeotropic and planar aligned films are highly birefringent and can show similar behavior under crossed polarizers. Moreover, in homeotropically aligned films of nontilted smectic phases, such as SmA, SmB, when the optical axis of the mesogen molecules are oriented parallel to the viewing direction (out of plane to the substrate), films appear isotropic when rotated under cross polarizers. However, confocal microscopy can be used to distinguish between homeotropic and planar alignments. A characteristic isogyre interference figure (commonly referred to as a "Maltese Cross") can be observed under convergent light, and remains, when a homeotropic film is rotated in the plane of the viewing stage. Isotropic films do not generate this interference effect. An example of the image observed by confocal microscopy exhibited by a homeotropic film of a diene RM containing the short phenyl-naphthalene core (structure shown in Table 10.1) is provided in Figure 10.3a. Another technique employed to identify the optical alignment director was optical retardation (Coates et al., 2000a, 2000b). When polarized light is incident on an optically birefringent film, the polarization phase is retarded, depending on the optic axes of the film. The variation of retardation with incident angle creates a characteristic retardation profile, which is dependent on the macroscopic alignment of the RM film (Coates et al., 2000a, 2000b). These profiles can be used to distinguish homeotropic from planar alignment. Examples of a retardation profiles obtained from a homeotropically aligned thin film of RM2 is illustrated in Figure 10.3b.

### 10.3.3  INTERMOLECULAR DISTANCES IN RM VERSUS NONREACTIVE ANALOGUE

Intermolecular distances in semiconducting films determine the efficiency of charge hopping between the molecules and distances in the range 3.5–4.5 Å are considered

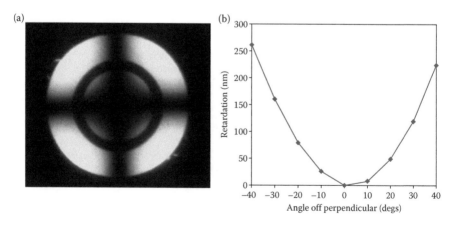

**FIGURE 10.3**  (a) "Maltese Cross" interference image observed on viewing a homeotropic film of RM2 (phenyl napthalene–diene RM) by confocal microscopy; (b) Retardation profiles of homeotropically aligned thin film of RM2. Data points are shown as circles, and the line is provided to guide the eye.

to be optimal for charge transport (Bredas et al., 2004; Coropceanu et al., 2007). Increase in intermolecular spacing leads to a reduction in hopping rates and consequently in lower charge mobility. As discussed in Section 10.2, addition of reactive groups has a number of effects on liquid crystal behavior of RMs compared to non-reactive analogues. The need to accommodate bulky end groups also leads to an increase of intermolecular distances in high order smectic phases. As observed by two-dimensional grazing incidence wide angle x-ray scattering (Sirringhaus et al., 2000a), in-plane characteristic distances (correlated with $\pi-\pi$ stacking) in homeotropic films of M2 are typically 4.0 and 4.6 Å. In the same setup, the data for a diene group RM version of the same molecular core (RM4) is somewhat different, revealing approximately 1 Å increase for in-plane distances (characteristic spacing 5.1 and 5.8 Å). Not surprisingly, mobilities of charge carriers are found to be lower in RM4 films compared to M2 as described in Section 10.4.

### 10.3.4 IMPACT OF CROSS-LINKING ON FILM MORPHOLOGY: XRD STUDIES

The impact of cross-linking on mesogen organization is related to the number density of the polymerizable unit, which can be minimized by increasing core and spacer lengths. However, if this density is too low, then cross-linking does not efficiently occur. When the polymerization occurs through a vinyl-type addition, but rather, involves a ring-opening mechanism, such as in the case for both oxetanes and epoxies, the effect of shrinkage on polymerization is further reduced. For example, on comparing a diene polymerization with an oxetane polymerization, it was observed that the carrier mobility of the diene mesogen decreases, whereas, the oxetane increases. The polymerization of a diene mesogen RM4 (Table 10.1) was observed by two-dimensional grazing incidence wide-angle x-ray scattering as shown in Figure 10.4. Dienes can undergo a self-initiated photopolymerization reaction resulting in a fused stiff bicyclopentane backbone polymer (Contoret et al., 2002; Hall et al., 2004). The prepolymerization film of RM4 exhibited many higher-order reflections, being the signature of a highly crystalline structure. Strong out-of-plane reflections indicate periodic lamella structure with characteristic distances of ~30 Å. Within the layers molecular cores are tilted in respect to the normal. Film appears to be consisting of crystallites, maintaining the same layered lamella structure and with common orientation of tilt angle within the crystallite. It is also evident that the crystallites have a high degree of orientation with a crystalline axis parallel to the film normal, that is, homeotropic alignment—otherwise the peaks would have appeared as circle segments. After polymerization, peaks appear to be in the same location, implying that the crystalline structure is unchanged. However, the angular spread has increased somewhat, suggesting that the polymerization has slightly disturbed the orientations of the crystallites or molecular periodicity has been disrupted by the formation of a stiff cyclic backbone. It is suggested that this reduction in degree of molecular order contributes to reduced charge mobility. In contrast, the cross-linked oxetane mesogen (RM5—Table 10.1) with larger domain sizes, exhibited an increase in hole mobility (confirmed by both time-of-flight (TOF) and FET results), most likely caused by closer $\pi-\pi$ distances of the cores and possibly a reduced grain-boundary effect.

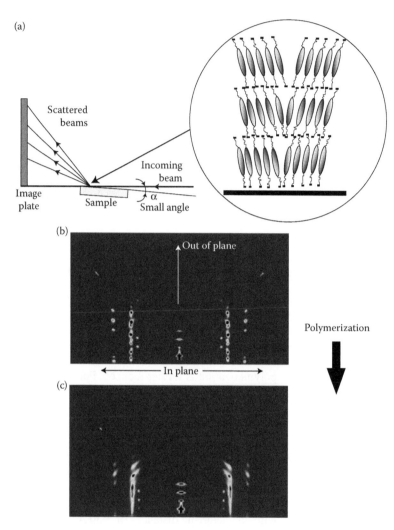

**FIGURE 10.4** Grazing incidence wide angle x-ray scattering on thin films of RM4 with diene polymerizable groups. (a) Schematic illustration of experimental setup showing the incoming beam, the sample at a small incidence angle $\alpha_i$, the scattered beams and the image plate for recording diffraction peaks. A schematic of homeotropically aligned molecular layers in a smectic phase is shown on a right-hand side. (b) An image obtained before polymerization. (c) An image obtained after polymerization.

## 10.3.5 Characterization of Photopolymerization Process

Photopolymerization can be quantitatively monitored by Fourier transform infrared spectroscopy (FTIR), by observing the disappearance of the C=C bond stretch vibration (McCulloch et al., 2003). Although vinyl bonds, with low dipole moments, do not exhibit particularly strong absorptions, the disappearance of the carbon–carbon double bond in a diene end-group polymerization, can be observed at $1641 \text{cm}^{-1}$ by

FTIR. A more qualitative verification that cross-linking has occurred, can be achieved by immersing the film, after cross-linking, in a solvent which dissolves the monomer, then drying the film and checking by either profilometry or UV–visible absorption spectrometer for any loss of film thickness.

A more detailed investigation of RM photopolymerization reactions can be conducted using Photo-DSC experiments (Thiem et al., 2005), provided that polymerization kinetic is fast enough to be detected with this type of equipment.

## 10.4  TIME OF FLIGHT MEASUREMENTS OF RMs

### 10.4.1  Technique Description

One of the most common techniques in studying fundamental charge transport properties of bulk organic semiconductors is a TOF method (LeBlanc, 1959; Kepler, 1960). It allows the mobility (defined as the carrier drift velocity, normalized to the electric field strength) to be determined by timing how long carriers take to cross a known distance (the sample length), as well as distinguishing between hole and electron transport. A typical schematic experimental arrangement is shown in Figure 10.5; the semiconductor (RM layer) is sandwiched between two electrodes (one of which is semitransparent) a distance $d$ apart. The RM samples are typically prepared by capillary filling these materials in isotropic phase into standard liquid crystal test cells with ITO electrodes coated with rubbed polyimide alignment layers. These cells are slowly cooled to the desired mesophase temperature to allow the formation of large liquid crystal domains.

The carriers are photogenerated by exciton dissociation in the vicinity of the semi-transparent electrode as a result of illumination by a short duration (typically $\sim$ ns) laser pulse of photon energy $h\omega$, and the polarity of the applied potential $\pm V$ determines the sign of carrier, which will drift through the bulk of the semiconductor. The relationship between the conduction current $i(t)$ and the signal voltage measured by the oscilloscope $v(t)$ is given by Equation 10.1 (Donovan et al., 2000):

$$i(t) = \frac{1}{R_L}\left\{v(t) + \tau\frac{dv(t)}{dt}\right\},\qquad(10.1)$$

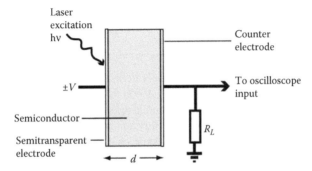

**FIGURE 10.5**  TOF sample schematic.

where $\tau$ is the electronic response time given by $\tau = C \sum R$ and $\sum R$ is the sum of all series resistances in the current loop, including $R_L$, any electrode resistances and the voltage supply output impedance. If the electronic response time is short compared to the times measured, the second term in the bracket can be neglected and Equation 10.1 simply reduces to Ohms law, $i(t) = (v(t)/R_L)$. This is called current mode TOF and is the most commonly used mode of this technique.

The mobility $\mu$ is calculated at a given electric field using Equation 10.2:

$$\mu = \frac{d^2}{Vt_t} \tag{10.2}$$

where $V$ is the voltage across the cell, $t_t$ is the transit time, and $d$ is sample thickness.

In the case of field independent mobility, a plot of drift velocity $d/t_t$ versus electric field strength $V/d$ will yield a straight line graph of gradient $\mu$.

## 10.4.2 MOBILITY-PHASE RELATIONSHIP

The molecular reorganization occurring in the formation of LC phases can have a dramatic effect on the charge transport. This is true for both discotic (Adam et al., 1994) and calamitic (Funahashi et al., 1997, 1999) systems, where carrier mobilities undergo sharp transitions at the phase boundaries. The overall pattern is one of increasing carrier mobility as the order of the phase is increased (usually carried out in cooling as the liquid crystal undergoes thermotropic transitions). The formation of a polycrystalline solid at low temperatures has a detrimental effect on charge transport as grain boundaries and other defects trap the charge carriers, usually making the measurement of mobility by TOF impossible as the charge trapping leads to featureless decaying photocurrent transients. We note that the underlying short range (short time) mobility in the crystalline phase may well be higher than that in the mesophases, but that the requirement for any reasonable range, fast transport in a molecular solid is that of a (reasonably) defect-free molecular crystal. An example of hole mobility parametric in temperature in a mesogen M5 DHTT2FPTT (thermotropic calamitic liquid crystal) is shown in Figure 10.6 while its structural formula is shown in Table 10.1. The mobility undergoes sharp transitions, from $2 \times 10^{-5}$ cm$^2$ V$^{-1}$s$^{-1}$ in the nematic phase to $1.5 \times 10^{-3}$ cm$^2$ V$^{-1}$s$^{-1}$ in the smectic C phase, then further increasing to $2 \times 10^{-2}$ cm$^2$ V$^{-1}$s$^{-1}$ in the smectic E phase before dropping to $3 \times 10^{-3}$ cm$^2$ V$^{-1}$s$^{-1}$ in the polycrystalline solid.

## 10.4.3 INFLUENCE OF REACTIVE END GROUPS

Although several groups have described polymerized reactive LC systems for charge transport (Farrar et al., 2002; Bleyl et al., 1997) the resulting films invariably show mobilities, which are lower than those found in the same mesophase of similar nonreactive LC systems. This reduction in mobility appears to be the result of the addition of reactive moieties to the end of the alkane chains even before any polymerization has taken place (and before the effect of any PI has to be taken into account). Researchers at the University of Hull (Woon et al., 2006) report different hole mobilities in two compounds that are identical except for the addition of a diene reactive

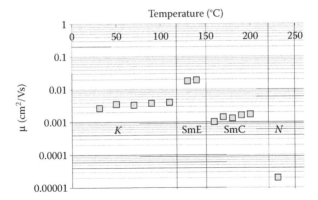

**FIGURE 10.6** Hole mobility parametric in temperature in mesogen M5 (DHTT2FPTT) measured by TOF.

end group, with values $\sim 2 \times 10^{-4}$ cm$^2$ V$^{-1}$ s$^{-1}$ for the reactive compound compared to $\sim 8 \times 10^{-4}$ cm$^2$ V$^{-1}$ s$^{-1}$ for the nonreactive compound, both at 100°C, nematic phase (Woon et al., 2006). They concluded that: "... some feature of the terminal group must be responsible for the significantly lower mobility value ..." A systematic study (Baldwin et al., 2007) involving a total of nine mesogenic compounds (consisting of three different aromatic cores, substituted in turn with a nonreactive alkane chain and two different reactive end-group-terminated chains) further showed that electron-poor and electron-rich end groups will affect hole and electron transport in different way.

Let us concentrate on two families of materials, those based on a quaterthiophene core (Table 10.1), and those based on a bis(4-heptylphenyl)-bithiophene core (Table 10.1), in each case consisting of a mesogen (M4, M2 M4, M2), a diene-substituted RM (RM8, RM4) and an oxetane-substituted RM (RM9, RM5). In each case the mesogen versions show fast ambipolar charge transport in the smectic mesophase (see TOF photocurrents Figures 10.7 and 10.8) and are therefore, ideal candidates for charge transport layer formation. A summary of the measured mesophase mobilities for the mesogens and nonpolymerized RMs is shown in Table 10.2.

Both diene substituted RMs show the expected reduction in hole mobility compared to the original LCs. In the case of M4 compared to RM8 by a factor of 14 and in the case of M2 compared to RM4 by a factor of 7. More dramatic drop is the reduction suffered by the electron mobility: in RM4 the reduction is by a factor of approximately 400 compared to the original LC and in RM8 the electrons trap before they can traverse the sample and the mobility is not even measurable. We can conclude that the diene end groups act as chemical traps for electrons. Moreover, reduction in charge carrier mobilities is also likely to be due to increased intermolecular distances in RM films compared to nonreactive analogues as described in Section 10.3.3.

The effect of the oxetane end group is somewhat less catastrophic: hole mobilities are reduced by a factor of 11 in RM9 compared to M4 and also by a factor of 5 in

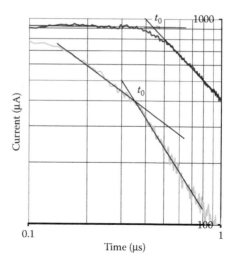

**FIGURE 10.7** Electron (lower curve, dispersive) and hole (upper curve, nondispersive) photocurrent transients across a 4.5 μm sample of M4 in the smectic G phase. Both photocurrents are under 7.5 V bias and yield $\mu_h = 6.3 \times 10^{-2}$ cm$^2$ V$^{-1}$ s$^{-1}$ and $\mu_e = 7.7 \times 10^{-2}$ cm$^2$ V$^{-1}$ s$^{-1}$.

RM5 compared to M2, and more importantly, the reduction suffered by the electron mobilities is not as dramatic as in the case of the dienes. Electron mobilities fall by a factor of 14 in RM9 compared to M4 and by a factor of 9 in RM5 compared to M2.

In summary, the addition of reactive end groups to the alkane chains will always reduce the charge carrier mobility. This is certainly true for holes (electron transport

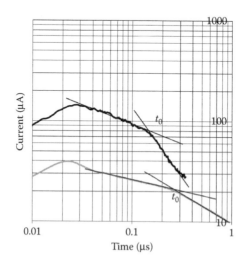

**FIGURE 10.8** Electron (upper curve) and hole (lower curve) photocurrent transients across a 4.9 μm sample of M2 in the smectic G phase. Both photocurrents are under 20 V bias and yield $\mu_h = 4.4 \times 10^{-2}$ cm$^2$ V$^{-1}$ s$^{-1}$ and $\mu_e = 7.1 \times 10^{-2}$ cm$^2$ V$^{-1}$ s$^{-1}$.

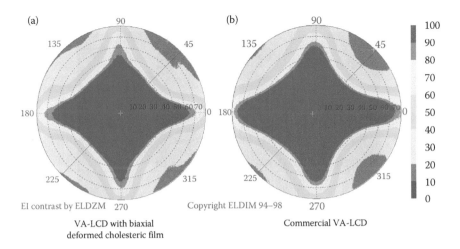

FIGURE 4.18 Comparison of 17-inch SXGA VA-LCD (Syncmaster 173P, *Samsung Electronics Co., Ltd.*) compensated with (a) biaxial deformed cholesteric film to (b) the product as it is. The biaxial deformed cholesteric film is attached only on the backlight side.

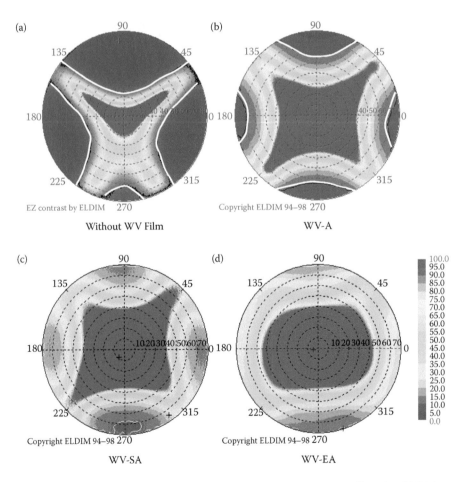

FIGURE 4.22 Viewing-angle performance of TN-LCDs (a) without WV film, (b) with WV-A, (c) with WV-SA, and (d) with WV-EA.

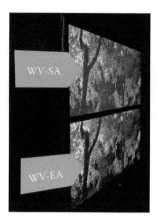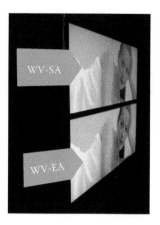

**FIGURE 4.24** Comparison of image quality of TN-LCDs between WV-SA and WV-EA at an oblique angle.

**FIGURE 5.3** LC polymer in a wedge cell polymerized under different conditions. The cells have planar alignment. A 25 mm × 25 mm LCP cell is seen between parallel and crossed polarizers on the top and bottom respectively. The quality of the cells is extremely sensitive on cross-linking conditions and amount of photoinitiator.

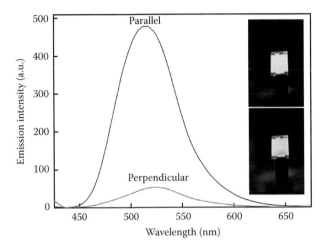

**FIGURE 6.6** Polarized emission of a film processed by *in situ* photopolymerization of a LC formulation of **18** (5%), **19** (85%), and **21** (10%) in a commercial glass cell for planar alignment. Emission was registered parallel and perpendicular to the preferred molecular orientation direction using polarized excitation.

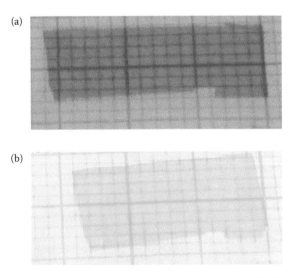

**FIGURE 7.8** Photographs of photodeformation of the azobenzene CLCP before UV light irradiation (a) and under UV light irradiation (b).

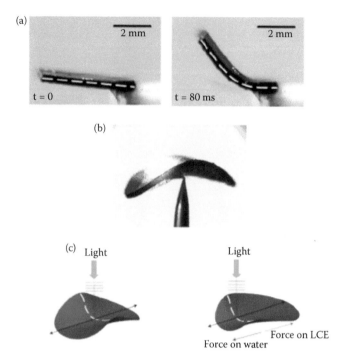

**FIGURE 7.10** Photomechanical response of the CLCP (a) and shape deformation of the CLCP upon exposure to 514 nm light (b). Schematic illustration is shown in (c) on the mechanism underlying the locomotion of the dye-doped CLCP.

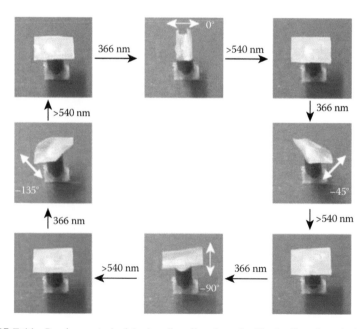

**FIGURE 7.11** Precise control of the bending direction of a film by linearly polarized light. Photographic frames of the film in different directions in response to irradiation by linearly polarized light of different angles of polarization (white arrows) at 366 nm, and bending flattened again by Vis light longer than 540 nm.

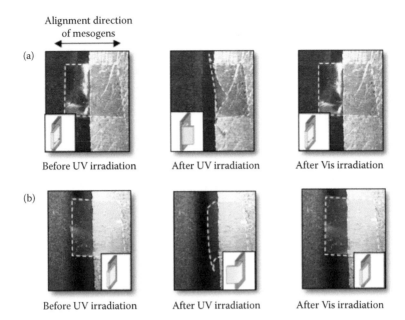

Alignment direction
of mesogens

(a)

Before UV irradiation     After UV irradiation     After Vis irradiation

(b)

Before UV irradiation     After UV irradiation     After Vis irradiation

**FIGURE 7.12** Photographs of the homogeneous film (a) and the homeotropic film (b) that exhibit photoinduced bending and unbending. The white dash lines show the edges of the films and the inset of each photograph is a schematic illustration of the film state.

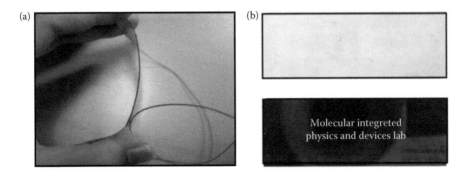

(a)        (b)

Molecular integreted
physics and devices lab

**FIGURE 8.29** A prototype of the flexible LC panel of 1.5 cm × 4 cm: (a) a bent state and (b) photographs showing a logo, which disappears at 0 V and appear at 10 V in a direct driving scheme.

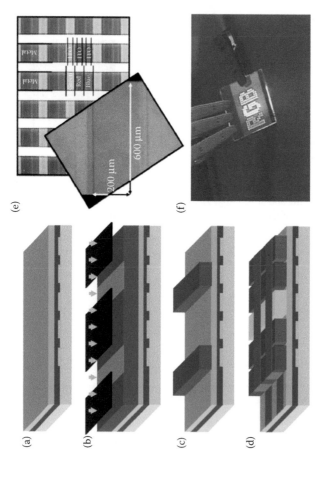

**FIGURE 11.13** (a)–(d) Schematic illustration of the direct lithography process (for explanations see text). (Reprinted by permission from Macmillan Publishers Ltd. Müller, C.D. et al., *Nature* **2003**, *421*, 829, copyright 2003.) (e) Microscopy image of the completed display taken through the glass substrate. The horizontal polymer stripes are parallel and well aligned with the underlying ITO anode stripes. The metal cathode columns are seen as vertical stripes. (the separating white stripes result from the microscope backlight overexposing the camera). The inset shows a single RGB triple at a higher magnification. (f) Photograph of a RGB OLED device. Dimensions of the glass substrate are 25 × 25 mm. (Gather, M.C. et al., *Adv. Funct. Mater.* **2007**, *17*, 191. Copyright Wiley-VCH Verlag GmbH & Co. KGaA. Reproduced with permission.)

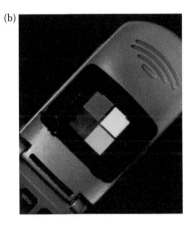

**FIGURE 11.21** (b) A prototype OLED with a red, green, and blue pixel on the same substrate, fabricated by spin coating on a PEDT film covering a patterned ITO substrate. The pixels were defined by irradiation at 325 nm through a mask. Nonirradiated material was removed by washing with chloroform. (Aldred, M.P. et al., *Adv. Mater.* **2005**, *17*, 1368. Copyright Wiley-VCH Verlag GmbH & Co. KGaA. Reproduced with permission.)

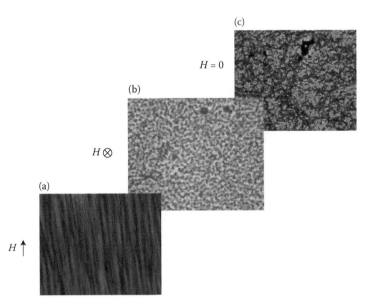

**FIGURE 18.10** Observation under the optical microscope of the texture of a ferrogel film formed in the presence (a, b) or in the absence (c) of a magnetic field. The magnetic field is applied in a direction parallel (a) or perpendicular (b) to the plane of the sample. Size of the pictures: 170 μm×130 μm. (Collin, D. et al. *Macromol. Rapid Commun.*, 24, 737, 2003. Copyright Wiley-VCH Verlag GmbH & Co. KGaA. Reproduced with permission.)

(a)

(a)
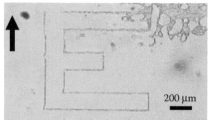

(b)
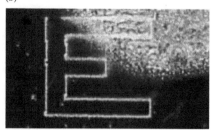

200 μm

**FIGURE 19.10** Letter "E" written by two-photon photolithography in the nematic phase (63°C) of an aligned sample under confocal microscope (sample cell thickness: 10 μm; microscope objective: numerical aperture 0.45, magnification 50). Femtosecond laser (80 MHz at 780 nm) was used as excitation beam for two-photon polymerization (500 mW/cm$^2$). The observation was made between uncrossed polarizers (a) and crossed polarizers (b), at $T = 81.5°C$ where the polymerized part ("E" letter) was in the nematic state, while the unpolymerized part (dark background in (b)) passed already in the isotropic state. The width of the E-letter is 800 μm. The resolution of two-photon beam is 7 μm. The nematic director orientation is indicated by the arrow. (From Sungur, E. et al., *Opt. Express*, 15, 6784–6789, 2007. With permission from Optical Society of America.)

(a)
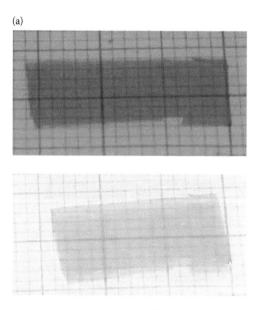

**FIGURE 19.13** (a) Elastomer E25% azo before UV irradiation (top) and under UV irradiation (bottom) (irradiation time: 130 s). (Background is a graduated paper.) (Li, M.-H. et al., *Adv. Mater.*, 15, 569–572, 2003. Copyright Wiley-VCH Verlag GmbH & Co. KGaA. Reproduced with permission.)

**TABLE 10.2**
**Summary of Materials Mesophase Mobilities**

| Material | $\mu_h$ (cm$^2$ V$^{-1}$ s$^{-1}$) | $\mu_e$ (cm$^2$ V$^{-1}$ s$^{-1}$) | Phase |
|---|---|---|---|
| M4 | $6.3 \times 10^{-2}$ | $7.7 \times 10^{-2}$ | SmG |
| RM8 | $4.6 \times 10^{-3}$ | N/A | SmG |
| RM9 | $5.8 \times 10^{-3}$ | $5.4 \times 10^{-3}$ | SmB |
| M2 | $4.4 \times 10^{-2}$ | $7.1 \times 10^{-2}$ | SmG |
| RM4 | $6.6 \times 10^{-3}$ | $1.7 \times 10^{-4}$ | SmG |
| RM5 | $9.5 \times 10^{-3}$ | $7.5 \times 10^{-3}$ | SmB |

has not been studied) in the case of acrylate end groups, although the occurrence of uncontrolled thermal cross-linking complicates matters as it is not clear whether a mesophase is formed. In the case of dienes and oxetanes, the end groups do not inhibit reproducible mesophase formation, but still affect the transport. Diene end groups always have a dramatic effect on electron transport, and a reduced effect on hole transport, whereas oxetane end groups appear to effect both hole and electron transport the least.

## 10.4.4 Cross-Linked Systems

Reactive mesogens have been successfully cross-linked and their performance as charge transport layers measured (Bleyl et al., 1997; Farrar et al., 2002; Kreouzis et al., 2005). The resulting films have carrier mobilities related to the mesophase in which they were formed. These range from $\mu_h \sim 4 \times 10^{-5}$ cm$^2$ V$^{-1}$ s$^{-1}$ for films formed in the nematic phase (Farrar et al., 2002) to $\mu_h \sim 10^{-5}$ cm$^2$ V$^{-1}$ s$^{-1}$ for those formed from a columnar discotic phase (Bleyl et al., 1997) and $\mu_h \sim 10^{-3}$–$10^{-2}$ cm$^2$ V$^{-1}$ s$^{-1}$ for those formed in a smectic mesophase (Baldwin et al., 2007), and so qualitatively follow the previously discussed mobility-phase relationship. Cross-linking within the correct mesophase is vital as films deliberately formed within a phase possessing undesirable transport properties (e.g., isotropic) result in poorly transporting layers.

It is worth noting that the transport properties of cross-linked films do not exactly mimic those of the original unreacted RM mesophase. These differences can be due to structural changes as a result of the reaction and to the presence of any PI used in order to carry out the photopolymerization. As demonstrated by Farrar at al (Farrar et al., 2002), diene group RM with a nematic phase showed an improvement in charge-carrier mobility after it has been polymerized. This is attributed to conformational changes within the film due to the formation of a rigid backbone. The PI, however, appears to have a largely negative effect on charge transport, and this is expected as it essentially constitutes the presence of an impurity. This may influence both the molecular packing within the mesophase (Bleyl et al., 1997) or act as a chemical charge carrier trap in itself.

Returning to the RMs previously shown (namely RM9, RM5, and RM4) the differences between the cross-linked film and its mesophase counterpart can be clearly seen. Typical room temperature photocurrent transients for cross-linked RM9 are shown in Figures 10.9a (holes) and 10.9b (electrons) yielding the relevant carrier mobilities. Table 10.3 lists the electron and hole mobilities for the unreacted materials and cross-linked films obtained by TOF. In the case of RM9 there is a small reduction in both hole and electron mobilities; this cannot be due to trapping from PI residue, as the same PI used in the RM5 reaction resulted in no decrease in hole and electron mobility (in fact, an increase), and is probably due to the molecular packing being affected in RM9 but not in RM5. The modest improvement in both electron and hole mobility found in cross-linked RM5 shows that the cationic PI used in oxetane cross-linking does not adversely affect transport in this system. This is in contrast to the cross-linked RM4 results, where a free-radical PI is used, and its trapping effect on holes is evident by the (one order of magnitude) drop in mobility. The molecular packing in cross-linked RM4 does not appear affected, however, as the electron mobility actually increased post reaction, suggesting that the diene moieties present before the reaction took place, act as electron traps.

In summary, films made from cross-linked RMs can have different charge transport properties to those occurring in the mesophase of an unreacted RM. The presence of PI, necessary for the cross-linking reaction, can act as an impurity, further complicating matters.

### 10.4.5  WEAK TEMPERATURE AND FIELD-DEPENDENT TRANSPORT

In contrast to polymer semiconductors, which typically display strongly field and temperature-dependent mobility (Arkhipov et al., 2007), cross-linked RMs show very weak field and temperature varying behavior. The field independence of the charge transport, in particular, seems an almost universal feature in such systems (Farrar et al.,

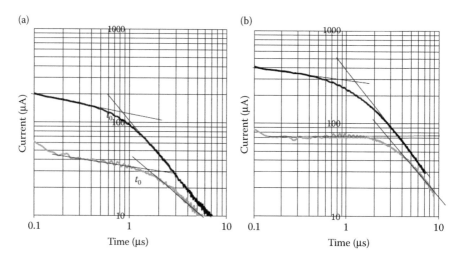

**FIGURE 10.9**  (a) Hole photocurrents in a 2.45 μm thick film of cross-linked RM9 (lower curve 21 V and upper curve 41 V). (b) Electron photocurrents in a 2.45 μm thick film of cross-linked RM9 (lower curve 21 V and upper curve 41 V).

**TABLE 10.3**
**Comparison between Mesophase RM and Polymerized**
**RM Mobilities**

| Material | $\mu_h$ (cm$^2$ V$^{-1}$ s$^{-1}$) | $\mu_e$ (cm$^2$ V$^{-1}$ s$^{-1}$) |
|---|---|---|
| RM9 | $5.8 \times 10^{-3}$ | $5.4 \times 10^{-3}$ |
| polymerized RM9 | $1.6 \times 10^{-3}$ | $1.1 \times 10^{-3}$ |
| RM5 | $9.5 \times 10^{-3}$ | $7.5 \times 10^{-3}$ |
| polymerized RM5 | $1.6 \times 10^{-2}$ | $2.8 \times 10^{-2}$ |
| RM4 | $6.6 \times 10^{-3}$ | $1.7 \times 10^{-4}$ |
| polymerized RM4 | $7.1 \times 10^{-4}$ | $7.5 \times 10^{-4}$ |

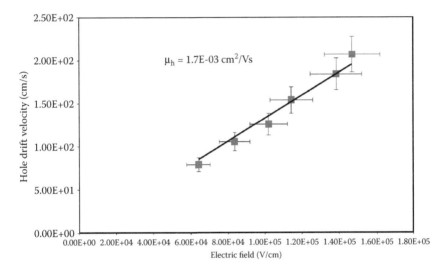

**FIGURE 10.10** Hole drift velocity versus electric field in cross-linked RM9 at room temperature.

2002; Bleyl et al., 1997; Baldwin et al., 2007). This is clearly shown in Figure 10.10, which shows a plot of hole drift velocity, $v_d = d/t_t$, versus electric filed strength, $E = V/d$, in cross-linked RM9 ; the straight line fit through the origin is indicative of field -independent mobility and yields $\mu_h = 1.7 \times 10^{-3}$ cm$^2$ V$^{-1}$ s$^{-1}$ over the range of fields studied. The hole mobility measured in cross-linked RM9 was also found to be temperature independent in the range 20°C to 145°C (Baldwin et al., 2007).

## 10.5   FIELD-EFFECT TRANSISTORS BASED ON RMS

Inorganic FETs are "building blocks" of modern integrated circuits and active-matrix backplanes in liquid crystal televisions and monitors. The same way organic FETs are envisioned to be the fundamental components for pixel switching in flexible displays (Granmar et al., 2005; Yagi et al., 2008), large-area wireless power-transmission sheets

(Sekitani et al., 2007), chemical and biological sensors (Dodabalapur, 2006; Wang et al., 2006), large area lightweight sheet image scanners (Someya et al., 2005), and pressure sensors for robotic skin (Manunza et al., 2006; Someya et al., 2004). In all these applications electrical performance (charge mobility, on/off current ratio), stability (or sensitivity to particular chemicals), and processability are key parameters that need to be improved to realize the enormous potential of organic semiconductor materials.

Contrary to TOF cells, FETs are "interfacial" devices (Singh et al., 2006; Park et al., 2007) where very thin, only few monolayers thick (Alam et al., 1997; Roichman et al., 2004), conducting channel between source and drain electrodes is modulated by the gate potential as shown schematically in Figure 10.11. As charge transport occurs at semiconductor–insulator interface, surface properties play major role in FET behavior. Organic film morphology, including molecular alignment and orientation, in general, is different from that of the bulk TOF devices based on the same RM materials. Injection from metal electrodes (typically gold) could limit current in an FET if there is substantial Schottky barrier between metal work-function and RM charge transport level energy, which can roughly be described as laying close to the molecular HOMO level.

### 10.5.1 TRANSISTOR PREPARATION AND CHARACTERIZATION

Most common types of RM transistors are shown in Figure 10.11. These are fabricated on highly doped crystalline silicon substrates. Typical transistor consists of a RM thin film coated on top of gate insulator and contacted by metal source and drain electrodes (Shkunov et al., 2003, 2004; McCulloch et al., 2006). The spacing between these electrodes defines the transistor channel length, which is varied between 2.5 and 80 $\mu$m. When bias is applied to the gate contact, charges are induced in the transistor channel and its conductivity is controlled by the gate bias. Since RM semiconducting layers are not intentionally doped, charges need to be injected from the metal electrodes. Poor physical contact between organic layer and injecting electrodes as well as high Schottky barrier can result in significant contact resistance effects in transistor current–voltage characteristics. In operation, transistor current is carried in a thin region (approximately 5 nm) next to the gate insulator (Roichman et al., 2004; Alam et al., 1997) and is very sensitive to both molecular order and quality

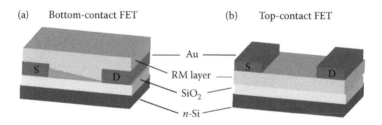

(a)    Bottom-contact FET                    (b)    Top-contact FET

————— Au —————
————— RM layer —————
————— SiO$_2$ —————
————— $n$-Si —————

**FIGURE 10.11** Schematic cross section of: (a) Bottom-gate bottom-contact FET. Charge accumulation layer in saturation regime is shown in gray between source and drain electrodes. (b) Bottom-gate top-contact FET.

of semiconductor–dielectric interface. Hopping dominated transport in RM films is influenced by charge density induced by the gate in the FET channel, resulting in effective gate–voltage dependence of charge carrier mobility (Horowitz, 2007). For a simplified analysis of FETs this dependence can be left out, especially if mobility is only weakly dependant on the gate bias. Qualitative behavior of organic transistors can be described using inorganic thin film transistor models (Sze, 1981), where source–drain current is expressed as

$$I_{sd}^{lin} = \frac{WC_0}{L}\mu^{lin}(V_g - V_0)V_{sd} \quad \text{linear regime} \tag{10.3}$$

$$I_{sd}^{sat} = \frac{WC_0}{2L}\mu^{sat}(V_g - V_0)^2 \quad \text{saturation regime} \tag{10.4}$$

where $W$ is the channel width, $L$ the channel length, $C_0$ the capacitance of unit area of insulating layer, $V_g$ the gate voltage, $V_{sd}$ is the source–drain voltage and $V_0$ is the transistor turn-on voltage.

In the linear regime, when $V_{sd} \ll (V_g - V_0)$, $I_{sd}^{lin}$ is directly proportional to $V_g$ (for a fixed $V_{sd}$) or to $V_{sd}$ (for a fixed $V_g$). In the saturation regime, where $V_{sd} > (V_g - V_0)$, $I_{sd}^{sat}$ is constant, independent of $V_{sd}$.

The transistor turn-on voltage can be related to the offset voltage at which current starts to rise sharply indicating creation of conducting channel between source–drain electrodes. $V_0$ can depend on various factors including unintentional doping of organic layer, trapped charges at the gate insulator and work function difference between the gate and source–drain contacts.

Field effect mobility is usually calculated in linear regime using Equation 10.5 assuming weak dependence of mobility on gate voltage:

$$\left(\frac{\partial I_{sd}^{lin}}{\partial V_g}\right)_{V_{sd}} = \frac{WC_0}{L}\mu^{lin}V_{sd} \tag{10.5}$$

In the presence of significant contact resistance, device mobility calculated in the saturation regime gives better indication of "real" material mobility (at corresponding change density) due to lowered contact resistance values at higher drain voltages (Horowitz, 2007). The expression for saturation mobility (Equation 10.6) can be easily derived for transistor transfer characteristic from Equation 10.4 using linear fit to the slope of $\sqrt{I_{sd}^{sat}}$ versus $V_g$

$$\mu^{sat} = 2\text{slope}^2\frac{L}{WC_0} \tag{10.6}$$

To fabricate RM FETs highly n-doped silicon wafers have been used as substrates. The gate insulator is a silicon oxide ($SiO_2$) layer, typically 230 nm thick, is thermally grown on silicon wafer that serves as a common gate electrode. Transistor source–drain gold contacts are either photolithographically defined on the $SiO_2$ layer, or deposited using thermal evaporator through a shadow mask. Prior to RM layer deposition, FET substrates are treated with silylating agents such as OTS or HMDS that render $SiO_2$ surface hydrophobic and as a result promote homeotropic orientation of RM molecules. Thin semiconductor films are then deposited by spin-coating a solution of a RM (typically 2–4 wt%) with an addition of 0.2–2 wt% (with respect to the RM monomer) of appropriate PI. To study the effect of PI on charge transport, some FET samples are prepared without any addition of PI and later used as reference. To achieve optimum molecular alignment and promote the growth of large domain samples are heated on precision temperature controlled hot stage above clearing point and then slowly cooled to the desired mesophase temperature (typically high order smectic phases). In some cases strong de-wetting of RM films was observed at the clearing point. For these samples heating was performed at lower temperatures (before reaching de-wetting conditions) followed by slow cooling to a particular mesophase.

Cross-linked samples are prepared by irradiation with filtered UV–VIS light using band-pass filters with transmission corresponding to the PI absorption band. The photo-cross-linking is conducted under nitrogen flow, at a temperature in which the RM film exhibited a desired mesophase. To remove unexposed areas, the film is then washed with a solvent. The electrical characterization of the transistor devices is carried out in a dry nitrogen atmosphere glove box, or under ambient air, using computer controlled Agilent 4155C Semiconductor Parameter Analyser. So far only p-type behavior is being observed for RM-based FETs.

## 10.5.2 MOLECULAR ALIGNMENT IN FET CHANNEL

As discussed in Sections 10.1 and 10.2, molecular alignment has profound effect on charge transport in RM films. Due to the in-plane direction of current flow in FET channel between source and drain electrodes at semiconductor–insulator interface two molecular orientations could provide optimum π-electron overlap: homeotropic films (Figure 10.2) with molecular axis orthogonal to the substrate, and planar aligned films with molecular axis in the plane of the substrate and also aligned parallel to the edge of source–drain electrodes. In both cases π–π stacking direction corresponds to charge transport direction.

### 10.5.2.1 Homeotropic Alignment of Smectic Phases

Smectic mesophases, especially higher order SmG, SmH, SmE represent the most ordered molecular layers (Gray et al., 1984). In thin film environment (few tens on nanometer thick films) RM homeotropic alignment prevailed over planar alignment possibly due to molecular side-chain interaction with low surface energy interfaces such as RM/air at the top of the film and RM/HMDS at the bottom of the film.

In all the RMs studied in this work only homeotropic alignment has been observed in smectic mesophases in FET channels. Typical RM film for tilted smectic phases such as SmC, SmG, SmH exhibited mosaic structure, where molecules in each domain had common director, but azimuthal angle of the director was different from domain to domain, when observed by polarized microscopy. Depending on the particular transistor electrode spacing and RM domain size, there were several domains in the FET channel, resulting in not a single, but multidomain charge transport between source–drain electrodes.

### 10.5.3    EFFECT OF PI ON CHARGE TRANSPORT

To study the effect of PI on charge transport in FETs a series of devices were made using RM4, where the amount of PI varied from 0.2 to 2 wt% (in respect to RM weight). FETs were measured in bottom-contact configurations (Figure 10.11a) before cross-linking the RM layer. As shown in Figure 10.12 for FETs with 0.2%, 0.5%, and 2%, transistor current was dropping in devices with increasing amount of PI from 6 µA for 0.2% PI content RM layer FET to 0.4 µA for 2% PI containing device. Corresponding device charge carrier mobility also reduced in proportion to the current. The drop in current is consistent with the suggestion that various impurities, including PI molecules can reduce transistor performance due to introduction of trap centers.

Reduced FET performance was consistently observed in both bottom- and top-contact FET when comparing RM layer with and without PI. Adding different PI (Irgacure 784 instead of IS12207) had the same effect on FET charge transport, resulting in approximately 5–50 times drop in current in devices containing 2% wt (Irgacure 784) compared with pristine RM FETs containing no PI.

### 10.5.4    EFFECT OF CROSS-LINKING ON FET PERFORMANCE

Cross-linking of polymerizable groups on RM molecules, results in a physical movement of these groups to accommodate the spatial decrease from van der Waals distances to that of covalent bonds. The effective shrinkage of the film and also the rigidity of the formed polymer backbone can have dramatic impact on charge transport in polymerized RM layers due to possible misalignment of π-conjugated cores in respect to each other and transport direction as well. For free-radical polymerized RMs (RM1, RM2, RM3, RM4, RM6, RM7, RM8—Table 10.1) the cross-linking always resulted in the decrease of charge-carrier mobility. This effect has also been

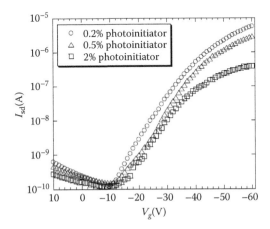

**FIGURE 10.12**    Transfer characteristics of bottom-contact RM4 FETs prepared with 0.2%, 0.5%, 2% PI (IS12207). Characteristics were measured in saturation regime at $V_{sd} = -60$ V. FET channel length $L = 10$ µm, and channel width $W = 1$ cm.

observed in thermally polymerized RM layers in FETs (Huisman et al., 2003). On the contrary, in cationically polymerized molecules (RM5, RM9, RM10, RM11—Table 10.1) mobility was increasing after the cross-linking process in full agreement with TOF results in Section 10.4.

### 10.5.4.1 FET Based on Diene-Group RMs

Diene-group RMs require free-radical polymerization reaction step to convert monomeric units to an insoluble polymer network. The diene-group RM layers were typically prepared by spin-coating 2 wt% solutions in butanone with an addition of 2 wt% PI Irgacure 784 (in respect to RM) or 0.5 wt% of PI IS12207 followed by photopolymerization within high-order smectic mesophase. The best FET results were obtained in bottom-gate top-contact geometry where source–drain electrodes were deposited through a shadow mask on top of RM layers. Results were obtained for two sets of initially identical samples where in the first set of RM films remained un-cross-linked and in the second set RMs have been photo cross-linked. The amount of PI was kept the same for both sets of samples. Figure 10.13 shows transfer characteristics for both cross-linked and un-cross-linked devices plotted as square root of saturated current versus gate voltage. Thin solid lines demonstrate good linear fit to the above threshold part of the curves, indicating constant mobility for these voltage ranges. In the polymerized FETs current was lowered to 0.6 μA compared to 1.7 μA for unpolymerized sample, and corresponding mobility dropped to $1.3 \times 10^{-3}$ cm²/V s in contrast to $3 \times 10^{-3}$ cm²/V s.

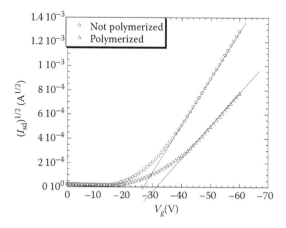

**FIGURE 10.13** Square root of saturated current transfer characteristics of top-contact RM4 FETs prepared with 0.5% PI (IS12207). Open circles: un-cross-linked RM layer FET. Open triangles: RM layer was crosslinked in SmG phase at 150°C by exposing the film to 18 mW/cm² filtered UV–VIS light (350–500 nm) for 15 min. Characteristics were measured at room temperature at $V_{sd} = -60$ V. FET channel length $L = 80\,\mu$, and channel width $W = 5$ mm. Thin lines are linear fits of $\sqrt{I_{sd}^{sat}}$ versus $V_g$.

For a number of FET devices polymerization step was typically accompanied by two to eight times reduction of mobility, in some cases reaching 30, depending on PI and polymerization conditions.

### 10.5.4.2  FET Based on Oxetane-Group RMs

Oxetane-group RMs were cross-linked using cationic polymerization reaction using triarylsulfonium hexafluoroantimonate salts as PIs. 10-biphenyl-4-yl-2-isopropyl-9-oxo-9H-thioxanthen-10-ium hexafluorphosphate (Omnicat 550 by IGM Resins) has also been successfully used as cationic PI. The FET fabrication procedure was in general similar to that of diene-group RMs devices, but with different UV exposure conditions. Due to faster kinetic of cationic polymerization reaction the illumination time was shortened to 5 min, and UV filter center wavelength was fixed at 308 nm to coincide with PI absorption band and at the same time with the minimum of RM absorption in this spectral region. Performance of un-cross-linked oxetane RM devices was low with mobility reaching only $4 \times 10^{-5}$ cm$^2$/V s and a large hysteresis ($\sim$15 V) for forward and reverse transfer scans. After cross-linking FET characteristics improved with mobility increasing five fold to $2 \times 10^{-4}$ cm$^2$/V s and no hysteresis in transfer scans. Figure 10.14 shows FET transfer characteristics in saturation regime for sample before and after polymerization. Low charge carrier mobility and strong hysteresis for unpolymerized RM films could be attributed to the presence of cationic PI. Due to its ionic nature, charges could be trapped at PI molecules, resulting in low mobility, and ionic species can move in RM film under the influence of the drain bias causing significant hysteresis effect. Following cross-linking step, PI decomposes and

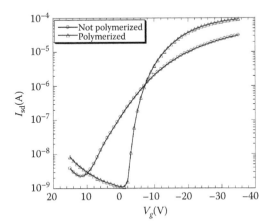

**FIGURE 10.14**  Transfer characteristics of bottom-contact RM9 FETs. Open circles: un-cross-linked device. Open triangles: cross-linked RM layer FET (cross-linking conditions: PI UVI 6976, photopolymerized in SmG phase at 100°C; by exposed to UV light (narrow band-pass filter centered at 308 nm) at 4 mW/cm$^2$ intensity for 5 min. Characteristics were measured in saturation regime at $V_{sd} = -18$ V at room temperature. FET channel length $L = 10$ µm, and channel width W = 31.2 cm.

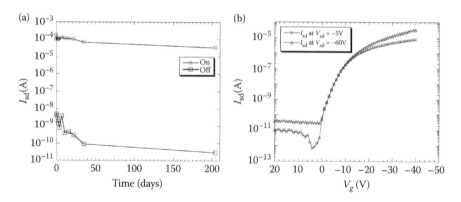

**FIGURE 10.15** Ambient stability of bottom-contact un-cross-linked RM4 FETs (a) On/Off ratio versus time in ambient air: circles–On current, squares–Off current (b) Transfer characteristics after 204 days exposure to air. Open circles: linear regime ($V_{sd} = -5$ V); open triangles: saturation regime ($V_{sd} = -60$ V). FET channel length $L = 10$ μm, and channel width W = 2 cm.

its fragments are likely to get immobilized within polymerized matrix, resulting in improved FET performance.

### 10.5.5 AMBIENT STABILITY OF RM FETs

Environmental stability of RM based FETs was studied by monitoring device performance in ambient air at fixed time intervals using bottom-gate bottom-contact device architecture. RM layer remained unprotected and no encapsulation layers have been used. Immediately after exposure to air, transistors were measured at shorter intervals, usually in tens of minutes, then hours, and later at longer time delays (days). RM4 FETs demonstrated the best ambient stability with only minor degradation of transistor characteristics. Transistor On/Off ratio defined as $I_{on}/I_{off}$, (where $I_{on}$ is the on current at highest gate bias, and $I_{off}$ is the off current at a gate bias, where device is switched off) was plotted as a function of time as shown in Figure 10.15a. More than 200 days after exposure to air RM4 transistors demonstrated excellent characteristics with On/Off ratio as high as $10^6$, turn-on voltage at 0 Volts, and also well-defined linear and saturation regimes (Figure 10.15b). Due to high ionization potential of 5.45 eV for RM4 core, electrochemical oxidation of this semiconducting molecule was minimal, resulting in high ambient stability of FET devices.

## 10.6  CONCLUDING REMARKS

Development of solution-processable organic semiconductors progressed substantially in the last few years. In organic electronic applications performance of semiconducting component is as important as processability of the whole device. In this respect, conjugated RMs are compatible with solution coating and printing technologies due to substantially higher solubility in typical organic solvents compared to nonreactive analogues. Moreover, excellent quality thin RM semiconducting films can be obtained using solution coating methods. The RM layers can be photopolymerized and thus create stable insoluble molecular networks, effectively working as negative photoresist. The morphology of these networks preserves structure of the films before polymerization, including the organization of higher-order smectic liquid crystal phases.

Due to higher degree of molecular order in smectic phases, such as SmB, SmG, SmH, charge carrier mobility is higher than in lower order mesophases, and is substantially higher than in isotropic phase, indicating that LC ordering is preferential for charge transport.

As measured by TOF technique, bulk charge transport in polymerized RM films is ambipolar and also is temperature independent for a very wide temperature range. The mobilities in access of $2 \times 10^{-2}$ cm$^2$/V s for both holes and electrons have been demonstrated for these cross-linked films.

In FET devices only p-type transport has been observed. Mobility of holes in FETs based on unpolymerized RM films is comparable to TOF with the values reaching $10^{-2}$ cm$^2$/V s. However, in polymerized RM FETs hole mobility is lower approximately by a factor of 10 reaching $10^{-3}$ cm$^2$/V s.

In summary, efficient bulk mobility makes RM approach highly suitable for charge transport layers in organic light-emitting diodes and solar cells. FETs still require some optimization with the potential to exceed mobility shown by TOF method. Yet, FET devices demonstrate excellent environmental stability for completely unprotected RM layers. Cross-linked RM films are impervious to processing solvents thus allowing multilayer device fabrication.

## ACKNOWLEDGMENTS

Authors would like to thank all the Merck synthetic chemists for the synthesis of the compounds discussed in the chapter, particularly Dr. Martin Heeney and Dr. Weimin Zhang. Prof. John Goodby (University of York, UK) is acknowledged for valuable discussions on identification of mesophases and Prof. Peter Strohriegl (Universität Bayreuth, Germany) for stimulating consultations on polymerization aspects of RMs.

## REFERENCES

Adam, D., P. Schuhmacher, J. Simmerer, et al. 1994. Fast photoconduction in the highly ordered columnar phase of a discotic liquid-crystal. *Nature* 371(6493):141–143.

Alam, M. A., A. Dodabalapur, and M. R. Pinto. 1997. A Two-dimensional simulation of organic transistors. *IEEE Trans. Electron Devices* 44(8):1332–1337.

Arkhipov, V. I., I. I. Fishchuk, A. Kadashchuk, and H. Bässler, eds. 2007. *Charge Transport in Neat and Doped Random Organic Semiconductors*. Eds., G. Hadziioannou and G. G. Malliaras. 2 vols. Vol. 1, *Semiconducting Polymers*. New York, NY: Wiley.

Baldwin, R. J., T. Kreouzis, M. Shkunov, et al. 2007. A comprehensive study of the effect of reactive end groups on the charge carrier transport within polymerized and nonpolymerized liquid crystals. *J. Appl. Phys.* 101(2):10.

Bleyl, I., C. Erdelen, K. H. Etzbach, et al. 1997. Photopolymerization and transport properties of liquid crystalline triphenylenes. *Mol. Cryst. Liq. Cryst. Sci. Technol., Sect. A - Mol. Cryst. Liq. Cryst.* 299:149–155.

Brabec, C. J. 2004. Organic photovoltaics: technology and market. *Sol. Energy Mater. Sol. Cells* 83 (2–3):273–292.

Bredas, J. L., D. Beljonne, V. Coropceanu, and J. Cornil. 2004. Charge-transfer and energy-transfer processes in pi-conjugated oligomers and polymers: A molecular picture. *Chem. Rev.* 104(11):4971–5003.

Broer, D. J. 1995. Creation of supramolecular thin film architectures with liquid-crystalline networks. *Mol. Cryst. Liq. Cryst. Sci. Technol.*, Sect A 261:513–523.

Broer, D. J., J. Boven, G. N. Mol, and G. Challa. 1989. *In situ* photopolymerization of oriented liquid-crystalline acrylates 3. Oriented polymer networks from a mesogenic diacrylate. *Makromol. Chem.-Macromol. Chem. Phys.* 190(9):2255–2268.

Broer, D. J., G. Challa, and G. N. Mol. 1991. *Macromol. Chem.* 192:59.

Broer, D. J., H. Finkelmann, and K. Kondo. 1988. *In situ* photopolymerization of an oriented liquid-crystalline acrylate. *Makromol. Chem.-Macromol. Chem. Phys.* 189 (1):185–194.

Broer, D. J., J. Lub, and G. N. Mol. 1995. Wide-band reflective polarizers from cholesteric polymer networks with a pitch gradient. *Nature* 378(6556):467–469.

Carrasco-Orozco, M., C. T. Wing, M. O'Neill, et al. 2006. New photovoltaic concept: Liquid-crystal solar cells using a nematic gel template. *Adv. Mater. (Weinheim, Ger.)* 18 (13):1754–1758.

Coates, D., O. Parri, M. Verrall, K. Slaney, and S. Marden. 2000a. Aligned LC. *Macromol. Symp.* 154:59.

Coates, D., O. Parri, M. Verrall, K. Slaney, and S. Marden. 2000b. Polymer films derived from aligned and polymerised reactive liquid crystals. *Macromol. Symp.* 154:59–71.

Contoret, A. E. A., S. R. Farrar, M. O'Neill, and J. E. Nicholls. 2002. The photopolymerization and cross-linking of electroluminescent liquid crystals containing methacrylate and diene photopolymerizable end groups for multilayer organic light-emitting diodes. *Chem. Mater.* 14:1477–1487.

Coropceanu, V., J. Cornil, D. A. daSilvaFilho, et al. 2007. Charge transport in organic semiconductors. *Chemical Reviews* 107(4):926–952.

Dimitrakopoulos, C. D. and D. J. Mascaro. 2001. Organic thin-film transistors: A review of recent advances. *IBM J. Res. Dev* 45(1):11–27.

Dodabalapur, A. 2006. Organic and polymer transistors for electronics. *Mater. Today* 9 (4):24–30.

Donovan, K. J. and T. Kreouzis. 2000. Deconvolution of time of flight photocurrent transients and electronic response time in columnar liquid crystals. *J. Appl. Phys.* 88 (2):918–923.

Farrar, S. R., A. E. A. Contoret, M. O'Neill, et al. 2002. Nondispersive hole transport of liquid crystalline glasses and a cross-linked network for organic electroluminescence. *Phys. Rev. B* 66(12):5.

Funahashi, M. and J. Hanna. 1997. Fast ambipolar carrier transport in smectic phases of phenylnaphthalene liquid crystal. *Appl. Phys. Lett.* 71(5):602–604.

Funahashi, M. and J. I. Hanna. 1999. Carrier transport in calamitic mesophases of liquid crystalline photoconductor 2-phenylnaphthalene derivatives. *Mol. Cryst. Liq. Cryst. Sci. Technol., Sect. A Mol. Cryst. Liq. Cryst.* 331:2369–2376.

Granmar, M. and A. Cho. 2005. Technology—Electronic paper: A revolution about to unfold? *Science* 308(5723):785–786.

Gray, G. W. and J. W. G. Goodby. 1984. *Smectic Liquid Crystals*. Glasgow: Leonard Hill.

Hall, A. W., M. J. Godber, K. M. Blackwood, P. E. Y. Milne, and J. W. Goodby. 2004. The photoinitiated cyclopolymerization of dienes in the creation of novel polymeric systems and three-dimensional networks. *The Roy. Soc. Chem.* 14:2593–2602.

Horowitz, G., ed. 2007. *Organic Thin-Film Transistors*. Edited by G. Hadziioannou and G. G. Malliaras. 2 vols. Vol. 2, *Semiconducting Polymers*: Wiley.

Huisman, B.-H., J. P. V. Josue, N. Wim, L. Johan, and ten H. Wolter. 2003. Oligothiophene-based networks applied for field-effect transistors. *Adv. Mater.* 15(23):2002–2005.

Katz, H. E., J. G. Laquindanum, and A. J. Lovinger. 1998. Synthesis, solubility, and field-effect mobility of elongated and oxa-substituted alpha, omega-dialkyl thiophene oligomers. Extension of "polar intermediate" synthetic strategy and solution deposition on transistor substrates. *Chem. Mater.* 10 (2):633–638.

Kepler, R. G. 1960. Charge carrier production and mobility in anthracene crystals. *Phys. Rev.* 119 (4):1226–1229.

Kreouzis, T., R. J. Baldwin, M. Shkunov, et al. 2005. High mobility ambipolar charge transport in a cross-linked reactive mesogen at room temperature. *Appl. Phys. Lett.* 87(17):172110.

LeBlanc, O. H. 1959. Electron drift mobility in liquid n-hexane. *J. Chem. Phys.* 30(6): 1443–1447.

Lestel, L., G. Galli, M. Laus, and E. Chiellini. 1994. Thermal and dynamic-mechanical properties of new chiral smectic networks. *Poly. Bull.* 32(5–6):669–674.

Lub, J., P. van de Witte, C. Doornkamp, J. P. A. Vogels, and R. T. Wegh. 2003. Stable photopatterned cholesteric layers made by photoisomerization and subsequent photopolymerization for use as color filters in liquid-crystal displays. *Adv. Mater.* 15(17):1420–1425.

Manunza, I., A. Sulis, and A. Bonfiglio. 2006. Pressure sensing by flexible, organic, field effect transistors. *Appl. Phys. Lett.* 89(14):143502-1–143502-3.

McCulloch, I., C. Bailey, K. Genevicius, et al. 2006. Designing solution-processable air-stable liquid crystalline crosslinkable semiconductors. *Philos. Trans. R. Soc., A* 364 (1847):2779–2787.

McCulloch, I., M. Coelle, K. Genevicius, et al. 2008. Electrical properties of recative liquid crystal semiconductors. *Jpn. J. Appl. Phys.* 47(1):488–491.

McCulloch, I., W. Zhang, M. Heeney, et al. 2003. Polymerizable liquid crystalline organic Semiconductors and their fabrication in organic field effect transistors. *J. Mater. Chem.* 13(10):2436–2444.

Mushrush, M., A. Facchetti, M. Lefenfeld, H. E. Katz, and T. J. Marks. 2003. Easily processable phenylene-thiophene-based organic field-effect transistors and solution-fabricated nonvolatile transistor memory elements. *J. Am. Chem. Soc.* 125 (31):9414–9423.

O'Neill, M. and S. M. Kelly. 2003. Liquid crystals for charge transport, luminescence, and photonics. *Adv. Mater.* 15(14):1135–1146.

Park, Y. D., J. A. Lim, H. S. Lee, and K. Cho. 2007. Interface engineering in organic transistors. *Materials Today* 10(3):46–54.

Podzorov, V., S. E. Sysoev, E. Loginova, V. M. Pudalov, and M. E. Gershenson. 2003. Single-crystal organic field effect transistors with the hole mobility approx. 8 cm$^2$/Vs. *Appl. Phys. Lett.* 83(17):3504–3506.

Roichman, Y., Y. Preezant, and N. Tessler. 2004. Analysis and modeling of organic devices. *Phys. Stat. Sol. a-Appl. Res.* 201(6):1246–1262.

Sekitani, T., M. Takamiya, Y. Noguchi, et al. 2007. A large-area wireless power-transmission sheet using printed organic transistors and plastic MEMS switches. *Nat. Mater.* 6 (6):413–417.

Shkunov, M., W. Zhang, C. Bailey, et al. 2004. Self-assembled liquid crystalline solution processable semiconductors. *SPIE-The Internat. Soc. Opt. Eng.* 5464:60–71.

Shkunov, M. N., Z. Weimin, G. David, et al. 2003. New liquid crystalline solution processible organic semiconductors and their performance in field effect transistors. *SPIE-The International Society for Optical Engineering* 5217 (Organic Field Effect Transistors II):181–192.

Singh, T. B. and N. S. Sariciftci. 2006. Progress in plastic electronics devices. *Ann. Rev. Mater. Res.* 36:199–230.

Sirringhaus, H., P. J. Brown, R. H. Friend, et al. 2000a. Microstructure-mobility correlation in self-organised, conjugated polymer field-effect transistors. *Syn. Metals* 111:129–132.

Sirringhaus, H., R. J. Wilson, R. H. Friend, et al. 2000b. Mobility enhancement in conjugated polymer field-effect transistors through chain alignment in a liquid-crystalline phase. *Appl. Phys. Lett.* 77(3):406–408.

Someya, T., Y. Kato, S. Iba, et al. 2005. Integration of organic FETs with organic photodiodes for a large area, flexible, and lightweight sheet image scanners. *IEEE Trans. Electron Dev.* 52(11):2502–2511.

Someya, T., T. Sekitani, S. Iba, et al. 2004. A large-area, flexible pressure sensor matrix with organic field-effect transistors for artificial skin applications. *Proc. Natl. Acad. Sci. U.S.A.* 101(27):9966–9970.

Sze, S. M. 1981. *Physics of Semiconductor Devices*. 2nd edition. New York, NY: John Wiley & Sons.

Thiem, H., P. Strohriegl, M. Shkunov, and I. McCulloch. 2005. Photopolymerization of reactive mesogens. *Macromol. Chem. Phys.* 206(21):2153–2159.

van de Craats, A. M., N. Stutzmann, O. Bunk, et al. 2003. Meso-epitaxial solution-growth of self-organizing discotic liquid-crystalline semiconductors. *Adv. Mater.* 15(6):495–499.

Wang, L., D. Fine, D. Sharma, L. Torsi, and A. Dodabalapur. 2006. Nanoscale organic and polymeric field-effect transistors as chemical sensors. *Anal. Bioanal. Chem.* 384(2): 310–321.

Whitehead, K. S., M. Grell, D. D. C. Bradley, M. Jandke, and P. Strohriegl. 2000. Highly polarized blue electroluminescence from homogeneously aligned films of poly (9,9-dioctylfluorene). *Appl. Phys. Lett.* 76(20):2946–2948.

Wittmann, J. C. and P. Smith. 1991. Highly oriented thin-films of poly(tetrafluoroethylene) as a substrate for oriented growth of materials. *Nature* 352(6334):414–417.

Woon, K. L., M. P. Aldred, P. Vlachos, et al. 2006. Electronic charge transport in extended nematic liquid crystals. *Chem. Mater.* 18(9):2311–2317.

Yagi, I., N. Hirai, Y. Miyamoto, et al. 2008. A flexible full-color AMOLED display driven by OTFTs. *J. Soc. Inform. Disp.* 16(1):15–20.

Yoshimoto, N. and H. Jun-Ichi. 2002. A novel charge transport material fabricated using a liquid crystalline semiconductor and crosslinked polymer. *Adv. Mater. (Weinheim, Ger.)* 14(13–14):988–991.

# 11 Reactive Mesogens in Organic Light-Emitting Devices

*Peter Strohriegl*

## CONTENTS

## 11.1 INTRODUCTION

The focus of this chapter is the application of reactive mesogens in organic light-emitting devices (OLEDs). OLEDs have attracted enormous interest since 1986 when C. Tang and S. vanSlyke described the first thin layer OLED with high brightness and a low operating voltage [1]. The basic structure and the materials used in C. Tang's OLED are shown in Figure 11.1. Two different organic materials have been used, the electron conductor and green emitter tris(8-hydroxyquinoline)aluminum (Alq$_3$) and the hole conducting aromatic amine 1,1-bis[(di-4-tolylamino)phenyl]cyclohexane (TAPC). Both materials are deposited on top of an indium tin oxide (ITO) glass by vacuum evaporation.

Meanwhile, OLEDs have reached a high level of perfection and exhibit high efficiencies. This goal has been reached by optimizing the single layers and by adding additional layers to the OLEDs. State-of-the-art OLEDs exhibit up to seven layers, which are responsible for hole injection, hole transport, light emission, electron transport, and electron injection, respectively. Additional hole- or electron-blocking layers are also frequently used. In Section 11.2, a brief introduction into OLED materials will be given. For a detailed description of the different layers and the materials used the reader is referred to a number of excellent books and review articles on OLEDs [2–6].

Today, the major applications of OLEDs are displays in MP3 players, cellular phones, and digital cameras. Such displays consist of single pixels in the three basic

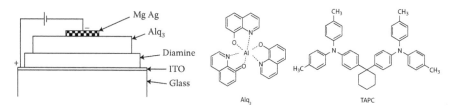

**FIGURE 11.1** Basic structure and materials used in the first efficient thin layer OLED. (Adapted from Tang, C. and vanSlyke, S.A., *Appl. Phys. Lett.* **1987**, *51*, 913.)

colors red, green, and blue (RGB). In the displays that have been commercialized until today, the pixels are defined by vacuum evaporation of low molar mass materials like Alq3 and TAPC through shadow masks. If polymers are used to fabricate OLED displays, the pixels are usually made by various printing techniques, but with these techniques it is still more difficult to fabricate OLED displays. In the last few years, the field of white phosphorescent OLEDs for lighting has emerged as a second upcoming of OLED application.

## 11.2 OLED MATERIALS: A SHORT OVERVIEW

There are two classes of materials from which OLED devices can be made, low molar mass compounds, which are vacuum evaporated [1] and polymers, which are manufactured from solution [7]. Figure 11.2 schematically shows a multilayer OLED fabricated by vacuum evaporation of small molecules. The OLED consists of a transparent, conductive ITO-electrode, a hole injection and a hole transport layer (HTL), an emission layer (EL), a hole blocking layer (HBL), an electron transport and an electron injection layer (EIL), and a metal cathode. Not all layers shown in Figure 11.2 have to be necessarily present in OLEDs, but over the years it turned out that a stack like it is shown in Figure 11.2 or similar setups lead to bright and highly efficient OLEDs.

In this chapter the materials used in the different layers will be introduced. As a first layer on top of the conductive ITO a hole injection layer (HIL) is often used. This layer can be either an evaporated small molecule, for example, copper phthalocyanine, or a conductive polymer like poly(3,4-ethylenedioxythiophene) (PEDT) shown in Figure 11.3 [8]. The role of the HIL is to match the highest occupied molecular

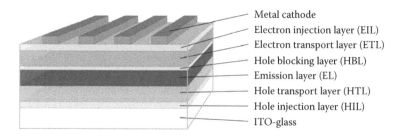

Metal cathode
Electron injection layer (EIL)
Electron transport layer (ETL)
Hole blocking layer (HBL)
Emission layer (EL)
Hole transport layer (HTL)
Hole injection layer (HIL)
ITO-glass

**FIGURE 11.2** Scheme of a multilayer OLED.

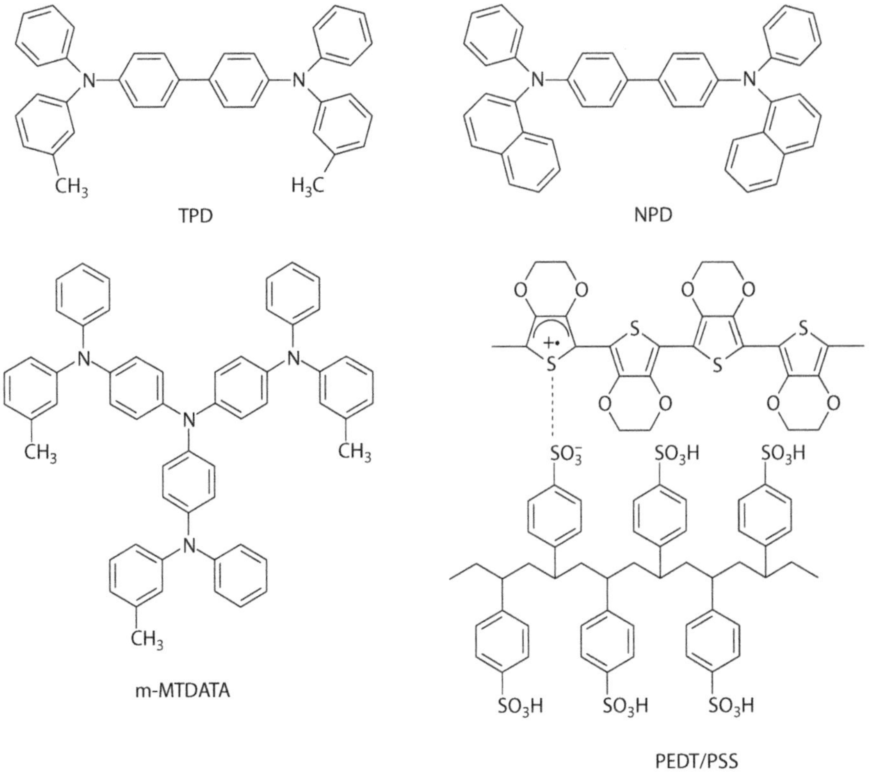

**FIGURE 11.3** Examples of typical hole transport materials used in OLEDs and molecular structure of the conductive polymer PEDT/PSS.

orbital (HOMO) levels of the ITO and the aromatic amines, which are normally used as hole transport materials and, especially in the case of PEDT, to smoothen the ITO surface.

Aromatic amines are the most common hole transport materials in OLEDs. In Figure 11.3 a number of aromatic amines are shown. $N,N'$-diphenyl-$N,N'$- bis(3-methylphenyl)-1,1'biphenyl-4,4'-diamine (TPD) is known for a long time and has been used in photoconductor drums before it became a popular hole conducting material in OLEDs [9]. Since TPD has a relatively low glass transition temperature ($T_g$) of 60°C it has been replaced by $N,N'$-bis(1-naphthyl)-$N,N'$-diphenyl-(1,1'-biphenyl)-4,4'diamine (NPD) with a $T_g$ of 98°C. An important aspect that affects the performance of an OLED is the morphological stability of the organic thin film layers. The star shaped aromatic amine 4,4',4''-tris(3-methylphenylphenylamino)triphenylamine (m-MTDATA) forms a stable amorphous phase and functions as an excellent hole transport material. m-MTDATA is an example for a group of materials often called "molecular glasses" which was introduced by Shirota [4,10].

In Figure 11.4 some common emitter materials are shown. Dimethylquinacridone and 4-(dicyanomethylene)-2-methyl-6-(4-dimethylaminostyryl)-4H-pyran (DCM)

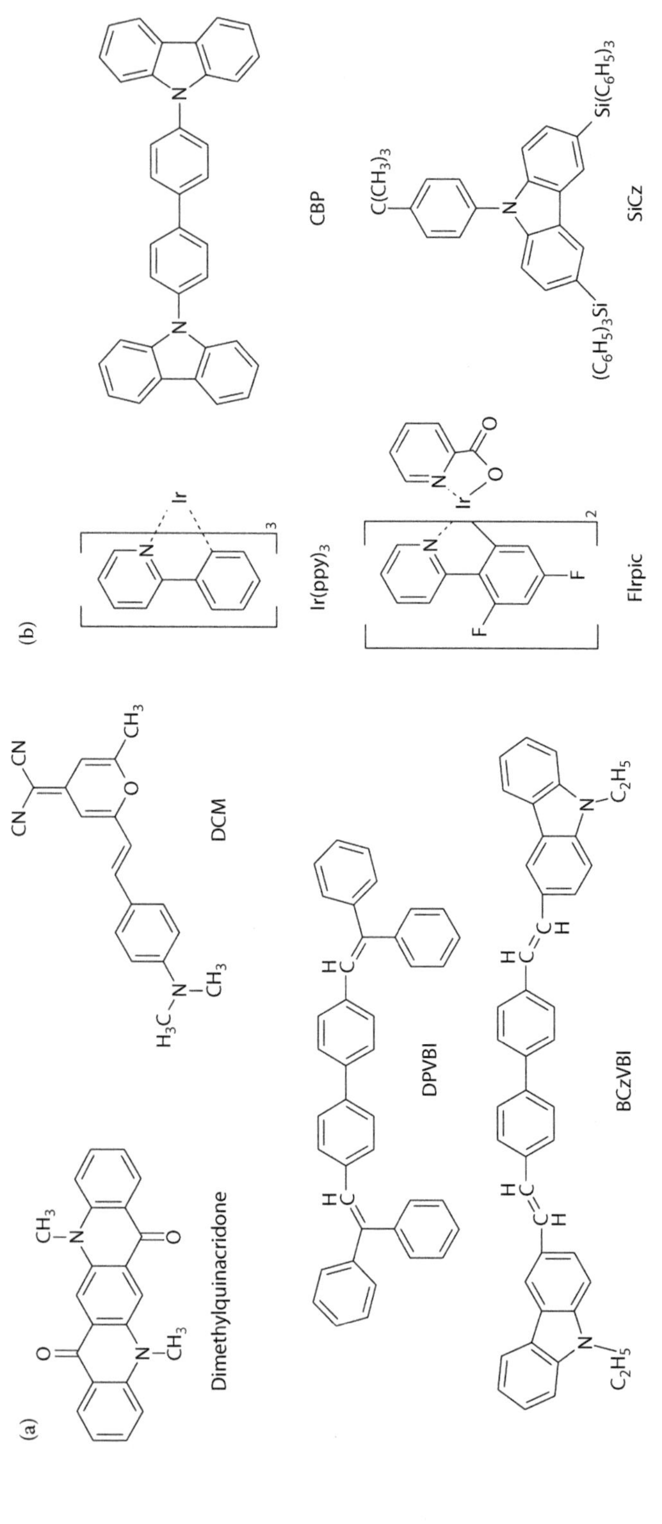

**FIGURE 11.4** Chemical structures of some common fluorescent (a) and phosphorescent (b) emitters and matrix materials.

are fluorescent green and red emitters, which are doped into suitable matrix materials like Alq$_3$. 4,4′-Bis(2-(9-ethylcarbazole-3-yl))ethylen-1-yl)biphenyl (BCzVBI) is an efficient fluorescent blue emitter and is used in a matrix of the distyrylarylene DPVBI (4,4′-bis(2,2-diphenylethylen-1-yl)biphenyl) [11]. In fluorescent OLEDs the internal quantum efficiency is limited to 25% due to spin statistics. With phosphorescent emitters both singlet and triplet excitons contribute to the electroluminescence (EL) and hence the theoretical limit of the quantum efficiency rises to 100%. In 1999 Forrest and Thompson introduced the green phosphorescent emitter Tris(phenylpyridyl)iridium (Ir(ppy)$_3$), which is used as a dopant in a 4,4′-Bis(carbazol-9-yl)biphenyl (CBP) matrix [12]. CBP has a high (singlet ground state-first excited triplet state) $S_0$–$T_1$ band gap of 2.55 eV and is well suited as matrix material for Ir(ppy)$_3$. FIrpic (Iridium-bis(4,6-difluorophenyl)-pyridinato-N,C$^{2'}$)picolinate) is a phosphorescent greenish-blue emitter. For such emitters, matrix materials with a $S_0$–$T_1$ band gap of almost 3 eV like 3,6-(triphenylsilyl)-9-(4-$tert$. butylphenyl)carbazole (SiCz) [13] are necessary to ensure an efficient energy transfer from the host material to the emitter.

Figure 11.5 shows the structures of typical electron conductors. Alq$_3$, which can act both as an electron conductor and emitter has been described already in C. Tang's first paper on OLEDs and is still in use today. Besides metal complexes electron deficient heterocycles are frequently used as electron conductors. In Figure 11.5 the triazole derivative TAZ (3-(4-biphenylyl)-4-phenyl-5-$tert$-butylphenyl-1,2,4-triazole), the oxadiazole PBD (2-biphenyl-5-(4-$tert$-butylphenyl)-1,3,4-oxadiazole), and the phenylquinoxaline TPBI (1,3,5-tris (1-phenyl-1$H$-benzimidazol-2-yl)benzene) serve as typical examples. 2,9-Dimethyl-4,7-diphenyl-1,10-phenanthroline (Bathocuproine) (BCP) has a very low lying HOMO level and is often used as a HBL.

In Figure 11.6 the chemical structures of a number of polymers used in OLEDs is shown. EL in polymers was discovered in poly(1,4-phenylenevinylene) (PPV) in 1990 [7]. Since PPV itself is completely insoluble and thin films are only accessible by thermal conversion of a soluble polyelectrolyte precursor, a number of soluble PPV-derivatives have been developed. Among these are poly-(2-methoxy-5-(2-ethylhexyloxy)-1,4-phenylenevinylene) (MeH-PPV) [14] and a number of soluble PPVs developed by Covion (Merck Organic Semiconductors) [15]. With PPV emitters the green and the red part of the visible spectrum are accessible. The best known blue emitters are polyfluorene homo- and copolymers [16,17].

Polymer OLEDs usually have much less layers compared to OLEDs prepared by vacuum evaporation. This is due to the fact that the deposition of a second polymer layer often leads to problems since the first polymer is partially dissolved. This problem can in some cases be overcome by orthogonal solvents, which means that the solvent for the second layer is a nonsolvent for the first one. A good example is poly(3,4-ethylenedioxythiophene/polystyrene sulfonic acid (PEDT/PSS), which is used as aqueous suspension and does not dissolve in common organic solvents so that other conjugated polymers can be deposited on top of PEDT/PSS. Nevertheless, orthogonal solvents are often not available for a pair of conjugated polymers and this makes the preparation of multilayer OLEDs difficult. One alternative for multilayer

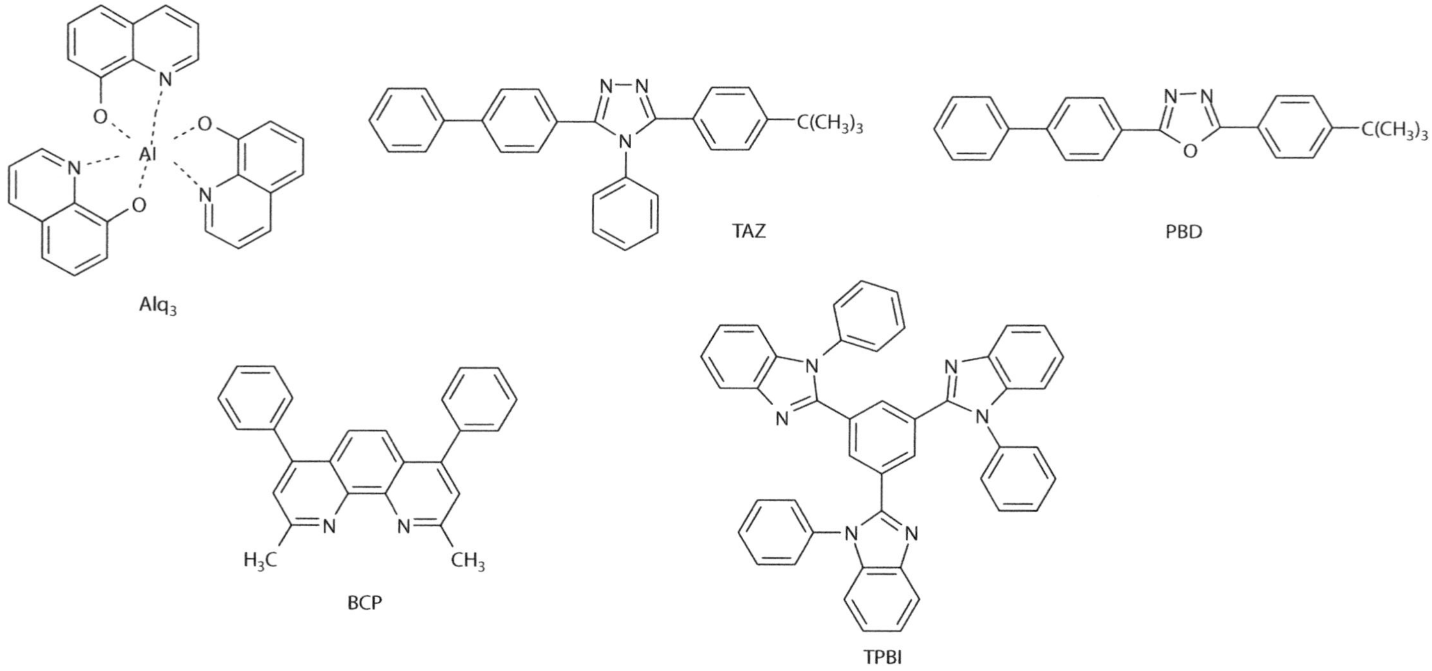

**FIGURE 11.5** Chemical structures of typical electron conducting and hole blocking materials.

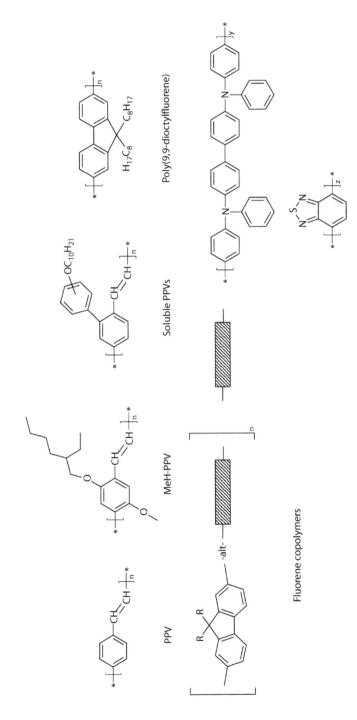

**FIGURE 11.6** Chemical structures of conjugated polymers frequently used in OLEDs.

OLEDs are conjugated polymers with photocrosslinkable units. These polymers are spin coated and then photochemically crosslinked, which makes the layers insoluble so that the next polymer layer can be coated on top. Photocrosslinking is discussed in detail in Section 11.5.

## 11.3   WHY LIQUID CRYSTALS? WHY REACTIVE MESOGENS?

Liquid crystals (LC) nowadays are the most versatile display technology. The LC-displays manufactured today range from small, low-cost displays for wristwatches and calculators to large active matrix TV screens with sizes of more than 1 m.

This gives rise to the question if the huge amount of knowledge on the self-assembling properties of liquid crystals can be advantageously used in the upcoming field of OLED displays. To my opinion there are *two major aspects*, which make the use of oriented liquid crystalline materials in OLEDs attractive.

The first is the fact that OLEDs in which rod-like emitters are aligned parallel to each other directly emit polarized light. This may lead to large area polarized light sources for applications like liquid crystal display (LCD) backlights [18]. Spatial patterning of the polarization direction, which will be discussed later in this review can form the basis of three-dimensional displays [19].

The second important property of well-aligned LC-films from conjugated materials is their high charge-carrier mobility. Mobilities up to $10^{-1}$ cm$^2$/V s have been demonstrated in highly ordered discotic [20] and smectic [21] LC-phases. The high carrier mobilities of liquid crystalline polymers [22] and of networks from reactive mesogens [23] are of particular interest for the development of organic field-effect transistors (OFETs). This aspect will be covered in a separate chapter of this book and is not in the focus of this review.

The use of reactive mesogens in optoelectronic applications offers an additional benefit compared to low molar mass liquid crystals and LC-polymers. Well ordered, densely crosslinked polymer networks are obtained by polymerization of reactive mesogens. The LC-phase in such networks is stable over a broad temperature range up to the thermal decomposition temperature. Furthermore, the LC-networks become completely insoluble upon crosslinking. This allows the preparation of multilayer devices using solution based processes. In many cases the crosslinking of the reactive mesogens is carried out by a photochemical reaction of acrylate or other reactive groups in the molecule (Figure 11.7). If a photomask is applied in addition, the material behaves like a conventional negative photoresist. By dissolving the noncrosslinked parts of the film, high-resolution patterns are obtained. Resolutions down to 1 μm are easily obtained [24], which is much smaller compared to the resolution of usual printing techniques.

The above-mentioned advantages of liquid crystals compared to isotropic materials and the benefits from crosslinking reactive mesogens will serve as an outline of this review. Section 11.4 deals with the orientation of liquid crystals in OLEDs and describes the work on polarized OLEDs. In Section 11.5 the use of photocrosslinked layers in OLEDs will be reported. In Section 11.6 both aspects are combined and the use of crosslinked layers in polarized OLEDs is discussed.

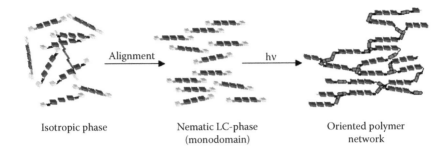

Isotropic phase      Nematic LC-phase      Oriented polymer
(monodomain)      network

**FIGURE 11.7** Photolithographic patterning of reactive mesogens.

## 11.4 POLARIZED OLEDs

The self-organization of liquid crystalline materials has been used to design polarized OLEDs. The underlying principle is rather simple. If the emitters in an OLED are arranged parallel to each other in an LC-monodomain, the device will directly emit linearly polarized light. Polarized OLEDs could be particularly useful as backlights for conventional LCDs because they make a polarizer with its absorptive losses redundant.

Such an opportunity was first discussed by Dyrekelev [25] who used mechanical stretching of a bithiophene polymer for the orientation and obtained an OLED with an EL polarization ratio ($EL_{||}/EL_{\perp}$) of 2.4. The method of stretch alignment, however, is not suitable for technical application in OLEDs. Different methods like mechanical alignment, Langmuir–Blodgett deposition and liquid crystalline self-organization have been proposed to align anisotropic molecules in OLEDs and are described in detail in a 1999 review article by M. Grell and D.D.C. Bradley [26]. Among these methods the alignment of liquid crystalline molecules to large monodomains has turned out to be most attractive.

From the materials investigated fluorene containing polymers and low molar mass compounds have attracted a lot of interest. This is due to the ability of many fluorene based materials to form mesophases in conjunction with the strong blue fluorescence of the fluorene chromophor [27]. Dialkylated polyfluorenes like poly(9,9-di(2-ethylhexyl))fluorene (PF 2/6) exhibit liquid crystalline phases. M. Grell et al. showed that the parallel orientation of the liquid crystalline polyfluorenes in an OLED directly leads to the emission of linearly polarized light [28]. In their OLED setup they used rubbed polyimide as orientation layer. Due to the high electrical resistivity the polyimide had to be doped with an aromatic amine to ensure hole transport through the orientation layer. They achieved polarization ratios of 14:1 if EL is measured parallel and perpendicular to the rubbing direction (refer Figure 11.8).

We have described the orientation of poly-(2,7-(9,9-dioctyl)fluorene) with a rubbed PPV layer (Figure 11.9) [29,30]. Since PPV is a conjugated polymer itself the problems with the extremely good insulator polyimide, which is not well suited as a HTL in OLEDs, are not present in PPV-OLEDs. With an OLED setup consisting of ITO, a 30 nm thick rubbed PPV orientation layer, an annealed 70 nm poly-2,7-(9,9-dioctyl)fluorene (PFO) layer, and a calcium top electrode, we obtained blue EL with an orientation ratio of 25:1 parallel and perpendicular to the rubbing direction of the PPV. If polythiophene is deposited on top of oriented PPV and subsequently

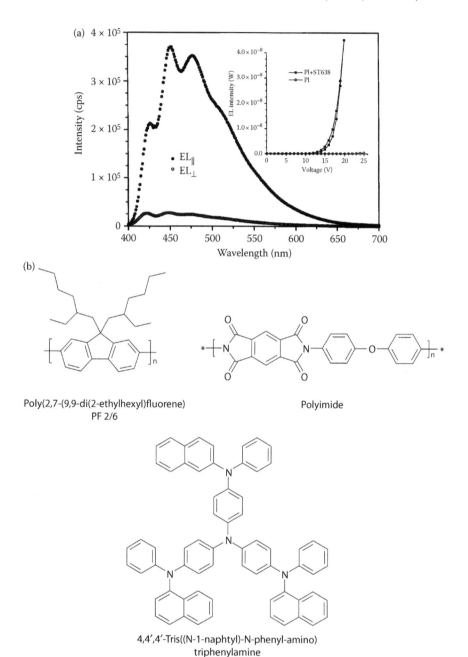

**FIGURE 11.8** (a) EL spectra of a thermally aligned device (ITO/17% aromatic amine in 83% polyimide (rubbed)/PF2/6/Ca/Al with polarizer detection parallel or orthogonal to the rubbing direction. The EL polarization ratio for the 477 nm peak is 15. Drive voltage was 19 V. (b) Chemical structures of the materials. (Grell, M. et al., *Adv. Mater.* **1999**, *11*, 671. Copyright Wiley-VCH Verlag GmbH & Co. KGaA. Reproduced with permission.)

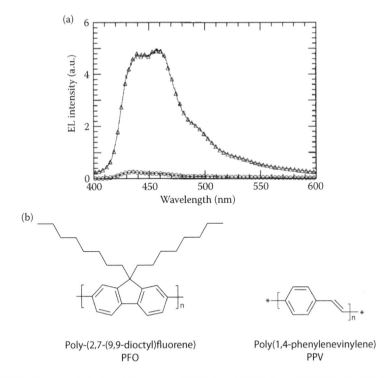

Poly-(2,7-(9,9-dioctyl)fluorene)
PFO

Poly(1,4-phenylenevinylene)
PPV

**FIGURE 11.9** (a) Polarized EL from an ITO/rubbed PPV/aligned PFO/Ca OLED. Spectra are shown for light polarized parallel (open triangles) and perpendicular (open circles) to the rubbing direction. Polarization ratio is 25:1. (b) Chemical structures of PFO and PPV. (Reprinted with permission from Whitehead, K.S. et al., *Appl. Phys. Lett.* **2000**, *76*, 2946. Copyright 2000, American Institute of Physics.)

rubbed, it is also possible to prepare OLEDs with two orthogonally polarized emitting layers [31].

Neher et al. used a doped photo-orientation layer containing azo-groups in polyfluorene OLEDs and reached orientation ratios up to 14:1 [32].

Based on these stimulating results on polyfluorenes, a number of groups synthesized low-molar-mass fluorene model compounds with a different number of fluorene units. The first paper about defined fluorene oligomers is from Klaerner et al., who described the synthesis of a mixture of oligomers and their separation through high-pressure liquid chromatography (HPLC) [33]. The synthesis of monodisperse oligofluorenes with up to seven fluorene units by repetitive Suzuki and Yamamoto coupling reactions was described in detail recently [34].

Fluorene oligomers with up to 12 units were synthesized by Geng et al. [35,36]. Cullingan et al. [37] succeeded in making linearly polarized OLEDs from these oligomers. The fluorene dodecamer F(MB)10F(EH)2 (Figure 11.10) has a $T_g$ of 123°C and exhibits a broad nematic mesophase up to the thermal decompositon at 375°C.

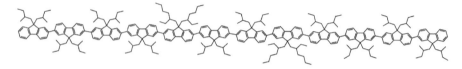

**FIGURE 11.10** Molecular structure of the fluorene decamer F(MB)10F(EH)2. (Adapted from Chen, A.C.A. et al., *Adv. Mater.* **2004**, *16*, 783.)

From this fluorene dodecamer polarized OLEDs were fabricated (Figure 11.11). Uniaxial molecular alignment was accomplished by spin casting F(MB)10F(EH)2 on top of a rubbed PEDT/PSS conductive alignment layer with subsequent thermal annealing. In an OLED with an additional electron conducting layer and a LiF/MgAg electrode a peak polarization ratio of 31 and a luminance yield of 1.1 cd/A were obtained. In 2004 the concept has been extended to polarized green and red OLEDs obtained by doping red and green emitting rod-like oligomers into a blue fluorene matrix [38]. Förster energy transfer leads to polarized emission with polarization ratios up to 26 and luminance yields up to 6.4 cd/A. In two recent papers [39,40], fluorene oligomers with both electron-donating arylamine and electron-accepting triazine units and their performance in polarized OLEDs are described.

The results discussed above show that from both low molar mass fluorene model compounds and high molecular weight polyfluorenes OLEDs with reasonably high polarization ratios can be fabricated. A number of orientation layers with hole-transporting properties have been developed, for example, rubbed polyimide doped with an aromatic amine, rubbed PPV, and rubbed PEDT. In addition to these layers, where orientation is achieved by mechanical forces, photo-orientation layers look very promising for OLED applications [32,41], since the orientation is achieved by irradiation with linearly polarized light and mechanical rubbing is not necessary. A further option of photo-orientation layers is their ability to create sets of OLED pixels with orthogonal polarization. This may lead to 3D-displays and is discussed in Section 11.6 in more detail.

The work on polarized emission described above has been carried out with polyfluorenes or low molar mass fluorene model compounds. None of the materials described above contains reactive end groups. By the introduction of polymerizable moieties the materials can be crosslinked and permanently fixed into dense polymer networks. This is the topic of the next section.

## 11.5 OLEDs WITH CROSSLINKED TRANSPORT AND EMISSION LAYERS

In the last section, the use of aligned LCs as emission layer in polarized OLEDs has been discussed in detail. The second prominent feature of reactive mesogens is the presence of reactive groups, which can be readily crosslinked. Crosslinking is usually carried out by illumination and leads to completely insoluble polymer networks. If a photomask is used in the crosslinking process the material behaves like a negative photoresist and can be easily patterned (Figure 11.13).

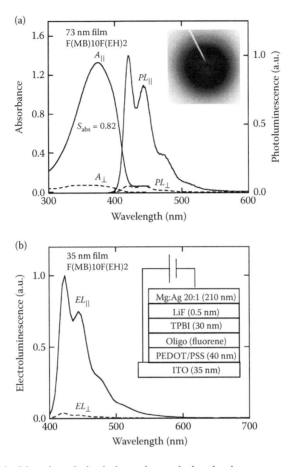

**FIGURE 11.11** Linearly polarized absorption and photoluminescence spectra (excitation wavelength 370 nm) of an OLED containing a 73 nm thick F(MB)10F(EH)2 film, with an inset showing the electron diffraction pattern (a). Device structure and polarized EL spectra of an OLED containing a 35 nm thick F(MB)10F(EH)2 film (b). (Culligan, S.W. et al., *Adv. Mater.* **2003**, *15*, 1176. Copyright Wiley-VCH Verlag GmbH & Co. KGaA. Reproduced with permission.)

Although OLEDs prepared by solution based processes have seen tremendous advances over the years, the pixilation of RGB spots in the emissive layer still remains one of the key challenges for the production of full color displays. Various printing techniques, especially ink-jet printing, have received much attention in recent years, mostly because of their potential for cost-effective production. Nevertheless, there are still a number of problems to be solved before printing of conjugated polymers can be used as a routine method to manufacture OLEDs. One problem is that printing often has to be done on prepatterned substrates, which are usually fabricated by conventional lithographic techniques.

In this section the crosslinking of photoactive organic materials is described as an alternative method for patterning RGB pixels in OLEDs. The discussion is not limited to reactive mesogens but also includes polymers and oligomers with photocrosslinkable units.

Probably the most recognized example for photopatternable OLED materials comes from a collaboration of the groups of K. Meerholz and O. Nuyken [42]. A number of RGB emitting spiro copolymers have been prepared, the structure of which is shown in Figure 11.12. The polymers are prepared by Suzuki through the cross coupling of a diborolane (monomer 1) and mixtures of various comonomers (monomer 2). One of the comonomers contains two photoactive oxetane units and serves as a crosslinker. The other comonomers of which only some are shown in Figure 11.12 are used for color tuning.

From these polymers pixelated OLED displays are made and shown in Figure 11.13. (a) On top of an ITO line structure, homogenous layers of PEDT/PSS and a crosslinkable hole-transport material are spin-coated and crosslinked. (b) Subsequently, a solution of the blue-light-emitting polymer is deposited and exposed to UV light through an aligned shadow mask. (c) After a soft-curing step, the noncrosslinked parts are dissolved in an organic solvent. (d) Repeating this procedure for green and red results in parallel stripes of blue-, green-, and red-emitting polymers. The device is completed by evaporating a metal cathode through a shadow mask.

The photopatterning of the spirobifluorene-*co*-fluorene copolymers described above is probably the most prominent but not the first example for the use of photocrosslinking in materials for optical applications. The basic idea goes back to D.J. Broer who developed thermally stable *passive* optical devices such as optical retarders [44] and efficient reflective polarizers with tunable absorption wavelength from mixtures of a cholesteric diacrylate and a nematic monoacrylate [45] (Figure 11.14).

This approach leads to densely crosslinked, thermally stable polymer networks and is very promising for the use in multilayer OLEDs where the underlying polymer layer is in many cases dissolved when a second polymer is spun cast on top. This problem can be circumvented by spin coating a polymer layer and subsequent photocrosslinking.

From 1997 the first research papers on crosslinkable HTLs appeared in the literature. Li et al. described a polymethacrylate terpolymer with blue-emitting distyrylbenzene, electron conducting oxadiazole and photocrosslinkable cinnamoyl units [46]. The polymer could be crosslinked by UV-irradiation but suffered from bleaching of the distyrylbenzene chromophores. Bellman et al. described a series of polynorbonenes with pendant triarylamine groups. The remaining double bonds in the polynorbonene backbone can be crosslinked by UV irradiation, but the efficiency of the OLEDs decreased after crosslinking [47]. Thermal crosslinking of oligotriarylamines with styrene endgroups [48] and the thermal crosslinking of triphenylamines using silane chemistry [49] have both been reported.

In 1999 Bacher et al. described the photopolymerization of triphenylenes with one, two, and three acrylate units [50]. Their paper is one of the first examples in which photopatterning of OLED materials with a photomask has been successfully demonstrated.

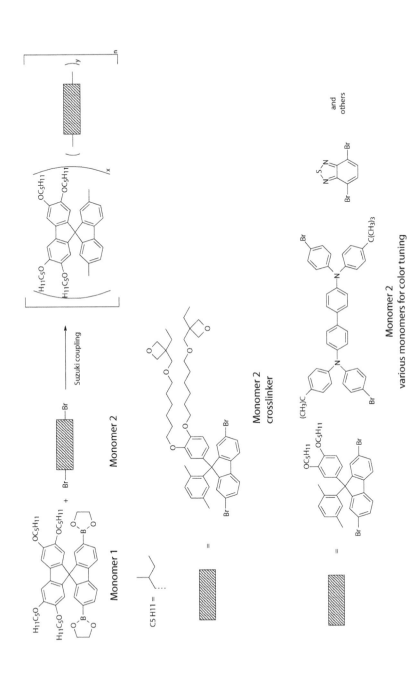

**FIGURE 11.12** Synthesis of crosslinkable spirobifluorene-*co*-fluorene copolymers. (Adapted from Mueller, C.D. et al., *Nature* 2003, 421, 829.)

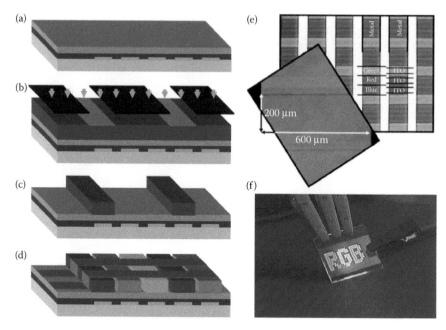

**FIGURE 11.13** **(See color insert following page 304.)** (a)–(d) Schematic illustration of the direct lithography process (for explanations see text). (Reprinted by permission from Macmillan Publishers Ltd. Müller, C.D. et al., *Nature* **2003**, *421*, 829, copyright 2003.) (e) Microscopy image of the completed display taken through the glass substrate. The horizontal polymer stripes are parallel and well aligned with the underlying ITO anode stripes. The metal cathode columns are seen as vertical stripes. (the separating white stripes result from the microscope backlight overexposing the camera). The inset shows a single RGB triple at a higher magnification. (f) Photograph of a RGB OLED device. Dimensions of the glass substrate are 25 × 25 mm. (Gather, M.C. et al., *Adv. Funct. Mater.* **2007**, *17*, 191. Copyright Wiley-VCH Verlag GmbH & Co. KGaA. Reproduced with permission.)

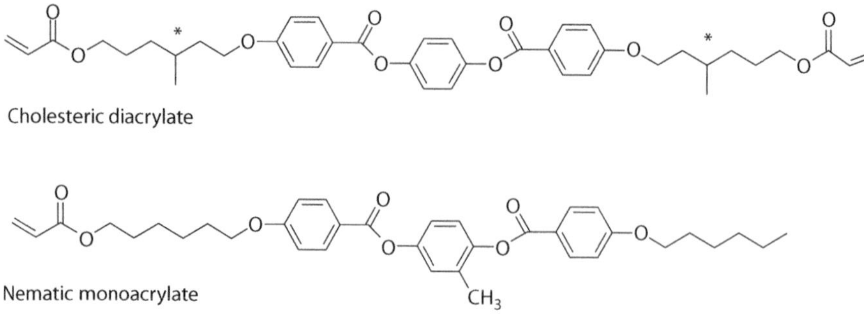

Cholesteric diacrylate

Nematic monoacrylate

CH₃

**FIGURE 11.14** Chemical structure of a cholesteric diacrylate and a nematic monoacrylate used for wide-band reflective polarizers. (Adapted from Broer, D.J. et al., *Nature* 1995, 378, 467.)

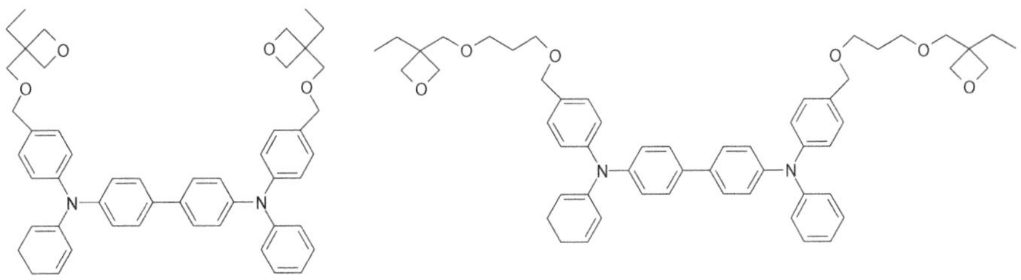

**FIGURE 11.15** Photocrosslinkable TPD derivatives with cationically polymerizable oxetane units. (Adapted from Bayerl, M.S. et al., *Macromol. Rapid Commun.* 1999, 20, 224.)

In the same year Bayerl et al. prepared TPD-based HTLs with pendant oxetane units that can be cationically photopolymerized [51] (Figure 11.15). Over the years, the material set with photocrosslinkable oxetane groups has been extended from hole transporting aromatic amines [51,52] to the multifunctional spirobifluorene copolymers described before [47,48] and to efficient electrophosphorescent diodes with iridium complex emitters [53].

We have published a number of papers on photocrosslinkable fluorene containing reactive mesogens and polymers. In 2001, we presented a number of acrylate functionalized fluorene trimers and pentamers and an oligomer [54]. Since this is one of the early publications on reactive mesogens which can be oriented and subsequently photopolymerized it is described in the next section. The monodomain alignment and subsequent photopolymerization of the acrylate functionalized fluorenes are described in more detail in Ref. [55]. In recent papers, we focused on the lithographic patterning of fluorene oligomers with pendant acrylate units and could demonstrate that resolutions of 1 μm are easily fabricated when the fluorene oligomers is processed as negative photoresist [24,56] (Figure 11.16).

## 11.6  POLARIZED OLEDs WITH CROSSLINKED EMISSION LAYERS

In the last two sections the orientation of liquid crystalline OLED-emitters, which leads to polarized emission and the patterning of OLED displays by crosslinking photopolymerizable OLED materials has been discussed. With reactive mesogens it is possible to combine both methods. So, reactive mesogens are aligned in their LC-phase in the first step. Subsequently, the ordered LC-phase is permanently fixed by photopolymerization (Figure 11.7). If a photomask is used in this process, the crosslinked layer can be patterned like a negative photoresist as described before (Figure 11.13). In this section work on reactive mesogens, which involves both orientation and crosslinking is reviewed.

One of the first reports on orientation and subsequent crosslinking of a reactive mesogen for OLED applications has been published by Bacher et al. [57] who synthesized the conjugated bisstilbene with two polymerizable acrylate groups shown in Figure 11.17. The reactive mesogen was oriented by heating into the LC phase and subsequently thermally crosslinked at 175°C.

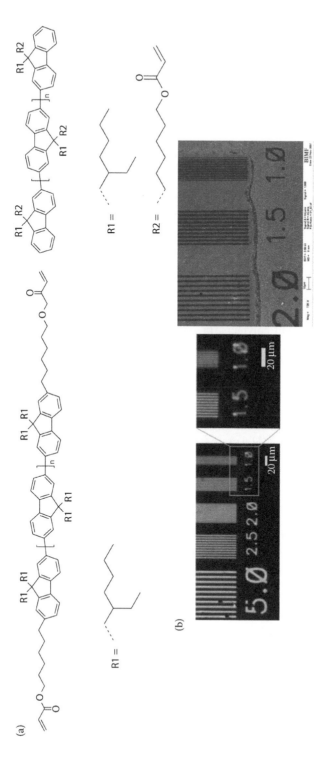

**FIGURE 11.16** (a) Chemical structures of photocrosslinkable fluorene oligomers (Scheler, E., Bauer, I., and Strohriegl, P. *Macromol. Symp.* **2007,** 254, 203. Copyright Wiley-VCH Verlag GmbH & Co. KGaA. Reproduced with permission.) (b) Fluorescence image (left) and SEM image (right) of microstructures prepared from the fluorene oligomers. (Scheler, E. and Strohriegl, P. *J. Mater. Chem.* **2009,** 19, 3207–3212. Reproduced by permission of The Royal Society of Chemistry.)

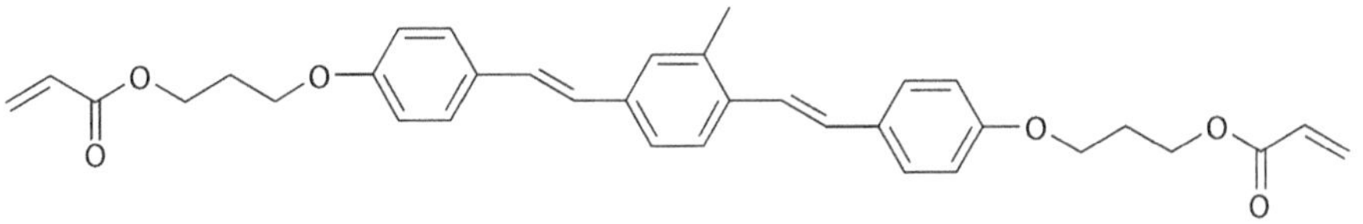

**FIGURE 11.17** Chemical structure of the conjugated bisstilbene with polymerizable acrylate units. (Adapted from Bacher, A. et al., *J. Mater. Chem.* 1999, 9, 2985.)

In 2001, we have described the synthesis of fluorene trimers, pentamers and oligomers with pendant acrylate units for the first time [54] (Figure 11.18). The fluorene reactive mesogens exhibit broad mesophases. For example, the pentamer has a nematic phase between −10°C and 123°C, which makes it ideally suited for orientation experiments. In a 25 nm thick layer on polyimide the pentamer exhibits a photoluminescence orientation ratio of 25. In a 90 nm thick layer this ratio decreases to 9. This result is important for the design of polarized OLEDs since it implies a maximum orientation close to the alignment layer/fluorene interface. If the recombination of electrons and holes can be confined to this zone, high polarization ratios can be expected. This might be the reason for the high polarization ratio of 25 that we observed in an OLED with a rubbed PPV orientation layer and a polyfluorene emitter [29].

The conversion of macroscopically oriented reactive mesogens to an intractable polymer network has been described by the group of S. Kelly and M. O'Neill in Hull in 2000 [41]. In their paper they already mentioned the possibility of pixel formation by photopatterning. For their experiments they used the reactive mesogen with a fluorene/thiophene core and photocrosslinkabele pentadiene end groups shown in Figure 11.19. The reactive mesogen was oriented by annealing on top of a photo-orientation layer consisting of a polymer with coumarine side groups and doped with an aromatic amine to ensure hole transport and subsequently crosslinked by UV-irradiation. A polarization ratio of 10 and a brightness of 60 cd/m$^2$ were achieved in an OLED.

The synthesis of a number of reactive mesogens with methacrylate and different dienes as polymerizable groups and their polymerization behavior is described in detail in [58]. It turned out that dienes as shown in Figure 11.19 require about 30 times longer irradiation for crosslinking compared to methacrylates, but photodegradation is claimed to be less pronounced with the diene reactive mesogens compared to methacrylates.

The Hull group published a number of papers on reactive mesogens for OLED applications over the years. The work until 2003 was reviewed by M. O'Neill and S. Kelly in Ref. [59].

Recent work of S. Kelly et al. showed the use of acrylate and diene photopolymerizable end groups in the α, ω-positions of mesogens containing fluorene and thiophene units [60,61]. The three different reactive mesogens shown in Figure 11.20 were subsequently deposited. Using a photomask the red, green, and blue areas of the OLED shown in Figure 11.21 could be defined by photochemical crosslinking and removal of the noncrosslinked parts by washing with chloroform. The green reactive mesogen

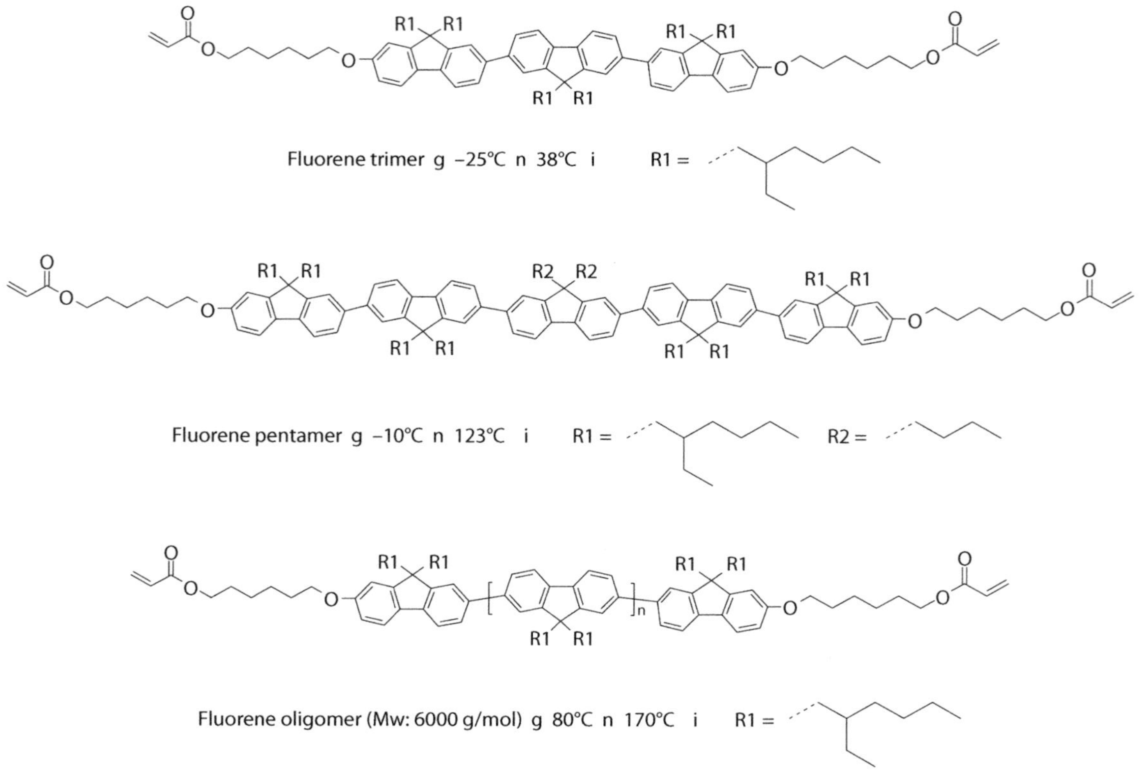

**FIGURE 11.18**  Chemical structure of a fluorene trimer, a pentamer, and an oligomer with pendant photocrosslinkable acrylate groups.  (Adapted from Jandke, M. et al., *SPIE Proc.* 2001, 4105, 338.)

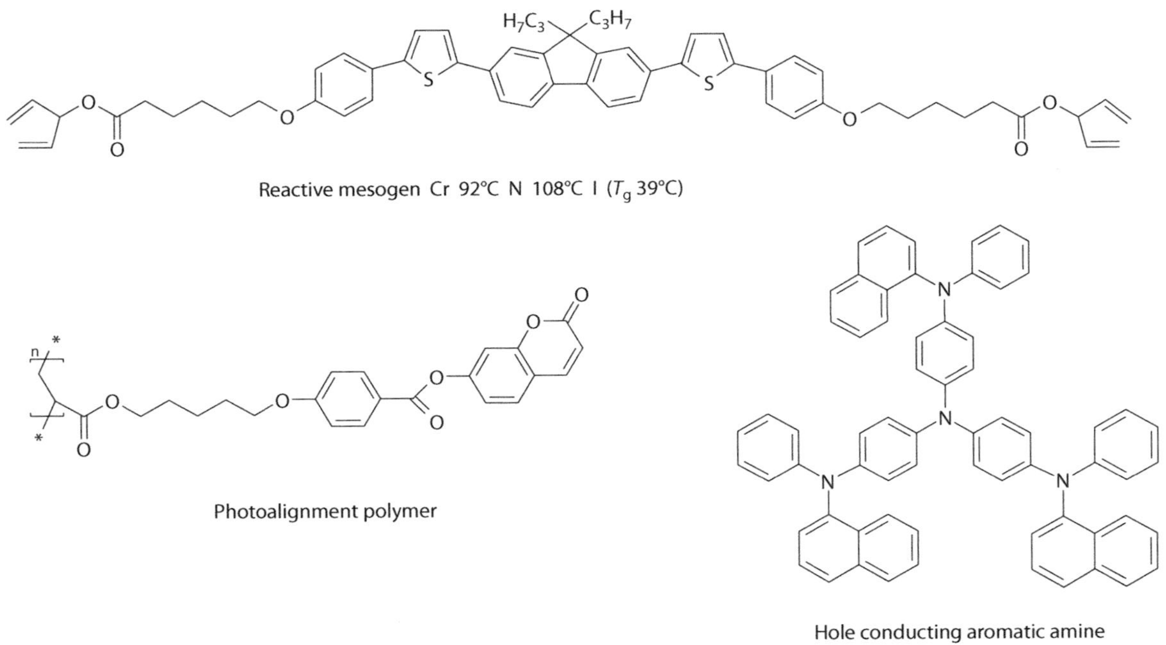

Reactive mesogen Cr 92°C N 108°C I ($T_g$ 39°C)

Photoalignment polymer

Hole conducting aromatic amine

**FIGURE 11.19** Chemical structure of the reactive mesogen, the photoalignment polymer and the hole conducting aromatic amine used for polarized OLEDs. (Adapted from Contoret, A.E.A. et al., *Adv. Mater.* 2000, 13, 971.)

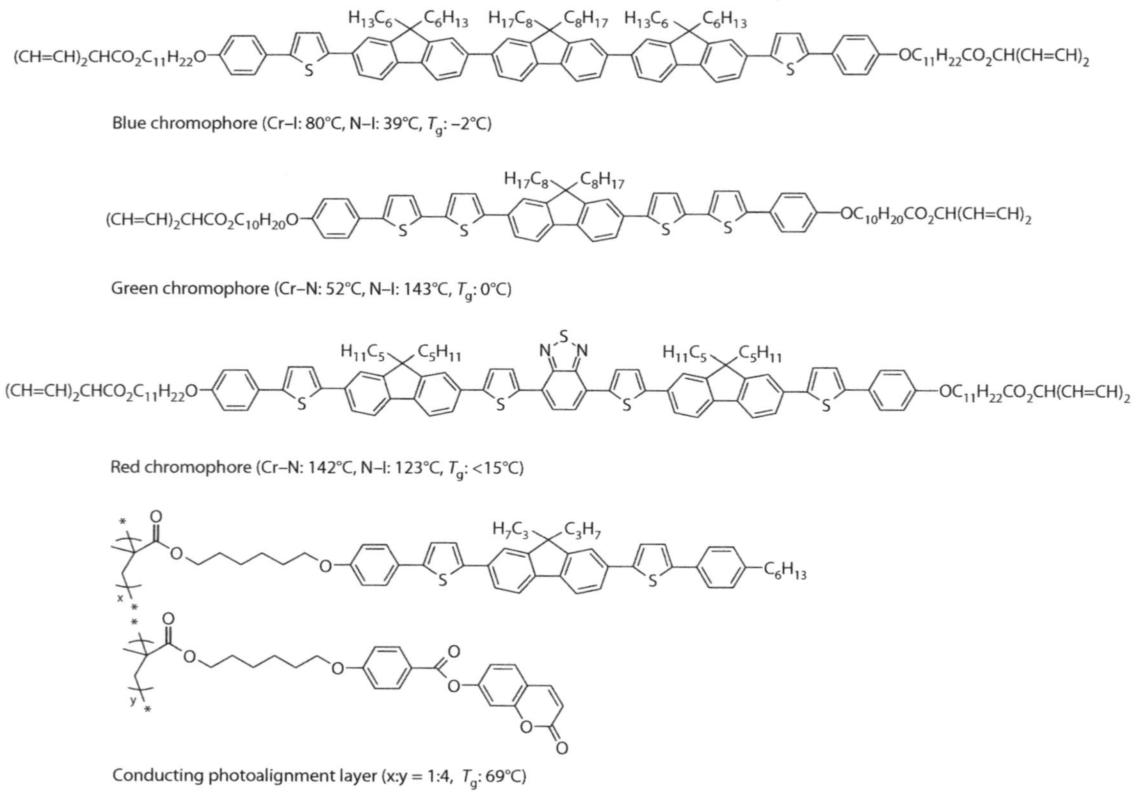

**FIGURE 11.20** Chemical structures and phase transition temperatures of the blue, green, and red emitting reactive mesogens and the conducting photoalignment layer. Cr–N represents the melting of the crystalline state to the nematic phase, N–I represents the transition from the nematic phase to the isotropic liquid. Cr–I is the transition directly from the crystalline state to the isotropic liquid. $T_g$: glass transition temperature. (Adapted from Aldred, M.P. et al., *Chem. Mater.* 2004, 16, 4928.)

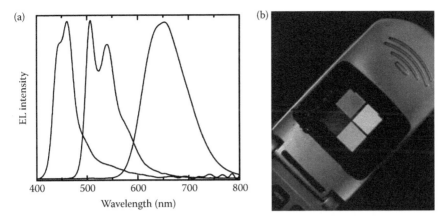

**FIGURE 11.21   (See color insert following page 304.)** (a) Normalized EL spectra from the red, green, and blue OLEDs. (b) A prototype OLED with a red, green, and blue pixel on the same substrate, fabricated by spin coating on a PEDT film covering a patterned ITO substrate. The pixels were defined by irradiation at 325 nm through a mask. Nonirradiated material was removed by washing with chloroform. (Aldred, M.P. et al., *Adv. Mater.* **2005**, *17*, 1368. Copyright Wiley-VCH Verlag GmbH & Co. KGaA. Reproduced with permission.)

has also been coated on top of a tailor-made photo-orientation layer [62] containing both charge transporting groups and photo-orienting coumarine units. After thermal alignment, an EL polarization ratio of 13:1 is obtained (Figure 11.22). The photograph in Figure 11.22 shows that polarization patterns are obtained if the photoalignment

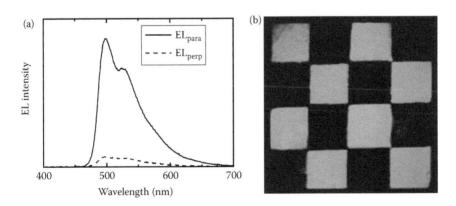

**FIGURE 11.22**   (a) Polarized EL spectrum from an OLED of the photochemically crosslinked green reactive mesogen (Figure 11.20) on a thin, crosslinked film of the orienting polymer previously irradiated with polarized UV light. (b) Photoluminescence of the green emitter deposited onto a photoalignment layer previously oriented in a chess pattern by exposing adjacent squares with orthogonally polarized light. Viewed through a polarizer, the chess pattern of the green squares in which the emitter molecules are aligned parallel to the polarizer and the black, orthogonally polarized squares is observed. (Aldred, M.P. et al., *Adv. Mater.* **2005**, *17*, 1368. Copyright Wiley-VCH Verlag GmbH & Co. KGaA. Reproduced with permission.)

layer is structured with a photomask. The pattern was obtained by exposing adjacent squares to orthogonally polarized light. The PL image was viewed through a polarizer and shows that in the green and in the dark regions the emitter molecules are orthogonally aligned. The viewing of patterned, polarized emission through a pair of orthogonal polarizers creates 3D-effects as each eye sees a different set of pixels. So creating photoalignment with two different polarizations of the pixels in combination with self-emitting OLEDs may pave the way to 3D-displays.

In three recent papers, the Hull group has extended the number of available reactive mesogens [63–65] and investigated the phase behavior of the materials and their transport properties [66] in detail.

## 11.7 CONCLUSIONS AND OUTLOOK

In this chapter the work on reactive mesogens for OLED applications has been summarized. The major question that has to be answered before reactive mesogens can be applied in commercial displays is the advantages these materials have compared to the current state-of-the-art low molar mass compounds and polymers described in Section 11.2. To my opinion there are mainly three points which make reactive mesogens attractive for OLED applications:

1. The parallel alignment of calamitic LCs lead to polarized emission.
2. The formation of insoluble, densely crosslinked polymer networks by photochemical polymerization of thin layers of reactive mesogens with polymerizable groups.
3. The possibility of pattern formation if a mask is used during the photopolymerization of reactive mesogens. In this case, the material behaves like a negative photoresist.

Polarized OLEDs may become attractive as backlights for LCD displays, since one polarizer, which causes huge absorption losses can be omitted. Nevertheless, the efficiency of polarized OLEDs from reactive mesogens has to be increased before they can compete with current light sources, for example, fluorescent lamps or inorganic high brightness LEDs. In recent years, the efficiency of white OLEDs has rapidly increased and they are now considered as efficient white-light sources for large area lighting. In such OLEDs phosphorescent metal complexes are used as emitters, and it would be an interesting topic to look for rod-like phosphorescent emitters which can help to increase the efficiency of polarized OLEDs.

The formation of crosslinked, insoluble polymer networks by irradiation of thin films of reactive mesogens allows the formation of multilayer OLED displays by solution processing. Due to problems with dissolving the underlying polymer layer when a second polymer is spin coated on top, efficient multilayer OLEDs are up to now a domain of vacuum-deposited small molecules.

The third important advantage of reactive mesogens is their ability to form high-resolution color pixels for OLED displays if they are irradiated through a photomask. In the irradiated parts densely crosslinked, completely insoluble films are formed.

The noncrosslinked materials in the nonexposed areas of the sample can be subsequently washed away. This technique in which the conjugated reactive mesogens are applied like a conventional negative photoresist is widely used in the semiconductor industry to fabricate integrated circuits with feature sizes of less the 100 nm. With reactive mesogens, feature sizes of 1 μm have been already demonstrated, and it can be expected that smaller feature sizes are accessible. It should be noted here that feature sizes in the submicron range are not crucial for pixels in OLED displays, but the possibility to create submicron patterns of conjugated organic materials with the well-established photoresist technique will probably have a huge impact on future applications of these materials, for example, in the field of sensors in which organic materials are combined with conventional silicon electronics.

If both peculiarities of reactive mesogens, polarized emission and pattern formation are combined, OLED displays in which two sets of pixels emit with orthogonal polarization can be realized if an additional photoorientation layer is used. This may pave the way for 3D-OLED displays and shows that the reactive mesogens described in this chapter are promising materials for future OLED displays.

## 11.8  ABBREVIATIONS

| | |
|---|---|
| Alq$_3$ | Tris(8-hydroxyquinoline)aluminum |
| BCP | 2,9-Dimethyl-4,7-diphenyl-1,10-phenanthroline (Bathocuproine) |
| BCzVBI | 4,4'-Bis(2-(9-ethylcarbazole-3-yl))ethylen-1-yl)biphenyl |
| CBP | 4,4'-Bis(carbazol-9-yl)biphenyl |
| DCM | 4-(Dicyanomethylene)-2-methyl-6-(4-dimethylaminostyryl)-4H-pyran |
| DPVBI | 4,4'-Bis(2,2-diphenylethylen-1-yl)biphenyl |
| EIL | Electron injection layer |
| EL | Electroluminescence |
| EL | Emission layer |
| ETL | Electron transport layer |
| FIrpic | Iridium-bis(4,6-difluorophenyl)-pyridinato-N,C$^{2'}$)picolinate |
| HBL | Hole blocking layer |
| HIL | Hole injection layer |
| HOMO | Highest occupied molecular orbital |
| HPLC | High pressure liquid chromatography |
| HTL | Hole transport layer |
| Ir(ppy)$_3$ | Tris(phenylpyridyl)iridium |
| ITO | Indium tin oxide |
| LCD | Liquid crystal display |
| MeH-PPV | Poly-(2-methoxy-5-(2-ethylhexyloxy)-1,4-phenylenevinylene) |
| m-MTDATA | 4,4',4''-Tris(3-methylphenylphenylamino)triphenylamine |
| NPD | N,N'-bis(1-naphthyl)-N,N'-diphenyl-(1,1'-biphenyl)-4,4'diamine |
| OFET | Organic field effect transistor |
| OLED | Organic light emitting device |

| | |
|---|---|
| PBD | 2-Biphenyl-5-(4-*tert*-butylphenyl)-1,3,4-oxadiazole |
| PEDT | Poly(3,4-ethylenedioxythiophene) |
| PEDT/PSS | Poly(3,4-ethylenedioxythiophene/polystyrene sulfonic acid) |
| PF 2/6 | Poly-2,7-(9,9-di(2-ethylhexyl))fluorene |
| PFO | Poly-2,7-(9,9-dioctyl)fluorene |
| PPV | Poly(1,4-phenylenevinylene) |
| RGB | Red, green, blue |
| $S_0$ | Singlet ground state |
| SiCz | 3,6-(Triphenylsilyl)-9-(4-*tert*. butylphenyl)carbazole |
| $T_1$ | First excited triplet state |
| TAPC | 1,1-Bis[(di-4-tolylamino)phenyl]cyclohexane |
| TAZ | 3-(4-Biphenylyl)-4-phenyl-5-*tert*-butylphenyl-1,2,4-triazole |
| $T_g$ | Glass transition temperature |
| TPBI | 1,3,5-Tris(1-phenyl-1*H*-benzimidazol-2-yl)benzene |
| TPD | N,N'-diphenyl-N,N'- bis(3-methylphenyl)-1,1'biphenyl-4,4'-diamine |

## REFERENCES

1. Tang, C. and vanSlyke, S.A., *Appl. Phys. Lett.* **1987**, *51*, 913.**Q8**
2. Chen, C.H., Shi, J., and Tang, C.W., *Macromol. Symp.* **1997**, *125*, 1.
3. Kraft, A., Grimsdale, C., and Holmes, A.B., *Angew. Chem. Int. Ed.* **1998**, *37*, 402.
4. Shirota, Y., *J. Mater. Chem.* **2000**, *10*, 1.
5. Nalwa, H.S. and Rohwer, L.S. (Eds.), *Organic Light Emitting Diodes.* American Sci. Publ., Valencia, California, 2003.
6. Müllen, K. and Scherf, U. (Eds.), *Organic Light Emitting Devices.* Wiley VCH, Weinheim, Germany, 2006.
7. Burroughes, J.H., Bradley, D.D.C., Brown, A.R., Marks, R.N., Mackay, K., Friend, R.H., Burns, P.L., and Holmes, A.B., *Nature* **1990**, *347*, 539.
8. Groenendaal, L.B., Jonas, F., Freitag, D., Pielartzik, H., and Reynolds, J.R., *Adv. Mater.* **2000**, *12*, 481.
9. Borsenberger, P.M. and Weiss, D.S. (Eds.), *Organic Photoreceptors for Xerography.* Marcel Dekker Inc., New York, 1999.
10. Strohriegl, P. and Grazulevicius, J.V., *Adv. Mater.* **2002**, *14*, 1439.
11. Hosokawa, C., Higashi, H., Nakamura, H., and Kusumoto, T., *Appl. Phys. Lett.* **1995**, *67*, 3853.
12. Baldo, M.A., Lamansky, S., Burrows, P.E., Thompson, M.E., and Forrest, S.R., *Appl. Phys. Lett.* **1999**, *75*, 4.
13. Tsai, M.H., Lin, H.W., Su, H.C., Ke, T.H., Wu, C.C., Fang, F.C., Liao, Y.L., Wong, K.T., and Wu, C.I., *Adv. Mater.* **2006**, *18*, 1216.
14. Braun, D. and Heeger, A.J., *Appl. Phys. Lett.* **1991**, *58*, 1982.
15. Spreitzer, H., Becker, H., Kluge, E., Kreuder, W., Schenk, H., Demandt, R., and Schoo, H., *Adv. Mater.* **1998**, *10*, 1340.
16. Bernius, M.T., Inbasekaran, M., O'Brien, J., and Wu, W., *Adv. Mater.* **2000**, *12*, 1737.
17. Scherf, U. and List, E.J.W., *Adv. Mater.* **2002**, *14*, 477.
18. Grell, M. and Bradley, D.D.C., *Adv. Mater.* **11**, *895*, 1999.
19. O'Neill, M. and Kelly, S.M., *Ekisho* **2005**, *9*, 9.

20. Adam, D., Schuhmacher, P., Simmerer, J., Haeussling, L., Siemensmeyer, K., Etzbach,K. H., Ringsdorf, H., and Haarer, D., *Nature* **1994**, *371*, 141.

21. Funahashi, M. and Hanna, J., *Appl. Phys. Lett.* **2000**, *76*, 2574.

22. Sirringhaus, H., Wilson, R.H., Friend, R.H., Inbasekaran, M., Wu, W., Woo, E.P., Grell, M., and Bradley, D.D.C., *Appl. Phys. Lett.* **2000**, *77*, 406.

23. McCulloch, I., Zhang, W., Heeney, M., Bailey, C., Giles, M., Graham, D., Shkunov, M., Sparrowe, D., and Thierney, S., *J. Mater. Chem.* **2003**, *13*, 2436.

24. Scheler, E., Bauer, I., and Strohriegl, P., *Macromol. Symp.* **2007**, *254*, 203.

25. Dyrekelev, P., Berggren, M., Inganäs, O., Andersson, M.R., Wennerström, O., and Hjertberg, T., *Adv. Mater.* **1995**, *7*, 43.

26. Grell, M. and Bradley, D.D.C. *Adv. Mater.* **1999**, *11*, 895.

27. Grice, A.W., Bradley, D.D.C., Bernuis, M.T., Inbasekaran, M., Wu, W.W., and Woo, E.P. *Appl. Phys. Lett.* **1998**, *73*, 629.

28. Grell, M., Knoll, W., Lupo, D., Meisel, A., Miteva, T., Neher, D., Nothofer, H.-G., Scherf, U., and Yasuda, A. *Adv. Mater.* **1999**, *11*, 671.

29. Whitehead, K.S., Grell, M., Bradley, D.D.C., Jandke, M., and Strohriegl, P. *Appl. Phys. Lett.* **2000**, *76*, 2946.

30. Jandke, M., Strohriegl, P., Gmeiner, J., Brütting, W., and Schwoerer, M., *Synth. Metals* **2000**, *111–112*, 177.

31. Bolognesi, A., Botta, C., Facchinetti, D., Jandke, M., Kreger, K., and Strohriegl, P., *Adv. Mater.* **2001**, *13*, 1072.

32. Sainova, D., Zen, A., Nothofer, H.-G., Asawapirom, U., Scherf, U., Hagen, R., Bieringer, T., Kostromine, S., and Neher, D. *Adv. Funct. Mater.* **2002**, *12*, 49.

33. Klaerner, G. and Miller, R.D. *Macromolecules* **1998**, *31*, 2007.

34. Jo, J., Chi, C., Höger, S., Wegner, G., and Yoon, D.Y. *Chem. Eur. J.* **2004**, *10*, 2681.

35. Geng, Y., Cullingham, S.W., Trajkovska, A., Wallace, J.U., and Chen, S.H., *Chem. Mater.* **2003**, *15*, 542.

36. Geng, Y., Chen, A.C.A., Ou, J.J., Chen, S.H., Klubek, K., Vaeth, K., and Tang, C.W., *Chem. Mater.* **2003**, *15*, 4352.

37. Culligan, S.W., Geng, Y., Chen, S.H., Klubek, K., Vaeth, K.M., and Tang, C.W., *Adv. Mater.* **2003**, *15*, 1176.

38. Chen, A.C.A., Culligan, S.W., Geng, Y., Chen, S.H., Klubek, K.P., Vaeth, K.M., and Tang, C.W., *Adv. Mater.* **2004**, *16*, 783.

39. Chen, A.C.A., Wallace, J.U., Wei, S.K.-H., Zeng, L., Chen, S.H., and Blanton, T.N., *Chem. Mater.* **2006**, *18*, 204.

40. Chen, A.C.A., Wallace, J.U., Klubek, K.P., Madaras, M.B., Tang, C.W., and Chen, S.H., *Chem. Mater.* **2007**, *19*, 4043.

41. Contoret, A.E.A., Farrar, S.R., Jackson, P.O., Khan, S.M., May, L., O'Neill, M.O., Nicholls, J.E., Kelly, S.M., and Richards, G.J., *Adv. Mater.* **2000**, *13*, 971.

42. Müller, C.D., Falcou, A., Reckefuss, N., Rojahn, M., Wiederhirn, V., Rudati, P., Frohne, H., Nuyken, O., Becker, H., and Meerholz, K., *Nature* **2003**, *421*, 829.

43. Gather, M.C., Köhnen, A., Falcou, A., Becker, H., and Meerholz, K., *Adv. Funct. Mater.* **2007**, *17*, 191.

44. Broer, D.J., Boven, J. Mol, G.N., and Challa, G., *Makromol. Chem.* **1989**, *190*, 2255.

45. Broer, D.J., Lub, J., and Mol, G.N., *Nature* **1995**, *378*, 467.

46. Li, X.C., Yong, T.M., Grüner, J., Holmes, A.B., Moratti, S.C., Cacialli, F., and Friend, R.H., *Synth. Metals* **1997**, *84*, 437.

47. Bellmann, E., Shaheen, S.E., Thayumanavan, S., Barlow, S., Grubbs, R.H., Marder, S.R., Kippelen, B., and Peyghambarian, N., *Chem. Mater.* **1998**, *10*, 1668.

48. Chen, J.P., Klaerner, G., Lee, J.-I., Markiewicz, D., Lee, V.Y., Miller, R.D., and Scott, J.C., *Synth. Metals* **1999**, *107*, 129.

49. Li, W., Wang, Q., Cui, J., Chou, H., Shaheen, S.E., Jabbour, G.E., Anderson, J. et al., *Adv. Mater.* **1999**, *11*, 731.

50. Bacher, A., Erdelen, C.H., Paulus, W. Ringsdorf, H., Schmidt, H.W., and Schuhmacher, P., *Macromolecules* **1999**, *32*, 4551.

51. Bayerl, M.S., Braig, T., Nuyken, O., Mueller, C.D., Gross, M., and Meerholz, K., *Macromol. Rapid Commun.* **1999**, *20*, 224.

52. Mueller, C.D., Braig, T., Nothofer, H.G., Arnoldi, M., Gross, M., Scherf, U., Nuyken, O., and Meerholz, K., *Chem. Phys. Chem.* **2000**, *1*, 207.

53. Yang, X., Müller, D.C., Neher, D., and Meerholz, K., *Adv. Mater.* **2006**, *18*, 948.

54. Jandke, M., Hanft, D., Strohriegl, P., Whitehead, K., Grell, M., and Bradley, D.D.C., *SPIE Proc.* **2001**, 4105, 338.

55. Thiem, H., Jandke, M., Hanft, D., and Strohriegl, P, *Macromol. Chem. Phys.* **2006**, *207*, 370

56. Scheler, E. and Strohriegl, P., *J. Mater. Chem.* **2009**, 19, 3207.

57. Bacher, A., Bentley, P.G., Bradley, D.D.C., Douglas, L.K., Glarvey, P.A., Whitehead, K.S., and Turner, M.L., *J. Mater. Chem.* **1999**, *9*, 2985.

58. Contoret, A.E.A., Farrar, S.R., O'Neill, M., Nicholls, J.E., Richards, G.J., Kelly, S.M., and Hall, W.A., *Chem. Mater.* **2002**, *14*, 1477.

59. O'Neill, M. and Kelly, S.M., *Adv. Mater.* **2003**, *15*, 1135.

60. Aldred, M.P., Eastwood, A.J., Kelly, S.M., Vlachos, P., Contoret, A.E.A., Farrar, S.R., Mansoor, B., O'Neill, M., and Tsoi, W.C. *Chem. Mater.* **2004**, *16*, 4928.

61. Aldred, M.P., Contoret, A.E.A., Farrar, S.R., Kelly, S.M., Mathieson, D., O'Neill, M., Tsoi, W.C., and Vlachos, P. *Adv. Mater.* **2005**, *17*, 1368.

62. Aldred, M.P., Vlachos, P., Contoret, A.E.A., Farrar, Chung-Tsoi, W., Mansoor, B., Woon, K.L., Hudson, R., Kelly, S.M., and O'Neill, M., *J. Mater. Chem.* **2005**, *15*, 3208.

63. Aldred, M.P., Vlachos, P., Dong, D., Kitney, S.P., Chung-Tsoi, O'Neill, M.W., and Kelly, S.M., *Liquid Crystals.* **2005**, *32*, 951.

64. Aldred, M.P., Eastwood, A.J., Kitney, S.P., Richards, G.J., Vlachos, P., Kelly, S.M., and O'Neill, M., *Liquid Crystals.* **2005**, *32*, 1251.

65. Aldred, M.P., Carrasco-Orozoco, Contoret, A.E.A., Dong, D., Farrar, S.R., Kelly, S.M., Kitney, S.P., Mathieson, D., O'Neill, M., Chung Tsoi, W., and Vlachos, P., *Liquid Crystals.* **2006**, *33*, 459.

66. Woon, K.L., Aldred, Vlachos, P., Mehl, G.H., Stirner, T., Kelly, S.M., and O'Neill, M., *Chem. Mater.* **2006**, *18*, 2311.

# Part II

---

*Weakly Cross-Linked Systems:*
*Liquid Crystal Elastomers*

# 12 Physical Properties of Liquid Crystalline Elastomers

*Eugene M. Terentjev*

## CONTENTS

## 12.1   INTRODUCTION

In ordinary elastic solids deformations are created by relative movement of the same atoms (or molecules) that form the bonded low-symmetry lattice. Hence, when the deformation is small, the lattice symmetry is preserved and one obtains an ordinary elastic response (although often anisotropic). There is a classical elastic response found in glasses as well (either isotropic or anisotropic), where in place of the crystalline lattice recording the preferred position of atoms they are confined by constraints of local cages. Either way, the elements of the body "know" their positions and the system responds with an increase of elastic energy when these elements are displaced; large deformations destroy the lattice (or cages) integrity and simply break the material.

In contrast, in elastomers and gels the macroscopic elastic response arises from the entropy change of polymer chains on relative movement of their cross-linked end-points, which are relatively far apart. What happens to chain segments (monomer moieties) on a smaller length scale is a relatively independent matter and such weakly cross-linked network behaves as a liquid on length scales smaller than the end-point separation of strands. For instance, liquid crystalline order can be established within these segments, if they possess sufficient molecular anisotropy, and its director can rotate, in principle, independently of deformation of the cross-linking points. Such an internal degree of freedom within, and coupled to the elastic body constitutes what is known as the Cosserat medium: The relative movement of cross-linking points provides elastic strains and forces, while the director rotation causes local torques and couple-stresses–both intricately connected in the overall macroscopic response of the body. However, the physics of liquid crystalline elastomers (LCE) is much richer than of notional Cosserat solids because (again due to the entropic nature of long polymer chains connecting the cross-linking points) rubbers are capable of very large shear deformations (being at the same time essentially incompressible). Hence, one expects a variety of unique physical properties, especially in the region of large deformations. Indeed, some such properties have been reported in recent years.

It is important to realize that a unique physical system, such as LCE and gels, should be looked upon from the point of view of equally unique applications. Because of their slow dynamics and high fields required to overcome the elastic resistance, LCE are poor for electro-optical display devices, which is one of the main thrusts in conventional liquid crystals. Because of their chemical complexity, they are not optimal for shoe soles and tennis balls (although for high-tech rubber tires this is perhaps an open question). What LCE seem to be made for is medium providing the mechanical actuation in response to changes of temperature, irradiation by light, exposure to specific solvents and even electric field. Unlike other famous shape-memory systems, the shape changes (or exerted forces) are fully reversible since they originate from equilibrium properties of the material itself. Another prospective area is rubbers with cholesteric order, which form a photonic bandgap system, the properties of which can be continuously changed by mechanical deformation. Other potential applications could be based on manipulation of axis of optical bire-fringence by mechanical means. Unusual nonsymmetric elasticity, with very low shear modulus and anomalously high damping is another characteristic physical property still waiting for an appropriate application. Piezoelectric and nonlinear optic properties of LCE, again in contrast to traditional crystalline and ceramic materials, allow large deformations and manipulation of polarization by mechanical means.

After the concept of nematic elastomers was put forward by de Gennes in 1975[1] and the first side-chain liquid crystalline polymer was cross-linked into elastomer by Finkelmann in 1981,[2] the initial research has been mainly focusing on synthetic and characterization work. The reviews[3–6] give a comprehensive picture of that period. In recent years the emphasis has been gradually shifting toward studies of new physical properties and the geography of research into LCE has been significantly broadened.

## 12.1.1 SYNTHESIS

For many years the prevailing type of LCE-forming materials was the side-chain liquid crystalline polymer. After the early work of Ringsdorf and Finkelmann, poly-acrylate backbones with a number of mesogenic pendants have been used by different groups[7–10] to produce a variety of LCE. However, it has been quickly recognized that polyacrylate-based polymer chains have certain practical disadvantages, in par-ticular the high glass transition $T_g \geq 50°C$ and low backbone anisotropy. Side-chain liquid crystalline polymers based on siloxane backbone have shown more dramatic mechanical properties due to a much higher chain anisotropy and are conveniently liquid crystalline at room temperature (with $T_g \leq 5°C$). Methods of cross-linking have varied from chemical, using copolymerization with a small proportion of reac-tive groups on a chain and adding di- or tri-functional cross-linking agents,[11,12] to radiation processes using UV light with photoinitiators[13] or even gamma-radiation.[14] In this context one has to mention an important parallel development of densely cross-linked anisotropic networks (mainly, although not exclusively, based on poly-merization of mesogenic diacrylates);[15,16] such systems exhibit glass-like elasticity and their internal degree of freedom are immobile (which makes them no less useful in appropriate applications).

Much less synthetic work has been done on networks of main-chain mesogenic polymers, apart from the long history of mesogenic epoxy resins prepared over the years by Carfagna et al.[17] and Ober et al.[4] More recently, however, there has been a noticeable surge of activity in this area. An interesting group of rigid-rod poly-mers, somewhat reminiscent of the activity in stiff-chain Kevlar-type fibers,[18] but showing remarkable liquid crystalline properties in a swollen state, was prepared by Zhao et al.[19] (polyisocyanate chains were cross-linked into networks by hydrosi-lation reaction). The Cornell group has produced lightly cross-linked (i.e., rubbery without swelling) nematic and smectic elastomers based on the rod-like mesogenic monomers connected through flexible spacers.[20,21] Finkelmann et al. have prepared another group of nematic main-chain elastomers based on semiflexible polyether chains.[22] Another source of (mainly photo-cross-linkable) main-chain polymers has been developed by Zentel et al.[23] More recently there has been a widespread move toward main-chain liquid crystalline polymers (and their cross-linked elastomers) which have the "traditional" aromatic rod-like units separated by the flexible spacers made of siloxane chains of different length.[24–26] The main advantage achieved here is the lowering of the glass transition temperature so that the rubber–elastic region spans over room temperatures and below. In all these new reports the synthetic work was accompanied by important physical experiment, characterizing the stress–strain behavior and the stress-induced alignment. One expects, and indeed finds dramatic elastic effects (in the range of strains up to 3–400%) due to the high chain anisotropy of main-chain mesogenic polymers.

Another interesting lyotropic LCE system (showing the mesogenic behavior in response to changes in solvent concentration, water in this case) has been prepared in Freiburg.[27] The material was based on the lamellar phase of sidechain polysurfactant. This work explored a concept of permanently tethering the layers through cross-linking across them either through the hydrophobic polymer backbone or hydrophilic side chains.

## 12.1.2   PHYSICAL PROPERTIES

One of the main difficulties preventing a widespread effort in experimental studies and applications of LCE has been their sparse availability. Because a complicated synthesis is required, the cross-linked (or cross-linkable) liquid crystalline polymers are not available commercially and only produced in the few research laboratories mentioned in the previous section. Nevertheless, the last several years has seen a substantial increase in experimental research. One factor contributing to this increase is the growing ability of synthetic groups to perform sophisticated physical experiments: many of the remarkable physical properties have been discovered in the same laboratories where the materials have been prepared.[28-30] Another factor is the increasing ease of preparing basic LCE systems, which allows some more traditional physics research groups to enter the field.[31-34] Below we shall examine some of the most important recent discoveries related to structure and mechanical response, photonic and polarization properties, and dynamical effects.

In the next section we shall review the main features of Cosserat-like nematic rubber elasticity, using the continuum nonlinear description to illustrate the similarities and contrasts with the conventional elasticity of crystalline solids. Although instructive, the highly complex nonlinear Cosserat elasticity is not always needed to describe the relevant physical properties of LCE. The approximation of linear anisotropic elasticity, assuming only the small deformations, may not always be relevant for a rubber capable of large strains. However, it is fully applicable to a number of important physical effects naturally involving small deformations, for example, acoustic waves, or thermal fluctuations. After all, statistics demonstrate that 95% of all modern applications of any kind of rubber involve its deformation below 5%.

The following sections describe recent advances in several key areas of physical properties of LCE. The theoretically predicted effects of soft elasticity, a remarkable phenomenon when there is no rubber–elastic energy in response to certain sets of strains, have been confirmed by several experimental findings. The role of quenched random disorder in forming the equilibrium polydomain textures and in controlling the slow dynamics of deformations in LCE has become clearer in the last few years, making an attractive parallel with a number of glass systems. The additional one-dimensional translational symmetry breaking in smectic or lamellar elastomers and gels makes them an equally puzzling and provocative physical system, with a characteristic two-dimensional entropic rubber elasticity within the layers and a very rigid solid-like response to deformations along the layer normal. Finally, much progress has been made recently in theoretical and experimental studies of piezoelectric effects in chiral LCE and we shall review the main points of principle in this area.

## 12.2   NEMATIC RUBBER ELASTICITY

Rubber–elastic response to deformations of a polymer network stems from the entropy change when the number of conformations allowed for the chains is reduced on stretching their end-to-end distance terminated by network cross-links. Within the simplest affine deformation approach one regards the change in each chain end-to-end vector $R$ as $R' = \underline{\underline{\lambda}} \cdot R$, when a deformation characterized by a deformation gradient tensor $\lambda_{ij}$ is applied to the whole sample.[1] Assuming the chain connecting the two cross-links is long enough, the Gaussian approximation for the number of its conformations $W(R)$ gives for the free energy (per chain): $F_{ch} = -kT \ln W(R') \simeq (kT/a^2N)[\underline{\underline{\lambda}}^T \cdot \underline{\underline{\lambda}}]_{ij}R_iR_j$, where $a$ is the step length of the chain random walk and $N$ the number of such steps. In order to find the total free energy of all chains affinely deforming in the network, one needs to add the contributions $F_{ch}(R)$ with the statistical weight to find a chain with a given initial end-to-end distance $R$ in the system. This procedure, called the quenched averaging, produces the average $\langle R_iR_j \rangle \simeq (1/3)a^2N\delta_{ij}$ in $F_{ch}$. The resulting rubber-elastic free energy (per unit volume) is $F_{el} = (1/2)n_ckT(\underline{\underline{\lambda}}^T : \underline{\underline{\lambda}})$, with $n_c$ a number of chains per unit volume of the network. This is a remarkably robust expression, with many seemingly relevant effects, such as the fluctuation of cross-linking points, only contributing a small quantitative change in the prefactor. The value of the rubber modulus is found on expanding the deformation gradient tensor in small strains, say for a small extension, $\lambda_{zz} = 1 + \varepsilon$, and obtaining $F_{el} \simeq (3/2)n_ckT\varepsilon^2$. This means the extension (Young) modulus $E = 3n_ckT$; the analogous construction for a small simple shear will give the shear modulus $G = n_ckT$, exactly a third of the Young modulus as required in an incompressible medium. This shear modulus, having its origin in the entropic effect of reduction of conformational freedom on polymer chain deformation, is usually so much smaller than the bulk modulus (determined by the enthalpy of compressing the dense polymer liquid) that the rubber is considered as deforming at constant volume. This constraint leads to the familiar rubber–elastic expression $F_{el} = (1/2)n_ckT(\lambda^2 + 2/\lambda)$ where one has assumed that the imposed extension $\lambda_{zz} = \lambda$ is accompanied by the symmetric contraction in both transverse directions, $\lambda_{xx} = \lambda_{yy} = 1/\sqrt{\lambda}$ due to the incompressibility.

## 12.2.1 THE TRACE FORMULA

When the chains forming the rubbery network are liquid crystalline, their end-to-end distance distribution becomes anisotropic. The much more complicated case of smectic/lamellar ordering is a subject of a special article in this collection. In the case of a simple uniaxial nematic one obtains $\langle R_\| R_\| \rangle = (1/3)\ell_\| L$ and $\langle R_\perp R_\perp \rangle = (1/2)\ell_\perp L$, with $L = aN$ the chain contour length and $\ell_\|/\ell_\perp$ the ratio of average chain step lengths along and perpendicular to the nematic director. In the isotropic phase one recovers $\ell_\| = \ell_\perp = a$. The uniaxial anisotropy of polymer chains has a principal axis along the nematic director $n$, with a prolate ($\ell_\|/\ell_\perp > 1$) or oblate ($\ell_\|/\ell_\perp < 1$) ellipsoidal conformation of polymer backbone. The ability of this principal axis to rotate independently under the influence of network strains makes the rubber elastic response nonsymmetric (see Ref. 35 for the original molecular theory of nematic rubber elasticity), so that

$$F_{el} = \frac{1}{2}G\mathrm{Tr}(\underline{\underline{\lambda}}^T \cdot \underline{\underline{\ell}}^{-1} \cdot \underline{\underline{\lambda}} \cdot \underline{\underline{\ell}}_0) + \frac{1}{2}\tilde{B}(\mathrm{Det}[\underline{\underline{\lambda}}] - 1)^2 \qquad (12.1)$$

with $\underline{\underline{\ell}}$ the uniaxial matrices of chain step-lengths before ($\underline{\underline{\ell}}_0$) and after the deformation to the current state: $\ell_{ij} = \ell_\perp\delta_{ij} + [\ell_\| - \ell_\perp]n_in_j$. The last term, the additional bulk-modulus contribution independent of the configurational entropy of polymer chains, is determined by molecular forces resisting the compression of a molecular liquid, $\tilde{B} \sim 10-100$ GPa, much greater than the typical value of rubber shear modulus $G \sim 0.1-1$ MPa. This large energy consequence constrains the value of the strain determinant, $\text{Det}[\underline{\underline{\lambda}}] \approx 1$ (which in other words means that the material is physically incompressible). Note that the deformation gradient tensor $\underline{\underline{\lambda}}$ does no longer enter the elastic energy in the symmetric combination $\underline{\underline{\lambda}}^T \cdot \underline{\underline{\lambda}}$, but is "sandwiched" between the matrices $\underline{\underline{\ell}}$ with different principal axes. This means that antisymmetric components of strain will now have a nontrivial physical effect, in contrast to isotropic rubbers and, more crucially, to elastic solids with uniaxial anisotropy. There, the anisotropy axis is immobile and the response is anisotropic but symmetric in stress and strain. The uniqueness of nematic rubbers stems from the competing microscopic interactions and the difference in characteristic length scales: The uniaxial anisotropy is established on a small (monomer) scale of nematic coherence length, while the strains are defined (and the elastic response is arising) on a much greater length scale of polymer chain end-to-end distance.

The "Trace formula" (Equation 12.1) has proven very successful in describing many physical properties of nematic elastomers. One of the most important consequences of coupling the nematic order to the shape of an elastic body (and perhaps the most relevant for applications) is the effect of spontaneous uniaxial extension/contraction. It is a very simple phenomenon, pictorially illustrated in transformation between states (a) and (b) of Figure 12.1. Mathematically, it is straightforward to obtain from Equation 12.1 for the fixed director orientation $n_z = 1$ and $\text{Det}[\underline{\underline{\lambda}}] = 1$, that

$$F_{el} = \frac{1}{2}G\left(\lambda^2 \frac{\ell_\parallel^{(0)}}{\ell_\parallel} + \frac{2}{\lambda}\frac{\ell_\perp^{(0)}}{\ell_\perp}\right), \tag{12.2}$$

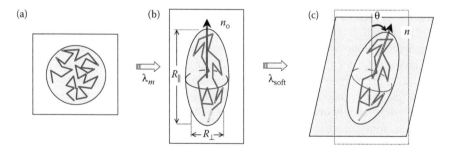

**FIGURE 12.1**  Relation between the equilibrium chain shape and deformations in nematic LCE. When the network of initially isotropic chains forming a spherical gyration shape, (a) is brought into the nematic phase, (b) a corresponding spontaneous deformation of the sample occurs in proportion to the backbone anisotropy, $\lambda_m = (\ell_\parallel/\ell_\perp)^{-1/3}$. An example of soft deformation is given in (c) when rotating anisotropic chains can affinely accommodate all strains (a combination of compression along the initial director, extension across it and a shear in the plane of director rotation), not causing any entropic rubber-elastic response.

minimizing which one obtains the equilibrium uniaxial extension $\lambda$ along $n_z$. In the case when the reference state $\underline{\underline{\ell}}_0$ is isotropic, $\ell_\parallel^{(0)} = \ell_\perp^{(0)} = a$, this spontaneous extension along the director takes the especially simple form:

$$\lambda_m = (\ell_\parallel / \ell_\perp)^{1/3}. \qquad (12.3)$$

This key relation has been obtained long before the Trace formula itself.[36,37] It lies at the heart of nematic elastomers performing as thermal actuators or artificial muscles.[10,31] A large amount of work has been recently reported on photo-actuation of nematic elastomers containing azobenzene moieties[38–40]—in all cases the effect is due to the changing of underlying nematic order parameter $Q$, affecting the magnitude of the ratio $(\ell_\parallel / \ell_\perp)$ and thus the length of elastomer sample or the exerted force if the length is constrained.[41] A plot demonstrating the equilibrium relation between the spontaneous uniaxial extension of a monodomain nematic elastomer, $\lambda$, and the underlying order parameter $Q$ (which may be changed by temperature[101] or by photo-isomerization[41]) is shown in Figure 12.2. One should add that it is rare that reports show such a sharp transition region, this requires a very particular preparation of uniaxial samples with isotropic genesis and no network entanglements.

## 12.2.2 BEYOND THE IDEAL MOLECULAR MODEL

However useful, the concise expression (Equation 12.1) does arise from a specific molecular model based on the ideal Gaussian statistics of polymer chains between cross-links and, as such, may seem oversimplified. It turns out that the symmetry features, correctly captured in the Trace formula, are more important than various corrections and generalizations of the model. In particular, a much more complex theory

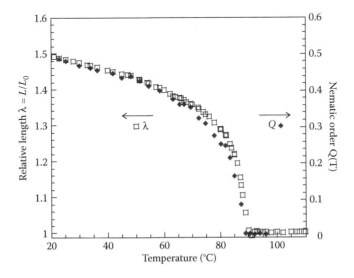

**FIGURE 12.2** The simultaneous plot of uniaxial extension and the nematic order illustrating their direct correlation. (Adapted from Clarke, S. M. et al. 2001. *Phys. Rev. Lett.*, **86**, 4044.)

of nematic rubber elasticity taking into account chain entanglements (which have to play a very important role in cross-linked networks, where the constraint release by diffusion is prohibited), developed in Ref. 42 shows the same key symmetries:

$$F_{el} = \frac{2}{3}G\frac{2M+1}{3M+1}\mathrm{Tr}(\underline{\underline{\lambda}}^T \cdot \underline{\underline{\ell}}^{-1} \cdot \underline{\underline{\lambda}} \cdot \underline{\underline{\ell}}_0) + \frac{3}{2}G(M-1)\frac{2M+1}{3M+1}\left(\overline{|\underline{\underline{\ell}}^{-1/2} \cdot \underline{\underline{\lambda}} \cdot \underline{\underline{\ell}}_0^{1/2}|}\right)^2$$

$$+ G(M-1)\ln|\underline{\underline{\ell}}^{-1/2} \cdot \underline{\underline{\lambda}} \cdot \underline{\underline{\ell}}_0^{1/2}|, \tag{12.4}$$

where $M$ is a number of entanglements on a chain of length $N$ ($M = 1$ in the ideal Trace formula), and the notation $\overline{|\ldots|}$ denotes an orientational average of the matrix ... applied to an arbitrary tube segment (see Ref. 42 for detail). The important feature of the complicated expression (Equation 12.4) is the "sandwiching" of the deformation gradient tensor $\underline{\underline{\lambda}}$ between the two orientational tensors defined in the current and the reference states, respectively. The continuum model of fully nonlinear elasticity is also quite complicated in its applications, but the fundamental requirements that any elastic free energy expression must be (1) invariant with respect to body rotations of the reference state and the current state, separately, and (2) reduce to the isotropic expression $\mathrm{Tr}(\underline{\underline{\lambda}}^T \cdot \underline{\underline{\lambda}})$, requires it to have the form

$$F_{el} \propto Tr(\underline{\underline{\lambda}}^T \cdot [\underline{\underline{\ell}}]^{-m} \cdot \underline{\underline{\lambda}} \cdot [\underline{\underline{\ell}}_0]^m), \tag{12.5}$$

where the power is only allowed to be $m = 1, 2, 3$. Once again, one recognizes the key symmetry of separating the two powers of deformation gradient tensor by the orientation tensors defined in the current and reference frames, respectively.

The similarities and contrasts with the conventional uniaxial elasticity are more apparent on the simple level of linear continuum elasticity, when only small deformations and small director rotations are considered. Introducing the deformation tensor $u_{ij}$ as the gradient of local displacement vector $\upsilon$ ($u_{ij} = \nabla_i u_j$, with the deformation gradient $\lambda_{ij} = \delta_{ij} + u_{ij}$ taken to be incompressible, with $\mathrm{Det}[\underline{\underline{\lambda}}] = 1$), one obtains

$$F_{el} = C_1(\mathbf{n}\underline{\tilde{\varepsilon}}\mathbf{n})^2 + 2C_4[n \times \underline{\tilde{\varepsilon}} \times \mathbf{n}]^2 + 4C_5(\mathbf{n} \cdot \underline{\tilde{\varepsilon}} \times \mathbf{n})^2$$

$$+ \frac{1}{2}D_1[(\mathbf{\Omega} - \omega) \times \mathbf{n}]^2 + D_2\mathbf{n}\underline{\tilde{\varepsilon}}[(\mathbf{\Omega} - \omega) \times \mathbf{n}] + \frac{1}{2}K(\nabla\mathbf{n})^2, \tag{12.6}$$

where $\underline{\tilde{\varepsilon}}$ is the symmetric traceless part of deformation tensor, $(1/2)(\underline{u} + \underline{u}^T) - (1/3)\underline{\delta}\,\mathrm{div}\,u$, the only relevant strain in the conventional linear elastic theory.[2] A full incompressibility, assumed here for simplicity, has rendered irrelevant two additional terms in (12.6), which are proportional to $\mathrm{Tr}(\underline{\varepsilon})$. Two vectors, $\mathbf{\Omega}$ and $\omega$, describe the rotational contributions of deformation ($\mathbf{\Omega} = (\overline{1}/2)\mathrm{curl}\,\underline{u}$, the antisymmetric part of $\underline{u}$) and the nematic director ($\omega = [n \times \delta n]$). Characteristically, the elastic energy depends on the relative rotation combinations $[(\mathbf{\Omega} - \omega) \times \mathbf{n}]$ reflecting the independently mobile degrees of freedom in this Cosserat medium. The unit vector $\mathbf{n}$ in Equation 12.6 represents the axis of uniaxial anisotropy before deformation (i.e., in the reference state) and should be regarded fixed at this level of approximation, except

in the traditional nematic Frank-elasticity contribution, the last term in the continuum free energy density (Equation 12.6), schematically presented in the one-constant
approximation. Clearly, only the uniform *relative rotation* between the nematic and
rubber-elastic subsystems contributes to the free energy (Equation 12.6) through the
coupling terms $D_1$ and $D_2$, first written down phenomenologically by de Gennes.[43]
Applying a particular molecular model, the Trace formula (Equation 12.1), one finds
the values of relevant constants:

$$C_1 = 2C_4 = G; \quad C_5 = \frac{1}{8}G\left(\frac{\ell_\parallel^2 + \ell_\perp^2}{\ell_\parallel \ell_\perp}\right)^2$$

$$D_1 = G\left(\frac{\ell_\parallel^2 - \ell_\perp^2}{\ell_\parallel \ell_\perp}\right)^2; \quad D_2 = -G\frac{\ell_\parallel^2 - \ell_\perp^2}{\ell_\perp^2}. \tag{12.7}$$

Appropriately, the relative rotation coupling constants must vanish in the isotropic
phase, at $\ell_\parallel = \ell_\perp$. Models of nematic polymer chains relate the average backbone
anisotropy $(\ell_\parallel/\ell_\perp - 1)$ to the local nematic order parameter $Q$, defined as the thermodynamic average of mesogenic monomer long axes $\langle(3/2)(u \cdot n)^2 - (1/2)\rangle$. One then
finds that in the ideal model $D_1$ must scale as $Q^2$ because the adverse effect on relative
rotation $\sim (\mathbf{\Omega}-\boldsymbol{\omega})^2$ cannot depend on the sign of $Q$, that is, on whether the polymer
chains are prolate or oblate. It has been discovered that, in some systems far from the
ideal rotationally-invariant case described by the Trace formula (Equation 12.1), the
constant $D_1$ has also a contribution $\sim \alpha|Q|$ with a small prefactor $\alpha$ that is a measure
of the so-called "semisoftness".[44] In contrast, $D_2$ must scale as $Q$ in the leading
approximation because the sign of director rotation induced by symmetric shear $\underline{\underline{\varepsilon}}$ is
different in prolate and oblate elastomers.

Two main consequences of the coupling between the elastic modes of polymer
network and the rotational modes of the nematic director are the reduction of the
effective elastic response and the adverse effect on the director fluctuations. The first
effect has received the name "soft elasticity" and is the result of the special symmetry
of the coupling between the orientational modes and deformation gradients. It is easy
to see Ref. 46 that there is a whole continuous class of deformations described by
the form

$$\underline{\underline{\lambda}} = \underline{\underline{\ell}}_\theta^{1/2} \cdot \underline{\underline{V}} \cdot \underline{\underline{\ell}}_0^{-1/2}, \tag{12.8}$$

with $\underline{\underline{V}}$ an arbitrary unitary matrix, which leave the elastic free energy $F_{\text{el}}$ at its
constant minimum value (for an incompressible system). Remarkably, this remains
true whether one examines the Trace formula (Equation 12.1), or any other expression
above, (Equations 12.4 or 12.5), as long as they respect the correct symmetries of the
two independent degrees of freedom. Figure 12.1c illustrates one example of such
soft deformation.

In the linear continuum theory one arrives at the same conclusion after integrating
out (minimizing over) the director fluctuations $\delta n$ in the expression $F_{\text{el}}$ in Equation 12.6. In some cases this may even result in the total elimination of elastic response,
for example, the renormalized shear modulus $\tilde{C}_5 = C_5 - D_2^2/8D_1 \rightarrow 0$, and reflects

the ability of anisotropic polymer chains to rotate their long axis to accommodate some imposed elastic deformations without changing their shape. If one instead chooses to focus on the director modes in a nematic elastomer with a fixed (constrained) shape, the coupling terms $D_1$ and $D_2$ provide a large energy adverse effect for uniform director rotations $\delta\boldsymbol{n}$ (with respect to the elastically constrained network). This adverse effect, which appears as a mass term in the expression for mean-square director fluctuation $\langle|\delta n_q|^2\rangle \simeq (k_B T/V)/(Kq^2 + \tilde{D})$ with $K$ the Frank constant, results in the suppression of fluctuations and the related scattering of light from a nematic elastomer.[45] In contrast to optically turbid ordinary liquid nematics where light is scattered by long-wavelength director fluctuations, the aligned monodomain nematic rubber is totally transparent. However, when the elastic deformations in the network are not constrained and are free to relax, there are certain combinations of polarization and wave vectors of director fluctuations (corresponding to the soft deformation modes), for which the "effective mass" $\tilde{D}$ vanishes and the fluctuation spectrum should appear as in ordinary liquid nematics.[46]

### 12.2.3  SOFT ELASTICITY AND STRIPE DOMAINS

It is natural to expect that if a sample of monodomain, uniformly aligned nematic elastomer (which usually implies that it has been cross-linked in the aligned nematic phase[8,11]) is stretched along the axis perpendicular to the nematic director $\boldsymbol{n}_0$, the director will switch and point along the axis of uniaxial extension. The early theory (ignoring the effect of soft elasticity)[35] has predicted, and the experiment on polyacrylate LCE[47] confirmed that this switching may occur in an abrupt discontinuous fashion when the natural long dimension of anisotropic polymer chains can fit into the new shape of the sample, much extended in the perpendicular direction. However, the same experiment performed on a polysiloxane LCE[28] has shown an unexpected stripe domain pattern. Further investigation has proven that the nematic director rotates continuously from $\boldsymbol{n}_0$ toward the new perpendicular axis, over a substantial range of deformations, but the direction of this rotation alternates in semiregular stripes of several microns width oriented along the stretching direction, Figure 12.3a. Later the same textures have been observed by other groups and on different materials, including polyacrylates,[12,48] although there also have been reports confirming the single director switching mode.[49]

Theoretical description of stripe domains[50] has been very successful and has led to several long-reaching consequences. First of all, this phenomenon has become a conclusive proof of the soft elasticity principles. Therefore, the nematic elastomers have been recognized as part of a greater family of elastic systems with microstructure (shape memory alloys being the most famous member of this family to date[51]), all exhibiting similar phenomena. Finally, the need to understand the threshold value of extension $\lambda_c$ has led to deeper understanding of internal microscopic constraints in LCE and resulted in introduction of the concept of semi-softness: a class of deformations that has the "soft" symmetry, as in Equation 12.8, penalized by a small but physically distinct elastic energy due to such random constraints.

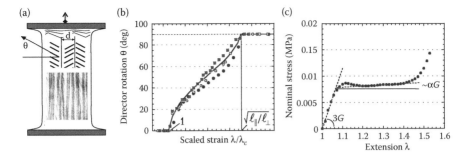

**FIGURE 12.3** Stripe domains in nematic LCE: (a) The picture of alternating domains with $\pm\theta$ on stretching the sample perpendicular to the nematic director $n_0$; the inset shows the microscopic image of stripes between crossed polars; (b) The director rotation angle $\theta(\lambda)$ in each domain—the theoretical curve Equation 12.9 and several different experiments collapsed on it by rescaling of variables; (c) An associated stress–strain curve showing the soft elastic plateau during the period of director rotation, between the threshold $\lambda_c$ and the completion of rotation at $\lambda_c\sqrt{\ell_\parallel/\ell_\perp}$. Note a small hump in the threshold region, which is due to an extremely slow stress relaxation at the moment when the stripe domains are being formed by negotiating between local shears and mechanical constraints.

The main result of theoretical model[50] gives the director angle variation with strain,

$$\theta(\lambda) = \pm\arcsin\left[\frac{\ell_\parallel}{\ell_\parallel - \ell_\perp}\left(1 - \frac{\lambda_c^2}{\lambda^2}\right)\right]^{1/2} \tag{12.9}$$

with only one free parameter, the threshold strain $\lambda_c$. In fact, if one uses the pure Trace formula (Equation 12.1), the threshold is not present at all, $\lambda_c = 1$. The backbone chain anisotropy $\ell_\parallel/\ell_\perp$, which enters the theory, is an independent experimentally accessible quantity related, for example, to the spontaneous shape change of LCE on heating it into the isotropic phase, $\lambda_m \approx (\ell_\parallel/\ell_\perp)^{1/3}$ in Figure 12.1. This allowed the data for director angle, obtained in different experiments on many different materials, to be collapsed onto the same master curve, Figure 12.3b, spanning the whole range of nonlinear deformations.

The physical reason for the stretched LCE to break into stripe domains with the opposite director rotation $\pm\theta(\lambda)$ becomes clear when one recalls the idea of soft elasticity.[46,52] The polymer chains forming the network are anisotropic, in most cases having the average shape of uniaxial prolate ellipsoid, see Figure 12.1. If the assumption of affine deformation is made, the strain applied to the whole sample is locally applied to all network strands. The origin of (entropic) rubber elasticity is the corresponding change of shape of the chains, away from their equilibrium shape frozen at network formation, which results in the reduction in entropy and rise in the elastic free energy. However, the nematic (e.g., prolate anisotropic) chains may find another way of accommodating the deformation: If the sample is stretched perpendicular to the director *n* (the long axis of chain gyration volume), the chains may rotate their *undeformed* ellipsoidal shapes—thus providing an extension, but necessarily in combination with simple shear—and keep their entropy constant and elastic free energy zero! This, of course, is unique to nematic elastomers: isotropic chains (with spherical shape) have to deform to accommodate any deformation. It is also important that with deformations causing no elastic energy, there is no need for the material to change its preferred nematic order parameter $Q$, so the ratio $\ell_\parallel/\ell_\perp$ remains constant for the duration of soft deformations (this is not the case when, e.g., one stretches the elastomer along the director axis). The physical explanation of stripe domains is now clear: the stretched nematic network attempts to follow the soft deformation route to reduce its elastic energy, but this requires a global shear deformation which is prohibited by rigid clamps on two opposite ends of the sample, Figure 12.3a. The resolution of this macroscopic constraint is to break into small parallel stripes, each with a completely soft deformation (and a corresponding shear) but with the sign of director rotation (and thus the shear) alternating between the stripes. Then there is no global shear and the system can lower its elastic energy in the bulk, although it now has to accept the consequence for domain walls and for the nonuniform region of deformation near the clamps. The balance of gains and losses determines the domain size $d$.

The argument above seems to provide a reason for the threshold strain $\lambda_c$, which is necessary to overcome the barrier for creating domain walls between the "soft" stripes. However, it turns out that the numbers do not match. The energy of domain walls must be determined by the gradient Frank elasticity of the nematic (with the Frank elastic constant independently measured, $K \sim 10^{-11}$N) and thus should be very small, since the characteristic length scale of nematic rubbers is $\xi = \sqrt{K/G} \sim 10^{-9}$m. Hence, the threshold provided by domain walls alone would be vanishingly small whereas, most experiments have reported $\lambda_c \sim 1.1$ or more. This mystery has led to a whole new concept of what is now called semisoftness of LCE. The idea is that, due to several different microscopic mechanisms,[53] a small addition to the classical nematic rubber–elastic free energy is breaking the symmetry required for the soft deformations:

$$F_{el} \approx \frac{1}{2}G[\text{Tr}(\underline{\underline{\lambda}}^T \cdot \underline{\underline{\ell}}_\theta^{-1} \cdot \underline{\underline{\lambda}} \cdot \underline{\underline{\ell}}_0) + \alpha(\delta n + n \cdot \underline{\underline{\lambda}} \times \delta n)^2] \qquad (12.10)$$

(usually $\alpha \ll 1$). The soft-elastic pathways are still representing the low-energy deformations, but the small price $\sim \alpha G$ provides the threshold for stripe domain formation ($\lambda_c \sim 1 + \alpha$) and also makes the slope of the stress–strain soft-elastic plateau small but nonzero, Figure 12.3c. Compare this with the ordinary extension modulus of rubber before and after the region of stripe domains. Regardless of "small complications" of semisoft corrections, the main advance in identifying the whole class of special low-energy soft deformations in LCE and proving their existence by direct experiment is worth noting. Recently, a number of fundamental papers have established very general rules of soft and semisoft deformations, as well as the symmetry-based structure of nematic–elastic energy.[54,55]

## 12.3 RANDOM DISORDER IN LCE

It has been recognized long ago that, if no special precautions are taken to preserve the monodomain director alignment in the network, the liquid crystal elastomers always form with a highly disordered director texture. In LCE literature such a texture has been historically called "polydomain" although this might be a misleading term: the system does not have pronounced uniform domains separated by accentuated walls, as for instance in ferromagnets or polycrystalline solids. The disordered texture may appear similar to liquid crystalline polymers and low-molar mass liquid crystals, where the quenching from isotropic to the nematic phase results in a dense defect texture (called *Schlieren* in nematics), however, the defect texture is not an equilibrium and coarsens with time. This evolution of topological defect structure has led to the famous analogy with the cosmological theory of the early universe undergoing symmetry-breaking transitions.[56] It has been exhaustively investigated in liquid crystals as well as in other ordered media[57] and also reviewed in the context of mixtures.[58]

The major difference with cross-linked liquid crystal elastomers is that their disordered director textures represent a true thermodynamic equilibrium. If such a material is made uniformly aligned by application of an external field or mechanical stretching (what we call polydomain–monodomain transition below), it always returns to its disordered state after the external aligning influence is removed. If a disordered, polydomain nematic rubber is heated to the isotropic phase, on return to the nematic state the texture is reestablished with the same characteristic length scale (the "domain size," $\xi_D$). This feature is preserved for the networks that are cross-linked in the high-temperature isotropic state and then cooled into the nematic (or other) liquid crystal phase. Therefore, it is not the case that certain topological defects are permanently cross-linked into the network—the texture forms spontaneously: the system "knows" the precise length scale, that is, the size of the regions in which it can afford having the aligned director, and even how this size reversibly changes with temperature.[59] Many experimental observations report that this domain size $\xi_D$ is between a fraction of and a few microns, so each small aligned region contains many cross-link sites (the average distance between cross-links is only a few nanometers in a typical rubber).

A concept of quenched sources of random disorder and their influence on the overall structure has been put forward a long time ago and fruitfully explored in many areas of modern physics, in particular, spin glasses, superconductor vortex lattices and neural networks.[60,61] The basic result is the destruction of long-range correlations in the order parameter, which nevertheless remains locally strong and well defined. In other words, the system is well-ordered locally but breaks into uncorrelated regions on a large scale and has no overall macroscopic order or alignment. Applying the knowledge accumulated in this field to nematic systems, one must identify quenched sources of random orientational disorder. For example, when liquid crystals are confined in silica gels,[62] the source of disorder stems from the director anchoring on random interfaces. Such sources of disorder are strong and have its own (rather large) characteristic size.

In nematic elastomers, the microscopic cross-links could be the entities carrying the quenched orientational disorder. Figure 12.4a illustrates the point: even for idealized point cross-links one can always identify a direction of anisotropy, which is quenched since the cross-link is not totally free to rotate under thermal noise. In practice, cross-linking moieties are always anisotropic and frequently deliberately made of mesogenic rods themselves. Freezing their orientation by linking to several tethered chains would restrict their orientational freedom, but their coupling to the surrounding mesogenic units would remain in force. In crude terms, one can say that each such cross-link

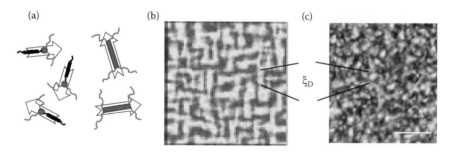

**FIGURE 12.4** (a) Schematic of the orientational effect of network cross-links in nematic LCE and the translational effect in smectic LCE. (b) A computer-simulated image of the polydomain nematic elastomer; the contrast represents the difference in director orientation and indicates the length scale $\xi_D$, as calculated by N. Uchida.[63] (c) A typical LCE polydomain texture image between crossed polars (the bar $= 10\,\mu$m). (Adapted from Elias, F. et al., 1999. *Europhys. Lett.*, **47**, 442.)

adds an energy $\sim -(1/2)g\mathbf{k} \cdot \underline{\underline{Q}} \cdot \mathbf{k}$ to the system Hamiltonian, where $g$ is the coupling strength, $\mathbf{k}$ the local orientation of the source and $Q_{ij}$ is the nematic order parameter at this point in space.[64] Adding all the cross-links together, introducing their continuum density $\rho(\mathbf{r}) = \sum_x \delta(\mathbf{r} - \mathbf{r}_x)$ and separating the magnitude of local nematic order from its principal direction $\mathbf{n}$, one obtains the *random field* contribution to the nematic free energy density

$$F_{RF} = -\int \frac{1}{2} g Q \rho(\mathbf{r})(\mathbf{k} \cdot \mathbf{n})^2 d^3\mathbf{r}. \tag{12.11}$$

Here both the density of sources and their orientation are the randomly quenched variables. When points are uniformly distributed in space their density adopts a Gaussian distribution

$$P[\rho(\mathbf{r})] \simeq \exp\left\{ -\int d^3\mathbf{r} \frac{[\rho - n_c]^2}{2n_c} \right\}, \tag{12.12}$$

where $n_c$ is the mean density of cross-links in the system. As the defects are at the same time randomly oriented in three dimensions, we have the probability of their orientation as simply $P[\mathbf{k}] = 1/4\pi$. Obviously, when a monodomain elastomer is prepared by cross-linking in the aligned state, the distribution $P[\mathbf{k}]$ would become increasingly biased, so the cross-links would then support the equilibrium macroscopic alignment, as opposed to disrupting it in the disordered case. This is how the monodomain, well-aligned elastomers are produced.

## 12.3.1 Phase Transitions with Quenched Disorder

How can one describe nematic ordering in a system with quenched disorder, when the system will remain misaligned on a macroscopic scale? The same question can be asked in many other systems, the most famous being the case of spin glass where magnetic ordering occurs in a lattice doped with impurity atoms that maintain rigid, and random, orientation of their magnetic moments ("random-anisotropy magnets"). One has to make a firm distinction between the state of macroscopic alignment, which one would detect in experiments measuring birefringence, dichroism, or wide-angle x-ray scattering accessing the volume $\gg \xi_D$, and the state of local ordering of mesogenic units, which is detected by powder-averaged nuclear magnetic resonance or electron spin resonance techniques as well as by its thermal signature. The difference is as between the "ensemble average" of rods over the large volume at a given moment of time, and the "time average" of orientations of an individual rod. These two averages are only different in nonergodic systems, but of course the quenched random disorder leads to nonergodic, glassy behavior.

Simulations[65] and continuum model[66] of quenched disorder predict different regimes of the nematic transition. The scalar amplitude of the nematic order parameter $Q$ is homogeneously set in the system, while the local principal axis of $\underline{\underline{Q}}$ (the director $n$) follows an equilibrium disordered texture with the characteristic size of correlated regions $\xi_D$. The continuum Landau–de Gennes free energy density gets an additional $Q$-dependent term. In the limit of small $Q$ and weak disorder, this renormalization leads to

$$F \approx \frac{1}{2}AQ^2 - \frac{1}{3}BQ^3 + \frac{1}{4}CQ^4 + \frac{Vn_c}{40\pi^2 a^3}\frac{g^2(kT)^2 Q^2}{K^3}, \qquad (12.13)$$

where $n_c$ is the density of cross-links, $a$ is the small-distance length scale cutoff (crudely, a mesogen size), and $V$ the system volume. Recall that $g$ is a measure of disorder energy per cross-link. Since the Frank constant is $K \propto Q^2$ at $Q \ll 1$, this additional term scales as $Q^{-4}$ and evidently prevents the system from ever achieving a state with $Q = 0$. Nevertheless, depending on the strength of disorder, one may still see a sharp transition between the free energy minima, or a continuous transition that may resemble supercritical behavior. As the disorder strength increases the discontinuous jump decreases and eventually disappears, however, even below the critical point one is left with small residual paranematic order above $T_{ni}$. Recently a detailed NMR study of the local order parameter $Q$ in the misaligned polydomain elastomers has been reported.[67] It has demonstrated that a critical continuous phase transition takes place, Figure 12.5 maintaining the analogy with well-known studies of spin glasses and the first-order phase transition with quenched impurities.

The suppression of thermal fluctuations by network cross-links is a well-established effect in all kinds of LCE.[45] In the case of smectic elastomers, the pinning of layers in a well-aligned monodomain system leads to effectively long-range order in a one-dimensional stack of layers.[68] In contrast, a similar x-ray scattering experiment[69] on smectic rubber, which was aligned and cross-linked in the nematic phase and later cooled down to the smectic-A state, has reported the opposite trend: the scattering from the layer stack was very diffuse, with broad peaks and clearly improper order in equilibrium.

Although the conventional liquid nematic–smectic A transition could be continuous according to symmetry arguments, director fluctuations change this picture and induce a first-order phase transition. This important result is due to Halperin et al.[70]

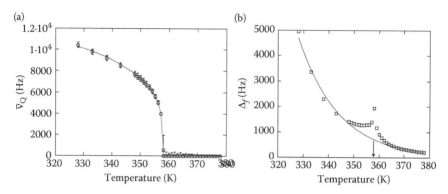

**FIGURE 12.5** (a) The NMR line splitting, proportional to the scalar magnitude of local order parameter $Q(T)$, and (b) the line width, which is a measure of mean-square fluctuations near the transition suggest a line of continuous critical transitions. (Adapted from Feio, G. et al., 2008. *Phys. Rev. B*, **78**, 020201.)

The quenched pinning of layers by the cross-link placed randomly with respect to the smectic density modulation changes this conclusion again.[71] Figure 12.6 shows three possible ways the cross-links can be established in the resulting smectic structure. The monodomain genesis (A), which has been the case in classical experiments[29,68,72] results in cross-links underpinning the layers and suppressing their thermal fluctuations. The isotropic genesis (C) results in a strong polydomain structure where the locally uniaxial orientation of layers is disordered in the manner resembling the short-range spin-glass correlations, often referred to as "Bragg glass."[73,74] In between, lies the uniaxial genesis (B) where the cross-links are established in the uniaxially aligned state (nematic or due to external fields) but the smectic layers form with quenched

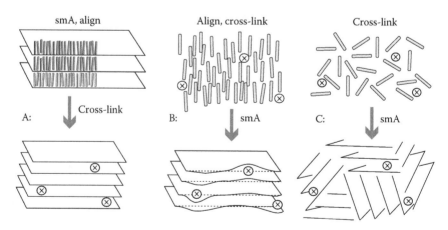

**FIGURE 12.6** Three ways of obtaining smectic elastomers, in decreasing order of resulting alignment. Procedure A results in a true monodomain smectic elastomer. Following the procedure B one obtains a uniaxially aligned elastomer with layer frustration, while the procedure C leads to a true polydomain smectic elastomer.

frustrations, leading to nonthermal narrowing of correlation range and resulting in broadening of x-ray lines.

Similar to Equation 12.11 one has to write the coupling energy

$$F_{RF} = \int \frac{1}{2} g |\psi(\mathbf{r})| \rho(\mathbf{r}) \cos[q_0(z - \mathbf{u}(\mathbf{r}))] d^3 r, \qquad (12.14)$$

where $\psi = |\psi| e^{i\Phi(\mathbf{r})}$ is the smectic order parameter, $q_0 = 2\pi/d$ is the equilibrium smectic layer periodicity, $z$ is the current position of the layer and $\mathbf{u}$ is the elastic displacement of the medium reflecting the movement of cross-links. The mean-field correction to the Landau–de Gennes free energy of the nematic–smectic transition takes the form

$$F(\psi) \approx \frac{1}{2} A |\psi|^2 + B(\psi) + \frac{1}{4} C |\psi|^4, \qquad (12.15)$$

where the additional $B$-term has a complicated dependence on the local smectic order:

$$B(\psi) \approx -\frac{kT}{12\pi K^{3/2}} \left[ g_\perp q_0^2 |\psi|^2 + \frac{4\tilde{C}_5 D_1^2}{4\tilde{C}_5 D_1 + (D_1 + D_2)^2} \right]^{3/2}.$$

Here $g_\perp q_0^2 |\psi|^2$ is the classical smectic term[70] proportional to the layer compression constant $B \sim 10^6$ Ps, and $\tilde{C}_5 = C_5 - D_1^2/8D_2$ is the familiar measure of softness in the nematic phase at higher temperature. In the ideally soft case, as well as in the uncrosslinked material, $\tilde{C}_5 = 0$ and one recovers the fluctuation-driven Halperin–Lubensky–Ma first-order phase transition. The main effect of quenched disorder on the nematic–smectic transformation in an elastomer network is a simple renormalization of the transition temperature and, depending on the nature of the nematic elastomer (degree of its softness), a change in the order of the transition. However, on further cooling below a certain crossover temperature,[71] the strength of layer coupling to the network cross-links increases and the "classical" smectic state changes to a glassy structure. As in polydomain nematic LCE, there is a characteristic length scale $\xi_D$ above in which smectic layers are no longer correlated.

## 12.3.2 CHARACTERISTIC DOMAIN SIZE

In the continuum theory, any deviation of nematic director from a uniform orientation is penalized by Frank elasticity. The random field energy density, Equation 12.11, has to be added to it and together these two terms capture the main features of systems with quenched orientational disorder. For the free energy containing these two terms one can now apply the famous Imry–Ma estimate.[75] For a texture with the correlation distance $\xi_D$ we can estimate the gradient $\nabla n$ as $\sim 1/\xi_D$, immediately getting an estimate for the Frank elastic energy (per domain of the size $\xi_D$): $F_{Fr} \sim KV\xi_D^{-2} = K\xi_D$, when the volume is $V \sim \xi_D^3$ in the normal three dimensions. The number of sources of random anisotropy $N_x$ in such a domain is proportional to the domain volume, $N_x \simeq n_c \xi_D^3$. To minimize the free energy the system will tend to have nematic director parallel to the direction most of the cross-links (vectors $\mathbf{k}_i$) within this volume are parallel too. To visualize this picture, one can take all $\boldsymbol{k}_i$ and make a chain from them by connecting them head to tail. The end-to-end vector of such a "chain" will give the preferred alignment direction for the nematic director. This is the same construction as that of an ordinary random walk, and we are looking for the net value of the orientational field. The mean square length of this end-to-end vector will be proportional to $N_x^{1/2} = n_c^{1/2} \xi_D^{3/2}$. This value indicates the magnitude of the mean field of quenched sources, averaged over the chosen domain volume. Thus the system will gain random-field energy of the order of $F_{RF} \simeq -g(n_c \xi_D^3)^{1/2}$ by aligning in this direction. Hence, the domain free energy can be estimated as: $F \simeq K\xi_D - gn_c^{1/2}\xi_D^{3/2}$. Minimizing this with respect to the domain size we obtain the characteristic length

$$\xi_D \sim K^2/n_c g^2. \tag{12.16}$$

This scaling argument is known to be very robust and is confirmed by explicit calculations within several different models by many authors over the years. It shows that no matter how weak the random disorder is, it will eventually achieve and disprove the correlation of director orientation, dividing the system into the misaligned regions of size $\xi_D$.

One needs to emphasize that, similar to spin glasses and other physical systems with quenched disorder, the structure is not one of uniformly aligned domains with sharp boundaries (as the word "polydomain" might imply). The length $\xi_D$ represents the characteristic size of correlated regions within which the local nematic director $n$ is more or less aligned (or in the analogous smectic LCE the region of layer correlations). The local order parameter $Q$ remains homogeneous, but the average alignment is absent (and so the birefringence or x-ray measurements will show zero macroscopic order). It is important to realize that $\xi_D$ is much larger than the average distance between cross-linking points (which is measured in few nanometers) and thus the sources of disorder can be treated as a continuous mean field, as opposed to the case of large-scale constraints in silica gel or outer porous systems.

At this time, there are only few direct measurements of domain size of polydomain nematic elastomers. It appears that in various different systems and materials this length scale is, roughly, between 0.5 and a few microns. Accordingly, such materials scatter light very strongly between the misaligned regions of very high birefringence. Figure 12.7a shows an example of light-scattering pattern, which allows for extracting the characteristic size from the reciprocal space, and also remarkably agrees with the simulation prediction of a rectangular domain arrangement shown in Figure 12.4b. Figure 12.7b shows how size $\xi_D$ evolves with temperature, as the local nematic order $Q$ decreases. It turns out[77] that director correlations decay differently depending how the separation $|r - r'|$ compares with the length scale $\xi_D$:

$$\langle \boldsymbol{n}(0) \cdot \boldsymbol{n}(r) \rangle \sim \begin{cases} \exp(-r/\xi_D), & \text{for } r \ll \xi_D \\ (r/\xi_D)^{-\upsilon}, & \text{for } r \gg \xi_D \end{cases}. \tag{12.17}$$

Far from the isotropic phase, and when the disorder is sufficiently strong, the expression (Equation 12.16) is a valid estimate for $\xi_D$. However, near the transition (at

(a)    (b)

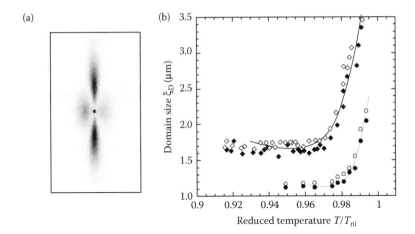

**FIGURE 12.7**    (a) Light-scattering image of polydomain LCE between crossed polars; note the 4-fold symmetry of scattering, which agrees with model predictions of Figure 12.4b. (b) The texture size $\xi_D$ from two different systems: siloxane side-chain LCE[76] and main-chain LCE.[59] In both plots, filled symbols show $\xi_D$ on heating toward $T_{ni}$; open symbols on cooling, indicating the equilibrium nature of $\xi_D$. The lines show a fit Equation 12.18.

$Q \ll 1$) as well as for very weak disorder this size takes on a different scaling limit:

$$\xi_D \sim (K^2/n_c g^2)e^{2kTa/\pi^2 K} \propto Q^2 e^{1/Q^2}. \tag{12.18}$$

This estimate is taken under the assumption that $g \propto Q$, see Equation 12.11 and an established estimate for the Frank constant $K \propto (kT/a)Q^2$, where $a$ is the continuum-limit cutoff scale, of the order of the mesogen size. Evidently this "domain size" $\xi_D$ diverges as $Q \to 0$, as the experimental data in Figure 12.7b demonstrates.

### 12.3.3 Polydomain–Monodomain Transition

Another fundamental feature of disordered, polydomain nematic elastomers, also important practically, is their alignment under applied uniaxial stress. Again, there is a direct analogy with alignment of random-anisotropy spin glasses in a strong magnetic field, and frustrated vortex lattices under the influence of underlying crystalline order. The basic result says that, on increasing the external field, the globally misaligned texture orders in the direction of the field—by rotating the individual domains toward a common axis (rather than, for instance, changing of relative size of correctly and wrongly aligned domains),[78] Figure 12.8a. As the domains become more aligned, the overall order can be characterized by a macroscopic alignment parameter $S_{macro} = Q(1/V) \int d^3 r ((3/2)\cos^2\theta - (1/2))$, where $Q$ is the (homogeneous) local nematic order and $\theta$ the angle between a local director orientation and the direction of applied stress.

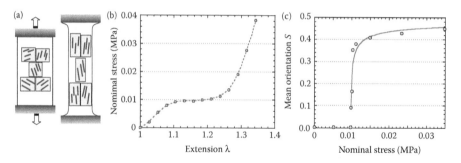

**FIGURE 12.8** (a) A sketch of the polydomain–monodomain transition under stress. (b) Stress–strain curve for the polydomain–monodomain transition in polysiloxane side-chain LCE; the stress value on the soft plateau is the critical stress $\sigma_c$ in $S(\sigma)$. (c) The onset of macroscopic alignment $S_{macro}(\sigma)$ during the polydomain–monodomain transition, fitted with Equation 12.19.

The case of elastomers is special because when each nematic domain attempts to rotate its director, the associated local deformations need to be elastically compatible. Of course, the reason for each region of uniformly aligned director to rotate toward the extension axis is the attempt to follow the low-energy soft deformation path. This is the reason for an extended soft plateau in stress–strain curves during the polydomain–monodomain transition, Figure 12.8b. The width of this plateau is different from that in the stripe-domain case in monodomain LCE, Figure 12.3c, but is also determined by the local anisotropy $\ell_{\parallel}/\ell_{\perp}$ and has been seen to stretch to over 200% in main-chain nematic LCE. The structural change occurring during this transition under an imposed uniaxial stress (or extension) is illustrated by the rise in the macroscopic alignment parameter $S$ shown in Figure 12.8c; this data is obtained from wide-angle x-ray scattering. The trend is evident: initially there is no macroscopic alignment, but after a threshold stress $\sigma_c$ (which is temperature dependent and takes very different values in LCE materials with different $\ell_{\parallel}/\ell_{\perp}$) the rise in alignment is very rapid. Characteristically, the $S(\sigma)$ plot saturates at a value $S = Q \sim 0.5$–$0.6$, which one would have expected of the ordinary nematic order parameter.

Theoretical calculations have shown how the macroscopic order increases in the system under increasing external stress. The transition point at a critical stress $\sigma_c$ is poorly described by the theories, but the subsequent evolution of $S = S(\sigma)$ is predicted as a sharp approach to saturation[78]:

$$S \simeq s_0 + (Q - s_0)\exp\left\{-\frac{const}{(\sigma - \sigma_c)^{1/2}}\right\}. \tag{12.19}$$

The small constant $s_0$ accounts for the jump at the vicinity of the transition point, which is only described by a more complicated analysis of replica symmetry breaking or functional renormalization group. At $\sigma \to \infty$, the macroscopic order $S$ tends to $Q$, the thermodynamic value of the nematic order parameter in a fully aligned elastomer. Modeling also shows that the equilibrium domain wall width $w$ decreases with growing applied deformation $\lambda$[64]:

$$w = \frac{\xi_D}{\lambda - 1}\left(\sqrt{\ell_\parallel/\ell_\perp} - 1\right) \tag{12.20}$$

(the fact that $w \to \infty$ at $\lambda = 1$ simply reflects the fact that there are no domain "walls" in equilibrium, the relative director distortions increasing continuously with distance). This feature of domain wall localization is very similar to the narrowing interface width of stripe domains.

## 12.4   CHIRALITY AND CHOLESTERIC ELASTOMERS

When chiral molecular moieties (or dopants) are added to nematic polymers, they form cholesteric phases. Cross-linked networks made from such polymers accordingly have a spontaneously twisted director distribution. Their director distribution being periodic with characteristic length scale in the optical range, they display their inhomogeneous modulation in brilliant colors of selective reflection. Being macroscopically noncentrosymmetric, such elastic materials possess correspondingly rare physical properties, such as piezoelectricity.

As with nematic LCE, in order to access many unique properties of cholesteric elastomers one needs to form well-aligned monodomain samples, which would remain ordered due to cross-linking. As with nematic systems, the only way to achieve this is to introduce the network cross-links when the desired phase is established by either external fields or boundary conditions—otherwise a polydomain elastomer would result. As with nematics, there are certain practical uses of the cholesteric polydomain state—for instance, the applications of stereo-selective swelling (see below) may work even better in a system with micron-size cholesteric domains.[79]

There are two established ways to create monodomain cholesteric elastomer films with the director in the film plane and the helical axis uniformly aligned perpendicular to the film. Each has its advantages, but also unavoidable drawbacks. Traditionally, a technique of photopolymerization (and network cross-linking when di-functional molecules are used, such as di-acrylates) was used in cells where the desired cholesteric alignment was achieved and maintained by proper boundary conditions. Many elaborate and exciting applications have been developed using this approach.[80,81] However, it is well known that the influence of surface anchoring only extends a maximum of few microns into the bulk of a liquid crystal (unless a main-chain mesogenic polymer, with dramatically increased planar anchoring strength is employed).[82] Therefore, only very thin cholesteric LCE films can be prepared by this surface-alignment and photopolymerization technique.

A different approach to preparing monodomain cholesteric elastomers has been developed by Kim and Finkelmann,[83] utilizing the two-step cross-linking principle originally used in nematic LCE. A sample of cholesteric polymer gel was prepared, cross-linked only partially, and deposits it on a substrate to de-swell. Since the area in contact with the substrate remains constrained, the sample cannot change its two lateral dimensions in the plane to accommodate the loss of volume; only the thickness can change. These constraints in effect apply a large uniaxial compression, equivalent to a biaxial extension in the flat plane, which serves to align the director in the plane and the helical axis perpendicular to it. After this is achieved, the full cross-linking of the network is initiated and the well-aligned free-standing cholesteric elastomer results. This is a delicate technique, which requires a careful balance of cross-linking/deswelling rates, solvent content, and temperature of the phases. In principle it has no significant restrictions on the resulting elastomer area or thickness. However, there is a limitation of a different nature: one cannot achieve too long cholesteric pitch, for example, in the visible red or infrared range. The reason is the required high strength of chiral twisting at the stage of biaxial extension on deswelling: With no or weak twisting strength the material is close to the ordinary nematic and although the symmetric biaxial extension does confine the director to its plane, it does nothing to counter the quenched disorder of the establishing cross-links and the resulting propensity to form a two-dimensional (planar) polydomain state. Only at sufficiently high chiral twisting power the cholesteric propensity overcomes this planar disorder—and the resulting cholesteric pitch will end up relatively short. Another unavoidable drawback of this two-step cross-linking technique is the disordering effect of the portion of cross-links established at the first stage of gel formation (one has to have them to sustain elastic stress and prevent flow on deswelling), so the optical quality of samples prepared in this way is always worse than that in thin films photopolymerized in a perfect surface-aligned cholesteric state.

Cholesteric elastomers and gels respond to imposed mechanical deformations in different ways. Bending of flexible free-standing cholesteric LCE has been used to great effect.[84] The more mechanically straightforward idea is to compress the elastomer along the helical axis, which is easiest to achieve by imposing a symmetric biaxial extension in the director plane.[85] This leads to the affine contraction of the helical pitch and the corresponding dramatic shift of the selective reflection band, see Figure 12.9. If a more traditional uniaxial stretching of elastomer fills is applied, with the initial helical pitch perpendicular to the direction of stretching, the modified director distribution is usually no longer simply helical although the characteristic length scales also affinely scale with sample dimensions. On deformation the director texture remains periodic along the former helical pitch, but becomes a nonchiral stepwise modulation,[86,87] leading to new photonic bandgaps and selective reflection in both right- and left-handed circular polarized light, Figure 12.9. Laser emission has been predicted and observed in cholesteric systems when light was excited in bulk, for example, by a photo-dye stimulated by a pumping laser emitting near the bandgap edge. The ability to continuously change the position of bandgap by mechanically deforming cholesteric elastomers has led to an attractive application opportunity of tunable lasers.[84,85,88]

Another possibility is topological imprinting of helical director distribution. Being anchored to the rubber matrix due to cross-linking, the director texture can remain macroscopically chiral even when all chiral dopants are removed from the material.[89,90] Internally stored mechanical twists can cause cholesteric elastomers to interact selectively with solvents according to its imprinted handedness. Separation of racemic mixtures by stereo-selective swelling is an interesting application possibility.

The lack of centrosymmetric invariance in chiral elastic materials leads to interesting and rare physical properties,[91] in particular, piezoelectric effect and nonlinear

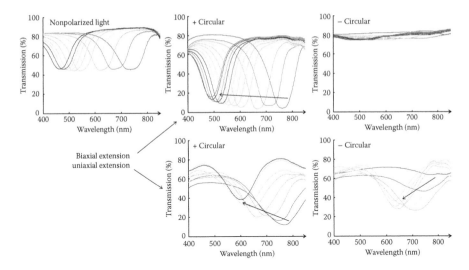

**FIGURE 12.9** Selective reflection of light by cholesteric elastomer shifts under deformation. The top row of plots shows the results for a biaxial extension in the plane, leading to uniaxial compression along the helix from 1 to 0.65, separating the transmission of right- and left-handed circular polarized light. The bottom row of plots shows the analogous results for a uniaxial extension in the plane, from 1 to 1.6. The key difference to note is the emergence of the matching selective reflection for the opposite chirality of light.

optical properties. Such characteristics of elastomers with a chiral smectic C* order (the ferroelectric liquid crystalline elastomers, FLCE) have been studied with some intensity for several years now. After the permanently monodomain (fixed by cross-linking) free-standing samples of FLCE were prepared by Finkelmann and Zentel groups,[92,93] several useful experiments targeting various electromechanical and electro-optical properties, in particular—the piezoelectricity and the nonlinear optical response, have been reported in recent years.[94,95] Clearly, the prospect of important applications will continue driving this work. In this short review we shall concentrate on the analogous effect in chiral nematic LCE, which do not possess a spontaneous polarization.

With much higher symmetry of nematic and cholesteric elastomers (the point group $D_\infty$ in a chiral material, in contrast to a very low symmetry: $C_2$ plus the translational effect of layers in ferroelectric smectic C*), there is a real possibility to identify microscopic origins of piezoelectricity in amorphous polymers or indeed elastomers, if one aims to have a equilibrium effect in a stress-resistant material. The piezoelectric effect in a uniform chiral nematic LCE has been described phenomenologically[91] and by a fully nonlinear microscopic theory similar in its spirit to the basic Trace formula for nematic rubber elasticity.[96] All experimental research so far has concentrated on the helically twisted cholesteric elastomers.[32,97,98] However, the cholesteric texture under the required shear deformation,[99] Figure 12.10a, will produce highly nonuniform distortions giving rise to the well-understood flexoelectric effect and masking the possible chiral piezoelectricity. In this context one has to mention a new development of nematic elastomers where the mesogenic groups are made of bent-core molecules,[100] where one expects to find an anomalously high flexoelectric response.

The uniform linear piezoelectricity, that is, the polarization induced by a uniform strain, Figure 12.10b, (with the small deformation $\varepsilon = \lambda - 1$), is unknown in completely amorphous polymers and rubbers. Even the well-known PVDF polymer-based piezoelectric has its response due to inhomogeneity: crystalline regions affected by macroscopic deformation. The molecular theory[96] has examined the effect of chirality in molecular structure of chain monomers and the bias in their short-axis alignment when the chains are stretched at an angle to the average nematic director $\mathbf{n}$. If the monomers possess a transverse dipole moment, this bias leads to macroscopic polarization:

$$\mathbf{P} \simeq -\frac{1}{2}(n_c \Delta)\underline{\underline{\varepsilon}} : (\underline{\underline{\lambda}}^T \cdot \underline{\underline{\ell}}_{\Theta}^{-1} \cdot \underline{\underline{\lambda}} \cdot \underline{\underline{\ell}}_0). \qquad (12.21)$$

Here, as in the underlying rubber elasticity, $n_c$ is the number of network strands per unit volume (the cross-linking density) and the parameter $\Delta$ is the measure of monomer chirality with the transverse dipole moment $d_t$.[96] Compare this with the original Trace formula of Equation 12.1, to note the characteristic "sandwiching" of strain between the two different orientation tensors. This expression involves the full deformation gradient tensor $\underline{\underline{\lambda}}$ and therefore, can describe large deformations of a chiral nematic rubber. When shear deformations are small, the linear approximation of Equation 12.21 gives, for a symmetric shear,

$$\mathbf{P} \simeq \gamma[\mathbf{n} \times (\underline{\underline{\varepsilon}} \cdot \mathbf{n})],$$

with the linear coefficient $\gamma = \partial P/\partial \varepsilon \approx -(1/2)n_c \Delta (\ell_\parallel^2 - \ell_\perp^2)/\ell_\parallel \ell_\perp$ clearly proportional to the degree of local nematic order through the chain anisotropy $\ell_\parallel - \ell_\perp$.

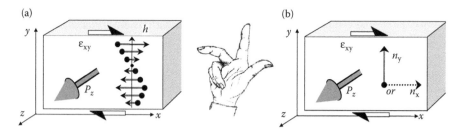

**FIGURE 12.10** Piezoelectric effects in chiral elastomers. (a) Polarization induced by the shear applied to helically twisted textures (flexoelectric effect); (b) Polarization induced by shearing a uniformly aligned chiral nematic, with the director either along x- or y- axes (the true piezoelectric effect $\mathbf{P} = \gamma[\mathbf{n} \times (\underline{\underline{\varepsilon}} \cdot \mathbf{n})]$). The sketch illustrates the chiral geometry that produces a polarization vector along $z$ due to the shear and anisotropy axis lying in the $x - y$ plane: (1) In continuum elasticity this tensor is often called $\underline{\underline{F}} = \nabla f$, a gradient of the geometrical mapping $f$ from the reference to the current state of deformation; (2) The symbolic second-rank tensor $[\mathbf{n} \times \underline{\underline{\varepsilon}} \times \mathbf{n}]$ is the shorthand for the matrix product $A_{\alpha\beta} = \epsilon_{\alpha jk} n_j \varepsilon_{kl} n_m \epsilon_{ml\beta}$ (symmetric, $A_{\alpha\beta} = A_{\beta\alpha}$). Similarly, the symbolic vector $(\mathbf{n} \cdot \underline{\underline{\varepsilon}} \times \mathbf{n})$ is the shortform for the matrix product $n_k \varepsilon_{kl} n_m \epsilon_{ml\alpha}$.

Piezoelectricity in amorphous rubbers is not only interesting from the point of view of fundamentals of chiral polymer random walks and symmetry breaking. On the practical side, due to the rubber modulus much lower than in ordinary solid piezo-electrics (typically $G \sim 10^5 \text{Pa}$), the relevant coefficient $d = \partial P / \partial \sigma = \gamma / G$ is much higher than the corresponding response to stress in, for instance, quartz or PVDF. The corresponding low mechanical impedance should make the piezoelectric rubber attractive for many energy transducing applications.

## 12.5  A LOOK INTO THE FUTURE

This short review examines some of the recent and relevant findings about the new class of materials—LCE and gels. Nematic rubbers have already proved themselves an unusual and exciting system, with a number of unique optical and mechanical properties, indeed a real example of the Cosserat media with couple-stress elasticity. Polymer networks with cholesteric order are an equally provocative system promising new physical properties.

The basic molecular theory of nematic rubber appears to be exceptionally simple in its foundation and does not involve any model parameters apart from the backbone polymer chain anisotropy $\ell_{\parallel} / \ell_{\perp}$, which can be independently measured. This is a great advantage over many other soft condensed-matter systems requiring complicated, sometimes ambiguous theoretical approaches. Of course, in many real situations and materials one finds a need to look deeper into the microscopic properties; an example of this is the "semisoftness" of nematic networks. Most of these systems are char-acterized by nonuniform deformations: Even in the simplest experimental setup a large portion of the sample near the clamps is subjected to nonuniform strains and, therefore, responds with a nonuniform director field.

Looking into the future, many challenging and fundamental problems in this field are still outstanding. The field of dynamics and relaxation in rubbery networks, although not young by any means, is still not offering an unambiguous physical picture of stress relaxation. Adding the liquid crystalline order, we find an additional (director) field undergoing its own relaxation process and coupled to that of an elastic network.[101] In the particular case of polydomain (i.e., randomly disordered in equi-librium) elastomers, we can identify a hierarchical sequence of physical processes in the underlying network (above its $T_g$) and the superimposed glass-like nematic order.[102] This leads to a particularly slow relaxation, but much remains to be done to understand the physics of such complex systems.

The general problem of dynamic mechanical properties, rheology, and relaxation in Cosserat-like incompressible solids, also characterized by the effect of soft elasticity, brings to mind a number of possible applications. An example would be the selective attenuation of certain acoustic waves, with polarization and propagation direction satisfying the condition for softness, essentially leading to an acoustic filtering system. Another example of application of soft elasticity, also related to the prob-lem of relaxation, is the damping of shear vibrations in an engineering component when its surface is covered by a layer of nematic or smectic rubber, particularly aligned to allow the director rotation and softness.

Another very important area of applications is based on polarizational properties of materials with chirality. Most nonlinear optical applications (which have a great technological potential) deal with elastomers in the ferroelectric smectic C* phase. The low symmetry (in particular, chirality) and large spontaneous polarization of C* smectics have a strong effect on the underlying elastic network, and vice versa. This also returns to the general problem of mechanical properties of smectic rubbers and gels. In conclusion, after the initial period of exploration and material synthesis, LCE, in all their variety, now offer themselves as an exciting ground for both fundamental research and for technology.

Smectic and lamellar LCE and gels are a subject of another article in this collection. This is a much more recent area of study, with further theoretical and experimental work required to underpin their dramatically anisotropic and nonlinear mechanical properties combining a two-dimensional rubber-elastic response and a solid-like properties in the third direction.

## ACKNOWLEDGMENTS

I am grateful to a number of colleagues for their help and stimulating input that contributed to this review, in particular the motivation provided by Dirk J. Broer and Mark Warner and the original data of Yoshi Hirota and Francesca Serra used in some illustrations.

## REFERENCES

1. de Gennes, P. G. 1975. Reflections on a type of nematic polymers, *C. R. Acad. Sci. B,* **281,** 101.
2. Finkelmann, H., Koch, H. J., and Rehage, G. 1981. Investigations on liquid crystalline polysiloxanes 3. Liquid crystalline elastomers—A new type of liquid crystalline materials, *Macromol. Rapid Commun.,* **2,** 317.
3. Gleim, W. and Finkelmann, H. 1989. Side chain liquid crystal elastomers. In: *Side-Chain Liquid Crystal Polymers,* ed. C. B. McArdle (Blackie & Sons, Glasgow), p. 287.
4. Barclay, G. G. and Ober, C. K. 1993. Liquid crystalline and rigid-rod networks, *Prog. Polymer Sci.,* **18,** 899.
5. Warner, M. and Terentjev, E. M. 1996. Nematic elastomers—A new state of matter? *Prog. Polymer Sci.,* **21,** 853.
6. Brand, H. R. and Finkelmann, H. 1998. Physical properties of liquid crystalline elastomers. In: *Handbook of Liquid Crystals,* eds. D. Demus, Goodby, J., Gray, G. W., Spiess, H. W., and Vill, V. (Wiley-VCH) Vol. 3, Chapter V, pp. 277–302.
7. Zentel, R. and Reckert, G. 1986. Liquid crystalline elastomers based on liquid crystalline side group, main chain and combined polymers, *Macromol. Chem.,* **187,** 1915.
8. Legge, C. H., Davis, F. J., and Mitchell, G. R. 1991. Memory effects in liquid crystal elastomers, *J. Phys. II,* **1,** 1253.
9. Zubarev, E. R., Talroze, R. V., Yuranova, T. I., Vasilets, V. N., and Plate, N. A. 1996. Influence of network topology on phase behavior of acrylate-based LC elastomers, *Macromol. Rapid Commun.,* **17,** 43.
10. Thomsen, D. L., Keller, P., Naciri, J., Pink, R., Jeon, H., Shenoy, D., and Ratna, B. R. 2001. Liquid crystal elastomers with mechanical properties of a muscle, *Macromolecules* **34,** 5868.

11. Küpfer, J. and Finkelmann, H. 1991. Nematic liquid single crystal elastomers, *Makromol. Chem., Rapid Commun.*, **12**, 717.

12. Kundler, I. and Finkelmann, H. 1998. Director reorientation via stripe-domains in nematic elastomers: Influence of cross-link density, anisotropy of the network and smectic clusters, *Macromol. Chem. Phys.*, **199**, 677.

13. Brehmer, M., Zentel, R., Wagenblast, G., and Siemensmeyer, K. 1994. Ferroelectric liquid-crystalline elastomers, *Macromol. Chem. Phys.*, **195**, 1891.

14. Zubarev, E. R., Yuranova, T. I., Talroze, R. V., Plate, N. A., and Finkelmann, H. 1998. Influence of network topology on polydomain-monodomain transition in side chain LC elastomers with cyanobiphenyl mesogens, *Macromolecules* **31**, 3566.

15. Broer, D. J., Hikmet, R. A. M., and Challa, G. 1989. *In-situ* photopolymerization of oriented liquid-crystalline acrylates: Influence of a lateral methyl substituent on polymer network properties of a mesogenic diacylate, *Macromol. Chem.*, **190**, 3201.

16. Hikmet, R. A. M., Lub, J., and Broer, D. J. 1991. Anisotropic networks formed by photopolymerization of liquid-crystalline molecules, *Adv. Mater.*, **3**, 392.

17. Carfagna, C., Amendola, E., and Giamberini, M. 1997. Liquid crystalline epoxy based thermosetting polymers, *Prog. Polym. Sci.*, **22**, 1607.

18. Ciferri, A. and Ward, I. M. (eds.) 1979. *Ultra-High Modulus Polymers* (Applied-Science: London).

19. Zhao, W. Y., Kloszkowski, A., Mark, J. E., Erman, B., and Bahar, I. 1996. Main-chain lyotropic liquid-crystalline elastomers. 2. Orientation and mechanical properties of polyisocyanate films, *Macromolecules*, **29**, 2805.

20. Ortiz, C., Kim, R., Rodighiero, E., Ober, C. K., and Kramer, E. J. 1998. Deformation of a polydomain, liquid crystalline epoxy-based thermoset, *Macromolecules*, **31**, 4074.

21. Shiota, A. and Ober, C. K. 1998. Smectic networks obtained from twin lc epoxy monomer: Mechanical deformation of smectic networks, *J. Polym. Sci. B - Polymer Phys.*, **36**, 31.

22. Bergmann, G. H. F., Finkelmann, H., Percec, V., and Zhao, M. Y. 1997. Liquid crystalline main-chain elastomers, *Macromol. Rapid Commun.*, **18**, 353.

23. Beyer, P., Terentjev, E. M., and Zentel, R. 2007. Monodomain liquid crystal main chain elastomers by photocrosslinking, *Macromol. Rapid Commun.*, **28**, 1485.

24. Rousseau, I. and Mather, P. T. 2003. Memory effect exhibited by smectic-C liquid crystalline elastomers, *J. Am. Chem. Soc.*, **125**, 15300.

25. Ren, W., McMullan, P. J., and Griffin, A. C. 2008. Poisson's ratio of monodomain liquid crystalline elastomers, *Macromol. Chem. Phys.*, **209**, 1896.

26. Komp, A., Rühe, J., and Finkelmann, H. 2005. A versatile preparation route for thin free-standing liquid single crystal elastomers, *Macromol. Rapid Commun.*, **26**, 813.

27. Fischer, P. and Finkelmann, H. 1998. Lyotropic liquid-crystalline elastomers, *Progr. Colloid Polym. Sci.*, **111**, 127.

28. Kundler, I. and Finkelmann, H. 1995. Strain-induced director reorientation in nematic liquid single crystal elastomers, *Macromol. Rapid Commun.*, **16**, 679.

29. Nishikawa, E. and Finkelmann, H. 1997. Orientation behavior of smectic polymer networks by uniaxial mechanical fields, *Macromol. Chem. Phys.*, **198**, 2531.

30. Ortiz, C., Ober, C. K., and Kramer, E. J. 1998. Stress relaxation of a main-chain, smetic, polydomain liquid crystalline elastomer, *Polymer*, **39**, 3713.

31. Tajbakhsh, A. R. and Terentjev, E. M. 2001. Spontaneous thermal expansion of nematic elastomers, *Euro. Phys. J. E*, **6**, 181.

32. Chang, C.-C., Chien, L.-C., and Meyer, R. B. 1997. Piezoelectric effects in cholesteric elastomer gels, *Phys. Rev. E*, **55**, 534.

33. Chang, C.-C., Chien, L.-C., and Meyer, R. B. 1997. Electro-optical study of nematic elastomer gels, *Phys. Rev. E*, **56**, 595.

34. Verduzco, R., Meng, G. N., Kornfield, J. A., and Meyer, R. B. 2006. Buckling instability in liquid crystalline physical gels, *Phys. Rev. Lett.*, **96**, 147802.

35. Bladon, P., Terentjev, E. M., and Warner, M. 1994. Deformation-induced orientational transitions in liquid crystal elastomers, *J. Phys. II*, **4**, 75.

36. Abramchuk, S. S. and Khokhlov, A. R. 1987. Molecular theory of high elasticity of polymer networks with accounting for orientational ordering of segments, *Doklady Akad. Nauk SSSR (Doklady-Phys. Chem.)*, **297**, 385.

37. Warner, M., Gelling, K. P., and Vilgis, T. A. 1988. Theory of nematic networks, *J. Chem. Phys.*, **88**, 4008.

38. Finkelmann, H., Nishikawa, E., Pereira, G. G., and Warner, M. 2001. A new opto-mechanical effect in solids, *Phys. Rev. Lett.*, **87**, 015501.

39. Hogan, P. M., Tajbakhsh, A. R., and Terentjev, E. M. 2002. UV manipulation of order and macroscopic shape in nematic elastomers, *Phys. Rev. E*, **65**, 041720.

40. Li, M. H., Keller, P., Li, B., Wang, X. G., and Brunet, M. 2003. Light-driven side-on nematic elastomer actuators, *Adv. Mater.*, **15**, 569.

41. Cviklinski, J., Tajbakhsh, A. R., and Terentjev, E. M. 2002. UV isomerisation in nematic elastomers as a route to photo-mechanical transducer, *Eur. Phys. J. E*, **9**, 427.

42. Kutter, S. and Terentjev, E. M. 2001. Tube model for the elasticity of entangled nematic rubbers, *Eur. Phys. J. E*, **6**, 221.

43. de Gennes, P. G. 1980. Weak nematic gels. In: *Liquid Crystals of One and Two-Dimensional Order*, eds. W. Helfrich and G. Heppke (Springer-Verlag, Berlin), p. 231.

44. Verwey. G. C. and Warner, M. 1995. Soft rubber elasticity, *Macromolecules*, **28**, 4303.

45. Schönstein, M., Stille, W., and Strobl, G. 2001. Effect of the network on the director fluctuations in a nematic side-group elastomer analysed by static and dynamic light scattering, *Eur. Phys. J. E*, **5**, 511.

46. Olmsted, P. D. 1994. Rotational invariance and goldstone modes in nematic elastomers and gels, *J. Phys. II*, **4**, 2215.

47. Mitchell, G. R., Davis, F. J., and Guo, W. 1993. Strain-induced transitions in liquid-crystal elastomers, *Phys. Rev. Lett.*, **71**, 2947.

48. Talroze, R. V., Zubarev, E. R., Kuptsov, S. A., Merekalov, A. S., Yuranova, T. I., Plate, N. A., and Finkelmann, H. 1999. Liquid crystal acrylate-based networks: Polymer backbone—LC order interaction, *Reactive & Functional Polymers*, **41**, 1.

49. Roberts, P. M. S., Mitchell, G. R., and Davis, F. J. 1997. A single director-switching mode for monodomain liquid crystal elastomers, *J. Phys. II*, **7**, 1337.

50. Verwey, G. C., Warner, M., and Terentjev, E. M. 1996. Elastic instability and stripe domains in liquid crystalline elastomers, *J. Phys. II*, **6**, 1273.

51. Bhattacharya, K. 2003. *Microstructure of Martensite. Why It Forms and How It Gives Rise to the Shape-Memory Effect* (Oxford University Press, Oxford).

52. Warner, M., Bladon, P., and Terentjev, E. M. 1994. "Soft elasticity" deformation without resistance in liquid crystal elastomers, *J. Phys. II*, **4**, 93.

53. Verwey, G. C. and Warner, M. 1997. Compositional fluctuations and semisoftness in nematic elastomers and nematic elastomers cross-linked by rigid rod linkers, *Macromolecules,* **30**, 4189 and 4196.

54. Ye, F. F., Mukhopadhyay, R., Stenull, O., and Lubensky, T. C. 2007. Semisoft nematic elastomers and nematics in crossed electric and magnetic fields, *Phys. Rev. Lett.*, **98**, 147801.

55. Biggins, J. S., Terentjev, E. M., and Warner, M. 2008. Semisoft elastic response of nematic elastomers to complex deformations, *Phys. Rev. E*, **78**, 041704.

56. Chuang, I., Turok, N., and Yurke, B. 1991. Late-time coarsening dynamics in a nematic liquid crystal, *Phys. Rev. Lett.*, **66**, 2472.

57. Volovik, G. E. 2003. *The Universe in a Helium Droplet* (Oxford University Press, Oxford).

58. Bray, A. 2002. Theory of phase-ordering kinetics, *Adv. Phys.*, **51**, 481.

59. Elias, F., Clarke, S. M., Peck, R., and Terentjev, E. M. 1999. Equilibrium textures in main-chain liquid crystalline polymers, *Europhys. Lett.*, **47**, 442.

60. Mezard, M., Parisi, G., and Virasoro, M. A. 1987. *Spin Glass Theory and Beyond* (World Scientific, Singapore).

61. Dotsenko, V. S. 1994. *Introduction to the Theory of Spin-Glasses and Neural Networks*, (World Scientific, Singapore).

62. Bellini, T., Clark, N. A., Muzny, C. D., Wu, L., Garland, C. W., Shaeffer, D. W. and Olivier, B. J. 1992. Phase behavior of the liquid crystal 8CB in a silica aerogel, *Phys. Rev. Lett.*, **69**, 788.

63. Uchida, N. 2000. Soft and nonsoft structural transitions in disordered nematic networks, *Phys. Rev. E*, **62**, 5119.

64. Fridrikh, S. V. and Terentjev, E. M. 1999. Polydomain-monodomain transition in nematic elastomers, *Phys. Rev. E*, **60**, 1847.

65. Pasini, P., Skacej, G., and Zannoni, C. 2005. A microscopic lattice model for liquid crystal elastomers, *Chem. Phys. Lett.*, **413**, 463.

66. Petridis, L. and Terentjev, E. M. 2006. Nematic-isotropic transition with quenched disorder, *Phys. Rev. E*, **74**, 051707.

67. Feio, G., Figueirinhas, J. L., Tajbakhsh, A. R., and Terentjev, E. M. 2008. Critical fluctuations and random-anisotropy glass transition in nematic elastomers, *Phys. Rev. B*, **78**, 020201.

68. Wong, G. C. L., de Jeu, W. H., Shao, H., Liang, K. S., and Zentel, R. 1997. Induced long-range order in crosslinked "one-dimensional" stacks of fluid monolayers, *Nature*, **389**, 576.

69. Muresan, A. S., Ostrovskii, B. I., Sanchez-Ferrer, A., Finkelmann, H., and de Jeu, W. H. 2005. Main-chain smectic liquid-crystalline polymers as randomly disordered systems (Rapid Note), *Eur. Phys. J. E*, **19**, 385.

70. Halperin, B. I., Lubensky, T. C., and Ma S-K, 1974. First-order phase transitions in superconductors and smectic-A liquid crystals, *Phys. Rev. Lett.*, **32**, 292.

71. Olmsted, P. D. and Terentjev, E. M. 1996. Mean-field nematic-smectic-A transition in a random polymer network, *Phys. Rev. E*, **53**, 2444.

72. Brodowsky, H. M., Terentjev, E. M., Kremer, F., and Zentel, R. 2002. Induced roughness in thin films of smectic C* elastomers, *Europhys. Lett.*, **57**, 53.

73. Le Doussal, P. and Giamarchi, T. 1998. Moving glass theory of driven lattices with disorder, *Phys. Rev. B*, **57**, 11356.

74. Radzihovsky, L. and Toner, J. 1999. Smectic liquid crystals in random environments, *Phys. Rev. B*, **60**, 206.

75. Imry, Y. and Ma, S.-K. 1975. Random-field instability of the ordered state of continuous symmetry, *Phys. Rev. Lett.*, **35** 1399.

76. Clarke, S. M., Nishikawa, E., Finkelmann, H. and Terentjev, E. M. 1997. Light-scattering study of random disorder in liquid crystalline elastomers, *Macromol. Chem. Phys.*, **198**, 3485.

77. Petridis, L. and Terentjev, E. M. 2006. Quenched disorder and spin-glass correlations in XY nematics, *J. Phys. A: Math. Gen.*, **39**, 9693.

78. Fridrikh, S. V. and Terentjev, E. M. 1997. Order-disorder transition in an external field in random ferromagnets and nematic elastomers, *Phys. Rev. Lett.*, **79**, 4661.

79. Courty, S., Tajbakhsh, A. R., and Terentjev, E. M. 2003. Phase chirality and stereo-selective swelling of cholesteric elastomers, *Eur. Phys. J. E*, **12**, 617.

80. Broer, D. J., Lub, J., and Mol, G. N. 1995. Wide-band reflective polarizers from cholesteric polymer networks with a pitch gradient, *Nature* **378**, 467.

81. Hikmet, R. A. M. and Kemperman, H. 1998. Electrically switchable mirrors and optical components made from liquid-crystal gels, *Nature* **392**, 476.

82. Terentjev, E. M. 1995. Density functional model of anchoring energy at a liquid crystalline polymer—solid interface, *J. Phys. II*, **5**, 159.

83. Kim, S. T. and Finkelmann, H. 2001. Cholesteric liquid single-crystal elastomers (LSCE) obtained by the anisotropic deswelling method, *Macromol. Rapid Commun.*, **22**, 429.

84. Matsui, T., Ozaki, R., Funamoto, K., Ozaki, M., and Yoshino, K. 2002. Flexible mirror-less laser based on a freestanding film of photopolymerized cholesteric liquid crystal, *Appl. Phys. Lett.*, **81**, 3741.

85. Finkelmann, H., Kim, S. T., Munoz, A., Palffy-Muhoray, P., and Taheri, B. 2001. Tunable mirrorless lasing in cholesteric liquid crystalline elastomers, *Adv. Mater.*, **13**, 1069.

86. Warner, M., Terentjev, E. M., Meyer, R. B., and Mao, Y. 2000. Untwisting of a cholesteric elastomer by a mechanical field, *Phys. Rev. Lett.*, **85**, 2320.

87. Cicuta, P., Tajbakhsh, A. R., and Terentjev, E. M. 2002. Evolution of photonic structure on deformation of cholesteric, *Phys. Rev. E*, **65**, 051704.

88. Schmidtke, J., Stille, W., and Finkelmann, H. 2003. Defect mode emission of a dye doped cholesteric polymer network, *Phys. Rev. Lett.*, **90**, 083902.

89. Mao, Y. and Warner, M. 2000. Theory of chiral imprinting, *Phys. Rev. Lett.*, **84**, 5335.

90. Courty, S., Tajbakhsh, A. R., and Terentjev, E. M. 2003. Stereo-selective swelling of imprinted cholesteric networks, *Phys. Rev. Lett.*, **91**, 085503.

91. Terentjev, E. M. 1993. Phenomenological theory of non-uniform nematic elastomers: Free energy of deformations and electric-field effects, *Europhys. Lett.*, **23**, 27.

92. Finkelmann, H., Benne, I., and Semmler, K. 1995. Smectic liquid single crystal elastomers, *Macromol. Symp.*, **96**, 169.

93. Gebhard, E. and Zentel, R. 1998. Freestanding ferroelectric elastomer films, *Macromol Rapid Commun.*, **19**, 341.

94. Brehmer, M., Zentel, R., Giesselmann, F., Germer, R., and Zugemaier, P. 1996. Coupling of liquid crystalline and polymer network properties in LC- elastomers, *Liq. Cryst.*, **21**, 589.

95. Lehmann, W., Gattinger, P., Keck, M., Kremer, F., Stein, P., Eckert, T., and Finkelmann, H. 1998. interferometer, *Ferroelectrics*, **208**, 373.

96. Terentjev, E. M. and Warner, M. 1999. Piezoelectricity of chiral nematic elastomers, *Euro. Phys. J. B*, **8**, 595.

97. Meyer, W. and Finkelmann, H. 1990. Piezoelectricity of cholesteric elastomers, *Macromol. Chem. Rapid Commun.*, **11**, 599.

98. Valerien, S. U., Kremer, F., Fischer, E. W., Kapitza, H., Zentel, R., and Poths, H. 1990. Experimental proof of piezoelectricity in cholesteric and chiral smectic C*-phases of LC-elastomers. *Macromol. Chem. Rapid Commun.*, **11**, 593.

99. Pelcovits, R. A. and Meyer, R. B. 1995. Piezoelectricity of cholesteric elastomers, *J. Phys. II*, **5**, 877.

100. Chambers, M., Verduzco, R., Gleeson, J. T., Sprunt, S., and Jakli, A. 2009. Calamitic liquid-crystalline elastomers swollen in bent-core liquid-crystal solvents, advanced materials, *Adv. Mater.*, **21**, 1622.
101. Clarke, S. M., Tajbakhsh, A. R., Terentjev, E. M., and Warner, M. 2001. Anomalous viscoelastic response of nematic elastomers, *Phys. Rev. Lett.*, **86**, 4044.
102. Clarke, S. M. and Terentjev, E. M. 1998. Slow stress relaxation in randomly disordered nematic elastomers and gels, *Phys. Rev. Lett.*, **81**, 4436.

# 13 Lagrange Elasticity Theory of Liquid Crystal Elastomers

*Tom C. Lubensky and Olaf Stenull*

## CONTENTS

## 13.1  INTRODUCTION

Liquid crystal elastomers [1] are fascinating materials that combine the elastic properties of rubber with the orientational properties of liquid crystals [2,3]: like rubber
but unlike a fluid they have the capability of supporting shear, yet unlike conventional
rubbers, they exhibit the optical anisotropy of a liquid crystal. Typical liquid crystal
elastomers are prepared by weakly cross-linking side-chain [4] or main-chain [5] liquid crystal polymers. A dilute concentration of cross links has a negligible or very
small effect on the interactions favoring the formation of liquid crystalline phases.
Thus, liquid crystalline elastomers can be expected to exhibit phase behavior similar to
that of uncross-linked polymers or small-molecule systems, and indeed they do [1].
They exhibit the same phases—nematic, cholesteric, smectic-$A$ (Sm$A$), smectic-$C$
(Sm$C$)—as do small-molecule liquid crystals. There are, however, important differences between liquid crystalline elastomers and polymers. Most importantly, as
highlighted above, elastomers support shear. In response to external stress, they exhibit
reversible strain rather than flow. Strain couples with nematic order and vice versa, so
the imposition of stress modifies the magnitude and direction of nematic order and the
imposition of orienting external fields or boundary conditions induces strain. These
couplings along with a small rubber-like shear rigidity are responsible for most of the
remarkable properties of liquid crystalline elastomers, including large thermo-elastic
effects (see, e.g., [1,6,7]), soft [8,9], and semisoft [10–12] elasticity, periodic shear
modulation in an external electric field [13], strain-induced pitch change in cholesteric
elastomers, and more.

Any theoretical description of liquid crystalline elastomers must not only treat
elasticity, but also, either explicitly or at least implicitly, liquid crystalline order, and
the coupling between them. Currently, there are two widely used approaches to the
development of theories of liquid crystalline elastomers that differ in their treatment
of elastic energies. One approach, developed by Warner and Terentjev and coworkers
[1,14] generalizes the classical theory of rubber elasticity [15] to include the effects of
orientational anisotropy on the random walks of constituent polymer links. This "neoclassical" model treats large strains with the same ease as they are treated in rubbers
in the absence of orientational order, it provides direct estimates of the magnitudes of
elastic energies (a shear modulus of order $10^6$ Pa), and it is characterized by a small
number of parameters. It has been enormously successful in explaining various properties of liquid crystalline elastomers. As in the theory of classical rubber elasticity,
the bulk modulus $B$ is two or three orders of magnitude greater than the shear modulus,
and an incompressibility constraint is generally imposed. A second approach is to use
Lagrangian elasticity theory [16] to describe the elastic properties of the elastomer.
This theory is expressed in terms of the nonlinear Cauchy–Saint–Venant nonlinear
strain tensor $u_{ij}$ and, in general, includes nonlinear terms beyond quadratic order in
the strain. The latter property provides for a realistic and fairly simple description of
the transition in elastomers from the isotropic to the nematic state and from the Sm$A$
to the Sm$C$ state in terms of strain alone without the explicit introduction of nematic
degrees of freedom. Elastic constants and higher-order couplings are regarded as phenomenological parameters whose magnitudes are determined either by experiment
or through a more microscopic theory as are symmetry-permitted couplings between

strain and liquid crystalline degrees of freedom. As we will discuss, care must be taken in both approaches in the treatment of couplings between liquid crystalline order and strain.

This chapter will discuss the second approach almost exclusively, though, when appropriate, it will refer to the first one. It will focus mostly on the unusual phenomenon of soft elasticity in nematic and SmC elastomers and on semisoft elasticity in nematic elastomers. Soft elasticity is a direct consequence of the spontaneous breaking of a continuous symmetry and associated Goldstone mode in these phases. In the nematic phase of traditional fluid liquid crystals, rigid rotations cost no energy, but spatially inhomogeneous rotations cost an energy proportional to the spatial gradient of the rotation angle as described by the Frank free energy. In nematic elastomers, nematic order can develop along any direction of the originally isotropic elastomer. Distortions that merely rotate this direction of order cost no energy. The result is that the shear modulus for shears in planes containing the nematic anisotropy axis is zero. In addition, nonlinear strains, induced for example, by stress perpendicular to the anisotropy axis that describe macroscopic rotations of the anisotropy direction cost no energy and give rise to a stress–strain curve in which the stress is zero up to a critical strain and then rises above that strain. SmC liquid crystals exhibit a similar soft elasticity in response to appropriate stresses. Section 13.3 describes soft elastic response in both nematic and SmC liquid crystals and will describe in detail the shape changes of macroscopic samples in soft regime where the stress in response to strain is zero.

Soft elasticity is a property of an idealized aligned monodomain material formed by a spontaneous symmetry-breaking transition. In practice, such materials cannot be produced in the laboratory because random microscopic-scale inhomogeneities in the rubber create random aligning fields that inhibit the formation of monodomain order. Essentially monodomain order can be frozen in by multistage cross-linking procedures involving aligning stresses. These methods have been developed for and successfully applied to both nematic [17] and smectic elastomers [18–22]. When applied to a nematic, this process creates a true uniaxial solid, with five independent and nonzero elastic constants that exhibit linear–strain relations at small strain. At a critical strain, however, the stress–strain curve of such systems can undergo an abrupt decrease in slope to a nearly flat plateau that terminates at a second critical strain at which there is an abrupt increase in this slope. In Section 13.4, we will discuss the simplest theory of this semisoft response. In this theory, which is based upon a purely elastic model in which nematic degrees of freedom have been integrated out, the semisoft response is a consequence of the existence of a symmetry-breaking biaxial phase when external stress perpendicular to the initial anisotropy direction is equal to the internal aligning stress that freezes in nematic order. Though physical soft systems do not exist, their stress–strain response closely approximates that of semisoft systems when the stress or field that freezes-in nematic alignment approaches zero. Thus, soft systems provide an interesting and well-controlled limit of physically realizable semisoft systems. Also, they are of great pedagogical value because the consequences of broken symmetry and Goldstone modes are more easily understood in them than in semisoft systems.

The elasticity of liquid crystalline phases provides a rich and exciting basis for theoretical and experimental research. A related and likewise interesting subject is that of the phase transitions between the various phases. Section 13.5 considers Landau theories in terms of both orientation and strain for the isotropic-to-nematic ($I$–$N$) and the SmA-to-SmC transitions. In addition to just shedding light on these transitions, the Landau theories allow us to understand the soft elasticity of the emergent nematic and SmC phases from a perspective different from that in Section 13.3.

The discussions in Sections 13.4–13.5 rely on a common formalism that is developed in detail in Section 13.2. This section discusses the difference between symmetry operations that involve changes in the final or target-space coordinates of the sample and those that involve operations on the coordinates in the original unstretched sample or reference space, and it identifies the vector and tensor fields that transform under these operations. It emphasizes in particular that the usual Cauchy–Saint–Venant strain tensor $u_{ij}$ transforms as tensor with respect to reference-space operations but as a scalar with respect to target-space operations whereas, the usual nematic director $n_i$ transforms as a vector with respect to target-space operations and as a scalar with respect to reference–space operations. This means that contractions of the form $n_i u_{ij} n_j$ are not scalars and cannot appear in a phenomenological free-energy density. We show how the polar decomposition theorem of linear algebra [23] can be used to convert $n_i$ into a reference–space vector $\tilde{n}_i$ and allow us to construct scalars of the form $\tilde{n}_i u_{ij} \tilde{n}_j$. Section 13.2 also presents a general discussion of the different kinds of stress tensors and of a number of model elastic energies.

Liquid crystalline elastomers necessarily exhibit local random stresses and orientation fields that disfavor the formation of long-range order in single domain samples. The theoretical treatment of these stresses and fields is complicated and still not completely understood. We will not consider them in this chapter.

## 13.2  MODELS

Lagrange elasticity theory facilitates a phenomenological description of slowly varying distortions of an elastic medium or body from its equilibrium configuration. In this section we review the basic elements of this formalism to provide essential background information, to establish notation, and to explain concepts that are important for studying liquid crystal elastomers theoretically. Liquid crystalline degrees of freedom are *not* a subject of classical Lagrange theory. Since they are, however, evidently important for liquid crystal elastomers, ways for combining or coupling elastic and liquid crystalline degrees of freedom had to be found to provide for a Lagrangian description of these materials. We review how this coupling can be achieved. A brief review of model elastic energies liquid crystal elastomers in their isotropic, nematic and smectic phases completes this section.

### 13.2.1  Basics of Lagrange Elasticity Theory

#### 13.2.1.1  Reference and Target Spaces

In the Langrangian formalism mass points in an undistorted medium, which occupy some region of the three-dimensional Euclidean space $\mathcal{E}$, are labeled by vectors $\mathbf{x} = (x, y, z)$. We refer to the space defined by $\mathbf{x}$ as the reference space $S_R$, cf. Figure 13.1. Upon distortion of the medium, a point originally at $\mathbf{x}$ is mapped to a new position

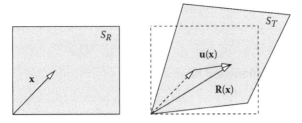

**FIGURE 13.1** Reference and target configurations.

$\mathbf{R}(\mathbf{x}) = (R_x(\mathbf{x}), R_y(\mathbf{x}), R_z(\mathbf{x}))$ in $\mathcal{E}$. We refer to the space defined by $\mathbf{R}$ as the target space $S_T$. It is useful and customary to introduce a displacement (phonon) field $\mathbf{u}(\mathbf{x})$ that measures the deviation of $\mathbf{R}(\mathbf{x})$ from $\mathbf{x}$,

$$\mathbf{R}(\mathbf{x}) = \mathbf{x} + \mathbf{u}(\mathbf{x}). \tag{13.1}$$

Both $\mathbf{x}$ and $\mathbf{R}(\mathbf{x})$ can be decomposed into components along the standard orthonormal basis $\{\mathbf{a}_i | i = x, y, z\}$ of $\mathcal{E}$:

$$\mathbf{x} = x_i \mathbf{a}_i, \quad \mathbf{R}(\mathbf{x}) = R_i(\mathbf{x}) \mathbf{a}_i. \tag{13.2}$$

Here and in what follows, we use the summation convention on repeated indices unless we indicate otherwise. We will also use the convention that indices from the middle of the alphabet run over all space coordinates, $i, j, k = x, y, z$. When discussing uniaxial phases, we choose our coordinate system so that the $z$-axis is along the uniaxial direction. Indices from the beginning of the alphabet $a, b, c$, run over $x$ and $y$ only, that is, over directions perpendicular to the anisotropy axis.

Though reference- and target-space vectors both exist in $\mathcal{E}$, they transform under distinct and independent transformation operations. Let $\underline{\underline{O}}_R$ and $\underline{\underline{O}}_T$ denote, respectively, matrices describing transformations (which we will take mostly to be rotations but which could include reflections and inversions as well) in the reference and target spaces; then under these transformations, $\mathbf{x} \rightarrow \mathbf{x}' = \underline{\underline{O}}_R \mathbf{x}$ and $\mathbf{R}(\mathbf{x}) \rightarrow \mathbf{R}'(\mathbf{x}) = \underline{\underline{O}}_T \mathbf{R}(\mathbf{x})$, or in terms of components relative to the $\mathbf{a}$-basis

$$R_i'(\mathbf{x}) = O_{T,ij} R_j(\mathbf{x}), \tag{13.3a}$$

$$x_i' = O_{R,ij} x_j. \tag{13.3b}$$

Unless otherwise specified, we will view $\underline{\underline{O}}_R$ and $\underline{\underline{O}}_T$ as operators that rotate vectors rather than coordinate systems.

We will usually represent the reference-space points $\mathbf{x}$ in terms of their coordinates relative to the basis $\{\mathbf{a}_i\}$. We will, however, find it useful on occasion to consider orthonormal bases locked to the reference medium and to represent reference-space vectors relative to them. The initial basis $\{\tilde{\mathbf{e}}_i | i = x, y, z\}$ is identical to the basis $\{\mathbf{a}_i | i = x, y, z\}$, and $\mathbf{x} = x_i \mathbf{a}_i \equiv x_i \tilde{\mathbf{e}}_i$. Under rotations of the body basis, $\mathbf{x} = x_i' \tilde{\mathbf{e}}_i' = x_i \mathbf{e}_i$, where

$$\tilde{\mathbf{e}}_i' = O_{R,ij} \tilde{\mathbf{e}}_j \tag{13.4}$$

and $x_i' = O_{R,ij}x_j$. As we shall discuss in more detail in Section 13.2.2 associated with each reference-space vector, there is a target-space vector of the same length. Thus, associated with the reference basis $\{\tilde{e}_i\}$, there is a target basis $\{e_i\}$.

### 13.2.1.2   Deformations and Strains

Distortions that vary slowly on a length scale set by microscopic length of the reference medium, for example, the average distance between the cross links in cross-linked liquid crystal polymers, are described by the Cauchy deformation tensor $\underset{=}{\Lambda}$ with components

$$\Lambda_{ij} = \partial R_i/\partial x_j \equiv \partial_j R_i = \delta_{ij} + \eta_{ij}, \tag{13.5}$$

where $\delta_{ij}$ is the Kronecker symbol, that is, the $\delta_{ij}$ are the components of the unit matrix $\underset{=}{\delta}$, and the $\eta_{ij} = \partial_j u_i$ are the components of the displacement gradient tensor $\underset{=}{\eta}$. The Cauchy deformation tensor transforms under the operations of Equation 13.3 according to

$$\underset{=}{\Lambda}(x) \rightarrow \underset{=}{O}_T \underset{=}{\Lambda}(x') \underset{=}{O}_R^{-1} \tag{13.6}$$

that is, the right subscript transforms under target-space rules and the left under reference-space rules.

The elastic energy of a distorted medium relative to its reference configuration depends on how much the distance between nearby points changes in the distortion,

$$(d\mathbf{R})^2 - (d\mathbf{x})^2 = 2u_{ij}\,dx_i dx_j, \tag{13.7}$$

where $u_{ij}$ are the components of the Cauchy–Saint–Venant [16,24] nonlinear strain tensor defined by

$$u_{ij}(\mathbf{x}) = \frac{1}{2}(g_{ij} - \delta_{ij}) \tag{13.8a}$$

$$= \frac{1}{2}(\partial_i u_j + \partial_j u_i + \partial_i u_k \partial_j u_k), \tag{13.8b}$$

with $g_{ij}$ being the components of the metric tensor

$$\underset{=}{g} = \underset{=}{\Lambda}^T \underset{=}{\Lambda}. \tag{13.9}$$

The strain $\underset{=}{u}$ is a tensor in $S_R$, that is, it transforms like a tensor under reference-space operations,

$$\underset{=}{u}(\mathbf{x}) \rightarrow \underset{=}{O}_R \underset{=}{u}(\mathbf{x}') \underset{=}{O}_R^{-1}. \tag{13.10}$$

This expression applies both to physical transformations of reference-space vectors under $x_i' = O_{R,ij}^{-1}x_j$ or under changes of basis described by Equation 13.4 under which $\Lambda_{ij} \rightarrow \Lambda_{ij}' = \partial R_i/\partial x_j' = \Lambda_{ik}O_{R,kj}^{-1}$. Under global rotations in $\underset{=}{O}_T$ in $S_T$, on the other hand, $\underline{u}$ is invariant. In other words, in $S_T$, $\underline{u}$ is a scalar. As we will discuss in the following, this property makes $\underline{u}$ a convenient variable for formulating elastic energies.

Liquid crystal elastomers are, like conventional rubbers, essentially incompressible, and it is important, at least for certain problems, to implement this incompressibility into theoretical description. Under a deformation $\underline{\underline{\Lambda}}$, an element of volume $d^3x$ in $S_R$ is transformed into an element of volume $d^3R = \det \underline{\underline{\Lambda}} d^3x$. Thus, a volume $V_0 = \int d^3x$ in $S_R$ becomes a volume $V = \int d^3x \det \underline{\underline{\Lambda}}$ in $S_T$, or

$$V/V_0 = \det \underline{\underline{\Lambda}} \approx 1 + u_{ii}, \qquad (13.11)$$

when $\underline{\underline{\Lambda}}$ is spatially constant. The final form is valid for small strains. Thus, when strains are small, one can approximate the true incompressibility constraint $\det \underline{\underline{\Lambda}} = 1$ by setting $u_{ii} = 0$.

### 13.2.1.3 Lagrange Elastic Energies

If there are no external fields, as we shall assume unless mentioned otherwise, the elastic energy of a body cannot depend on its orientation or location in physical space, and thus elastic energies have to be invariant under rigid rotations and translations in $S_T$. Moreover, elastic energies must be invariant under the under symmetry operations of $S_R$ of the form $\mathbf{x} \to \mathbf{x}' = \underline{\underline{O}}_R^{-1}\mathbf{x} + \mathbf{b}$, where $\mathbf{b}$ is a constant vector and $\underline{\underline{O}}_R$ is a matrix associated with some symmetry element of the reference space. For example, for a uniaxial elastomer the elastic energy must be invariant under global rotations about the anisotropy axis. Overall elastic energies are invariant under transformations of the form

$$\mathbf{R}(\mathbf{x}) \to \mathbf{R}'(\mathbf{x}') = \underline{\underline{O}}_T \mathbf{R}(\underline{\underline{O}}_R^{-1}\mathbf{x} + \mathbf{b}) + \mathbf{X}, \qquad (13.12)$$

where $\underline{\underline{O}}_T$ is an arbitrary target-space rotation matrix and $\mathbf{X}$ is a constant displacement vector. In what follows, we will generally ignore the displacements $\mathbf{X}$ and $\mathbf{b}$. The use of $\underline{\underline{O}}_R^{-1}$, in Equation 13.12 rather than $\underline{\underline{O}}_R$ is a matter of convention. Recall that the strain tensor $\underline{u}$ incorporates rotation invariance in $S_T$ by construction. Thus, $\underline{u}$ is a natural variable for elasticity theory; in many cases, elastic energies can be most conveniently formulated by using $\underline{u}$.

Now, let us turn to the form of elastic free energies $F$ in Lagrange theory, which can be expressed as an integral over the reference volume of the free-energy density $f(\mathbf{x})$:

$$F = \int d^3x f(\mathbf{x}). \qquad (13.13)$$

To harmonically order in the nonlinear strain tensor, these Lagrangian densities can be written in the form

$$f = \frac{1}{2}K_{ijkl}u_{ij}u_{kl}, \qquad (13.14)$$

where $K_{ijkl}$ are the components of the elastic constant or elastic modulus tensor. Additional linear "stress"-like contributions to $f$ are possible, however, they usually can be removed by the redefinition of $u_{ij}$ (unless they stem from transverse internal random stresses). The elastic modulus tensor has inherent symmetry, $K_{ijkl} = K_{klij} = K_{jikl} = K_{ijlk}$, and it has to reflect the symmetries of $S_R$.

An isotropic elastomer is invariant under arbitrary rotations in $S_R$, and hence $K_{ijkl} = \lambda \delta_{ij} \delta_{kl} + \mu (\delta_{ik} \delta_{jl} + \delta_{il} \delta_{kj})$ such that

$$f_{\text{iso}} = \frac{1}{2} \lambda u_{ii}^2 + \mu u_{ij} u_{ij} = \frac{1}{2} B u_{ii}^2 + \mu \hat{u}_{ij} \hat{u}_{ij}. \tag{13.15}$$

$\lambda$ and $\mu$ are the so-called Lamé coefficients with $\mu$ the shear modulus. The elastic constant $B = \lambda + (2/3)\mu$ is the bulk modulus describing dilation or expansion of the bulk volume, and

$$\hat{u}_{ij} = u_{ij} - \frac{1}{3} \delta_{ij} u_{kk} \tag{13.16}$$

defines the traceless part of the strain tensor.

For a uniaxial elastomer, for example, a nematic or a SmA elastomer, $K_{ijkl}$ has to reflect $D_{\infty h}$ symmetry. Then, there are, in general, five independent elastic constants producing

$$f_{\text{uni}} = \frac{1}{2} C_1 u_{zz}^2 + C_2 u_{zz} u_{ii} + \frac{1}{2} C_3 u_{ii}^2 + C_4 \hat{u}_{ab}^2 + C_5 u_{az}^2, \tag{13.17}$$

where $\hat{u}_{ab} = u_{ab} - (1/2)\delta_{ab} u_{cc}$ are the components of the two-dimensional symmetric, traceless strain tensor. The elastic constant $C_1$ describes dilation or compression along $z$ and $C_3$ describes dilation or compression of the bulk. $C_2$ couples these two types of strains. $C_4$ and $C_5$, respectively, describe shears in the plane perpendicular to the anisotropy axis and in the planes containing it.

SmC elastomers possess monoclinic or $C_{2h}$ symmetry. In this case, unless soft-elasticity plays a role, there are 13 independent elastic constants implied in $K_{ijkl}$ and

$$
\begin{aligned}
f_{C_{2h}} = {} & \frac{1}{2} C_{xyxy} u_{xy}^2 + \frac{1}{2} C_{yzyz} u_{yz}^2 + C_{xyyz} u_{xy} u_{yz} + \frac{1}{2} C_{zzzz} u_{zz}^2 \\
& + \frac{1}{2} C_{xzxz} u_{xz}^2 + C_{zzxx} u_{zz} u_{xx} + C_{zzyy} u_{zz} u_{yy} + \frac{1}{2} C_{xxxx} u_{xx}^2 + \frac{1}{2} C_{yyyy} u_{yy}^2 \\
& + C_{xxyy} u_{xx} u_{yy} + C_{xxxz} u_{xx} u_{xz} + C_{yyxz} u_{yy} u_{xz} + C_{zzxz} u_{zz} u_{xz}.
\end{aligned}
\tag{13.18}
$$

### 13.2.1.4 Stresses

Next, we briefly review external stresses and various versions of the stress tensor. The internal force on a small volume element centered at point $\mathbf{x}$ can only be transmitted across the surfaces of that element. Thus internal force density $\mathbf{f}(\mathbf{x})$ with components $\mathbf{f}_i(\mathbf{x})$ must be the divergence of a stress tensor:

$$f_i = \partial_j \sigma_{ij}^I. \tag{13.19}$$

Here $\sigma_{ij}^I$ is the first Piola–Kirchhoff stress tensor, often refereed to as the engineering stress $\sigma_{ij}^{eng}$; it is the force per unit area of in $S_R$. $\mathbf{f}_i$ is a vector in $S_T$. Thus, $\sigma_{ij}^I$, like the deformation tensor, is a mixed tensor whose left and right indices belong to $S_T$ and $S_R$, respectively, and it is not necessarily a symmetric tensor. $\sigma_{ij}^I$ is the stress tensor associated with internal forces, and it is only nonzero in the interior $D$ of a physical sample; it is zero at the outer boundary $\partial D$ of the sample. Now, consider the work done by the internal force in displacing points at $\mathbf{x}$ a distance $\delta\mathbf{u}$:

$$\delta W = \int_D d^3x f_i \delta u_i = \int_{\partial D} dS_j \sigma_{ij}^I \delta u_i - \int_D d^3x \sigma_{ij}^I \partial_j \delta u_i$$

$$= - \int_D d^3x \sigma_{ij}^I \delta \Lambda_{ij}. \tag{13.20}$$

The change in free energy is $-\delta W$, and

$$\sigma_{ij}^I = \partial f / \partial \Lambda_{ij}. \tag{13.21}$$

From the theoretical physicist's standpoint, it is often more convenient to work with the second Piola–Kirchhoff stress tensor defined by

$$\sigma_{ij}^{II} = \partial f / \partial u_{ij}. \tag{13.22}$$

The second Piola–Kirchhoff stress is often called the thermodynamic stress. It cannot, however, be expressed as the force per unit area in any space. It is symmetric by construction, and like the strain tensor $u_{ij}$, it is a tensor in $S_R$. Working in theory with $\sigma_{ij}^{II}$ instead of the experimentally more relevant $\sigma_{ij}^I$ poses no problem because one can easily convert between the two exploiting their relation

$$\sigma_{ij}^I = \Lambda_{ik}\sigma_{kj}^{II}. \tag{13.23}$$

For completeness, let us also mention the Cauchy stress $\sigma_{ij}^C$ which is defined as the force per unit area in $S_T$. The work done by $\sigma_{ij}^C$ is

$$\delta W = - \int d^3R \sigma_{ij}^C \frac{\partial u_i}{\partial R_j}, \tag{13.24}$$

from which it follows that

$$\sigma_{ij}^C = \frac{1}{\det\underline{\underline{\Lambda}}} \Lambda_{ik}\sigma_{kl}^{II}\Lambda_{lj}^T. \tag{13.25}$$

Note that the Cauchy stress tensor is symmetric as it must be.

## 13.2.2 COUPLING ELASTIC AND LIQUID CRYSTALLINE DEGREES OF FREEDOM

In traditional uncross-linked liquid crystals, there is no reference space, and all physical fields like the smectic layer-displacement field $U$, the layer normal $\mathbf{N}$, the Frank director $\mathbf{n}$, and the Maier–Saupe–de Gennes order parameter $\underline{\underline{Q}}$ are defined at target-space points $\mathbf{R}$, and they transform as scalars, vectors, and tensors under rotations in $S_T$. Moreover, external fields, for example, electric or magnetic, define a direction in $S_T$ and are therefore, target-space vectors. In the Lagrangian theory of elasticity, on the other hand, fields are defined at reference-space points $\mathbf{x}$, and they transform into themselves under the symmetry operations of that space. To develop a comprehensive theory of liquid-crystalline elastomers, it is necessary to combine target-space liquid crystalline and external fields and reference-space elastic variables to produce scalars that are invariant under arbitrary rotations in $S_T$ and under symmetry operations of $S_R$. It is, therefore, necessary to be able to express vectors and tensors in either space [25].

To be more specific, let $\mathbf{b}$ be a target-space vector, which by definition transforms under rotations to $\mathbf{b}' = \underline{\underline{O}}_T \mathbf{b}$, and let $\tilde{\mathbf{b}}$ be a reference-space vector, which transforms to $\tilde{\mathbf{b}}' = \underline{\underline{O}}_R \tilde{\mathbf{b}}$. Recall that both reference- and target-space vectors exist in the same physical Euclidean space $\mathcal{E}$. Therefore, both $\mathbf{b}$ and $\tilde{\mathbf{b}}$ can be expressed in terms of components $b_i$ and $\tilde{b}_i$ relative to a fixed orthonormal basis $\{e_i | i = x, y, z\}$ with $\mathbf{e}_i \cdot \mathbf{e}_j = \delta_{ij}$:

$$\mathbf{b} = b_i \mathbf{e}_i, \quad \tilde{\mathbf{b}} = \tilde{b}_i \mathbf{e}_i, \tag{13.26}$$

There must be a transformation that converts a given reference-space vector to a target-space vector and vice versa while preserving length. This transformation is provided by the deformation matrix $\underline{\underline{\Lambda}}$ and the matrix polar decomposition theorem [23], which states that any nonsingular square matrix can be expressed as the product of a rotation matrix and a symmetric matrix. If $\tilde{\mathbf{b}}$ is a reference-space vector, then $\underline{\underline{\Lambda}}\tilde{\mathbf{b}}$ is a target-space vector that transforms under $\underline{\underline{O}}_T$ but does not change under $\underline{\underline{O}}_R$. The transformation $\tilde{\mathbf{b}} \rightarrow \underline{\underline{\Lambda}}\tilde{\mathbf{b}}$, however, does not preserve length. To construct a transformation that does, we simply multiply $\underline{\underline{\Lambda}}$ by the square root of the metric tensor to get

$$\underline{\underline{O}} = \underline{\underline{\Lambda}}\, \underline{\underline{g}}^{-1/2}. \tag{13.27}$$

This operator clearly satisfies $\underline{\underline{O}}^T \underline{\underline{O}} = \underline{\underline{O}}\,\underline{\underline{O}}^T = \underline{\underline{\delta}}$ and $\det \underline{\underline{O}} = 1$, and it is thus a length-preserving rotation matrix. Equation 13.27, which can be recast in the form $\underline{\underline{\Lambda}} = \underline{\underline{O}}\underline{\underline{g}}^{1/2}$ is simply a restatement of the polar decomposition theorem because $\underline{\underline{g}}$ is a symmetric matrix. To first order in $\eta_{ij}$, $O_{ij}$ reduces to the standard expression for an infinitesimal local rotation of an elastic body through an angle $\Omega = (1/2)\nabla \times \mathbf{u}$,

$$O_{ij} = \delta_{ij} - \epsilon_{ijk}\Omega_k + \cdots \tag{13.28a}$$

$$= \delta_{ij} - \eta_{A,ij} + \cdots, \tag{13.28b}$$

where $\epsilon_{ijk}$ is the Levi–Civita tensor, and where $\eta_{A,ij} = (1/2)(\eta_{ij} - \eta_{ji})$ is the antisymmetric part of the displacement gradient tensor. Equipped with $\underline{\underline{O}}$ we can convert (or rotate) any reference-space vector $\tilde{\mathbf{b}}$ to a target space vector $\mathbf{b}$ through

$$\mathbf{b} = \underline{\underline{O}} \cdot \tilde{\mathbf{b}} \tag{13.29}$$

and a target-space vector to a reference space vector through

$$\tilde{\mathbf{b}} = \underline{\underline{O}}^T \cdot \mathbf{b}, \tag{13.30}$$

that is, $b_i = O_{ij}\tilde{b}_j$ and $\tilde{b}_i = O_{ij}^T b_j$.

An important property of $\underline{\underline{O}}$ is that it reduces to the unit matrix when $\underline{\Lambda}$ is symmetric, that is, under pure shear transformations, target- and reference-space vectors are identical. Thus, if a reference-space vector is known (calculated, e.g., by minimizing a free energy that depends only on reference-space vectors and tensors), the associated target-space vector is obtained by rotating the reference-state vector by the operator $\underline{\underline{O}}$, which is the same operator that rotates the pure shear configuration to the target-space configuration described by $\underline{\Lambda}$. Figure 13.2 depicts the effect on an initial reference state unit vector of a symmetric shear and then a subsequent rotation to a final target state.

An alternative interpretation of the relation between $\mathbf{b}$ and $\tilde{\mathbf{b}}$ follows from

$$\mathbf{b} = b_i\mathbf{e}_i = O_{ij}\tilde{b}_j\mathbf{e}_i \equiv \tilde{b}_j\mathbf{t}_j, \tag{13.31}$$

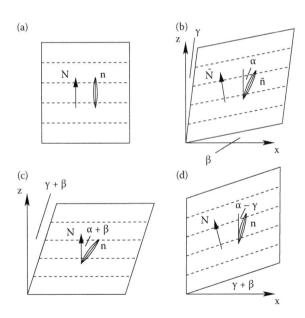

**FIGURE 13.2** Schematic representation of distortions in the $xz$-plane induced by the SmA-to-SmC transition: (a) undistorted SmA phase, (b) SmC phase with a symmetric deformation tensor $\underline{\Lambda}$, (c) SmC phase with $\Lambda_{xz} > 0$ and $\Lambda_{zx} = 0$, and (d) SmC phase with $\Lambda_{xz} = 0$ and $\Lambda_{zx} > 0$.

with

$$\mathbf{t}_j = O_{ji}^T \mathbf{e}_j = g_{jk}^{-1/2} \frac{\partial \mathbf{R}}{\partial x_k}, \tag{13.32}$$

where we used $\Lambda_{kl}^T \mathbf{e}_l = (\partial R_l / \partial x_k) \mathbf{e}_l = \partial \mathbf{R} / \partial x_k$. The set of vectors $\{\mathbf{t}_i | i = x, y, z\}$ forms an orthonormal target-space basis in the tangent space of the deformed medium (recall that $\partial \mathbf{R} / \partial x_j$ is a tangent-space vector). Thus, $\tilde{b}_i$ represents the components of the target-space vector $\mathbf{b}$ relative to the orthonormal tangent-space basis defined by $\mathbf{t}_i$.

Having the approach based on the polar decomposition theorem at our disposal, we can readily construct couplings between liquid crystalline or external fields and the Lagrangian strain tensor that have the required symmetry properties, that is, that are rotationally invariant in $S_T$ and transform under the point group of $S_R$. As a first example, let us consider an isotropic elastomer in an external electric field $\mathbf{E}$. Because the material is polarizable, the electric field couples to strain. Contributions to the total elastic energy that model this effect have to be a scalar under simultaneous rotations of the electric field and the strain in reference space. Contractions of the components of $\mathbf{E}$ with the stress–strain components $u_{ij}$ do not produce a scalar because $E_i$ and $u_{ij}$ transform under different operators. To create scalar contractions, we can convert $\mathbf{E}$ to a reference-space vector by $\tilde{\mathbf{E}} = \underline{\underline{O}}^{-1} \mathbf{E}$. The leading order coupling of $\mathbf{E}$ to strain is thus proportional to the scalar

$$\tilde{E}_i u_{ij} \tilde{E}_j = E_i \upsilon_{ij} E_j, \tag{13.33}$$

where $\upsilon_{ij}$ being the components of the so-called left Cauchy–Green strain tensor

$$\underline{\underline{\upsilon}} = \underline{\underline{O}} \underline{\underline{u}} \underline{\underline{O}}^T = \frac{1}{2} (\underline{\underline{\Lambda}} \underline{\underline{\Lambda}}^T - \underline{\underline{\delta}}), \tag{13.34}$$

which, of course, transforms as a tensor in $S_T$ and as a scalar in $S_R$.

As a second example, let us consider the coupling of strain and the Maier–Saupe–de Gennes order parameter $\underline{\underline{Q}}$, which is a tensor in $S_T$ that we can convert to a reference-space tensor by $\tilde{\underline{\underline{Q}}} = \underline{\underline{O}}^{-1} \underline{\underline{Q}} \underline{\underline{O}}$. Modeling the phase transition from an isotropic to a nematic elastomer, we need to devise couplings that are rotationally invariant in $S_R$ and $S_T$. The leading couplings of this type are $u_{ij} \tilde{Q}_{ij} = \upsilon_{ij} Q_{ij}$ and $u_{kk} \tilde{Q}_{ij} \tilde{Q}_{ji} = \upsilon_{kk} Q_{ij} Q_{ji}$.

As our third and perhaps most important example, let us consider the coupling, the Frank director and strain. To apply our formalism, it is useful to represent the director as

$$\mathbf{n} \equiv (\mathbf{c}, n_z), \quad n_z = \sqrt{1 - c_a^2}, \tag{13.35}$$

where $\mathbf{c}$ is the so-called c-director. Using the polar-decomposition technique, we can convert $\mathbf{n}$ to a reference-space vector via $\tilde{\mathbf{n}} = \underline{\underline{O}}^{-1} \mathbf{n}$ with

$$\tilde{\mathbf{n}} \equiv (\tilde{\mathbf{c}}, \tilde{n}_z), \quad \tilde{n}_z = \sqrt{1 - \tilde{c}_a^2}. \tag{13.36}$$

In a uniaxial elastomer, for example, this approach leads straightforwardly to $\tilde{n}_a u_{az}$ as the leading coupling term.

### 13.2.3  MODEL ELASTIC ENERGIES

Liquid crystal elastomers exhibit a wide variety of interesting, often remarkable phenomena. Several theoretical models have been devised to predict or explain some of these phenomena. The level of detail of these models does vary depending on what kind of questions they are supposed to address. For example, one can describe the $I-N$ and the SmA-to-SmC phase transitions in elastomers in terms of strain-only models, which do not explicitly contain any liquid crystalline degrees of freedom. These models, of course, have nothing to say about quantities like the director. More comprehensive theories involve strains and liquid crystalline degrees of freedom; we have just learned in Section 13.2.2 how to model couplings between these kinds of variables. The point of view adopted in Section 13.2.2 is that of the famous Landau theory, that is, one constructs combinations of the relevant variables that are compatible with symmetry and then write a model free energy by collecting all terms to a certain order. An alternative approach is to start with microscopic models and then extract a total free energy from these, as we did for smectic elastomers in Refs. [26,27]. This latter approach has the virtue that it can provide an estimate of the values of the elastic constant featured in the model. In a plain Landau expansion the elastic constants are generically only phenomenological parameters whose value is not known. On the other hand, deriving a free energy from microscopic models is usually much more involved than writing down a Landau expansion right away. To keep our arguments as simple as possible, we will therefore, restrict ourselves here to models in the spirit of Landau expansions. Moreover, we will focus on models with strain and liquid crystalline degrees of freedom and we will limit our considerations to three concrete examples: a model of nematic elastomers with strain and director and models, respectively, for the $I-N$, and SmA-to-SmC phase transitions in elastomers.

#### 13.2.3.1  Model of the *I–N* Transition

A generic model elastic energy density of $I-N$ transition will consist of the isotropic elastic energy density $f_{iso}$ of Equation 13.15, a contribution $f_Q$ describing nematic-orientational ordering, and a nemato-elastic contribution $f_{coupl}$ coupling the strain tensor and the nematic-order parameter,

$$f_{I-N} = f_{iso} + f_Q + f_{coupl}. \tag{13.37}$$

Near the $I-N$ transition we can use for $f_Q$ the usual Landau–de Gennes free energy density [2]:

$$f_Q = \frac{1}{2}r\mathrm{Tr}\underline{\underline{Q}}^2 - w\mathrm{Tr}\underline{\underline{Q}}^3 + \upsilon(\mathrm{Tr}\underline{\underline{Q}}^2)^2 \tag{13.38a}$$

$$= \frac{1}{2}r\mathrm{Tr}\underline{\underline{\tilde{Q}}}^2 - w\mathrm{Tr}\underline{\underline{\tilde{Q}}}^3 + \upsilon(\mathrm{Tr}\underline{\underline{\tilde{Q}}}^2)^2. \tag{13.38b}$$

Terms containing gradients of $\underline{\underline{Q}}$ could also be added, but they are beyond the scope of this review and they are inconsequential for the mean-field consideration to be presented in Section 13.5. A simple phenomenological coupling energy density can be constructed along the lines of Section 13.2.2:

$$f_{\text{coupl}} = -2t\hat{u}_{ij}\tilde{Q}_{ij} - su_{kk}\tilde{Q}_{ij}\tilde{Q}_{ji}$$
$$= -2t\text{Tr}\underline{\hat{u}}\underline{\tilde{Q}} - s\text{Tr}\underline{u}\text{Tr}\underline{\tilde{Q}}^2. \tag{13.39}$$

This form reflects the key features of strain-orientational coupling, that is, that the development of orientational order will drive an anisotropic distortion and induce a small volume change.

Despite its simplicity, $f_{I-N}$ captures the essentials of the $I-N$ transition in mean field theory. We will return to $f_{I-N}$ in Section 13.5, where we review its mean-field analysis.

### 13.2.3.2  Model of Nematic Elastomers with Strain and Director

A simple phenomenological model elastic energy density of nematic elastomers that features strain and director can be written as the sum of three parts,

$$f_{\text{nem}} = f_{\text{uni}} + f_{\text{tilt}} + f_{\text{coupl}}. \tag{13.40}$$

$f_{\text{uni}}$ is the uniaxial elastic energy density of Equation 13.17. In an equilibrium nematic elastomer the director prefers alignment along the uniaxial direction. The simplest phenomenological model for the energy cost that produced when $\mathbf{n}$ tilts away from the preferred uniaxial axis $\mathbf{n}_0 = \mathbf{e}_z$ is

$$f_{\text{tilt}} = \frac{1}{2}D_1[1 - (\mathbf{n} \cdot \mathbf{n}_0)^2] = \frac{1}{2}D_1\tilde{c}_a^2, \tag{13.41}$$

where in the second equality we used the relation $\mathbf{n} \cdot \mathbf{n}_0 = \tilde{\mathbf{n}} \cdot \tilde{\mathbf{e}}_z$. $f_{\text{coupl}}$, finally, couples $\mathbf{n}$ and the strain tensor $\underline{u}$; we have already learned to construct the terms entering $f_{\text{coupl}}$ in Section 13.2.2. At leading order,

$$f_{\text{coupl}} = D_2\tilde{c}_a u_{az}. \tag{13.42}$$

Other contributions to $f$ are conceivable, such as the well-known Frank energy describing deviations from homogeneous director alignment. To keep our reasoning as simple as possible, we will neglect these contributions here.

When the combinations $\tilde{c}_a^2$ and $\tilde{c}_a u_{az}$ are expanded with the help of Equation 13.28, they reproduce, at leading order, those in the original de Gennes theory [28]. Linearized deviations of the target-space director from its equilibrium $\mathbf{n}_0$ can be expressed as $\delta\mathbf{n} = \omega \times \mathbf{n}_0$, where $\omega$ is a rotation angle. Then $\delta\tilde{\mathbf{n}} = \tilde{\mathbf{c}} = (\omega - \Omega) \times \mathbf{n}_0$ and, for example, $\tilde{c}_a u_{az} \rightarrow u_{az}[(\omega - \Omega) \times \mathbf{n}_0]_a$.

Having $f_{\text{nem}}$, a worthwhile question, whose answer reveals some interesting physics, is the following: How can we retrieve from Equation 13.40 an elastic energy density in terms of strain only? We can do so by minimizing over $\tilde{c}_a$. To facilitate this procedure, often referred to as "integrating out" the director, it is convenient to recast $f_{nem}$ as

$$f_{nem} = \frac{1}{2}C_1 u_{zz}^2 + C_2 u_{zz} u_{ii} + \frac{1}{2}C_3 u_{ii}^2 + C_4 \hat{u}_{ab}^2 + C_5^R u_{az}^2 + \frac{1}{2}D_1 \left[\tilde{c}_a + \frac{D_2}{D_1} u_{az}\right]^2,$$

(13.43)

which demonstrates that the director will relax locally to $\tilde{c}_a = -(D_2/D_1)u_{az}$, which eliminates the dependence of $f_{nem}$ on the director. This relaxation leads to a reduction of the shear modulus for shears in planes containing the anisotropy axis, that is, $C_5$ is renormalized to

$$C_5^R = C_5 - \frac{1}{2}D_2^2/D_1.$$

(13.44)

In an ideal soft-nematic elastomer the contributions $C_5$ and $\frac{1}{2}D_2^2/D_1$ are equal so that $C_5^R = 0$. In a semisoft nematic, $C_5^R$ is nonzero but small.

### 13.2.3.3  Model of the SmA-to-SmC Transition

In principle, smectic elastomers can be produced either by cooling nematic elastomers through the transition to the smectic phase or by cross linking in the smectic phase. Cross linking in the smectic phase tends to lock the smectic layers to the cross-linked network [29]. Without this lock-in, the phase of the smectic mass-density-wave can translate freely relative to the elastomer as it can in smectics in aerogels [30]. To keep our discussion as simple as possible, we will not consider in the following the case of cross linking in the nematic phase, and we take the lock-in of the smectic layers and the elastic matrix as given.

In the SmA phase, the director is parallel to both the layer normal $\mathbf{N}$, whose components are given by

$$N_i = \frac{\Lambda_{zi}^{-1}}{g_{zz}^{-1}},$$

(13.45)

and the anisotropy axis $\mathbf{e}_z$, which are parallel to each other. Thus, there are terms in the free-energy density proportional to $(\tilde{\mathbf{N}} \cdot \tilde{\mathbf{n}})^2$ and $(\tilde{\mathbf{e}}_z \cdot \tilde{\mathbf{n}})^2$, which when combined yield a term proportional to $\tilde{c}_a^2$, whose coefficient vanishes at the SmA–SmC transition, and higher order terms involving the strain and strain-director coupling. To stabilize the SmC phase, higher order terms in $(\tilde{\mathbf{N}} \cdot \tilde{\mathbf{n}})^2$ and $(\tilde{\mathbf{e}}_z \cdot \tilde{\mathbf{n}})^2$ must be added. The final free-energy density up to inconsequential higher order terms is thus

$$f_{A-C} = f_{uni} + f_{tilt} + f_{coupl},$$

(13.46)

where $f_{uni}$ is once again the uniaxial energy density (Equation 13.17) and, different from what we had above,

$$f_{tilt} = \frac{1}{2}r_{\tilde{c}}\tilde{c}_a^2 + \frac{1}{4}g(\tilde{c}_a^2)^2,$$

(13.47)

$$f_{coupl} = \lambda_1 \tilde{c}_a^2 u_{zz} + \lambda_2 \tilde{c}_a^2 u_{bb} + \lambda_3 \tilde{c}_a \hat{u}_{ab} \tilde{c}_b + \lambda_4 \tilde{c}_a u_{az} \tilde{n}_z.$$

(13.48)

If $\tilde{c}_a$ is integrated out of $f$, the result is identical to harmonic order to $f_{uni}$ with $C_5$ renormalized to $C_5^R = C_5 - \lambda_4^2/(2r_{\tilde{c}})$. When $\lambda_1 = \lambda_2 = \lambda_3 = 0$, this model is equivalent to that studied in Ref. [31] when polarization is ignored. One can include additional terms, like $-B_1 u_{zz} u_{za}^2$, $B_2 (u_{za}^2)^2$ and $\lambda_5 u_{zz} u_{az} \tilde{c}_a$, for example, if one seeks to describe the analog in smectic elastomers of the Helfrich–Hurault effect in uncross-linked smectics [32–35]. For simplicity, we will neglect here any of such higher order terms. We will revisit $f$, when we study the SmA-to-SmC transition in mean-field theory in Section 13.5.

## 13.3   SOFT ELASTICITY

### 13.3.1   ORIGIN AND CONCEPT

Symmetries, long-wavelength excitations, and their interplay are fundamental to the understanding of condensed-matter systems [36]. The famous Goldstone theorem [37, 38] states that ordered thermodynamic phases that break a continuous symmetry have low-energy excitations, referred to as *soft* or Goldstone modes that move the system among a continuum of degenerate ground states.

The role of the Goldstone theorem in the context of liquid crystal elastomers was first discussed by Golubović and Lubensky [39]. They considered uniaxial solids that form spontaneously from an isotropic phase. The epitome of such a system is an idealized nematic elastomer that is produced when an isotropic liquid crystal elastomer undergoes a phase transition to the nematic phase in which the continuous-rotational symmetry of the isotropic phase is broken. As a consequence of the Goldstone theorem, this idealized nematic phase exhibits soft elasticity characterized by the vanishing of the shear modulus $C_5$ and by a stress–strain curve for strains $u_{xx}$ (or $u_{yy}$) and stresses $\sigma_{xx}$ (or $\sigma_{yy}$) perpendicular to the anisotropy axis in which strains up to a critical value are produced at zero stress, *cf.* Figure 13.4.

Arguments in the spirit of Golubović and Lubensky also apply to smectics. When an ideal SmC elastomer is formed from a SmA elastomer by cooling through the SmA-to-SmC phase transition, rotational invariance in the smectic planes is spontaneously broken. As a result, like ideal nematic elastomers, these ideal SmC elastomers exhibit soft elasticity with vanishing shear moduli and unconventional stress–strain curves [26,40].

Monodomain samples of nematic or SmC elastomers cannot be produced without locking in a preferred anisotropy direction, usually by the Küpfer–Finkelmann procedure [17] in which a first cross linking in the absence of aligning stress is followed by a second or more cross-linking steps with stress. This process introduces a mechanical aligning field $h$, analogous to an external electric or magnetic field, and lifts the value of the soft elastic moduli from zero. For fields $h$ that are not too large, however, these moduli are predicted to remain small and the corresponding elastomers are expected exhibit semisoft elasticity [1,10].

Strictly speaking, the genuine Goldstone arguments for soft elasticity apply only for $h = 0$. The fact that this assumption is violated in all monodomain samples that have been produced thus far, has apparently led to some disagreement over the value an applicability of Goldstone arguments to real systems. Our point of view is that the ideal case $h = 0$ can be viewed as an approximation to the real case $h > 0$ provided that $h$ is small. In our eyes, soft elasticity is a first approximation to the semisoft elasticity of real samples. Indeed, the symmetry arguments leading to softness can be generalized systematically to $h > 0$, as was done very recently for nematics [41]. These generalized arguments predict semisoftness that reduces to softness for $h \to 0$ and thereby underscore the validity of the concept of soft elasticity as a first approximation to semisoftness. We will return to semisoftness in Section 13.4.

In the remainder of this section we will review the symmetry arguments leading to softness in ideal nematic and SmC elastomers. It is important to note that these arguments exclusively rely on the existence of a broken-symmetry state with the macroscopic symmetry of the nematic and the SmC phase, respectively. They do not depend on the detailed form of any elastic energy density such as Equations 13.17 and 13.18 and, in particular, they are not restricted to small strains.

### 13.3.2 SOFTNESS OF NEMATIC ELASTOMERS

When an ideal nematic elastomer forms spontaneously from an isotropic elastomer upon cooling through the $I-N$ transition, the anisotropy direction $\mathbf{n}_0$ of the resulting uniaxial phase is arbitrary, see Figure 13.3. Let us characterize an elastomer formed in this way by its equilibrium strain tensor

$$\underline{\underline{u}}^0 = \frac{1}{2}(\underline{\underline{\Lambda}}^{0T}\underline{\underline{\Lambda}}^0 - \underline{\underline{\delta}}),$$

(13.49)

where $\underline{\underline{\Lambda}}^0$ is the equilibrium deformation tensor defined by $\Lambda_{ij}^0 = \partial R_i^0 / \partial x_j$ and $\mathbf{R}^0(\mathbf{x}) \equiv \mathbf{x}'$ is the equilibrium position of the point $\mathbf{x}$. The full deformation tensor relative to reference space points $\mathbf{x}$ can be expressed as a product of the deformation tensor $\Lambda'_{ik} = \partial R_i / \partial x'_k$ relative to the new reference space $S'_R$ defined by the equilibrium points $\mathbf{x}'$ and $\Lambda_{kj}^0$:

$$\Lambda_{ij} = \frac{\partial R_i}{\partial x_j} = \frac{\partial R_i}{\partial x'_k}\frac{\partial R_k^0}{\partial x_j} = \Lambda'_{ik}\Lambda_{kj}^0.$$

(13.50)

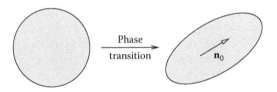

**FIGURE 13.3** Spontaneous symmetry breaking at the $I-N$ phase transition. The direction of the equilibrium director $\mathbf{n}_0$ is arbitrary.

Strains $\underline{\underline{u}}'$ relative to $S_R'$ are then linearly proportional to the deviation $\delta\underline{\underline{u}} = \underline{\underline{u}} - \underline{\underline{u}}^0$ of the strain $\underline{\underline{u}}$ from the equilibrium strain:

$$\underline{\underline{u}}' = \frac{1}{2}(\underline{\underline{\Lambda}}'^T\underline{\underline{\Lambda}}' - \underline{\underline{\delta}}) = \frac{1}{2}(\underline{\underline{\Lambda}}^{0T})^{-1}(\underline{\underline{\Lambda}}^T\underline{\underline{\Lambda}} - \underline{\underline{\Lambda}}^{0T}\underline{\underline{\Lambda}}^0)(\underline{\underline{\Lambda}}^0)^{-1}$$

$$= \frac{1}{2}(\underline{\underline{\Lambda}}^{0T})^{-1}\delta\underline{\underline{u}}(\underline{\underline{\Lambda}}^0)^{-1}. \tag{13.51}$$

Because the anisotropy direction in the ordered state is arbitrary, the state with strain $\underline{\underline{u}} = \underline{\underline{O}}_R\underline{\underline{u}}^0\underline{\underline{O}}_R^{-1}$ relative to $S_R$ must have the same energy as the state with strain $\underline{\underline{u}}^0$. Thus, there is no energy difference between the strains $\underline{\underline{u}}$ and $\underline{\underline{u}}^0$ and no energy cost associated with the strain

$$\underline{\underline{u}}' = (\underline{\underline{\Lambda}}^{0T})^{-1}\left[\underline{\underline{O}}_R\underline{\underline{u}}^0\underline{\underline{O}}_R^{-1} - \underline{\underline{u}}^0\right](\underline{\underline{\Lambda}}^0)^{-1} \tag{13.52}$$

relative to $S_R'$. The strain $\underline{\underline{u}}'$ describes the Goldstone modes of the spontaneously ordered nematic elastomer.

We now consider a specific realization of $\underline{\underline{O}}_R$. Let us choose our $z$-direction in $S_R$ along $\tilde{n}_0$ so that the equilibrium deformation tensor is diagonal with diagonal elements $\Lambda_{xx}^0 = \Lambda_{yy}^0 = \Lambda_\perp^0$ and $\Lambda_{zz}^0 = \Lambda_{||}^0$, and let us rotate counterclockwise about the $y$ direction, that is, $\underline{\underline{O}}_R = \underline{\underline{O}}_{R,y}$ with

$$\underline{\underline{O}}_{R,y} = \begin{pmatrix} \cos\vartheta & 0 & \sin\vartheta \\ 0 & 1 & 0 \\ -\sin\vartheta & 0 & \cos\vartheta \end{pmatrix}, \tag{13.53}$$

where $\vartheta$ is an arbitrary rotation angle. Using this $\underline{\underline{O}}_R$ in Equation 13.52, we see that $\underline{\underline{u}}'(\vartheta)$ is a nontrivial strain even though it merely describes a rigid rotation in $S_R$:

$$\underline{\underline{u}}'(\vartheta) = \frac{r-1}{4}\begin{pmatrix} 1 - \cos 2\vartheta & 0 & r^{-1/2}\sin 2\vartheta \\ 0 & 0 & 0 \\ r^{-1/2}\sin 2\vartheta & 0 & -r^{-1}(1 - \cos 2\vartheta) \end{pmatrix}, \tag{13.54}$$

where $r$ denotes the anisotropy ratio of the uniaxial state,

$$r = (\Lambda_{||}^0)^2/(\Lambda_\perp^0)^2. \tag{13.55}$$

This anisotropy ratio must not be confused with the parameter $r$ of the Landau–de Gennes free-energy density, Equations 13.38 and 13.88.

To harmonic order in the strain, the free energy density of the uniaxial state formed through spontaneous symmetry breaking from an isotropic state must have the form of Equation 13.17. To preserve stability and global rotational invariance in $S_R$ of the parent isotropic state, it must also have higher order terms in the strain. For infinitesimal rotation angle $\vartheta$, the strain $\underline{\underline{u}}'(\vartheta)$ has nonzero components, namely

$$u_{xz}'(\vartheta) = u_{zx}'(\vartheta) = \frac{r-1}{2\sqrt{r}}\vartheta. \tag{13.56}$$

Thus, the shear modulus $C_5$ must vanish in an ideal nematic elastomer, whose harmonic elastic energy is, therefore, characterized by only four elastic constants,

$$f_{\text{nem}}^{\text{soft}} = \frac{1}{2}C_1(u'_{zz})^2 + C_2 u'_{zz}u'_{ii} + \frac{1}{2}C_3(u'_{ii})^2 + C_4(\hat{u}'_{ab})^2. \tag{13.57}$$

Consequently, shear strains $u'_{az}$ in planes containing the anisotropy direction cost no elastic energy and cause no restoring forces.

The existence of zero-energy strains that reproduce rotations in $S_R$ has consequences reaching further then just the softness with respect to shear strains $u'_{xz}$, that is, depending on the experimental boundary conditions, extensional or compressional strains $u'_{xx}$ or $u'_{zz}$ can also be soft. If the boundary conditions are such that no relaxation of strains is allowed, then strains $u'_{xx}$ and $u'_{zz}$ will cost an elastic energy proportional to $(u'_{xx})^2$ and $(u'_{zz})^2$, respectively. If, however, strain relaxation is allowed and one imposes, for example, $u'_{xx}$ with the right sign, then $u'_{zz}$ and $u'_{xz}$ can relax under the right circumstances to produce the zero-energy strain of Equation 13.54, that is, to make $u'_{xx}$ a soft deformation.

To discuss this in more detail, let us assume for concreteness that the anisotropy of the sample is positive, $r > 1$. Then, this relaxation is possible only for $u'_{xx} > 0$ (extension perpendicular to the uniaxial direction $\mathbf{n}_0$), $u'_{zz} < 0$ (compression along $\mathbf{n}_0$), and either positive or negative $u'_{xz}$. Let us consider here as an example a stretch along $x$, that is, $u'_{xx} > 0$. Comparison with Equation 13.54 shows that, if strain relaxation is allowed and $u'_{zz}$ and $u'_{xz}$ relax to

$$u'_{zz} = -r^{-1}u'_{xx}, \tag{13.58a}$$

$$u'_{xy} = \pm\sqrt{\frac{u'_{xx}(r - 1 - 2u'_{xx})}{2r}}, \tag{13.58b}$$

then $u'_{xx}$ is converted into a zero-energy rotation through an angle

$$\vartheta = \pm\sin^{-1}\sqrt{\frac{2u'_{xx}}{r - 1}}. \tag{13.59}$$

Thus, $u'_{xx}$ costs no elastic energy as long as $0 < u'_{xx} < (r - 1)/2$.

When $u'_{xx}$ is increased from zero, $\vartheta$ increases from zero until it reaches $\pi/2$ at $u'_{xx} = (r - 1)/2$ [and $u'_{zz} = (r^{-1} - 1)/2$, $u'_{xz} = 0$] at which point, the deformation tensor defined by $\Lambda'_{0ij} = \partial R_i/\partial x'_j$ is

$$\underline{\underline{\Lambda}}'_0 = \begin{pmatrix} r^{1/2} & 0 & 0 \\ 0 & 1 & 0 \\ 0 & 0 & r^{-1/2} \end{pmatrix}, \tag{13.60}$$

which leads via $\Lambda_{ij} = \Lambda_{ik}^0 \Lambda'_{0kj}$ to an overall deformation

$$\underline{\underline{\Lambda}} = \begin{pmatrix} \Lambda_{zz}^0 & 0 & 0 \\ 0 & \Lambda_{yy}^0 & 0 \\ 0 & 0 & \Lambda_{xx}^0 \end{pmatrix} \tag{13.61}$$

relative the original uniaxial state. Thus, at $\vartheta = \pi/2$, $\Lambda_{ij}$ is identical to $\Lambda_{ij}^0$ except with $\Lambda_{xx}^0$ and $\Lambda_{zz}^0 = \sqrt{r}\Lambda_{xx}^0$ interchanged, that is, the $x$ and $y$ axes have been interchanged in going from $\vartheta = 0$ to $\vartheta = \pi/2$. In the process, the director rotates from being parallel to the $z$-axis to being parallel to the $x$-axis.

For $u'_{xx} > (r-1)/2$, there is no real solution for $\vartheta$, and a further increase in $u'_{xx}$, measured by $\delta u'_{xx} = u'_{xx} - (r-1)/2$, which stretches the system along the space-fixed $x$-axis, cost energy. Since the anisotropy axis is now along the $x$- rather than the $z$-axis, this stretching costs the same energy as it would have cost to stretch the original system with anisotropy axis along the space-fixed $z$-axis by the same amount. The $xx$-component of the strain relative to the state with $\vartheta = \pi/2$ is $\Delta u'_{xx} = (\Lambda'_{0xx})^{-2}\delta u'_{xx} = r^{-1}\delta u'_{xx}$. Thus, because the $z$- and $x$-axes have been interchanged, the free energy as a function of $u'_{xx}$ is

$$f_{nem}^{soft} = \begin{cases} 0 & \text{for } \delta u'_{xx} < 0, \\ \frac{1}{2}Y_z(\delta u'_{xx})^2 & \text{for } \delta u'_{xx} > 0. \end{cases} \tag{13.62}$$

where

$$Y_z = \frac{1}{r}\left\{C_1 - \frac{C_2^2}{C_3 + C_4}\right\} \tag{13.63}$$

is the Young's modulus for stretching along the anisotropy axis along $x$ (originally along $z$). Consequently, the engineering stress is to leading order

$$\sigma_{yy}^{eng} = \frac{\partial f_{nem}^{soft}}{\partial \Lambda'_{xx}} = \begin{cases} 0 & \text{for } \Lambda'_{xx} < \sqrt{r} \\ Y_z \Lambda'_{0xx}\delta u'_{xx} & \text{for } \Lambda'_{xx} > \sqrt{r} \end{cases}. \tag{13.64}$$

For $\Lambda'_{xx}$ near $\Lambda'_{0xx}$, $\delta u'_{xx} \approx \Lambda'_{0xx}(\Lambda'_{xx} - \Lambda'_{0xx})$, and $\sigma_{xx}^{eng} \approx Y_z(\Lambda'_{xx} - \Lambda'_{0xx})$. Figure 13.4 depicts the dependence of $\sigma_{xx}^{eng}$ on $\Lambda'_{xx}$. From $\Lambda'_{xx} = 0$ up to a critical deformation $\Lambda'_{0xx} = \sqrt{r}$ the stress is zero. Above the critical deformation, $\sigma_{xx}^{eng}$ grows linearly from zero.

### 13.3.3   SOFTNESS OF SMECTIC-C ELASTOMERS

As mentioned above, when a SmC elastomer forms through a phase transition from a SmA elastomer, rotational invariance in the plane of the smectic layers is spontaneously broken. Hence, the Goldstone theorem mandates the existence of a soft mode in such an ideal SmC elastomer. To analyze the nature of this soft mode, we will use a somewhat different starting point from that in Section 13.3.2 in that we first consider soft deformations [1,9,25] rather than soft strains. First, this is interesting in its own right. Second, this will set the stage for a comparison of the predictions on the softness of SmC elastomers made, respectively, by Lagrange theory and the neoclassical model, cf. Chapter 17 of this book.

Let us first determine the general form of soft deformations. The equilibrium or "ground state" deformation tensor $\underline{\underline{\Lambda}}^0$, which can be cast with the appropriate choice

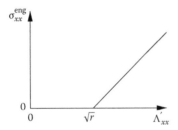

**FIGURE 13.4** Schematic plot (arbitrary units) of the engineering stress $\sigma_{xx}^{eng}$ versus the deformation $\Lambda'_{xx}$ for a soft nematic elastomer with an equilibrium director along $z$ and positive anisotropy, $r > 0$. Up to a critical deformation $\Lambda'_{0xx} = \sqrt{r}$ the sample responds to the deformation merely by rotating the director and consequently the stress is zero. Above $\Lambda'_{0xx}$ the sample stretches along the new direction $x$ of the director and $\sigma_{xx}^{eng}$ grows linearly for small $\Lambda'_{xx} - \Lambda'_{0xx}$.

of coordinates such that the SmA-to-SmC transition amounts to simple shear in the $xz$-plane,

$$
\underline{\underline{\Lambda}}^0 = \begin{pmatrix} \Lambda_{xx}^0 & 0 & \Lambda_{xz}^0 \\ 0 & \Lambda_{yy}^0 & 0 \\ 0 & 0 & \Lambda_{zz}^0 \end{pmatrix}, \tag{13.65}
$$

maps points in $S_R$ to points in $S_T$ via $\mathbf{R}(\mathbf{x}) = \underline{\underline{\Lambda}}^0\mathbf{x}$ as depicted in Figure 13.5. Rotational invariance about the $z$-axis in $S_R$ ensures that $\mathbf{R}(\underline{\underline{O}}_{R,z}^{-1}\mathbf{x}) = \underline{\underline{\Lambda}}^0\underline{\underline{O}}_{R,z}^{-1}\mathbf{x}$, where

$$
\underline{\underline{O}}_{R,z} = \begin{pmatrix} \cos\vartheta & -\sin\vartheta & 0 \\ \sin\vartheta & \cos\vartheta & 0 \\ 0 & 0 & 1 \end{pmatrix} \tag{13.66}
$$

describes a state with equal energy, that is, an alternative ground state. In other words, a deformation described by

$$
\underline{\underline{\bar{\Lambda}}}^0 = \underline{\underline{\Lambda}}^0\underline{\underline{O}}_{R,z}^{-1} \tag{13.67}
$$

has the same energy as one described by $\underline{\underline{\Lambda}}^0$. Any deformation $\underline{\underline{\Lambda}}$ relative to the original reference system can be expressed in terms of a deformation $\underline{\underline{\Lambda}}'$ relative to the reference system obtained from the original reference system via $\underline{\underline{\Lambda}}^0$ through the relation

$$
\underline{\underline{\Lambda}} = \underline{\underline{\Lambda}}'\underline{\underline{\Lambda}}^0. \tag{13.68}
$$

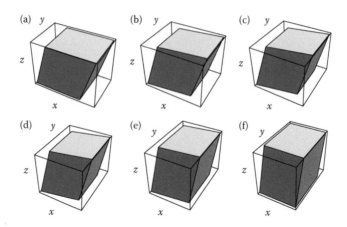

**FIGURE 13.5**  Effect of deformations $\underline{\tilde{\Lambda}}'$ as given in Equation 13.72 for a series of values of $\vartheta$ between (a) 0 and (f) $\pi/2$ with $\vartheta$ increased in steps of $\pi/10$. In the process a parallelepiped-shaped sample with initial shear in the $xz$-plane is transformed into a parallelepiped with shear in the $yz$-plane that appears as if the original parallelepiped had been rotated by $\pi/2$ about the $z$-axis.

Thus, choosing $\underline{\underline{\Lambda}} = \underline{\tilde{\Lambda}}^0$, we find that the deformation tensor

$$\underline{\underline{\Lambda}}' = \underline{\underline{\Lambda}}^0 \underline{\underline{O}}_{R,z}^{-1} (\underline{\underline{\Lambda}}^0)^{-1} \tag{13.69a}$$

$$= \begin{pmatrix} \cos\vartheta & r_\perp^{1/2}\sin\vartheta & s[1-\cos\vartheta] \\ -r_\perp^{1/2}\sin\vartheta & \cos\vartheta & sr_\perp^{-1/2}\sin\vartheta \\ 0 & 0 & 1 \end{pmatrix}, \tag{13.69b}$$

describes a zero-energy deformation of the reference state represented by $\underline{\underline{\Lambda}}^0$. The parameters $s = \Lambda_{xz}^0/\Lambda_{zz}^0$ and $r_\perp = (\Lambda_{xx}^0/\Lambda_{yy}^0)^2$ describe the shear of the sample and its anisotropy in the $xy$-plane, respectively. Further rotations of $\underline{\underline{\Lambda}}'$ in $S_T$, of course, do not change the energy, and the most general soft deformation tensor is

$$\underline{\tilde{\Lambda}}' = \underline{\underline{O}}_T \underline{\underline{\Lambda}}' \tag{13.70}$$

where $\underline{\underline{O}}_T$ is an arbitrary target-space rotation matrix. Of particular interest to our discussion of response to imposed strain, which we present shortly, will be soft strains with a vanishing $xy$ component. To construct such a soft deformation tensor, we rotate through an angle $\omega$ about the $z$ axis,

$$\underline{\tilde{\Lambda}}' = \underline{\underline{O}}_{T,z}(\omega)\underline{\underline{\Lambda}}'. \tag{13.71}$$

Then, the condition $\tilde{\Lambda}'_{xy} = 0$ is satisfied when the target-space and reference-space rotation angles are related through $\tan \omega = r_\perp^{1/2} \tan \vartheta$ in which case

$$\tilde{\underline{\underline{\Lambda}}}' = g(\vartheta) \begin{pmatrix} 1 & 0 & -s[1 - \cos \vartheta] \\ \frac{1}{2} r_\perp^{-1/2}(r_\perp - 1) \sin 2\vartheta & \cos^2 \vartheta + r_\perp \sin^2 \vartheta & \frac{1}{2} r_\perp^{-1/2} s[-(r_\perp - 1) \sin 2\vartheta + 2r_\perp \sin \vartheta] \\ 0 & 0 & 1 \end{pmatrix},$$
(13.72)

where $g(\vartheta) = [1 + (r_\perp - 1) \sin^2 \vartheta]^{-1/2}$. When $\vartheta = \pi/2$, then

$$\tilde{\underline{\underline{\Lambda}}}'_0 \equiv \tilde{\underline{\underline{\Lambda}}}'(\vartheta = \pi/2) = \begin{pmatrix} r_\perp^{-1/2} & 0 & -sr_\perp^{-1/2} \\ 0 & r_\perp^{1/2} & s \\ 0 & 0 & 1 \end{pmatrix},$$
(13.73)

which corresponds to an overall deformation

$$\underline{\underline{\Lambda}} = \tilde{\underline{\underline{\Lambda}}}'_0 \underline{\underline{\Lambda}}^0 = \begin{pmatrix} \Lambda^0_{yy} & 0 & 0 \\ 0 & \Lambda^0_{xx} & \Lambda^0_{xz} \\ 0 & 0 & \Lambda^0_{zz} \end{pmatrix},$$
(13.74)

that is, to a shear deformation of the original SmA in the $yz$- rather than the $xz$-plane. Figure 13.5 shows the effect of deformations $\tilde{\underline{\underline{\Lambda}}}'$ for a series of values of $\vartheta$ between 0 and $\pi/2$. In the course of this deformation, the $x$- and the $y$-axis interchange their roles.

The soft elasticity of ideal SmC elastomers can also be studied in the framework of the neoclassical model, as has been done by Adams and Warner (AW) [42]. At this point it is worthwhile to compare our findings to those of AW. Our soft deformation tensor (Equation 13.71) is, up to differences in notation, identical to the soft deformation tensor found by AW (see the last equation in the appendix of Ref. 42). The same holds true for the change in the director associated with these soft deformations (which we will not discuss further to save space). However, whereas the AW derivation emphasizes geometric constraints, ours emphasis is that softness arises from invariances with respect to reference-space rotations and the independent nature of reference- and target-space rotations. It should be noted that unlike the AW derivation, ours does not impose incompressibility; rather the incompressibility condition of the soft deformation arises naturally from Equation 13.69.

To discuss the implication of the rotational invariance on the Lagrange elastic energy, we now switch from deformations to strains. In terms of the soft deformation, the general form of the soft strain tensor is given by

$$\underline{\underline{u}}' = \frac{1}{2} \left[ (\underline{\underline{\Lambda}}')^T \underline{\underline{\Lambda}}' - \underline{\underline{\delta}} \right] = \frac{1}{2} \left[ (\tilde{\underline{\underline{\Lambda}}}')^T \tilde{\underline{\underline{\Lambda}}}' - \underline{\underline{\delta}} \right]$$
(13.75)

independent of target-space rotations. Note that Equations 13.75 and 13.52 are equivalent, as can be checked straightforwardly by inserting Equation 13.69a or

Equation 13.70 into Equation 13.75. Inserting Equation 13.69b, or Equation 13.72 into 13.75 yields explicit expressions for the components of the soft strain $\underline{\underline{u'}}$ in terms of the rotation angle $\vartheta$. Here, however, this tensor is considerably more complicated than its counterpart (Equation 13.54) for a soft nematic elastomer, and thus, we refrain from stating it. We restrict ourselves to noting that four of its components are nonzero for infinitesimal $\vartheta$, namely,

$$u'_{xy}(\vartheta) = u'_{yx}(\vartheta) = \frac{r_\perp - 1}{2\sqrt{r_\perp}}\vartheta, \tag{13.76a}$$

$$u'_{yz}(\vartheta) = u'_{zy}(\vartheta) = \frac{s}{2\sqrt{r_\perp}}\vartheta. \tag{13.76b}$$

To ensure that these infinitesimal strains do not cost elastic energy the following combination of elastic constants in the elastic energy density (Equation 13.18) has to vanish:

$$C_{xyxy}(r_\perp - 1)^2 + 2C_{xyyz}s(r_\perp - 1) + C_{yzyz}s^2 = 0. \tag{13.77}$$

This equation is justified for

$$C_{xyxy} = \bar{C}\cos^2\theta, \tag{13.78a}$$

$$C_{xyyz} = \bar{C}\cos\theta\sin\theta, \tag{13.78b}$$

$$C_{yzyz} = \bar{C}\sin^2\theta, \tag{13.78c}$$

with the angle $\theta$ given by

$$\theta = \tan^{-1}\left(\frac{1 - r_\perp}{s}\right). \tag{13.79}$$

Thus, the elastic energy density of a soft SmC elastomer is

$$f^{soft}_{C_{2h}} = \frac{1}{2}\bar{C}\left[\cos\theta\, u'_{xy} + \sin\theta\, u'_{yz}\right]^2 + \text{remaining terms}, \tag{13.80}$$

where the remaining terms are identical to the last 10 terms in Equation 13.18.

Evidently, soft elasticity in a SmC elastomer is more complex than in a nematic elastomer. To derive the physical implications of this softness, we can proceed two ways. In the first derivation, we recast the first three terms in the elastic energy density (Equation 13.80) by exploiting the fact that $\cos\theta u'_{xy} + \sin\theta u'_{yz}$ can be viewed as the dot product of the "vectors" $\vec{v} = (u'_{xy}, u'_{yz})$ and $\vec{e}_1 = (\cos\theta, \sin\theta)$. Thus, the first term in Equation 13.80,

$$\frac{1}{2}\bar{C}\left[\cos\theta u'_{xy} + \sin\theta u'_{yz}\right]^2 = \frac{1}{2}\bar{C}(\vec{e}_1 \cdot \vec{v})^2, \tag{13.81}$$

is independent of $\vec{e}_2 \cdot \vec{v}$, where $\vec{e}_2 = (-\sin\theta, \cos\theta)$ is the vector perpendicular to $\vec{e}_1$. Thus, distortions of the form $-\sin\theta u'_{xy} + \cos\theta u_{yz} = \vec{e}_2 \cdot \vec{v}$, that is, distortions with $\vec{v}$

along $\vec{e}_2$, cost no elastic energy. A manifestation of this softness is that certain stresses cause no restring force and thus lead to large deformations. To find these stresses we take the derivative of the elastic energy density (Equation 13.80) with respect to $u'_{xy}$ and $u'_{yz}$ which tells us that

$$-\sin\theta\,\sigma^{II}_{xy} + \cos\theta\,\sigma^{II}_{zy} = \vec{e}_2 \cdot \vec{w} = 0, \qquad (13.82)$$

where are $\sigma^{II}_{ij}$ components of the second Piola–Kirchhoff stress tensor defined in Equation 13.22, and where we have introduced the "vector" $\vec{w} = (\sigma^{II}_{xy}, \sigma^{II}_{zy})$ in the $xz$-plane. Equation 13.82 means that there are no restoring forces for external forces in the $xz$-plane along $\pm\tilde{e}_2 \equiv \pm(-\sin\theta, 0, \cos\theta)$ applied to opposing surfaces with normal along $\pm\tilde{e}_y$, and there no restoring forces for external forces along $\pm\tilde{e}_y$ applied to surfaces with normal along $\pm\tilde{e}_2$. These kinds of stresses are depicted in Figure 13.6.

A second derivation of the softness of the SmC phase involves a transformation to a rotated coordinate system, as described in Equation 13.4, in which the first term in Equation 13.80 is diagonal. The components $x''_{i'}$ of a reference-space vector $\mathbf{x}'$ expressed with respect to a rotated basis $\{\tilde{\mathbf{e}}'_{i'} | i' = x', y', z'\}$, where $\tilde{\mathbf{e}}'_{i'} = O_{R,y;i'j}\tilde{\mathbf{e}}_j$ with $\underset{=R,y}{O}$ describing a counterclockwise rotation of the reference-space basis about the $y$-axis through the angle $\theta$ (given by Equation 13.53 with $\vartheta$ replaced by $\theta$), are simply $x''_i = O_{R,y;ij}x'_j$. The components of the strain matrix expressed in the rotated basis are $u''_{i'j'} = \tilde{\mathbf{e}}'_{i'} \cdot \underline{u}' \cdot \tilde{\mathbf{e}}'_{j'} = O_{R,y;i'i}u_{ij}O^T_{R,y;jj'}$, from which we obtain through Equation 13.10 that $u''_{x',y'} = \cos\theta\,u'_{xy} + \sin\theta\,u'_{yz}$ and $u''_{y',z'} = -\sin\theta\,u'_{xy} + \cos\theta\,u'_{yz}$. Thus, taking $\theta$ to be the angle appearing in Equation 13.18 and dropping the double-prime from the strains, we obtain

$$f^{\text{soft}}_{C_{2h}} = \frac{1}{2}C_{x'y'x'y'}(u_{x'y'})^2 + \frac{1}{2}C_{z'z'z'z'}(u_{z'z'})^2 + \frac{1}{2}C_{x'z'x'z'}(u_{x'z'})^2 + C_{z'z'x'x'}u_{z'z'}u_{x'x'}$$

$$+ C_{z'z'y'y'}u_{z'z'}u_{y'y'} + \frac{1}{2}C_{x'x'x'x'}(u_{x'x'})^2 + \frac{1}{2}C_{y'y'y'y'}(u_{y'y'})^2 + C_{x'x'y'y'}u_{x'x'}u_{y'y'}$$

$$+ C_{x'x'x'z'}u_{x'x'}u_{x'z'} + C_{y'y'x'z'}u_{y'y'}u_{x'z'} + C_{z'z'x'z'}u_{z'z'}u_{x'z'}. \qquad (13.83)$$

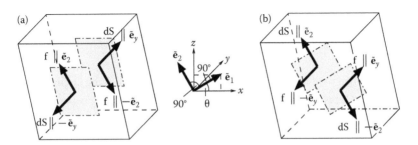

**FIGURE 13.6** Soft distortions of SmC elastomers. There are no restoring forces (a) for external forces in the $xz$-plane along $\pm\tilde{e}_2$ applied to opposing surfaces with normal along $\pm\tilde{e}_y$, and (b) for external forces along $\pm\tilde{e}_y$ applied to surfaces with normal along $\pm\tilde{e}_2$.

for the elastic energy density of SmC elastomers in the rotated coordinates. Note that, in this coordinate system, the elastic constants $C_{x'y'y'z'}$ and $C_{y'z'y'z'}$ are zero, and the elastic energy does not depend at all on $u_{y'z'}$. $C_{x'y'x'y'}$ is identical to $\bar{C}$ and $C_{y'y'y'y'} = C_{yyyy}$. The remaining new elastic constants are nonvanishing conglomerates of the elastic constants defined by Equation 13.18 and sines and cosines of $\theta$. The vanishing of $C_{x'y'y'z'}$ and $C_{y'z'y'z'}$ means that SmC elastomers are soft with respect to shears in the $y'z'$-plane. If one can cut a rectangular sample with faces perpendicular to the $\tilde{\mathbf{e}}'_{i'}$, then there are no restoring forces for external forces along $\pm\tilde{\mathbf{e}}'_{z'}$ applied to opposing surfaces with normal along $\pm\tilde{\mathbf{e}}'_{y'}$, and for external forces along $\pm\tilde{\mathbf{e}}'_{y'}$ applied to surfaces with normal along $\pm\tilde{\mathbf{e}}'_{z'}$.

As we just have seen, the softness of SmC elastomers with respect to shear is rather intricate: either one has to apply a specific linear combination of shear or one has to use a specific coordinate system. Both depend on the angle $\theta$, which in turn depends on temperature. Thus, it will be difficult if not impossible to realize soft shears in SmC elastomers, experimentally. This brings about an important question: are ideal SmC elastomers, like nematic elastomers, soft with respect to plain extensional or compressional strains? If so, there is hope that semisoft elasticity in SmC elastomers can be realized in experiments. To address this question, we assume, for the sake of the argument, positive anisotropy in the $xy$-plane, $r_\perp > 1$. We consider extensional strains along the $y$-axis, $u'_{yy} > 0$, as a specific example, and we assume that the remaining strain components have freedom to relax to their equilibrium values in the presence of $u'_{yy} > 0$. A straightforward but slightly tedious calculation reveals that $u'_{yy}$ is converted into a zero-energy rotation through an angle $\vartheta$, if the remaining components relax to $u'_{xx} = -r_\perp^{-1}u'_{yy}$ and so on. When $u'_{yy}$ is increased from zero to $u'_{yy} = u^C_{yy} \equiv (r_\perp - 1)/2$, $\vartheta$ grows from zero to $\pi/2$ and the state of the elastomer, originally described by the equilibrium strain tensor $\underline{u}^0$ associated with the equilibrium deformation tensor $\underline{\underline{\Lambda}}^0$, Equation 13.65, is changed without costing elastic energy to the strain tensor associated with the deformation tensor of Equation 13.73.

In this process, the shape of the sample changes as depicted in Figure 13.5. As already discussed, the configuration at $\vartheta = \pi/2$ describes a sample in which the SmA phase was sheared in the $yz$- rather than the $xz$ plane. Thus, further increase in $u'_{yy}$ beyond $u^C_{yy}$ is equivalent to increasing $u'_{xx}$ beyond zero in the original sample sheared in the $xz$-plane. As a consequence, the engineering stress, is given at leading order by

$$\sigma_{yy}^{\text{eng}} = \begin{cases} 0 & \text{if } \tilde{\Lambda}'_{yy} < \sqrt{r_\perp}, \\ Y_x\tilde{\Lambda}'_{0yy}\left(\tilde{\Lambda}'_{yy} - \tilde{\Lambda}'_{0yy}\right) & \text{if } \tilde{\Lambda}'_{yy} > \sqrt{r_\perp}. \end{cases} \quad (13.84)$$

Therefore, when plotted as a function of the deformation $\tilde{\Lambda}'_{yy}$, the engineering stress $\sigma_{yy}^{\text{eng}}$ for a SmC elastomer has the same form as the corresponding curve for a nematic elastomer, *cf.* Figure 13.4.

## 13.4 SEMISOFT ELASTICITY

As discussed in detail in Section 13.3, Goldstone arguments for soft response predict that certain shear moduli vanish in ideal nematic and SmC elastomers as a

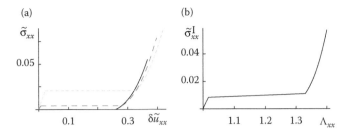

**FIGURE 13.7** (a) Soft (full line) and semisoft (dashed and dotted lines) stress–strain curves at $\tilde{r} = 0.08$ with $\tilde{h} = 0, 0.8\tilde{h}_c, 4\tilde{h}_c$, respectively. (b) Semisoft curve of $\sigma_{xx}^{\sim l}$ as a function of $\Lambda_{xx}$ at $\tilde{r} = 0.08$ and $\tilde{h} = 2\tilde{h}_c$, where we have set $\upsilon = w$.

consequence of *spontaneous* symmetry breaking at the $I{-}N$ and $A{-}C$ transitions, respectively. Monodomain samples of nematic or SmC elastomers, however, cannot be produced without applying some mechanical aligning field $h$ which imprints into the material preferred anisotropy directions. For example, nematic elastomers prepared in this way have a nonzero $C_5$ and are thus simply uniaxial solids with a linear stress–strain relation at small strain. For fields $h$ that are not too large, however, they are predicted to exhibit semisoft elasticity [1,10] in which the nonlinear stress–strain curve exhibits a flat plateau at finite stress as shown in Figure 13.7. Measured stress–strain curves in appropriately prepared nematic samples unambiguously exhibit the characteristic semisoft plateau [11,12]. In this section we discuss the semisoft elasticity of nematic elastomers in terms of a simple model which explicitly accounts for the presence of a mechanical aligning field. Similar, yet because the lower symmetry is more involved, arguments can be applied to study of semisoftness in SmC elastomers [43]. To our knowledge, experiments seeking to detect semisoftness in SmC elastomers have not been reported to date. We are optimistic, however, that this predicted property can be measured if the corresponding experiments are done. For simplicity, we will focus in the following exclusively on semisoftness in nematics.

The Goldstone argument for soft response predicts $C_5 = 0$ in the nematic phase, making reasonable conjectures that $C_5$ should remain small at finite $h$ when semisoft response is expected and that semisoft response might not exist at all in the supercritical regime [44] beyond the mechanical critical point (with $h = h_c$) terminating the paranematic($PN$)–nematic($N$) coexistence line [45]. There is now strong evidence [46,47] that samples prepared with the Küpfer–Finkelmann technique are supercritical. In addition, $C_5$ measured in linearized rheological experiments is not particularly small [47]. These results have caused some to doubt the interpretation of the measured stress–strain plateau in terms of semisoft response [48].

In this section we clarify the nature of semisoft response by reviewing arguments developed in Ref. [41]. First, we discuss simple symmetry arguments that allow us to understand main features of the semisoft stress–strain curves, Figure 13.7, without much effort. These arguments culminate in a so-called Ward identity that provides a rigorous basis for semisoft elasticity beyond mean-field theory. We then consider

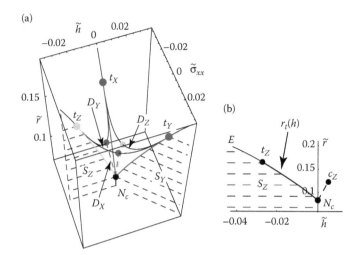

**FIGURE 13.8** Phase diagrams (a) in the $\tilde{h} - \tilde{\sigma}_{xx} - \tilde{r}$ space showing the $S_Y$ and $S_Z$ ($S_X$ hidden)$CC$ and the $D_X, D_Y$, and $D_Z$ $DC$ surfaces along with the tricritical points $t_X, t_Y, t_Z$ and (b) in the $\tilde{r} - \tilde{h}$ plane ($\tilde{\sigma}_{xx} = 0$) showing the first-order uniaxial $PN$-$N$ coexistence line $N_c c_Z$, the mechanical critical point $c_Z$, and the $S_Z$ surface terminated by the line $\tilde{r}_t(\tilde{h})$ with respective first- and second-order segments $N_c t_Z$ and $t_Z E$ meeting at the tricritical point $t_Z$.

the simplest or minimal model, which is formally equivalent to the Maier–Saupe–de Gennes model [2] for nematic liquid crystals, that exhibits semisoft response. We review the global mean-field phase diagram, Figure 13.8, for this model. It turns out that semisoft response is associated with biaxial phases that spontaneously break rotational symmetry, and that semisoft response exists well into the supercritical regime. Figure 13.7 shows calculated stress–strain curves for $h = 0.8h_c$ and $h = 4h_c$ that clearly exhibit semisoft behavior both for $h < h_c$ and in the supercritical regime with $h > h_c$. Though the minimal model provides a robust description of semisoft response, we will briefly discuss changes in this response that extensions of the minimal model can bring about.

### 13.4.1 A WARD IDENTITY

A generic model elastic energy density for a nematic elastomer with an internal aligning field is of the form

$$f(\underline{u}) = f_{\text{iso}}(\underline{u}) + f_{\text{ani}}(\underline{u}). \tag{13.85}$$

Here, $f_{\text{iso}}$ is the elastic energy density of an isotropic elastomer, which is to harmonic order given by Equation 13.15. However, our arguments to follow do not rely on the harmonic approximation and we can think of $f_{\text{iso}}$ as including isotropic terms to arbitrary order. The anisotropic term arising from the imprinting process has the form of an imposed stress: $f_{\text{ani}}(\underline{u}) = -\text{Tr}\underline{\underline{h}}\,\underline{u}$. Without loss of generality and in accord with our conventions used throughout this review, we choose the aligning field to favor stretching along the $z$-axis, $h_{ij} = \tilde{h}\tilde{e}_{zi}\tilde{e}_{zj} \cdot f_{iso}(\underline{u})$ is invariant under rotations of $\underline{u}$, that is, under $\underline{u} \to \underline{\underline{O}}_R \underline{u} \underline{\underline{O}}_R^{-1}$ where $\underline{\underline{O}}_R$ is any reference-space rotation matrix. Thus

$$f(\underline{\underline{O}}_R \underline{u} \underline{\underline{O}}_R^{-1}) = f_{iso}(\underline{u}) - \mathrm{Tr}\underline{h} \, \underline{\underline{O}}_R \underline{u} \underline{\underline{O}}_R^{-1}, \qquad (13.86)$$

for any $\underline{\underline{O}}_R$ and, in particular, for one describing an infinitesimal rotation by $\vartheta$ about the $y$-axis with components $O_{R,y:ij} = \delta_{ij} + \epsilon_{yij}\vartheta$ (the linearized version of Equation 13.53), where $\epsilon_{ijk}$ is the Levi–Civita antisymmetric tensor. Equating the term linear in $\vartheta$ in $f(\underline{\underline{O}}_R \underline{u} \underline{\underline{O}}_R^{-1})$ to that of $\mathrm{Tr}\underline{h} \, \underline{\underline{O}}_R \underline{u} \underline{\underline{O}}_R^{-1}$ yields the Ward identity

$$\sigma_{xz}^{II}(u_{zz} - u_{xx}) = (\sigma_{xx}^{II} - \sigma_{zz}^{II} - h)u_{xz}. \qquad (13.87)$$

This identity applies for any $f$ of the form of Equation 13.85 so long as $f_{ani}$ is linear in $\underline{u}$. Despite the arguments leading to it are simple, this Ward identity allows us to understand fundamental aspects of semisoft stress–strain curves. To see this, we note that in the experimental geometry behind Figure 13.7, stress is applied along the $x$ direction, and $\sigma_{xz}^{II} = \sigma_{zz}^{II} = 0$ but $\sigma_{xx}^{II} > 0$. Thus, either $u_{xz} = 0$ or $\sigma_{xx}^{II} = h$ for any nonzero $u_{xz}$ is in full agreement with the stress–strain curves shown in Figure 13.7a.

In physical experiments, the engineering stress, $\sigma_{ij}^{I}$, or the Cauchy stress, $\sigma_{ij}^{C}$ (as in [1,12]), rather than $\sigma_{ij}^{II}$ is externally controlled. The $\sigma_{xx}^{I} - \Lambda_{xx}$ stress–strain curve is easily obtained from the $\sigma_{xx} - u_{xx}$ curve using $\sigma_{xx}^{I} = \Lambda_{xx}\sigma_{xx}$ and $\Lambda_{xx} = \sqrt{1 + 2u_{xx}}$. These two curves are similar, but the flat plateau in the $\sigma_{xx}^{I} - \Lambda_{xx}$ curve rises linearly with $\Lambda_{xx}$ as shown in Figure 13.7b, and there is a unique value of $\Lambda_{xx}$ for each value of $\sigma_{xx}^{I}$.

### 13.4.2 SEMISOFTNESS BEYOND THE MECHANICAL CRITICAL POINT: PHASE DIAGRAM OF NEMATIC ELASTOMERS

A minimal model for semisoftness in nematic elastomers can be constructed as follows. First, one imposes the constraint $\mathrm{Tr}\underline{u} = 0$, enforcing incompressibility at small but not large $\underline{u}$, rather than the full nonlinear incompressibility constraint $\det\underline{\underline{\Lambda}} = [\det(\underline{\delta} + 2\underline{u})]^{1/2} = 1$ that more correctly describes nematic elastomers, whose bulk moduli are generally orders of magnitude larger than their shear moduli. The theory thus depends only on the traceless part $\hat{\underline{u}}$ of the strain tensor, see Equation 13.16. For the anisotropy energy density, in particular, this limit means that $f_{ani} \to -h\hat{u}_{zz}$. Second, one chooses the Landau–de Gennes form [2] for $f_{iso}$ [cf. Equation 13.38]:

$$f_{iso}(\hat{\underline{u}}) = \frac{1}{2}r\mathrm{Tr}\hat{\underline{u}}^2 - w\mathrm{Tr}\hat{\underline{u}}^3 + \upsilon(\mathrm{Tr}\hat{\underline{u}}^2)^2, \qquad (13.88)$$

where $w > 0$ and where $r = a(T - T^*)$ with $T$ the temperature and $T^*$ the temperature at the metastability limit of the $PN$ phase. In the isotropic phase with $\hat{\underline{u}} = \underline{0}$, $r = 2\mu$, where $\mu$ is the $T$-dependent shear modulus. Third, one models the presence of an external second Piola–Kirchhoff stress by adding to the free energy density $f$ an external-stress part $f_{\text{ext}} = -\sigma_{xx}^{II}\hat{u}_{xx}$. In the following, we will suppress the super-script $II$ of $\sigma_{xx}^{II}$ for notational simplicity. Overall, the minimal model is defined by the Gibbs free energy density

$$g(\hat{\underline{u}}, h, \sigma_{xx}, r) = f_{\text{iso}}(\hat{\underline{u}}, r) + f_{\text{ani}}(\hat{\underline{u}}, h) + f_{\text{ext}}(\hat{\underline{u}}, \sigma_{xx}). \tag{13.89}$$

Note that this model is formally equivalent to that for a nematic liquid crystal in crossed electric and magnetic fields, $\mathbf{E} = E\mathbf{e}_z$ and $\mathbf{H} = H\mathbf{e}_x$, in which $\hat{u}_{ij} \leftrightarrow Q_{ij}$, $h \leftrightarrow \frac{1}{2}\Delta\epsilon E^2$, and $\sigma_{xx} \leftrightarrow (1/2)\chi_i H^2$, where $\Delta\epsilon$ and $\chi_i$ are, respectively, the anisotropic parts of the dielectric tensor and the magnetic susceptibility. In the following, we will often express quantities in reduced form: $\tilde{u}_{ij} = (\upsilon/w)\hat{u}_{ij}$, $\tilde{r} = r\upsilon/w^2$, $\tilde{h} = h\upsilon^2/w^3$, $\tilde{\sigma}_{ij} = \sigma_{ij}\upsilon^2/w^3$, $\tilde{C}_5 = C_5\upsilon/w^2$, and similarly for other elastic moduli.

We begin our discussion of the global phase diagram [49] with the $\sigma_{xx} = 0$ plane, which we will refer to as the Z-plane because the anisotropy field $h$ favors uniaxial order along the $z$-axis. The $h \geq 0$ half of this plane exhibits the familiar nematic clearing point $N_c$ at $(\tilde{r}_N, \tilde{h}_N) = (1/12, 0)$ and the PN–N coexistence line terminating at the mechanical critical point $(\tilde{r}_c, \tilde{h}_c) = (1/8, 1/192)$. Throughout the $h > 0$ half-plane, there is prolate uniaxial order with $\hat{u}_{ij} = S(n_i n_j - (1/3)\delta_{ij})$ with $S > 0$ and the Frank director $\mathbf{n}$ along $\mathbf{e}_z$. In the $N$ phase at $h = 0$ and $r < r_N$, $\mathbf{n}$ can point anywhere on the unit sphere. Negative $h$ induces oblate rather than prolate uniaxial order along $\mathbf{e}_z$ and $S = -S' < 0$ at high temperature. When $h < 0$ is turned on for $r < r_N$ at which nematic order exists at $h = 0$, $\mathbf{n}$ aligns in the two-dimensional $xy$-plane. This creates a biaxial environment and biaxial rather than uniaxial order. Since $\mathbf{n}$ can point anywhere in the $xy$-plane, the biaxial state at $h < 0$ exhibits a spontaneously broken symmetry. There must be a transition along a line $r = r_t(h)$ between the high-temperature oblate uniaxial state and the low-temperature biaxial state, which exists throughout the $S_Z$ surface shown in Figure 13.8. This transition is first order at small $|h|$ because the PN–N transition is first order at $h = 0$ and second order at larger $|h|$, and there is a tricritical point [50,51] $t_Z$ at $(\tilde{r}_t, \tilde{h}_t) = (21/128, 27/1024)$ separating the two behaviors as shown in Figure 13.8b. A continuum of biaxial states coexists on $S_Z$. We will refer to such surfaces as $CC$ surfaces and ones on which a discrete set of states coexist as $DC$ surfaces.

The full phase diagram reflects the symmetries of $g$. The $x$- and $z$-directions are equivalent in $f_{\text{iso}}$, and the $\sigma_{xx} = 0$ and the $h = 0$ planes are symmetry equivalent. These planes are also equivalent (apart from stretching) to the vertical plane with $\sigma_{xx} = h$, but with positive and negative directions interchanged. To see this, we note that $\hat{u}_{zz} + \hat{u}_{xx} = -\hat{u}_{yy}$ and $h\hat{u}_{zz} + \sigma_{xx}\hat{u}_{xx} = -h\hat{u}_{yy}$ when $h = \sigma_{xx}$. Thus, the phase structure of the Z-plane is replicated in the X-plane ($h = 0$) and the Y-plane ($\sigma_{xx} = h$) with respective preferred uniaxial order along $\mathbf{e}_x$ and $\mathbf{e}_y$, critical points $c_X$ and $c_Y$, biaxial coexistence surfaces $S_X$ and $S_Y$, and tricritical points $t_X$ and $t_Y$.

To fill in the 3D phase diagram, we consider perturbations away from the X-, Y-, and Z-planes. Turning on $\sigma_{xx}$ converts the PN–N coexistence line into a DC surface $D_Z$, on which two discrete in general biaxial phases coexist. Turning on $\sigma_{xx}$ near the $S_Z$ surface favors alignment of the biaxial order along $\mathbf{e}_x$ when $\sigma_{xx} > 0$ and along $\mathbf{e}_y$ when $\sigma_{xx} < 0$. Thus, $\sigma_{xx}$ is an ordering field for biaxial order whereas, a linear combination of $h$ and $\sigma_{xx}$ acts as a nonordering field. The topology of the phase diagram near $t_Z$ is that of the Blume–Emery–Griffiths model [52] with $DC$ surfaces $D_X$ and $D_Y$ emerging from the first-order line $N_c t_Z$ terminating $S_Z$. The $D_X$ and $D_Y$ surfaces terminate, respectively, on the critical lines $N_c t_X$ and $N_c t_Y$ in the $X$- and $Y$-planes. The surfaces $D_X$, $D_Y$, and $D_Z$ form a cone with vertex at $N_c$.

Before considering the $\sigma_{xx}$–$u_{xx}$ stress–strain curve, it is useful to look more closely at elastic response in the vicinity of the $Z$-plane and the nature of order in the $Y$-plane. Throughout the $h > 0$ $Z$-plane, the equilibrium state is prolate uniaxial with order parameter $S = S_0$, and thus strains $\hat{u}^0_{zz} = (2/3)S_0 = -2\hat{u}^0_{xx} = -2\hat{u}^0_{yy}$. We are primarily interested in shears in the $xz$-plane and the response to an imposed $\sigma_{xx}$ with no additional stress along $z$. In this case $\delta u_{zz} \equiv \hat{u}_{zz} - \hat{u}^0_{zz}$ will relax to an imposed $\delta u_{xx}$, and the free energy of harmonic deviations from equilibrium can be written as $\delta f = (1/2)C_3(\delta u_{xx})^2 + (1/2)C_5(\delta u_{xz})^2$. The modulus $C_3$ gives the slope of $\sigma_{xx}$ versus $\delta u_{xx}$, and $C_5$ is measured in linearized rheology experiments [47,53,54]. $C_3$ and $C_5$ are easily calculated as a function of $r$ and $h$. In reduced units, the ordered pair $(\tilde{C}_3, \tilde{C}_5)$ takes on the value $(1/8, 1/6)$ just above $N_c(\tilde{r} = \tilde{r}^+_N)$, $(3/8, 0)$ just below $N_c(\tilde{r} = \tilde{r}^-_N)$, $(0, 1/12)$ at the critical point, and $(57/112, 1/12)$ in the supercritical regime at $(\tilde{r}, \tilde{h}) = (\tilde{r}_c, 2\tilde{h}_c)$. To keep our arguments as simple as possible, we will measure here elastic distortions using $\delta u_{ij}$ rather than the strain $u'_{ij}$ relative to the reference space $S'_R$ defined by the equilibrium configuration at any given $T$, see Equation 13.51.

On the $h > 0$, $Y$-plane, there is oblate uniaxial order aligned along the $y$-direction at high $T$ and biaxial order at low $T$. A convenient representation of the tensor-order parameter is

$$\hat{\underline{u}} = \begin{pmatrix} \frac{1}{3}S' - \eta_1 & 0 & \eta_2 \\ 0 & -\frac{2}{3}S' & 0 \\ \eta_2 & 0 & \frac{1}{3}S' + \eta_1 \end{pmatrix}, \tag{13.90}$$

where $S' > 0$. The vector $\vec{\eta} \equiv (\eta_1, \eta_2) \equiv \eta(\cos 2\theta, \sin 2\theta)$ is the biaxial order parameter, which is nonzero on the $S_Y$ surface. We define the equilibrium values of $S'$ and $\eta$ in the biaxial phase to be $S'_0$ and $\eta_0$, respectively. Energy in this phase is independent of the rotation angle $\theta$. Away from the $Y$-plane, $f_{ani} + f_{ext} = -(1/3)(h + \sigma_{xx})S' + (\sigma_{xx} - h)\eta_1$. Thus, $\sigma_{xx} < h$ favors $\eta_1 > 0$ and $\sigma_{xx} > h$ favors $\eta_1 < 0$, implying that $\vec{\eta} = (\eta_0, 0)$ (or $\theta = 0$) at $\sigma_{xx} = h^-$ and $\vec{\eta} = (-\eta_0, 0)$ (or $\theta = (\pi/2))\vec{\eta} = (-\eta_0, 0)$ (or $\theta = \frac{\pi}{2}$) at $\sigma_{xx} = h^+$. These considerations imply that the modulus $C_5$ is zero at $\sigma_{xx} = h^{\pm}$ because $C_5 = \partial^2 f / \partial u^2_{xz}|_{u_{xz} \to 0} = (2\eta_0)^{-2}\partial^2 f / \partial \theta^2|_{\theta \to 0} = 0$.

We can now construct the $\sigma_{xx}$–$u_{xx}$ stress–strain curves. At $\sigma_{xx} = 0$, $\hat{u}_{xx} = \hat{u}^0_{xx}$; as $\sigma_{xx}$ is increased from zero, $\delta u_{xx}$ grows with initial slope $1/C_3$ until $\sigma_{xx} = h^-$ at which point, $\delta u_{xx} = (1/3)S'_0 - \hat{u}^0_{xx} - \eta_0$. At $\sigma_{xx} = h$, further increase of $\delta u_{xx}$ to a maximum of $(1/3)S'_0 - \hat{u}^0_{xx} + \eta_0$ produces a zero-energy rotation of $\vec{\eta}$ to yield $\delta u_{xx} = (1/3)S'_0 - \hat{u}^0_{xx} - \eta_0 \cos 2\theta$ and a nonzero shear $u_{xz} = \eta_2 = \eta_0 \sin 2\theta$. The growth of $\eta_2$ from zero is induced by the vanishing of $C_5$ at $\sigma_{xx} = h^{\pm}$ and its becoming negative for $|\eta_1| < \eta_0$. Thus, the characteristic semisoft plateau is a consequence of $C_5$'s vanishing at $\sigma_{xx} = h$ and not at $\sigma_{xx} = 0$. Measurements of $C_5$ at $\sigma_{xx} = 0$

do not provide information about what happens at $\sigma_{xx} = h$. For $\sigma_{xx} > h$, $\delta u_{xx}$ again grows with $\sigma_{xx}$. Figure 13.7 shows stress–strain curves for different values of $\tilde{h}$. Thus, semisoft response is associated with the $S_Y$ surface, which exists at $r$ and $h$ well into the supercritical regime.

Thus far, we have focused on the effects of an external second Piola–Kirchhoff stress $\sigma_{xx}$. Switching to the physical first Piola–Kirchhoff stress as described above, the $S_Y$ surface in the $r - h - \sigma_{xx}$ phase diagram would open into a finite volume biaxial region in the $r - h - \sigma_{xx}^I$ phase diagram with a particular value of $\bar{\eta}$ at each point in it. The phase diagram in the $h - \sigma_{xx}^I$ plane for $r_c < r < r_t$ is similar to that in Figure 13.9b.

We can now consider modifications of the minimal model. A simple modification is to replace the constraint $\text{Tr}\underline{u} = 0$ with the real volume constraint $\det \underline{\underline{\Lambda}} = 1$. This replacement does not change the validity of the Ward identity and the resulting phase diagram has the same structure as that for $\text{Tr}\underline{u} = 0$ but with different boundaries for the $CC$ and $DC$ surfaces. In particular, the mechanical critical point is at $(\tilde{r}_c, \tilde{h}_c) = (0.1279, 0.0052)$ and the tricritical point is at $(\tilde{r}_t, \tilde{h}_t) = (0.1900, 0.0247)$. Other modifications of the minimal model replace $f_{ani}$ with nonlinear functions of $u_{zz}$. Modifications of this kind can spread the $CC$ surface $S_Y$ into a finite volume or convert it to a $DC$ surface, as shown in Figure 13.9. If $f_{ani} = -hu_{zz}^2$, two states coexist, whereas with other forms that might arise in a hexagonal lattice, three or more discrete states might coexist. When $S_Y$ is a $DC$ surface, rather than exhibiting a homogeneous rotation of the biaxial order parameter (if boundary conditions are ignored) in response to an imposed $u_{xx}$, samples will break up into discrete domains of the allowed states. In other words, their response to external stress will be martensitic [55] rather than semisoft.

The neoclassical model, see Chapter 17 of this book, can also be discussed in our language. The free energy of this model is a function of $\underline{\underline{\Lambda}}$ and $\underline{\underline{Q}}$. It consists of an isotropic part, invariant under simultaneous rotations of $\underline{\underline{\Lambda}}$ and $\underline{\underline{Q}}$ in the target space

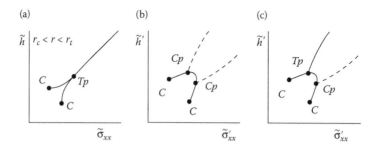

**FIGURE 13.9** Schematic phase diagrams in the $h$(or $h'$)-stress plane. The points $Tp$, $C$ and $Cp$ are, respectively, triple points, liquid–gas-like critical points, and critical endpoints. (a) Diagrams for the minimal model, where all transitions are first order; (b) and (c) Phase diagrams for more general $f_{ani}$ or $f_{ext}$ in which the first-order line from $S_Y$ is replaced by a surface terminated by two second-order (dashed) lines or one first-order line and one second-order line. $h'$ and $\sigma_{xx}'$ are, respectively, the generalized aligning field and generalized external stress resulting from the more general $f_{ani}$ or $f_{ext}$.

and under rotations of $\underline{\underline{\Lambda}}$ in the reference space, and a semisoft anisotropic energy [10], which is effectively nonlinear in the strain, that breaks rotational symmetry in the reference space. The phase diagram of this model is similar to that of the minimal model in the space of $r - h - \sigma_{xx}^l$. In it, semisoft behavior also persists above the mechanical critical point [56].

In our discussion, we have ignored boundary conditions and random stress, both of which can modify stress–strain curves. When Frank elastic energies are ignored, detailed calculations of domain structure induced by boundary conditions reproduce soft and semisoft response [57,58]. Small isotropic randomness appears not to affect soft response, but large randomness does [59]. The Lagrange approach discussed here should provide a basis for further study of randomness.

## 13.5 LANDAU THEORIES

In this section we will review mean-field theories for the $I-N$ and the $A-C$ phase transitions. As vantage points, we will use the Landau-type models for these transitions discussed in Section 13.2.3. The aim is, on one hand, to elucidate the $I-N$ and the $A-C$ transitions in elastomers and, on the other hand, to understand the elasticity of the emergent nematic and SmC phases from a perspective different from that in Section 13.3.

### 13.5.1 NEMATIC ELASTOMERS

#### 13.5.1.1 The I–N Transition

To study the $I-N$ transition, it is useful to recast the model elastic-energy density (Equation 13.37) as a sum of two terms

$$f_{I-N} = f_{I-N}^{(1)} + f_{I-N}^{(2)}, \tag{13.91}$$

with one of the terms, $f_{I-N}^{(2)}$, depending on $\tilde{\underline{\underline{Q}}}$ only:

$$f_{I-N}^{(1)} = \frac{1}{2} B \left[ \text{Tr}\underline{\underline{u}} - \alpha \text{Tr}\tilde{\underline{\underline{Q}}}^2 \right]^2 + \mu \text{Tr} \left[ \underline{\underline{u}} - \beta \tilde{\underline{\underline{Q}}} \right]^2, \tag{13.92a}$$

$$f_{I-N}^{(2)} = \frac{1}{2} r^R \text{Tr}\tilde{\underline{\underline{Q}}}^2 - w \text{Tr}\tilde{\underline{\underline{Q}}}^3 + v^R \left( \text{Tr}\tilde{\underline{\underline{Q}}}^2 \right)^2. \tag{13.92b}$$

In Equation 13.92a we have introduced the dimensionless parameters $\alpha = s/B$ and $\beta = t/\mu$, which should not be confused with the angles $\alpha$ and $\beta$ defined in Figure 13.2. $r^R$ and $v^R$ are renormalized versions of the corresponding original parameters, $r^R = r - 2t^2/\mu$ and $v^R = v - s^2/(2B)$. Equation 13.92a makes it immediately clear that the equilibrium values of $\text{Tr}\underline{\underline{u}}$ and $\hat{\underline{\underline{u}}}$ are given by

$$\text{Tr}\underline{\underline{u}}^0 = \alpha \text{Tr}(\tilde{\underline{\underline{Q}}}^0)^2, \tag{13.93a}$$

$$\hat{u}_{ij}^0 = \beta \tilde{Q}_{ij}^0. \tag{13.93b}$$

The equilibrium values $\tilde{\underline{Q}}^0$ the order parameter tensor is determined by the minima of $f_{I-N}^{(2)}$. In the nematic phase,

$$\tilde{Q}_{ij}^0 = S\left(\tilde{n}_i^0 \tilde{n}_j^0 - \frac{1}{3}\delta_{ij}\right), \tag{13.94}$$

with the scalar order parameter $S$ satisfying the equation of state

$$r^R S - w S^2 + \frac{8}{3}\upsilon^R S^3 = 0. \tag{13.95}$$

In accord with the conventions used throughout our review, we choose the $z$-axis in $S_R$ to lie along the uniaxial direction, $\tilde{\mathbf{e}}_z = \tilde{\mathbf{n}}^0$. With this setting we find by exploiting the definition of $\hat{\underline{u}}$ and Equation 13.93 that the equilibrium strain tensor in the nematic phase is diagonal to diagonal-elements

$$u_{xx}^0 = u_{yy}^0 = -\frac{1}{3}\beta S + \frac{2}{9}\alpha S^2, \tag{13.96a}$$

$$u_{zz}^0 = \frac{2}{3}\beta S + \frac{2}{9}\alpha S^2. \tag{13.96b}$$

The strain $\underline{u}^0$ provides a complete description of the macroscopic equilibrium state after the phase transition to the nematic state, but it provides no information about a sample's specific orientation in space. The latter information is contained in the equilibrium Cauchy deformation tensor $\underline{\Lambda}^0$ which is related to $\underline{u}^0$ through Equation 13.49. Recall that $\underline{\Lambda}^0$ is not uniquely determined by $\underline{u}^0$ since rotations in $S_T$ change $\underline{\Lambda}^0$ but do not change $\underline{u}^0$. Because $\underline{u}^0$ is diagonal, it is natural in the present case not to rotate the strain after the transition. Then, $\underline{\Lambda}^0$ is also diagonal with diagonal elements given by

$$\Lambda_\perp^0 \equiv \Lambda_{xx}^0 = \Lambda_{yy}^0 = \sqrt{1 - \frac{2}{3}\beta S + \frac{4}{9}\alpha S^2}, \tag{13.97a}$$

$$\Lambda_\perp^0 = \Lambda_{zz}^0 = \sqrt{1 + \frac{4}{3}\beta S + \frac{4}{9}\alpha S^2}. \tag{13.97b}$$

Having the equilibrium deformation tensor, we can easily establish the connection of the anisotropy ratio (Equation 13.55) and the scalar order parameter,

$$S = \frac{(\Lambda_\perp^0)^2}{2\beta}(r-1), \tag{13.98}$$

that is, $S$ is indeed a direct measure of the anisotropy of the nematic phase as it should. Once again, we have to be careful here not to confuse the anisotropy ratio $r$ and the parameter $r$ in the Landau–de Gennes energy density.

## 13.5.1.2 Elasticity of the Nematic Phase

To determine the elastic properties of the new state, we expand $f_{I-N}$ in the deviations $\delta \underline{u} = \underline{u} - \underline{u}^0$ and $\delta \tilde{\underline{\underline{Q}}} = \tilde{\underline{\underline{Q}}} - \tilde{\underline{\underline{Q}}}^0$ from equilibrium. This can be done most conveniently by following the standard treatments of the $I-N$ transition, that is by introducing a complete set of five orthonormal symmetric-traceless matrices $\underline{\underline{I}}^n$ satisfying $I_{ij}^n I_{ij}^m = \delta^{nm}$ and expressing $\delta \hat{\underline{u}}$ and $\delta \tilde{\underline{\underline{Q}}}$ as $\delta \hat{\underline{u}} = \sum_{n=0}^{4} \delta u_n \underline{\underline{I}}^n$ and $\delta \tilde{\underline{\underline{Q}}} = \sum_{n=0}^{4} \delta \tilde{Q}_n \underline{\underline{I}}^n$, respectively. Explicit expressions for the matrices $\underline{\underline{I}}^n$ can be found for example in Appendix A of Ref. [25]. The $\delta \tilde{Q}_n$ have the following physical interpretation: $\delta \tilde{Q}_0$ describes deviations in the magnitude of uniaxial order, $\delta \tilde{Q}_1$ and $\delta \tilde{Q}_2$ measure the two independent components of incipient biaxial ordering, and $\delta \tilde{Q}_3$ and $\delta \tilde{Q}_4$ describe rotations of the uniaxial direction. Expressed in terms of the new variables, the elastic energy density changes to harmonic order

$$\delta f_{I-N} = \frac{B}{2} \left[ \text{Tr} \underline{u} - \alpha' \text{Tr} \delta \tilde{Q}_0 \right]^2 + \mu \sum_{n=0}^{4} \left[ \delta u_n - \beta \delta \tilde{Q}_n \right]^2,$$

$$+ \frac{1}{2} A_1 (\delta \tilde{Q}_0)^2 + \frac{1}{2} A_2 \left[ (\delta \tilde{Q}_1)^2 + (\delta \tilde{Q}_2)^2 \right],$$

(13.99)

where $\alpha' = \sqrt{8/3} S \alpha$ and where $A_1$ and $A_2$ are elastic constants composed of the elastic constants of Equation 13.92b and $S$. Note that the role played by the rotation modes $\delta \tilde{Q}_3$ and $\delta \tilde{Q}_4$ is different from that played by the longitudinal mode $\delta \tilde{Q}_0$ and the biaxial modes $\delta \tilde{Q}_1$ and $\delta \tilde{Q}_2$: whereas the former appear only in the form of complete squares, the latter also appear in the $A_1$ and $A_2$ terms, which are referred to in Landau theory as *mass* terms. Owing to these terms, fluctuations in $\delta \tilde{Q}_0, \delta \tilde{Q}_1$ and $\delta \tilde{Q}_2$ are small and hence these variables can be integrated out without loosing important physics. This procedure yields after some algebra

$$\delta f_{I-N} = \frac{1}{2} B_1 (\delta u_{zz})^2 + B_2 \delta u_{zz} \delta u_{aa} + \frac{1}{2} B_3 (\delta u_{aa})^2$$

$$+ B_4 (\delta u_{ab})^2 + 2\mu [\delta u_{az} - \beta S \delta \tilde{c}_a]^2,$$

(13.100)

where we have exploited that $\delta \tilde{Q}_{az} = S \delta \tilde{c}_a$. $B$'s in Equation 13.100 are conglomerates of the original elastic constants and $S$ whose details are not important here.

We could continue our discussion here using $\tilde{c}_a$. To foster a comparison with existing theories on the elasticity of nematic elastomers, we find it useful at this point to switch to the physical c-director. Moreover, as discussed earlier, the strain $\delta \underline{u}$ describes distortions relative to the new uniaxial reference state measured in the coordinates of the original isotropic state, and it is customary and more intuitive to express the elastic energy in terms of a strain $\underline{u}'$, Equation 13.51, relative to the new reference space $S_R'$. Switching to the new strain, we have to be careful because the transformation from $S_R$ to $S_R'$ affects the rotation operator (Equation 13.27). When expressed relative to $S_R'$ as opposed to $S_R$, the leading contributions to $\underline{\underline{Q}}$ depend no longer only on the antisymmetric part of the displacement gradient tensor, *cf.* Equation

13.28b, but rather, there are additional contributions proportional to the symmetric part of the displacement gradient tensor (the linear part of the strain tensor). Taking this subtlety into account, one obtains

$$\delta\tilde{c}_a = \delta c_a - \eta'_{A,az} - \frac{1-\sqrt{r}}{1+\sqrt{r}}u'_{az}, \qquad (13.101)$$

to linear order in $\delta c_a$ and $\eta'_{ij} = \partial'_j u'_i$. Here and in the remainder of our discussion of the Landau theory of the $I$–$N$ transition it is understood that strains are linearized, that is, $u'_{ij} = \eta'_{S,ij}$, where $\underline{\underline{\eta}}'_S$ is the symmetric part of displacement gradient tensor with respect to $S'_R$. The final form of $\delta f_{I-N}$ can be written as

$$\delta f_{I-N} = \frac{1}{2}C_1(u'_{zz})^2 + C_2 u'_{zz}u'_{aa} + \frac{1}{2}C_3(u'_{aa})^2$$
$$+ C_4(u'_{ab})^2 + \frac{1}{2}\mu'[u'_{az} - \beta'(\delta\tilde{c}_a - \eta'_{A,az})]^2, \qquad (13.102)$$

where $\mu' = (1+r)^2(\Lambda_\perp^0)^4\mu, \beta' = (r-1)/(r+1), C_1 = B_1(\Lambda_\parallel^0)^4$ and so on. Several salient features of Equation 13.102 are worth pointing out: (1) Its first four terms coincide with the nematic elastomer energy density $f_{\text{nem}}^{\text{soft}}$ of Equation 13.57 up to a trivial redefinition of the elastic constants $C_1$, $C_2$ and $C_3$ (in Equation 13.102 we use $u_{aa}$ as opposed to the $u'_{ii}$ used in Equation 13.57). (2) It demonstrates that $\delta\tilde{c}_a$ will relax to $\delta c_a = \eta'_{A,az} + (\beta')^{-1}u'_{az}$. After this relaxation, $\delta f_{I-N}$ is independent of $u'_{az}$, and hence, the elastomer is soft with respect to shear in planes containing the director, as expected from the arguments of Section 13.3.2. (3) It is identical to the result of Ref. [25], where the $I$–$N$ transition in elastomers has been studied in terms of the nematic order parameter $\underline{\underline{Q}}$ and the left Cauchy–Green strain tensor $\underline{\underline{v}}$. (4) It is also identical in form to Equation 13.43 with $C_5^R = 0$ up to the aforementioned trivial redefinition of $C_1$, $C_2$, and $C_3$. Thus, we can make the following identifications: $\mu' = D_2^2/D_1$ and $\beta' = -D_1/D_2$. These identifications lead to the relations $D_1/\mu' = (\beta')^2$ and $D_2/\mu' = 2\beta'$ that are identical (up to differences in notation) to well-known relations derived by Olmsted [9] for the rotationally invariant neoclassical model of rubber elasticity. These relations are required by rotational symmetry, and Lagrange theory precisely reproduces them.

### 13.5.2 SMECTIC-$C$ ELASTOMERS

#### 13.5.2.1 The SmA-to-SmC Transition

We can analyze the $A$–$C$ transition in a similar way as the $I$–$N$ transition. We complete the squares involving the strains and the director-strain couplings. The resulting elastic energy density is once more a sum of two terms,

$$f_{A-C} = f_{A-C}^{(1)} + f_{A-C}^{(2)}, \qquad (13.103)$$

where

$$f_{A-C}^{(1)} = \frac{1}{2}C_1 w_{zz}^2 + C_2 w_{ii} w_{zz} + \frac{1}{2}C_3 w_{ii}^2 + C_4 w_{ab}^2, \tag{13.104}$$

is quadratic in the shifted strains

$$w_{zz} = u_{zz} - \sigma \tilde{c}_c^2, \tag{13.105a}$$

$$w_{ii} = u_{ii} - \tau \tilde{c}_c^2, \tag{13.105b}$$

$$w_{ab} = \hat{u}_{ab} - \omega(\tilde{c}_a \tilde{c}_b - \frac{1}{2}\delta_{ab}\tilde{c}_c^2), \tag{13.105c}$$

$$w_{az} = u_{az} - \rho \tilde{c}_a, \tag{13.105d}$$

and where

$$f_{A-C}^{(2)} = \frac{1}{2}r_{\tilde{c}}^R \tilde{c}_c^2 + \frac{1}{4}g^R (\tilde{c}_c^2)^2, \tag{13.106}$$

depends on $\tilde{c}$ only. The coefficients in Equation 13.105 are given by

$$\begin{pmatrix} \sigma \\ \tau \end{pmatrix} = \frac{-1}{C_1 C_3 - C_2^2} \begin{pmatrix} C_3\lambda_1 - C_2\lambda_2 \\ C_1\lambda_2 - C_2\lambda_1 \end{pmatrix}, \tag{13.107}$$

$\omega = -\lambda_3/(2C_4)$ and $\rho = -\lambda_4/(2C_5)$. The renormalized elastic constants $r_{\tilde{c}}^R$ and $v^R$ in Equation 13.106 read $r_{\tilde{c}}^R = r_{\tilde{c}} - \lambda_4^2/(2C_5)$ and $g^R = g - 2\sigma^2 C_1 - 4\sigma\tau C_2 - 2\tau^2 C_3 - 2\omega^2 C_4 + 4\rho^2 C_5$.

Next, we minimize $f_{A-C}$ to assess the equilibrium states. With our coordinate system in $S_R$ chosen so that $\tilde{c}$ aligns along $x$, we obtain readily from Equation 13.106 that $\tilde{c}_y^0 = 0$ for any value of $r_{\tilde{c}}^R$, and

$$S \equiv \tilde{c}_x^0 = \sin\alpha = \begin{cases} 0 & \text{for } r_{\tilde{c}}^R > 0, \\ \pm\sqrt{-r_{\tilde{c}}^R/g^R} & \text{for } r_{\tilde{c}}^R < 0, \end{cases} \tag{13.108}$$

where $\alpha$ is the angle that the reference-space director makes with the $z$-axis. The full reference space nematic director is thus

$$\tilde{\mathbf{n}}^0 = (\sin\alpha, 0, \cos\alpha). \tag{13.109}$$

Note that this corresponds to a counterclockwise rotation through $\alpha$ about the $y$-axis of the original director $\tilde{\mathbf{n}} = (0, 0, 1)$ in the SmA phase. The director (Equation 13.109) is also the target space director under a symmetric deformation tensor $\underline{\underline{\Lambda}}^0$ as shown in Figure 13.2b. The components of the equilibrium strain tensor then follow from

Equation 13.106 as

$$u^0_{xx} = \frac{1}{2}(\tau + \omega - \sigma)S^2, \tag{13.110a}$$

$$u^0_{yy} = \frac{1}{2}(\tau - \omega - \sigma)S^2, \tag{13.110b}$$

$$u^0_{zz} = \sigma S^2, \tag{13.110c}$$

$$u^0_{xz} = u^0_{zx} = \rho S, \tag{13.110d}$$

and zero for the remaining components.

Once again, we have to choose our coordinate system in $S_T$. As in Section 13.3.3 we choose this system so that the $A$–$C$ transition amounts to the simple shear shown in Figure 13.2c with $\Lambda^0_{xz} > 0$, and $\Lambda^0_{zx} = 0$ such that the equilibrium deformation tensor has the form shown in Equation 13.65. With this choice,

$$\Lambda^0_{xx} = \sqrt{1 + (\tau + \omega - \sigma)S^2}, \tag{13.111a}$$

$$\Lambda^0_{yy} = \sqrt{1 + (\tau + \omega - \sigma)S^2}, \tag{13.111b}$$

$$\Lambda^0_{xz} = \frac{2\rho S}{\sqrt{1 + (\tau + \omega - \sigma)S^2}}, \tag{13.111c}$$

$$\Lambda^0_{zz} = \sqrt{1 + 2\sigma^2 S^2 + \frac{4\rho^2 S^2(1 - S^2)}{1 + (\tau + \omega - \sigma)S^2}}. \tag{13.111d}$$

Knowing $\tilde{c}^0$ and $\underline{\underline{\Lambda}}^0$ we can discuss what happens in the SmC phase to the layer normal, the director, and the uniaxial anisotropy axis. Under the simple shear (Equation 13.111), $(\Lambda^0_{zi})^{-1} = (\Lambda^0_{zz})^{-1}\delta_{zi}$, and hence

$$\mathbf{N}^0 = (0, 0, 1). \tag{13.112}$$

Thus, as expected, the shear deformation induced by the transition to the SmC phase slides the smectic layers parallel to each other. In this geometry, it does not rotate the layer normal. Since $\mathbf{N}$ is parallel to the $z$-axis under simple shear, the angle between the layer normal and the nematic director $\mathbf{n}$ is the angle that the director makes with the $z$ axis under simple shear. This angle is simply $\Theta = \alpha + \beta$, where $\beta$ is the angle through which the sample has to be rotated to bring the symmetric-shear configuration to the simple-shear configuration. Under symmetric shear, the symmetric deformation tensor is given by

$$\underline{\underline{\Lambda}}_S = \underline{\underline{g}}^{1/2} = (\underline{\underline{\delta}} + 2\underline{\underline{u}})^{1/2} \tag{13.113}$$

In order to calculate $\beta$, we need the symmetric equilibrium deformation tensor $\underline{\underline{\Lambda}}^0_S$, given by Equation 13.113 with $\underline{\underline{g}}$ replaced by $\underline{\underline{g}}^0 = \underline{\underline{\Lambda}}^{0T}\underline{\underline{\Lambda}}^0$. In terms of the components of $\underline{\underline{\Lambda}}^0_S$, $\beta = \tan^{-1}[(1 + 2u^0)^{1/2}]_{zx}/[(1 + 2u^0)^{1/2}]_{zz} \approx u^0_{xz}$. Tedious but straightforward algebra verifies that the simple-shear deformation tensor $\underline{\underline{\Lambda}}^0$, whose components are given by Equation 13.111, satisfies $\underline{\underline{\Lambda}}^0 = \underline{\underline{O}}_y(\beta)(\underline{\underline{g}}^0)^{1/2}$, where $\underline{\underline{O}}_y(\beta)$ is the matrix for a counterclockwise rotation about the $y$-axis, which is shown in Figure 13.2, through $\beta$, as given by the right-hand side of Equation 13.53 with $\vartheta$ replaced by $\beta$. The angle between $\mathbf{n}$ and $\mathbf{N}$, which is equivalent to the angle between $\mathbf{n}$ and the $z$-axis, is $\Theta = \alpha + \beta = [1 - \lambda_4/(2C_5)]S$. The uniaxial anisotropy vector $\tilde{\mathbf{e}}$ becomes $\mathbf{e} = (\sin\beta, 0, \cos\beta)$. Note finally that the angle $\gamma$ in Figure 13.2 is $\gamma = \tan^{-1}[(1 + 2u^0)^{1/2}]_{xz}/[(1 + 2u^0)^{1/2}]_{zz} \approx u^0_{xz}$. Thus, at lowest order $\gamma = \beta$. They differ, however, at higher order in $S$. The mechanical tilt angle $\phi \equiv \Lambda^0_{xz}/\Lambda^0_{zz}$ is given in terms of the angles defined in Figure 13.2 by $\phi = \beta + \gamma \approx 2\beta$. Thus, the spontaneous mechanical tilt of the sample, as described by $\phi$, and the tilt of the mesogens, as described by $\Theta$, are not equal.

### 13.5.2.2 Elasticity of the SmC Phase

To analyze the elastic properties of the SmC phase we expand the elastic energy density $f_{A-C}$ about the SmC equilibrium state. Expansion of $f^{(1)}_{A-C}$ to harmonic order results in

$$\delta f^{(1)}_{A-C} = \frac{1}{2}C_1(\delta w_{zz})^2 + C_2\delta w_{ii}\delta w_{zz} + \frac{1}{2}C_3(\delta w_{ii})^2 + C_4(\delta w_{ab})^2 + C_5(\delta w_{az})^2$$

(13.114)

with the composite strains

$$\delta w_{zz} = \delta u_{zz} - 2\sigma S\delta\tilde{c}_x, \tag{13.115a}$$

$$\delta w_{ii} = \delta u_{ii} - 2\tau S\delta\tilde{c}_x, \tag{13.115b}$$

$$\delta w_{xx} = -\delta w_{yy} = \frac{1}{2}(\delta u_{xx} - \delta u_{yy}) - \omega S\delta\tilde{c}_x, \tag{13.115c}$$

$$\delta w_{xy} = -\delta w_{yx} = \delta u_{xy} - \omega S\delta\tilde{c}_y \tag{13.115d}$$

$$\delta w_{xz} = \delta w_{zx} = \delta u_{xz} - \rho(1 - \frac{3}{2}S^2)\delta\tilde{c}_x, \tag{13.115e}$$

$$\delta w_{yz} = \delta w_{zy} = \delta u_{yz} - \rho(1 - \frac{1}{2}S^2)\delta\tilde{c}_y. \tag{13.115f}$$

The expansion of $f^{(2)}_{A-C}$ is particularly simple. It leads to

$$\delta f^{(2)}_{A-C} = g^R S^2(\delta\tilde{c}_x)^2. \tag{13.116}$$

A glance at Equations 13.114, 13.115, and 13.116 shows that the two components of the c-director, $\delta\tilde{c}_x$ and $\delta\tilde{c}_y$, play qualitatively different roles. Whereas, $\delta\tilde{c}_y$ appears only in the composite strains (Equation 13.115), the component $\delta\tilde{c}_x$ also appears in Equation 13.116. In the spirit of Landau theory of phase transitions, the term $g^R S^2(\delta\tilde{c}_x)^2$ makes $\delta\tilde{c}_x$ a massive variable. $\delta\tilde{c}_y$, on the other hand, is massless. Since $\delta\tilde{c}_x$ is massive, the softness of the SmC phase that we expect from what we have learned in Section 13.3.3 cannot come from the relaxation of $\delta\tilde{c}_x$. Rather it has to result from the relaxation of $\delta\tilde{c}_y$. Anticipating this relaxation of $\delta\tilde{c}_y$, we rearrange $\delta f_{A-C}^{(2)}$ so that $\delta\tilde{c}_y$ appears only in one place. Then we combine the two contributions $\delta f_{A-C}^{(1)}$ and $\delta f_{A-C}^{(2)}$ and integrate the massive variable $\delta\tilde{c}_x$. For details on these steps we refer to Ref. [26].

Our final step in deriving the elastic energy density is to change from $\delta\underline{u}$ to the strain $\underline{\underline{u}}'$ defined in Equation 13.51, which takes us to

$$\delta f_{A-C} = f_{C_{2h}}^{\text{soft}} + \Delta \left[ \delta\tilde{c}_y + \Lambda_{yy}^0 \frac{(2\Lambda_{xx}^0 C_4 \Pi + \Lambda_{xz}^0 C_5 \Xi)u_{xy}' + \Lambda_{zz}^0 C_5 \Xi u_{yz}'}{\Delta} \right]^2, \tag{13.117}$$

where $\Pi = -\omega S$, $\Xi = -\rho(1 - S^2/2)$, and $\Delta = 2C_4\Pi^2 + C_5\Xi^2$. $f_{C_{2h}}^{\text{soft}}$ is exactly of the same form as the result stated in Equation 13.80. Equation 13.117 shows clearly that $\delta\tilde{c}_y$ can relax locally to

$$\delta\tilde{c}_y = -\Lambda_{yy}^0 \frac{(2\Lambda_{xx}^0 C_4 \Pi + \Lambda_{xz}^0 C_5 \Xi)u_{xy}' + \Lambda_{zz}^0 C_5 \Xi u_{yz}'}{\Delta} \tag{13.118}$$

which eliminates the dependence of the elastic energy density on the linear combination of strains appearing on the right-hand side of Equation 13.118. In other words, Landau theory produces

$$\delta f_{A-C} = f_{C_{2h}}^{\text{soft}}, \tag{13.119}$$

in absolute agreement with the symmetry arguments presented in Section 13.3.3.

## 13.6 SUMMARY

In this chapter, we have discussed the Lagrange elasticity theory of liquid crystal elastomers. Rather than giving an exhaustive overview of this topic, we have explained in detail the general approach and then discussed in some depth the unusual, related phenomena of soft elasticity in nematic and in SmC phases of elastomers and semisoft elasticity in nematic elastomers.

Lagrange elasticity has the reputation of being cumbersome. Indeed, it takes some time and effort to get used to thinking in terms of reference and target spaces. Dealing properly with strains, stresses, and deformations, which, depending on whether one considers the reference or the target space, behave as tensors or scalars or even have a mixed behavior, requires some training. Also, notation tends to be heavy, and there are many indices involved. Despite these difficulties, Lagrange elasticity allows for an elegant and powerful theoretical description of liquid crystalline elastomers. What makes it so appealing from the viewpoint of theoretical physicists is that it offers a clear and systematic way to account for the complicated macroscopic symmetries of liquid crystal elastomers. When combined with the polar decomposition theorem, it allows treating liquid crystal and elastic degrees of freedom on equal footing and thus opens the door to a theoretical understanding of the interaction between elasticity and liquid crystalline order.

One of the virtues of Lagrange elasticity is that it provides for a general approach that can be used to study numerous topics in liquid crystal elastomers in a common framework. To give a flavor for its generality, we would like to give a few examples of topics that we could have included in the chapter but have omitted them because of space constraints. Lagrange elasticity has been used to investigate the low-frequency, long-wavelength dynamics of nematic [60–63] and smectic [64,65] elastomers. It has been used, as has the neoclassic model [66], to study Sm*A* elastomers under strain along the layer normal [27]. It has been employed to analyze the electrostriction effect in chiral Sm*A*\* elastomers in lateral electric fields [27]. Lagrange elasticity provided the basis to study anomalous elasticity in nematic elastomers [67–70], where thermal fluctuations lead to a renormalization of the elastic behavior whereby certain elastic constants become dependent on length scale and also form universal Poisson ratios. It provided a route for studying the phase diagrams and the elasticity of nematic [71,72] and smectic [73] elastomer membranes, and so on [1].

Undoubtedly, there is an abundance of interesting phenomena in liquid crystal elastomers, resulting from the interaction between elasticity and liquid crystalline order. Considerable achievements have already been made in this field, experimentally and theoretically, and it is almost certain, that the future holds more important discoveries. Lagrange elasticity theory will play a significant role in this quest.

## REFERENCES

1. M. Warner and E. M. Terentjev, *Liquid Crystal Elastomers, International Series of Monographs on Physics* (Oxford University Press, Oxford, 2003).
2. P. G. de Gennes and J. Prost, *The Physics of Liquid Crystals* (Oxford University Press, Oxford, 1993).
3. S. Chandrasekhar, *Liquid Crystals* (Cambridge University Press, Cambridge, England, 1992).
4. H. Finkelmann, H. J. Kock, and G. Rehage, *Makromol. Chem.-Rapid Commun.* **2**, 317, 1981.
5. R. Zentel, *Angew. Chem.* **101**, 1437, 1989.
6. A. Greve, H. Finkelmann, and M. Warner, *Eur. Phys. J. E* **5**, 281, 2001.
7. S. M. Clarke, A. Hotta, A. R. Tajbakhsh, and E. M. Terentjev, *Phys. Rev. E* **64**, 061702, 2001.
8. M. Warner, P. Bladon, and E. M. Terentjev, *J. Phys. II* **4**, 93, 1994.
9. P. D. Olmsted, *J. Phys. II (France)* **4**, 2215, 1994.
10. G. Verway and M. Warner, *Macromolecules* **28**, 4303, 1995.
11. J. Küpfer and H. Finkelmann, *Macromol. Chem. Phys.* **195**, 1353, 1994.
12. M. Warner, *J. Mech. Phys. Solids* **47**, 1355, 1999.
13. E. M. Terentjev, M. Warner, R. B. Meyer, and J. Yamamoto, *Phys. Rev. E* **60**, 1872, 1999.
14. P. Blandon, E. Terentjev, and M. Warner, *J. Phys. II (France)* **4**, 75, 1994.
15. L. R. G. Treloar, *The Physics of Rubber Elasticity* (Clarendon Press, Oxford, 1975).
16. L. D. Landau and E. M. Lifshitz, *Theory of Elasticity*, 3rd ed. (Pergamon, New York, 1986).
17. J. Küpfer and H. Finkelmann, *Makromol. Chem.-Rapid Commun.* **12**, 717, 1991.

18. M. Brehmer, A. Wiesemann, R. Zentel, K. Siemensmeyer, and G. Wagenblast, *Polymer Preprints* **34**, 708, 1993.

19. M. Brehmer, R. Zentel, G. Wagenblast, and K. Siemensmeyer, *Macromol. Chem. Phys.* **1995**, 849, 1994.

20. I. Benné, K. Semmler, and H. Finkelmann, *Macromol. Rapid Commun.* **15**, 295, 1994.

21. K. Hiraoka and H. Finkelmann, *Macromol. Rapid Commun.* **22**, 456, 2001.

22. K. Hiraoka, W. Sagano, T. Nose, and H. Finkelmann, *Macromolecules* **38**, 7352 , 2005.

23. R. A. Horn and C. R. Johnson, *Topics in Matrix Analysis* (Cambridge University Press, New York, 1991).

24. A. Love, *A Treatise on the Mathematical Theory of Elasticity* (Dover Publications, New York, 1944).

25. T. C. Lubensky, R. Mukhopadhyay, L. Radzihovsky, and X. J. Xing, *Phys. Rev. E* **66**, 011702, 2002.

26. O. Stenull and T. C. Lubensky, *Phys. Rev. E* **74**, 051709, 2006.

27. O. Stenull and T. C. Lubensky, *Phys. Rev. E.* **76**, 011706, 2007.

28. P. G. de Gennes, in *Polymer Liquid Crystals*, A. Ciferri, W. Krigbaum, and W. Helfrich (Eds.) (Springer, Berlin, 1982), p. 231.

29. T. C. Lubensky, E. M. Terentjev, and M. Warner, *J. Phys. II (France)* **4**, 1457, 1994.

30. L. Radzihovsky and J. Toner, *Phys. Rev. B* **60**, 206, 1999.

31. E. M. Terentjev and M. Warner, *J. Phys. II (France)* **4**, 849, 1994.

32. W. Helfrich, *J. Chem. Phys.* **55**, 839, 1971.

33. W. Helfrich, *Appl. Phys. Lett.* **17**, 531, 1970.

34. J. P. Hurault, *J. Chem. Phys.* **59**, 2086, 1973.

35. N. Clark and R. Meyer, *Appl. Phys. Lett.* **22**, 493, 1973.

36. P. M. Chaikin and T. C. Lubensky, *Principles of Condensed Matter Physics* (Cambridge University Press, Cambridge, 1995).

37. Y. Nambu, *Phys. Rev. Lett.* **4**, 380, 1960.

38. J. Goldstone, *Nuovo Cimento* **19**, 155, 1961.

39. L. Golubovic and T. C. Lubensky, *Phys. Rev. Lett.* **40**, 2631, 1989.

40. O. Stenull and T. C. Lubensky, *Phys. Rev. Lett.* **94**, 018304, 2005.

41. F. Ye, R. Mukhopadhyay, O. Stenull, and T. C. Lubensky, *Phys. Rev. Lett.* **98**, 147801, 2007.

42. J. M. Adams and M. Warner, *Phys. Rev. E* **72**, 021798, 2005.

43. O. Stenull and T. C. Lubensky, unpublished.

44. O. Stenull and T. C. Lubensky, *Eur. Phys. J. E* **14**, 333, 2004.

45. P. D. Gennes, *C.R. Acad. Sci. Ser. B* **281**, 101, 1975.

46. A. Lebar, Z. Kutnjak, S. Zumer, H. Finkelmann, A. Sanchez-Ferrer, and B. Zalar, *Phys. Rev. Lett.* **94**, 197801, 2005.

47. D. Rogez, G. Francius, H. Finkelmann, and P. Martinoty, *Eur. Phys. J. E* **20**, 369, 2006.

48. H. R. Brand, H. Pleiner, and P. Martinoty, *Soft Matter* **2**, 182, 2006.

49. B. J. Frisken, B. Bergersen, and Palffy-Muhoray, *Mol. Cryst. Liq. Cryst.* **148**, 45, 1987.

50. C.-P. Fan and M. J. Stephen, *Phys. Rev. Lett.* **25**, 500, 1970.

51. R. G. Priest, *Phys. Lett. A* **47**, 475, 1974.

52. M. Blume, V. Emory, and R. Griffiths, *Phys. Rev. A* **4**, 1071, 1971.

53. A. Hotta and E. M. Terentjev, *Eur. Phys. J. E* **10**, 291, 2003.

54. E. M. Terentjev, A. Hotta, S. M. Clarke, and M. Warner, *Philos. Trans. R. Soc. Lond. A* **361**, 653, 2003.

55. K. Bhattacharya, *Microstructure of Martensite: Why It Forms and How It Gives Rise to the Shape-Memory Effect* (Oxford University Press, New York, 2003).

56. F. Ye and T. C. Lubensky (2007). Unpublished.

57. S. Conti, A. DeSimone, and G. Dolzmann, J. *Mech. Phys. of Solids* **50**, 1431, 2002.
58. S. Conti, A. DeSimone, and G. Dolzmann, *Phys. Rev. E* **66**, 061710, 2002.
59. N. Uchida, *Phys. Rev. E.* **62**, 5199, 2000.
60. H. R. Brand and H. Pleiner, *Physica A* **208**, 359, 1994.
61. E. M. Terentjev, I. V. Kamotski, D. D. Zakharov, and L. J. Fradkin, *Phys. Rev. E* **66**, 052701(R), 2002.
62. L. J. Fradkin, I. V. Kamotski, and D. D. Terentjev, E. M. Zakharov, *Proc. R. Soc. Lond. A* **459**, 2627, 2003.
63. O. Stenull and T. C. Lubensky, *Phys. Rev. E* **69**, 051801, 2004.
64. O. Stenull and T. C. Lubensky, *Phys. Rev. E* **73**, 030701(R), 2006.
65. O. Stenull and T. C. Lubensky, *Phys. Rev. E* **74**, 031711, 2007.
66. J. M. Adams and M. Warner, *Phys. Rev. E* **71**, 011703, 2005.
67. X. Xing and R. Radzihovsky, *Phys. Rev. Lett.* **90**, 168301, 2003.
68. X. Xing and R. Radzihovsky, *Europhys. Lett.* **61**, 769, 2003.
69. O. Stenull and T. C. Lubensky, *Europhys. Lett.* **61**, 776, 2003.
70. O. Stenull and T. C. Lubensky, *Phys. Rev. E* **69**, 021807, 2004.
71. X. Xing, R. Mukhopadhyay, T. C. Lubensky, and L. Radzihovsky, *Phys. Rev. E* **68**, 021108, 2003.
72. X. Xing and L. Radzihovsky, *Phys. Rev. E* **71**, 011802, 2005.
73. O. Stenull, *Phys. Rev. E* **75**, 051702, 2007.

# 14 Orientational Order and Paranematic–Nematic Phase Transition in Liquid Single Crystal Elastomers

## *Nuclear Magnetic Resonance and Calorimetric Studies*

*Boštjan Zalar, Zdravko Kutnjak,*
*Slobodan Žumer, and Heino Finkelmann*

**CONTENTS**

## 14.1  INTRODUCTION

One of the major open problems in the physics of liquid single crystal elastomers
(LSCE) [1] is the nature of the paranematic–nematic phase transition, particularly
in view of the still open debate [2–6] on their mechanical properties, related to the
question whether the elasticity in these materials is of the soft-, semisoft-, or nonsoft-
type [7]. Conventional nematogenic liquid crystals (LCs) exhibit a first-order phase
transition, easily identifiable by a discontinuous jump (Figure 14.1a) of the nematic
order parameter (OP) $S(T)$ at the clearing temperature $T_{NI}$, observed in birefringence
measurements or deuteron nuclear magnetic resonance (DNMR). In LSCEs, on the
contrary, the onset of the nematic order is continuous, with three characteristic temperature regions [8,9]: high-temperature paranematic, intermediate, and low-temperature
nematic (Figure 14.1b). At high temperatures, in the so-called paranematic (PN) phase,
the nematic (N) order, characterized by the nematic OP denoted by $S$, is small but
nonzero, and monotonously increases with decreasing temperature. In the intermediate transition region, close to $T_{NI}$ of a low-molar-mass mesogenic component, which
typically constitutes more than 90% of the total LSCE mass, a substantially higher
slope in $S(T)$ is detected, resulting in a large jump $\delta S_{N-PN}$ of about 0.5 in the OP
within a relatively narrow temperature interval $\delta T_{N-PN}$ (typically a few K). The width
of this interval strongly depends on the type of LSCE network and its preparation conditions. In some cases, it can span several tens of K. In the low-temperature nematic
region, $S$ becomes saturated.

The coupling between the orientational order of the mesogen and the cross-linked
polymer network results in a temperature-dependent changes [10] of LSCE specimen
length, $l = l[S(T)]$, with the strain $\lambda(T) = l(T)/l(S = 0)$ and elastic deformation
$e(T) = \lambda(T) - 1$, which directly mimics the nematic order, that is, $e(T) \propto S(T)$. In

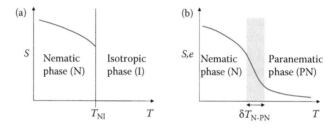

**FIGURE 14.1**  (a) A typical discontinuous temperature profile of $S$, found in conventional
low-molar-mass LCs and linear LC polymers; (b) Continuous profile of the nematic OP $S(T)$,
or equivalently, profile of the elastic deformation $e(T) \propto S(T)$, of a LSCE. The intermediate
N–PN transition region is shown with a shaded background.

view of the huge application potential of these materials for actuator and biomimetic devices (like artificial muscles) [11], the understanding of the physical mechanisms leading to the smearing out of the $S(T)$ profile seems to be of an extreme importance. Specifically, this knowledge could contribute to the optimization of $\delta T_{\text{N–PN}}$ and $\delta S_{\text{N–PN}}$, or, equivalently, $\delta T_{\text{N–PN}}$ for various applications, from on–off-type mechanical triggers with narrow $\delta T_{\text{N–PN}}$ to sensors and micro-motion actuators with broader $\delta T_{\text{N–PN}}$.

Two radically different descriptions of the N–PN phase transition in LSCE elastomers were proposed in the literature:

1. The first scenario relies on the assumption that LSCE's are inherently heterogeneous materials due to the composite nature of their structure at the nanoscopic level [12]. Any local variations in the concentrations of the three main constituents, the polymer backbone, the mesogenic or nonmesogenic cross-linkers, and the mesogen, can result in a distribution of the intermolecular coupling coefficients. This introduces a certain degree of glass-like behavior to the thermodynamics of the system, with all the difficulties related to the calculation of the partition function, even in the mean-field approximation. In order to overcome these troubles, yet on account of disregarding the nanoscopic collective disorder, the heterogeneity can be introduced at the phenomenological level, by simply dividing the system into microdomains, each with a well-defined set of Landau–deGennes (LdG) free energy expansion coefficients. The average OP temperature profile of the system, $\langle S(T) \rangle \propto e(T)$, is then calculated as a superposition of profiles $S(T)$ arising from the individual microdomains, with weight factors corresponding to the distribution of the LdG expansion coefficients over microdomains. The experimentally observed smooth $S(T)$ behavior in the transition region can then be readily reproduced theoretically with a proper selection of the distribution functions. Macroscopically, such a heterogeneous system then exhibits a smeared N–I transition, with a broad coexistence region of both phases and spatially inhomogeneous $S$ (Figure 14.2a). The intermediate transition region is also distinguished by a nonzero latent heat, relaxed over a broad temperature interval. However, in many conventional LSCE networks with broad $\delta T_{\text{N–PN}}$, $S$ remains nonzero even at temperatures more than 20 or 30 K above the nominal $T_{\text{N–PN}}$ (roughly the temperature of the maximal $S(T)$ slope). Since it is rather implausible that chemical heterogeneity could result in such a strong upward shift of $T_{\text{N–PN}}$ in any domain, a different physical mechanism must also be considered;

2. The behavior of $S(T)$ in LSCE can be attributed to the supercritical character of the N–PN transition [13]. It is a well-known fact that a linear coupling of the nematic OP with a conjugate internal or external field $g$, accounted for by the free energy term $-gS$, can drive the transition to a supercritical regime, characterized by zero latent heat and continuous $S(T)$ profile (Figure 14.2b). This occurs whenever $g$ exceeds a critical value $g_c$. Recently, high magnetic fields have been successfully used to suppress the first-order character of the

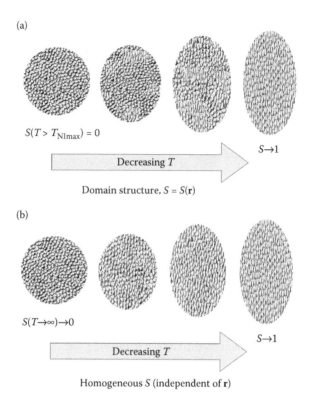

(a)

$S(T > T_{\text{NImax}}) = 0$

$S \to 1$

Decreasing $T$

Domain structure, $S = S(\mathbf{r})$

(b)

$S(T \to \infty) \to 0$

$S \to 1$

Decreasing $T$

Homogeneous $S$ (independent of $\mathbf{r}$)

**FIGURE 14.2** Schematic representation of the temperature evolution of the orientational order of mesogenic molecules in a conventional side-chain LSCE network. An effective, that is, time-averaged molecular shape is shown. (a) "Heterogeneous" scenario with spatially inhomogeneous nematic order of domains. There are no domains with very small nematic order, since the phase transition is first order (discontinuous); (b) "Supercritical" scenario with spatially homogeneous nematic order. In this case, $S$ can be arbitrarily small at high temperatures in the PN state. In real systems, domains with small but nonzero $S$ are clearly detected in DNMR experiment, as well as the coexistence of PN and N phases. This speaks in favor of combined heterogenous–supercritical mechanism of the orientational freezing transition in conventional LSCEs.

N–I transition in polymer LCs [14]. In LSCE, $g = g_{int} + g_{exg}$ is the mechanical stress field that consists of the internal, monodomain-state maintaining field $g_{int}$, imprinted into the system during the two-step "Finkelmann cross-linking procedure" [15] and the external field $g_{ext}$, applied by straining the sample.

As evident from above, (1) the "heterogeneous" and (2) the "supercritical" scenarios both provide for qualitatively satisfactory description of the OP temperature profile in LSCEs. It is anticipated that in real systems, the two mechanisms go hand in hand. The remaining question is which one prevails and whether the properties of LSCEs can be tailored so that either the heterogeneous or the supercritical nature is promoted.

In order to distinguish between the two scenarios of the N–PN transition, Selinger et al., have exploited the fact that in the N–PN transition region, the two models predict slightly mismatching $S(T)$ curves [12]. From the analysis of the relative mechanical strain versus $T$ curves, $e(T)$, in the samples exposed to static external mechanical stress of various amplitudes $g_{ext}$, it is concluded that the smooth $S(T)$ profile is predominantly determined by the heterogeneity of the sample, which can be modeled by a Gaussian distribution of clearing temperatures $T_{N–PN}$. By extrapolating the $g/g_c$ versus $g_{ext}$ experimental curves to $g_{ext} = 0$, it is estimated that the internal field is far below the critical value $(g_{int}/g_c \sim 0.2 - 0.5)$. Although, the above study represents the first serious attempt at discerning the nature of the N–PN transition in LSCEs, the applied experimental approach has some shortcomings: (1) The resolution is rather low, since the experimental error of the $e(T)$ points is of the same order of magnitude as the difference between the theoretical best-fit $e(T)$ points of the "heterogeneous" and "supercritical" scenarios, respectively; (2) The primary OP, $S(T)$, should be analyzed, rather than the secondary OP $e(T)$; and (3) Application of external stress is mandatory to facilitate the measurement of internal field $g_{int}$, which is an intrinsic property of the system; one needs to look for alternative routes of determining $g_{int}$, in the absence of $g_{ext}$.

Only recently, this ambiguity has been rather well resolved by combining nuclear magnetic resonance and high-resolution calorimetry [16,17]. It is the purpose of this chapter to discuss the experimental approach used in these studies.

In Section 14.2, we will describe two case studies to which the above-mentioned experimental methodology has been applied. The basics of DNMR and the method of associating the disorder in the nematic OP and domain misalignments with DNMR line-shape moments are discussed in Section 14.3. The impact of molecular diffusion on the DNMR spectrum is also discussed in this section. Section 14.4 copes with theoretical modeling using LdG approach and its use in determining the character of the N–PN transition. In Section 14.5 we introduce the concept of "smeared criticality." The potential of high-resolution calorimetry is demonstrated in Section 14.6. In Section 14.7 we relate DNMR and ac calorimetry experimental methods. Section 14.7 is devoted to conclusions.

## 14.2  CASE STUDIES OF REPRESENTATIVE LSCE SYSTEMS

We shall only consider conventional side-chain LSCEs, prepared by the "Finkelmann procedure." LSCE materials based on poly[oxy(methylsilylene)] were synthesized as described in Reference [15], typically in the form of $(l_a = 0.3-0.5 \, \text{mm}) \times (l_b = 5-8 \, \text{mm}) \times (l_c = 20-50 \, \text{mm})$ stripes so that the average nematic director $\bar{\mathbf{n}}$ was pointing along the longest edge $\mathbf{c}$ of length $l_c$ (Figure 14.3a). During the second cross-linking step, nematic domains were aligned by loading the specimen with a standard stress of about $5-10 \, \text{mN/mm}^2$.

To acquire DNMR spectra with an optimal filling factor in the NMR detection coil, LSCE stripes were cut into smaller pieces with $l'_c$. A few of them, typically $N = 5$, were stacked together to assemble a measurement specimen in the form of a block with dimensions $Nl_a \times l_b \times l'_c$ (Figure 14.3b). The specimen was fitted into the coil of the NMR probehead equipped with goniometer that provides for setting an arbitrary

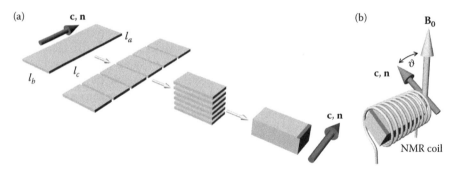

**FIGURE 14.3** Typical LSCE sample geometry (a) and preparation of the sandwich sample for angular-dependent DNMR measurements (b).

orientation of the sample with respect to external magnetic field $B_0$. The orientation is specified by the tilt angle $\vartheta\angle(\mathbf{c}, \mathbf{B_0})$. For calorimetry studies, only a single "small" piece of dimensions $l_a \times l_b \times l_c'$ was used.

Two groups of samples were prepared in order to investigate the nature of the PN-N phase transition: Samples, doped with different concentrations of low-molar-mass LC (Section 14.2.1) and samples containing different concentrations of cross-linking molecules (Section 14.2.2). We will show that in both cases, by altering the dopant or cross-linker concentrations, respectively, subcritical and supercritical phase-transition behavior can be promoted or suppressed.

### 14.2.1 DOPING WITH LOW-MOLAR-MASS NEMATOGEN

Samples in the first group were doped with octylcyanobiphenyl (8CB), deuterated at the two α-positions in the hydrocarbon chain (8CB–$\alpha d_2$). Doping was performed by initially swelling the elastomer in the controlled molarity solution of 8CB in cyclohexane, to which toluene was subsequently added in small steps to progressively and nondestructively swell the sample to about 400% of its initial volume. After being soaked for a few hours at $T = 330$ K, samples were deswollen to their initial volume by drying in the rotavapor pumping system at the same temperature. In this way, LSCEs are weakly swollen with low-molecular-weight mesogen [18] of a concentration $x_{LC}$. We shall denote this class of samples with LSCE-$x_{LC}$.

It is important to recognize the twofold convenience of swelling with 8CB–$\alpha d_2$: (1) It provides for selective DNMR sensitivity, not available in original LSCEs where only low natural abundance deuterons are found at any proton position. In low concentrations, 8CB–$\alpha d_2$ molecules probe the orientational order of the LSCE mesogenic components [19]; (2) 8CB–$\alpha d_2$ is itself a nematogen with a first-order N–I transition at $T_{NI}(8CB) \approx 315$ K. The increase of $x_{LC}$ should, apart from the drop in the N–PN transition temperature, $T_{N-PN}(x_{LC} \neq 0) < T_{N-PN}(x_{LC} = 0) \approx 357$ K, result in the change of the N–PN transition character from the supposedly supercritical regime for $x_{LC} = 0$ toward the critical or below critical regime for high enough $x_{LC}$. In order to achieve this, one must select an LSCE network, which is intrinsically supercritical.

## 14.2.2    IMPACT OF CROSS-LINKING DENSITY

The internal mechanical field $g_{int}$, conjugate to the nematic OP $S$, is strongly related to the type and density of cross-links [20]. On increasing the concentration of cross-linking molecules, $x_{c\text{-}l}$, the temperature profile of the OP $S$ exhibits an increasingly smeared N–PN transition [21]. It is therefore expected that for large enough $x_{c\text{-}l}$, the system will be pushed beyond a critical point, given that it is subcritical in the low $x_{c\text{-}l}$ limit. Furthermore, local variations in the cross-linking density may lead to quenched randomness, macroscopically manifested as random field-induced smearing of criticality described by a distribution of internal mechanical fields [22].

A convenient choice to prove the above predictions are side-chain LSCEs, cross-linked with the rod-like 4-(undec-10-enyloxy)phenyl-4-(undec-10-enyloxy)benzoate (V6) and the trifunctional point-like 1,3,5-tris(undec-10-enyloxy)cyclohexane (V3). The samples are denoted as V6-LSCE-$x_{c\text{-}l}$ and V3-LSCE-$x_{c\text{-}l}$, respectively. The polymer backbone consists of siloxane units and 4-but-3-enyl-benzoic acid 4-methoxy-phenyl ester was used as a mesogen (Figure 14.4).

Let us note that both V6- and V3-type samples were doped with 8CB–$\alpha$d$_2$, in order to render them DNMR-sensitive. However, $x_{LC}$ was low enough so that it did not alter the intrinsic N–PN transition behavior. This is in contrast to the case described in the previous subsection where the presence of 8CB–$\alpha$d$_2$ in higher concentrations significantly changes the transition behavior.

## 14.3    DNMR AND MOLECULAR ORIENTATIONAL ORDER

Here, we describe a recently developed analytical method, which overcomes the above limitations and allows for a clear-cut discrimination among subcritical, critical, and supercritical N–PN transition nature. The method is based on the analysis of the temperature profiles of the first and second moments ($M_{\nu,1}$ and $M_{\nu,2}$) of DNMR spectra of deuterated mesogenic molecules. It utilizes the fact that $M_{\nu,2}$ contains information on the second moment $M_{S,2}$ of the distribution function $w_S(S)$ of the nematic OP. Within the ideal "supercritical" scenario, $M_{S,2} = 0$ at any temperature since the order is homogeneous and thus described by a single OP value $S$. In the "heterogeneous" case, on the other hand, $M_{S,2}$ is nonzero and strongly temperature dependent. It is vanishingly small, far up in the PN phase and deep in the N phase, whereas it exhibits a pronounced maximum in the N–PN transition temperature range where small variations of the parameters of the model result in large variations of $S$. This anomaly, characteristic for the "heterogeneous" scenario and absent in the "supercritical" scenario, is directly reflected in the experimentally observable anomaly in $M_{\nu,2}(T)$. The intensity of the $M_{\nu,2}(T)$ maximum is therefore a direct measure of the degree of inherent heterogeneity.

DNMR spectrum of an 8CB–$\alpha$d$_2$ molecule confined to a LSCE nematic or PN domain with OP $S$ is a doublet of sharp resonance peaks that can be written in terms of two shifted Dirac $\delta$–functions,

$$f(\nu; S, \theta) = 1/2\{\delta[\nu + 3/4\nu_Q SP_2(\cos\theta)] + \delta[\nu - 3/4\nu_Q SP_2(\cos\theta)]\}. \quad (14.1)$$

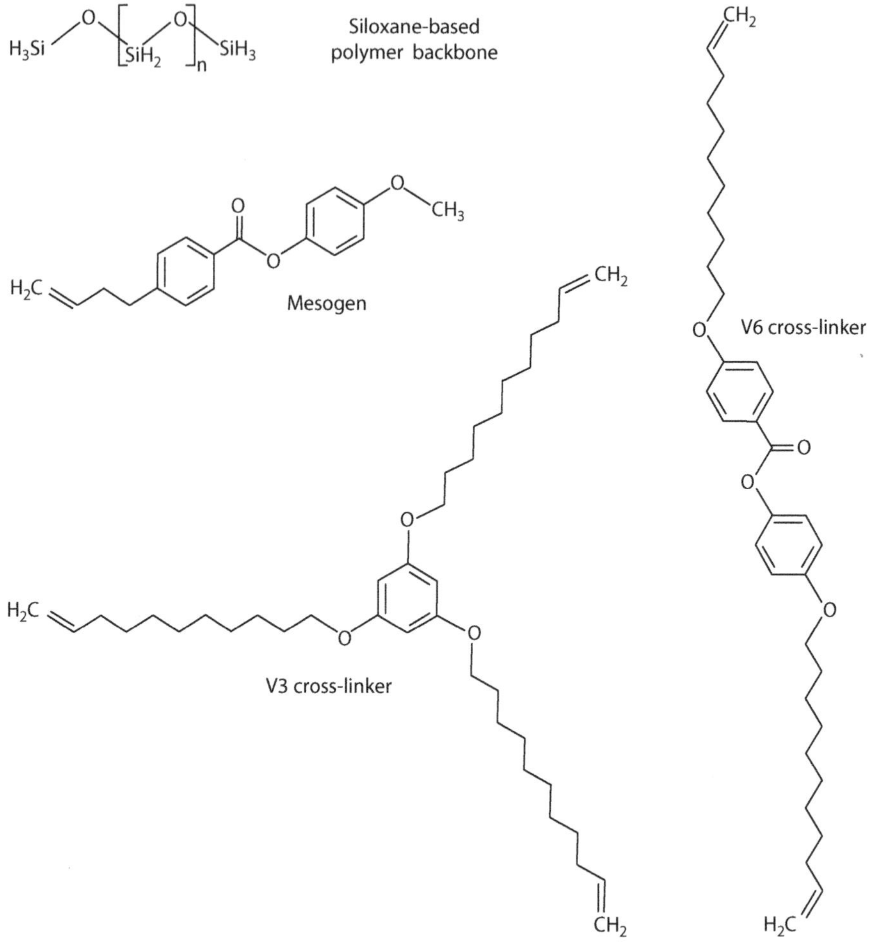

**FIGURE 14.4** Building blocks of a conventional side-chain LSCE network.

Here, the frequency offset $\nu$ is measured with respect to the Larmor frequency $\nu_L$ and $\nu_Q \approx 60\,\text{kHz}$ is the quadrupole coupling constant, averaged over fast reorientations about the 8CB long molecular axis, of the two indistinguishable deuterons at the $\alpha$-position of the 8CB hydrocarbon chain [23]. $P_2$ is the Legendre polynomial of order 2, whereas $\theta$ denotes the angle between the domain director $\mathbf{n}$ and the magnetic field $\mathbf{B}_0, \theta\angle(\mathbf{n}, \mathbf{B}_0)$. $\mathbf{B}_0$ is the vertical magnetic field of the superconducting magnet. In the experiments discussed in this paper, $\mathbf{B}_0 = 9\,\text{T}$, corresponding to deuteron Larmor frequency $\nu_L = 58.34\,\text{MHz}$. All DNMR spectra were taken using the $(\pi/2)_x - \tau - (\pi/2)_y$—"solid echo" pulse sequence with quadrature detection scheme [24] on cooling the sample from $T = 380\,\text{K}$.

Representative temperature dependences of the LSCE-$x_{LC}$ DNMR spectra are shown in Figure 14.5. Obviously, the two doublet lines are far from being $\delta$-like. Their considerable line widths reflect a substantial amount of inherent disorder, present in

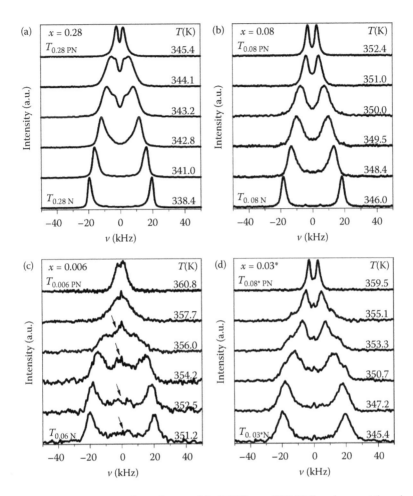

**FIGURE 14.5** Temperature dependences of the LSCE-$x_{LC}$ DNMR line shapes at the orientation $\vartheta = 0$ for concentrations $x_{LC} = 0.28$ (a), 0.08 (b), 0.006 (c), and 0.03* (d). The comparison between samples of different $x_{LC}$ clearly demonstrates the broadening of the N–PN transition temperature interval $\delta T_{N-PN}$ and smearing of the transition on decreasing $x_{LC}$. Arrows mark the background originating from natural abundance deuterons. The asterisk denotes the sample, cross-linked more completely in the first step (longer time was allowed for the cross-linking), whereas in the second step it was submitted to a higher stress of about 20 mN/mm². In this way, stronger internal field $g_{int}$ was expected to be locked in the system (pronounced supercritical behavior).

conventional LSCEs, a fact that apparently speaks in favor of the "heterogeneity" scenario. Similar behavior is found in the V6-LSCE-$x_{c-l}$ samples (Figure 14.6), except for the fact that here an increasing $x_{c-l}$ (promotion of supercriticality) has an opposite effect from increasing $x_{LC}$ in LSCE-$x_{LC}$ samples (suppression of supercriticality). Nevertheless, in a real LSCE network, one has to allow for a combined heterogeneity-supercriticality scenario, that is, for a "smeared" N–PN transition nature.

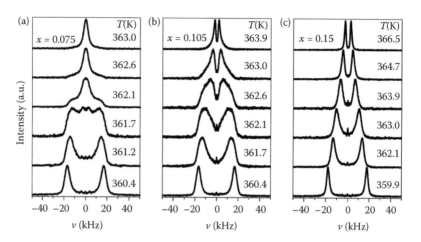

**FIGURE 14.6** Temperature dependences of the LSCE-$x_{c-1}$ DNMR line shapes at the orientation $\vartheta = 0$ for cross-linking concentrations $x_{c-1} = 0.075$ (a), $0.105$ (b), and $0.15$ (c). At low $x_{c-1}$, the subcritical nature is demonstrated through the coexistence of N and PN phases, whereas no such coexistence is present in the supercritical, high $x_{c-1}$ samples.

### 14.3.1 Broadening of DNMR Lines and Nematic Disorder

Let us observe that in real samples the domains are never ideally aligned, that is, $\mathbf{n} \neq \bar{\mathbf{n}} \| \mathbf{c} \Rightarrow \theta \neq 0$. Moreover, different domains can exhibit different orientational order $S$. These facts can be taken into account by introducing a distribution function $w(S, \cos\theta, \phi; \cos\vartheta)$. $\theta$ and $\phi$ are spherical coordinates of the domain director $\mathbf{n}$ in the laboratory frame $\mathbf{XYZ}$ with $\mathbf{Z} \| \mathbf{B}_0$. It is plausible to assume that the distributions of $S$ and $\mathbf{n}$ are uncorrelated, $w(S, \cos\theta, \phi; \cos\vartheta) = w_S(S)w_\mathbf{n}(\cos\theta, \phi; \cos\vartheta)$. The cumulative DNMR spectrum is then given by

$$F(v; \cos\vartheta) = \int\int\int w_S(S)w_\mathbf{n}(\cos\theta, \phi; \cos\vartheta)f(v; S, \theta)\, dS\, d\cos\theta\, d\phi. \qquad (14.2)$$

The distribution function of domain (nematic) directors $w_\mathbf{n}$ must possess cylindrical symmetry about the sample's uniaxiality axis $\mathbf{c}$. Consequently, at the specimen orientation $\vartheta = 0$ where $\mathbf{c}$ matches $\mathbf{B}_0$, $w_{\mathbf{n},0} = w_\mathbf{n}(\vartheta = 0)$ can be expanded in terms of Legendre polynomials $P_l(\cos\theta)$ as

$$w_{\mathbf{n},0}(\cos\theta) = (4\pi)^{-1}\sum_{l=0}^{\infty}(2l+1)\mathcal{S}_l P_l(\cos\theta). \qquad (14.3)$$

Here $\mathcal{S}_l \equiv \overline{P_l(\cos\theta)} = 2\pi \int w_{\mathbf{n},0}(\cos\theta)P_l(\cos\theta)\, d\cos\theta$ are the nematic director OPs [25]. The spectrum is then calculated as a superposition of two mirrored components,

$F(v) = 1/2[\tilde{F}(v) - \tilde{F}(-v)]$, with

$$\tilde{F}(v; \cos \vartheta) \equiv \frac{2}{9v_Q} \sum_{l=0}^{\infty} (2l+1) P_l(\cos \vartheta) S_l \int \frac{w_S(S) P_l[\chi(v,S)]}{S\chi(v,S)} \, dS, \quad (14.4a)$$

$$\chi(v,S) \equiv \sqrt{\left(1 + \frac{8v}{3v_Q S}\right)/3}. \quad (14.4b)$$

The above inhomogeneously broadened DNMR spectrum $F(v)$ has a relatively complex shape, determined by distributions of $S$ and $\theta$. For any robust DNMR line-shape analysis, description of the spectra in terms of a small number of experimentally accessible parameters, rather than in terms of $F(v)$, is mandatory.

## 14.3.2 DNMR Spectral Moments and LSCE OPs

The shape of the DNMR spectrum in an "inhomogeneous" LSCE is rather complex and depends on the shape of the nematic OP distribution function $w_S(S)$ as well as on the director OPs $\{S_l\}$. The analysis can be greatly simplified by considering the moments of the DNMR spectrum. Specifically, the first moment

$$M_{v,1}(\cos \vartheta) \equiv \int v\tilde{F}(\cos \vartheta, v) \, dv = \langle v \rangle \quad (14.5a)$$

and the second moment

$$M_{v,2}(\cos \vartheta) \equiv \int [v - M_{v,1}(\cos \vartheta)]^2 \tilde{F}(\cos \vartheta, v) \, dv = \langle v^2 \rangle - \langle v \rangle^2 \quad (14.5b)$$

are easily integrable from $F(v; \cos \vartheta)$:

$$M_{v,1}(\cos \vartheta) = \frac{3}{4} v_Q M_{S,1} S_2 P_2(\cos \vartheta), \quad (14.6a)$$

$$M_{v,2}(\cos \vartheta) = \frac{9}{16} v_Q^2 \left\{ (M_{S,2} + M_{S,1}^2) \left[ \frac{18}{35} S_4 P_4(\cos \vartheta) + \frac{2}{7} S_2 P_2(\cos \vartheta) + \frac{1}{5} \right] \right. \quad (14.6b)$$

$$\left. - M_{S,1}^2 S_2^2 P_2^2(\cos \vartheta) \right\}.$$

The first and the second DNMR line shape moments are therefore simple functions of the first and the second moments of $w_S(S)$,

$$M_{S,1} \equiv \int S \, w_S(S) \, dS = [S]_{av}, \quad (14.7a)$$

and

$$M_{S,2} \equiv \int [S - M_{S,1}(\cos \vartheta)]^2 w_S(S) \, dS = [S^2]_{av} - [S]_{av}^2, \quad (14.7b)$$

as well as of the nematic director, equivalently orientational OPs $S_2$ and $S_4$. In an inhomogeneous system, it is $[S]_{av}$ that assumes the role of nematic OP, whereas $[S^2]_{av} - [S]^2_{av}$ measures the width of its distribution. In the specific case of ideal domain alignment, that is, for $S_l = 1$, the spectral second moment directly probes the disorder in $S$, as $M_{v,2}(\cos \vartheta) = 9/16\, v_Q^2\, M_{S,2}\, P_2^2(\cos \vartheta)$. Parameters $M_{S,1}, M_{S,2}, S_2$, and $S_4$ can be in principle determined from the $\cos \vartheta$—dependences of spectral moments given by Equation 14.6. However, such an approach does not work in practice, since we have assumed that $M_{v,1}$ and $M_{v,2}$, or equivalently $\langle v \rangle$ and $\langle v^2 \rangle$, are calculated from and $\tilde{F}(v)$ and not from $F(v)$. For $\langle v^2 \rangle$ the result is the same in both cases. But in order to calculate $\langle v \rangle$ one must first decompose the spectrum $F(v)$ into its mirrored components $\tilde{F}(v)$ and $\tilde{F}(-v)$, since the first moment of $F(v)$ is zero. This is only possible, but subject to an increase in the experimental error of $\langle v \rangle$, if the two components are well separated. Such is the case at the orientation $v = 0$.

### 14.3.3 TRANSLATIONAL DIFFUSION AND MOTIONAL NARROWING OF INHOMOGENEOUS SPECTRAL LINES

Dynamical processes like fluctuations of the nematic or director OP or (restricted) molecular diffusion can substantially alter the spectrum. This problem is particularly acute with guest 8CB molecules that are free to diffuse over the LSCE network. In order to determine the "static" parameters of the LSCE network, $w_S(S)$ and $\{S_l\}$, without having to consider dynamical aspects, one must probe spectral parameters that are independent of the motion of deuteron spins.

DNMR line shape does not satisfy this condition, as it is strongly modified by the so-called RMTD (rotations mediated by translational displacements) mechanism. This mechanism is effective in systems with small-enough nematic domains. Specifically, spectral averaging will take place (Figure 14.7) whenever the characteristic RMTD frequency $v_{RMTD} = 6D/(2\pi \langle r^2 \rangle)$ becomes comparable to or exceeds the rigid-lattice spectral line width $\Delta v$, $v_{RMTD} \gtrsim \Delta v$. The latter is $\Delta v \approx 9v_Q S/8$ in poly-domain samples with homogeneous $S$ and $\Delta v \approx 3v_Q \sqrt{M_{S,2}}/4$ in perfectly aligned LSCEs with inhomogeneous $S$. If one approximates the 8CB–$\alpha d_2$ diffusion coefficient with its bulk nematic phase value $D = 6 \times 10^{-11}\, m^2/s$ and the domain size with $\langle r^2 \rangle^{1/2} \approx 1\, \mu m$, the above condition is indeed satisfied for any feasible value of $S$ or $M_{S,2}$.

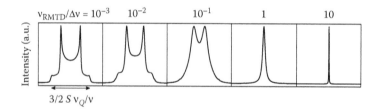

**FIGURE 14.7**  Theoretical motionally averaged DNMR spectra of a polydomain LC elastomer for different values of the ratio $v_{RMTD}/\Delta v$.

Contrary to DNMR spectral shapes, line shape moments $M_{v,1}$ and $M_{v,2}$ are "constants of motion" [27], that is, they are invariant to the RMTD process. Therefore, these experimentally determinable DNMR line shape parameters are exactly what we were looking for: they can be used to extract information on the average nematic OP $[S]_{av}$, its disorder $[S^2]_{av} - [S]_{av}^2$, and the degree of domain misalignment $S_2$, and $S_4$, without the need to consider motional averaging.

Extreme care should, however, be taken when calculating the line-shape moments. According to above conclusions, spectra of Figure 14.7 all have identical second moments. This is a bit counterintuitive, particularly in view of a Lorentzian-shaped spectra in the fast motional limit, $v_{RMTD} \gg \Delta v$, since it is a known fact that Lorentzian lines have infinite second moment. The discrepancy is resolved by noting that fast motional limit lines are Lorentzians only in the vicinity of the "isotropic" spectral peak. If the non-Lorentzian wings of the spectrum are properly taken into account, a correct second moment is obtained. Unfortunately, the wings have vanishing intensity and extend to $\pm\infty$ in the case of extreme motional narrowing, thus preventing experimental determination of the second moments. In the intermediate ($v_{RMTD} \approx \Delta v$) and slow-motion $v_{RMTD} < \Delta v$ regime, on the other hand, this problem is not present, so that the moments can be calculated from the experimental spectra with sufficient confidence.

## 14.4 LANDAU–DE GENNES PHENOMENOLOGICAL DESCRIPTION AND MODELING OF OPs

Let us now theoretically model $M_{v,1}$ and $M_{v,2}$ by assuming that the director disorder is quenched, that is, that the distribution $w_\mathbf{n}(\cos\theta, \phi; \cos\vartheta)$ and consequently $\{S_l\}$ are all independent of temperature. A further simplification is made by approximating the director distribution function by a spherical Gaussian [28] $w_{\mathbf{n},0}(\cos\theta) \propto \exp[(1 - \cos^2\theta)/(2\cos^2\theta \tan^2\sigma_\theta)]$, so that all director OPs can be expressed in terms of Gaussian error functions of a single parameter $\sigma_\theta$ that describes the angular dispersion of domain orientations about the specimen symmetry axis $\mathbf{c}$. Furthermore, the "heterogeneity" scenario is implicated by a Gaussian distributions of nominal transition temperatures $T^*$, $w_{T^*}(T^*)$, and of internal mechanical fields $g$, $w_g(g)$, with mean values $\bar{T}^*$ and $\bar{g}$ and dispersions $\sigma_{T^*}$ and $\sigma_g$, respectively [12]:

$$w_S(S; T, \bar{T}^*, \sigma_{T^*}, a, b, c, \bar{g}, \sigma_g) = \frac{1}{2\pi\sigma_{T^*}\sigma_g} \int \exp\left[-\frac{(T^* - \bar{T}^*)^2}{2\sigma_{T^*}^2} - \frac{(g - \bar{g})^2}{2\sigma_g^2}\right]$$

$$\delta[S - S_{LdG}(T, T^*, a, b, c, g)] \, dT^* dg. \qquad (14.8)$$

We have replaced $g_{int}$ with $g$ since none of the samples considered here was loaded externally ($g_{ext} = 0$). The above relation can be generalized to an arbitrary distribution of LdG free energy functional parameters.

The nematic OP $S_{LdG}(T, T^*, a, b, c, g)$, obtained from the minimization of the LdG free energy

$$F(S, T^*, a, b, c, g) = F_0 + \tfrac{a}{2}(T - T^*)S^2 + \tfrac{b}{3}S^3 + \tfrac{c}{4}S^4 - gS, \qquad (14.9)$$

is discontinuous at the transition (clearing) temperature $T_{N-PN} = T^* + 2b^2/(9ac) + 3cg/(ab)$ for internal field values $g$ that are below the critical value $g_c = -b^3/(27c^2)$, whereas it becomes a continuous function of $T$ for $g \geq g_c$ [7]. Equation 14.8 is used to calculate the nematic order moments $M_{S,1}(T, \bar{T}^*, \sigma_{T^*}, a, b, c, \bar{g}, \sigma_g)$ and $M_{S,2}(T, \bar{T}^*, \sigma_{T^*}, a, b, c, \bar{g}, \sigma_g)$. These are subsequently inserted into Equation 14.6 to obtain the theoretical spectral moments $M_{v,1}(\cos\vartheta, T; \bar{T}^*, \sigma_{T^*}, a, b, c, \bar{g}, \sigma_g, \sigma_\theta)$ and $M_{v,2}(\cos\vartheta, T; \bar{T}^*, \sigma_{T^*}, a, b, c, \bar{g}, \sigma_g, \sigma_\theta)$. Calculated this way, the moments encompass and probe the three basic structural characteristics of LSCEs: the heterogeneity, quantified by $\sigma_{T^*}$ and $\sigma_g$, the average internal mechanical field $\bar{g}$, and the misalignment of domains, quantified by $\sigma_\theta$. These values can be estimated from the simultaneous fits (with a single set of parameters for both $M_{v,1}$ and $M_{v,2}$) of these theoretical moments to the experimental temperature profiles.

The $w_S(T)$ profiles corresponding to the disorder in $T^*$ are different from the profiles corresponding to the disorder in $g$. This is so since changes in $T^*$ merely result in a shift of $T_{N-PN}$ while the shape of $S_{LdG}(T)$ is preserved, whereas changes in $g$ also alter the temperature profile of $S_{Ldg}(T)$. Exploiting this fact we find that high-temperature tails of $M_{2,v}(T)$ can be reproduced more perfectly by distributed internal fields $g$ rather than distributed transition temperatures $T^*$ (Figure 14.8). The disorder in $S$ therefore mainly arises from the disorder in local mechanical fields $g$.

Figure 14.9 shows the temperature evolution of the spectral moments in the LSCE-$x_{LC}$ sample group. As anticipated, the nominal transition temperature $\bar{T}_{N-PN}$ increases with decreasing 8CB–$\alpha d_2$ concentration $x_{LC}$ and approaches its limiting value $\bar{T}_{N-PN}(x = 0) \approx 357$ K. $\bar{T}_{N-PN}$ also coincides with the temperature of the $M_{v,2}(T)$ maximum in all samples. This provides for a direct experimental determination of $\bar{T}_{N-PN}$. The relatively high values of $S_2$ confirm that the samples retain their monodomain state even after being treated with low-molecular-mass mesogen.

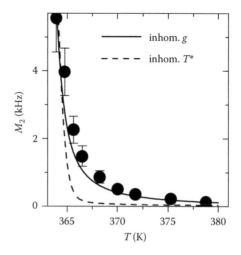

**FIGURE 14.8** High-temperature tails of the in $M_{v,2}(T)$ in V6-LSCE-0.105, fitted with a distribution of $g$ (solid line) and $T^*$ (dashed line).

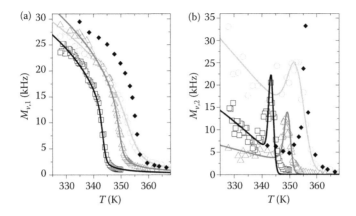

**FIGURE 14.9** Temperature dependences of (a) the first and (b) the second moments of LSCE-$x_{LC}$ DNMR line shapes at the orientation $\vartheta = 0$ for concentrations $x_{LC} = 0.28$ ($\square$), 0.08 ($\triangle$), 0.006 ($\diamond$), and 0.03* ($\bigcirc$). Best fits with the nematic OP distribution function $w_S(S)$ (Equation 14.8) are shown as solid lines.

We stress that it is the second moment of the spectrum, $M_{v,2}$, and not the first moment, $M_{v,1}$, which distinguishes between the "heterogeneous" (large $\sigma_{T*}$ or $\sigma_g$, $\bar{g} < g_c$) regime and the "supercritical" (small $\sigma_{T*}$ or $\sigma_g$, $\bar{g} \geq g_c$) regime (Figure 14.10). The $M_{v,1}(T)$ curves for "heterogeneous," "supercritical," or combined scenario are practically indistinguishable in both sample groups (LSCE-$x_{LC}$ and LSCE-$x_{c-1}$). The suppression of supercritical nature on increasing $x_{LC}$ and decreasing $x_{c-1}$ is clearly

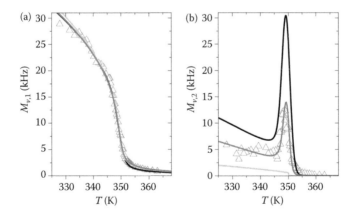

**FIGURE 14.10** Best simultaneous (a) $M_{v,1}(T)$ and (b) $M_{v,2}(T)$ fits of theoretical model of Section 14.4 to experimental data (black line) for LSCE-0.08. Also shown are the forced $\sigma_{T*} = 0$ fit (light gray solid line) and forced $g = g_c$ fit (dark gray line). The three theoretical $M_{v,1}(T)$ curves are almost identical and cannot be used to distinguish among "heterogeneous", "supercritical," or combined scenario of the N–PN transition. It is the $M_{v,2}(T)$ that are extremely sensitive to the changes in the model parameter. In LSCE-0.08, by far the best matching is obtained with $g/g_c = 1.5$, a fact, which clearly supports the supercritical scenario.

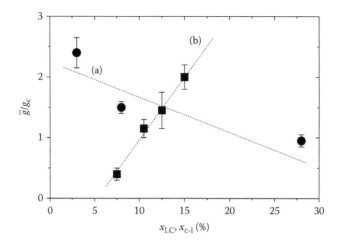

**FIGURE 14.11** (a) Suppression of supercritical behavior, evidenced by decreasing $\bar{g}/g_c$ on increasing the LC dopant concentration $x_{LC}$ in LSCE-$x_{LC}$; (b) promotion of supercritical behavior, evidenced by increasing $\bar{g}/g_c$ on increasing the cross-linking concentration $x_{c-l}$ in V6-LSCE-$x_{c-l}$.

evidenced from the experimentally determined dependence of $\bar{g}/g_c$ on the respective $x$ (Figure 14.11).

The above analytical method also clearly reveals that conventional LSCEs are not ideally aligned systems. Domain misalignment parameter $\sigma_\theta$ is typically between 10° and 15°, corresponding to nematic director orientational order of $S_2 \approx 0.9$.

## 14.5   SMEARED CRITICALITY IN LSCEs

A particularly notable property of the investigated LSCE systems is their inherent heterogeneity. As demonstrated above, the heterogeneity is characterized either by $\sigma_{T^*}$ or $\sigma_g$, that is, by the dispersion of the transition temperature $T^*$ or the internal mechanical field $g$ about their respective mean values $\bar{T}^*$ or $\bar{g}$. Due to a large number of fitting parameters, it is somewhat ambiguous to determine both $\sigma_{T^*}$ and $\sigma_g$ concurrently. Nevertheless, a reliable estimate of the heterogeneity is obtained by assuming either $\sigma_{T^*} = 0$ or $\sigma_g = 0$. With homogeneous $T^*$, the heterogeneity in $g$ is typically estimated at $\sigma_g \approx 0.5 - 0.8\, g_c$ [17]. This corresponds to $\sigma_{T^*} \approx 1–3$ K in the homogenous $g$ limit. As already discussed above, description in terms of homogeneous $T^*$ and heterogeneous $g$ seems to be more appropriate, for it predicts, in accordance with experimental data, a slower high-temperature decay of $M_{v,2}(T)$. Let us observe that $\sigma_g$ values in LSCE-$x_{c-l}$ systems with high $x_{c-l}$ are not much smaller than the mean values $\bar{g}$, so that $\bar{g} - \sigma_g < g_c$ can easily be satisfied. Consequently, the tails of the $g$-distribution function $w_g(g)$ extend below the critical value $g_c$ even if $\bar{g} > g_c$. The opposite is also true: in low $x_{c-l}$ systems, $w_g(g)$ extends beyond the critical value $g_c$ even if $\bar{g} < g_c$. Even though the average response of the system is supercritical, the local response in some domains can be critical and supercritical; vice

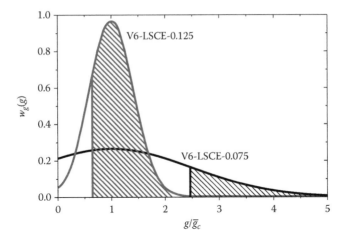

**FIGURE 14.12** Smeared criticality in LSCEs from the point of view of distribution function $w_g(g)$ of internal mechanical field $g$. The depicted distributions are calculated with the fit values of $\bar{g}$ and $\sigma_g$ for an effectively subcritical V6-LSCE-0.075 and effectively supercritical V6-LSCE-0.125. Shaded areas represent the supercritical parts of the respective distributions $w_g(g)$. (Adapted from Cordoyiannis, G. et al., *Phys. Rev. Lett.*, 99, 197801, 2007.)

versa, in an effectively subcritical system, there are domains that respond in a critical or supercritical manner (Figure 14.12). We shall term such a behavior "smeared criticality."

## 14.6  CALORIMETRIC INVESTIGATIONS OF LSCE NETWORKS

This Section describes the high-resolution calorimetric technique and its role in describing and characterizing LSCE phase behavior. The emphasis will be given on high-resolution ac and relaxation or the so-called nonadiabatic calorimetric techniques (Sections 14.6.1 and 14.6.2), which have been well established in elucidating the nature of LC phase transitions [29]. Section 14.6.3 deals with the calorimetric results obtained for N–PN phase conversion, both in LSCE-$x_{LC}$ and LSCE-$x_{c-1}$ systems.

### 14.6.1  ac CALORIMETRY

Heat-capacity data were taken by a computerized calorimeter. Description of the technique is extensively given in [29,30]. The calorimeter is capable of automated operation in either ac or relaxation modes in a temperature range from 80 to 470 K. LSCE sample was carefully wrapped in a Teflon tape and then placed in a sealed silver cell. The Teflon tape was used in order to prevent LSCE sample to stick to the cell surface, which might be a source of additional uncontrolled stress. Same approach was used in the DNMR experiments. Before the measurement, samples were annealed by cycling several times across the N–PN transition temperature. The sample cell is thermally linked to a temperature-stabilized bath (within 0.1 mK) by support wires and by air. The thermal link can be represented by a thermal resistivity $R$ typically between 200 and 250.

The classical high-resolution method of measuring heat capacity and enthalpy in various soft matter systems is ac calorimetry. Since it was discovered in 1960s by Kraftmakher [31], it was successfully applied for studying LCs [32]. In fact, most knowledge about the $C_p$ behavior in the vicinity of the LC phase transitions was acquired by the high-resolution ac calorimetry.

In the ac mode, an oscillating heat $P_{ac}e^{i\omega t}$ is supplied to the sample by a thin resistive heater. The frequency $\omega = 0.0767\,s^{-1}$ is chosen so that $1/\omega \gg \tau_{int}$. Here $\tau_{int}$ is the characteristic time for thermal diffusion in the sample cell. The temperature oscillations $T_{ac} = P_{ac}/(1/R + i\omega C)$ of the sample are detected by the small bead 1 M$\Omega$ thermistor. The complex heat capacity $C = C'(\omega) - iC''(\omega)$ is in most cases purely real, frequency-independent quantity $C = C'(0)$. Because of the anomalous response in the presence of latent heat, the ac mode does not provide quantitative value of the latent heat in the case of the first-order phase conversion. However, the phase shift $\tan(\Phi) = 1/\omega RC' + C'/C''$ of $T_{ac}$ can be useful in detecting the presence of the latent heat because it is exhibiting the anomalous peak within the two-phase coexistence range [33].

The data were taken on heating the annealed sample from the N to PN phase with the heating rates between 700 and 850 mK/h. The typical amplitudes of $T_{ac}$ were between 30 and 60 mK. The masses of LSCE samples were between 20 and 40 mg. The heat capacity of the empty cell and Teflon tape was later subtracted from the $C_p$ data. The net heat capacity $C_p$, obtained in this way, was divided by the mass of LSCE sample in order to obtain the specific heat capacity $C_p$ in J/gK.

### 14.6.2    RELAXATION CALORIMETRY

As mentioned in the previous section, the ac mode does not provide for quantitative value of the latent heat. Nevertheless, it can very precisely measure heat capacity $C_p$ as a function of temperature. The integral of $C_p(T)$ measures the continuous enthalpy changes $\delta H = \int C_P(T)\,dT$. Another method is thus necessary in order to elucidate the discontinuous enthalpy changes, that is, the latent heat. Such method, which was successfully employed in various systems, is relaxation calorimetry, or the so-called nonadiabatic scanning calorimetry [30]. It has recently been successfully applied in the detection of a critical point in ferroelectric relaxors [34].

This method was used for the first time in 1974 to study nematic to smectic-A transition [30] and its improved version was recently revived by Ema [30,35].

In the improved version of the relaxation method, the heater power supplied to the cell is linearly ramped [30]. Typically, after the bath temperature is stabilized to very high precision, the heating/cooling ramps are performed on the sample. During the heating run, $P = 0$ for $t < 0$, $P = \dot{P}t$ for $0 \le t \le t_1$, and $P = P_0 = \dot{P}t_1$ for $t > t_1$. For $t \le 0$, the initial sample temperature is equal to the bath temperature $T_B$. For $t \gg t_1$ the sample temperature eventualy reaches a plateau $T(\infty) = T_B + RP_0$. Here typically $t_1 \sim 480\,s$ during which about 1500 sample temperature $T(t)$ data points were taken. In the cooling run the heater power profile is reversed: $P = P_0$ for $t \le 0$, $P = P_0 - \dot{P}t$ for $0 \le t \le t_1$, and $P = 0$ for $t > t_1$. Here the initial temperature is $T(\infty)$ and the final temperature is equal to $T_B$. The effective heat capacity is calculated from $C_{eff}(T) = dH/dT = [P - (T - T_B)/R]/(dT/dt)$, where $R = (T(\infty) - T_B)/P_0$ and $P$ is the power at some time $0 \leq t \leq t_1$, corresponding to the sample temperature $T$ between the bath temperature $T_B$ and $T(\infty)$. The sample temperature heating/cooling rate $dT/dt$ is calculated over a short time interval centered at $t$. Except for a brief period of time just after $t = 0$ and for $t > t_1$, the temperature rate $dT/dt$ is nearly linear. The total enthalpy $\Delta H$ variation obtained in the ramping step is calculated as $\Delta H(T) = \int_{T_B}^{T} C_{eff}(T)dT$. After the completed heating/cooling cycle the bath temperature is shifted to a new value and the whole procedure is repeated. The typical heating/cooling rate of $dT/dt \sim 7.5$ K/h was used in the relaxation mode, with typical ramping steps $T(\infty) - T_B$ of about 0.7 K. The data obtained in the relaxation mode have slightly lower signal-to-noise ratio in comparison to the ac mode. In spite of that, the advantage of this mode, known also as nonadiabatic scanning mode, is a much better sensitivity to the latent heat than the ac mode, thus enabling the estimation of the discontinuous jump in the total enthalpy in the presence of latent heat. In practice, in addition to the determination of $C_p(T)$, estimation of the total enthalpy $\Delta H(T)$ and the contiunuous variations of the enthalpy $\delta H(T)$, the comparison of the enthalpy integrals $\delta H(T) = \int_{T_B}^{T} C_p(T)\,dT$ and $\Delta H(T) = \int_{T_B}^{T} C_{eff}(T)\,dT$ obtained in the ac and relaxation mode, respectively, provides for a quantitative estimation of the latent heat value $L = \Delta H - \delta H$.

## 14.6.3 CALORIMETRIC RESULTS

The heat capacity data can be used to locate the phase transition and provide for important information about the critical behavior in its vicinity. The calorimetric results show the existence of a vapor–liquid-type critical point both in LSCE-$x_{LC}$ and LSCE-$x_{c-l}$ systems.

### 14.6.3.1 Critical Point in LSCEs Doped with Low-Molar-Mass Nematogen

The effective specific heat of the LSCE-0.006 and LSCE-0.28 samples was measured across the N–PN phase transition by employing the ac and relaxation techniques simultaneously in order to detect and quantify the amount of latent heat, if present. Figure 14.13 shows the effective heat capacity determined in the weakly doped LSCE-0.006. Both data sets obtained in ac (open circles) and relaxation run (solid circles) perfectly match, thus overruling the presence of the latent heat. The smooth, rounded continuous evolution of the heat capacity anomaly suggests a noncritical N–PN evolution, that is, a supercritical behavior. Figure 14.14 shows the effective heat capacity determined in LSCE-0.28. A comparison of Figures 14.13 and 14.14 shows that the heat capacity anomaly in LSCE-0.28 is much more pronounced and sharper than in LSCE-0.006. Indeed, data sets obtained in the ac (open circles) and relaxation run (solid circles) do not match, indicating the presence of the latent heat. Although the amount of latent heat is relatively small ($L = 0.24$ J/g) in comparison wit the total enthalpy ($\Delta H = 2.91$ J/g), the nonzero latent heat clearly demonstrates that the N–PN transition is driven across the critical point to become weakly first order. This implies the existence of a vapor–liquid-type critical point in LSCE systems [16].

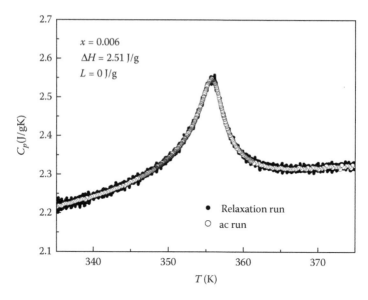

**FIGURE 14.13**   Specific heat temperature dependence obtained from the ac run (open circles) and relaxation run (solid circles) in LSCE-0.006. No difference between the relaxation and ac data excludes the presence of the latent heat.

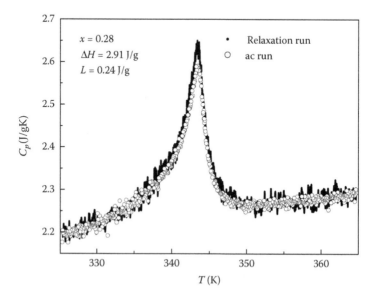

**FIGURE 14.14**   Specific heat temperature dependence obtained from the ac run (open circles) and relaxation run (solid circles) in LSCE-0.28. The difference between the relaxation and ac data indicates nonzero latent heat.

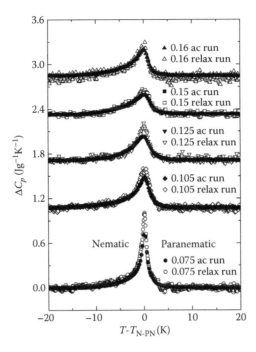

**FIGURE 14.15**  Temperature dependence of ac (solid symbols) and relaxation mode (open symbols) $C_p$ data obtained for different concentrations of V6 cross-linkers. Circles stand for $x = 0.075$, rhombs for $x = 0.105$, down-triangles for $x = 0.125$, squares for $x = 0.15$ and up-triangles for $x = 0.16$.

### 14.6.3.2 Critical Point in LSCEs with Different Cross-Linking Density

Figure 14.15 shows the $C_p$ data for several concentrations of the rod-like cross-linker V6. Data sets corresponding to different concentrations have been shifted for clarity along the $C_p$-axis. At low cross-linking concentrations $x_{c-l}$, the data of the ac and relaxation modes differ significantly. This demonstrates a significant presence of the latent heat, which is released in a few K broad coexistence range, indicating a broad distribution of the internal mechanical fields. With increasing cross-linking concentration, the latent heat gradually decreases and eventually vanishes at a critical concentration $x_{c-l}^{crit} \approx 0.15$, beyond which the transition becomes continuous. This behavior is in accordance with the scenario in which the first-order transition line is terminated in an isolated critical point.

In Figure 14.16, the $C_p$ data are plotted for the point-like V3 cross-linker. A small difference between the ac and relaxation data can be observed only at low cross-linking concentrations. Latent heat rapidly vanishes with increasing $x_{c-l}$. At the critical concentration $x_{c-l}^{crit} \approx 0.10$, transition becomes continuous in similar fashion as in the case of V6-LSCE-$x_{c-l}$ samples. A small amount of latent heat observed in V3 samples demonstrates that these samples are already very close to the critical point.

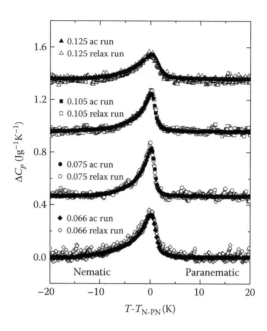

**FIGURE 14.16** Temperature dependence of ac (solid symbols) and relaxation mode (open symbols) $C_p$ data obtained for different concentrations of V3 cross-linkers. Rhombs stand for $x = 0.066$, circles for $x = 0.075$, squares for $x = 0.105$, and up-triangles for $x = 0.125$.

## 14.7   RELATING DNMR AND ac-CALORIMETRY RESULTS

The latent heat $L$ is related to the discontinuous jump in the nematic OP at the transition temperature $T_{N-PN}$ through [36]

$$L = \frac{\alpha}{2\rho} T_{N-PN}(S_N^2 - S_{PN}^2). \tag{14.10}$$

$S_N$ and $S_{PN}$ are the nematic OPs of the coexisting N and PN phase, respectively, at the transition temperature $T_{N-PN}$, whereas $\rho$ is the density of the LSCE network ($\rho \approx 1\ \text{mg/mm}^3$). The disorder in a heterogeneous system, which can be adequately modeled with a Gaussian like $w_g(g)$, results in a smearing of transition temperatures $T_{N-PN}(g)$. Equation 14.10 can then be generalized into

$$L = \frac{\alpha}{2\rho} \int w_g(g) T_{N-PN}(g)[S_N^2(g) - S_{PN}^2(g)]\, dg. \tag{14.11}$$

$L$ is therefore determinable from DNMR experiments, independently from calorimetric experiments. However, this is not a direct measurement of $L$, like in the case of ac calorimetry, since all $g$-dependant quantities in the integrand of Equation 14.11 need to be modeled theoretically (see Section 14.4).

Calorimetric and DNMR $L(x)$ data for the LSCE-$x_{c-1}$ specimen are summarized in Figure 14.17. The error bars take into account the data scattering in the heat capacity

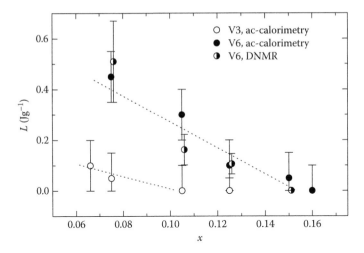

**FIGURE 14.17**   The cross-linking concentration dependence of the latent heat of the N–PN transition for the V6 and V3 cross-linked LSCEs as determined from calorimetric and DNMR experiments.

experiment and the cumulative errors of all fitting parameters in DNMR experiments. The two independent experimental methods yield consistent results, which undoubtedly reveal that the N–PN transition is driven toward the critical point and beyond with increasing cross-linking concentration $x_{c\text{-}l}$.

## 14.8   CONCLUSIONS

In this chapter, we have shown that the internal stress field $g$ and consequently the critical properties of the PN–N transition in LSCEs can be controlled in a systematic way by varying the concentration of the cross-linking molecules [17].

DNMR and ac calorimetry results clearly demonstrate that conventional side-chain LSCE networks are neither purely subcritical and heterogeneous, nor purely supercritical and homogeneous. A certain degree of disorder, that is, heterogeneity, in the nematic OP is inherently present in these systems. This disorder is accompanied by disorder in the orientation of nematic domains, which is however rather weak. The influence of heterogeneity on the nature of the N–PN transition is best summarized as "smeared criticality." This term is used to describe a thermodynamic system with coexisting subcritical, critical, and supercritical regions (domains, clusters). A straightforward manifestation of the smeared criticality is the presence of latent heat in many LSCE systems that exhibit supercritical effective thermodynamic response.

LSCEs can be tailored to exhibit thermodynamic response spanning from subcritical to supercritical. However, classification in terms of "subcritical," "critical", or "supercritical" is only applicable, due to smeared criticality, in describing the average, effective response. The conversion from the average subcritical behavior into the average supercritical behavior is provoked by increasing the concentration of cross-linking molecules (e.g., V6-LSCE-$x_{c-1}$ and V3-LSCE-$x_{c-1}$ networks). A reverse conversion is promoted by doping the systems with low-molar-mass nematic mesogens (e.g., LSCE-$x_{LC}$ networks).

DNMR methodology of determining the moments of the nematic OP distribution function by measuring the moments of the DNMR spectral line shapes, in combination with ac calorimetry, is not restricted solely to LSCEs. It is applicable to LCs in restricted geometries [37] in general, since in these systems, confinement is considered phenomenologically as a source of randomness that results in a distribution of nematic OP values. In this sense, confined LCs are similar to LSCEs where locally inhomogeneous mechanical fields $g$ assume the role of random fields, giving rise to smeared criticality. The DNMR analytical approach, described in this paper, is similar to the one used in the studies of disordered ferroelectrics to determine the moments of the distribution function of local OPs [38,39].

## REFERENCES

1. Brand, H. R., and Finkelmann, H., *Handbook of Liquid Crystals, Vol. 3: High Molecular Weight Liquid Crystals,* Eds Demus, D., Goodby, J., Gray, G. W., Spiess, H. -W., and Vill, V. (Wiley-VCH, Weinheim, 1998), p. 277.
2. Martinoty, P., Stein, P., Finkelmann, H., Pleiner, H., and Brand, H. R., Mechanical properties of monodomain side chain nematic elastomers, *Eur. Phys. J. E*, 14, 311, 2004.
3. Terentjev, E., and Warner, M., Commentary on "Mechanical properties of monodomain side chain nematic elastomers" by P. Martinoty, P. Stein, H. Finkelmann, H. Pleiner and H.R. Brand, *Eur. Phys. J. E*, 14, 323, 2004.
4. Martinoty, P., Stein, P., Finkelmann, H., Pleiner, H., and Brand, H., Reply to the Commentary by E. M. Terentjev and M. Warner on "Mechanical properties of monodomain side chain nematic elastomers", *Eur. Phys. J. E*, 14, 329, 2004.
5. Stenull, O., and Lubensky, T., Commentary on "Mechanical properties of monodomain side chain nematic elastomers" by P. Martinoty et al., *Eur. Phys. J. E*, 14, 333, 2004.
6. Martinoty, P., Stein, P., Finkelmann, H., Pleiner, H., and Brand, H., Reply to the Commentary by O. Stenull and T.C. Lubensky on "Mechanical properties of monodomain side chain nematic elastomers", *Eur. Phys. J. E*, 14, 339, 2004.
7. Warner, M., and Terentjev, E. M., *Liquid Crystal Elastomers* (Clarendon Press, Oxford, 2003).
8. Küpfer, J., Nishikawa, E., and Finkelmann, H., Densely crosslinked liquid single-crystal elastomers, *Polym. Adv. Technol.*, 5, 110, 1994.
9. Disch, S., Schmidt, C., and Finkelmann, H., Nematic elastomers beyond the critical-point, *Macromol. Rapid Commun.*, 15, 303, 1993.
10. Küpfer, J., and Finkelmann, H., Liquid-crystal elastomers–influence of the orientational distribution of the cross-links on the phase-behavior and reorientation process, *Macromol. Chem.*, 195, 1353, 1994.
11. Hébert, M., Kant, R., and de Gennes, P. -G., Dynamics and thermodynamics of artificial muscles based on nematic gels, *J. Phys. I France*, 7, 909, 1997.
12. Selinger, J. V., Jeon, H. G., and Ratna, B. R., Isotropic–nematic transition in liquid-crystalline elastomers, *Phys. Rev. Lett.*, 89, 225701-1, 2002.
13. de Gennes, P. -G., Réflexions sur un type de polymères nématiques, *C. R. Acad. Sc. Paris B*, 281, 101, 1975.

14. Boamfa, M. I., Viertler, K., Wewerka, A., Stelzer, F., Christianen, P. C. M., and Maan, J. C., Magnetic-field-induced changes of the isotropic–nematic phase transition in side-chain polymer liquid crystals, *Phys. Rev. E*, 67, 050701-1, 2003.
15. Küpfer, J., and Finkelmann, H., Nematic liquid single-crystal elastomers, *Makromol. Chem. Rapid Commun.*, 12, 717, 1991.
16. Lebar, A., Kutnjak, Z., Žumer, S., Finkelmann, H., Sánchez-Ferrer, A., and Zalar, B., Evidence of supercrtitical behavior in liquid single crystal elastomers, *Phys. Rev. Lett.*, 94, 197801, 2005.
17. Cordoyiannis, G., Lebar, A., Zalar, B., Žumer, S., Finkelmann, H., and Kutnjak, Z., Criticality controlled by cross-linking density in liquid single-crystal elastomers, *Phys. Rev. Lett.*, 99, 197801, 2007.
18. Yusuf, Y., Ono, Y., Sumisaki, Y., Cladis, P. E., Brand, H. R., Finkelmann, H., and Kai S., Swelling dynamics of liquid crystal elastomers swollen with low molecular weight liquid crystals, *Phys. Rev. E*, 69, 021710-1, 2004.
19. Severing, K., and Saalwächter, K., Biaxial nematic phase in a thermotropic liquid-crystalline side-chain polymer, *Phys. Rev. Lett.*, 92, 125501-1, 2004.
20. Verwey, G. C., and Warner, M., Nematic elastomers cross-linked by rigid rod linkers, *Macromolecules*, 30, 4196, 1997.
21. Greve, A., and Finkelmann, H., Nematic elastomers: The dependence of phase transformation and orientation processes on crosslinking topology, *Macromol. Chem. Phys.*, 202, 2926, 2001.
22. Petridis, I., and Terentjev, E. M., Nematic–isotropic transition with quenched disorder, *Phys. Rev. E*, 74, 051707, 2006.
23. Zalar, B., Lavrentovich, O. D., Zeng, H., and Finotello, D., Deuteron NMR investigation of a photomechanical effect in a smectic-A liquid crystal, *Phys. Rev. E*, 62, 2252, 2000.
24. Davis, J. H., Jeffrey, K. R., Bloom, M., Valic, M. I., and Higgs, T. P., Quadrupolar echo deuteron magnetic-resonance spectroscopy in ordered hydrocarbon chains, *Chem. Phys. Lett.*, 42, 390, 1976.
25. Boeffel, C., and Spiess, H. W., Highly ordered main chain in a liquid-crystalline side-group polymer, *Macromolecules*, 21, 1626, 1988.
26. Vilfan, M., Apih, T., Gregorovič, A., Zalar, B., Lahajnar, G., Žumer, S., Hinze, G., Böhmer, R., and Althoff, G., Surface-induced order and diffusion in 5CB liquid crystal confined to porous glass, *Mag. Res. Imag.*, 19, 433, 2001.
27. Cowan, B., *Nuclear Magnetic Resonance and Relaxation* (Cambridge University Press, Cambridge, 1997).
28. Zalar, B., Blinc, R., Albert, W., and Petersson, J., Discrete nature of the orientational glass ordering in $Na_{1-x}K_xCN$, *Phys. Rev. B*, 56, R5709, 1997.
29. Kutnjak, Z., Kralj, S., Lahajnar, G., and Žumer, S., Calorimetric study of octyl-cyanobiphenyl liquid crystal confined to a controlled-pore glass, *Phys. Rev. E*, 68, 021705, 2003.
30. Yao, H., Ema, K., and Garland, C. W., Nonadiabatic scanning calorimeter, *Rev. Sci. Instrum.*, 69, 172, 1998.
31. Kraftmakher, Y. A., Modulation method for measuring specific heat, *Zh. Prikl. Mekh. Tekh. Fiz.* (in Russian), 5, 176, 1962.
32. Garland, C. W., Calorimetric studies, in *Liquid Crystals: Experimental Study of Physical Properties and Phase Transitions*, edited by S. Kumar (Cambridge University Press, Cambridge, 2001), p. 240.
33. Djurek, D., Baturic-Rubcic, J., and Franulovic, K., Specific-heat critical exponents near the nematic–smectic-A phase transition, *Phys. Rev. Lett.*, 33, 1126, 1974.

34. Kutnjak, Z., Petzelt, J., and Blinc, R., The giant electromechanical response in ferroelectric relaxors as a critical phenomenon, *Nature*, 441, 956, 2006.

35. Ema, K., Uematsu, T., Sugata, A. and Yao, H., Complex calorimeter with AC- and relaxation-mode operation, *Jpn. J. Appl. Phys.*, 32, 1846, 1993.

36. Jamee, P., Pitsi, G., and Thoen, J., Systematic calorimetric investigation of the effect of silica aerosils on the nematic to isotropic transition in heptylcyanobiphenyl, *Phys. Rev. E*, 66, 021707, 2002.

37. Finotello, D., Phase transitions in restricted geometries, in *Liquid Crystals in Complex Geometries Formed by Polymer and Porous Networks*, Eds G. P. Crawford and S. Žumer (Taylor & Francis, London, 1996), p. 325.

38. Blinc, R., Dolinšek, J., Pirc, R., Tadić, B., Zalar, B., Kind, R., and Liechti, O., Local-polarization distribution in deuteron glasses, *Phys. Rev. Lett.*, 63, 2248, 1989.

39. Blinc, R., Dolinšek, J., Gregorovič, A., Zalar, B., Filipič, C., Kutnjak, Z., Levstik, A., and Pirc, R., Local polarization distribution and Edwards–Anderson order parameter of relaxor ferroelectrics, *Phys. Rev. Lett.*, 83, 424, 1999.

# 15 Computer Simulations of Liquid Crystal Polymeric Networks and Elastomers

*G. Skačej and Claudio Zannoni*

## CONTENTS

## 15.1 INTRODUCTION

Weakly cross-linked polymeric networks—elastomers—with embedded liquid crystalline units represent a novel class of functional soft materials. They couple the elastic properties of rubbers with the orientational anisotropy of liquid crystals [1]. Accordingly, liquid crystal elastomers (LCE) are characterized by pronounced responsiveness to external stimuli such as temperature variation, application of external fields, mechanical or electric, or irradiation with ultraviolet (UV) light, that can be exploited, for example, for the construction of new actuator and sensor devices [2]. The design of these devices relies on the (at least qualitative) understanding of structural features and system behavior on the microscopic level. On the fundamental side, irregular cross-links in a LCE network result in quenched disorder similar to that observed in spin glasses with magnetic impurities [3] or random anisotropy [4]. Hence, LCE have become interesting both from theoretical and application point of view. Theoretically, LCE have been described at the continuum level by anisotropic rubber elasticity [1] and by phenomenological Landau-type approaches [5–7]. Finite-element continuum computer simulations have been used to study various phenomena in macroscopic LCE samples [8]. Along with the existing continuum descriptions that are based on a specific free energy expression, microscopic molecular-level or coarse-grained computer simulation studies can provide significant complementary insight

**451**

into LCE behavior. Unfortunately, the modeling and simulation of these complex systems is not straightforward and a number of attempts have been made so far in the field of microscopic computer simulation of LCE. According to the level of detail treated in the modeling, they can be classified into three groups: (1) lattice models, (2) anisotropic beads and spring models, and (3) atomistic models.

1. Lattice models are coarse-grained and thus typically contain only the very essential features of LCE [9–13]: The orientational degrees of freedom for the liquid crystal units embedded into a continuous polymeric matrix. Computationally, they are less expensive so simulation samples can be made large enough even for a qualitative prediction of selected experimental observables.
2. Anisotropic beads and spring models consist, for example, of Gay–Berne particles linked into a network [14,15]; they contain both orientational and translational degrees of freedom for all network components and are hence more realistic, but also computationally more demanding. First stress–strain experiments on rather small samples have already been carried out.
3. Atomistic models contain details within mesogenic and polymer units. As such at present they do not allow for a study of an extended polymeric network but rather focus on intramolecular details such as UV light-induced conformation change. Ideally, in this hierarchy the high-level-of-detail approaches should provide information on parameters entering the coarse-grained models that, in turn, should then be used in large-scale simulations for the prediction of experimental observables. This approach has been developed until now only for relatively simple polymeric systems (see, e.g., Ref. [16]).

In this chapter we are going to focus on lattice modeling where most progress has been made until now.

## 15.2 LATTICE MODELING

### 15.2.1 UNIAXIAL ELASTOMER

Lattice modeling refers to a class of coarse-grained nonphenomenological studies considering the lowest possible degree of microscopic detail. Following the approach presented in Refs. [10–12], there are three main contributions to the Hamiltonian of the system: (1) rubber elasticity of polymer chains, (2) interactions between mesogenic units (anisotropic van der Waals or steric), and (3) the strain-alignment coupling. In the modeling, polymeric chains are replaced by a stretchable elastic lattice, and the mesogenic material within one unit cell by a rotor/particle, that is, a unit vector $\mathbf{u}_i$ representing a cluster of $\sim 10^2$ close-packed liquid crystal molecules (see Figure 15.1).

1. The rubber elastic part of the pseudo-Hamiltonian for a uniaxial deformation by a factor of $\lambda$ (Figure 15.1) can be written (for noninteracting chains,

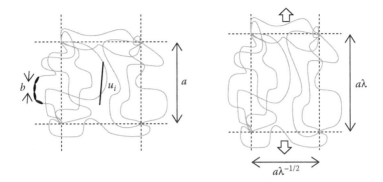

**FIGURE 15.1** Coarse-grained lattice model for LCE: deformable unit cell. (Reprinted from P. Pasini, G. Skačej, and C. Zannoni, *Chem. Phys. Lett.* **413**, 463, 2005. Copyright 2005 with permission from Elsevier.)

assuming deformation affinity and sample incompressibility) as

$$\mathcal{H}_e = N k_B T \alpha \left( \lambda^2 + \frac{2}{\lambda} \right), \tag{15.1}$$

where $N$ is the number of lattice unit cells and $\alpha = 3a^2/2Mb^2$ is a constant related to the elastic modulus of the sample. (Here $a$ stands for the average distance between cross-links, $M$ for the average monomer number between two cross-links, and $b$ for monomer size.)

2. Interactions between mesogenic units are modeled by

$$\mathcal{H}_n = - \sum_{\langle i<j \rangle} \epsilon_{ij} P_2(\mathbf{u}_i \cdot \mathbf{u}_j), \tag{15.2}$$

where $P_2(x) \equiv (1/2)(3x^2 - 1)$. The above sum goes over nearest-neighbor cells $i$ and $j$, assuming periodic boundary conditions, $\epsilon_{ij}$ being the corresponding interaction strengths. For $\epsilon_{ij} = \epsilon > 0$, $\mathcal{H}_n$ promotes parallel alignment of $\mathbf{u}_i$ as in a liquid crystal [17,18]. On the other hand, for $\epsilon_{ij}$ sampled from a distribution with positive and negative $\epsilon_{ij}$, a highly frustrated system is obtained. The first case corresponds to a regular sample cross-linked in an aligned state, while the second is more irregular and similar to a spin glass [3]. Here a reduced temperature scale is conveniently introduced as $T^* = k_B T/\epsilon$.

3. The coupling between strain and alignment of liquid crystalline units is modeled through a mechanical aligning field, in analogy with the mean-field Maier–Saupe theory for ordinary nematics [19]. For a uniaxial stretch along the $z$-axis by a factor of $\lambda$ the pseudo-Hamiltonian can be written as

$$\mathcal{H}_c = -k_B T \chi Q(\lambda) \sum_{i=1}^{N} P_2(\mathbf{u}_i \cdot \mathbf{z}), \tag{15.3}$$

where $\chi$ is a coupling constant, $\mathbf{z}$ denotes a unit vector parallel to the $z$-axis, and $Q(\lambda)$ is associated with the mechanical field strength. Here $Q(\lambda)$ is introduced as a quadrupolar average measuring the anisotropy of the polymer chain end-to-end tensor distribution, where this distribution is represented by a uniaxial ellipsoid $\varepsilon$ obtained from an isovolume deformation of a unit sphere by a factor of $\lambda$ along the $z$-axis. Denoting with $\theta$ the polar angle measured with respect to $\mathbf{z}$, and with $\Omega$ the corresponding solid angle, the integration over $\varepsilon$ gives $Q(\lambda) = (4\pi)^{-1} \int_\varepsilon P_2 (\cos\theta)\, d\Omega = (3\lambda^3/2)L(\lambda^3-1)-1/2$, where

$$L(\zeta) = \begin{cases} \left(\text{atanh}\sqrt{-\zeta} - \sqrt{-\zeta}\right)/\sqrt{-\zeta^3}, & -1 < \zeta < 0, \\ 0, & \zeta = 0, \\ \left(\sqrt{\zeta} - \arctan\sqrt{\zeta}\right)/\sqrt{\zeta^3}. & \zeta > 0; \end{cases} \tag{15.4}$$

The dependence $Q(\lambda)$ is shown in Figure 15.2 as inset. In "positive" (e.g., main chain) materials with $\chi > 0$ stretching ($\lambda > 1$) favors $\mathbf{u}_i$ alignment along $z$, while compression results in alignment perpendicular to $z$. (Vice versa for "negative" materials with $\chi < 0$, such as some side-chain materials.) Then, the total Hamiltonian of the system is given by $\mathcal{H} = \mathcal{H}_e + \mathcal{H}_n + \mathcal{H}_c$. A similar Hamiltonian has been derived by Selinger and Ratna [13], however, starting from continuum neoclassical rubber theory [1].

To explore the behavior of such model elastomers, large-scale constant-stress ($\sigma$) Monte Carlo (MC) simulations have been performed. The analysis presented in Ref. [10] has followed the standard MC algorithms [18,20,21] and has been performed for sample sizes $N = 30^3$ or $50^3$, with $\chi = 0.5$, $a = 4.6$ nm, $b = 1$ nm, and $M = 100$ (yielding $\alpha \approx 0.3$). The stress–strain and temperature scan experiments were implemented as a series of MC stretch/release and heating/cooling runs, respectively, with at least 60/66 MC kcycles for equilibration/production for given parameter values.

Stress–strain isotherms $\lambda(\sigma^*)$ (where $\sigma^* = \sigma a^3/\epsilon$ stands for reduced engineering stress) for a regular sample with $\epsilon_{ij} = \epsilon > 0$ and $50^3$ unit cells are shown in Figure 15.2 (a). From equilibrium $\lambda$ values at zero stress one can see that the isotropic–nematic (IN) transition is observed at $T^* = 1.141 \pm 0.003$, which is slightly higher than in an ordinary nematic due to the stabilizing effect of the mechanical field. Stretching the sample in the nematic phase (curves $a$ and $b$ in Figure 15.2), a Hookean behavior is seen, with an estimated Young modulus of $E \sim 52$ kPa (assuming that $\epsilon \approx 0.023$ eV.) When the stretching starts, on the other hand, from the isotropic phase, there is a stress-induced first-order aligning transition, seen also experimentally [7], at a $T^*$-dependent value of $\sigma^*$, accompanied by observable stretch/release hysteresis. Above $T^* \sim 1.150$, however, the discontinuity in the aligning transition disappears and the $\lambda(\sigma^*)$ isotherms become continuous. In between these two behavior types there is a critical point at approximately $\sigma_c^* = 0.021 \pm 0.002$, $T_c^* = 1.149 \pm 0.002$. Such critical behavior has been predicted by de Gennes [22], and is similar also to the behavior of ordinary nematics liquid crystals exposed to, for example, an external electric field [23].

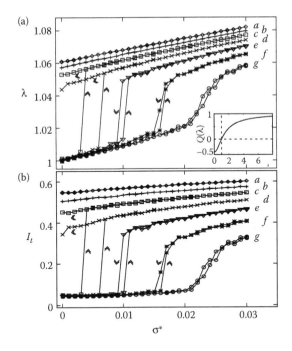

**FIGURE 15.2** Regular sample stretched: stress–strain isotherms (a) and light transmittance (b). Curves $a$–$g$ correspond to $T^* = 1.135$–$1.15$, with a step size of $0.0025$. Inset: $Q(\lambda)$ dependence. Arrows denote the scan direction. (Reprinted from P. Pasini, G. Skačej, and C. Zannoni, *Chem. Phys. Lett.* **413**, 463, 2005. Copyright 2005 with permission from Elsevier.)

Upon stretching, the nematic director aligns along $\mathbf{z}$ and the sample becomes birefringent. (Birefringence measurements have been used to explore the optical behavior of LCE under stress [1,7,24].) A measure of macroscopic degree order is given by $P_2^z = \langle P_2(\mathbf{u}_i \cdot \mathbf{z}) \rangle$ (averaged over lattice sites and MC cycles) or, alternatively, by the simulated light intensity when a beam, directed along the $y$-axis, passes through a sample placed between crossed polarizers at $\pm 45°$ from $\mathbf{z}$. In this case, light transmittance is given by [25]

$$I_t = \sin^2(\delta/2), \tag{15.5}$$

with $\delta = 2\pi d \Delta n/\lambda_0$ being the phase difference between the ordinary and extraordinary polarization accumulated on passing through a sample; here $d$ is the sample thickness, $\lambda_0$ light wavelength in vacuum, and $\Delta n$ the refractive index anisotropy. For low $\Delta n$, one has $\Delta n \propto P_2^z$, and, consequently, $I_t$ reflects orientational ordering in the sample. Light propagation within each pixel in the $xz$-plane was modeled using the Jones matrix formalism [25], neglecting diffraction and scattering, assuming that local optical axes coincide with $\mathbf{u}_i$. Before averaging the transmitted light intensity over pixels to obtain $I_t$, $L = 10$ light passes through the sample were allowed to yield $d \approx 2.3\ \mu$m. Other parameters: $\Delta n = 0.2175$ (for a perfectly ordered sample), and $\lambda_0 = 632.8$ nm. The $I_t(\sigma^*)$ curves shown in Figure 15.2 reveal a good correlation with the stress–strain isotherms. In an isotropic sample almost no light is transmitted, while a nonzero $I_t$ is a signature of ordering in the system.

Turn now to the effects of cross-link inhomogeneity and to temperature scan experiments. We have studied $N = 30^3$ samples of four different LCE types, A–D, with increasing degree of quenched disorder. Sample A corresponds to a regularly cross-linked LCE with $\epsilon_{ij} = \epsilon$, as before. For the B, C, and D types, $\epsilon_{ij}$ was sampled from a Gaussian distribution of width $\epsilon$, centered at $\langle \epsilon_{ij} \rangle \approx 0.8\epsilon, 0.5\epsilon$, and 0, respectively. Figure 15.3 shows the temperature dependences of the average elongation $\lambda$ and the degree of order for samples A–D, for $\sigma^* = 0$ and $\sigma^* = 0.1$ (A and B only). Cooling isotropic samples A–C, at a given temperature mesogenic units align, for example, at $T_A^* \approx 1.15 \pm 0.01$ for unstressed sample A. If, on the other hand, the cooling is carried out under stress, for $\sigma^* = 0.1 > \sigma_c^*$ the sample becomes supercritical and the IN transition is smeared out [26]. In samples B and C, the IN transition shifts towards lower temperatures, with $T_B^* \approx 0.87 \pm 0.01$ and $T_C^* \approx 0.49 \pm 0.01$. The highly frustrated sample D, however, remains macroscopically disordered even as $T^* \to 0$, as suggested by $P_2^z \approx 0$ for all $T^*$. To check for local ordering at low $T^*$, a "local" order parameter $S_L$ can be introduced, which is calculated as follows. For each unit cell the MC cycle-averaged ordering matrix $Q_i = (1/2)\langle 3\mathbf{u}_i \otimes \mathbf{u}_i - I \rangle$ is calculated and diagonalized. Then the largest eigenvalue $S_i$ and the local director $\mathbf{n}_i$ are identified, and finally $S_L = N^{-1} \sum_{i=1}^{N} S_i$. $S_L$ takes the role of the Edwards–Anderson order parameter used for detecting local order in spin glasses [27]. The temperature dependence of $S_L$ indeed shows local ordering in the system below $T^* \sim 0.2$: orientational fluctuations of $\mathbf{u}_i$ are frozen-in along the local $\mathbf{n}_i$, which is analogous to glassy states observed in magnetic systems. The variance of $S_L$ peaks at $T_g^* \approx 0.1$ and locates the transition to a locally ordered glassy state. Note that sample C (with $\langle \epsilon_{ij} \rangle \neq 0$) exhibits both the IN and the glass transition and thus qualitatively covers the elastomer behavior over a wide temperature range. Nematic alignment is accompanied by a spontaneous macroscopic deformation of the sample. For $\chi > 0$ the sample is elongated on cooling. The deformation magnitude (here $\sim 10\%$) increases with increasing $\chi$. In sample D for $\sigma^* = 0$ there is no macroscopic alignment and therefore no extension. The transmittance $I_t(T^*)$, Figure 15.3, agrees with the behavior of $P_2^z$. Here $L = 15$ and $d \approx 2.07\ \mu\text{m}$.

While light transmission methods measure the overall degree of orientational order, $^2$H NMR can be used to probe local order in LCE [1,28] and hence distinguish between isotropic, aligned, and glassy states. Quadrupolar interactions in deuterated mesogenic units result in a frequency splitting [29]

$$\omega_Q^j = \pm \delta \omega_Q P_2 (\mathbf{u}_j \cdot \mathbf{b}),\tag{15.6}$$

where the unit vector $\mathbf{b}$ determines the orientation of the NMR spectrometer magnetic field and $\delta \omega_Q$ is a coupling constant. To calculate NMR spectra, for $\mathbf{b} \| \mathbf{z}$ the free induction decay signal

$$G(t) = \left\langle \exp \left( i \int_0^t \omega_Q^j(t')\, dt' \right) \right\rangle_j \tag{15.7}$$

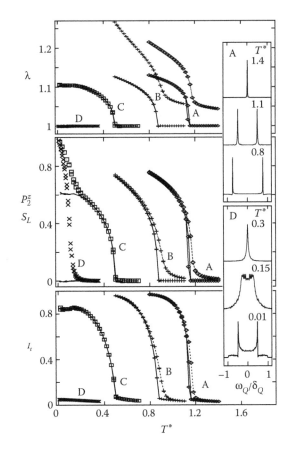

**FIGURE 15.3**   Samples A–D: temperature scans. Sample length $\lambda$ (top), $P_2^z$ (lines) and $S_L$ (symbols) order parameters (center), and transmittance $I_t$ (bottom). No stress and $\sigma^* = 0.1$ (solid and dotted lines, respectively). Insets: $^2$H NMR spectra for $\sigma^* = 0$; top: sample A, bottom: sample D. (Reprinted from P. Pasini, G. Skačej, and C. Zannoni, *Chem. Phys. Lett.* **413**, 463, 2005. Copyright 2005 with permission from Elsevier.)

was generated and Fourier-transformed [30]. Here $\langle \ldots \rangle_j$ stands for ensemble averaging; diffusion is neglected. Typical line shapes for samples A and D are shown as insets in Figure 15.3. Below $T_{IN}^*$ regular sample A gives double-peaked spectra, with peaks at $\omega_Q = \pm \delta \omega_Q P_2^z$. Glassy sample D below $T_g^*$, however, yields a powder pattern, with a width proportional to $S_L$.

Note that the random-bond quenched disorder mimicking the effect of local chemical inhomogeneity has also been studied by Selinger and Ratna, focusing on the width of the IN phase transition in elastomers [6,13]. At the same time, they also considered random-field quenched disorder, representing the presence of orientationally frozen-in anisotropic cross-linkers in the elastomer sample. Both mechanisms are seen to broaden the IN transition. Random-field quenched disorder was also studied in a two-dimensional unstretchable lattice elastomer by Yu et al. [9], with a special emphasis on long-range correlations in the nematic director orientation.

## 15.2.2 BIAXIAL ELASTOMER

In nematic elastomers, liquid crystal units usually exhibit uniaxial orientational ordering. However, a possibility of biaxial ordering in ordinary thermotropic nematics has been predicted both theoretically [31] and from MC simulations [32,33]. Moreover, recently there have been several experimental observations of biaxial order in thermotropic nematics [34–37]. Biaxial order has also been observed in $^2$H-NMR studies of nematic liquid-crystalline side-chain polymers [38,39]. Therefore, one might soon expect the synthesis of biaxial nematic elastomers capable of showing biaxial orientational order and, as a result, biaxial elastic deformations in a monodomain sample. This section will be dealing with an extension of coarse-grained lattice models to biaxial elastomers, as presented in Ref. [12]. Like in the uniaxial case, the Hamiltonian of the system can be split into three contributions: the rubber elastic part $\mathcal{H}_e$, the nematic part $\mathcal{H}_n$, and the strain-alignment coupling part $\mathcal{H}_c$.

In case of biaxial deformations, Equation 15.1 has to be rewritten as

$$\mathcal{H}_e = Nk_BT\alpha(\lambda_x^2 + \lambda_y^2 + \lambda_z^2). \tag{15.8}$$

Here it has been again assumed that the biaxial deformation is homogeneous and affine, and that it changes the sides of each unit cell from $\{a, a, a\}$ into $\{\lambda_x a, \lambda_y a, \lambda_z a\}$. To meet the incompressibility constraint, one has $\lambda_x\lambda_y\lambda_z = 1$.

The particles representing close-packed clusters of mesogenic molecules now become biaxial objects. The orientation of each particle is hence described by an orthonormal triad $\{s_j, t_j, u_j\}$, where $t_i$ and $u_i$ correspond to the short and long axis of the cluster, respectively. Assuming that the interaction between the molecular clusters is mainly of dispersive origin, the corresponding Hamiltonian reads [32]

$$\mathcal{H}_n = -\epsilon\sum_{\langle ij \rangle}\{R_{00}^2(\omega_{ij}) + 2v[R_{02}^2(\omega_{ij}) + R_{20}^2(\omega_{ij})] + 4v^2R_{22}^2(\omega_{ij})\}, \tag{15.9}$$

where $v$ is the cluster biaxiality parameter, $\omega_{ij}$ stands for the relative orientation of the neighboring particles $i$ and $j$, defined by the three Euler angles, $\alpha_{ij}, \beta_{ij}$, and $\gamma_{ij}$. $R_{mn}^2$ are combinations of Wigner functions and are given by [40]

$$R_{00}^2(\omega_{ij}) = \frac{3}{2}\cos^2\beta_{ij} - \frac{1}{2}, \tag{15.10}$$

$$R_{02}^2(\omega_{ij}) = \sqrt{\frac{3}{8}}\sin^2\beta_{ij}\cos 2\gamma_{ij}, \tag{15.11}$$

$$R_{20}^2(\omega_{ij}) = \sqrt{\frac{3}{8}}\sin^2\beta_{ij}\cos 2\alpha_{ij}, \tag{15.12}$$

$$R_{22}^2(\omega_{ij}) = \frac{1}{4}(\cos^2\beta_{ij} + 1)\cos 2\alpha_{ij}\cos 2\gamma_{ij} - \frac{1}{2}\cos\beta_{ij}\sin 2\alpha_{ij}\sin 2\gamma_{ij}. \tag{15.13}$$

For $v = 0$ $\mathcal{H}_n$ reduces to the previous expression for uniaxial particles, Equation 15.2, while $0 < v < 1/\sqrt{6}$ and $v > 1/\sqrt{6}$ represent prolate (rod-like) and oblate (plate-like) biaxial particles, respectively. Depending on temperature and $v$, an ensemble of biaxial particles can be found in the isotropic ($I$), nematic ($N$), or biaxial ($B$) phase. In the $I$ phase, no orientational order is present. In the $N$ phase, for prolate particles their long axes $\mathbf{u}_i$ align ($N_+$ phase), while oblate particles align along their short axes ($N_-$ phase). In the biaxial phase, all three particle axes are aligned [32]. Again, periodic boundary conditions are assumed.

Finally, to describe the strain-alignment coupling the concept of uniaxial mechanical field needs to be generalized to the biaxial case. Following the mean-field treatment of biaxial orientational ordering in liquid crystals proposed by Straley [31] and the approach adopted in the previous section for uniaxial particles, $\mathcal{H}_c$ is given by

$$\mathcal{H}_c = -k_B T \chi \sum_{i=1}^{N} [r_{00}^2(\lambda) R_{00}^2(\omega_i) + 2r_{20}^2(\lambda) R_{20}^2(\omega_i)], \qquad (15.14)$$

where $\chi$ is a coupling constant and $\omega_i$ are the Euler angles for a relative orientation of $i$th particle and $\{\mathbf{x}, \mathbf{y}, \mathbf{z}\}$—an orthonormal triad defining the principal axes of the biaxial sample deformation. Further, $r_{mn}^2(\lambda)$ are the deformation-dependent quantities that measure the anisotropy of the polymer chain end-to-end tensor distribution. Like in the uniaxial case this distribution is represented by an ellipsoid $\varepsilon$ obtained by deforming a unit sphere at constant volume by the factors $\lambda_x$, $\lambda_y$, and $\lambda_z$ along $\mathbf{x}$, $\mathbf{y}$, and $\mathbf{z}$, respectively. Denoting with $\theta$ and $\phi$ the polar and azimuthal angles measured with respect to $\mathbf{z}$ and $\mathbf{x}$, respectively, and with $\Omega$ the corresponding solid angle, the $r_{mn}^2$s are defined as

$$r_{00}^2(\lambda) = (4\pi)^{-1} \int_{\varepsilon(\lambda)} \left( \frac{3}{2} \cos^2 \theta - \frac{1}{2} \right) d\Omega, \qquad (15.15)$$

$$r_{20}^2(\lambda) = (4\pi)^{-1} \int_{\varepsilon(\lambda)} \sqrt{\frac{3}{8}} \sin^2 \theta \cos 2\phi \, d\Omega \qquad (15.16)$$

and are calculated numerically in every simulation step. Having avoided terms that depend on $\gamma_i$ in $\mathcal{H}_c$, it is assumed that only the orientation of the particle long axis, $\mathbf{u}_i$, is actually coupled to the polymer network. Then, biaxial sample deformations are invoked if biaxiality in $\mathbf{u}_i$ fluctuations is present, and vice versa. Like for the uniaxial coupling the sign and the magnitude of the parameter $\chi$ depend on the specific complex architecture of the elastomeric material [10].

Again, we did a large set of constant-force MC simulations to study the behavior of our biaxial model system. The sample size was set to $N = 50^3$ unit cells. Then, we performed several heating and cooling MC run cascades, with at least 70/66 MC kcycles for equilibration/production. In the cascades, structural transitions between the $I, N_+, N_-$, and $B$ phases can be detected by calorimetry through anomalies in the heat capacity of the system, $C_V$. Reduced $C_V$ is calculated as

$$c_V^* = \frac{C_V}{Nk_B} = \frac{\langle U^2 \rangle - \langle U \rangle^2}{N(k_B T)^2}, \qquad (15.17)$$

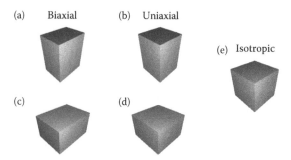

**FIGURE 15.4** Schematic depiction of sample shape (reflecting the symmetry of the polymer chain end-to-end tensor distribution) for different types of orientational ordering. In all cases the nematic director is vertical. (a,b) Positive materials with prolate mesogens, as well as negative materials with oblate mesogens. (c,d) Positive materials with oblate mesogens, or negative materials with prolate mesogenic units. (e) Undeformed sample, no orientational order. (Reprinted with permission from G. Skačej and C. Zannoni, *Eur. Phys. J.* E **25**, 181, 2008. Copyright 2008 by EDP Sciences.)

where $U$ is the internal energy and $\langle\ldots\rangle$ denotes averaging over MC cycles. Alternatively, since orientational order couples to elastic deformation, the average simulation box sides, $\lambda_x$, $\lambda_y$, and $\lambda_z$, are also relevant observables.

Heating/cooling runs have been performed for different values of the biaxiality parameter $v$. The parameter $\alpha$ introduced in Equation 15.1 was set to 0.3 as for uniaxial particles. The deformation types observed during the cooling runs, are shown schematically in Figure 15.4. The temperature dependences of $c_V^*$ and of $\lambda_\mu$ ($\mu = x, y, z$) for a positive material with $\chi = 0.5$ are shown in Figure 15.5. Setting $v = 0$, the interaction between mesogenic units becomes uniaxial as in Ref. [10]. On cooling from the isotropic phase the specific heat $c_V^*$ (Figure 15.5, top) peaks at $T^* \approx 1.141 \pm 0.003$ [10], which is a signature of the weakly first-order $IN_+$ phase transition. The particle alignment is accompanied by a deformation of about 8% along the director (Figure 15.5, bottom) and a lateral contraction of about 4% (at constant volume). Cooling to even lower $T^*$, the deformation increases together with the degree of nematic order.

Experimentally, biaxial orientational order in nematic polymers has been detected by $^2$H NMR [38,39]. We therefore used the simulation output to predict $^2$H NMR spectra of biaxial elastomers, assuming in the analysis that the electric field gradient (EFG) tensor of the carbon–deuteron bond is effectively uniaxial, with the symmetry axis of the tensor parallel to the long particle axis. Then, one can apply the same methodology for the calculation of spectra as in the uniaxial case. In each run, we have simultaneously calculated three spectra, with the spectrometer magnetic field **b** directed along **x**, **y**, and **z**, to easily detect any biaxiality in ordering. Translational diffusion was neglected. Like in the uniaxial case, the duration of one NMR cycle, $2\pi/\delta\omega_Q$, was set to 1024 MC cycles. For smoothening, a convolution with a Lorentzian kernel was applied. Figure 15.6, left, shows the corresponding $^2$H NMR spectra, with superimposed curves for different spectrometer field orientations: $\mathbf{b}\|\mathbf{x}, \mathbf{y}$, or $\mathbf{z}$, that is, parallel or perpendicular to the nematic director **n**. In the isotropic phase a single line

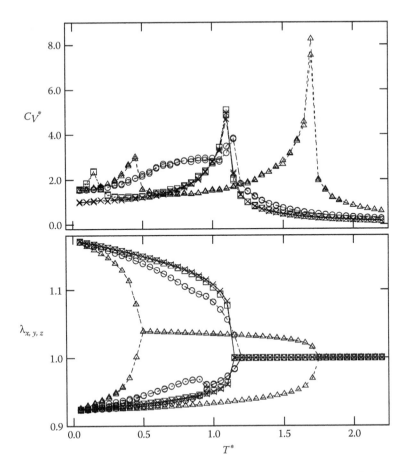

**FIGURE 15.5**  Temperature dependence of specific heat (top) and average sample dimensions (bottom) for different values of the biaxiality parameter: $v = 0$ ($\times$), $v = 0.2$ ($\square$), $v = 1/\sqrt{6} \approx 0.4082$($\circ$), and $v = 0.6$($\triangle$). Positive material, $\chi = 0.5$. (Reprinted with permission from G. Skačej and C. Zannoni, *Eur. Phys. J. E* **25**, 181, 2008. Copyright 2008 by EDP Sciences.)

at zero quadrupolar splitting can be seen. In the nematic phase, however, a doublet appears at $\pm\Delta\omega$ and $\pm\Delta\omega/2$ for $\mathbf{b}\|\mathbf{n}$ and $\mathbf{b}\perp\mathbf{n}$, respectively, where $\Delta\omega$ denotes the maximum frequency splitting observed for a given degree of order. The spectra for $\mathbf{b}\perp\mathbf{n}$ overlap as long as ordering is uniaxial. Decreasing $T^*$, $\Delta\omega$—proportional to the degree of nematic ordering—increases.

Weakly biaxial prolate particles with $v = 0.2$ exhibit biaxial order, however only for $T^* < 0.2$. This can be seen from a second peak in the temperature dependence of $c_V^*$ in Figure 15.5 that corresponds to the nematic–biaxial ($N_+B$) transition. The $N_+B$ transition is a second-order transition: the $c_V^*$ peak is not as sharp as the one at $T^* \approx 1.15$ attributed to the $IN_+$ transition. For $v = 0.2$ the biaxiality is weak and essentially undetectable from asymmetries in lateral sample contraction or in $^2$H NMR spectra for $\mathbf{b}\perp\mathbf{n}$ (Figure 15.6).

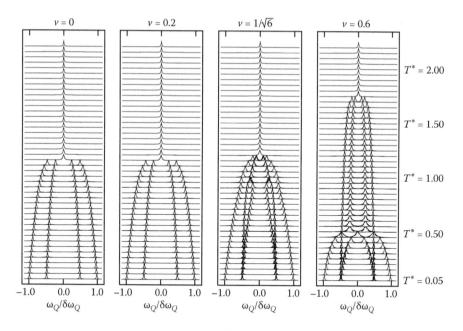

$v = 0$  $v = 0.2$  $v = 1/\sqrt{6}$  $v = 0.6$

$T^* = 2.00$

$T^* = 1.50$

$T^* = 1.00$

$T^* = 0.50$

$T^* = 0.05$

$-1.0 \quad 0.0 \quad 1.0$  $-1.0 \quad 0.0 \quad 1.0$  $-1.0 \quad 0.0 \quad 1.0$  $-1.0 \quad 0.0 \quad 1.0$
$\omega_Q/\delta\omega_Q$  $\omega_Q/\delta\omega_Q$  $\omega_Q/\delta\omega_Q$  $\omega_Q/\delta\omega_Q$

**FIGURE 15.6** Temperature dependence of $^2$H NMR spectra for different values of $v$. The spectrometer magnetic field was directed along **x**, **y**, and **z**; the resulting spectral sets are superimposed. Positive material, $\chi = 0.5$. (Reprinted with permission from G. Skačej and C. Zannoni, *Eur. Phys. J. E* **25**, 181, 2008. Copyright 2008 by EDP Sciences.)

In the ordinary nematic, the stability temperature range of the biaxial phase is widest at the limit between prolate and oblate particles (for $v = 1/\sqrt{6} \approx 0.4082$) where there is a direct second-order isotropic–biaxial (IB) transition [32] at the Landau point. Performing a cooling run with our model system for $v = 1/\sqrt{6}$ this direct IB transition splits into a pair of near transitions: $IN_+$ at $T^* \approx 1.17 \pm 0.01$ and $N_+B$ at $T^* \approx 0.96 \pm 0.01$, see Figure 15.5, bottom. The corresponding $c_V^*$ peak turns out to be extremely broad. The biaxial asymmetry is clearly visible also from $^2$H NMR spectra: the two **b**⊥**n** spectral sets do not overlap any longer, except in the narrow temperature range where ordering is uniaxial. More detailed temperature scans show that in comparison with the ordinary nematic the stability range of the $N_+$ phase formed by rod-like mesogenic particles has slightly increased at the expense of the $N_-$ phase formed by plate-like particles (Figure 15.7). This is a consequence of the mechanical field presence: as the particle alignment couples to strain exclusively through the orientation of the particle long axis $\mathbf{u}_i$—see Equations 15.14 through 15.16—the mechanical field stabilizes the alignment of $\mathbf{u}_i$ rather than that of the short axes $\mathbf{t}_i$. The new position of the Landau point is estimated to be $v \approx 0.414 \pm 0.001$.

For oblate particles, the isotropic phase on cooling transforms into the negative nematic phase, where the short particle axes are aligned. While for prolate particles the $IN_+$ transition temperature is essentially $v$-independent, this is not the case for oblate particles. For $v = 0.6$ the $IN_-$ transition takes place at $T^* \approx 1.75$, while the $N_-B$ transition can be observed at $T^* \approx 0.50$ (Figure 15.5). In the $N_-$ phase the sample

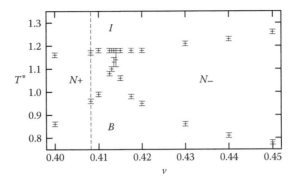

**FIGURE 15.7**   Orientational phase diagram of a biaxial elastomer (positive material, $\chi = 0.5$). Note the shift of the Landau point from $v = 1/\sqrt{6} \approx 0.4082$ (dashed line) to a slightly higher value. (Reprinted with permission from G. Skačej and C. Zannoni, *Eur. Phys. J.* E **25**, 181, 2008. Copyright 2008 by EDP Sciences.)

contracts along the director and expands laterally. The degeneracy of the lateral sample sides is lifted in the biaxial phase. The $^2$H NMR spectra show two pairs of peaks: one at $\pm\Delta\omega/2$ and one at $\pm\Delta\omega/4$, corresponding to **b**∥**n** and **b**⊥**n**, respectively. (**n** here is assigned to the average orientation of the short molecular axes $\mathbf{t}_i$ and the EFG tensor principal axis is directed along the long axes $\mathbf{u}_i$ that are assumed to move fast on the $^2$H NMR time scale.) Again, the peaks at $\pm\Delta\omega/4$ split once biaxial order is obtained.

Let us also comment on negative (e.g., end-on side chain) materials. The orientational ordering behavior (and, consequently, the resulting $^2$H NMR spectra) is essentially the same as that for positive materials. The main difference appears in average sample dimensions: for example, in the $N_+$ phase a positive material expands along **n**, while a negative material contracts. In the $N_-$ phase a positive material contracts along **n**, and vice versa for a negative material. Again, in the $B$ phase, the symmetry of the two lateral components (perpendicular to **n**) is broken. See also Figure 15.4.

### 15.2.3   EXTERNAL FIELD

The orientation of mesogenic units can be controlled by an external (electric or magnetic) field. Hence, at least in principle, the application of a strong enough stimulus can result—through strain-alignment coupling—in an induced elastic deformation of the elastomeric material. In general, a rather strong field is required to invoke deformations; however, particular (semi)soft deformation modes at low free energy cost [1,41,42]—like director rotation accompanied by shear—may facilitate the deformation process even in confined samples. Experimentally, director response to a normal electric field has been observed in a slab of cross-linked nematic gel, and was seen to be accompanied by shear deformation [43,44]. The corresponding threshold was characterized by critical electric field strength rather than by critical voltage as in ordinary liquid crystals. This electromechanical Fréedericksz effect has been explained

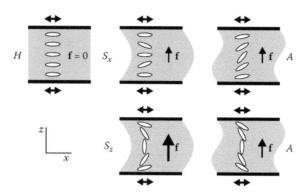

**FIGURE 15.8** Schematic depiction of configurations in a planar elastomeric slab exposed to a normal external field **f**, with anchoring along $x$. In absence of field the homogeneous configuration $H$ is observed. In weak fields the antisymmetric configuration $A$ (two versions) and the symmetric configuration $S_x$ (reported in [1,44]) are seen for weakly and strongly anisotropic networks, respectively. At high field strengths the symmetric $S_z$ structure sets in. (Reprinted with permission from G. Skačej and C. Zannoni, *Eur. Phys. J. E* **20**, 289, 2006. Copyright 2006 by EDP Sciences.)

using the continuum neoclassical rubber elastic theory [44]. Here the same problem is addressed, starting from a shearable lattice model, again performing large-scale MC simulations. Apart from the Fréedericksz transition studied in [44], additional structural transitions are observed at field strengths well above the Fréedericksz threshold. Some of the possible structures are shown in Figure 15.8.

At this point the previous lattice modeling has to be extended to treat shear deformations as well. Upon deformation, the displacement of every lattice point is given by the displacement field $\mathbf{v}_i$ ($i = 1 \ldots N$). Suppose now that the deformation takes the $i$th lattice point located initially at $\mathbf{r}_i$ to the new coordinates $\mathbf{r}'_i = \mathbf{r}_i + \mathbf{v}_i$ so that $\mathbf{r}'_i = \lambda_i \mathbf{r}_i$, where $\lambda_i = \partial \mathbf{r}'_i / \partial \mathbf{r}_i$ is the deformation tensor. Its components are

$$\lambda_i^{jk} = \delta^{jk} + \frac{\partial v_i^j}{\partial r_i^k}, \tag{15.18}$$

where $\delta^{jk}$ denotes the Kronecker symbol, $v_i^j$ stands for the $j$th component of $\mathbf{v}_i$, and $r_i^k$ represents the $k$th Cartesian coordinate. Then the rubber elastic contribution to the pseudo-Hamiltonian is given by [1,10]

$$\mathcal{H}_e = k_B T \alpha \sum_{i=1}^{N} [\text{Tr}(\lambda_i \lambda_i^T) - 3], \tag{15.19}$$

where the undeformed sample contribution has been subtracted.

The interaction between mesogenic units (assumed uniaxial here) is again given by $\mathcal{H}_n$, Equation 15.2. Further, as mesogens are anisotropic dielectrics, they align in an external electric field. The interaction of mesogenic clusters $\mathbf{u}_i$ with the field is modeled through

$$\mathcal{H}_f = -\epsilon\eta \sum_{i=1}^{N} P_2(\mathbf{u}_i \cdot \mathbf{f}), \tag{15.20}$$

where $\mathbf{f}$ is a unit vector defining the field orientation and $\eta = \epsilon_a\epsilon_0 a^3 E^2/3\epsilon$ is the field coupling constant; here $E$ denotes the field strength, $\epsilon_a$ the dielectric anisotropy of the material, and $\epsilon_0$ the dielectric constant of vacuum. For $\eta > 0$, $\mathbf{u}_i$ align parallel to $\mathbf{f}$. Taking $\epsilon_a \sim 6$, $a \sim 4.6$ nm, and $\epsilon \sim 0.023$ eV, $\eta = 0.1$ corresponds to $E \sim 14.4$ V/$\mu$m. For simplicity, in the following analysis the external field will be assumed homogeneous throughout the sample. This assumption is actually accurate only for elastomers with rather low dielectric anisotropy.

The strain-alignment coupling now has to be adapted to also consider shear deformations. This is accomplished by coupling the local ordering matrix $Q_i$ for mesogenic units with the local gyration tensor $G_i$ for polymer chains. Denoting with $\mathbf{q}_n$ the end-to-end vector for the $n$th polymer chain in an undeformed sample, close to the $i$th lattice point after an affine deformation one has $\mathbf{q}'_n = \lambda_i\mathbf{q}_n$. The local gyration tensor is then given by

$$G_i = \langle \mathbf{q}'_n \otimes \mathbf{q}'_n \rangle_i = \frac{Ma^2}{3}\lambda_i\lambda_i^T, \tag{15.21}$$

where $\langle \cdots \rangle_i$ stands for averaging over polymer chains [45]. Here it has been assumed that the gyration tensor for undeformed polymer chains is isotropic, that is, $\langle \mathbf{q}_n \otimes \mathbf{q}_n \rangle_i = (Ma^2/3)I$. Then, any deformation-induced anisotropy of $G_i$ results in an aligning tendency for mesogenic units, and an appropriate scalar invariant is given by $\text{Tr}(Q_iG_i) \propto \text{Tr}(\lambda_i\lambda_i^T Q_i)$ [1,5,22]. This corresponds to a generalization of the mechanical field concept to cases where this field is inhomogeneous and its direction is not fixed. The strain-alignment coupling is then described by

$$\mathcal{H}_c \propto -k_BT \sum_{i=1}^{N} \text{Tr}(\lambda_i\lambda_i^T Q_i). \tag{15.22}$$

For small deformations one can neglect terms quadratic in $\partial v_i^k/\partial r_i^l$, which also ensures sample stability: the coupling contribution should never overwhelm the positive elastic energy. (We will assume that this is qualitatively accurate even outside the small deformation limit. Note that considering linear terms only is compatible with the coupling given by Equation 15.3 for weak deformations.)

The elastomer slab was sandwiched between two parallel plates lying in $xy$-planes at $z = \pm d_z/2$. The external field was applied normal to the plates, that is, along the $z$-axis. The orientations of vectors $\mathbf{u}_i$ in the $z = \pm d_z/2$ layers were fixed along the $x$-axis to mimic anchoring to solid sample walls, while at the remaining simulation box sides in $x$ and $y$ directions periodic boundary conditions were assumed. Before every external field strength scan, a run in absence of the field ($\eta = 0$) was carried out to fix the simulation box dimensions (including $d_z$): in case of nematic ordering the system (assumed incompressible) is elongated along the director, $x$-axis, and contracted laterally. Then, to simulate a switching experiment, a cascade of runs with increasing/decreasing $\eta$ was performed. For each $\eta$ value, 70 MC kcycles were typically performed both for equilibration and for production. Different sample sizes (from $16^3$ to $40^3$ unit cells) were considered. In the MC evolution, plane shift trial moves have been introduced to facilitate the occurrence of shear. These moves consist of picking one of the $xy$ lattice planes (perpendicular to the sample normal) at random and generating a random in-plane shift of the corresponding lattice points.

To monitor the degree of orientational ordering with respect to external field direction $\mathbf{f}$, the order parameter

$$P_2^z = \langle P_2(\mathbf{u}_i \cdot \mathbf{f}) \rangle \tag{15.23}$$

was calculated, where $\langle \cdots \rangle$ denotes averaging over lattice sites and MC cycles. $P_2^z$ defined in this way is sensitive to both director reorientations and variations in the degree of order, but the latter should be negligible in the present case. Hence, below and far above the switching threshold negative ($H$ structure) and positive ($S_z$ structure) values of $P_2^z$ are expected, respectively. Further, to monitor the field-induced deformation, we calculated the overall displacement

$$\mathbf{V} = \langle \mathbf{v}_i - \mathbf{v}_i^0 \rangle, \tag{15.24}$$

where $\mathbf{v}_i^0$ is the deformation field prior to field application, and $\langle \cdots \rangle$ has the same meaning as above. In the chosen geometry, one expects displacements along the $x$ axis; hence $V_y \approx 0$ and $V_z = 0$. Undeformed ($H$) as well as deformed antisymmetric samples ($A$) give $V_x \approx 0$, while $V_x \neq 0$ is a clear signature of symmetrically deformed samples ($S_x$ and $S_z$; see Figure 15.8). Note that shear deformations with negative and positive $x$ components of the displacement $\mathbf{v}_i - \mathbf{v}_i^0$ are degenerate. Both are seen in our simulations.

The director structures can be visualized by means of polarizing microscopy, performing a simulated experiment where polarized light is shone along the $y$-axis, and the sample is placed between crossed polarizers at $\pm\pi/4$ from the $z$-axis. (Note that experimentally switching in elastomers has so far been studied by conoscopy-type experiments where the light is actually shone along the sample normal $z$ [43], which is easier to implement than the setup proposed here.) The intensity transmittance for the chosen setup is given by

$$I_t = \sin^2(\delta/2) \cos^2 2\phi \tag{15.25}$$

and depends on the angle $\phi$ between the director and the $x$-axis [25]. Consequently, the spatial dependence of $\phi$ results in characteristic intensity patterns for various director fields shown in Figure 15.8. In the simulation, light propagation was again modeled through the Jones matrix formalism, however, omitting the intensity averaging in the $xz$-plane. The effective sample thickness was set to $\sim$2.2 $\mu$m in the undeformed case; it was tuned so as to give $\sin^2(\delta/2)$ rather close to 1; this provides the highest possible dark/bright contrast in calculated patterns.

As a complement, it is possible to use an alternative set-up where the angle between the polarizer and the analyzer is $\pi/4$ rather than $\pi/2$. Then, for the polarizer at $\pi/4$ with respect to the $x$-axis (measured counter-clockwise) and the analyzer along $x$, the transmittance is given by

$$I'_t = \frac{1}{2}[1 + \sin^2(\delta/2) \sin 4\phi]. \tag{15.26}$$

The different odd–even symmetry of $I_t$ and $I'_t$ with respect to the central $z = 0$ plane facilitates the recognition of structures with different symmetries in the director profile $\phi(z)$.

All simulations were run at $T^* = 1.0$, that is, well inside the nematic phase [10]. Consider now a positive elastomeric material with a strain-alignment coupling strength equivalent to three times the value used in Section 15.2.1. As the director was fixed by anchoring, there is a spontaneous sample elongation along $x$, the easy axis. This elongation establishes a mechanical field parallel to $x$ and is accompanied by lateral contraction, as required by the incompressibility constraint. The equilibrium box shape depends somewhat on the sample dimension; on average, however, the initial cubic unit cells deform into boxes with approximate dimensions $1.44a \times 0.85a \times 0.82a$. There is weak systematic biaxiality in the lateral contraction that stems from the inequivalence of the $y$ and $z$ directions due to confinement. For $\eta = 0$ the system is on average homogeneous and thus corresponds to the $H$-structure.

Starting the external field scan with $\eta > 0$, the mismatch between the mechanical and the external field (directed along $x$ and $z$, respectively) below the switching threshold results merely in a slight decrease in the degree of ordering, while the director remains homogeneous. Then, at the threshold, the director reorients, which is accompanied by a spontaneous elastic shear deformation resulting in the $S_x$ structure; see Figures 15.9 and 15.10. Here the alignment along $x$ at $z = 0$ seems to be stabilized by the strong enough mechanical field. Note that the wavelength of the director field distortion and of the accompanying elastic shear deformation equals $d_z$ and $2d_z$, respectively, where $d_z$ is the distance between the confining plates. (For the Fréedericksz transition in ordinary nematics the corresponding director field wavelength equals $2d_z$.) The $H \rightarrow S_x$ transition is easily detected from the $P_2^z(\eta)$ and $V_x(\eta)$ dependences shown in Figure 15.9: as the mesogens align towards the field direction, an accompanying shear deformation results in an overall distortion along the $x$-axis. Within the current scan resolution, there is no observable hysteresis for the $H \leftrightarrow S_x$ switching. Moreover, the switching threshold decreases with increasing sample size. The dependence is less pronounced in large samples where the gradient terms related to surface boundary conditions become less important. In very large samples the threshold seems to approach 0, which is in agreement with the fact that the coupling term in the Hamiltonian, $\mathrm{Tr}(\lambda_i \lambda_i^T Q_i)$, allows ideally soft deformations [1]. On the other hand, experiments have shown that the $H \rightarrow S_x$ switching in elastomers is actually characterized by a nonzero threshold even in large samples [43, 44], but in that case the existence of a threshold could be attributed to the intrinsic semisoftness of the elastomer (not included in our modeling).

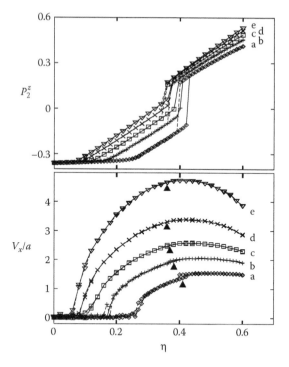

**FIGURE 15.9** An example of external field strength scans. $P_2^z(\eta)$ (top) and $V_x(\eta)$ (bottom) dependences are plotted for different sample sizes: $16^3$ (a), $20^3$ (b), $24^3$ (c), $30^3$ (d), and $40^3$ (e). Solid and dashed lines correspond to increasing and decreasing $\eta$, respectively. The arrows indicate the approximate position of the $S_x \leftrightarrow S_z$ transition. (Reprinted with permission from G. Skačej and C. Zannoni, *Eur. Phys. J. E* **20**, 289, 2006. Copyright 2006 by EDP Sciences.)

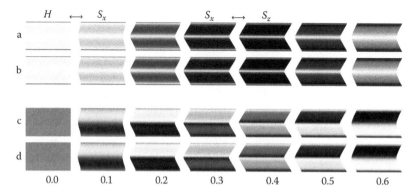

**FIGURE 15.10** Patterns of transmitted light intensity with visualized shear deformation; $N = 40^3$. Classical (a,b) and alternative (c,d) polarizer set-up. Scans with increasing (a,c) and decreasing (b,d) field strength, with the corresponding $\eta$ values reported below. (Reprinted with permission from G. Skačej and C. Zannoni, *Eur. Phys. J. E* **20**, 289, 2006. Copyright 2006 by EDP Sciences.)

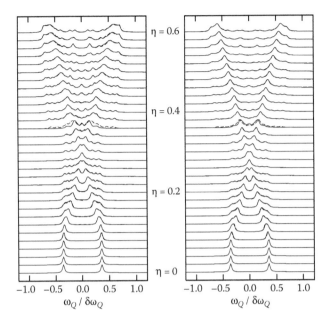

**FIGURE 15.11** External field-induced switching as seen by $^2$H NMR: spectra for $N = 30^3$ (left) and $N = 40^3$ (right). (Reprinted with permission from G. Skačej and C. Zannoni, *Eur. Phys. J. E* **20**, 289, 2006. Copyright 2006 by EDP Sciences.)

In rather strong external fields a second transition is observed, resulting in the symmetric $S_z$ structure. Here the wavelengths for director and elastic distortion both equal $2d_z$, and the director is aligned along $z$ even in the central layer of the sample. The $S_x \rightarrow S_z$ transition results in a jump of $P_2^z$ becoming less and less pronounced with increasing sample size, while $V_x$ is not very sensitive to this transition (Figure 15.9). As for progressive alignment along $z$ the occurrence of shear becomes less favorable, the effective deformation $V_x$ decreases with increasing $\eta$. Hence, $V_x$ peaks at a certain value of $\eta$, which may be important for the optimization of the actuation process. Unlike for the $H \rightarrow S_z$ transition the $S_x \rightarrow S_z$ threshold seems to be fairly sample size-independent because it involves only a thin wall between domains of opposite shear. In large enough samples the thickness of the domain wall (a few lattice spacings) becomes sample size-independent, too.

The calculated optical patterns (Figures 15.10), as well as $^2$H NMR spectra (Figure 15.11), all reveal a rather gradual character of the switching above the $H \leftrightarrow S_x$ threshold. The uniform transmittance patterns observed at zero-field become more structured once the sample is distorted. Note that while both the classical and the alternative polarizing microscope setup are able to detect the $H \leftrightarrow S_x$ transition, only the alternative setup sees the $S_x \leftrightarrow S_z$ transition. Further, the predicted $^2$H NMR spectra (Figure 15.11) mainly consist of doublets. Therefore, since for **b**$\|$**f** one has $\omega_Q \propto P_2(\cos \phi)$ there seems to be no sizable distribution of director orientation in our samples: the mesogenic units seem to be well aligned either along the external or the strain-induced mechanical field. The only exceptions are the thin subsurface regions where anchoring effects become important, as well as the also thin shear domain wall(s). Below the threshold the sample is aligned along the $x$-axis (i. e., perpendicular to **b**), with two well-resolved spectral peaks at a splitting $\mp\delta\omega_Q P_2^z/2$, and above it massive field-induced alignment in the $z$-direction (coinciding with **b**) is observed, eventually yielding a two-peak spectrum with a splitting equal to $\pm\delta\omega_Q P_2^z$. Residual surface-induced alignment along the $x$-axis is visible even above the threshold. $^2$H NMR is unable to detect the $S_x \leftrightarrow S_z$ transition as it only involves a thin layer between the domains of positive and negative shear and the fraction of the reoriented elastomeric material is too small to be detectable within the bulk response of the system.

If the anisotropy (i.e., the strain-alignment coupling constant) of the elastomer is decreased, the switching behavior becomes somewhat more complex and, especially in smaller samples, new asymmetric (A) structures (Figure 15.8) may appear in the switching sequence. In addition, the switching process is now typically accompanied by pronounced hysteresis and, moreover, asymmetric structures yield almost no net displacement with $V_x \approx 0$. Therefore, their potential for use in external field-induced actuation seems to be rather limited.

## 15.3 CONCLUSIONS AND OUTLOOK

The first steps in computer simulations of LCEs have been done over the past few years, resorting mainly to coarse-grained lattice models. This class of models contains no translational degrees of freedom and elasticity is reproduced by deforming the lattice as a whole. The mesogenic units embedded into the lattice are subject to a deformation-dependent mechanical aligning field. While the mesogen–mesogen interactions are described sufficiently well within lattice models, the rubber elasticity and the strain-alignment coupling are typically treated in a very simplified manner. Such simplicity on the one hand allows studies of samples large enough for the prediction of selected experimental observables, while on the other it also poses limitations in predictive power of lattice modeling, especially in considering inhomogeneous deformations. Therefore, more insight should be obtained from off-lattice modeling, albeit on considerably smaller samples, and the output linked to lattice modeling in an appropriate way. We envisage that this will be one of the main directions for development in the field of LCE simulations in the next few years.

## REFERENCES

1. M. Warner and E. M. Terentjev, *Liquid Crystal Elastomers* (Oxford University Press, Oxford, 2003).
2. M. Hébert, R. Kant, and P. G. de Gennes, Dynamics and thermodynamics of artificial muscles based on nematic gels, *J. Phys. I* **7**, 909, 1997.
3. D. Sherrington and S. Kirkpatrick, Solvable model of a spin-glass, *Phys. Rev. Lett.* **35**, 1792, 1975.
4. R. Harris, M. Plischke, and M. J. Zuckermann, New model for amorphous magnetism, *Phys. Rev. Lett.* **31**, 160, 1973.
5. N. Uchida and A. Onuki, Elastic interactions in nematic elastomers and gels, *Europhys. Lett.* **45**, 341, 1999.

6. J. V. Selinger, H. G. Jeon, and B. R. Ratna, Isotropic-nematic transition in liquid-crystalline elastomers, *Phys. Rev. Lett.* **89**, 225701, 2002.
7. W. Kaufhold, H. Finkelmann, and H. R. Brand, Nematic elastomers, 1. Effect of the spacer length on the mechanical coupling between network anisotropy and nematic order, *Makromol. Chem.* **192**, 2555, 1991.
8. S. Conti, A. DeSimone, and G. Dolzmann, Semisoft elasticity and director reorientation in stretched sheets of nematic elastomers, *Phys. Rev. E* **66**, 061710, 2002.
9. Y.-K. Yu, P. L. Taylor, and E. M. Terentjev, Exponential decay of correlations in a model for strongly disordered 2D nematic elastomers, *Phys. Rev. Lett.* **81**, 128, 1998.
10. P. Pasini, G. Skačej, and C. Zannoni, A microscopic lattice model for liquid crystal elastomers, *Chem. Phys. Lett.* **413**, 463, 2005.
11. G. Skačej and C. Zannoni, External Field-induced switching in nematic elastomers: A Monte Carlo study, *Eur. Phys. J. E* **20**, 289, 2006.
12. G. Skačej and C. Zannoni, Biaxial Liquid-crystal elastomers: A lattice model, *Eur. Phys. J. E* **25**, 181, 2008.
13. J. V. Selinger and B. R. Ratna, Isotropic-nematic transition in liquid-crystalline elastomers: Lattice model with quenched disorder, *Phys. Rev. E* **70**, 041707, 2004.
14. R. Berardi, D. Micheletti, L. Muccioli, M. Ricci, and C. Zannoni, A computer simulation study of the influence of a liquid crystal medium on polymerization, *J. Chem. Phys.* **121**, 9123, 2004.
15. D. Micheletti, L. Muccioli, R. Berardi, M. Ricci, and C. Zannoni, Effect of nanoconfinement on liquid-crystal polymer chains, *J. Chem. Phys.* **123**, 224705, 2005.
16. G. Milano and F. Müller-Plathe, Mapping atomistic simulations to mesoscopic models: A systematic coarse-graining procedure for vinyl polymer chains, *J. Phys. Chem. B* **109**, 18609, 2005.
17. P. A. Lebwohl and G. Lasher, Nematic-liquid-crystal order. A monte carlo calculation, *Phys. Rev. A* **6**, 426, 1972.
18. U. Fabbri and C. Zannoni, A monte carlo investigation of the Lebwohl-Lasher lattice model in the vicinity of its orientational phase transition, *Mol. Phys.* **58**, 763, 1986.
19. W. Maier and A. Saupe, Eine einfache molekularstatistische Theorie der nematischen kristallinflüssigen Phase, *Z. Naturforsch. A* **14**, 882, 1959; *Z. Naturforsch. A* **15**, 287, 1960.
20. P. Pasini and C. Zannoni, *Advances in the Computer Simulations of Liquid Crystals*, Kluwer, Dordrecht, 2000.
21. G. Raos and G. Allegra, Mesoscopic bead-and-spring model of hard spherical particles in a rubber matrix. I. Hydrodynamic reinforcement, *J. Chem. Phys.* **113**, 7554, 2000.
22. P. G. de Gennes, Réflexions sur un type de polymères nématiques, *C. R. Acad. Sc. Paris.* **281**, 101, 1975.
23. I. Lelidis and G. Durand, Electric-field-induced isotropic-nematic phase transition, *Phys. Rev. E* **48**, 3822, 1993.
24. J. Küpfer and H. Finkelmann, Nematic liquid single crystal elastomers, *Macromol. Chem. Rapid Commun.* **12**, 717, 1991.
25. P. J. Collings and J. S. Patel, *Handbook of Liquid Crystal Research* (Oxford University Press, New York, 1997).
26. A. Lebar, Z. Kutnjak, S. Žumer, H. Finkelmann, A. Sánchez-Ferrer, and B. Zalar, Evidence of supercritical behavior in liquid single crystal elastomers, *Phys. Rev. Lett.* **94**, 197801, 2005.
27. S. F. Edwards and P. W. Anderson, Theory of spin glasses, *J. Phys. F: Met. Phys.* **5**, 965, 1975.

28. S. Disch, C. Schmidt, and H. Finkelmann, Nematic elastomers beyond the critical point, *Macromol. Rapid Commun.* **15**, 303, 1994.

29. R. Y. Dong, *Nuclear Magnetic Resonance of Liquid Crystals* (Springer Verlag, New York, 1994).

30. C. Chiccoli, P. Pasini, G. Skačej, S. Žumer, and C. Zannoni, NMR spectra from Monte Carlo simulations of polymer dispersed liquid crystals, *Phys. Rev. E* **60**, 4219, 1999.

31. J. P. Straley, Ordered phases of a liquid of biaxial particles, *Phys. Rev. A* **10**, 1881, 1974.

32. F. Biscarini, C. Chiccoli, P. Pasini, F. Semeria, and C. Zannoni, Phase diagram and orientational order in a biaxial lattice model: A monte carlo study, *Phys. Rev. Lett.* **75**, 1803, 1995.

33. R. Berardi and C. Zannoni, Do thermotropic biaxial nematics exist? A Monte Carlo study of biaxial gay–berne particles, *J. Chem. Phys.* **113**, 5971, 2000.

34. B. R. Acharya, A. Primak, and S. Kumar, Biaxial nematic phase in bent-core thermotropic mesogens, *Phys. Rev. Lett.* **92**, 145506, 2004.

35. K. Neupane, S. W. Kang, S. Sharma, D. Carney, T. Meyer, G. H. Mehl, D. W. Allender, S. Kumar, and S. Sprunt, Dynamic light scattering study of biaxial ordering in a thermotropic liquid crystal, *Phys. Rev. Lett.* **97**, 207802, 2006.

36. L. A. Madsen, T. J. Dingemans, M. Nakata, and E. T. Samulski, Thermotropic biaxial nematic liquid crystals, *Phys. Rev. Lett.* **92**, 145505, 2004.

37. J. L. Figueirinhas, C. Cruz, D. Filip, G. Feio, A. C. Ribeiro, Y. Frère, T. Meyer, and G. H. Mehl, Deuterium NMR investigation of the biaxial nematic phase in an organosiloxane tetrapode, *Phys. Rev. Lett.* **94**, 107802, 2005.

38. K. Severing and K. Saalwächter, Biaxial nematic phase in a thermotropic liquid-crystalline side-chain polymer, *Phys. Rev. Lett.* **92**, 125501, 2004.

39. K. Severing, E. Stibal-Fischer, A. Hasenhindl, H. Finkelmann, and K. Saalwächter, Phase biaxiality in nematic liquid crystalline side-chain polymers of various chemical constitutions, *J. Chem. Phys. B* **110**, 15680, 2006.

40. M. E. Rose, Elementary theory of angular momentum, *Elementary Theory of Angular Momentum* (Wiley, New York, 1957).

41. L. Golubović and T. C. Lubensky, Nonlinear elasticity of amorphous solids, *Phys. Rev. Lett.* **63**, 1082, 1989.

42. P. D. Olmsted, Rotational invariance and goldstone modes in nematic elastomers and gels, *J. Phys. II France* **4**, 2215, 1994.

43. C.-C. Chang, L.-C. Chien, and R. B. Meyer, Electro-optical study of nematic elastomer gels, *Phys. Rev. E* **56**, 595, 1997.

44. E. M. Terentjev, M. Warner, R. B. Meyer, and J. Yamamoto, Electromechanical fredericks effects in nematic gels, *Phys. Rev. E* **60**, 1872, 1999.

45. P. D. Olmsted and S. M. Milner, Strain-induced nematic phase separation in polymer melts and gels, *Macromolecules.* **27**, 6648, 1994.

# 16 Electromechanical Effects in Swollen Nematic Elastomers

*Kenji Urayama and Toshikazu Takigawa*

## CONTENTS

## 16.1 INTRODUCTION

The coupling of rubber elasticity with orientational order is one of the most important characteristics of liquid crystal elastomers (LCEs). This coupling effect strongly correlates the macroscopic shape with the orientational order, and results in unique stimuli–response behavior of nematic elastomers [1,2]. A change in orientational order caused by external fields such as temperature variation and light irradiation drives a macroscopic deformation. Electric field is also an external field that can switch the director of liquid crystal (LC) molecules [3]. The director of many low-molecular-mass LCs (LMMLCs) is readily realigned by low electric fields because of their large dielectric anisotropy and nematic interaction. The resultant significant change in optical birefringence (electro-optical effect) leads to various industrial devices, such as, LC displays. What happens in nematic elastomers under electric fields? We expect not only electro-optical effect but macroscopic deformation induced by mesogen realignment because nematic elastomers behave as solids without flowing due to the cross links. Para-electric nematic elastomers in the neat state, however, need unusually high electric fields to exhibit a finite distortion because their high elastic moduli act as a strong resistance to dielectric forces [4]. Swelling of nematic elastomers by some

473

appropriate LMMLCs reduces the network modulus with neither decreasing dielectric anisotropy nor nematic interaction [5]. Several studies observed a finite distortion in such swollen nematic elastomers under moderate electric fields [6–11]. Some other researchers examined the electro-optical effects in the nematic networks formed in the cells in the presence of large quantities of LMMLCs [12–14]. However, none of them observed the simultaneous optical and mechanical effects under electric fields. In the present paper, on the basis of our recent results, we review (1) the role of the signs of dielectric anisotropy of the mesogens in electrical deformation; and (2) the simultaneous electro-optical and mechanical effects observed in swollen nematic elastomers under unconstrained geometry.

## 16.2 EXPERIMENTS

### 16.2.1 PREPARATION OF POLYDOMAIN LCE

Side-chain type-polydomain LCEs having positive or negative dielectric anisotropy (P-POLY-LCE or N-POLY-LCE, respectively) were prepared by polymerizing the mesogenic monomers MP or MN (Figure 16.1) with 1,6-hexanediol diacrylate (cross linker), respectively [9]. The cross-linker concentration in the feed was 1 mol%. The polymerization was performed in a glass capillary with a diameter of several hundred micron at 80°C where the reactant mixture was in the isotropic state. Azoisobuthyl-nitrile (AIBN) was used as the catalyst. The resultant cylindrical polydomain LCEs separated from the glass capillary were immersed in toluene to wash out the unreacted materials. After drying, P-POLY-LCE and N-POLY-LCE were allowed to swell at a certain temperature in the LMM LC 5CB or SN (Figure 16.1) with positive or negative dielectric anisotropy, respectively, until the swelling equilibrium was achieved. The swelling exhibited no anisotropy because of the polydomain texture. The degree of swelling ($Q$) defined by the ratio of the volumes in the dry and swollen states was

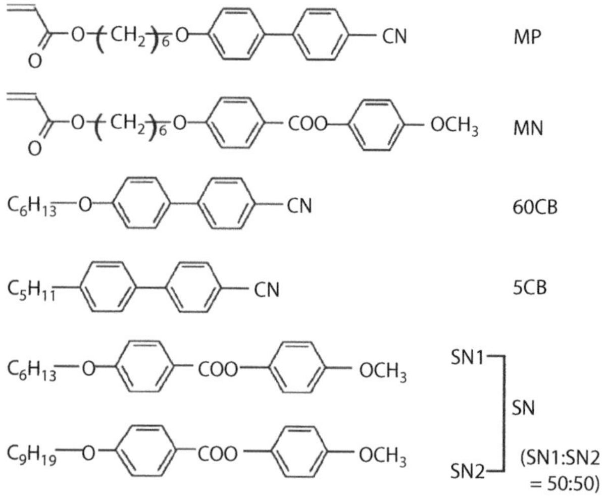

**FIGURE 16.1** Chemical structures of the employed reactive and nonreactive LCs.

evaluated using $Q = (d_s/d_0)^3$ where $d_s$ and $d_0$ are the diameters in the swollen and dry states.

## 16.2.2 PREPARATION OF MONODOMAIN LCE

Dielectrically positive or negative monodomain LCE samples (P-MONO-LCE or N-MONO-LCE, respectively) were prepared by photocrosslinking the monomer MP or MN in the glass cell whose surfaces were coated with uniaxially rubbed polyimide layer [15]. 1,6-Hexanediol diacrylate and Irgacure 784 were employed as the cross linker and photoinitiator, respectively. The nonreactive LMMLC 6OCB (Figure 16.1) or SN-1 was mixed with MP or MN in the molar ratio 50:50 to broaden the temperature range of the nematic phase, respectively. The photocrosslinking was performed at a temperature where the reactant nematic mixture exhibits a monodomain homogeneous texture. The cell gap was 20 or 25 μm. The resulting gel films were carefully separated from the glass substrates. After the washing procedures, the dry P-MONO-LCE and N-MONO-LCE films were immersed in 5CB and SN, respectively. The swelling was considerably anisotropic, and the anisotropy was characterized by the dimensional ratios $\lambda$ along the axes parallel ($\parallel$) and perpendicular ($\perp$) to the director before and after swelling. The reference state for $\lambda$ is the dry and isotropic states. The degree of swelling $Q$ was calculated by $Q = \lambda_{\parallel}^2 \lambda_{\perp}$ owing to the uniaxial anisotropy. The characteristics of the monodomain LCE samples are tabulated in Table 16.1.

## 16.2.3 MEASUREMENTS

Dimensional changes of the polydomain LCE samples under DC electric fields were measured by optical microscopy using the geometry of Figure 16.2. The cylindrical samples with a diameter of ca. 1 mm were not constrained by the Pt electrodes with the gap of ca. 2 mm. The cell was filled with the swelling solvent.

The electromechanical and optical effects of the monodomain LCE samples under AC electric fields were observed using the geometry of Figure 16.3. The electro-optical effect was examined by measuring the transmittance through crossed polarizers using a He–Ne laser. The specimen was placed such that the initial director was at an angle of 45° relative to the crossed polarizers. The swollen specimens with a thickness of ca. 30 μm were placed between the two transparent indium tin oxide (ITO) electrodes

## TABLE 16.1
## Sample Characteristics

| Sample | $C_x$ (mol%) | $Q$ | $d_f{}^a$ (μm) | $d_p{}^b$ (μm) |
|---|---|---|---|---|
| P-MONO-LCE | 3 | 5.5 | 26 | 20 |
| | 7 | 4.1 | 34 | 25 |
| | 14 | 2.0 | 23 | 25 |
| N-MONO-LCE | 5 | 3.0 | 21 | 20 |

[a] Film thickness in the swollen state.
[b] Cell thickness at sample preparation.

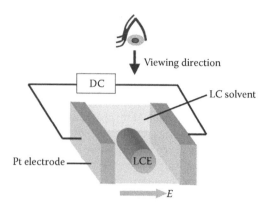

**FIGURE 16.2** Measurements of electrical deformation of swollen cylindrical polydomain LCEs.

with a gap of 40 μm. In this geometry, the specimens have no mechanical constraint from the electrodes. The cell was filled with transparent silicone oil, a nonsolvent for the specimens. The electrically induced deformation was observed with a polarizing microscope. The sample dimensions along (*x*) and normal (*y*) to the initial director were measured as a function of the voltage amplitude and frequency of the AC fields.

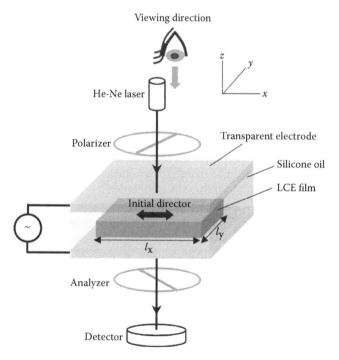

**FIGURE 16.3** Measurements of electro-optical and electromechanical effects for swollen monodomain LCEs.

The strain ($\gamma_i$) in the $i$-direction ($i = x, y, z$) is defined by $\gamma_i = (l_i - l_{i0})/l_{i0}$ where $l_i$ and $l_{i0}$ are the dimensions in the deformed and undeformed states, respectively.

## 16.3 ROLE OF DIELECTRIC ANISOTROPY IN ELECTRICAL DEFORMATION

Figure 16.4 shows the strain in the field direction as a function of field strength for the swollen P-POLY-LCE and N-POLY-LCE [9]. In the high-temperature isotropic state, the strains for both specimens are negative; the samples are compressed in the field direction independently of the signs of dielectric anisotropy. The compressive strain is proportional to the square of field strength in accordance with the familiar electro-strictive behavior. This shows that the LCEs in the isotropic phase behave as conventional para-electric materials. In contrast, the effect of the sign of dielectric anisotropy is prominent in the low-temperature nematic phase. The dielectrically positive and negative LCEs are stretched and compressed in the field direction, respectively. The macroscopic stretching direction accords with the direction of electrical mesogen realignment that is governed by the signs of dielectric anisotropy: The dielectrically positive and negative mesogens realign parallel and normal to the field direction [3]. The large elongation strain in the field direction for P-POLY-LCE shows that the distortion induced by the mesogen reorientation dominates the overall deformation, whereas the contribution of the electro-strictive effect is minor in the nematic phase. The degree of the induced strain increases with the field strength, and it reaches almost 20% at high fields for P-POLY-LCE. An increase in the cross-linker amount reduces the field-induced strain [9], because it increases the network modulus, and decreases the degree of swelling. A decrease in the degree of nematicity (controlled by the amount of mesogenic molecules in the networks) also reduced the electrical strain [9]. The results in Figure 16.4 show that (1) the direction of electrically induced stretching in the low-temperature nematic phase is variable according to the signs of dielectric anisotropy, and that (2) for dielectrically positive LCEs, it is also switchable depending on the temperature (i.e., nematic or isotropic phase).

## 16.4 MACROSCOPIC DEFORMATION COUPLED TO ELECTRO-OPTICAL EFFECT

The correlation between the mesogen reorientation and macroscopic strain is clearly observed in monodomain LCEs having global director as a pronounced electro-optical effect accompanying the macroscopic deformation, as shown in Figure 16.5 [16]. Figure 16.5 shows the response of a swollen P-MONO-LCE to the electric field ($z$-axis) normal to the initial director ($x$-axis). The significant change in birefringence regarding the $x$–$y$ plane stems from the director realignment along the field axis. Simultaneously, anisotropic deformation occurs: The film shrinks in the $x$-direction by ca. 10% with almost no dimensional change in the $y$-direction. The film is stretched by ca. 10% in the field ($z$) direction because this material is dielectrically positive and incompressible. When the imposed field is removed, the specimen recovers the initial shape and mesogen orientation.

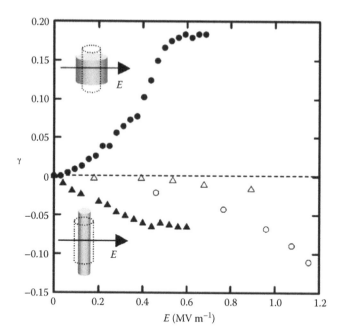

**FIGURE 16.4** Strain in the field direction as a function of field strength for P-POLY-LCE (circular) and N-POLY-LCE (triangular) in the low-temperature nematic (filled symbols) and high-temperature isotropic phases (open symbols): $Q = 15$ (nematic) and 17 (isotropic) for P-POLY-LCE; $Q = 8.9$ (nematic) and 33 (isotropic) for N-POLY-LCE. (The data are reproduced from Urayama, K. et al., *Phys. Rev. E*, 71, 051713, 2005.)

Figure 16.6 illustrates the strains in the $x$ and $y$ directions and the effective birefringence ($\Delta n_{\mathrm{eff}}$) in the $x$–$y$ plane as a function of voltage amplitude ($V_0$) for the swollen P-MONO-LCE [16]. These data were obtained in the steady state at each $V_0$. It should be noted that $V_0$ is the apparent amplitude between the electrodes, not the effective amplitude acting on the specimens. The sample cell consists of the two layers, that is, the elastomer and surrounding silicone oil. The effective field acting on the elastomer depends on the thickness and effective dielectric constant of each layer. Sufficiently high fields result in a marked reduction in $\Delta n_{\mathrm{eff}}$ (ca. 95% of the initial value), that is, almost full (90°) rotation of director toward the field direction. The simultaneous strain in the $x$-direction exceeds 10% while the strain in the $y$-direction is almost zero even at high fields. The agreement of the $V_0$ dependence of $\gamma_x$ and $\Delta n_{\mathrm{eff}}/\Delta n_{\mathrm{eff}}^\circ$ indicates the strong correlation between the degree of director rotation and macroscopic strain. Figure 16.7 illustrates the deformation coupled to the director rotation. The distortion dominantly occurs in the plane regarding the director rotation. The deformation mode driven by the full director rotation is pure shear. This anisotropic deformation is also characterized in terms of the Poisson's ratio ($\mu$) when we regard the $z$-axis as the uniaxial stretching direction: $\mu_{xz} = 1$ and $\mu_{yz} = 0$. This is in contrast to $\mu_{xz} = \mu_{yz} = 1/2$ for conventional elastomers.

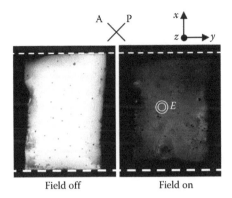

Field off                      Field on

**FIGURE 16.5** Electromechanical effect coupled to electro-optical effect for the swollen P-MONO-LCE ($C_x = 7$ mol%) with homogeneous mesogen alignment ($x$-axis) driven by the AC field ($z$-axis) of $V_0 = 750$ V and $f = 1$ kHz. The gap between electrodes and film thickness are 40 and 34 μm, respectively. The specimen is not constrained by transparent electrodes, and the cell is filled with transparent silicone oil (nonsolvent for LCEs). The white dashed lines are guidelines for eyes, and A and P denote the optical axes of the analyzer and polarizer, respectively. The film shrinks in the $x$-direction by ca. 10% but with almost no dimensional change in the $y$-direction.

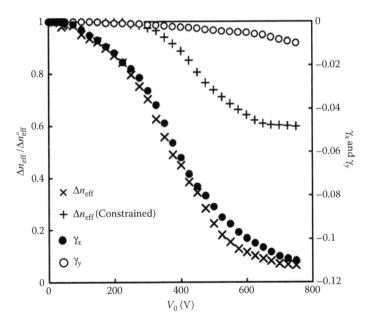

**FIGURE 16.6** Effective birefringence and strains as a function of voltage amplitude of imposed AC fields ($f = 1$ kHz) for the swollen P-MONO-LCE with $C_x = 7$ mol%. (The data are reproduced from Urayama, K., Honda, S., and Takigawa, T., *Macromolecules*, 38, 3574, 2005; Urayama, K., Honda, S., and Takigawa, T., *Macromolecules*, 39, 1943, 2006.)

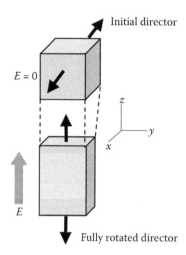

**FIGURE 16.7**  Characteristic deformation induced by a 90° rotation of director.

The presence of the strong coupling between the macroscopic shape and orientational order is well demonstrated by the same experiment using the constrained geometry where the film specimen is effectively sandwiched by electrodes [17]. This constrained geometry prohibits the strain in the field direction ($\gamma_z = 0$). As shown in Figure 16.6, the reduction in $\Delta n_{\text{eff}}$ for the constrained sample remains at ca. 40% even at high fields. This clearly indicates that the mechanical constraint also suppresses the director rotation, and that the full director rotation never occurs in the constrained geometry. The mode of the induced deformation in the constrained geometry is an interesting issue. Some types of inhomogeneous deformation are expected to occur as a result of the coupling of the rubber elasticity and Frank elasticity. The effects of the constraint condition and material parameters on the deformation type were discussed [14,18,19].

Figure 16.8 displays the electro-optical and mechanical effects for P-MONO-LCE with various amounts of cross linker ($C_x$) [17]. An increase in $C_x$ reduces the degree of swelling and increases the elastic modulus. Both the electro-optical and mechanical effects become smaller as $C_x$ increases. The strain for the elastomer of $C_x = 3\,\text{mol}\%$ (the lowest $C_x$) reaches ca. 17 % at high fields. The thicknesses of the elastomers with various $C_x$ are different (Table 16.1), and thus the effective field strengths for each specimen are not identical. However, it does not influence the qualitative conclusion about the $C_x$ dependence of the electro-optical and mechanical effects.

The rotation angle of director ($\theta$) is estimated from the data of $\Delta n_{eff}$ on the basis of the relation for the uniform systems with uniaxial optical anisotropy ($n_x \geq n_y = n_z$):

$$\sin^2\theta = \frac{n_{y0}^2}{n_{x0}^2 - n_{y0}^2} \left\{ \frac{n_{x0}^2}{n_x(\theta)^2} - 1 \right\} \qquad (16.1)$$

where $n_{i0}$ and $n_i(\theta)$ ($i = x,y$) is the refractive index along the $i$-axis at $\theta = 0°$ and $\theta = \theta$. We can safely assume $n_{y0} \gg \Delta n_{\text{eff}}$ and $n_{y0} \gg \Delta n_{\text{eff}}^{\circ}$ because $\Delta n_{\text{eff}}^0 (= n_{x0} - n_{y0})$ of

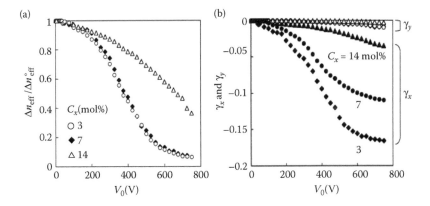

**FIGURE 16.8** (a) Effective birefringence and (b) strains as a function of voltage amplitude of imposed AC fields for the swollen P-MONO-LCEs with various $C_x$. (The data are reproduced from Urayama, K., Honda, S., and Takigawa, T., *Macromolecules*, 39, 1943, 2006.)

our samples is on the order of $10^{-2}$ whereas the refractive index normal to the long axis for 5CB is ca. 1.5 at 25°C. This condition simplifies Equation 16.1 as

$$\sin^2 \theta \approx 1 - \frac{\Delta n_{\text{eff}}(\theta)}{\Delta n_{\text{eff}}^0} \tag{16.2}$$

Figure 16.9 illustrates $\gamma_x$ and $\gamma_y$ as a function of $\sin^2 \theta$ estimated from $\Delta n_{\text{eff}}$ for the swollen P-MONO-LCE. It should be noted that $\sin^2 \theta = 1$ corresponds to the full rotation of director. The data of $\gamma_x$ for the samples with various $C_x$ appear to fall on the straight lines while $\gamma_y$ is almost zero independently of $\theta$. These characteristic

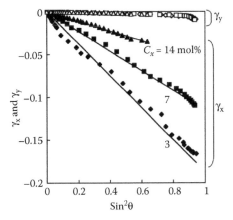

**FIGURE 16.9** Strains as a function of $\sin^2 \theta$ where $\theta$ is the rotation angle of director for the swollen P-MONO-LCEs with various $C_x$. (The data are reproduced from Urayama, K., Honda, S., and Takigawa, T., *Macromolecules*, 39, 1943, 2006.)

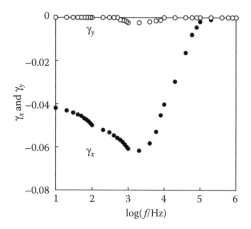

**FIGURE 16.10**   Strains as a function of the frequency of imposed AC fields ($V_0 = 400\,\text{V}$) for the swollen P-MONO-LCE with $C_x = 7\,\text{mol}\%$. (The data are reproduced from Urayama, K., Honda, S., and Takigawa, T., *Macromolecules*, 38, 3574, 2005.)

$\theta$ dependence of $\gamma_x$ and $\gamma_y$ ($\gamma_x(\theta) \sim \sin^2 \theta$ and $\gamma_y(\theta) \approx 0$) accords with the picture of soft deformation for sufficiently thin films driven by the full director rotation [17,20,21]. The detailed discussion is given in Ref. [17]. A recent experiment using the polarized Fourier transform infrared spectroscopy has also confirmed the linear relation $\gamma_x \sim \sin^2 \theta$ [22]. The response times to field-on and field-off for the electro-optical and electromechanical effects have also been examined as a function of $V_0$ in the recent experiments [23]. An mpeg movie for the electromechanical effect is available in the supporting information of Ref. [23]. A simple model has been proposed to describe the static and dynamic aspects of this phenomenon [23].

Figure 16.10 shows the strains, a function of frequency ($f$), of the imposed AC field for the P-MONO-LCE with $C_x = 7\,\text{mol}\%$ at $V_0 = 400\,\text{V}$ [16]. No appreciable response is observed at the high frequencies of $f > 10^5\,\text{Hz}$ because the field frequency is too high to induce a dipolar reorientation of the mesogens, as known in the electric-field response of LMMLCs. At the frequencies of $f < 10^4\,\text{Hz}$, the electromechanical effect becomes pronounced. The strain exhibits a maximum peak around $f = 2 \times 10^3\,\text{Hz}$. Appreciable $f$ dependence at low frequencies will be due to a reduction in effective field strengths stemming from a finite ionic current of impurities in the system.

Figure 16.11 displays $\gamma_x$ and $\gamma_y$ as a function of $V_0$ for the swollen N-MONO-LCE with $C_x = 5\,\text{mol}\%$ under the geometry of Figure 16.3. The orientational order of the dielectrically negative mesogens initially aligned along the $x$-axis further increases in the same direction under the electric fields normal to the initial director. This induces a stretching of the film along the initial director, that is, positive and negative values in $\gamma_x$ and $\gamma_y$, respectively. The electro-strictive effect leads to compression along the field axis, that is, positive values in both $\gamma_x$ and $\gamma_y$. In the $x$-direction, both effects result in positive (elongation) strains. In contrast, these two effects act on the strain along the $y$-direction in the opposite ways. The positive $\gamma_y$ in the experiments show that the

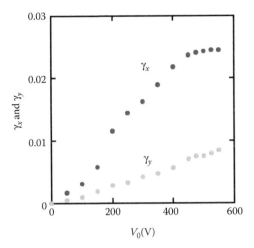

**FIGURE 16.11**   Strains as a function of voltage amplitude of imposed AC fields ($f = 600\,\text{Hz}$) for the swollen N-MONO-LCE with $C_x = 5\,\text{mol}\%$.

strain by the mesogen reorientation is not as large as the electrostriction. The strain in the $x$-direction lies below 3% even at high fields. These are because the degrees of negative dielectric anisotropy of the constituent mesogens in the swollen N-MONO-LCE are not large; In addition, in this case, the mesogen reorientation occurs in the same direction as the initial alignment, not a 90° switching. The induced distortion by the mesogen reorientation is much smaller than that in Figure 16.6. The difference in the major origin between $\gamma_x$ and $\gamma_y$ is also recognized in the $f$ dependence of $\gamma_x$ and $\gamma_y$ for the swollen N-MONO-LCE in Figure 16.12. The $f$ dependence of $\gamma_x$ is similar to that for the P-MONO-LCE. In contrast, $\gamma_y$ is almost independent of $f$.

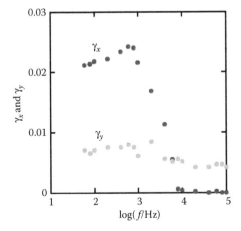

**FIGURE 16.12**   Strains as a function of the frequency of imposed AC fields ($V_0 = 500\,\text{V}$) for the swollen N-MONO-LCE with $C_x = 5\,\text{mol}\%$.

## 16.5   SUMMARY

The nematic elastomers swollen by LMMLCs exhibit a pronounced distortion in fast response to electric fields. The electrically induced reorientation of the mesogens drives a macroscopic deformation. The stretching direction is controllable by the signs of dielectric anisotropy of the mesogens: The dielectrically positive or negative LCE is elongated parallel or normal to the field direction. In the case of monodomain LCEs having global director, a significant change in birefringence coupled to deformation is also observed. In the loosely cross-linked LCEs, sufficiently high fields achieve the almost full rotation of director along the field axis and lead to the large strain more than 15%. These pronounced electromechanical effects coupled to electro-optical effects provide not only a basis to understand the characteristic deformation driven by director rotation in LCEs but also a potential of LCEs for electrically driven soft actuators.

## ACKNOWLEDGMENTS

The authors thank Mr. H. Kondo, Mr. S. Honda, and Dr. Y. O. Arai for their cooperation in the experiments. The authors appreciate Nissan Chemical Industries for providing the polyimide solution. This work was partly supported by a Grant-in-Aid on Priority Area "Soft Matter Physics" (No. 19031014) and a Grant-in-Aid for Scientific Research (B) (No. 16750186) from the Ministry of Education, Culture, Sports, Science and Technology (MEXT) of Japan. This research was also supported in part by the Global COE Program "International Center for Integrated Research and Advanced Education in Materials Science" (No. B-09) of MEXT of Japan, administrated by the Japan Society for the Promotion of Science. This work was also supported by a grant from the Murata Science Foundation.

## REFERENCES

1. Warner, M. and Terentjev, E. M. *Liquid Crystals Elastomers,* London: Clarendon Press, 2003.
2. Urayama, K. Selected issues in liquid crystal elastomers and gels, *Macromolecules*, 40, 2277, 2007.
3. de Gennes, P. G. and Prost, J., *The Physics of Liquid Crystals*, 2nd ed. New York, NY: Oxford University Press, 1993.
4. Terentjev, E. M., Warner, M., and Bladon, P., Orientation of nematic elastomers and gels by electric-fields, *J. Phys. II*, 4, 667, 1994.
5. Urayama, K., Arai, Y. O., and Takigawa, T., Anisotropic swelling and phase behavior of monodomain nematic networks in nematogenic solvents, *Macromolecules,* 38, 5721, 2005.
6. Zentel, R. Shape variation of cross-linked liquid-crystalline polymers by electric-fields, *Liq. Cryst.,* 1, 589, 1986.
7. Kishi, R., Suzuki, Y., Ichijo, H., and Hirasa, O. Thermotropic liquid-crystalline polymer gels.1. Electrical deformation of thermotropic liquid-crystalline polymer gels, *Chem. Lett.*, 2257, 1994.
8. Barnes, N. R., Davis, F. J., and Mitchell, G. R., Molecular switching in liquid-crystal elastomers, *Mol. Cryst. Liq. Cryst.*, 168, 13, 1989.

9. Urayama, K., Kondo, H., Arai, Y. O., and Takigawa, T., Electrically driven deformations of nematic gels, *Phys. Rev. E*, 71, 051713, 2005.

10. Yusuf, Y., Huh, J. H., Cladis, P. E., Brand, H. R., Finkelmann, H., and Kai, S., Low-voltage-driven electromechanical effects of swollen liquid-crystal elastomers, *Phys. Rev. E*, 71, 061702, 2005.

11. Huang, C., Zhang, Q. M., and Jakli, A. Nematic anisotropic liquid-crystal gels—Self-assembled nanocomposites with high electromechanical response, *Adv. Funct. Mater.*, 13, 525, 2003.

12. Hikmet, R. A. M. and Boots, H. M. J., Domain-structure and switching behavior of anisotropic gels, *Phys. Rev. E*, 51, 5824, 1995.

13. Chang, C. C., Chien, L. C., and Meyer, R. B., Electro-optical study of nematic elastomer gels, *Phys. Rev. E*, 56, 595, 1997.

14. Terentjev, E. M., Warner, M., Meyer, R. B., and Yamamoto, J., Electromechanical Fréedericksz effects in nematic gels, *Phys. Rev. E*, 60, 1872, 1999.

15. Urayama, K., Arai, Y. O., and Takigawa, T., Volume phase transition of monodomain nematic polymer networks in isotropic solvents accompanied by anisotropic shape variation, *Macromolecules*, 38, 3469, 2005.

16. Urayama, K., Honda, S., and Takigawa, T., Electrooptical effects with anisotropic deformation in nematic gels, *Macromolecules*, 38, 3574, 2005.

17. Urayama, K., Honda, S., and Takigawa, T., Deformation coupled to director rotation in swollen nematic elastomers under electric fields, *Macromolecules*, 39, 1943, 2006.

18. Muller, O. and Brand, H. R., Undulation versus Frederiks instability in nematic elastomers in an external electric field, *Eur. Phys. J. E*, 17, 53, 2005.

19. Skacej, G. and Zannoni, C., External field-induced switching in nematic elastomers, a Monte Carlo study, *Eur. Phys. J. E*, 20, 289, 2006.

20. Olmsted, P. D. Rotational invariance and goldstone modes in nematic elastomers and gels, *J. Phys. II*, 4, 2215, 1994.

21. Verwey, G. C. and Warner, M., Soft rubber elasticity, *Macromolecules*, 28, 4303, 1995.

22. Fukunaga, A., Urayama, K., Koelsch, P., and Takigawa, T., Electrically driven director-rotation of swollen nematic elastomers as revealed by polarized Fourier transform infrared spectroscopy, *Phys. Rev. E.*, 79, 051702, 2009.

23. Fukunaga, A., Urayama, K., Takigawa, T., DeSimone, A., and Teresi, L., Dynamics of electro-opto-mechanical effects in swollen nematic elastomers, *Macromolecules*, 41, 9389, 2008.

# 17 Smectic Elastomers

*Mark Warner*

## CONTENTS

## 17.1   INTRODUCTION

Liquid crystal elastomers (LCEs) are lightly cross-linked polymer networks with mesogenic order [1]. Being lightly linked they are capable, as classical elastomers are, of huge deformations. They can be elongated by factors of six or more, that is, of 600% strain, or sheared by of order one or more. Their deformability reflects the fact that elastomers are really polymeric liquids with high molecular mobility but where the cross-links between chains that hold the rubber network together do not allow macroscopic flow. The linking is light in the case of elastomers where deformability is great. Such linking leads to low mechanical moduli of typically $10^5 - 10^6$ Pa. Since these networks have conventional, liquid-like values for their bulk moduli that are much higher than the shear moduli, it follows that they change shape at constant volume. For small distortions, the sum of their Poisson ratios governing their size change in the two perpendicular directions must accordingly be 1. Classical (isotropic) elastomers are the same, and their Poisson ratios are both 1/2. We shall find smectic elastomers also conserve volume, but their Poisson ratios are most complex because their shape changes are dominated by the underlying layering of the smectic phase.

The most common mesogenic order in LCEs is nematic where the (typically) constituent rod-like molecules are orientationally ordered with respect to their ordering direction, the director $n$. We shall assume the reader is familiar with nematics, but point out that in elastomers the molecular mobility means that the director can be rotated by mechanical stresses or strains, with the same ease as it is rotated by electric fields in liquid nematics. Director rotation with respect to the solid matrix is not apparently found in nematic glassy networks. The reduced molecular mobility also leads to lower extensibility and much higher moduli ($\sim 10^9$ Pa).

Nematic elastomers have been much more widely studied than their layered (smectic) counterparts. They display several remarkable phenomena which rest upon two fundamental effects [1]. First, nematic polymer chains are (typically) elongated by the orientational order of their constituent rods. Since the chains determine the mechanical shape of the network, as they elongate on changing from the isotropic to the nematic phase, there are associated spontaneous strains (see Figure 17.1).

The elongation can be of several hundred per cent. There are vivid demonstrations of strips of nematic elastomers lifting weights, though they will probably find application ultimately in the micro to nano scale. Their specific strengths are much the same as human muscle.

The second effect is that response to mechanical strains is at much lower energy cost, at constant temperature and thus nematic order, by rotating the director if that is possible [2,3]. Simply put, the rubber tries to put its longer dimension (that along the director) along the direction of imposed elongation, or along the extension diagonal if shear is imposed (see Figure 17.2).

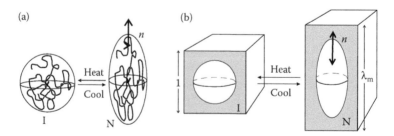

**FIGURE 17.1**   (a) The characteristic shape of polymer chains in the isotropic and nematic states. The rods that are ordering have been suppressed to show the backbone shape elongating. (b) The chains that form the solid network cause it to elongate by a factor of $\lambda_m$ as they are elongated by the orientational order. (Adapted from M. Warner and E.M. Terentjev. *Liquid Crystal Elastomers*. Oxford University Press (revised, paperback edition), 2007 by permission of Oxford University Press.)

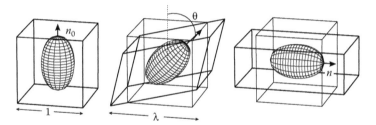

**FIGURE 17.2** A block of nematic rubber initially of unit dimensions accommodates a stretch perpendicular to the director by rotating the distribution of chains. Directing the long dimensions of chains toward the stretch direction minimizes their elongation away from their natural shape and thus keeps the free energy low. One can see that other sympathetic deformations, contraction along the original director and shears, accompany the imposed elongation. (Adapted from M. Warner and E.M. Terentjev. *Liquid Crystal Elastomers*. Oxford University Press (revised, paperback edition), 2007 by permission of Oxford University Press.)

Thus, an elongation perpendicular to the director will induce it to rotate and the stress generated can be orders of magnitude smaller than if one were to extend along the director. These are large amplitude effects, in fact comparable to the spontaneous extensions mentioned above. They are harder to observe in conventional rheology experiments where the strains imposed might be $10^{-5}$ times smaller, and certainly smaller than possible thresholds against such rotations. From these two effects follow the remarkable response of nematic elastomers to light and the fascinating opto-mechanical properties of cholesterics.

Smectic elastomers are no less remarkable in their phenomena. We shall find spontaneous distortions (shears in this case) in smectic elastomers, and much reduced elastic costs of distortion if the director and the layers can be induced to rotate. Indeed mechanical instabilities result when they are stretched along their layer normal, a direction where their modulus is initially very high. In fact, smectics essentially form two-dimensional elastomers. They can be soft ferroelectrics with a continuous manifold of polarization states and they too can respond to light. We review their physical properties and then introduce a model for their large distortions and rotations. One can take their rubber elasticity to be essentially that of the underlying nematic phase, but severely constrained by their layer structure, which can have a much higher rigidity than the rubber matrix in which the layers are embedded.

Nematic elastomers have been extensively reviewed, see for instance [1], and these phenomena put in the context of order change and director rotation, and also of nonlinear elasticity theory. Smectic elastomers have only more recently been studied in the same depth and there are accordingly fewer reviews of the new phenomena that have arisen and of the nonlinear elasticity theory required to describe them. One relatively recent review [1] from 2007 discusses experimental and theoretical advances in smectic elastomers to that date at some length. We give here a more compressed introduction and refer the reader to [1] for a fuller and more technical treatise.

We now briefly review the smectic liquid phase before discussing the weak solid forms of smectics.

## 17.2   SMECTIC LIQUID CRYSTALS

Smectic liquid crystals have an orientational order as nematics do, but additionally have a limited degree of spatial ordering in the form of layering. From the symmetry point of view there is a great variety of possible phases, combining one-dimensional layered order with various degrees of structure and various types of alignment of mesogenic groups. The word smectic derives from the Greek for soapy, an attribute that the phases have from the flow properties associated with layers sliding largely intact over each other. The simplest smectic order, smectic A (denoted by SmA), has the molecular anisotropy coaxial with the layer normal, that is, the nematic director (which is the principal axis of uniaxial optical birefringence) is locked perpendicular to the smectic layers (Figure 17.3).

Smectic layering arises often from the drive of the aliphatic tails segregating from their conjugated, rod-like, mesogenic cores by forming the layered morphology seen in Figure 17.3a.

We shall be interested in the polymer forms of smectics where, since both the spacers and the main chain often tend to be immiscible with the conjugated side chains rods [4] (Figure 17.3b) microphase separation is even more likely. These complex molecules are cross-linked to form smectic elastomers and rod-like cross-linkers can also be important (Figure 17.3c).

Smectic C (SmC) liquids have the nematic director tilted with respect to the layer normal. An additional degree of freedom then exists—rotation of the director on a cone about the layer normal. This will have important mechanical ramifications in elastomeric SmC. The role of chirality is also important since it leads to ferroelectricity in SmC liquids. The corresponding solids are soft, noncrystalline ferroelectrics that can be mechanically manipulated like no other ferro-elastic materials.

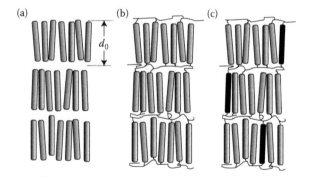

**FIGURE 17.3**   Schematic of mesogenic group arrangements in phases of SmA symmetry: (a) a classical smectic liquid, (b) a smectic side-chain polymer with the backbone confined between the layers, and (c) a smectic elastomer with cross-linking groups incorporated into the smectic layers. (Adapted from M. Warner and E.M. Terentjev. *Liquid Crystal Elastomers.* Oxford University Press (revised, paperback edition), 2007 by permission of Oxford University Press.)

## 17.2.1 SMA LIQUIDS

Smectic or lamellar order is characterized by a one-dimensional wave of density or composition most simply represented by its fundamental $\rho(z) \approx \rho_0 + |\psi| \cos(q_0 z + \Phi)$. The amplitude of the fundamental density modulation is $|\psi|$ and typically acts as the SmA order parameter. Experimentally, smaller higher harmonics can also be observed. The layer normal is along $z$ and $\Phi(\mathbf{r})$ is an arbitrary phase, which thus describes the distortion of the layers. The symmetry of the SmA phase is that of the point group $D_{\infty h}$, that is, of a cylinder. This symmetry and that of the phase from which the smectic arises (either isotropic or nematic) determines the order of the phase transformation [5]. In elastomers the transformations are, however, high influenced by the network.

When the smectic order is well established, the phase of layer modulation can be written as $\Phi = -q_0 v(\mathbf{r})$ with $v(\mathbf{r})$ describing the displacement of layers relative to their original position. (We will denote the displacement of the background solid matrix by $\mathbf{u}(\mathbf{r})$.) Note that $v$ is not a vector but only the component of such along the layer normal ($z$ or $\mathbf{n}_0$ for SmA): displacements in the layer plane have no meaning for smectic liquid crystals. The continuum description of smectic and lamellar phases uses the gradients of layer displacement $v(\mathbf{r})$ as effective distortion fields. The elastic free energy density must be invariant under the symmetry transformations of the phase, $D_{\infty h}$ (or the point group $D_\infty$, a cylinder with a screw thread, if noncentrosymmetric molecules form the smectic phase).

Rotations of the layers in the fluid state cannot be relevant to the free energy. For instance rotation of $\theta$ about the $y$-axis relates to a gradient of layer displacement as $\tan \theta = -\partial v / \partial x$, see Figure 17.4.

Thus the corresponding harmonic form $(\partial v / \partial x)^2 + (\partial v / \partial y)^2 \equiv (\nabla_\perp v)^2$ does not enter $F_{\text{SmA}}$. Layer rotations will no longer be irrelevant in smectic elastomers. Altering the equilibrium layer spacing $d_0 = 2\pi / q_0$ causes a layer strain $(d - d_0)/d_0 = \partial v / \partial z$ (Figure 17.4b) that costs a free energy measured by the layer compression constant $B$, of dimensions of energy density (Equation 17.1). At leading order, the smectic

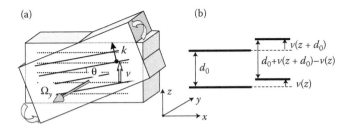

**FIGURE 17.4** (a) Rotation of a system of layers about the $y$-axis gives $\tan \theta = -\partial v / \partial x$. The layer normal $\mathbf{k}$ then develops an $x$-component $\delta k_x \approx -\partial v / \partial x$ or more generally the component of the normal in the original perpendicular plane is $\delta \mathbf{k} = -\nabla_\perp v$. (b) dilation of smectic layers: neighboring layers at $z$ and $z + d_0$ suffer differing displacements $v$ and hence acquire a new separation $d \approx d_0(1 + \partial v / \partial z)$. (Adapted from M. Warner and E.M. Terentjev. *Liquid Crystal Elastomers*. Oxford University Press (revised, paperback edition), 2007 by permission of Oxford University Press.)

elastic free energy density takes the form

$$F_{SmA} = \frac{1}{2}B\left(\frac{\partial v}{\partial z}\right)^2 + \text{curvature terms.} \qquad (17.1)$$

The definitive source for the smectic energy and its origins is the definitive book by de Gennes and Prost [6]; there is a recent, briefer review in the context of elastomers [1]. For instance, there is a relation between the two elastic moduli of a layered phase and the layer spacing: the ratio $K/B$ has a dimensionality of length squared, this length being of the order of the equilibrium layer spacing $d_0$. The layer compression modulus is roughly $B \simeq K_1/d_0^2 \sim 1.7 \times 10^6$ J/m$^3$, but significantly higher values are found in many smectic elastomers. $B$ turns out to dominate the properties of many smectic elastomers.

### 17.2.2  THE LAYER TILT INSTABILITY of SMECTICS

Stretch deformation along the layer normal is resisted by the cost $B$ of increasing layer separation. SmA liquids find a cheaper way of elongating, by rotating their layers. Simple geometry shows that layers, of intrinsic separation $d_0$, rotated by an angle $\theta$ then present a longer distance $d_0/\cos\theta$ along the stretch direction, as indicated in the marker bars of Figure 17.5a or more graphically where the two depictions meet.

This geometrical mechanism for avoiding strain is associated with many authors including Helfrich and Hurault who discovered it in cholesterics and with Clark and Meyer who discovered it in smectics [7]. We shall refer to it as the Clark–Meyer–Helfrich–Herault (CMHH) effect. The rotation angle is related to the layer displacement by $\tan\theta = -\partial v/\partial x$ when rotation is about the y-axis (see Figure 17.4). Thus the apparent spacing in the stretch direction is

$$d = d_0\sqrt{1 + \tan^2\theta} \simeq d_0(1 + \frac{1}{2}(\partial v/\partial x)^2)$$

for small angles, and with nonlinear effects entering at higher angles [8]. However, in the presence of boundaries at $z = 0$ and $z = L_z$ (for instance, the glass plates that

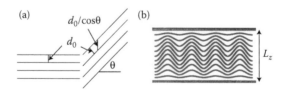

**FIGURE 17.5**   (a) When a system of layers is rotated it presents an effectively larger laying spacing with respect to the original (stretch) direction, $z$. (b) To avoid collision with the bounding plates, layers must adopt undulatory rotations. The no displacement boundary condition at the plates suppresses undulations in their vicinity. (Adapted from M. Warner and E.M. Terentjev. *Liquid Crystal Elastomers.* Oxford University Press (revised, paperback edition), 2007 by permission of Oxford University Press.)

contain the liquid and which are being separated), the layers cannot simply rotate and avoid colliding with the plates, $v = -x \tan \theta$. Thus, the rotation must be periodic to avoid the walls (see Figure 17.5b). Such a solution involves bending the layers which, as we saw in Equation 17.1, incurs essentially a splay cost. The outcome is a compromise between the advantage to rotating layers and the concomitant cost of bending them.

As the amplitude of undulations increases one must consider nonlinear corrections to this analysis [8]. The threshold is small in liquids because the Frank costs are small compared with layer deformations, but in elastomers the strain thresholds will turn out to be much larger. CMHH undulations can be relieved by inserting replacement layers by dislocation propagation. In elastomers this apparently does not happen since the undulatory state is long-lived.

### 17.2.3   SmC Liquids

SmC is a special phase among all the varieties of liquid crystals because of the spontaneous ferroelectricity in the chiral C* -smectics [9]. This class of materials has attracted great industrial and scientific interest. The key monographs on the subject of ferroelectric liquid crystals [5,6] give detailed accounts of symmetries and the resulting physical effects. The basic feature of SmC structure is the tilt of the average molecular axis (the director $n$) with respect to the layer normal, while maintaining a disordered (liquid) placement of rods within layers. The tilt angle $\theta$ is often regarded as the order parameter of A–C phase transition (Figure 17.6a).

Tilt is a good scalar measure, although strictly the order parameter is more complicated—an in-plane vector $\xi$ with components $(-n_z n_y, n_z n_x) = 1/2 \sin 2\theta$ $(-\sin \phi, \cos \phi)$ [5]. When the material is noncentrosymmetric (chiral), the resulting SmC* structure becomes helically twisted in equilibrium (Figure 17.6b) in such a way that the director preserves the constant tilt angle $\theta$ but rotates around a cone, in the azimuthal plane, as one travels along the layer normal, $\phi = qz$. The low symmetry of such a phase, with the layer normal $k$ and the director $n$ at an increasing angle to

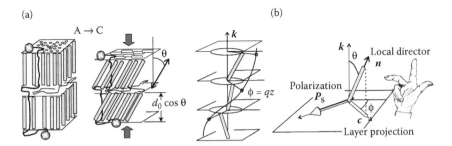

**FIGURE 17.6**   (a) A sketch of the SmA and SmC phases, indicating the tilt $\theta$ and the associated layer compression. (b) The helically twisted structure of chiral SmC*; the coordinates show the relative orientation of $n$, its projection to the layer plane $c$ and spontaneous polarization $P_s$ for right-handed chirality. (Adapted from M. Warner and E.M. Terentjev. *Liquid Crystal Elastomers*. Oxford University Press (revised, paperback edition), 2007 by permission of Oxford University Press.)

each other, allows spontaneous electric polarization to develop in the direction perpendicular to both, that is $P$ is along $k \times n$. Note that in centrosymmetric, nonchiral SmC such polarization cannot develop since the directions of $P$ and $-P$ cannot be distinguished. Ferroelectricity in chiral SmC* liquid crystals is "improper"—in contrast to classical ("proper") ferroelectrics, where the vector of spontaneous polarization is the genuine order parameter, here the tilt is in effect the order parameter. Unlike in classical ferroelectrics, where there is generally a discrete manifold of preferred axes, there is a continuum of polarization directions.

In equilibrium, the C*-smectic is helically twisted and so its polarization rotates in the layer plane. In order to obtain a uniformly polarized material one needs to unwind the helix—the classical way of doing so is forming the so-called bookshelf geometry of SmC* confined between two boundary layers (Figure 17.9e). If the director is in the plane of the sample, the polarization is perpendicular to this plane. By applying an external electric field across such cell one can switch the polarization from "up" to "down," by rotating the director azimuthally around the cone to its 180°-opposite position. Since this is the basis of some advanced liquid crystal displays [10], a huge literature exists on this subject.

## 17.3 SMECTIC ELASTOMERS AND SAMPLE PREPARATION

Several classes of synthetic polymers exhibit SmA order [11]. Historically, the most common are the side-chain mesogenic polymers based on siloxane [–O–Si–] and

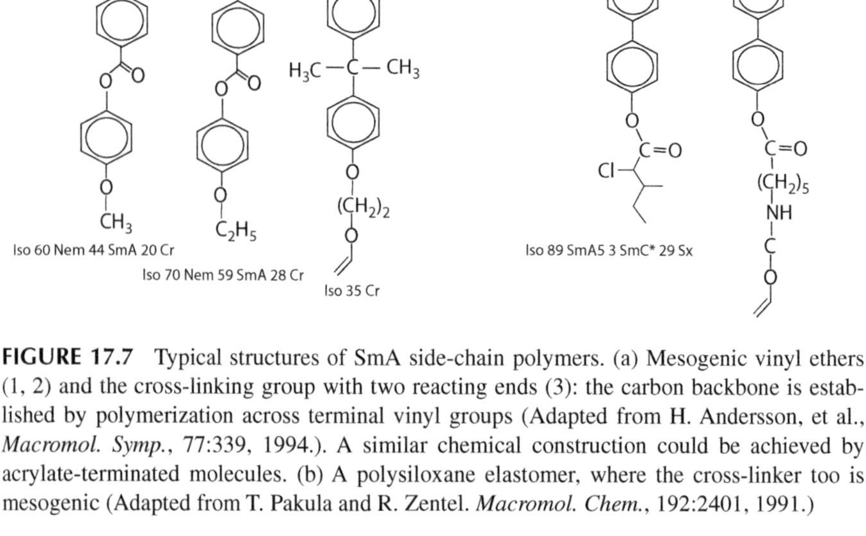

**FIGURE 17.7** Typical structures of SmA side-chain polymers. (a) Mesogenic vinyl ethers (1, 2) and the cross-linking group with two reacting ends (3): the carbon backbone is established by polymerization across terminal vinyl groups (Adapted from H. Andersson, et al., *Macromol. Symp.*, 77:339, 1994.). A similar chemical construction could be achieved by acrylate-terminated molecules. (b) A polysiloxane elastomer, where the cross-linker too is mesogenic (Adapted from T. Pakula and R. Zentel. *Macromol. Chem.*, 192:2401, 1991.)

aliphatic carbon [–C–C–] chains, with rod-like molecular moieties attached to one of the reacting bonds of Si or C atoms. Exploiting naturally smectic polymers was the first route to smectic elastomers [12], as described in review articles [13,14]. The recognized method to obtain smectic rather than nematic materials is to increase the length of flexible spacer between the mesogenic side groups and the backbone. Long spacers (note $(CH_2)_{11}$ in Figure 17.7) provide sufficient mobility for the backbone to aggregate in a microphase separated way between the lamellar regions with higher concentration of rod-like mesogenic groups. This, combined with the orientational order of the rods, leads to smectic layer formation.

Smectic elastomer preparation also has to take into account cross-link size and its relation to layers. In fact, the sketch in Figure 17.3c illustrates that one needs to be careful in selecting the size of cross-linking molecular groups. From the geometric point of view, their length must equal an integer multiple of the smectic layer spacing $d_0$. A "point-like" cross-link (or a flexible chain of the same nature as the backbone— which in some sense accounts for two "point-like" cross-links) would bind backbones confined within one interlayer plane. A rod-like cross-link such as bi-functional groups shown in Figures 17.3c and 17.7a would bind backbones across one layer if its length is $\approx d_0$. The cross-linking group shown in Figure 17.7b reacts with a vinyl bond of another such group from a different chain—the resulting length of the binding aggregate becomes $\approx 2d_0$. Anything in between would create a strong distortion of the local smectic order near an incommensurate cross-link and, therefore, would depress the existence of the phase. In any event, cross-linking polymers in the smectic phase creates a local dependence between the cross-link and the layer position and their relative movement along the layer normal should be difficult. We shall soon return to this effect, crucial to the understanding of smectic rubber elasticity. Note that the position of cross-links has no effect in a homogeneous nematic phase: one needs to break the translational symmetry of the mesophase in order to experience such a coupling.

As with nematic networks, the conditions of cross-linking define the texture of the resulting mesophase. When the network is formed in the isotropic phase, or no special aligning procedure is applied, the liquid crystalline elastomer invariably forms poly-domains with a very small characteristic texture. Polydomain nematic and smectic elastomers and gels strongly scatter light and, thus, appear opaque. However, when the final cross-linking of the network is performed in an aligned state (whether in a mesophase or in the isotropic state, for instance, by applying a stress or a strong magnetic field, which is then frozen in by cross-linking) a monodomain liquid crystalline phase results (see Figure 17.8).

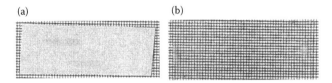

(a)      (b)

**FIGURE 17.8** (a) Polydomain (opaque) and (b) monodomain (transparent, cross-linked in the aligned state) smectic elastomers. (Photographs: with thanks to Professor H. Finkelmann.)

As in nematic elastomers, smectic monodomains are transparent since they do not scatter light because of long wavelength director fluctuations being suppressed by the polymer network.

### 17.3.1 SmA Elastomers

Prolate nematic polymer chains are best aligned by uniaxial extension. By contrast smectic or lamellar polymers usually align their layers on uniaxial compression. One can understand this from a sketch of backbone conformation, for example in Figure 17.9a.

Elastic deformation of a network is transmitted through this backbone. The uniaxial compression, accompanied by symmetric biaxial extension in the perpendicular direction due to incompressibility, results in the appropriate oblate configuration promoting a monodomain SmA alignment. The resulting monodomain sample has its smectic

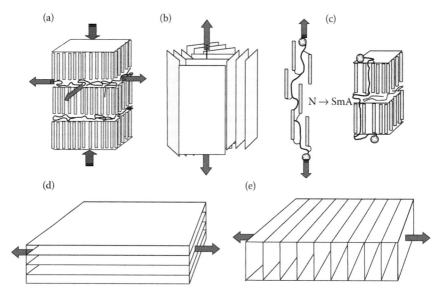

**FIGURE 17.9** The effects of deforming the SmA elastomer network: (a) uniaxial compression leads to the symmetric extension of backbones in the layer plane; (b) uniaxial extension in the smectic phase leads to the alignment of layer planes, but not their normals (only the layer plane positions are shown in this sketch); (c) uniaxial stretching during cross-linking in the nematic phase aligns the director $n$. On subsequent transition into the SmA this orientation is preserved and the layers are formed by re-arranging (partial flattening of) the backbone. Associated uniaxial contraction occurs at the N–A transition; see Figure 17.12. (d) the macroscopic arrangements of layers with respect to sample shape arising in case (a). The bookshelf geometry (e) arises from case (c). (Adapted from M. Warner and E.M. Terentjev. *Liquid Crystal Elastomers*. Oxford University Press (revised, paperback edition), 2007 by permission of Oxford University Press.)

layers parallel to its flat surface (Figure 17.9d). In practice an effective way of extending a weak, partially cross-linked gel is to deswell it while it is stuck to a substrate. Contraction is only possible in film thickness, while the other two dimensions, in not shrinking, are effectively stretched. When cross-linking and solvent removal are complete, a totally transparent monodomain SmA elastomer results [18]. Uniaxial compression is the same in effect as biaxial extension in an incompressible system.

Polydomain systems can be compression-aligned subsequent to their formation, but the effect is not permanent unless second-stage cross-linking is used to freeze in the established alignment. Simple application of uniaxial extension can result in layers oriented along the stretching direction, but with the layer normal randomly oriented in the perpendicular plane (Figure 17.9b). Moreover, in this situation one would still encounter internally quenched disorder from the random cross-links of the first stage.

When the liquid crystalline polymer, which will form the network also possesses a prolate nematic phase at higher temperatures, aligning this phase by uniaxial extension produces a monodomain nematic. Subsequent cooling into the SmA preserves this director orientation, with layers spontaneously forming in the plane perpendicular to the stretching axis (Figure 17.9c), thus providing an equally monodomain SmA elastomer. In a thin film the layers are in the bookshelf geometry with respect to the flat and long dimensions of the sample, as shown in Figure 17.9e. It has to be noted that, in some cases, no equilibrium nematic phase exists between the smectic and the isotropic states. However, if the cross-linking has been performed under a uniaxial load (which is one of the very few ways of producing monodomain textures), prolate nematic order may be induced in the resulting elastomer network just above the smectic transition point [19].

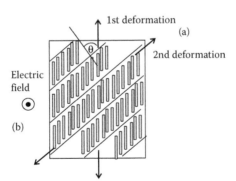

**FIGURE 17.10** Aligning a SmC (or C*) system before cross-linking, leading to the bookshelf geometry of the final elastomer. Method (a): a sequence of two mechanical deformations applied to a partially cross-linked gel at a prescribed angle to each other, followed by the final cross-linking. Method (b): aligning of a polymer melt by an external electric field, followed by UV-initiated cross-linking (evidently, this would only be applicable to ferroelectric C*-smectics). (Adapted from M. Warner and E.M. Terentjev. *Liquid Crystal Elastomers*. Oxford University Press (revised, paperback edition), 2007 by permission of Oxford University Press.)

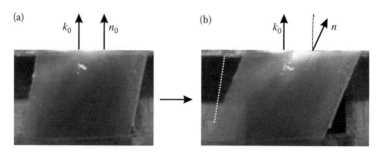

**FIGURE 17.11**   (a) A SmA elastomer (Adapted from K. Hiraoka et al., *Macromolecules*, 38:7352, 2005.). (b) Spontaneous shear $\lambda_{xz}$ in achieving the SmC state.

## 17.3.2   SMC AND FERROELECTRIC C* ELASTOMERS

The synthesis of SmC* polymers and their cross-linked elastomer networks was greatly motivated by their liquid analogues displaying ferroelectricity [12,20]. However, as with nematic elastomers, direct synthesis always results in a highly nonuniform texture with submicron size domains in equilibrium. Traditional alignment techniques of surface treatment, electric field, or shear flow often fail for polymers and are hopeless when applied to elastomers when already cross-linked.

Rapid progress in discovery and studies of novel physical properties of SmC* elastomers was achieved after effective methods of preparing monodomain aligned structures were developed. One route to monodomains follows the example of nematic and SmA networks by cross-linking under selective mechanical deformations. However, one mechanical stretching is not enough since the SmC phase is clearly biaxial. After an initial uniaxial stretch aligns the nematic director, the second stretch at an appropriate angle to $n$ is applied to align the layers (Figure 17.10a).

In effect after one stretch, one has a uniform director but with random layer normals. We discuss later the response of polydomain SmC elastomers to strain since strain is such an important route to monodomains. These stretches are generally performed while network cross-linking is in process, which then permanently fixes a well-aligned, untwisted bookshelf C*-smectic structure [21]. Monodomain SmC elastomers were also made by blowing a bubble of a smectic melt, which is thereby aligned by the accompanying biaxial elongational flow in the layer planes. Cross-linking of the bubble then follows [22]. The resulting balloons have most unusual elasticity. A third mechanical method [23] first obtains a monodomain SmA by stretching as in Figure 17.9c. A second cross-linking while imposing an in-plane shear on the still SmA elastomer imposes a preferred in-plane direction. Indeed, the SmA in Figure 17.11a shows a residual sheared shape. Cooling to the SmC* state gives rise to a spontaneous shear (Figure 17.11b), confirming mechanically the x-ray and optical evidence that a monodomain state has been achieved.

A perhaps more natural method of preparing monodomain ferroelectric elastomers is by aligning the polymer melt (in the C* phase) in a cell subjected to a high external electric field (Figure 17.10b). This creates the perfect bookshelf geometry, which is then cross-linked by a UV-initiated reaction (a noninvasive method of cross-linking at a fixed temperature) to form a permanent defect-free elastomer network [24]. In this approach the electric field aligns a relatively thick layer of material, but one may need to overcome the problem of removing the electrode plates from the surfaces (usually achieved by coating the electrodes with a sacrificial, e.g., water-soluble, layer). An alternative to electric alignment has been to use free-standing films of the corresponding polymer melt stretched on a rigid frame [25]. In this case a (very thin) film of a smectic polymer shows a very strong alignment of layers in the film plane, in the geometry of Figure 17.9d. If the cross-linking is performed in the SmA phase, the resulting film has been demonstrated to have a giant electrostriction because of the electroclinic effect, the field-induced A–C transformation [26,27]. When, after the drawing of a free-standing film in the C* phase, a lateral electric field is applied between two electrodes in the frame, a well-aligned molecular tilt is established and, after cross-linking, one obtains a ferroelectric film with polarization in the plane [28].

### 17.3.3 PHYSICAL PROPERTIES OF SmA ELASTOMERS

As smectic order develops, it changes the chain shape distribution and thus the macroscopic shape of the elastomer. Figure 17.12 shows the uniaxial thermal expansion on cooling the elastomer [19] into such a prolate "induced-nematic" (IN) state, which is followed by contraction in the SmA phase because of the flattening by layers of some portions of the otherwise stretched backbones.

This system was twice cross-linked under uniaxial extension, thereby remembering a preferred direction of elongation. But the elastomer has no intrinsic nematic phase.

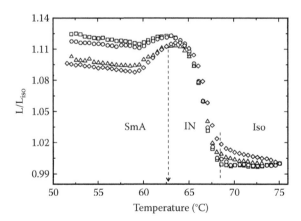

**FIGURE 17.12** Thermal expansion of a SmA elastomer with an associated prolate nematic phase at a higher temperature. The approximate smectic phase transition temperature (62°C) is marked by the arrow on the plot; above it one finds a prolate "IN" state, which eventually terminates at a true isotropic phase. Different data sets correspond to the same sample cooled or heated under an increasing load. (E. Nishikawa and H. Finkelmann. *Macromol. Chem. Phys.*, 200:312, 1999. Copyright Wiley-VCH Verlag GmbH & Co. KGaA. Reproduced with permission.)

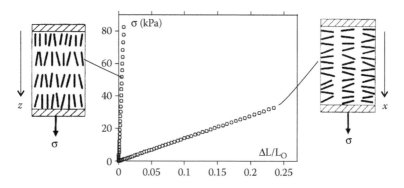

**FIGURE 17.13**  Stress against strain showing in-plane fluidity and parallel rigidity in a SmA elastomer. The Young modulus parallel and perpendicular to the layer normals differ very greatly—the rubber elasticity is two-dimensional. (E. Nishikawa, H. Finkelmann, and H.R. Brand. *Macromol. Rap. Commun.*, 18:65, 1997. Copyright Wiley-VCH Verlag GmbH & Co. KGaA. Reproduced with permission.)

As incipient layer formation and orientation develops, chains extend to a prolate shape, finally shrinking at a characteristic temperature where layer order becomes very high.

Stress, $\sigma$, accompanies elongational strain (Figure 17.13), from which a (Young) linear elastic modulus $E$ could be extracted: $\sigma = E\varepsilon$. First, for $\varepsilon_{zz}$ imposed parallel to the layer normal, a modulus of $E_{||} = 1.03 \times 10^7$ N/m² was measured. For $\varepsilon_{xx}$ imposed in-plane, a linear modulus $E_{\perp} = 1.4 \times 10^5$ N/m² emerged—about 75 times smaller than $E_{||}$. The physical reason for this one high modulus is clear: stretch along the layer normal attempts to alter the layer spacing which is resisted by the smectic ordering. This ordering acts on each side-chain monomer, rather than on chains as a whole as in rubber elasticity, and correspondingly dominates.

In many smectic elastomers, layers are constrained not to move relative to the rubber matrix. Deformations of a rubber along the layer normal are thus resisted by a layer spacing modulus, $B$, of the order of $10^2$ times greater than the shear modulus of the matrix. Distortions in plane, either extensions or appropriate shears, are simply resisted by the rubber matrix. Thus, SmA elastomers are rubbery in the two dimensions of their layer planes, but respond as hard conventional solids in their third dimension. Figure 17.13 shows this behavior. Such extreme mechanical anisotropy promises interesting applications.

Such high-elastic anisotropy is unparalleled even in strongly ordered nematic networks, where both Young moduli arising from the nematic elastomer free energy for stretch along and perpendicular to the director, $E_{||}$ and $E_{\perp}$, are of the same magnitude—roughly that of an ordinary rubber modulus $\mu$.

With such a high-elastic anisotropy, the shape of monodomain smectic rubber on being stretched ($\varepsilon_{xx}$) in the layer plane (Figure 17.14) does not look as surprising as it might have.

We know that all rubbers are physically incompressible, because the bulk modulus is many orders of magnitude greater than any of the rubber moduli. This usually means that rubber strips narrow on stretching. However, the sample in the picture,

**FIGURE 17.14**   A SmA elastomer showing no transverse contraction in the $z$-direction on being strained by $\sim$80% along $x$, in the layer plane. Layers are aligned with their normal along $z$. (E. Nishikawa, H. Finkelmann, and H.R. Brand. *Macromol. Rap. Commun.*, 18:65, 1997. Copyright Wiley-VCH Verlag GmbH & Co. KGaA. Reproduced with permission.)

although stretched by 80%, did not reduce in width at all. The reason, of course, is the elastic anisotropy $E_{\parallel} \gg E_{\perp}$. The contraction of the sample width in the direction visible in Figure 17.14 would mean the deformation along the layer normal $z$, which is penalized much more strongly. So, instead, the elastomer preserves its volume by reducing the thickness of the sample in Figure 17.13, which is another "weak," quasi-fluid direction in the layer plane.

The high modulus against extension along the layer normal in both liquid and rubbery smectics can be avoided by rotating the layers and thereby presenting along the extension direction a longer repeat interval (Figure 17.5a). This is the CMHH effect discussed above for liquid smectics. In rubber the elongation can then proceed by in-plane shear that we will see below in molecular and in continuum theory is cheaper than layer distortions. Rotation can, however, be incompatible with boundary constraints and layers have to have a periodic reversal in the sense of their rotation that is, they must buckle (see Figures 17.15a and b). The buckled layers now scatter light and are no longer transparent.

Nishikawa and Finkelmann (1999) have thoroughly investigated this effect of layer buckling under extension of monodomain smectic rubber along its principal axis—the layer normal (Figure 17.15b). At strains below threshold, the elastomer responded as a classical rubber, albeit with anomalously high Young modulus, *cf.* Figure 17.13. The onset of instability is easy to detect, both optically (because of strong light scattering on birefringence axis modulations, making the material opaque) and from the stress–strain variation (which develops a "soft" plateau because of internal relaxation of layer orientation); see Figure 17.16 and the original literature [29,19].

Chiral SmA* elastomers exhibit an electroclinic effect where an in-plane electric field induces a tilt of the director from the layer normal to the in-plane direction orthogonal to the field [30]. Where an associated mechanical distortion arises we have an inverse piezoelectric effect, possible in LCEs because now there is an elastic rigidity. We return to both these effects in reviewing SmC* elastomers where the deformations and polarization are natural rather than induced.

We have concentrated on bulk phases of strongly coupled elastomers. It is also possible to make free-standing, submicron thick elastomeric films on which one can conduct mechanical and electro-optical experiments. For instance one can induce

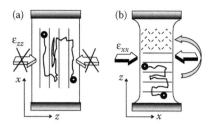

**FIGURE 17.15** A monodomain smectic elastomer with (a) elongational strain $\varepsilon_{xx}$ imposed along the layers (as in Figure 17.14) (b) elongational strain $\varepsilon_{zz}$ imposed along the layer normal. In the first case, the high-modulus resistance to layer compression prevents the sample from changing its perpendicular dimension. In the second case there is no such resistance and transverse contraction occurs, except at the clamps; see Figure 17.16a. However, at noninfinitesimal strains $\varepsilon_{zz}$, the smectic elastomer finds a lower energy deformation path, the layer buckling instability we have seen in liquid smectics (Figure 17.5). (Adapted from M. Warner and E.M. Terentjev. *Liquid Crystal Elastomers*. Oxford University Press (revised, paperback edition), 2007 by permission of Oxford University Press.)

compression of the smectic layers by applying an in-plane strain [31–33]. Evidently these elastomers do not have the strong coupling of the layers we have described thus far. Further, manipulating the micron-scale roughness of free-standing smectic films by mechanical stretching has also been reported [34].

### 17.3.3.1 SmC Elastomers

Liquid SmC has the director tilted by an angle θ from the layer normal. The solid, elastomeric analogues are also interesting because in their chiral forms they too are

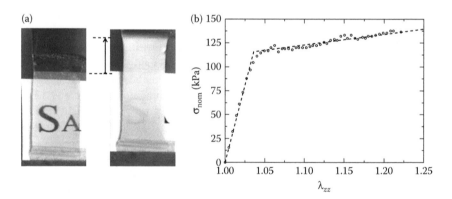

**FIGURE 17.16** Layer buckling (CMHH) instability on stretching the monodomain SmA elastomer along the layer normal. Images (a) show that the strip experiences the ordinary transverse contraction, but also becomes optically opaque. The stress–strain plateau (b) shows how internal layer rotations reduce the effective elastic modulus and also indicates the strain threshold eth $\varepsilon_{th} \sim 0.05$. Dotted line theory. (E. Nishikawa and H. Finkelmann. *Macromol. Chem. Phys.*, 200:312, 1999. Copyright Wiley-VCH Verlag GmbH & Co. KGaA. Reproduced with permission.)

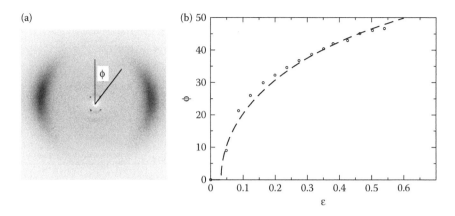

**FIGURE 17.17** (a) An x-ray scattering image illustrating the zigzag coarsened buckling of smectic layers, while the nematic alignment is left largely unchanged—vertical, as indicated by wide-angle arcs. Measuring the angle $\phi$ between the two pairs of smectic reflexes allows plotting the angle of local layer rotation $\phi(\varepsilon)$ against strain, (b). Dotted line theory. (E. Nishikawa and H. Finkelmann. *Macromol. Chem. Phys.*, 200:312, 1999. Copyright Wiley-VCH Verlag GmbH & Co. KGaA. Reproduced with permission.)

(improper) ferroelectrics. They are perhaps unique among solid ferroelectrics in that they are locally fluid-like, capable of huge distortions and with low moduli (about $10^4$ times smaller than crystalline materials or ceramics). Additionally their extra azimuthal degree of freedom of the tilted director, combined with their solidity, gives them the shear equivalent of the spontaneous distortion we saw in nematic elastomers.

Shear associated with the SmA to SmC transition [23] is shown in Figure 17.11. These authors characterize shear by the angle $\theta_E$ that the sample mechanically tilts through: $\tan \theta_E = \lambda_{xz}$ where $x$ is the in-plane tilt direction and $z$ that of the layer normal. They denote their molecular tilt, as measured by x-ray scattering, by $\theta_X$ (we simply use $\theta$). Its evolution as the C-phase is entered is shown in Figure 17.18a, while Figure 17.18b plots the dependence $\theta_E(\theta_X)$. It shows that mechanical tilt (i.e., shear $\lambda_{xz}$) is not identical to the molecular tilt [35].

Indeed there is a residual shear in the SmA, untilted phase from the two-step cross linking (see Section 17.3.2). The second stage was performed under an imposed shear, a memory of which is imprinted in the matrix, both in the mechanical shape and in the residual director tilt.

In SmC, director tilt has created a second special direction in addition to the layer normal. As explained for liquid SmC, for a chiral system, the existence of two noncolinear axes allows us to find a third polar vector, in this case the polarization, $P$, in the orthogonal direction. Thus, changes in the tilt magnitude or tilt direction should correspondingly change the magnitude or direction of $P$.

We can now understand how an electric field, $E$, applied in the layer planes of chiral SmA* can induce a polarization and, by inverse of the above connection between tilt and polarization, the field also induces a tilt. In effect a SmC* phase is being induced; this is the electroclinic effect [36,37]. The electroclinic coefficient is

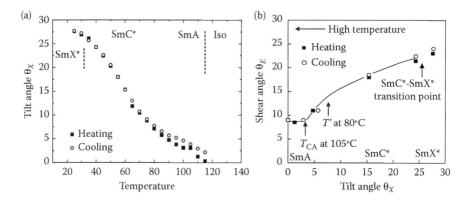

**FIGURE 17.18**   (a) Increase of molecular tilt $\theta_X$ with wearing temperature $T$ on entering the SmC phase. (Adapted from K. Hiraoka et al., *Macromolecules*, 38:7352, 2005.) and (b) Shear angle $\theta_E$ against molecular tilt.

defined as $\alpha = \partial\theta/\partial E$ and the values obtained in SmA* elastomers were in the range $\alpha = 0.045$ m/MV.

Since tilt and mechanical shear are related, then $E$ can as well induce shear $\lambda_{ck}$ (in a frame-independent form; $c$ is the c-director, the director projection on the layer plane, and $k$ the layer normal). This is the inverse piezoelectric effect, the direct effect being where an imposed $\lambda_{ck}$ induces a $P$ in the orthogonal $c \times k$ direction. The normal and inverse effects can be found in both SmA* and SmC* phases and tend to be largest near the A–C transition where the order is least rigid against changes of tilt [37].

When orientationally ordered rods tilt, they also affect the layer spacing. Naively one might expect in a tilted system that $d(\theta) = d_0 \cos\theta$. One can directly measure $d$ by x-rays as tilt is altered, for instance by electric fields. Alternatively, one can deduce there has been a layer spacing change by measuring possible deformation $\lambda_{kk}$ along the layer normal, assuming strong coupling between the layers and the rubber matrix, that is $\lambda_{kk} = d(\theta)/d_0$. In SmA* there is initially no tilt; the flatness of the cosine function around $\theta = 0$ means that $\lambda_{kk} \sim 1 - \theta^2/2$, that is strain, $\varepsilon$, is second order in $\theta$ and in this mode is contractile [38]. In SmA* therefore, the electrostriction resulting from field-induced layer tilt has a coefficient, $a$, defined by $\varepsilon = \lambda_{kk} - 1 = -aE^2$, that is connected to the electroclinic coefficient by $a = \alpha^2/2 - 10^{-15} m^2/V^2$ [37].

To summarize, one can induce $\lambda_{ck}$ and $\lambda_{kk}$ at linear order in $E$ in SmC*, but in SmA* $\lambda_{kk}$ is second order in tilt. This contrast is to be expected since for electrical effects one requires two directions, naturally provided by either shear or the tilted SmC* environment, but the two directions are not provided by compression or by a SmA* environment.

In SmC* where there is a preexisting tilt, $\theta_0$ say, then the connection between $\delta d/d_0$ and $\delta\theta$ is linear, that is $\delta d/d_0 \sim -\sin\theta_0\delta\theta$. Now we have a piezoelectric response to compression along the layer normal, or inversely a contraction along the layer normal in response to an in-plane field. Thus, we have two piezoelectric responses–to compression along the layer normal and to shear $\lambda_{ck}$, the latter being associated with much lower stresses since layer separation is not involved.

A final interesting possibility arises from analogous liquid smectic experiments, namely that of photo-ferroelectricity [39–41] where a SmC* phase has photoisomerizable elements that on photon absorption reduce the molecular tilt and hence, also reduce the electric polarization. The first steps to synthesizing photo-switchable SmC elastomers have been taken [42] where light can induce both mechanical tilt and an electrical response. Further recent related work examined the solidification of smectic liquid crystal phases with photosensitive gel forming agents [43] where irradiation causes the network to lose memory of its formation state—the cross-links are essentially dissolved and then remade on de-excitation.

## 17.4   A MOLECULAR MODEL OF SmA RUBBER ELASTICITY

Since smectic elastomers are rubbery, that is capable of large reversible deformations, their chains locally retain a liquid-like mobility and readily explore configurations. Being long, the chains have a Gaussian distribution of shapes, but one that is not necessarily isotropic. They are both orientationally ordered and subjected to spatial localization by layers. Being Gaussian, their conformations are characterized by a second moment proportional to a $\underline{\underline{\ell}}$ tensor. In Figures 17.1 and 17.2 the $\underline{\underline{\ell}}$ tensors are represented by spheres in the isotropic state and by spheroids in the nematic state. The mean square size of chains has an anisotropic form in such Gaussians that is, the average $\langle R_i R_j \rangle = (1/3)\ell_{ij}L$ where $L$ is the length of the chain. These are Gaussian forms familiar from simple polymers, but where the effective step lengths are different, parallel and perpendicular to the director, respectively taking the values $\ell_{||}$ and $\ell_{\perp}$. The ratio of these, $r = \ell_{||}/\ell_{\perp}$, characterizes the anisotropy of the distribution of chains and is the square of the ratio of the major to minor axes of the spheroids depicted in Figures 17.1 and 17.2. Specifically, the shape tensor is:

$$\underline{\underline{\ell}} = (\ell_{||} - \ell_{\perp})\boldsymbol{n}\,\boldsymbol{n} + \ell_{\perp}\underline{\underline{\delta}} \equiv \ell_{\perp}((r-1)\boldsymbol{n}\,\boldsymbol{n} + \underline{\underline{\delta}}) \tag{17.2}$$

where we shall drop the prefactor $\ell_{\perp}$ henceforth, because, as we shall see, it always appears with its inverse, and hence cancels, in free energy expressions.

We also require the deformation gradients $\underline{\underline{\lambda}}$. The diagonal elements of $\underline{\underline{\lambda}}$ are stretches and contractions, more precisely the relative changes in lengths of the relevant dimensions of the solid and are indicated for some directions in Figures 17.1 and 17.2—if the $z$-direction were chosen, then the relative length change would be $\lambda_{zz}$, and so on. Shears are specified by the off-diagonal elements $\lambda_{xz}$, and so on, where in this example $x$ would be the displacement direction and $z$ the gradient direction, that is, $\lambda_{xz} = \partial u_x/\partial z$. Small strains are denoted by $\varepsilon_{zz} = \lambda_{zz} - 1$, and so on, with symmetrized forms for shears $\varepsilon_{xz}$.

Given that network deformations are deforming Gaussian distributions of chains, there turns out to be [1] a rubbery part to their free energy density given by $(1/2)\mu\,\mathrm{Tr}(\underline{\underline{\ell}}_0 \cdot \underline{\underline{\lambda}}^{\mathrm{T}} \cdot \underline{\underline{\ell}}^{-1} \cdot \underline{\underline{\lambda}})$. This rubber elastic free energy is the simplest generalization of classical Gaussian rubber to the anisotropic case. It too is proportional to the square of the deformation gradients, but has encoded in it the original $(\underline{\underline{\ell}}_0)$ and current ($\underline{\ell}$) shapes and orientations of the network chains. We are taking the view that, because the underlying order is locally nematic, then the basic rubber free energy of smectics is also locally nematic-like, but with strong constraints offered by the layers of the phase.

We have seen that in smectic elastomers the layers have a profound effect on physical properties, and chains are constrained on an energy scale far greater than $\mu$. Extensions along the layer normal that would require cross-link translation with respect to the layers are penalized by a modulus $B$ comparable to or larger than that of the layer spacing modulus of smectic liquids. We shall assume that, in strongly coupled smectics, layers rigidly localize cross-links. Layers must then move affinely with the rubber matrix, that is, their orientation and spacing must deform as their local matrix does. We must then add to the rubber elastic deformation cost a penalty $1/2B(d/d_0 - 1)^2$ where $d$ and $d_0$ are the current and original layer spacings, an addition that can be explicitly derived, see Ref. [44]. Thus the overall smectic-A rubber energy density is

$$F_A = \tfrac{1}{2}\mu \text{Tr}(\underline{\underline{\ell}}_0 \cdot \underline{\underline{\lambda}}^T \cdot \underline{\underline{\ell}}^{-1} \cdot \underline{\underline{\lambda}}) + \tfrac{1}{2}B(d/d_0 - 1)^2 \qquad (17.3)$$

The second part of this energy may not be relevant to weakly coupled smectic elastomers where there is apparently no elastic signature of the smectic layer system [31]. Such elastomers could be adequately described by classical isotropic rubber elasticity as is appropriate for a nematic elastic matrix where the director cannot rotate (biaxial strains in the layer planes were examined). They have been prepared as very thin samples (a few hundreds of molecular layers) and it seems possible that layer numbers are not necessarily preserved under deformations.

In contrast, there is evidence [32] that homopolymer networks, where the smectogens are not diluted, experience a strong potential (suggested, as we will see, by their extreme Poisson ratios (0, 1)). Indeed, the overall $B$ in strongly coupled elastomers is much larger than that encountered in liquid smectics. Polymeric smectics evidently have a mechanism to give coherence to the movement of connected rods with respect to the layer potential and hence give more resistance to layer-spacing change, as well as there being a contribution to $B$ because of cross-linking. Overall, in Figure 17.13 one sees the much greater resistance to elongations along the layer normal than the rubbery response in-plane, the modulus for the former being around, or even greater than, $10^7$ Pa. Additionally, in Equation 17.3 we have assumed that the director remains at all times along the layer normal. Anchoring need not be rigid. We discuss below when it is weaker and shears can induce the director to rotate with respect to the layer normal—a "mechano-clinic" effect [30]. We shall concentrate on the remarkable, strongly coupled systems whereupon the geometry of the rotation and deformation of layers under $\underline{\underline{\lambda}}$ must be solved in order to give explicit form to the innocent-looking last term in the energy (17.3). The energy $F_A$ is highly nonlinear both because of incompressibility, $\text{Det}(\underline{\underline{\lambda}}) = 1$, and from the distorted layer spacing dependence $d(\underline{\underline{\lambda}})$ that we derive in the next section. Deformations are large and constraints are strong so nonlinearity must be confronted directly and molecular theory offers a direct route. However, since effects are complex, some guidance to their nature can be offered by linear continuum theory, especially since one can employ group theory to ensure that one encompasses all effects and that their symmetry is correct; see Refs. [30,38] for the complete SmA and SmC free linear continuum energies.

## 17.4.1 THE GEOMETRY OF AFFINE LAYER DEFORMATIONS

The layer normal (equivalent to the director in SmA) and the layer spacing transform as a result of layers affinely following the deformation $\underline{\underline{\lambda}}$. Since the layer normal is a unit vector and must remain so, the process is highly nonlinear. Two perpendicular unit vectors, $h$ and $m$ say, span a plane and their cross product is the layer normal, $n_0 = h \times m$, that defines the plane (Figure 17.19).

If $h$ and $m$ deform affinely with $\underline{\underline{\lambda}}$, they become the vectors $\underline{\underline{\lambda}} \cdot h$ and $\underline{\underline{\lambda}} \cdot m$ in the deformed plane and yield the new normal by again taking the cross product:

$$k = n = \frac{(\underline{\underline{\lambda}} \cdot h) \times (\underline{\underline{\lambda}} \cdot m)}{|(\underline{\underline{\lambda}} \cdot h) \times (\underline{\underline{\lambda}} \cdot m)|} \rightarrow \frac{\underline{\underline{\lambda}}^{-T} \cdot n_0}{|\underline{\underline{\lambda}}^{-T} \cdot n_0|} \tag{17.4}$$

where we continue dealing with strongly anchored SmA and where the layer normal and director are identical. In general $\underline{\underline{\lambda}} \cdot h$ and $\underline{\underline{\lambda}} \cdot m$ are no longer necessarily unit vectors or even still perpendicular to each other, thus the need for the normalizing denominator in Equation 17.4. The second form, one can readily show [44,1], is the result of layers deforming affinely with an *incompressible* medium, where $\underline{\underline{\lambda}}^{-T}$ is the inverse transpose of $\underline{\underline{\lambda}}$. In general vectors and planes embedded in a solid and deforming with it respond very differently. Figure 17.20 shows this vividly for vectors and planes of various orientations with respect to an imposed elongation in the $x$ direction. Each vector has a corresponding set of planes with their normal parallel to the vector—the vector and normal rotate in opposite senses when they are not aligned with a principal deformation direction.

It also follows from Equation 17.4 that the layer spacing becomes

$$d/d_0 = \frac{n_0 \cdot \underline{\underline{\lambda}}^{-1} \cdot \underline{\underline{\lambda}} \cdot n_0}{|\underline{\underline{\lambda}}^{-T} \cdot n_0|} \equiv \frac{1}{|\underline{\underline{\lambda}}^{-T} \cdot n_0|}. \tag{17.5}$$

Inserting the result, Equation 17.5, into the second part of the free energy (Equation 17.3) one obtains the highly nonlinear expression for the free energy density:

$$F_A = \tfrac{1}{2}\mu \operatorname{Tr}(\underline{\underline{\ell}}_0 \cdot \underline{\underline{\lambda}}^T \cdot \underline{\underline{\ell}}^{-1} \cdot \underline{\underline{\lambda}}) + \tfrac{1}{2}B\left(\frac{1}{|\underline{\underline{\lambda}}^{-T} \cdot n_0|} - 1\right)^2. \tag{17.6}$$

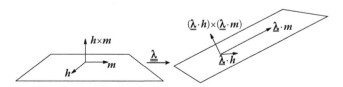

**FIGURE 17.19** The normal to the layer deforms as $\underline{\underline{\lambda}}^{-T}$ when the vectors in the layer deform according to $x \rightarrow \underline{\underline{\lambda}} \cdot x$.

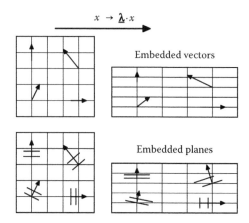

**FIGURE 17.20**  Vectors and sets of plans deforming affinely with the matrix in which they are embedded.

## 17.4.2  RESPONSE TO PRINCIPAL DEFORMATIONS

Despite the apparently highly tensorial and highly nonlinear character of $F_A$, one can analyze fundamental deformations, that is, those of Figure 17.21. None of these deformations are soft since director rotation is associated with deformations tightly constrained by layers.

### 17.4.2.1  Imposed In-Plane Extension $\lambda_{xx}$

We impose an extension $\lambda = \lambda_{xx}$ along one ($x$) of the equivalent in-plane directions (Figure 17.21a). Our notation is that the externally imposed component of $\underline{\underline{\lambda}}$ is written without its subscript index. The deformation gradient matrix and its inverse transpose take the form:

$$
\underline{\underline{\lambda}} = \begin{pmatrix} \lambda & 0 & 0 \\ 0 & \lambda_{yy} & 0 \\ 0 & 0 & \lambda_{zz} \end{pmatrix} ; \underline{\underline{\lambda}}^{-T} = \begin{pmatrix} \dfrac{1}{\lambda} & 0 & 0 \\ 0 & \dfrac{1}{\lambda_{yy}} & 0 \\ 0 & 0 & \dfrac{1}{\lambda_{zz}} \end{pmatrix}.
\tag{17.7}
$$

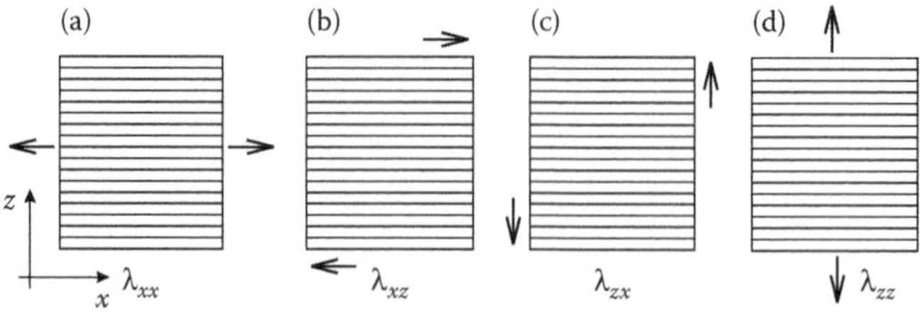

**FIGURE 17.21**  Imposed deformations treated in [44]: (a) stretch perpendicular to the layer normal, (b) shear of layers in their plane, (c) shear out of plane, and (d) stretch parallel to the layer normal, see [44].

As expected, the director is unchanged and the layer spacing changes simply with $\lambda_{zz}$:

$$\underline{\underline{\lambda}}^{-T} \cdot n_0 = (0,0,1/\lambda_{zz}) \Rightarrow n = (0,0,1) \quad \text{and} \quad d/d_0 = \lambda_{zz}. \tag{17.8}$$

Volume conservation is simply expressed by $1 = \lambda\lambda_{yy}\lambda_{zz}$. The shear $\lambda_{zx}$ is absent since it introduces torques from the change of shape in the presence of an $x$ component of force (from the $xx$-stress associated with imposing $\lambda_{xx}$) which tends to eliminate the distortion. The shear $\lambda_{xz}$ vanishes by the symmetry of this deformation.

Inserting $\underline{\underline{\lambda}}$ (with $\lambda_{zz} \rightarrow 1/\lambda \, \lambda_{yy}$) into the free energy results in:

$$F_A = \tfrac{1}{2}\mu \left\{ \lambda_{yy}^2 + \lambda^2 + \frac{1}{\lambda_{yy}^2\lambda^2} + b \left( \frac{1}{\lambda_{yy}\lambda} - 1 \right)^2 \right\}, \tag{17.9}$$

with the nondimensional layer compression ratio $b = B/\mu$. Minimization of this free energy with respect to $\lambda_{yy}$ gives:

$$\lambda^2\lambda_{yy}^4 - 1 = b(1 - \lambda\lambda_{yy}). \tag{17.10}$$

Our illustrations are for $b = 5, r = 2$. Material properties such as Poisson ratios, strain thresholds and moduli depend on this coupling and anisotropy; small $b$ amplifies thresholds in our illustrations.

Although a quartic in $\lambda_{yy}$, Equation 17.10 can be solved by treating it as a quadratic in $\lambda(\lambda_{yy})$ and plotting the result parametrically (Figure 17.22). It is clear that the limits (dashed lines) of large and small $b$ correspond respectively to the bounds $\lambda_{yy} = 1/\lambda$ (with no contraction along the layer normal, $\lambda_{zz} = 1$) and $\lambda_{yy} = \lambda_{zz} = 1/\sqrt{\lambda}$ (the classical limit with equal perpendicular contractions). The elastically isotropic material with a small ratio $B/\mu$ is still a SmA phase in the sense that the director is constrained to point along the layer normal and the elastomer could still be in the

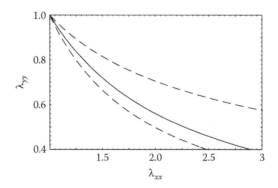

**FIGURE 17.22** In-plane contraction $\lambda_{yy}$ in response to an in-plane stretch $\lambda_{xx}$. $r = 2$ and $b = 5$. The dashed line comparisons $yy$-contractions are $1/\lambda_{xx}$ (lower) and $1/\sqrt{\lambda_{xx}}$ (upper). (Adapted from M. Warner and E.M. Terentjev. *Liquid Crystal Elastomers*. Oxford University Press (revised, paperback edition), 2007 by permission of Oxford University Press.)

strong coupling limit in that layer-matrix coupling is rigid; the layer strength is simply lower, as in [31] or one is near the SmA–N transition.

The Poisson ratios for $b \to 0$ and $b \gg 1$ reveal the isotropy and dimensional reductions implied in the limits of weak and strong layers. For small strains $\lambda_{yy} = 1 + \varepsilon$ and $\lambda = 1 + \omega$, Equation 17.10 gives the Poisson ratio in the $y$ direction, $v_y = -\varepsilon/\omega$, and that in the layer direction, $v_z$:

$$v_y = \frac{2+b}{4+b}; \quad v_z = \frac{2}{4+b}. \tag{17.11}$$

The crossover from Poisson ratios $(z, y) = (0, 1)$ to $(1/2, 1/2)$ is thus relatively slow. However, it is clear that for $b \sim 60$, as found in some experimental situations [19], the material is firmly in the $(0, 1)$ class. In this regime of large $b$, where layer compression is suppressed ($\lambda_{zz} \to 1$), the elastomer is like a classical 2-D rubber. Since there is no director rotation, $r$ does not enter into the elastic free energy density (Equation 17.9), which reduces to

$$F_A = \tfrac{1}{2}\mu\{\lambda^2 + \lambda_{yy}^2\} = \tfrac{1}{2}\mu\{\lambda^2 + 1/\lambda^2\} \tag{17.12}$$

as a consequence of higher layer stiffness and volume conservation. The Young modulus derived from Equation 17.12 is

$$E_\perp = \left.\frac{\partial^2 F}{\partial \lambda^2}\right|_{\lambda=1} = 4\mu, \tag{17.13}$$

rather than the value $3\mu$ that obtains for a classical isotropic rubber: dimensional constraints increase the effective response modulus.

### 17.4.2.2 Imposed In-Plane Shear $\lambda_{xz}$

Consider (Figure 17.21b) in-plane shear with deformation gradient matrix:

$$\underline{\underline{\lambda}} = \begin{pmatrix} \lambda_{xx} & 0 & \lambda \\ 0 & \lambda_{yy} & 0 \\ 0 & 0 & \lambda_{zz} \end{pmatrix}. \tag{17.14}$$

We suppress $\lambda_{zx}$ since typically the application of $\lambda_{xz}$ is with parallel plates that constrain the sample. Again, volume conservation is simply expressed $1 = \lambda_{xx}\lambda_{yy}\lambda_{zz}$ and eliminates $\lambda_{yy}$. The free energy is:

$$F_A = \tfrac{1}{2}\mu\left\{\lambda_{xx}^2 + \frac{1}{\lambda_{xx}^2\lambda_{zz}^2} + \lambda_{zz}^2 + r\lambda^2 + b(\lambda_{zz} - 1)^2\right\}. \tag{17.15}$$

The shear $\lambda$ and the $zz$-stretch do not couple and so the imposed shear cannot affect $\lambda_{zz}$, as expected in this geometry. Minimizing over $\lambda_{xx}$ and $\lambda_{zz}$ yields $\lambda_{zz} = \lambda_{xx} = 1$ and the free energy density becomes

$$F_A = \tfrac{1}{2}\mu\{3 + r\lambda^2\}. \tag{17.16}$$

The result is the same as for the simple shear in a nematic elastomer with unrotating director. The linear shear modulus is $r\mu$. The anisotropy enters in the classical way—for large $r$ the modulus becomes large because chains extend across many molecular shear planes. It turns out that this modulus is identical to that obtaining for $\lambda_{zz} \gg 1$ after the instability on stretching along the layer normal, because then deformation is largely through shears in the rotated planes (see Section 17.4.2.3).

It is this $xz$-shear of Figure 17.21b that would most directly induce director tilt with respect to the layer normal. A strong electro clinic response [26,27] suggests that a mechanoclinic should exist, after all electric fields are relatively very small compared with mechanical fields (seen by comparing elastic and electric field energy densities $\mu$ and $\Delta\varepsilon E^2$, respectively). Thus one can consider relaxing the strong anchoring condition of the director, both in nonlinear continuum theory [45] and in the molecularly based theory sketched here [46]. In the latter the deformations of Figure 17.21a, b, and d are evaluated while allowing varying degrees of director rotation. Experiments are difficult since shearing a flat sheet in its plane leads to buckling and wrinkling at 45 degrees to the shear directions that is, along the compression diagonal. Experimentally some degree of director rotation is observed [47] with which theory of $xz$-shear with finite director coupling agrees well [48] (see the latter for treatment of the wrinkling problem).

### 17.4.2.3  Imposed Out-of-Plane Shear $\lambda_{zx}$

The two principal deformations analyzed above have, characteristically, preserved the layer spacing $d_0$ and left the orientation of the layer normal intact. We now examine the other two deformations of Figure 17.21 which, with increasing complexity, affect these two smectic variables. Let us now impose the deformation of Figure 17.21(c) by placing the lower triangular element in the matrix $\underline{\underline{\lambda}}$:

$$\underline{\underline{\lambda}} = \begin{pmatrix} \lambda_{xx} & 0 & 0 \\ 0 & \lambda_{yy} & 0 \\ \lambda & 0 & \lambda_{zz} \end{pmatrix}. \tag{17.17}$$

As in the previous cases we must suppress the opposite shear, here $\lambda_{xz}$, because of the mechanical constraints implied in imposing $\lambda_{zx}$. From the geometry of this shear one expects the layers to be forced to rotate by an angle that we denote by $\phi$. Volume conservation gives $\lambda_{yy} = 1/\lambda_{xx}\lambda_{zz}$. In the rigid director-layer normal coupling limit, the director-layer normal and layer spacing evolve as:

$$k = n = \frac{1}{\sqrt{\lambda^2 + \lambda_{xx}^2}}(-\lambda, 0, \lambda_{xx}) \quad \text{and} \quad d/d_0 = \frac{\lambda_{xx}\lambda_{zz}}{\sqrt{\lambda_{xx}^2 + \lambda^2}}. \tag{17.18}$$

The free energy density is somewhat complicated in form since the director has rotated (which mixes up the elements of $\underline{\underline{\lambda}}$) but it can easily be minimized numerically [44]. A typical solution for the components of $\underline{\underline{\lambda}}$ and the corresponding layer spacing changes is illustrated in Figure 17.23.

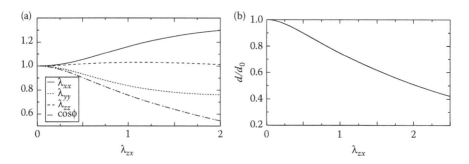

**FIGURE 17.23** (a) The relaxing components of the deformation tensor on imposing $\lambda_{zx}$ for $b = 5$ and $r = 2$; The associated director (layer normal) rotation is given by $\cos\phi$. (b) The associated contraction of layer spacing. (Adapted from M. Warner and E.M. Terentjev. *Liquid Crystal Elastomers*. Oxford University Press (revised, paperback edition), 2007 by permission of Oxford University Press.)

In this $zx$-deformation we are thus effectively compressing, as well as rotating the layers, which is now penalized by the large modulus $B$. Even for very large ratio $b$, the layer spacing eventually yields and begins to decrease.

### 17.4.2.4 Imposed Extension $\lambda_{zz}$ along the Layer Normal

The smectic planes are known experimentally, at first, to resist such extension with a high modulus and then complex rotational and shear instabilities arise that reduce the modulus (Figure 17.16). In this case (Figure 17.21d), the deformation tensor is

$$\underline{\underline{\lambda}} = \begin{pmatrix} \lambda_{xx} & 0 & 0 \\ 0 & \lambda_{yy} & 0 \\ \lambda_{zx} & 0 & \lambda \end{pmatrix} \quad \text{with} \quad \underline{\underline{\lambda}}^{-T} = \begin{pmatrix} \lambda_{yy}\lambda & 0 & 0 \\ -\lambda_{zx}\lambda_{yy} & \lambda_{xx}\lambda & 0 \\ 0 & 0 & \lambda_{xx}\lambda_{yy} \end{pmatrix}.$$

$$(17.19)$$

The shear $\lambda_{zx}$ induces layer rotation by an angle $\phi$ which, for purely geometrical reasons, presents a longer distance $d/\cos\phi$ along $z$ than the simple value $d$. This is the origin of CMHH instability for elongation along the layer normal in smectic liquid crystals (see Figure 17.5a).

The shear $\lambda_{xz}$ in the presence of a $z$-force of extension would lead to torques and is not observed in smectic elastomers [19] or in the analogous nematic elastomer geometries that involve simultaneous director rotation and shear. The volume constraint simply gives $\lambda_{yy} = 1/\lambda_{xx}\lambda$. Since $\underline{\underline{\lambda}}^{-T} \cdot k_0 = (-\lambda_{zx}\lambda_{yy}, 0, \lambda_{xx}\lambda_{yy})$, then the layer spacing and the director/layer normal become respectively:

$$d/d_0 = \frac{1}{\lambda_{yy}\sqrt{\lambda_{xx}^2 + \lambda_{zx}^2}} \quad \text{and} \quad k = \left( -\frac{\lambda_{zx}}{\sqrt{\lambda_{xx}^2 + \lambda_{zx}^2}}, 0, \frac{\lambda_{xx}}{\sqrt{\lambda_{xx}^2 + \lambda_{zx}^2}} \right). \quad (17.20)$$

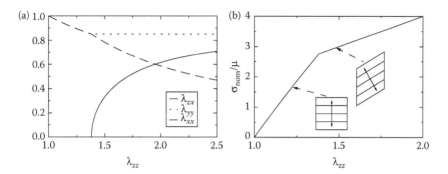

**FIGURE 17.24** (a) Deformation tensor components for an imposed stretch $\lambda_{zz} = \lambda$ parallel to the layer normal, for $b = 5$ and $r = 2$. The shear $\lambda_{zx}$ relaxes to an asymptote of $1/\sqrt{\lambda_c}$ for large $\lambda_{zz}$. (b) The nominal $zz$ stress, see [44].

The elastic free energy density is then:

$$F_A = \tfrac{1}{2}\mu \left\{ \lambda_{xx}^2 + \frac{1}{\lambda_{xx}^2 \lambda^2} + \lambda_{zx}^2 + \frac{(\lambda_{xx}^2 + r\lambda_{zx}^2)\lambda^2}{\lambda_{xx}^2 + \lambda_{zx}^2} + b\left( \frac{\lambda\lambda_{xx}}{\sqrt{\lambda_{xx}^2 + \lambda_{zx}^2}} - 1 \right)^2 \right\}.$$

(17.21)

This free energy can be analytically minimized [44] to fix the free components of the deformation tensor for a particular $b$ value (see Figure 17.24a).

There is a critical, threshold value of the elongation $\lambda = \lambda_c(b, r)$ when the layer rotation starts to occur. It is rather large because a relatively small ratio $b$ has been adopted for illustration. The threshold is to a stretched and sheared state given by the strain (Equation 17.19) and sketched in Figure 17.24b. The threshold exists independently of other possible causes of threshold, namely clamp constraints and microstructure, which arise in practical cases and which we discuss below. Frank elasticity that partly determines the fineness of microstructure, has not yet been invoked and will turn out to be largely irrelevant for the threshold.

Shear $\lambda_{zx}$ starts with a singular edge at the threshold and the transverse contraction $\lambda_{yy}$ thereafter, remains constant. This reminds one of the analogous response in nematic elastomers where $\lambda_{yy} = \text{const}$ was a key signature of soft (or semisoft) deformations. The accompanying stress also divides into two distinct regimes with a much higher modulus before the transition (Figure 17.24b) and the fit to experiment in Figure 17.16b. The layer rotation, given by $\mathbf{n}$ (Equation 17.20), has the same singular edge at $\lambda_c$. It is plotted against $\varepsilon = \lambda - 1$ in Figure 17.17b along with experiment [19] where $\lambda_c \sim 1.04$. There are strong geometric reasons why the shear $\lambda_{zx}$, the layer rotation and the softer elongation do not start immediately, but only onsets after a threshold, despite the cost of distortions being lower after the instability; see Ref. [1].

Before the layers start to rotate, $\lambda_{zx} = 0$, and the free energy density

$$F_A = \tfrac{1}{2}\mu \left\{ \lambda_{xx}^2 + \frac{1}{\lambda_{xx}^2 \lambda^2} + \lambda^2 + b(\lambda - 1)^2 \right\} \tag{17.22}$$

is minimal when $\lambda_{xx}^2 = \lambda_{yy}^2 = 1/\lambda$. Thus, for $\lambda < \lambda_c$, the two perpendicular directions are equivalent and the material has Poisson ratios $(1/2, 1/2)$ in the $(x, y)$ directions. The free energy density and nominal stress, $\sigma_{\text{nom}} = \partial F_A/\partial \lambda$, are

$$F_A = \tfrac{1}{2}\mu\{\lambda^2 + 2\lambda + b(\lambda - 1)^2\} \quad \text{and} \quad \sigma_{\text{nom}} = \mu\{\lambda - 1/\lambda^2 + b(\lambda - 1)\}. \tag{17.23}$$

Layer rotation starts at a threshold $\lambda_c$ that one can show is set by the tensile nominal stress in the unrotated and rotated phases being the same. The in-plane contraction $\lambda_{yy}$ is now constant, and volume conservation then requires $\lambda_{xx} = \sqrt{\lambda_c}/\lambda$. If deformations were small, then one could say that the material would now have Poisson ratios $(1, 0)$, to which we return when comparing with experiment.

Minimizing the free energy with respect to $\lambda_{zx}^2$ gives,

$$\lambda_{zx} = \pm\frac{1}{\sqrt{\lambda_c}}\sqrt{1 - \left(\frac{\lambda_c}{\lambda}\right)^2}; \lambda_{yy} = \frac{1}{\sqrt{\lambda_c}}; \lambda_{xx} = \frac{1}{\sqrt{\lambda_c}}\left(\frac{\lambda_c}{\lambda}\right). \tag{17.24}$$

The shear displays the singular edge seen in Figure 17.24a. Both signs of shear give the same dilation along $z$ at constant layer spacing. Both shears, and indeed all directions perpendicular to $z$ (not just $x$), are equivalently possibly and thus required in a description of any induced microstructure (Section 17.4).

The director (and thus layer) rotation can be derived from the explicit expression (Equation 17.20) for $\mathbf{n}$. For instance, the first component gives

$$\sin \phi = \sqrt{1 - (\lambda_c/\lambda)^2} \tag{17.25}$$

with the singular edge and distinctive form also seen in experiment (Figure 17.17b).

All components of induced deformation and the rotation of layers depend solely on (the imposed) $\lambda$ and (the observable) $\lambda_c$, not in any separate or detailed way on the smectic potential $b$ or the anisotropy $r$. Theory is very tightly constrained since there are no free parameters. It requires all relaxation–strain and rotation–strain relations (Equations 17.24 and 17.25) to be of the same form for all systems when $\lambda$ is reduced by $\lambda_c$.

We substitute $\lambda_{zx}$ back into (Equation 17.21) to obtain the free energy and thus the nominal stress for $\lambda > \lambda_c$:

$$F_A = \tfrac{1}{2}\mu \left\{ \frac{2}{\lambda_c} + \lambda_c^2 + r(\lambda^2 - \lambda_c^2) + b(\lambda_c - 1)^2 \right\} \quad \text{and} \quad \sigma_{\text{nom}} = \mu r \lambda. \tag{17.26}$$

Notice that the energy associated with layer spacing, the $b$ term, is now constant. The gradients of the nominal stresses Equations 17.23 and 17.26 give the effective Young moduli in each regime:

$$E_{\parallel} = \begin{cases} 3\mu + B & \lambda < \lambda_c \\ \mu r & \lambda > \lambda_c \end{cases} . \tag{17.27}$$

The layer spacing of the system as a function of the imposed stretch is

$$\frac{d}{d_0} = \frac{1}{\lambda_{yy}\sqrt{\lambda_{xx}^2 + \lambda_{zx}^2}} = \frac{\lambda\lambda_{xx}}{\sqrt{\lambda_{xx}^2 + \lambda_{zx}^2}} = \begin{cases} \lambda & \text{for } \lambda < \lambda_c \\ \lambda_c & \text{for } \lambda > \lambda_c \end{cases} . \tag{17.28}$$

Before layer rotation starts $\lambda_{zx} = 0$ and so the layer spacing increases as $d/d_0 = \lambda$. After layer rotation starts the layer spacing remains fixed; the only cost of deforming the system is that of shearing the rubber. This is because, as the layers rotate, the component of the force along the layer normals remains constant. Explicitly this component of nominal stress is $\sigma_{\text{nom}} \cos\phi$ which, by Equations 17.25 and 17.26, above the transition is $\mu r \lambda_c$ and thus constant. Further, because shear (as opposed to extension and contraction along principal directions) involves the chain anisotropy and it is the cost of this shear that is to be compared with that of layer dilation, we can understand that $r$ (as well as the relative modulus $B/\mu$) enters the expression for $\lambda_c$.

The continuity of the nominal stress with $\lambda$ that is, equating Equations 17.23 and 17.26, yields a cubic equation for the threshold elongation $\lambda_c$:

$$\lambda_c^3(r - b - 1) + b\lambda_c^2 + 1 = 0. \tag{17.29}$$

To obtain a threshold we require $b > r - 1$. For reduced layer moduli below this value, there is no instability: layer dilation is not significantly more costly than matrix distortion and it is no longer avoided by the intercession of instability.

In the physically important large $b$ limit,

$$\lambda_c = 1 + \frac{r}{b} + r(r - 3)\frac{1}{b^2} + O\left(\frac{1}{b^3}\right) \rightarrow r\mu/B \approx \lambda_c - 1. \tag{17.30}$$

The ratio of the linear response moduli of Equation 17.27 before and after the transition is thus simply related to $\lambda_c$ for large $b$, which provides another stringent constraint on theory. Experimentally one could obtain $\mu$ (actually $4\mu$; see Equation 17.12 and the text below the equation) from stretching the rubber in the layer plane, and the anisotropy of the polymers, $r$, from the Young modulus after the threshold (Equation 17.27). Thus the value of the threshold itself gives a direct measure of the layer compression constant $B$, which can then be compared with the modulus before threshold in (Equation 17.27). The large $B/\mu$ expression (Equation 17.30) for the threshold strain essentially agrees with the result of linear continuum mechanics [38,49] if one neglects Frank effects, which have a small effect on the threshold strain (although do influence the length scales of the subsequent microstructure). If one ignores the anisotropy of the underlying nematic network, then the factor $r$ is of course absent from an expression for the threshold [49].

### 17.4.2.4.1  CMHH Microstructure

When extension along the layer normal, $\lambda_{zz}$ becomes too costly we have seen that layers rotate instead of dilating further, as illustrated in the inset of Figure 17.24b. However, the sample must be clamped in order to apply a $z$-extensional force and thus the rotation cannot occur uniformly throughout the sample. It must vanish at $z = 0$ and $z = L_z$, that is at the clamps, and must alternate in the $x$-direction between the values $\pm\phi$ (Equation 17.25) sufficiently frequently that large layer translations are not built up which would then cost large elastic energies to satisfy the clamp constraints (see Figure 17.25).

This is a particularly simple example of microstructure or texture that we return to in considering SmC deformations. A system does not deform at low cost because it has textures, but rather textures allow the system to exploit an intrinsically lower cost deformation (here in-plane stretch and shear) while satisfying boundary conditions at variance with the geometry of preferred deformations.

On the other hand, very frequent alternation between $\pm\phi$ leads to the creation of many stripe interfaces (where a Frank elastic energy cost is paid). The resulting $x$-length scale and overall energy cost arises from optimizing the sum of these two energies. There will turn out to be significant differences from the standard CMHH instability to stripes (discussed under liquid smectics), where the thresholds are small and are set by the interplay between Frank and layer elasticity. Here the threshold to rotation was primarily set by the interplay between layer and rubber elasticity; see the threshold condition Equation 17.29.

Length scales emerge naturally from layer and matrix elastic moduli $B$ and $\mu$ competing with Frank elastic energies which, for simplicity, we represent by a single constant $K$. Minimizing over competing effects yields geometric quotients for the characteristic lengths. Rubber versus Frank effects gives $\xi = \sqrt{K/\mu} \sim 10^{-8}$m for the nematic penetration depth. It is a measure of how deeply a director variation can penetrate into the depth of a material while acting against the rubber elastic penalty for director rotation [1]. It determines stripe interfacial width and the seemingly instant coarsening in the analogous strain-induced microstructure of stripe domains in nematic elastomers [50]. Analogously, one defines the usual smectic penetration depth $\xi_{sm} = \sqrt{K/B} \simeq d_0 \leq 10^{-9}$m, which determines the penetration of distortion into a smectic structure. Note this length is shorter than layer spacing in elastomers, due to the fact that the layer compression constant $B$ is larger than in liquid smectic phases, where naturally $\sqrt{K/B} \approx d$. The length scale $\xi_{sm}$ is independent of rubber elasticity and its small value compared with that of nematic elastomers suggests that smectic microstructure should also be instantly coarsened. The geometric mean of the smectic and Frank scales gives a surface tension of interfaces, as the energy cost per unit area of stripe formation: $\gamma_{sm} = \sqrt{KB}$.

As in the analysis of stripe domains in nematic elastomers, the threshold $\lambda_{zz} = \lambda_c$ of a uniform system will be shifted very slightly by the addition of Frank effects to a higher $\lambda'_c \gtrsim \lambda_c$ at which point there will be a small jump to a finite $\phi > 0$. Microstructure to accommodate clamp constraints means there are spatial variations and thus a (small) Frank contribution to the energy. A little more strain must be imposed to overcome this additional cost. The stripe period in the $x$-direction can be

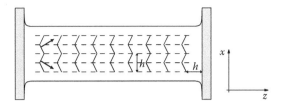

**FIGURE 17.25** The microstructure of a SmA elastomer loaded past the threshold stress. Stripes (dotted) of width $h$ are shown coarsened, the layer normals (arrows) being at $\pm\phi$ with respect to the extension axis, $z$, see [44].

estimated as

$$h \sim \sqrt{\frac{L_z \xi_{sm} B}{r\mu}} \, \frac{1}{(\lambda - \lambda_c)^{1/4}}. \qquad (17.31)$$

The period never diverges, since $\lambda \geq \lambda_c' \gtrsim \lambda_c$ (instant coarsening), and rapidly saturates to the value of a few $\mu$m for a sample of length $L_z \sim 10^{-2}$ m. Note that, in contrast to liquid smectics, in elastomers the long sample dimension is in the direction of the stretch, that is, the original layer normal. The resulting length scale would give the strong light scattering that is actually observed, such as in Figure 17.16a. However, other more complex calculations suggest there is possible 2/3 power scaling of microstructure with sample length [51].

### 17.4.2.4.2 Theory and Experiment

Three different types of experiments give insight into SmA elastomers–strain response, stress–strain, and rotation–strain measurements. We compare the theory with relevant experiment [19], principally the $\lambda_{zz}$ and $\lambda_{xx}$ distortions. The reader will see that many questions raised by the theory are as yet unanswered and thus a new experiment is suggested in the discussion below.

### 17.4.2.4.3 Strain Response

The Poisson ratios for imposed $\lambda_{zz}$ are isotropic and volume preserving, (1/2, 1/2), until the layer instability is reached. Thereafter, the transverse relaxation is apparently $(1/\sqrt{\lambda}, 1/\sqrt{\lambda})$ from closely inspecting the snapshot of large strain shown in Figure 17.16a. The authors do not give a functional dependence, but Figure 17.16a rules out the predicted monodomain post-threshold $xx$ and $yy$ responses of Equation 17.24 which, if strains were small would correspond to Poisson ratios (1, 0) respectively. This discrepancy is not surprising given that there is clearly not a monodomain after $\lambda_c$. Either layers are being destroyed, or they are rotating to *all* possible directions in the $xy$ plane and not just to the $x$-direction as in our monodomain analysis, see below.

Imposed in-plane stretches $\lambda_{xx}$ give predicted Poisson ratios $(\nu_y, \nu_z) = (2 + b/4 + b, 2/4 + b)$. Experiment apparently gives (1, 0) corresponding to $B/\mu \gg 1$ (see Figure 17.14). These Poisson ratios agree with the stress results that show that smectic order is much more rigid than rubber elastic effects.

#### 17.4.2.4.4 Stress–Strain Relations

Nominal stress against imposed strain ($\lambda_{zz} - 1$) along the initial layer normal direction in Figure 17.16b is fitted to Equations 17.23 and 17.26. The ratio of the slopes is $4.1 \times 10^{-2}$. Theory gives $\mu r/B$ for this ratio, whereupon one deduces $b \approx 25r$, which is evidently large. In this limit, Equation 17.30 predicts the direct connection $r/b = \lambda_c - 1$ for the threshold. This gives for the threshold strain $\varepsilon_c = \lambda_c - 1 \approx 4\%$ which is extremely close to that observed in Figure 17.16b.

In-plane stress and moduli in response to imposed $\lambda_{xx}$ were not reported by Nishikawa and Finkelmann. The in-plane Young modulus is in theory $E_\perp = 4\mu$ (Equation 17.13). It is known from other works to be comparable to the post-threshold modulus against stretches parallel to the layer normal, $E_{after} = r\mu$, in accord with theory (see Equation 17.26). However, the situation is potentially more subtle.

Many SmA elastomers that have been investigated are suspected to be de Vries phases, that is, where there is incipient SmC ordering. In this state the director tilt is not long-range correlated in its azimuthal direction. The signature of such local order is that the transition to the SmC state, with long-ranged order of tilting molecules, is not accompanied by a layer spacing change as expected from the transition from a standard SmA. Only small layer spacing changes have been observed even for induced tilts as large as 31° [52] and also in ferroelectric SmC* examples [53]. Applied strain $\lambda_{xx}$ in one in-plane direction could extend the correlation in tilt alignment and direct it along the strain, allowing the rubber to extend along $x$ at lower energy cost than $4\mu$. Tests of this type of response would be:

1. The observation of in-plane induced optical birefringence. While an untilting director remains anchored along the layer normal, the response should be that of a classical elastomer where stress induces very small birefringence compared with that in any liquid crystal system. In comparison a de Vries elastomer would have a huge birefringence response.
2. The ratio $E_{after}/E_\perp$ is predicted to be $r\mu/4\mu = r/4$. Departures from this ratio could be because of a low $E_\perp$ due to de Vries phase effects. However, to some extent de Vries effects should also intervene in $E_{after}$ since there is an element of in-plane stretch in the now rotated planes. The balance between the interventions of de Vries effects in the two moduli is not trivial since the component of stretch in-plane and the degree of shear acting both change with strain beyond the threshold for the $E_{after}$ case.

#### 17.4.2.4.5 Strain-Layer Rotation and X-Ray Scattering

Layer rotation against strain starts in a singular manner at a threshold $\lambda_c$ both in theory (Equation 17.25) and in an x-ray determination of layer orientation (Figure 17.17b). Agreement with $\phi(\lambda)$ is good, but a major problem of interpretation remains. As strain increases, the relative x-ray intensity associated with the rotating layer lines (that is, the ratio of wide angle to small angle scattering) diminishes sharply. Nishikawa and Finkelmann proposed that above $\lambda_c$ a diminishing fraction of the sample rotates while an increasing fraction of the layer system melts to a nematic state. One can, however, argue [1] that the smectic energy scale is high compared with the rubber elastic scale and that mechanically-induced melting is unlikely; see Ref. [44] for an estimate of the cost of "melting" in smectic elastomers. Investigation of the microstructures arising after the transition is required using 3-D techniques such as fluorescence confocal polarizing microscopy, that has been so effective in liquid crystals [54].

A more direct explanation [44] of the reduced intensity of x-ray scattering is that layers rotate their normals towards *all* directions perpendicular to the stretch along the original layer normal $\mathbf{n}_0$. Section 17.4.2.4 calculates the monodomain contraction and shear in the $x$ direction perpendicular to original layer normal $z$ (Equation 17.24) and the rotation of the normal toward $x$ (see Equation 17.25). But no direction perpendicular to the original director is privileged, in contrast to stripe formation in nematic elastomers. We must consider all other axes perpendicular to $\mathbf{n}_0$. This breakup of the sample into a microstructure of regions of tilted domains is cylindrically symmetric around the stretch axis. The regions that are tilted toward the x-ray beam no longer meet the Bragg condition for diffraction, and as a result do not contribute intensity to the observed scattered beam. One could suggest that the drop in x-ray intensity is simply a result of polydomain formation. Additionally, the overall Poisson ratios observed in the two, now equivalent directions perpendicular to the original layer normal are $(1/2, 1/2)$ rather than the monodomain values $(1, 0)$. In fact a 3-D x-ray scattering study on SmA* elastomers [55] undergoing a CMHH instability suggests the above explanation is correct. It has been confirmed in great detail by further very recent 3-D scattering [56] by rotation of the sample in the x-ray beam about the deformation axis.

## 17.5   A MODEL OF SmC RUBBER ELASTICITY

Molecular tilt leads to spontaneous shear in making the transition from the SmA to SmC phase (see Figure 17.11). The development of such shear has been described within nonlinear continuum theory [57] and within the current molecular theory [58]. Both methods draw a distinction between the molecular tilt and the shear angle, as required in experiment. The other challenge for theory is to describe the shear offset and the dependence of the spontaneous shear on the cross-linking history, that is, the dependence on the shear that was applied during the second-stage cross-linking and then allowed to relax when cross-linking was complete.

The most remarkable aspects of nematic rubber elasticity are the spontaneous distortions on entering the nematic phase and shape change at minimal free energy cost—soft elasticity. The two phenomena are intimately related: soft elasticity arises because of the degeneracy in the problem, that one can rotate the director independently of the body; the extent of the soft modes is given by the magnitude of the spontaneous distortion arising when symmetry is broken. In uniaxial, SmA networks the constraint of the director being perpendicular to the layers removed this freedom. In SmC we have this possibility—now the director is tilted one can move it independently of the body by rotating it on a cone about the layer normal, and with this rotation must come elastic distortion at no, or low elastic energetic cost. Such director rotations and deformations do not alter the layer spacings and thus these deformations do not transgress the rigid constraints acting on SmA. (In fact, *biaxial* SmA elastomers could also rotate their in-plane director about the layer normal and change shape without effecting layers. This route to elastic softness has been recognized theoretically [57, 59]; it is equivalent to the simplest soft modes of biaxial nematics.)

We again consider strongly coupled smectic elastomers that is, where the layer compression modulus $B$ is large and all deformations $\underline{\underline{\lambda}}$ affinely convect the layer system while respecting their spacing. In general the tilt angle, $\theta$, might be induced to change by stresses as in the weaker anchoring SmA case, but we take it to be fixed here. (A recent, nonlinear continuum theory of SmC elasticity [57] has relaxed the absolute rigidity of the tilt angle.) The difficulty arises in respecting the strong layer spacing constraint, that is the initially unit layer normal $k_0$ transforms to a new $k$, which remains a unit vector:

$$k = \underline{\underline{\lambda}}^{-T} \cdot k_0 \quad \text{with} \quad k_0^T \cdot \underline{\underline{\lambda}}^{-1} \cdot \underline{\underline{\lambda}}^{-T} \cdot k_0 = 1. \tag{17.32}$$

The transformation is the equivalent of Equation 17.4 for SmA rubber where the director and the layer normal were identical. In addition to not incurring any smectic penalty, the last part of the free energy (Equation 17.3), the deformations must also incur no nematic energy cost, the first (Trace) part. Such deformations are soft modes of the form:

$$\underline{\underline{\lambda}} = \underline{\underline{\ell}}_n^{1/2} \cdot \underline{\underline{W}} \cdot \underline{\underline{\ell}}_0^{-1/2}, \tag{17.33}$$

with the rotation $\underline{\underline{W}}$ being able to be chosen arbitrarily. These low-energy deformations are very important since, where possible, a sample will respond by changing its shape by selecting such a mode, or by adopting a microstructure that is a combination of such modes that when added together allow the satisfaction of boundary conditions at minimal cost.

Finding general soft modes (Equation 17.33) with the smectic constraint (Equation 17.32) is rather abstract; see Ref. [59] where they are given in their entirety. A more concrete, but very important, set of low free energy modes in SmC elastomers is extension in the layer planes perpendicular to the initial director, which is thereby induced to rotate without layer rotation or contraction of the layer spacing. The director rotation can then only be azimuthal, on the surface of a cone about the unchanging layer normal. Let us consider the specific coordinates shown in Figure 17.26.

Conceptually, one can think of the distortion being driven by an imposed extension in the $y$ direction, or by an imposed $xz$-shear, or by determining the rotation $\phi$ of the director in a cone about the layer normal (for instance by the application of an electric field in SmC* elastomers); we take a deformation matrix of the form:

$$\underline{\underline{\lambda}} = \begin{pmatrix} \lambda_{xx} & 0 & \lambda_{xz} \\ \lambda_{yx} & \lambda_{yy} & \lambda_{yz} \\ 0 & 0 & \lambda_{zz} \end{pmatrix}. \tag{17.34}$$

The components $\lambda_{xy}$ and $\lambda_{zy}$ are excluded because they deform the sample by translating the $y$ faces of the sample in the $\pm x$ and $\pm z$ directions. Any small $y$ forces associated with the $yy$ elongation would generate counter torques and quickly eliminate them. The $\lambda_{zx}$ component and (again) the $\lambda_{zy}$ component are excluded because they rotate the layer normal, which we want to remain fixed.

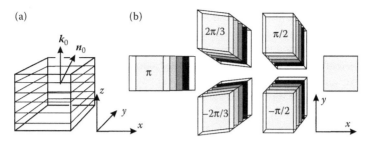

**FIGURE 17.26** (a) The undeformed SmC elastomer with layer normal $k_0 = z$ and with $n_0$ tilted toward the $x$ direction. (b) The shape changes of the elastomer viewed along the layer normal as the $c$-director advances (right to left) through $\phi = 0, \pm\pi/2, \pm 2\pi/3$, and $\pm\pi$. In this example the anisotropy and tilt are $r = 8, \theta = \pi/6$. Two alternative routes, $\pm\phi$ are shown—their combinations will be used in developing textures (microstructure) to achieve soft deformations in the presence of boundary conditions incompatible with such distortions. (Adapted from M. Warner and E.M. Terentjev. *Liquid Crystal Elastomers*. Oxford University Press (revised, paperback edition), 2007 by permission of Oxford University Press.)

In our coordinates, the director and the (unchanging) layer normal are

$$k_0 = (0, 0, 1); \quad n_0 = (\sin\theta, 0, \cos\theta) \rightarrow n = (\sin\theta\cos\phi, \sin\theta\sin\phi, \cos\theta)$$

where the tilt angle $\theta$ is typically around 20°. The layer normal, $k_0 = z$, cannot be moved by deformation tensors of the form (Equation 17.34) since it derives from $k = \underline{\underline{\lambda}}^{-T} \cdot k_0$. The soft mode, not rotating the layer normal and with rotation $\phi$ of the director about $k_0$, is [59,60]

$$\begin{pmatrix} a(\phi) & 0 & \dfrac{(r-1)\sin 2\theta}{2\rho}(-a(\phi) + \cos\phi) \\ \left(1 - \dfrac{\rho}{r}\right)\dfrac{\sin 2\phi}{2a(\phi)} & \dfrac{1}{a(\phi)} & \dfrac{(r-1)\sin 2\theta}{2\rho}\left(\sin\phi - \left(1 - \dfrac{\rho}{r}\right)\dfrac{\sin 2\phi}{2a(\phi)}\right) \\ 0 & 0 & 1 \end{pmatrix}$$

$$(17.35)$$

Anisotropy and tilt are encoded in the mode in the parameter $\rho = 1 + (r - 1)\cos^2\theta \le r$. The rotation angle appears in $a(\phi) = \left(1 - \dfrac{r-1}{r}\sin^2\theta\sin^2\phi\right)^{1/2} \le 1$. The deformation is pictured in Figure 17.26b for a sequence of rotations $\phi$. One clearly sees that $yx$ shear deforms the initially square $xy$ section shown on the right ($\phi = 0$) and that $xz$ and $yz$ shears give the lean of the layer stack seen at all other rotations. The final state ($\phi = \pi$) on the left is purely (negative) $xz$ shear resulting from a reversal of the tilted director. It is the maximum extent of soft shear and one thereby sees a connection to the spontaneous shear: consider Figure 17.27, which starts from Figure 17.27a to Figure 17.27b with a SmA elastomer spontaneously shearing by $\Lambda$ to the SmC state [23]; see also Figure 17.11.

The maximum extent of possible soft shear in going from Figure 17.27c to Figure 17.27d is $2\Lambda$ that is, twice the spontaneous shear on entering the SmC state. Figure 17.27d corresponds to the left-most state in Figure 17.26b. Two routes to maximum $xz$-shear are shown namely, positive and negative rotations $\pm\phi$ about the layer normal. We concentrate on one path here. The opposite one with some of the attendant shears of the opposite sign will be another, the canceling component of a texture that we consider subsequently.

The director rotation and the shears that develop as $\lambda_{yy}$ grows are shown in Figure 17.28. The relation $\lambda_{yy} = 1/a(\phi) \geq 1$ between extension and rotation $\phi$ can be read off the appropriate element of $\underline{\underline{\lambda}}$ in (Equation 17.35) and, using the definition of $a(\phi)$, can easily be inverted. For small $\phi$ and for $\phi \to \pi$ it is clearly a singular relation, $\phi \sim (\lambda_{yy} - 1)^{1/2}\sqrt{(2r/r - 1)}/\sin\theta$. The maximum soft extension occurs at $\phi = \pi/2$ and is $\lambda_{yy} = \lambda_m = \sqrt{r/\rho} \equiv (1 - (r-1)/r\sin^2\theta)^{1/2}$. The other components of $\underline{\underline{\lambda}}$ can also be easily read. A very subtle aspect of nonlinear elasticity is apparent: a maximal $yz$ shear of magnitude $\Lambda$ must develop when the director is tilted purely in the $y$ direction that is, when $\phi = \pi/2$. Inspection of Figure 17.28b shows that in fact this particular shear is apparently maximal for $\lambda_{yy} \sim 1.045$, and thus for a rotation $\phi \gtrsim \pi/4$.

This apparent contradiction is a consequence of geometrical nonlinearity of finite deformations, which must be compounded rather than added.

Very large strain experiments have still to be performed, both shear as sketched above, and experiments. This area remains one of exciting future studies and applications now that large SmC monodomains are accessible. Applications are especially attractive when chiral SmC* elastomers are considered. In the above discussion of soft $c$-director modes, the associated electric polarization was originally along the $y$ direction. On deformation, its $y$-projection varies as

$$P_y = P_0 \cos\phi = P_0\sqrt{((\lambda_m/\lambda_{yy})^2 - 1)/(\lambda_m^2 - 1)}. \qquad (17.36)$$

It is not singular at the start of the extension, at $\lambda_{yy}=1$, but does vary rapidly around the maximum $\lambda_m$ which may be of practical significance. Experiments probing the

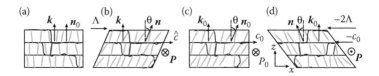

**FIGURE 17.27** (a) A SmA elastomer spontaneously shearing by $\Lambda$ on achieving the SmC state (b); see Figure 17.11 for a photo of an experiment. A regular shape (c) is cut out of this sheared elastomer and becomes the reference state for subsequent distortion. (d) its tilt direction reversed with an associated shear reversal of $-2\Lambda$. For the three shapes the polarization $P$ arising in the chiral (SmC*) case is shown in or out of the plane. (Adapted from M. Warner and E.M. Terentjev. *Liquid Crystal Elastomers*. Oxford University Press (revised, paperback edition), 2007 by permission of Oxford University Press.)

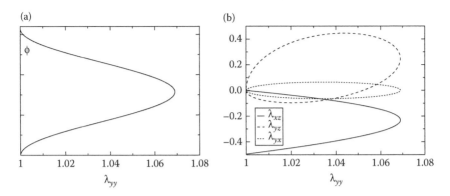

**FIGURE 17.28** (a) The director rotation in response to an imposed stretch $\lambda_{yy}$ perpendicular to the initial director. (b) The soft $yx$, $yz$, and $xz$ shears thereby induced by the imposed $yy$ extension. The anisotropy and tilt are $r = 2, \theta = \pi/6$. As discussed in the text, it would be possible to take another of the variable, for instance $\phi$ as the imposed quantity, see [59].

electrical response to this particular (extensional) mode have not yet been performed, but rather shear has been investigated.

### 17.5.1 DEFORMATION WITH TEXTURE IN SMC ELASTOMERS

We have seen in the CMHH undulatory instability of SmA elastomers (Figure 17.25) how soft routes to deformation can be found, even though they are in conflict with the macroscopic imposed strain, by the development of microstructure or texture. We now examine how textures can arise in SmC elastomers to lower the elastic energy of shape change.

A more naturally imposed deformation applies to a SmC elastomer than a $y$-extension or a rotation by $\phi$ of $\boldsymbol{n}$ about the layer normal might be an $xz$ in-plane shear. One could view the deformations as being simply those of the previous section, but using $\lambda_{xz}$ as a variable instead of $\lambda_{yy}$ or $\phi$, which played somewhat equivalent roles. The soft $\underline{\underline{\lambda}}$ would be identical and any accompanying polarization also. However, depending upon whether one has a slab or thin sheet geometry, that is, whether $z$ or $y$ is the smallest dimension of the sample in Figure 17.26a, one might have one or more elements of macroscopic shear suppressed by clamps, used to hold the sample, or by electrodes used to measure the electric polarization changes. In this event, the system will try to deform as softly as possible, albeit by the development of microstructure.

Take as an example an in-plane shear $\lambda_{xz}$ imposed by gripping a slab-like sample using rigid plates in the $xy$ planes bounding the top and bottom of the sample in the $z$ direction, and then moving the plates relatively in the $zx$ direction. We have seen that, in order to deform softly, the sample must rotate the director around on a cone, ideally from tilting in the $+x$ direction to the $-x$ direction, and thereby develops a $\lambda_{yz}$, among other shears. Suppose that a constraint on the sample is that the top plate may not move in the $y$ direction relative to the bottom plate. In that case the average $\lambda_{yz}$ shear must be suppressed; in other words an oscillating $\lambda_{yz}$ must develop preserving the

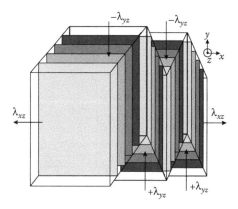

**FIGURE 17.29** The top and bottom plates are moved relatively in the $x$ direction to create a $\lambda_{xz}$ while perpendicular ($y$) displacement of the plates is forbidden. Alternating bands of $\pm\lambda_{yz}$ shears ensure this macroscopic boundary condition is respected. (Adapted from M. Warner and E.M. Terentjev. *Liquid Crystal Elastomers.* Oxford University Press (revised, paperback edition), 2007 by permission of Oxford University Press.)

zero $y$-displacement on the two plates. Figure 17.29 proposes one scheme as to how this can be done by alternating displacements (and hence shears) in the $y$ direction.

Although the elastic energy is lowered by the $yz$ shears, there are in fact other shears also in the soft mode and all of these except the $xz$ shear must be involved in laminates to avoid macroscopic distortion suppressed by clamps. Softer and more complicated modes of deformation and texture are discussed in [60] where detailed switching routes and the development of sheets of charge in SmC* elastomers are presented. For instance, the normal to the laminates rotates in the $zx$ plane as the director $\phi$ rotates about the layer normal (the latter remaining along the $z$ direction in this geometry). In technical terms, this arises from the requirement of "rank-one connectedness" between the deformation gradients of neighboring laminates that is, there is no geometric incompatibility that would tear the solid apart! A nontechnical explanation of this requirement is given in [60]. Since then particular laminates are oblique to the $z$-axis and thus cut through the layers, as the polarization rotates with $\phi$, then the laminates will also cut the polarization and the internal interfaces between laminates will be charged. Where the laminates meet the external ($yz$) surfaces, there will be surface charges of sign that alternates from laminate to laminate since the $x$ component of polarization that terminates or starts on this surface reverses too. Other laminates are known that do not cut the polarization and hence do not lead to internal charging [61]. As yet little is known of the structure of the laminates and, so far, no large amplitude shear experiments have been performed.

## 17.6  CONCLUSIONS

We have reviewed the physical properties of strongly coupled SmA and SmC elastomers. Because smectic layers resist distortion more strongly than the rubber matrix in which they are embedded, smectic elastomers have mechanical properties quite different from the nematic elastomers on which they are based. They are rubbery in two-dimensions and more solid-like in the third. Like classical and also nematic elastomers, they deform at constant volume (despite the higher energy scale offered by the layers) and are still capable of very large deformations. The latter property of large distortions demands that they be described by nonlinear elasticity, subject to the strong constraints of volume conservation and of resistance to layer spacing change.

Although great progress has been made in fabricating both SmA and SmC monodomains and performing experiments on them, our review has suggested a large number of future experiments.

There are rigid constraints on theories relating SmA moduli in different directions and before and after layer instabilities. Are these connections found? The role of de Vries order should also be clarified, and shear–optical methods of doing so are suggested. The fate of SmA elastomers, after their mechanical instabilities induced by strain along the layer normal, remains to be clarified. In particular what kinds of microstructure are induced? The description of SmC systems suggests experiment to determine the response to tensile and shear strains: Do systems break up into textures as the order evolves from the initial to final states? Chiral SmC* elastomers are ferroelectric. There is a great demand for mechanical experiments to examine the switching of polarization and to evaluate the potential of SmC* elastomers for large amplitude electromechanical and opto-mechanical actuation.

We have not reviewed the theory and experiment of the evolution of polydomains under applied strain. For SmC elastomers this is of particular importance because polydomains arise at the intermediate stages of multistage applied strains and cross linking. These are effective routes to monodomain production but we still await systematic studies of the evolution of smectic polydomains to the aligned state—see a theoretical study of this evolution [62]. Nor have we reviewed predictions [62] of the nonsoft evolution of SmC elastomers under imposed strain. The response is complex because it can be achieved partly by the intervention of director rotation that makes intermediate stages soft, but the largest deformations are inevitably hard. We are unaware of mechanical experiments thus far—the importance of such work may be that it will underpin studies of polydomain response.

## ACKNOWLEDGMENT

I am indebted to my collaborator Dr James Adams for insights into smectic elastomers and for a reading of this review in its draft stage.

## REFERENCES

1. M. Warner and E.M. Terentjev. *Liquid Crystal Elastomers*. Oxford University Press (revised, paperback edition), 2007.
2. J. Küpfer and H. Finkelmann. *Macromol. Chem. Phys.*, 195:1353, 1994.
3. I. Kundler and H. Finkelmann. *Macromol. Chem. Rap. Commun.*, 16:679, 1995.
4. V.P. Shibaev, S.G. Kostromin, and N.A. Plate. *Eur. Polym. J.*, 18:651, 1982.
5. S.A. Pikin. *Structural Transformations in Liquid Crystals*. Gordon & Breach, New York, 1991.

6. P.-G. de Gennes and J.P. Prost. *The Physics of Liquid Crystals*. Oxford University Press, Oxford, 1994.

7. N.A. Clark and R.B. Meyer. *Appl. Phys. Lett.*, 22:493, 1973.

8. S.J. Singer. *Phys. Rev. E*, 48:2796, 1993.

9. R.B. Meyer, L. Liebert, L. Strzelecki, and P. Keller. *J. Phys. Lett.*, 36:L69, 1975.

10. N.A. Clark and S.T. Lagerwall. *Appl. Phys. Lett.*, 36:899, 1980.

11. C.B. McArdle. *Side Chain Polymer Liquid Crystals*. Blackie & Sons, Glasgow, 1989.

12. H. Finkelmann, H.J. Kock, and G. Rehage. *Macromol. Rap. Commun.*, 2:317, 1981.

13. H. Finkelmann. *Adv. Polym. Sci.*, 60/61:99, 1984.

14. R. Zentel. *Angew. Chem. Adv. Mater.*, 101:1437, 1989.

15. H. Andersson, F. Sahlen, U.W. Gedde, and A. Hult. *Macromol. Symp.*, 77:339, 1994.

16. T. Pakula and R. Zentel. *Macromol. Chem.*, 192:2401, 1991.

17. K. Semmler and H. Finkelmann. *Polym. Adv. Tech.*, 5:231, 1994.

18. E. Nishikawa, J. Yamamoto, H. Yokoyama, and H. Finkelmann. *Macromol. Rap. Commun.*, 25:611, 2004.

19. E. Nishikawa and H. Finkelmann. *Macromol. Chem. Phys.*, 200:312, 1999.

20. H. Kapitza and R. Zentel. *Macromol. Chem.*, 189:1793, 1988.

21. I. Benné, K. Semmler, and H. Finkelmann. *Macromol. Rap. Commun.*, 15:295, 1994.

22. H. Schuring, R. Stannarius, C. Tolksdorf, and R. Zentel. *Macromolecules*, 34:3962, 2001.

23. K. Hiraoka, W. Sagano, T. Nose, and H. Finkelmann. *Macromolecules*, 38:7352, 2005.

24. M. Brehmer, R. Zentel, G. Wagenblast, and K. Siemensmeyer. *Macromol. Chem. Phys.*, 195:1891, 1994.

25. E. Gebhard and R. Zentel. *Macromol. Rapid Commun.*, 19:341, 1998.

26. W. Lehmann, H. Skupin, C. Tolksdorf, E. Gebhard, R. Zentel, P. Krüger, M. Lösche, and F. Kremer. *Nature*, 410:447, 2001.

27. C.M. Spillmann, B.R. Ratna, and J. Naciri. *Appl. Phys. Lett.*, 90:021911, 2007.

28. R. Zentel, E. Gebhard, and M. Brehmer. *Adv. Chem. Phys.*, 113:159, 2000.

29. E. Nishikawa, H. Finkelmann, and H.R. Brand. *Macromol. Rap. Commun.*, 18:65, 1997.

30. E.M. Terentjev and M. Warner. *J. de Phys. II*, 4:111, 1994.

31. R. Stannarius, R. Köhler, U. Dietrich, M. Löscher, C. Tolksdorf, and R. Zentel. *Phys. Rev. E*, 65:041707, 2002.

32. R. Stannarius, R. Köhler, M. Rössler, and R. Zentel. *Liquid Crystals*, 31:895, 2004.

33. R. Stannarius, V. Aksenov, J. Bläsing, A. Krost, M. Rössler, and R. Zentel. *Phys. Chem. Chem. Phys.*, 8:2293, 2006.

34. H.M. Brodowsky, E.M. Terentjev, F. Kremer, and R. Zentel. *Europhys. Lett.*, 57:53, 2002.

35. K. Hiraoka, Y. Uematsu, P. Stein, and H. Finkelmann. *Macromol. Chem. Phys.*, 203:2205, 2002.

36. S. Garoff and R.B. Meyer. *Phys. Rev. Lett.*, 38:848, 1977.

37. R. Köhler, R. Stannarius, C. Tolksdorf, and R. Zentel. *Appl. Phys.*, 80:381, 2005.

38. E.M. Terentjev and M. Warner. *J. de Phys. II*, 4:849, 1994.

39. A. Langhoff and F. Giesselmann. *J. Chem. Phys.*, 117:2232, 2002.

40. A. Langhoff and F. Giesselmann. *Chem. Phys. Chem.*, 3:424, 2002.

41. A. Saipa, M.A. Osipov, K.W. Lanham, C.H. Chang, D.M. Walba, and F. Giesselmann. *J. Mat. Chem.*, 16:1, 2006.

42. P. Beyer and R. Zentel. *Macromol. Rap. Commun.*, 26:874, 2005.

43. P. Deindorfer, A. Eremin, R. Stannarius, R. Davis, and R. Zentel. *Soft Matter*, 2:693, 2006.

44. J.M. Adams and M. Warner. *Phys. Rev. E*, 71:021708, 2005.

45. O. Stenull and T.C. Lubensky. *Phys. Rev. E*, 76:011706, 2007.

46. J.M. Adams, M. Warner, O. Stenull, and T.C. Lubensky. *Phys. Rev. E*, 78:011703, 2008.

47. D. Kramer and H. Finkelmann. *Phys. Rev. E*, 78:021704, 2008.
48. O. Stenull, T.C. Lubensky, J.M. Adams, and M. Warner. *Phys. Rev. E*, 78:021705, 2008.
49. J. Weilepp and H.R. Brand. *Macromol. Theor. Sim.*, 7:91, 1998.
50. H. Finkelmann, I. Kundler, E.M. Terentjev, and M. Warner. *J. Phys. II*, 7:1059, 1997.
51. R.V. Kohn and S. Müller. *Philos. Mag.*, 66:697, 1992.
52. M.S. Spector, P.A. Heiney, J. Naciri, B.T. Weslowski, D.B. Holt, and R. Shashidhar. *Phys. Rev. E*, 61:1579, 2000.
53. M. Rössle, R. Zentel, J.P.F. Lagerwall, and F. Giesselmann. *Liquid Crystals*, 31:883, 2004.
54. I.I. Smalyukh, S.V. Shiyanovkii, and O.D. Lavrentovitch. *Chem. Phys. Letts.*, 336:88, 2001.
55. C.M. Spillmann, J.H. Konnert, J.R. Deschamps, J. Naciri, and B.R. Ratna. *Chem. Mater.*, 20:6130, 2008.
56. C.M. Spillmann, J.H. Konnert, B.R. Ratna, J. Naciri, J. M. Adams, and J.R. Deschamps. *Phys. Rev. E*, 82:031705, 2010.
57. O. Stenull and T.C. Lubensky. *Phys. Rev. Lett.*, 94:018304, 2005.
58. J.M. Adams and M. Warner. *Phys. Rev. E*, 73:031706, 2006.
59. J.M. Adams and M. Warner. *Phys. Rev. E*, 72:011703, 2005.
60. J.M. Adams and M. Warner. *Phys. Rev. E*, 79:061704, 2009.
61. J.S. Biggins and K. Bhattacharya. *Phys. Rev. E*, 79:061705, 2009.
62. J.M. Adams and M. Warner. *Phys. Rev. E*, 77:021702, 2008.

# 18 Physical Properties of Magnetic Gels

*Helmut R. Brand, Philippe Martinoty, and Harald Pleiner*

## CONTENTS

## 18.1 INTRODUCTION AND HISTORICAL OVERVIEW

While liquid crystalline elastomers (LCEs) combining the properties of liquid crystals and of elastomers, had been synthesized first by Finkelmann's group over 25 years ago [1] (*cf.* Ref. [2] for a recent review of some macroscopic properties), the field of magnetic gels is much more recent. In the late 90's, Zrinyi's group reported the preparation of isotropic magnetic gels combining the properties of a gel with those of a magnetic liquid [3,4].

Since the late 90's, isotropic magnetic gels, which only react to magnetic field gradients, but not to homogeneous magnetic fields, have been investigated in detail synthetically, experimentally, and theoretically [5–25], where a large body of work has been accumulated in particular by Zrinyi's group. In particular, Ref. [5] gives a concise and clear overview of the early work on isotropic magnetic gels. Apart from combining ferrofluids and polymer networks, other groups have generated magnetic gels by embedding ferrite particles or $\gamma$-$Fe_2O_3$ particles of nanometer scale into a gel matrix [26–30]. When magneto-rheological fluids are combined with a polymer network, magneto-rheological gels result [31–37].

In 2003, the groups of Martinoty [38] and Zrinyi [39] pioneered the synthesis and characterization of uniaxial magnetic gels by generating the magnetic gel in an external magnetic field thus freezing-in the direction of the external magnetization. In contrast to isotropic magnetic gels, which respond only to field gradients, uniaxial magnetic gels can be oriented in a homogeneous magnetic field [38]. While the static properties of iron and magnetite particles containing magnetic gels have been elucidated in Zrinyi's group [39–44], Martinoty's group has focused on the investigation of the dynamic properties, in particular using piezorheometry. Very recently, Martinoty's group [45] has also studied the magnetic-field induced generation of a novel uniaxial magneto-rheological system—using a magneto-rheological fluid, which is solid-like in the field direction and liquid-like transverse to the direction of the external field. Correspondingly, the review and the analysis of the physical properties of uniaxial magnetic gels comprises the major part of this chapter.

We note that very recently the synthesis and characterization of thermo reversible ferrogels has also been reported [46]. In these systems the sol–gel transition can be observed as a function of temperature. It will be clearly very interesting to study the rheological behavior and the instabilities (e.g., the Rosensweig instability) in these systems using the temperature as a continuously variable parameter.

In the next section, we present a detailed analysis of the available experimental investigations of uniaxial magnetic gels followed by a section describing the macroscopic properties and novel cross-coupling effects for both, isotropic and uniaxial magnetic gels [18,47–49]. In this section we make use of the approach of hydrodynamics and macroscopic dynamics, which is valid in the low frequency, long wavelength limit (*cf.* [50] for a detailed exposition of the methods). In the last section, we discuss similarities and differences in the physical, in particular in the macroscopic properties between magnetic gels on the one hand and liquid crystalline gels and elastomers on the other.

## 18.2   EXPERIMENTAL ASPECTS

### 18.2.1   INTRODUCTION

Magnetic gels or elastomers are composite materials made up of magnetic particles embedded in a polymeric matrix. Like LCEs, they belong to the category of systems known as "intelligent." They are called uniaxial when the magnetic particles are oriented in a permanent way in a given direction. The interest of the latter systems is that they respond not only to gradients of the magnetic field, like super paramagnetic gels, but also to uniform magnetic fields.

In this section, we concentrate on the main experimental results obtained for uniaxial magnetic gels and elastomers. We address three fundamental problems raised by these materials: the mechanical anisotropy, the influence of an external magnetic field on the mechanical anisotropy, and the formation kinetics governing the properties of the final material. These topics will be discussed in the light of the experiments performed by the groups of Zrinyi and Martinoty. The static aspects, mainly studied by Zrinyi's group, and the dynamic aspects studied by Martinoty's group will be presented separately.

### 18.2.2   STATIC ASPECTS

In the experiments performed by Zrinyi's group [40–42], two different types of systems were investigated:

1. Magnetic elastomers composed of poly (dimethyl siloxane) (PDMS) networks filled either with carbonyl iron particles, or with $Fe_3O_4$ particles (magnetite particles). As shown in Figure 18.1, the carbonyl iron particles are polydisperse and have spherical shape with smooth surface, in contrast to the magnetite particles. The concentration of the filler particles in the elastomer varied from 10 to 40 wt.%.
2. Magnetic gels composed of poly(vinyl alcohol) (PVA) hydrogels filled with magnetite particles whose concentration in the gel was 1.2 wt.%.

The uniaxial samples were prepared by applying a uniform magnetic field during the formation process. To do that, the mixture containing the polymeric matrix and the magnetic particles was introduced into a cube-shape mold placed between the poles of an electromagnet. The formation of the uniaxial gel or elastomer proceeds in two steps. First, the embedded magnetic particles are aligned along a common direction, leading to fiber formation. The aligned particles are then locked by the chemical cross-linking, giving rise to a highly anisotropic material. For detail about the preparation of the sample see [40–42].

An example of the texture of the final material is given by observations using an optical microscope performed on a uniaxial magnetic PVA hydrogel prepared by Martinoty's group (see Section 18.2.3.4.2 dealing with dynamic aspects; Figure 18.10).

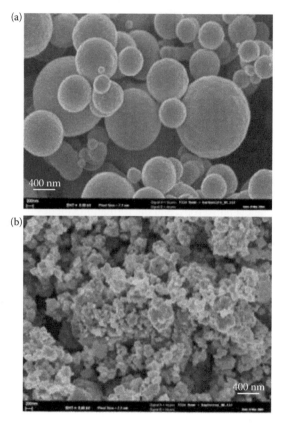

**FIGURE 18.1**   SEM pictures of the iron (a) and magnetite particles (b). The bar corresponds to 400 nm. (Reprinted from Varga, Z., Filipcsei, G., and Zrinyi, M., *Polymer,* 46, 7779, 2005. Copyright 2005, with permission from Elsevier.)

### 18.2.2.1  Shear Mechanical Anisotropy

In [40–42], the static shear modulus $G$ was determined by applying a unidirectional compression force on the sample and by measuring the resulting stress–strain relationship. $G$ is then deduced from the neo-Hookean law of rubber elasticity [51], which is given by

$$\sigma_n = G(\lambda - \lambda^{-2}) \qquad (18.1)$$

$\sigma_n$ is the nominal stress defined as the ratio $f/S$ where $f$ is the tensile force and $S$ the cross-sectional area of the sample measured in the undeformed state. $\lambda$ is the compression ratio $l/l_0$ where $l$ corresponds to the length of the sample measured in the direction of the force, and $l_0$ is the initial length associated with the undeformed state.

All the measurements were made at room temperature on samples containing a high concentration of magnetic particles varying from 10% to 40% in weight. The results obtained for a uniaxial PDMS sample filled with 40 wt% magnetite particles

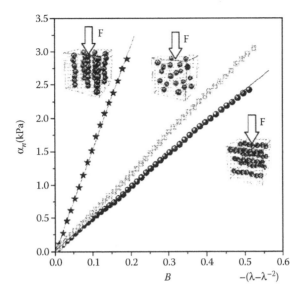

**FIGURE 18.2** Stress–strain behavior of PDMS elastomers filled with 40 wt.% magnetite particles for three different spatial distributions of the particles. The white arrows indicate the direction of the applied compression force. (Varga, Z. et al. *Macromol. Symp.*, 227, 123, 2005. Copyright Wiley-VCH Verlag GmbH & Co. KGaA. Reproduced with permission.)

and for its isotropic analog are shown in Figure 18.2. For each of the three distributions of the magnetic particles with respect to the applied force, the data clearly show that the stress–strain behavior obeys Equation 18.1. The significant differences in the slopes of the stress–strain curves demonstrate that $G$ is very sensitive to the spatial distribution of the magnetite particles. In particular, the uniaxial sample exhibits highly anisotropic properties with a ∼70% anisotropy in $G$. Similar types of behavior are observed for the other concentrations.

The situation is not the same when the PDMS elastomer is filled with iron particles. Two cases must be considered, according to whether the direction of the applied compression force is parallel or perpendicular to the fiber structure. For the perpendicular case, the stress–strain behavior roughly obeys the ideal behavior given by Equation 18.1. This is no longer the case when the direction of the compression force is parallel to the fiber structure, and a strong deviation from the ideal behavior is observed, as shown in Figure 18.3. This striking change in behavior for the parallel geometry between the iron-filled samples and the magnetite-filled samples indicates that the interaction of the particles with the polymeric matrix is different. As we will see below, the data obtained at large deformations lead to the same conclusion.

Since the elastic modulus is related to the degree of swelling, the anisotropy in the mechanical behavior can also be observed in swelling experiments. The data of Figure 18.4, taken for a PDMS elastomer filled with 40 wt.% magnetite particles, show that the swelling is anisotropic, indicating again the anisotropic nature of the material.

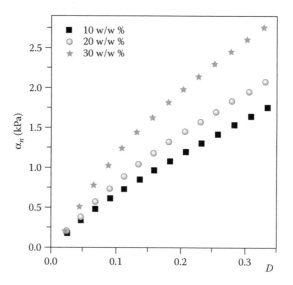

**FIGURE 18.3** Stress–strain behavior of a uniaxial magnetic PDMS elastomer filled with iron particles. The compression force is parallel to the direction of the fibers. The symbols indicate the concentrations of particles. A 400 mT uniform magnetic field was applied during the formation of the sample. $D = \lambda - \lambda^{-2}$. (Reprinted from Varga, Z., Filipcsei, G., and Zrinyi, M., *Polymer*, 46, 7779, 2005. Copyright 2005, with permission from Elsevier.)

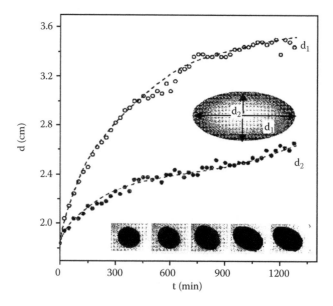

**FIGURE 18.4** Anisotropic swelling of a uniaxial magnetic PDMS gel filled with 40 wt.% magnetite particles in *n*-hexane. (Reprinted from Varga, Z., Filipcsei, G., and Zrinyi, M., *Polymer*, 46, 7779, 2005. Copyright 2005, with permission from Elsevier.)

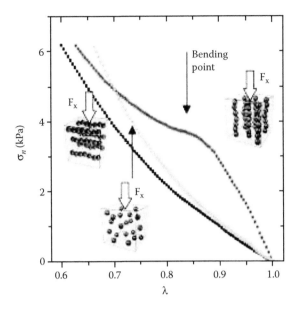

**FIGURE 18.5** Stress–strain curves of magnetite-loaded PVA gels for three different spatial distributions of the particles but with the same concentration of magnetite particles. The white arrows indicate the direction of the applied compression force. A 400 mT uniform magnetic field was applied during the formation of the sample. (Reprinted from Varga, Z., Filipcsei, G., and Zrinyi, M., *Polymer*, 46, 7779, 2005. Copyright 2005, with permission from Elsevier.)

All the previous experiments concern the behavior at small deformations. New effects appear at large deformations, as we will see now. Figure 18.5 shows the stress–strain curves taken for the uniaxial PVA gel loaded with magnetite particles, and for its isotropic analog. In all cases, it can be seen that the data taken at large deformations deviate from the linear behavior observed at small deformations. Particularly interesting is the situation where the applied force is parallel to the fibers. In that case a bending occurs below a critical strain. This effect, which reflects a stress induced softening, has been interpreted as a mechanical instability associated with the buckling of the fibers [41]. Similar types of behavior were observed for the PDMS elastomer loaded with magnetite particles.

Figure 18.6 shows the data obtained on the iron-filled PDMS elastomer. As it can be seen, the shape of the stress–strain curves is different from those observed in the systems filled with magnetite particles (PVA gel or PDMS elastomer). In the case where the applied force is parallel to the fibers, the compression of the sample induces a break point in the stress–strain curve instead of a softening. According to [41], this break point reflects the destruction of the fiber structure of the iron particles.

The difference in behavior between the magnetite-loaded and the iron-loaded systems can be explained by the fact that the iron particles do not interact strongly with the polymer network, contrary to the magnetite particles. As a result, the polymer network cannot prevent the break-up of the iron fiber structure under compression, while it prevents the destruction of the magnetite fibers. As stressed above, the nature

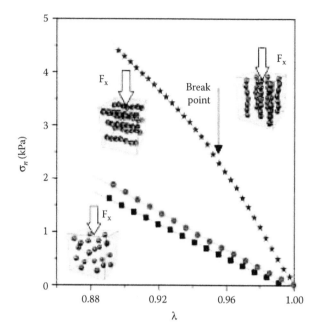

**FIGURE 18.6** Same as in Figure 18.5, but for iron-loaded PDMS elastomers. (Reprinted from Varga, Z., Filipcsei, G., and Zrinyi, M., *Polymer*, 46, 7779, 2005. Copyright 2005, with permission from Elsevier.)

of the magnetic particles (magnetite vs. iron), and probably their size (larger for the iron particles than for the magnetite particles), their shape and geometric features (spherical shape with smooth surface for the iron particles, irregular angular shape with sharp facets for magnetite particles) play a major role in the mechanical behavior of both systems.

### 18.2.2.2 Influence of an External Magnetic Field on the Behavior of the Shear Modulus *G*

Since the shape of the uniaxial elastomers is coupled to the magnetic field through the magnetic particles, it is interesting to investigate the influence of a uniform external field, $H$, on the behavior of $G$. This has been done in [42] on uniaxial (PDMS) elastomers filled with carbonyl iron particles. The experiments were performed as a function of particle concentration, intensity of the magnetic field and of the spatial distribution of the particles. By varying the direction of the magnetic field with respect to the direction of the fibers and to the direction of the compression stress, five different geometries were studied. In all cases an increase of $G$ was found with increasing $H$. However, this magnetic reinforcement effect is only significant when the applied field is parallel to the fibers. This is not too surprising since in this case the magnetic field reinforces the rigidity of the fibers. Typical results are shown in Figure 18.7 for the two geometries where the field is applied in a direction parallel to the fibers. It can

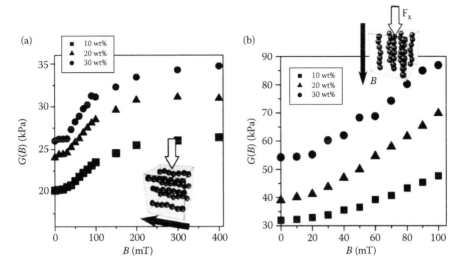

**FIGURE 18.7** Influence of a uniform magnetic field intensity on the elastic modulus of a uniaxial iron-loaded PDMS elastomer. The white and black arrows indicate the direction of the applied force and of the external magnetic field, respectively. (a) The force is applied perpendicularly to the applied magnetic field; (b) the applied force and the applied magnetic field are parallel. The symbols indicate the concentration of iron particles. (Reprinted from Varga, Z., Filipcsei, G., and Zrinyi, M., *Polymer*, 46, 7779, 2005. Copyright 2005, with permission from Elsevier.)

be seen that the largest increase in $G$ is observed when the magnetic field is parallel to both the particle alignment and the applied force.

It was found that the magnetically induced excess modulus, $G_M^E$, is proportional to $H^2$ at weak fields. To take into account this limiting behavior and the fact that $G_M^E$ approaches a saturation value at high fields, the $G_M^E$ data have been analyzed with the following equation:

$$G_M^E(H) = G_{M,\infty} \frac{H^2}{a_B + H^2} \tag{18.2}$$

where $a_B$ is a material parameter and $G_{M,\infty}$ the value at high field. This equation provides a good fit to the experimental data, as shown by Figure 18.8.

As will be shown in the next section this magnetic field-induced enhancement is not due to static magnetostriction effects, but rather associated with the magnetic field dependence of the elastic moduli.

### 18.2.3 DYNAMIC ASPECTS

#### 18.2.3.1 Introduction

Contrary to the static aspects, the dynamic aspects of uniaxial magnetic gels have practically never been studied, essentially because conventional rheometers do not

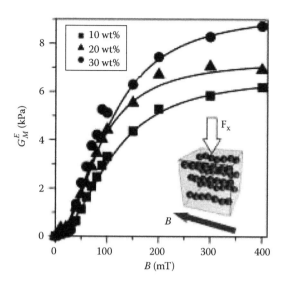

**FIGURE 18.8** The magnetically induced excess in the modulus as a function of the applied field for several uniaxial iron-loaded PDMS elastomers. The solid lines are fits to Equation 18.2. The symbols indicate the concentration in iron particles. (Reprinted from Varga, Z., Filipcsei, G., Zrinyi, M., *Polymer*, 47, 227, 2006. Copyright 2005, with permission from Elsevier.)

allow taking measurements in the presence of a magnetic field. The piezorheometer developed in Martinoty's group over the last few years makes it possible to perform these measurements. In this technique, the gels are formed directly in the measuring cell, which due to its reduced dimensions, can be placed in the air gap of an electromagnet. This makes it possible to follow the kinetics of the formation of the gel and to determine its frequency response in a wide frequency range. Both pieces of information are essential to optimize uniaxial magnetic gels for possible applications. Before discussing the experimental results, we first recall the expectations for the sol–gel transition, and the essential features of the piezorheometer.

### 18.2.3.2 The Sol–Gel Transition

Since the sol–gel transition is a connectivity transition, rheological experiments are one of the most powerful tools to determine the behavior of a gelling system. Here, we briefly recall the changes in behavior of the complex shear modulus $G = G' + iG''$, which occur when the system evolves from the sol phase to the gel phase.

At the beginning of the chemical reaction, in the sol phase, the low-frequency response of the system is liquid-like with

$$G' = \omega^2 \eta \tau \tag{18.3}$$

$$G'' = \omega \eta \tag{18.4}$$

where $\omega$ is the angular frequency, $\eta$ the viscosity, and $\tau$ a relaxation time. Equations 18.3 and 18.4 are valid for $\omega\tau \ll 1$ (hydrodynamic regime). During the gelation process, the number and size of the polymer clusters increase leading to an increase in $\eta$ and $\tau$. At the gel point, a first cluster extends across the entire sample. The steady-shear viscosity diverges to infinity, and the equilibrium modulus starts to rise to a finite value. Beyond the gel point the low-frequency response of the system is solid type with

$$G' = G_P \tag{18.5}$$

$$G'' = \omega\eta \tag{18.6}$$

where $G_P$ is a constant plateau value. In the gel phase, the network structure coexists with the remaining clusters, which do not yet extend across the sample. The size distribution of these clusters can be deduced from the high-frequency behavior of $G'$ and $G''$, which is given by the following power law.

$$G' \sim G'' \sim \omega^n \tag{18.7}$$

The low-frequency elastic response (i.e., $G_P$) of the gel continues to increase steadily until the chemical reaction is completed.

### 18.2.3.3 The Piezorheometer

The measurements of the complex shear modulus $G = G' + iG''$ as a function of frequency were taken with the piezorheometer used for studying different aspects of the shear mechanical properties of elastomers [52–59], polymers [60–62], and polyelectrolyte films [63]. As schematically shown in Figure 18.9, the piezorheometer is a plate–plate rheometer that uses piezoelectric ceramics vibrating in the shear mode to apply a small strain $\varepsilon$ to the sample and to measure the amplitude and the phase $\phi$ of the shear stress $\sigma$ transmitted through the sample. The complex shear modulus of the sample is given by the stress–strain ratio $G = \sigma/\varepsilon$. It can be determined for sample thicknesses ranging from 10 to 100 $\mu$m (for viscous liquids), over a wide frequency range ($10^{-2}$–$10^4$ Hz) and by applying very weak strains ($10^{-5}$–$10^{-2}$). The sample has an elastic response when the strain and the stress are in-phase ($\varphi = 0$), and a viscous response when the strain and the stress have a phase difference $\varphi$ of 90°. For a viscoelastic sample, the phase difference $\varphi$ is in between 0° and 90°. In practice, the sample is placed between two silica slides, respectively stuck to the emitting and to the receiving ceramics. The reliability of the setup was verified with liquids and soft solids (elastomers and gels), whose rheological properties are well known.

In the experiments presented here, $G$ was determined for frequencies ranging from 0.2 to $10^3$ Hz. The applied strain $\varepsilon$ was very small, $\sim 10^{-4}$, and the validity of the linear response was checked experimentally. Because of its small size, the cell was placed between the poles of an electromagnet during the growth of the gel and, once the gel was formed, under the objective of an optical microscope to check the organization of the magnetic particles. The experiments were performed with a homogeneous magnetic field (up to $\sim 1$ T) applied within the plane of the sample, in a direction parallel or perpendicular to the shear direction. Additional measurements were also taken without the magnetic field. All the experiments were performed at

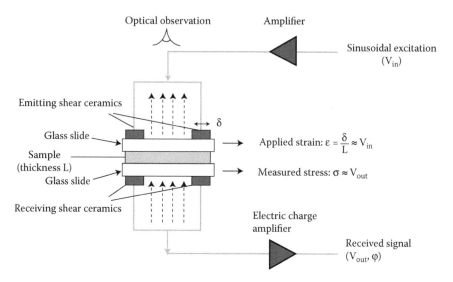

**FIGURE 18.9**   Schematic representation of the shear piezorheometer.

room temperature. The samples were films with ∼35 μm thickness and ∼2.5 cm² surface area. A PC monitoring the measuring system allowed the measurements to be carried out automatically.

### 18.2.3.4   Results

The system studied was a magnetic PVA hydrogel filled with ferrofluid particles [38].

#### 18.2.3.4.1   Preparation and Magnetic Properties of the Gel

These gels were prepared by introducing a magnetic fluid (M-300 iron oxide suspension from Sigma-Hi-Chemical, Japan) into a 7.5 wt.% aqueous solution of PVA (degree of hydrolyzation of 98–99%, $M_w = 31{,}000$–50,000 g/mol, from Aldrich Chemical Company). The cross-linking reaction was performed with glutardialdehyde (50 wt.% aqueous solution from Aldrich) at a pH of 1.5 (through addition of HCl 37% from Riedel–de Haën). The formation process of the magnetic gel (hereafter called ferrogel) is related to the competition between two antagonistic mechanisms associated with the fact that the ferrofluid has a pH of ∼ 8, and that the cross linker does not react above pH ∼ 2. Consequently, the lowering of the pH, which is necessary for the reticulation of the gel, induces a destabilization of the ferrofluid, leading to the formation of clusters of magnetic particles. This effect is compensated by the reticulation effect, which blocks the size of the clusters and their orientation in the direction of the magnetic field. As a result, the gel presents oriented fibers (length: several tens of microns, diameter: a few microns) perfectly visible with the optical microscope, as shows in Figure 18.10. For details about the preparation of the sample see [38].

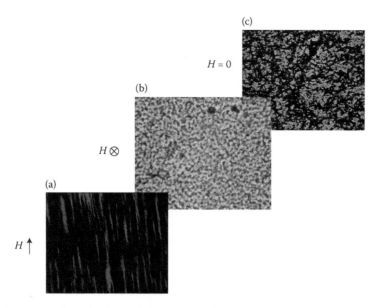

**FIGURE 18.10** **(See color insert following page 304.)** Observation under the optical microscope of the texture of a ferrogel film formed in the presence (a, b) or in the absence (c) of a magnetic field. The magnetic field is applied in a direction parallel (a) or perpendicular (b) to the plane of the sample. Size of the pictures: 170 $\mu$m × 130 $\mu$m. (Collin, D. et al. *Macromol. Rapid Commun.*, 24, 737, 2003. Copyright Wiley-VCH Verlag GmbH & Co. KGaA. Reproduced with permission.)

To gain insight into the magnetic nature of the gel, the orientation of a cylindrical gel (the analog of the planar sample studied in the piezorheometer) was studied as a function of the field amplitude. To do that, the gel was suspended in a homogenous magnetic field using a nonmagnetic wire of known torsional modulus. The rotation angle of the sample was then measured as a function of the field amplitude. The rotation angle is expected to be linear (ferromagnetic-like) or quadratic (paramagnetic-like) in the applied field strength. The data in Figure 18.11 show that the gel presents a ferromagnetic type response for weak fields, and that a paramagnetic contribution becomes dominant at larger fields, leading to substantial deviations from a linear law. This ferrogel is uniaxial because it can be oriented in a uniform magnetic field. Its axis of magnetization is defined by the direction of the magnetic field imposed during the reticulation.

### 18.2.3.4.2 Dynamical Shear Properties of the Gel

We start with the mechanical properties of the pure PVA gel. Its formation is illustrated in Figure 18.12a, which shows the variation of $G'$ as a function of time. The plotted values of $G'$ correspond to the hydrodynamic behavior characterized by the frequency-independent behavior of $G'$ shown in Figure 18.12b. The solid line in Figure 18.12a shows that the fit with a stretched exponential gives a good representation of the data.

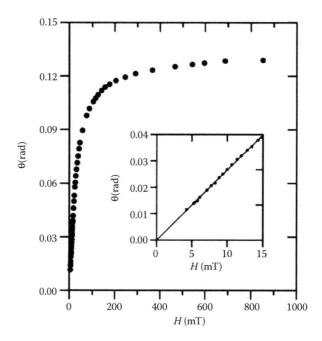

**FIGURE 18.11**   The rotation angle of a gel cylinder as a function of the magnetic field strength. The inset shows that the rotation angle is proportional to the field for small fields. (Collin, D. et al. *Macromol. Rapid Commun.*, 24, 737, 2003. Copyright Wiley-VCH Verlag GmbH & Co. KGaA. Reproduced with permission.)

The stretched exponential is given by Equation 18.8:

$$\frac{G'}{G'_\infty} = 1 - \exp\left(-\left[\frac{t-t_0}{\tau}\right]^x\right) \tag{18.8}$$

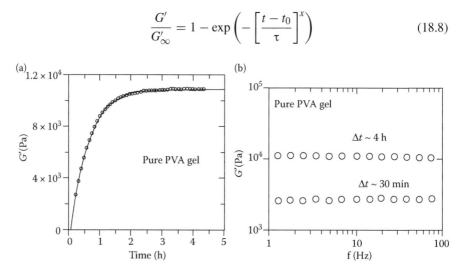

**FIGURE 18.12**   (a) variation of $G'$ as a function of time, for a pure PVA gel. (b) frequency-dependence of $G'$ for the same gel.

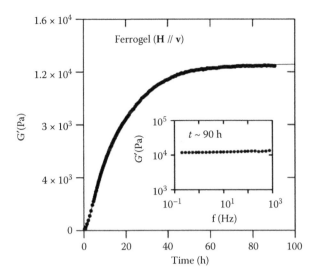

**FIGURE 18.13** Variation of $G'$ as a function of time for the uniaxial magnetic PVA hydrogel in the parallel orientation. A 200 mT uniform magnetic field was applied during the formation of the sample. The solid line is the fit with Equation 18.8. Inset: $G'$ as a function of frequency at $t \sim 90$ h. (Collin, D. et al. *Macromol. Rapid Commun.*, 24, 737, 2003. Copyright Wiley-VCH Verlag GmbH & Co. KGaA. Reproduced with permission.)

It contains four-fit parameters. The value $G'_\infty$ of $G'$ for long times, the time $t_0$ associated with the infinite cluster, the formation time $\tau$ of the network, and the exponent $x$ describing how stretched is the exponential. The resulting parameters are $G'_\infty = (1.091 \pm 0.001) \times 10^4$ Pa, $t_0 = 0.050 \pm 0.006$ h, $x = 1.00 \pm 0.01$ and $\tau = 0.585 \pm 0.007$ h.

A typical example showing the formation of a uniaxial ferrogel in a homogeneous field is given by Figure 18.13. The data correspond to the case where **H** is parallel to **v**. The field strength was 200 mT. The comparison with the time axis of Figure 18.12a immediately shows that the presence of magnetic particles slows down considerably the kinetics of formation of the gel.

As for the pure PVA gel, Equation 18.8 provides a good fit (solid line) to the experimental data with the following parameters, $G'_\infty = (1.261 \pm 0.002) \times 10^4$ Pa, $t_0 = 1.12 \pm 0.08$ h, $x = 1.01 \pm 0.01$ and $\tau = 16.87 \pm 0.10$ h. The inset in the figure shows that $G'$ is frequency independent at low frequencies, indicating that the data are in the hydrodynamic regime. Similar results were obtained for **H** perpendicular to **v**. However, it was not possible to determine whether the mechanical response of the gel is anisotropic or not, because the final value of $G'$ is very sensitive to slight differences in the concentration of the ingredients (glutardialdehyde and HCl), which are involved in the preparation of the gel. In any case, the anisotropy, if it exists, is very small. This indicates that the fibers seen in Figure 18.10 are not compact objects, but rather loosely packed objects.

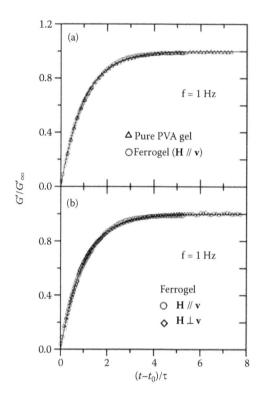

**FIGURE 18.14** Master curves for the pure PVA gel and the uniaxial magnetic PVA gel (a) in the parallel geometry, and for the uniaxial magnetic PVA gel in the parallel and perpendicular geometries (b). (Collin, D. et al. *Macromol. Rapid Commun.*, 24, 737, 2003. Copyright Wiley-VCH Verlag GmbH & Co. KGaA. Reproduced with permission.)

As shown in Figure 18.14a, a quantitative comparison between the data obtained for the pure PVA gel and for the ferrogel can be done by plotting the data for both gels as a function of the reduced variables $G'/G'_\infty$ and $(t - t_0)/\tau$. It can be seen that both curves perfectly superimpose, giving rise to a master curve, although the $\tau$ values differ by more than one order of magnitude.

Figure 18.14b shows a similar plot for the ferrogel data associated with the parallel and perpendicular geometries. Again all the data fall on the same master curve. For all the samples, the exponent $x$ remains around 1, indicating that these gels are characterized by a single timescale. The gel formation phenomenon is thus controlled by the same physics, regardless of whether the system is filled with magnetic particles or not.

Similar experiments are in progress on different systems for which the final material should show (or shows) a mechanical anisotropy. Also, experiments performed in the sol phase, similar to those reported in [45], should be interesting to give information about the precise nature of the interaction between the magnetic particles and the polymer network.

## 18.3   MACROSCOPIC STATIC AND DYNAMIC PROPERTIES*

### 18.3.1   STATICS AND THERMODYNAMICS

#### 18.3.1.1   Isotropic Ferrogels

The macroscopic description of a system starts with the identification of the relevant variables. Apart from the quantities that are related to local conservation laws, such as mass density $\rho$, momentum density $g$, energy density $\varepsilon$, and concentration $c$ of the swelling fluid, we consider the elastic strain $u_{ij}$ and the magnetization $M$ as additional variables. In a crystal the former is related to the broken translational symmetry due to the long range positional order, which gives rise to the displacement vector $\vec{u}$ as a hydrodynamic symmetry variable. Since, neither solid body translations nor rigid rotations give rise to elastic deformations, the strain tensor is used as a variable, which reads in linearized version $u_{ij} = 1/2(\nabla_i u_j + \nabla_j u_i)$. In amorphous solids, such as rubbers, gels, and so on, linear elasticity is still described by a second-rank, symmetric strain tensor. For a proper description of nonlinear elasticity, *cf.* [64]. The induced magnetization $M$ is a slowly relaxing variable in the super paramagnetic (isotropic) case.

Assuming local thermodynamic equilibrium, that is, all microscopic, fast relaxing quantities are already in equilibrium, we have the Gibbs relation

$$d\varepsilon = T d\sigma + \mu d\rho + \mu_c dc + \upsilon_i dg_i + H_i dB_i + h_i^M dM_i + \Psi_{ij} du_{ij}, \qquad (18.9)$$

relating all macroscopically relevant variables discussed above to the entropy density $\sigma$. $B$ is the magnetic induction field included here in order to accommodate the static Maxwell equations. In Equation 18.9 the thermodynamic quantities, chemical potential $\mu$, temperature $T$, relative chemical potential $\mu_c$, velocity $\upsilon_i$, elastic stress $\Psi_{ij}$, magnetic field $H_i$, and the magnetic molecular field $h_i^M$, are defined as partial derivatives of the energy density with respect to the appropriate variables [50].

To determine the thermodynamic conjugate variables we need an expression for the local energy density. This energy density must be invariant under time reversal as well as under parity and it must be invariant under rigid rotations, rigid translations, and be covariant under Galilei transformations. In addition to that the energy density must have a minimum, because there exists an equilibrium state for the gel. Therefore, the expression for the energy density needs to be convex. Taking into account these symmetry arguments we get [18]

$$\varepsilon = \varepsilon_0 + \frac{B^2}{2} - B \cdot M + \frac{\mu_{ijkl}}{2} u_{ij} u_{kl} - \frac{\gamma_{ijkl}}{2} M_i M_j u_{kl} + \frac{\alpha}{2} M_i^2 + \frac{\beta}{4}(M_i^2)^2$$
$$+ u_{ii}(\chi^\rho \delta\rho + \chi^\sigma \delta\sigma + \chi^c \delta c), \qquad (18.10)$$

where $\varepsilon_0$ is the energy density of a fluid binary mixture. Equation 18.10 explicitly contains the elastic and the magnetic energy, their cross coupling (the magnetoelastic energy) and bilinear couplings of compression with the scalar variables. To

---

* This section is based on Refs. [18,47].

discuss large elastic deformations (rubber elasticity) one should keep terms of higher order of $u_{ij}$, which are neglected here. The magnetostrictive coupling ($\sim \gamma_{ijkl}$) is cubic [65] and the $M^4$ contribution is kept in order to guarantee the thermodynamic stability. The elastic tensor $\mu_{ijkl}$ (and similarly $\gamma_{ijkl}$) takes the isotropic form $\mu_{ijkl} = \mu_1 \delta_{ij}\delta_{kl} + \mu_2(\delta_{ik}\delta_{jl} + \delta_{il}\delta_{jk} - (2/3)\delta_{ij}\delta_{kl})$, where $\mu_1$ is the compressibility and $\mu_2$ the shear modulus. The magnetoelastic energy is similar to that for ferromagnetic materials, where the compressional magnetostriction is neglected ($\gamma_1 = 0$) [65]. All static susceptibilities, such as the elastic and magnetoelastic moduli as well as those describing cross couplings between compression and the density, entropy density, and concentrations variations ($\chi^\rho$, $\chi^\sigma$, and $\chi^c$, respectively) can depend on $M^2$ and are thus magnetic field-strength dependent.

Using Equations 18.9 and 18.10, the magnetic Maxwell field $H_i$ is defined in the usual way

$$H_i = \left(\frac{\partial \varepsilon}{\partial B_i}\right)_{M,u_{ij},\ldots} = B_i - M_i, \tag{18.11}$$

while the magnetic molecular field $h_i^M$ reads

$$h_i^M = \left(\frac{\partial \varepsilon}{\partial M_i}\right)_{B,u_{ij},\ldots} = -B_i - \gamma_{ijkl}M_j u_{kl} + \alpha M_i + bM^2 M_i \tag{18.12}$$

Note that because of definition (Equation 18.11), it is not possible to have a direct coupling between the external field $\boldsymbol{B}$ and the strain; the deformation of the network is mediated by the magnetization through the coupling terms $\sim \gamma_{ijkl}$.

The elastic stress $\Psi_{ij}$ has the following form:

$$\Psi_{ij} = \left(\frac{\partial \varepsilon}{\partial u_{ij}}\right)_{M,B,\ldots} = \mu_{ijkl}u_{kl} - \frac{\gamma_{ijkl}}{2}M_k M_l + \delta_{ij}(\chi^\rho\delta\rho + \chi^\sigma\delta\sigma + \chi^c\delta c) \tag{18.13}$$

and depends on the magnetization.

### 18.3.1.2 Uniaxial Ferrogels

In uniaxial ferrogels there is a permanent magnetization, which defines a preferred direction and constitutes a spontaneously broken rotational symmetry. We will introduce a unit vector $m_i$ defined by $m_i = M_i/|\mathbf{M}|$ pointing to the direction of magnetization in analogy to the director $n_i$ in nematic liquid crystals. But there is a significant difference. While both are even under parity, the unit vector of magnetization $\mathbf{m}$ is odd under time reversal. This will permit static as well as dynamic couplings to other variables that are odd under time reversal. We can then define the transverse Kronecker tensor $\delta_{ij}^\perp = \delta_{ij} - m_i m_j$ and we have, together with the Levi–Cevità symbol $\varepsilon_{ijk}$, three invariants of the system, by which all the static material tensors (e.g., the elastic and the magnetostrictive tensor) and the transport tensors get their uniaxial form [47]. Rotations of $\boldsymbol{M}$ are hydrodynamic excitations of the system and, therefore, the energy density should be a function of gradients of $\boldsymbol{M}$, too. This involves replacing $h_i^M dM_i$ in the Gibbs relation (Equation 18.9) by $h_i^{'M} dM_i + \Phi_{ij}^M d(\nabla_j M_i)$ or using $h_i^M \equiv h_i^{'M} - \nabla_j \Phi_{ij}^M$. In addition, we consider a variable first introduced by de Gennes for LCEs [66] called relative rotation $\tilde{\Omega}_i$. This variable belongs to the class of slowly relaxing variables and describes the relative rotation between the polymer network and the orientation of the magnetization. It is defined linearly by $\tilde{\Omega}_i = \delta m_i - \Omega_i^\perp = \delta m_i - 1/2 m_j (\nabla_i u_j - \nabla_j u_i)$. Since $m_i$ is a unit vector, $\mathbf{m} \cdot \delta \mathbf{m} = 0$. This variable is odd under time reversal and even under parity. For a general, non-linear definition $cf.$ [67]. This degree of freedom requires the additional contribution $W_i d\tilde{\Omega}_i$ in the Gibbs relation. It should be noted, however, that for materials with an almost rigid coupling between the direction of $\mathbf{M}$ and the elastic network, this degree of freedom is clamped and less important.

In the uniaxial case the local energy density, Equation 18.10, has to be amended by the appropriate additional contributions [47]

$$\varepsilon^{uniax} = \frac{1}{2} K_{ijkl}(\nabla_i M_j)(\nabla_k M_l) + \frac{1}{2} D_1 \tilde{\Omega}_i \tilde{\Omega}_i + D_2(m_j \delta_{ik}^\perp + m_k \delta_{ij}^\perp) \tilde{\Omega}_i u_{jk} + c_{ijk} g_i \nabla_j M_k$$
$$+ \sigma_{ijk}^\sigma (\nabla_i M_j)(\nabla_k \delta \sigma) + \sigma_{ijk}^\rho (\nabla_i M_j)(\nabla_k \delta \rho) + \sigma_{ijk}^c (\nabla_i M_j)(\nabla_k \delta c) \quad (18.14)$$

containing a Frank-like magnetization–rotational energy, the relative rotations' energy and their energetic coupling to elastic shear, various couplings between magnetization rotation and density, temperature, and concentration gradients as well as the momentum density $g_i$. The latter coupling results in $g_i = \rho v_i - \rho c_{ijk} \nabla_j M_k$ and is similar to one of the couplings appearing in superfluid $^3$He-$A$ first introduced by Graham [68]. In this system one defines an axial vector $\mathbf{1}$ parallel to the direction of the net orbital momentum of the helium pairs. This vector does not commute with the total angular momentum vector and therefore, this variable breaks the continuous rotational symmetry spontaneously, similarly to the magnetization in our system.

The uniaxial form of the static material tensors in Equations 18.10 and 18.14 as well as the resulting expressions for the thermodynamic conjugate fields (like $h_i^M$, $W_i$, $\Psi_{ij}$ etc.) are listed in [47] and not repeated here.

## 18.3.2 DYNAMICS

To determine the dynamics of the variables we take into account that the conserved quantities obey a local conservation law while the dynamics of the other variables can be described by a simple balance equation where the counter term to the temporal change of the quantity is called a quasi current. As a set of dynamical equations we get [47]

$$\partial_t \rho + \nabla_i g_i = 0 \quad (18.15)$$

$$\partial_t \sigma + \nabla_i (\sigma v_i) + \nabla_i j_i^\sigma = \frac{R}{T} \quad (18.16)$$

$$\rho \partial_t c + \rho v_i \nabla_i c + \nabla_i j_i^c = 0 \quad (18.17)$$

$$\partial_t g_i + \nabla_i(\upsilon_j g_i + \delta_{ij}[p + B \cdot \mathbf{H}] + \sigma_{ij}^{th} + \sigma_{ij}) = 0 \tag{18.18}$$

$$\partial_t M_i + \upsilon_j \nabla_j M_i + (M \times \omega)_i + X_i = 0 \tag{18.19}$$

$$\partial_t u_{ij} + \upsilon_k \nabla_k u_{ij} + Y_{ij} = 0 \tag{18.20}$$

$$\partial_t \tilde{\Omega}_i + \upsilon_k \nabla_k \tilde{\Omega}_i + Z_i = 0 \tag{18.21}$$

where we have introduced the vorticity $\omega_i = (1/2)\varepsilon_{ijk}\nabla_j \upsilon_k$ and $\sigma_{ij}^{th} = -(1/2)(B_i H_j + B_j H_i) + (1/2)(\Psi_{jk}\epsilon_{ki} + \Psi_{ik}\epsilon_{kj})$ according to the requirement that the energy density should be invariant under rigid rotations [50]. The pressure $p = \partial(\int \varepsilon \, dV)/\partial V$ in Equation 18.18 is given by $p = -\varepsilon + \mu\rho + T\sigma + v \cdot g$.

For practical reasons we have used an entropy balance Equation 18.16 rather than the energy conservation law; both are not independent but related through the Gibbs relation (Equation 18.9). In the equation for the entropy density we introduced $R$, the dissipation function which represents the entropy production of the system. Due to the second law of thermodynamics $R$ must satisfy $R \geq 0$. For reversible processes this dissipation function is equal to zero while for irreversible processes it must be positive. In the following we will split the currents and quasi currents into reversible parts (denoted with a superscript $R$) and irreversible parts (denoted with a superscript $D$).

If we again make use of the symmetry arguments mentioned above and use Onsager's relations we obtain the following expressions for the reversible currents up to linear order in the thermodynamic forces [47]

$$j_i^{\sigma R} = -\kappa_{ij}^R \nabla_j T - D_{ij}^{TR} \nabla_j \mu_c + \xi_{ij}^{TR} \nabla_l \Psi_{jl} \tag{18.22}$$

$$j_i^{cR} = -D_{ij}^R \nabla_j \mu_c + D_{ij}^{TR} \nabla_j T + \xi_{ij}^{cR} \nabla_l \Psi_{lj} \tag{18.23}$$

$$\sigma_{ij}^R = -\Psi_{ij} - c_{ijk}^{RJ} h_k^M - \nu_{ijkl}^R A_{kl} + \xi_{ijk}^{\sigma R} W_k \tag{18.24}$$

$$Y_{ij}^R = -A_{ij} + \xi_{ijk}^{YR} W_k + \frac{1}{2}\lambda^M[\nabla_i(\nabla \times h^M)_j + \nabla_j(\nabla \times h^M)_i] - \frac{1}{2}\nabla_i(\xi_{jk}^R \nabla_l \Psi_{kl}$$
$$+ \xi_{jk}^{TR}\nabla_k T + \xi_{jk}^{cR}\nabla_k\mu_c) - \frac{1}{2}\nabla_j(\xi_{ik}^R\nabla_l\Psi_{kl} + \xi_{ik}^{TR}\nabla_k T + \xi_{ik}^{cR}\nabla_k\mu_c) \tag{18.25}$$

$$X_i^R = b_{ij}^R h_j^M + \lambda^M \epsilon_{ijk}\nabla_j\nabla_l\Psi_{kl} - c_{jki}^{RJ}A_{jk} + \xi_{ij}^{XR}W_j \tag{18.26}$$

$$Z_i^R = \tau_{ij}^R W_j - \xi_{ij}^{XR} h_j^M - \xi_{kli}^{\sigma R} A_{kl} - \xi_{kli}^{YR}\Psi_{kl} \tag{18.27}$$

In isotropic ferrogels the quasi current $Z_i^R$ is absent as well as the appropriate counter terms $\sim \xi_{ij}^{XR}, \xi_{kli}^{\sigma R}$, and $\xi_{kli}^{YR}$ in $X_i^R, \sigma_{ij}^R$, and $Y_{ij}^R$, respectively. These terms describe the reversible dynamic coupling of relative rotations to the magnetization, the momentum density and the network. The first coupling mediated by the tensor $\xi_{ij}^{XR}$ — is a new term that exists neither in nematic LCEs [69] nor in superfluid $^3$He-$A$, while the second coupling, $\xi_{ijk}^{\sigma R}$, already appeared in superfluid $^3$He-$A$, and the third one, $\xi_{ijk}^{YR}$, in nematic LCEs.

All the second rank tensors have to be odd under time reversal and are of the form $\alpha_{ij}^R = \alpha^R \epsilon_{ijk} m_k$ resembling the form of a Hall conductance. In the isotropic case,

$m_i$ has to be replaced by the induced magnetization or by the external field. These reversible linear field contributions to transport parameters or tensors are known even for isotropic fluids ($\kappa^R$, Righi–Leduc effect) and some of them have been discussed in detail for ferronematics [70] and isotropic ferrofluids [71]. The third and fourth rank material tensors are listed in Refs. [18,47]. Again, for isotropic ferrogels these reversible couplings are absent, if there is no external magnetic field.

We can use the dissipation function $R$ as a Liapunov functional to derive the irreversible currents and quasi currents. One can expand the function $R$ ($R/T$ is the amount of entropy produced within a unit volume per unit time) into the thermodynamic forces using the same symmetry arguments as in the case of the energy density. We obtain [47]

$$R = \frac{1}{2}\kappa_{ij}(\nabla_i T)(\nabla_j T) + D_{ij}^T(\nabla_i T)(\nabla_j \mu_c) + \xi_{ij}^T(\nabla_i T)(\nabla_k \Psi_{jk}) + \frac{1}{2}D_{ij}(\nabla_i \mu_c)(\nabla_j \mu_c)$$

$$+ \xi_{ij}^c(\nabla_i \mu_c)(\nabla_k \Psi_{jk}) + \frac{1}{2}\nu_{ijkl}A_{ij}A_{kl} + \xi_{ijk}^\sigma A_{ij}W_k + c_{ijk}^J A_{ij}h_k^M$$

$$+ \frac{1}{2}\xi_{ij}(\nabla_k \Psi_{ik})(\nabla_l \Psi_{jl}) + \frac{1}{2}b_{ij}h_i^M h_j^M + \frac{1}{2}\tau W_i W_i + \xi^X W_i h_i^M \qquad (18.28)$$

where we have introduced various second rank dissipative transport tensors describing heat conduction ($\kappa$), diffusion ($D$) and thermo-diffusion ($D^T$), elastic (or vacancy) diffusion ($\xi$) and appropriate cross couplings to temperature ($\xi^T$) and concentration gradients ($\xi^c$), and magnetization relaxation ($b$). These tensors are of the form $\alpha_{ij} = \alpha_{\parallel}m_i m_j + \alpha_{\perp}\delta_{ij}^\perp$, reducing to $\alpha_{ij} = \alpha\delta_{ij}$ in the isotropic case. The third rank tensor $\xi_{ijk}^\sigma = \xi^\sigma(m_i\epsilon_{jkl} + m_j\epsilon_{ikl})m_l$ vanishes in the isotropic case as well as $\tau$ and $\xi^X$ due to the lack of relative rotations. Only $c_{ijk}^J$ can exist in isotropic ferrogels, but is proportional to the square of the external magnetic field strength. The viscosity tensor $\nu_{ijkl}$ has the standard isotropic or uniaxial form.

To obtain the dissipative parts of the currents and quasi currents one has to take partial derivatives with respect to the appropriate thermodynamic forces, for example, $Y_{ij}^D = \partial R/\partial \Psi_{ij}$, $Z_i^D = \partial R/\partial W_i$, or $\sigma_{ij}^D = -\partial R/\partial(\nabla_j \upsilon_i)$. Again, explicit expressions are listed in [18,47].

## 18.4 EXAMPLES FOR NOVEL CROSS-COUPLING EFFECTS*

### 18.4.1 MAGNETOSTRICTION

#### 18.4.1.1 Static Elongation and Shear

For isotropic ferrogels we first discuss static elastic deformations in the presence of an external field. Magnetostriction leads to an anisotropic deformation. If then an external strain is applied, the stress–strain relation is more complicated than a simple Hooke's law.

In equilibrium, both the elastic stress equation 18.13 and the magnetic molecular field Equation 18.12 have to be zero. Without an external field or external strain

---

* This section is based on Refs. [18,47–49].

there is no magnetization and no strain in equilibrium. A finite external field, taken along the $z$-axis, $B = B_0 e_z$, induces an equilibrium magnetization ($M^0 = M_0 e_z$) and a nonzero strain $u_{ij}^0$ due to the magnetostriction effect. Neglecting the couplings of density, entropy, and concentration to the strain tensor, we have [18]

$$u_{xx}^0 = u_{yy}^0 = \frac{\mu_2 \gamma_1 - \mu_1 \gamma_2}{6\mu_1 \mu_2} M_0^2, \qquad (18.29)$$

$$u_{zz}^0 = \frac{\mu_2 \gamma_1 + 2\mu_1 \gamma_2}{6\mu_1 \mu_2} M_0^2, \qquad (18.30)$$

leading to the volume change $U^0 \equiv u_{xx}^0 + u_{yy}^0 + u_{zz}^0 = (\gamma_1/2\mu_1) M_0^2$. The magnetostrictive volume change of the ferrogel is determined by the bulk modulus $\mu_1$ and by the coefficient $\gamma_1$, which couples the trace of the stress tensor to the magnitude of the magnetization.

Magnetostriction is a well-known phenomenon in single- or polycrystalline ferromagnetic solids [72]. Ferrogels are isotropic and nonmagnetic without an external magnetic field and magnetostriction is then a nonlinear effect. The induced deformations, Equations 18.29 and 18.30, are of uniaxial symmetry and in this state the ferrogel behaves more like a uniaxial ferromagnet than an isotropic one.

This state is then disturbed by an external deformation $\Delta u_{ij}$ by some mechanical device. Due to the magnetostriction effect this gives rise to a change in the magnetization also. In the static limit the magnetic degree of freedom is still in equilibrium and the change of the magnetization can be obtained from the condition $h_i^M = 0$, Equation 18.12. The applied deformations give, directly by Hooke's law, and indirectly by the change of the magnetization, elastic stresses [18]

$$\Psi_{zz} = (\mu'' - \chi_0 \gamma''^2 M_0^2)\Delta u_{zz} + (\mu' - \chi_0 \gamma' \gamma'' M_0^2)(\Delta u_{xx} + \Delta u_{yy}), \qquad (18.31)$$

$$\Psi_{xx} = (\mu'' - \chi_0 \gamma'^2 M_0^2)\Delta u_{xx} + (\mu' - \chi_0 \gamma'^2 M_0^2)\Delta u_{yy} + (\mu' - \chi_0 \gamma' \gamma'' M_0^2)\Delta u_{zz}, \qquad (18.32)$$

$$\Psi_{zx} = 2(\mu_2 - \chi_0 \gamma_2^2 M_0^2)\Delta u_{zx}, \qquad (18.33)$$

$$\Psi_{xy} = 2\mu_2 \Delta u_{xy} \qquad (18.34)$$

with two additional relations similar to Equations 18.32 and 18.33, but with $x$ and $y$ subscripts interchanged. For the magnetic susceptibility, $\chi_0 = \chi(B_0)$, its value at the external field strength $B_0$ has to be taken while $\mu'' = \mu_1 + (4/3)\mu_2$ and $\mu' = \mu_1 - (2/3)\mu_2$ and similar abbreviations for the $\gamma's$ are used. Note that, even if the deformation does conserve the volume ($\Delta u_{xx} + \Delta u_{yy} + \Delta u_{zz} = 0$), the trace of the elastic stress tensor is not zero, but given by $\Psi_{kk} = -6\chi_0 \gamma_1 \gamma_2 M_0^2 \Delta u_{zz}$. Formulas (Equations 18.31 through 18.34) are applicable for small strains only, since it is based on Hooke's law, while for larger strains deviations from this law due to rubber elasticity are to be expected.

The effective elastic moduli, that is, the ratio between the measured stress and the applied strain, for example, $\Psi_{zz}/\Delta u_{zz}$ or $\Psi_x/\Delta u_x$, show a decrease as a function of the applied field due to magnetostriction. This is in contrast to the experiments shown in Figure 18.8, where an increase is found. Thus, for the materials discussed in Figure 18.8 magnetostriction is a small effect, completely dominated by the intrinsic field dependence of the material parameters, $\mu''(M_0) - \mu'' \sim M_0^2$. The latter effect is nonhydrodynamic and based on microscopic structure changes due to the external field. The proportionality factor can be obtained from experiments and is positive for the materials discussed here.

### 18.4.1.2    Propagation of Sound

Due to the presence of the permanent polymer network in ferrogels compared to ferrofluids, there are transverse as well as longitudinal sound eigenmodes. In this section we derive the longitudinal and the transverse sound of the system with an external magnetic field parallel to the $z$-axis. We neglect all diffusional processes connected, for example, with viscosity and heat conduction as well as their reversible counterparts. Only the relaxation of the magnetization field is kept.

Assuming a one-dimensional plane wave with space–time dependence $\sim \exp i(-\omega t + \mathbf{k} \cdot \mathbf{r})$ for all deviations $\delta u_{ij}, \delta M_i, \upsilon_i, \delta\rho$ from the equilibrium values, the linearized set of dynamic equations becomes an algebraic one. We consider sound in the two cases, where the external magnetic field and the equilibrium magnetization are either perpendicular or parallel to the wave vector. Field fluctuations $\delta B_i$ are fixed by the static Maxwell equations to $\delta B_i = \delta M_j(\delta_{ij} - k_i k_j k^{-2})$.

In the low-frequency limit ($\omega < 1/\tau_M$) below the magnetization relaxation frequency and for the external field being perpendicular to the wave vector the velocities of the longitudinal $c_l$ and the transverse sounds $c_{t1}, c_{t2}$ read [18]

$$c_l^2 = \frac{\tilde{\mu}(M_0)}{\rho} - \tilde{\mu}\frac{\mu_2\gamma_1 - \mu_1\gamma_2}{6\rho\mu_2\mu_1}M_0^2, \tag{18.35}$$

$$c_{t1}^2 = \frac{\mu_2(M_0)}{\rho} - \frac{\mu_2\gamma_1 - \mu_1\gamma_2}{6\rho\mu_1}M_0^2, \tag{18.36}$$

$$c_{t2}^2 = \frac{\mu_2(M_0)}{\rho} - \frac{2\mu_2\gamma_1 + 7\mu_1\gamma_2}{12\rho\mu_1}M_0^2, \tag{18.37}$$

while for a parallel field the transverse sound velocities are equal according to the effectively uniaxial symmetry due to the external field.

The sound speeds at low frequencies and zero field give information about the compressibility and the elastic moduli (bulk and shear). There is a dependence on $M_0^2$ due to intrinsic field dependence of the moduli as measured by static experiments. However, even if the static magnetostrictive effect is neglected (as done in Equations 18.35 through 18.37), there are additional contributions $\sim M_0^2$. They are of the bilinear $\gamma\mu$ type and a dynamic manifestation of magnetostriction emerging through the nonlinear elastic stress contribution to the stress tensor in $\sigma_{ij}^{th}$, which, however, is effectively linear due to the nonzero equilibrium strains. Therefore, the effective moduli measured by sound propagation in an external field are different from those given by the static elastic stress $\Psi_{ij}$ discussed in the preceding section. The coincidence of static linear elasticity and low-frequency sound speed is restored in the limit of vanishing magnetic field, when no magnetostrictive deformation is present and the additional contribution in the stress tensor $\sigma_{ij}^{th}$ is nonlinear and absent in the sound spectra. Of course, in this limit the sound spectra are isotropic as is the ferrogel.

The sound velocities change with an external magnetic field basically with the second power of the field, which is in accordance with experiments on longitudinal sound [10]. Whether the sound velocities are decreased or increased by the field cannot be established by general rules, since the signs of $\gamma_{1,2}$ are not fixed and can be material dependent. Measurements of transverse and longitudinal sound velocities in the different geometries will provide information on the magnitude and sign of the magnetostrictive and elastic moduli. As a first approximation the magnetostrictive volume change $(\sim \gamma_1 M_0^2 / \mu_1)$ can be neglected in those rubbers and only shape changes remain.

Damping of sound waves generally is rather weak and given by the imaginary part of the dispersion relation. In addition to the usual magnetic-field-independent sound damping due to viscosity and other diffusional processes there is a field-dependent sound damping in ferrogels. This is an effect of the reversible, dynamic coupling of the magnetization to flow, either phenomenological ($c_{ijk}^R(M)$ in Equation 18.26) or kinematic ($\epsilon_{ijk}M_j\omega_k$ in Equation 18.19) and its counterparts in the Navier–Stokes equation. For example, when the magnetic field is parallel (perpendicular) to the wave vector a field-dependent damping of longitudinal (transverse) sound reads, respectively [18],

$$\mathfrak{Im}(\omega_l) = -\frac{1}{2}\frac{[\chi_0\gamma'' - (2c_1^R + c_2^R)]^2}{\rho b}M_0^2 k^2, \tag{18.38}$$

$$\mathfrak{Im}(\omega_{t2}) = -\frac{1}{2}\frac{(\chi_0\gamma_2 - c_1^R + \frac{1}{2})^2}{\rho b}M_0^2 k^2, \tag{18.39}$$

the first of which can be related to the observed increase of the apparent viscosity due to the magnetic field [73]. In all cases, $\mathfrak{Im} < 0$, as it should be according to the second law of thermodynamics.

We will not investigate the high-frequency limit for $\omega > 1/\tau_M$, since for higher frequencies possible viscoelastic effects should also be taken into account, which goes beyond the scope of this review.

### 18.4.2 SHEAR-INDUCED MAGNETIZATION

Uniaxial ferrogels differ qualitatively from the isotropic ones by the macroscopic variables associated with relative rotations. These variables describe, as already mentioned, the relative rotations between the orientation of the magnetization and the polymer network. In this section we discuss an effect associated with these variables. We apply a constant shear flow and determine the change of magnetization. We assume that the direction of the permanent magnetization of the uniaxial ferrogel $m = e_x$ is parallel to the flow direction of an external stationary shear flow $S_{kl} = S^{ext}\delta_{ky}\delta_{lx}\nabla_k\upsilon_l$. Furthermore, we assume spatial homogeneity and consider only linear effects. Contributions due to magnetostriction effects are neglected. We do not apply an external magnetic field. Therefore, we can assume that the magnitude of the magnetization is not changed but only its direction $M = M_0(m + \delta m)$.

The resulting set of inhomogeneous algebraic equations for the four variables $\delta m_y$, $\delta m_z$, $u_{xy}$, and $u_{xz}$ has the following solutions [47]:

$$\delta m_y = \frac{b^R(\xi^X + 2\xi^{YR}c^J) - b_\perp(\xi^{XR} - 2\xi^{YR}c_1^{RJ})}{2\xi^{YR}(\alpha M_0 - B_0)(b_\perp^2 + b^{R2})}S^{\text{ext}} \tag{18.40}$$

$$\delta m_z = -\frac{b_\perp(\xi^X + 2\xi^{YR}c^J) + b^R(\xi^{XR} - 2\xi^{YR}c_1^{RJ})}{2\xi^{YR}(\alpha M_0 - B_0)(b_\perp^2 + b^{R2})}S^{\text{ext}} \tag{18.41}$$

For this experimental setup we thus predict a rotation of the magnetization within the shear plane, $\delta m_y \neq 0$, as well as out of the shear plane, $\delta m_z \neq 0$, which are both proportional to the applied external force. This effect is due to the variables associated with relative rotations, because all contributions are proportional to either $\xi^{XR}$, $\xi^X$, or $\xi^{YR}$, which represent the (reversible) dynamical coupling of relative rotations to the magnetization and the strain field, respectively. Reversible (superscript $R$) and irreversible transport parameters are systematically paired in the numerator. There are elastic deformations as well, not only in the shear plane, but also out of the shear plane, $u_{xz}$, which are constant and proportional to $S^{\text{ext}}$. For oscillatory shear $S^{\text{ext}} \to S^{\text{ext}} \exp(i\omega t)$ the same variables as in the static case will be excited with the same frequency, but will show some phase lag due to various dissipation effects. The change of the direction of the magnetization obtained should be observable by Hall probes, if the degree of freedom of relative rotations is operable and not clamped in the ferrogel used.

### 18.4.3    Surface Waves and Rosensweig Instability

#### 18.4.3.1    Surface Waves in Isotropic Ferrogels

Surface undulations of the free surface of viscous liquids are known to be able to prop-agate as gravity or capillary waves. In more complex systems like viscoelastic liquids or gels the transient or permanent elasticity allows for modified transverse elastic waves at free surfaces [74]. They are excited, for example, by thermal fluctuations or by imposed temperature patterns on the surface. In ferrofluids magnetic stresses at the surface come into play. In particular, in an external magnetic field normal to the surface there is a focusing effect on the magnetization at the wave crests of an undulating surface with the tendency to increase the undulations [75]. At a critical field strength no wave propagation is possible and the surface becomes unstable with respect to a stationary pattern of surface spikes (Rosensweig or normal field insta-bility). Combining the two aspects of elasticity and superparamagnetic response and using linearized dynamic equations and boundary conditions we get the general sur-face wave dispersion relation for ferrogels (in a normal external field), which contains as special cases those for ferrofluids and nonmagnetic gels and can be generalized to viscoelastic ferrofluids and magnetorheological fluids. A linear stability analysis reveals the threshold condition, above which stationary surface spikes grow. This crit-ical field depends on gravity, surface tension and on the elastic (shear) modulus of the gel, while the critical wavelength of the emerging spike pattern is independent of the latter. As in the case of ferrofluids neither the threshold nor the critical wavelength depends on the viscosity.

To derive the surface wave dispersion relation several approximations are made [48]. Of the various reversible and irreversible dynamic cross couplings between flow, elasticity and magnetization discussed above, we only keep those that are presumably the relevant ones for the present problem. In particular, we keep the magnetic Maxwell and the elastic and viscous contributions to the stress tensor. The magnetization $M$ is assumed to have relaxed to its static value given by the magnetic field and magnetostatics, div $B = 0$ and curl $H = 0$, can be applied. Global incompressibility, div $\upsilon = 0$, and incompressibility of the gel network, $u_{kk} = 0$, is employed and only the shear elastic modulus $\mu_2$ and the shear viscosity $\upsilon_2$ enter the stress tensor.

Neglecting the thermal degree of freedom, magnetostriction, and taking the simplest form for the dynamics of the elasticity we are left with the linearized dynamic equations

$$\frac{\partial}{\partial t}\rho\upsilon_i + \nabla_j\sigma_{ij} = -\rho g\delta_{iz} \tag{18.42}$$

$$\frac{\partial}{\partial t}\epsilon_{ij} - \frac{1}{2}(\nabla_i\upsilon_j + \nabla_i\upsilon_j) = 0. \tag{18.43}$$

where the gravitational force ($\sim g$) is acting along the negative $z$-axis.

We model our system by an originally flat surface $z = 0$ dividing the magnetic gel ($z < 0$) from vacuum ($z > 0$), where the applied external field ($B^{\text{vac}} = B_0 e_z = H^{\text{vac}}$) is normal to the surface.

At the free surface we need boundary conditions for our dynamic variables. First, there are the magnetic ones [76], the mechanical ones guaranteeing a stress-free surface, and the (linearized) kinematic one

$$e \times H = e \times H^{\text{vac}} \quad e \cdot B = e \cdot B^{\text{vac}} \tag{18.44}$$

$$e \times \sigma \cdot e = e \times \sigma^{\text{vac}} \cdot e \tag{18.45}$$

$$e \cdot \sigma \cdot e - e \cdot \sigma^{\text{vac}} \cdot e = \rho g\xi + \sigma\,\text{div}\,e \tag{18.46}$$

$$\frac{\partial}{\partial t}\xi = \upsilon_z \tag{18.47}$$

where $\xi(x, y, t)$ describes surface displacements. The unit vector $e$ is the surface normal, $e = \nabla(z - \xi)/|\nabla(z - \xi)|$, and div $e$ is twice the mean curvature. The vacuum stresses (superscript vac) are solely due to the magnetic fields (vacuum Maxwell stress tensor). The normal stress difference between the magnetic gel and the vacuum is balanced by gravity and the Laplace stress due to curvature of the surface and the surface tension $\sigma$.

The system of equations and boundary conditions (42–47) always has the trivial solution (ground state), where the surface is flat ($\xi = 0, e_0 = e_z$), flow and deformations are absent ($\upsilon = 0, u_{ij} = 0$), and the fields are constant ($M_0 = M_0 e_z$ with $M_0 = (1 - 1/\mu)B_0$, where $\mu$ is magnetic permeability).

We now allow for periodic surface undulations $\xi(x, y, t) = \hat{\xi} \exp(-ik_x x - ik_y y + i\omega t)$ with frequency $\omega = \omega_0 - i\lambda$ ($\omega_0$ and $\lambda$ real) and wave vector $\mathbf{k} = (k_x, k_y, 0)$ describing propagating and damped surface waves. For $\omega = 0$ a stationary spatial pattern is obtained. Generally $\omega$ is a complex function of $\mathbf{k}$. Deviations from the ground state of all the other variables have to be proportional to $\xi(x, y, t)$ and in a linear theory the amplitude $\hat{\xi}$ is undetermined.

To have a nontrivial solution for the resulting set of linear algebraic equations, the determinant of coefficients must vanish. This leads to the dispersion relation of surface waves for ferrogels [48]

$$\rho\omega^2 (2\tilde{\mu}_2(\omega)k^2 - \rho\omega^2) + \rho\omega^2 \left( \sigma k^3 + \rho g k + 2\tilde{\mu}_2(\omega)k^2 - \frac{\mu}{1+\mu} M_0^2 k^2 \right)$$

$$- 4\tilde{\mu}_2^2(\omega)k^4 \left[ 1 - \left( 1 - \frac{\rho\omega^2}{\tilde{\mu}_2(\omega)k^2} \right)^{1/2} \right] = 0 \qquad (18.48)$$

with the frequency dependent $\tilde{\mu}_2(\omega) \equiv \mu_2 + i\omega\nu_2$. In the absence of an external magnetic field ($M_0 = 0$) Equation 18.48 reduces to the dispersion relation for nonmagnetic gels [74]. It also contains, as a special case, the surface wave dispersion relation for ferrofluids (in an external field) by choosing $\tilde{\mu}_2 = i\omega\nu_2$. It can be generalized to viscoelastic ferrofluids, whose elasticity relaxes on a time scale $\tau^{-1}$, by replacing $\mu_2$ with $i\omega\tau\mu_2/(1 + i\omega\tau)$ [74], and to magnetorheological fluids by allowing $\mu_2, \nu_2$, and $\tau$ to be functions of the external field.

### 18.4.3.2   Rosensweig Instability

The effect of an external magnetic field perpendicular to the surface is a destabilizing one [75]. From Equation 18.48 it is evident that an external field leads to an effective reduction of the surface stiffness (provided by surface tension, gravity or elasticity) and decreases the frequency (squared) of the propagating waves in all regimes by $\sim M_0^2 k^2$. If the field is large enough, this reduction is the dominant effect and can lead to $\omega = 0$ and thus, to the breakdown of propagating waves. Indeed, Equation 18.48 can be slightly reinterpreted: It is an equation for that external field strength (or $M_0$), where a surface disturbance with wave vector $k$ and frequency $\omega_0$ relaxes to zero or grows exponentially for $\lambda$ negative or positive, respectively. For $\lambda = 0$ such a surface disturbance is marginally stable (or unstable) against infinitesimal disturbances. The function $M_0$ still depends on $\omega_0$ and $k$ and has to be minimized with respect to the latter quantity in order to get the true instability threshold.

For $\omega_0 = 0$ (stationary instability) the linear threshold condition is completely independent of $\nu_2$ and simplifies to [48]

$$M_0^2 = \frac{1+\mu}{\mu} \left( \sigma k + \frac{\rho g}{k} + 2\mu_2 \right). \qquad (18.49)$$

Minimizing with respect to $k$ leads to the critical wave vector $k_c = \sqrt{\rho g/\sigma}$ and the critical field $M_c^2 = (2/\mu)(1 + \mu)(\sqrt{\sigma\rho g} + \mu_2)$. Obviously, $k_c$ is identical to that in ferrofluids and not dependent on elasticity, but the critical field is enhanced by elasticity. The latter finding is no surprise, since elasticity increases the surface stiffness. For typical polymer gels with a shear elastic modulus of 1 kPa, the elastic contribution to $M_c$ exceeds the surface tension contribution roughly by a factor of 5 and elasticity is the dominant factor. Critical values of 100–200 Gauss for $M_0$ have to be expected for typical samples. The critical wavelength is in the range of 1 cm, which for surface waves lies in the elasticity dominated regime. Thus, if the system cannot choose the optimal (critical) wavelength, but is fixed to a prescribed one like in many surface wave-scattering experiments, the field necessary to destabilize the surface wave is about $M_c$ in the elastic regime and higher in the other ones, where $k > k_c$ or $k < k_c$. For very soft gels with $\mu_2 < 10$ Pa, the influence of the elasticity is rather negligible and ordinary ferrofluid behavior is found.

A linear theory can neither determine the actual spike pattern, nor the true nature of the instability (forward, backward, etc.). The standard weakly nonlinear (amplitude expansion) theory that provides suitable amplitude equations, by which these questions can be answered, is more complicated in the present situation due to two problems. First the driving force of the instability is manifest in the boundary conditions, but not in the bulk equations, and second the surface profile (the location where the boundary conditions have to be applied) changes with the order of the amplitude expansion. Thus, for ferrofluids a different path has been chosen [77,78]. Neglecting the viscosity (and all other dynamic processes) from the beginning, the system is Hamiltonian and its stability governed by a free energy, more precisely by the surface free energy, since the magnetic destabilization acts at the surface. We generalize this approach to (isotropic) ferrogels by taking into account in addition the elastic free energy [49]. The results have to be taken with the caveat that neglecting the viscosity is justified at the (linear) instability threshold, but is an unproven assumption for the nonlinear domain and for the pattern formation and selection processes.

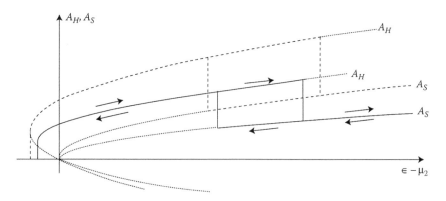

**FIGURE 18.15**   Qualitative sketch (not to scale) of the evolution of the amplitudes for squares ($A_S$) and hexagons ($A_H$).The dashed lines correspond to the case of a ferrofluid while the solid lines qualitatively describe the behavior for ferrogels with finite shear modulus $\mu_2$. The dotted lines represent the unstable branches. (Adapted from Bohlius, S., Pleiner, H., and Brand, H.R., *J. Phys. Cond. Matter*, 18, S2671, 2006.)

After some rather involved algebraic calculations the stability diagram for a square and a hexagon spike pattern is obtained (Figure 18.15) where the amplitudes are shown as function of the reduced external field strength ($\epsilon = \mu_2$ constitutes the linear threshold field). We note a decrease in size of the hysteretic region (for negative $\epsilon - \mu_2$) with increasing shear modulus. The second hysteretic region for the transition between squares and hexagons also shrinks with increasing $\mu_2$. This result should be experimentally detectable, at least qualitatively. This picture has recently been corroborated by deriving the amplitude equation for this problem [79].

## 18.5 A COMPARISON WITH THE PHYSICAL PROPERTIES OF LIQUID CRYSTALLINE GELS AND ELASTOMERS

Most of the other chapters of this book deal with the preparation, the characterization, and the physical properties of liquid crystalline gels and LCEs (*cf.* [80] for an early review covering the material until about 1998). Therefore, we briefly discuss here similarities and differences of these systems compared to the magnetic gels studied in this chapter.

First of all, one notices a big difference in the width of the hydrodynamic regime for water-based gels like PVA gels. For example, from the completely flat plateau value of $G'$ shown in the inset of Figure 18.13 we conclude that the hydrodynamic regime for this class of magnetic gels covers at least the frequency range from 100 mHz to 1 kHz that is four orders of magnitude. We note that the upper bound for the frequency of the hydrodynamic regime in magnetic gels is actually not known as yet, since the technical limitation of the piezorheometer used for these experiments set an upper limit of about 2 kHz. However, clearly the upper bound must be lower than that for a water- or acetone-based magnetic fluid, which is typically about 2 MHz. In contrast to PVA gels, PDMS gels have a much reduced hydrodynamic regime and show a cross-over to nonhydrodynamic behavior around 100 Hz. This can be traced back to the viscoelastic effects of the PDMS gels.

These observations should be compared with our knowledge about the hydrodynamic regime in the fields of liquid crystalline gels and elastomers. There we have shown (*cf.*, e.g., Ref. [57], that the cross-over from the hydrodynamic regime to the regime where more microscopic, nonhydrodynamic effects become important, takes place at about 10–100 Hz for a nematic elastomer as is evidenced by the frequency dependence of $G'$ and $G''$. We had earlier pointed out [53, 54], that for smectic A elastomers this cross-over arises for even lower frequencies ($\sim 1-10$ Hz), thus cutting down even further in frequency, the applicability of the hydrodynamic approach in these layered systems. In Refs. [53,54] this further reduction of the hydrodynamic regime has been linked to the formation of giant clusters involving the mesogenic side chains organized in layers.

From this comparison it emerges that magnetic gels have a comparable hydrodynamic range to LCEs, when they are based on typical polymers (PDMS), while there is a much larger hydrodynamic regime than in most LCEs when they are water-based like magnetic gels on PVA basis. We note that from the dynamic investigations of magnetic gels as well of LCEs and gels it emerges that piezorheometry is an ideally suited tool to probe the dynamic mechanical properties of both types of systems.

Another important issue, which has not yet been addressed for magnetic gels, but has turned out to be very important for LCEs is the question to what extent the preferred direction (the director in LCEs and the direction of the frozen-in magnetization in anisotropic magnetic gels) can act as an independent macroscopic variable.

In LCEs this topic is closely related to the importance of relative rotations between the director associated with nematic order and the rotations of the polymer network [66,69]. In nematic LCEs, the consequences of relative rotations for the onset of instabilities as well as for the reorientation behavior under an external force such as shear has been studied in quite some detail—compare, for example, Refs. [2,57,81,82].

For anisotropic magnetic gels the study of the question whether the direction of the frozen-in magnetization is a macroscopic variable is a challenge for future experimental investigations. In Sections 18.3 and 18.4 on the macroscopic properties and novel cross-coupling effects we have outlined several possibilities to couple variations of the preferred direction in magnetic gels. From the experimental investigations carried out so far there appear to be several features, which can possibly influence this question. As discussed in Section 18.2, the experimental results obtained under static external magnetic fields depend sensitively on several factors including the use of iron versus magnetite particles, on the concentration of magnetic particles, for example $\sim 2\%$ versus $\sim 30\%$, and also on the question whether one uses an aqueous or a polymeric "solvent." As a conclusion, we stress that the study of the interaction between the magnetic particles and the network clearly needs further investigations statically as well as dynamically.

One of the most remarkable features of nematic side chain LCEs is a plateau-type behavior of the stress–strain curve when a liquid single crystal elastomer (LSCE) is subjected to a mechanical stress perpendicular to the uniform director orientation of the LSCE [83]. Therefore, the question arises whether all or any of the anisotropic magnetic gels produced so far also show a plateau-like behavior in the stress–strain curve under analogous conditions as LSCEs.

In this connection it might help to summarize some of the key issues raised and discussed so far concerning the plateau observed for LSCEs. Different models have been put forward for LSCEs by the group of Warner and Terentjev. In their first model nematic LCEs are an anisotropic rubber and the director is not a macroscopic degree of freedom [84]. Here, the plateau in the stress-strain curve is interpreted to be the evidence for "soft elasticity," for which there is no other experimental evidence as yet. The concept of soft elasticity goes back originally to Golubovic and Lubensky [85], who developed it for a novel type of solids (not yet found experimentally). In such a solid one starts from an isotropic phase without an external field [85] and obtains spontaneously a preferred direction. Clearly this concept does not apply to the LSCEs studied experimentally by Küpfer and Finkelmann, since in their case the preferred direction is not at all spontaneous, but rather generated by the cross-linking carried out under an external mechanical stress. It has been shown recently [86], that one can obtain a plateau in the stress–strain curve without invoking the concept of soft elasticity. Thus, one arrives at the conclusion that there is no experimental evidence for "soft elasticity" in the LSCEs as they exist today.

Later on Terentjev and Warner acknowledged that due to the nonspontaneous preferred direction a modification of the concept of "soft elasticity" was necessary; they called it "semisoft elasticity." They state that the plateau observed in the stress–strain curve is related closely to "semisoft elasticity", since it exhibits the vanishing of the relevant effective elastic coefficient exactly at the beginning and end of the plateau. Such a soft mode behavior (resembling that at a second-order phase transition) is in marked contrast to the original "soft elasticity" picture featuring a Nambu-Goldstone scenario, where the relevant linear elastic coefficient is identically zero by symmetry. Nevertheless, soft mode behavior is not a signal of "semi-soft elasticity," but can also be obtained by conventional nonlinear elasticity in combination with nonlinear relative relations [87]. Finally, we point out that the experimental situation regarding the soft mode behavior is still controversial and not yet settled.

## ACKNOWLEDGMENTS

Helmut R. Brand acknowledges the Deutsche Forschungsgemeinschaft for partial support of his work through the Forschergruppe FOR 608 "Nichtlineare Dynamik komplexer Kontinua." Philippe Martinoty thanks the Federation de Recherche CNRS 2863 J. Villermaux for partial support. Both of us (Helmut R. Brand and Philippe Martinoty) acknowledge partial support of their work by PROCOPE 312/proms through the Deutscher Akademischer Austauschdienst and by PAI 02951XJ through the Ministère des Affaires Etrangères.

## REFERENCES

1. Finkelmann, H., Kock, H.J., and Rehage, H., Investigations of liquid crystalline polysiloxanes: 3. Liquid crystalline elastomers—a new type of liquid crystalline material, *Makromol. Chem. Rapid Commun.*, 2, 317, 1981.
2. Brand, H.R., Pleiner, H., and Martinoty, P., Selected macroscopic properties of liquid crystalline elastomers, *Soft Matter*, 2, 182, 2006.
3. Zrinyi, M., Barsi, L., and Büki, A., Deformation of ferrogels induced by nonuniform magnetic fields, *J. Chem. Phys.*, 104, 8750, 1996.
4. Barsi, L., Büki, A., Szabo, D., and Zrinyi, M., Gels with magnetic properties, *Progr. Colloid Polym. Sci.*, 102, 57, 1996.
5. Zrinyi, M., Magnetic field sensitive polymer gels, *Trends Polymer Sci.*, 5, 280, 1997.
6. Zrinyi, M., Barsi, L., and Büki, Ferrogel: A new magneto-controlled elastic medium, *Polymer Gels Networks*, 5, 415, 1997.
7. Zrinyi, M., Barsi, L., Szabo, D., and Kilian, H.-G., Direct observation of abrupt shape transition in ferrogels induced by nonuniform magnetic field *J. Chem. Phys.*, 106, 5685, 1997.
8. Szabo, D., Szeghy, G., and Zrinyi, M., Shape transtion of magnetic field sensitive polymer gels, *Macromolecules*, 31, 6541, 1998.
9. Zrinyi, M., Szabo, D., and Kilian, H.-G., Kinetics of the shape change of magnetic field sensitive polymer gels, *Polymer Gels Networks*, 6, 441, 1998.
10. Mitsumata, T., Ikeda, K., Gong, J.P., Osada, Y., Szabo, D., and Zrinyi, M., Magnetism and compressive modulus of magnetic fluid containing gels, *J. Appl. Phys.*, 85, 8451, 1999.

11. Li, S., John, V.T., Irvin, G.C., Rachakonda, S.H., McPherson, G.L., and O'Connor, C.J., Synthesis and magnetic properties of a novel ferrite organogel, *J. Appl. Phys.*, 85, 5965, 1999.

12. Lebedev, V.T., Torok, G., Cser, L., Buyanov, A.L., Revelskaya, L.G., Orlova, D.N., and Sibilev, A.I., Magnetic phase ordering in ferrogels under applied field, *J. Magn. Magn. Mater.*, 201, 136, 1999.

13. Zrinyi, M., Intelligent polymer gels controlled by magnetic fields, *Colloid Polym. Sci.*, 278, 98, 2000.

14. Xulu, P.M., Filipcsei, G., and Zrinyi, M., Preparation and responsive properties of magnetically soft poly (*N*-isopropylacrylamide) gels, *Macromolecules*, 33, 1716, 2000.

15. Babincova, M., Jeszcynska, D., Sourivong, P., Cicmanec, P., and Babinec, P., Superparamagnetic gel as a novel material for electromagnetically induced hyperthermia, *J. Magn. Magn. Mater.*, 225, 109, 2001.

16. Török, G., Lebedev, V.T., Cser, L., Kali, G., and Zrinyi, M., Dynamics of PVA-gel with magnetic macrojunctions, *Physica B*, 297, 40, 2001.

17. Raikher, Y.L. and Rusakov, V.V., Linear orientational magnetodynamics of a ferrogel, *Colloid J.*, 63, 607, 2001.

18. Jarkova, E., Pleiner, H., Müller, H.-W., and Brand, H.R., Hydrodynamics of isotropic ferrogels, *Phys. Rev. E*, 68, 041706, 2003.

19. Kato, N., Sakai, Y., Kagaya, D., and Sekiya, S., Magnetically isotropic–anisotropic change of thermally responsive polymer gels, *Jpn. J. Appl. Phys.*, 42, 102–103, 2003.

20. Hernandez, R., Sarafian, A., Lopez, D., and Mijangos, C., Viscoelastic properties of poly(vinyl alcohol) hydrogels and ferrogels obtained through freezing–thawing cycles, *Polymer*, 46, 5543, 2004.

21. Narita, T., Knaebel, A., Munch, J.-P., and Zrinyi, M., Candau, S.J., Microrheology of chemically crosslinked polymer gels by using diffusing-wave spectroscopy, *Macromol. Symp.*, 207, 17, 2004.

22. Teixeira, A.V. and Licinio, P., Local deformations of ferrogels induced by uniform magnetic fields, *J. Magn. Magn. Mater.*, 289, 126, 2005.

23. Wang, G., Tian, W.J., and Huang, J.P., Response of ferrogels to an AC magnetic field, *J. Phys. Chem. B*, 110, 10738, 2006.

24. Filipcsei, G., Szilagyi, A., Csetneki, I., and Zrinyi, M., Comparative study on the collapse transition of poly (*N*-isopropylacrylamide) gels and magnetic nanoparticles loaded poly (*N*-isopropylacrylamide) gels, *Macromol. Symp.*, 239, 130, 2006.

25. Liu, T.-Y., Hu, S.-H., Liu, T.-Y., Liu, D.-M., and Chen, S.-Y., Magnetic-sensitive behavior of intelligent ferrogels for controlled release of drug, *Langmuir*, 22, 5974, 2006.

26. Bentivegna, F., Ferré, J., Nyvlt, M., Jamet, J.P., Imhoff, D., Canva, M., Brun, A., Veillet, P., Visnovsky, S., Chaput, F., and Boilot, J.P., Magnetically textured $\gamma$–$Fe_2O_3$ nanoparticles in a silica gel matrix: Structural and magnetic properties, *J. Appl. Phys.*, 83, 7776, 1998.

27. Bentivegna, F., Nyvlt, M., F., Ferré, J., Jamet, J.P., Brun, A., Visnovsky, S., and Urban, R., Magnetically textured $\gamma$-$Fe_2O_3$ nanoparticles in a silica gel matrix: Optical and magneto-optical properties, *J. Appl. Phys.*, 85, 2270, 1999.

28. Uritani, M. and Hamada, A., A simple and inexpensive dot-blotter for immunoblotting, *Biochem. Education*, 27, 169, 1999.

29. Mitsumata, T., Furukawa, K., Juliac, E., Iwakura, K., and Koyama, K., Compressive modulus of ferrite containing polymer gels, *Int. J. Mod. Phys. B*, 16, 2419, 2002.

30. Mitsumata, T., Juliac, E., Furukawa, K., Iwakura, K., Taniguchi, T., and Koyoma, K., Anisotropy in longitudinal modulus of polymer gels containing ferrites, *Macromol. Rapid Commun.*, 23, 175, 2002.

31. Wilson, M.J., Fuchs, A., and Gordaninejad, F., Development and characterization of magnetorheological polymer gels, *J. Appl. Polym. Sci.*, 84, 2733, 2002.

32. Horvath, A.T., Klingenberg, D.J., and Shkel, Y.M., Determination of rheological and magnetic properties for magnetorheological composites via shear magnetization measurements, *Int. J. Mod. Phys. B*, 16, 2690, 2002.

33. Ginder, J.M., Clark, S.M., Schlotter, W.F., and Nichols, M.E., Magnetostrictive phenomena in magnetorheological elastomers, *Int. J. Mod. Phys. B*, 16, 2412, 2002.

34. Schlotter, W.F., Cionca, C., Paruchuri, S.S., Cunningham, J.B., Dufresne, E., Dierker, S.B., Arms, D., Clarke, R., Ginder, J.M. and Nichols, M.E., The dynamics of magnetorheological elastomers studied by synchrotron radiation speckle analysis, *Int. J. Mod. Phys. B*, 16, 2426, 2002.

35. Bellan, C. and Bossis, G., Field dependence of viscoelastic properties of MR elastomers, *Int. J. Mod. Phys. B*, 16, 2447, 2002.

36. Szabo, D. and Zrinyi, M., Nonhomogeneous, non-linear deformation of polymer gels swollen with magneto-rheological suspensions, *Int. J. Mod. Phys. B*, 16, 2616, 2002.

37. Davis, L.C., Model of magnetorheological elastomers, *J. Appl. Phys.*, 85, 3348, 1999.

38. Collin, D., Auernhammer, G.K., Gavat, O., Martinoty, P., and Brand, H.R., Frozen-in magnetic order in uniaxial magnetic gels: Preparation and physical properties, *Macromol. Rapid Commun.*, 24, 737, 2003.

39. Varga, Z., Feher, J., Filipcsei, G., and Zrinyi, M., Smart nanocomposite polymer gels, *Macromol. Symp.*, 200, 93, 2003.

40. Varga, Z., Filipcsei, G., Szilagyi, A., and Zrinyi, M., Electric and magnetic field-structured smart composites, *Macromol. Symp.*, 227, 123, 2005.

41. Varga, Z., Filipcsei, G., and Zrinyi, M., Smart composites with controlled anisotropy, *Polymer*, 46, 7779, 2005.

42. Varga, Z., Filipcsei, G., and Zrinyi, M., Magnetic field sensitive functional elastomers with tunable elastic modulus, *Polymer*, 47, 227, 2006.

43. Csetneki, I., Filipcsei, G., and Zrinyi, M., Smart nanocomposite polymer membranes with on/off switching control, *Macromolecules*, 39, 1939, 2006.

44. Hajsz, T., Csetneki, I., Filipcsei, G., Zrinyi, M., Swelling kinetics of anisotropic filler loaded PDMS networks, *Phys. Chem. Chem. Phys.*, 8, 977, 2006.

45. Auernhammer, G.K., Collin, D., and Martinoty, P., Viscoelasticity of suspensions of magnetic particles in a polymer: Effect of confinement and external field, *J. Chem. Phys.*, 124, 204907, 2006.

46. Lattermann, G. and Krekhova, M., Thermoreversible ferrogels, *Macromol. Rapid Commun.*, 27, 1373, 2006; 1968, 2006.

47. Bohlius, S., Brand, H.R., and Pleiner, H., Macroscopic dynamics of uniaxial magnetic gels, *Phys. Rev. E*, 70, 061411, 2004.

48. Bohlius, S., Brand, H.R., and Pleiner H., Surface waves and Rosensweig instability in isotropic ferrogels, *Z. Physik. Chemie*, 220, 97, 2006.

49. Bohlius, S., Pleiner, H., and Brand, H.R., Pattern formation in ferrogels: Analysis of the Rosensweig instability using the energy method, *J. Phys. Cond. Matter*, 18, S2671, 2006.

50. Pleiner, H. and Brand, H.R., Hydrodynamics and electrohydrodynamics of nematic liquid crystals, in *Pattern Formation in Liquid Crystals*, A. Buka and L. Kramer (Eds.), Springer, New York, pp. 15 ff 1996.

51. Treloar, L.R.G., *The Physics of Rubber Elasticity*, Oxford, Clarendon Press, 1949.

52. Gallani, J.-L., Hilliou, L., Martinoty, P., Doublet, F., and Mauzac, M., Mechanical behavior of side-chain liquid crystalline networks, *J. Phys. II (France)*, 6, 443, 1996.

53. Weilepp, J., Stein, P., Aßfalg, N., Finkelmann, H., Martinoty, P., and Brand, H.R., Rheological properties of mono-and polydomain liquid crystalline elastomers exhibiting a broad smectic-A phase, *Europhys. Lett.*, 47, 508, 1999.

54. Weilepp, J., Zanna, J.J., Aßfalg, N., Stein, P., Hilliou, L., Mauzac, M., Finkelmann, H., Brand, H.R., and Martinoty, P., Rheology of liquid crystalline elastomers in their isotropic and smectic-A state, *Macromolecules*, 32, 4566, 1999.

55. Stein, P., Aßfalg, N., Finkelmann, H., Martinoty, P., Shear modulus of polydomain, mono-domain and non-mesomorphic side-chain elastomers: Influence of the nematic order, *Eur. Phys. J. E*, 4, 255, 2001.

56. Zanna, J.J., Stein, P., Marty, J.D., Mauzac, M., and Martinoty, P., Influence of molecular parameters on the elastic and viscoelastic properties of side-chain liquid crystalline elastomers, *Macromolecules*, 35, 5459, 2003.

57. Martinoty, P., Stein, P., Finkelmann, H., Pleiner, H., and Brand, H.R., Mechanical properties of monodomain side-chain nematic elastomers, *Eur. Phys. J. E*, 14, 311, 329, and 339, 2004.

58. Rogez, D., Francius, F., Finkelmann, H., and Martinoty, P., Shear mechanical anisotropy of side chain liquid-crystal elastomers: Influence of sample preparation, *Eur. Phys. J. E*, 20, 369, 2006.

59. Rogez, D., Brandt, H., Finkelmann, H., and Martinoty, P., Shear mechanical properties of main chain liquid crystalline elastomers, *Macromol. Chem. Phys.*, 207, 735, 2006.

60. Martinoty, P., Hilliou, L., Mauzac, M., Benguigui, L., and Collin, D., Side-chain liquid-crystal polymers: Gel-like behavior below their gelation points, *Macromolecules*, 32, 1746, 1999.

61. Collin, D. and Martinoty, P., Dynamic macroscopic heterogeneities in a flexible linear polymer melt, *Physica A*, 320, 235, 2003.

62. Collin, D. and Martinoty, P., Commentary on Solid-like rheological response of non-entangled polymers in the molten state by H. Mendil et al., *Eur. Phys. J. E*, 19, 87–98, 2006.

63. Collin, D., Lavalle, P., Garza, J.M., Voegel, J.C., Schaaf, P., and Martinoty, P., Mechanical properties of cross-linked hyaluronic acid/poly-(L-lysine) multilayer films, *Macromolecules*, 37, 10195, 2004.

64. Temmen, H., Pleiner, H., Liu, M., and Brand, H.R., Convective nonlinearity in non-Newtonian fluids, *Phys. Rev. Lett.* 84, 3228, 2000.

65. Landau, L.D. and Lifshitz, I.M., *Electrodynamics of Continuous Media* (Addison-Wesley, Reading, MA, 1960).

66. De Gennes, P.G., Weak nematic gels, pp. 231ff in *Liquid Crystals of One- and Two-Dimensional Order*, Helfrich, W., Heppke, G., (Eds.), Springer, New York, NY, 1980.

67. Menzel, A., Pleiner, H., and Brand, H.R., Nonlinear relative rotations in liquid crystalline elastomers, *J. Chem. Phys.*, 126, 234901, 2007.

68. Graham, R., Hydrodynamics of He-3 in anisotropic A phase, *Phys. Rev. Lett.* 33, 1431, 1974.

69. Brand, H.R. and Pleiner, H., Electrohydrodynamics of nematic liquid crystalline elastomers, *Physica A*, 208, 359, 1994.

70. Jarkova, E., Pleiner, H., Müller, H.-W., and Brand, H.R., Macroscopic dynamics of ferronematics, *J. Chem. Phys.* 118, 2422, 2003.

71. Ryskin, A., Müller, H.-W., and Pleiner, H., Hydrodynamic instabilities in ferronematics, *Eur. Phys. J. E* 11, 389, 2003.

72. du Trémolet de Lacheisserie, E., *Magnetostriction—Theory and Applications of Magnetoelasticity* (CRC Press, Boca Raton, FL, 1993).

73. Nikitin, L.V., Mironova, L.S., Stepanov, G.V., and Samus, A.N., The influence of a magnetic field on the elastic and viscous properties of magnetoelastics, *Polym. Sci., Ser. A Ser. B* 43, 443, 2001.

74. Harden, J.L., Pleiner, H., and Pincus, P.A., Hydrodynamic surface modes on concentrated polymer solutions and gels, *J. Chem. Phys.* 94, 5208, 1991.

75. Rosensweig, R.E., *Ferrohydrodynamics*, Cambridge University Press, Cambridge, 1985.

76. Jackson, J.D., *Classical Electrodynamics*, 2nd edn., John Wiley & Sons, Inc., New York, NY, 1999.

77. Gailitis, A., Formation of hexagonal pattern on surface of a ferromagnetic fluid in an applied magnetic field, *J. Fluid Mech.* 82, 401, 1977.

78. Friedrichs, R. and Engel, A., Pattern and wavenumber selection in magnetic fluids, *Phys. Rev. E*, 64, 021406, 2001.

79. Bohlius, S., Brand, H.R., and Pleiner, H. Amplitude equation for the Rosensweig instability, *Progr. Theor. Phys. Suppl.* 175, 27, 2008.

80. Brand, H.R. and Finkelmann, H., Physical Properties of liquid crystalline elastomers, in *Handbook of Liquid Crystals*, J.W. Goodby et al (Eds.), Wiley, New York, NY, Vol.3, pp. 277ff (1998).

81. Mueller, O. and Brand, H.R., Undulation versus Frederiks instability in nematic elastomers in an external electric field, *Eur. Phys. J. E*, 17, 53, 2005.

82. Weilepp, J., and Brand, H.R., Director reorientation in nematic liquid single crystal elastomers, *Europhys. Lett*, 34, 495, 1996; *Europhys. Lett*, 37, 499, 1997.

83. Küpfer, J. and Finkelmann, H., Liquid crystal elastomers—influence of the orientational distribution of the cross-links on the phase behavior and reorientation processes, *Macromol. Chem. Phys.*, 195, 1353, 1994.

84. Warner, M. and Terentjev, E.M., *Liquid Crystal Elastomers*, Oxford University Press, Oxford, 2003.

85. Golubovic, L. and Lubensky, T.C., Nonlinear elasticity of amorphous solids, *Phys. Rev. Lett.*, 63, 1082, 1989.

86. Menzel, A.M., Pleiner, H., and Brand, H.R. On the nonlinear stress-strain behavior of nematic elastomers—materials of two coupled preferred directions, *J. Appl. Phys.*, 105, 013503, 2009.

87. Menzel, A.M., Pleiner, H., and Brand, H.R., Response of prestretched nematic elastomers to external fields, *Eur. Phys. J. E* 30, 371, 2009.

# 19 Side-On Nematic Liquid-Crystalline Elastomers for Artificial Muscle Applications

*Min-Hui Li and Patrick Keller*

## CONTENTS

## 19.1 INTRODUCTION

Artificial muscles are man-made materials that try to reproduce the two main characteristics of real muscle fibers, namely, elasticity and contractility. They respond to various external stimulations (ion concentration, electric field, temperature, light etc.) by a significant shape or size change. In addition to classical materials such as piezoelectric ceramics and shape memory alloys, polymer-based artificial muscles have become the most important muscle-like materials since the 1990s [1]. They offer operational similarity to biological muscles in response to external stimulation. They are resilient and damage tolerant and they exhibit large actuation strains (stretching, contraction, or bending). Many new polymer materials/approaches are under active investigation, including ferroelectric polymers, dielectric elastomers, liquid crystal (LC) elastomers and gels, ionic polymer gels, ionomeric polymer–metal composites,

565

conductive polymers, and carbon nanotubes [1 and references therein]. In addition to
the obvious attractiveness of such studies in basic science, artificial muscle systems
have many potential applications of great interest, including serving as the materials
foundation for fabrication of sensors, microrobots, micropumps, and actuators with
combinations of size, weight, and performance parameters beyond those currently
achievable. One remarkable advantage of these systems is that they offer the possibility of driving micromachines and nanomachines without the aide of motors and
gears.

In this chapter, we will discuss one particular kind of polymer-based artificial
muscles, the side-on LC elastomers. LC elastomers have been the subject of a number
of reviews [2–4]. They combine the properties of LC systems with orientational
ordering and those of polymer networks with rubbery elasticity [5]. LC elastomers as
artificial muscles were first proposed by de Gennes [6–7]. They can be divided into
two categories, depending on the LC phases exhibited by the systems, namely, nematic
LC elastomers [8–9] and smectic LC elastomers [10–12]. The actuation mechanism
is rather different for these two LC elastomers. We will focus here on the nematic
LC elastomers. The chapter is organized as follows. We will first describe briefly the
actuation mechanism (Section 19.2) and then present a series of self-assembled LC
elastomers developed in our group (Section 19.3). We will try to illustrate the bottom-
up design of stimuli responsive materials in which the overall material response
reflects the individual macromolecular response, using LC polymer as building block.
This approach is particularly suitable for the development of micro- or nano-sized
artificial muscles.

## 19.2   ACTUATION MECHANISM IN NEMATIC LC ELASTOMERS

LCs are typically fluids made of relatively stiff rod molecules with long-range orientational order. The simplest ordered phase is the nematic mesophase in which the
mean ordering direction of the rods is uniform [13]. Long polymer chains, which
incorporate rigid rod-like units called mesogens, can also order nematically and thus
form nematic polymers (Figure 19.1). In those nematic polymers, the conformation
of the macromolecular backbone is coupled with the orientational nematic order
[14–15]. This coupling is the strongest in the main-chain nematic polymer, where
mesogens are directly incorporated into the polymer backbone (Figure 19.1a), and
the weakest in end-on type of side-chain LC polymers where it is almost zero [15]
(Figure 19.1c). The polymer chains elongate when their mesogens orient in the
nematic phase [14,16,17], while in the isotropic phase, they recover a random coil

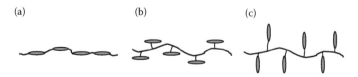

**FIGURE 19.1**   Main-chain (a) and side-chain [(b) side-on and (c) end-on] LC polymers. The
small elongated ellipsoid represents the rod-like mesogen.

conformation, driven by entropy [18]. A change of the average molecular shape is thus introduced, from elongated to spherical, when the nematic–isotropic phase transition takes place (Figure 19.2a). When the polymer chains are lightly linked together, a new kind of rubber, a LC elastomer, is obtained. It shows elasticity similar to that of conventional elastomers but with some specific properties [5]. If the LC elastomer is judiciously prepared so that all mesogens are oriented uniformly in the whole sample and thus, by coupling, all chains are elongated in the same direction, as a monodomain (called by Finkelmann as the liquid single crystal elastomer [19]), the individual polymer chain shape changes will then be translated to a macroscopic shape change of the elastomer sample at the nematic–isotropic transition (Figure 19.2b). De Gennes first proposed to use this principle to prepare artificial muscles [6,7,20]. Theoretical analysis of the dynamical and thermodynamical properties of the nematic LC elastomers predicted that nematic LC elastomers would contract at the nematic–isotropic transition similarly as real muscles. These theoretical works stimulated a large interest in artificial muscles based on LC elastomers. The uniqueness of the LCs lies in their responsiveness to various external stimuli such as temperature, light, electric field, magnetic field, and so on. Therefore, in principle, different stimuli-responsive artificial muscles could be prepared with LC elastomers. Thermo-responsive [8,9], photo-responsive [21–23], and electroresponsive [24–27] muscle-like materials have been reported. Generally, any polymer chain shape change

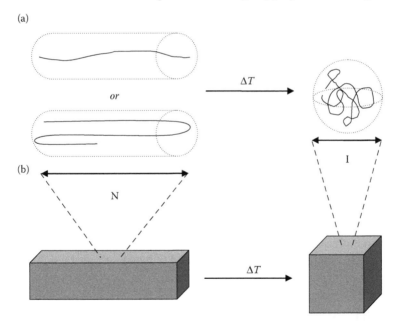

**FIGURE 19.2** (a) Conformations of main-chain LC polymers in the nematic (N) and isotropic (I) phases. In the nematic phase, depending on the chain molecular weight, two possible chain conformations (linear conformation and hairpin conformation) were predicted theoretically [14] and then observed experimentally by neutron scattering [17]. In the isotropic phase, a random coil (Gaussian-like) conformation was observed [18]. (b) Macroscopic shape change of the monodomain sample of main-chain LC elastomer at nematic–isotropic transition.

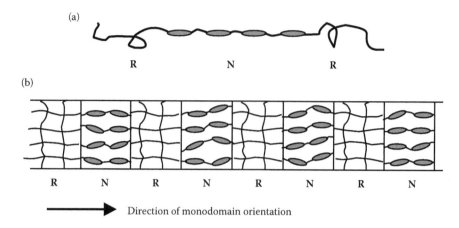

Direction of monodomain orientation

**FIGURE 19.3**   A striated artificial muscle based on a triblock copolymer **RNR** (a) in a lamellar phase with suitable cross-linking of the elastomer part **R** (b) [7].

induced by the disorganization or reorganization of LC mesogens in response to external stimulation can be used to produce muscle-like contraction in a monodomain sample. The response of the mesogens can result in the nematic-to-isotropic phase transition or the reorientation of the mesogen direction.

The examples of artificial muscles given above have uniform microstructure, where the cross-linking bridges are randomly distributed in the whole monodomain sample. De Gennes [7], additionally, proposed a composite structure based on the lamellar phase of a triblock copolymer **RNR** (**R** = classical elastomer, **N** = nematic polymer) (Figure 19.3). The cross-linking is only present in **R** parts. This striated structure should be mechanically more robust and the monodomain nematic ordering could be better preserved during the contraction/elongation cycles. From the biomimetic point of view, this structure is also more interesting, since biological structures are active and organized. This triblock composite structure mimics not only muscle function (contraction/elongation) but also, to a certain extent, the hierarchical and striated structure of real muscle [28]. It is also reminiscent of the functioning of primitive muscles for which contraction was related to a conformational change of proteins induced by calcium ions [29–31].

Motivated by de Gennes's artificial muscle models, our group since 1997 had started to develop artificial muscles with lamellar structure [32–34], as well as thermoresponsive and photoresponsive artificial muscles based on nematic polymers with random cross linking [9,22,35].

## 19.3   SELF-ASSEMBLED LC ELASTOMERS AS ARTIFICIAL MUSCLES

Three key characteristics are at the heart of the triblock model proposed by de Gennes [7]: lamellar phase, monodomain, and cross linking. The triblock copolymers **RNR** (Figure 19.3) should first self-assemble into a lamellar phase; the lamellar phase as well as the nematic mesophase in the sublayer N should be aligned as a monodomain; and finally the whole molecular organization should be "frozen" by cross-linking the flexible sublayers **R**. In his theoretical paper, de Gennes suggested to use a main-chain nematic polymer (Figure 19.1a) for the **N** block, since the order conformation coupling is the strongest in this case [15]. However, for practical synthesis considerations outlined below, we used instead a side-on nematic polymer (Figure 19.1b). The synthesis of block copolymers with well-defined structures and narrow molecular-weight distributions is a crucial step in the production of artificial muscles with lamellar structure, because only block copolymers with well controlled molecular weights and narrow molecular-weight distributions of each of their components will display very regular, long-range ordered lamellar structures. Most of main-chain LC polymers are synthesized by polycondensation reactions, which give polymers with fairly large polydispersity indexes [36] which will be detrimental to a good organization. Another point is relative to the copolymerization. The central blocks **N** (LC polymer, normally first synthesized) need to be functionalized at the two chain ends, which are necessary for the chain extension in order to get the triblock copolymer **RNR**. Again, a perfect control of the end groups in polycondensation reactions is hard to achieve. In contrast, polymerization techniques employed to prepare side-chain LC polymers (Figures 19.1b and 19.1c), such as living or controlled radical polymerization, ring-opening metathesis polymerization, or anionic polymerization, allow for the synthesis of macromolecules with narrow molecular weight distributions and functionalized chain ends. Therefore, we decided to use one particular class of side-chain LC polymer, the side-on nematic polymers (Figure 19.1b), in the synthesis of the triblocks **RNR**. Side-on nematic polymers were chosen because structural studies made by neutron scattering [15,37–39] demonstrated that their conformational behavior was similar to that of main-chain LC polymers. In the nematic phase, the backbone is strongly elongated in the direction of the nematic director due to the strong coupling between the liquid crystalline order and the chain conformation. Would the side-on nematic polymer chains, when attached together, behave actually as artificial muscles at the nematic–isotropic transition? This point had to be addressed before the tedious synthesis of triblock copolymers with side-on central block had to be attempted. In Section 19.3.1, we present results obtained for simple LC elastomers based on random cross-linked side-on nematic homopolymers. Their muscles-like properties were verified by using two different stimulations, temperature and light. Then we will describe the preparation of side-on LC triblock elastomers with the three key characteristics: lamellar phase, monodomain, and cross-linking in Section 19.3.2.

## 19.3.1 ARTIFICIAL MUSCLES BASED ON SIDE-ON NEMATIC LC HOMOPOLYMERS

### 19.3.1.1 Thermo-Responsive Systems

#### 19.3.1.1.1 Macroscopic Liquid Crystalline Actuators

On the basis of our previous conformational studies by neutron scattering [37,38], side-on monomers with the molecular structure shown in Figure 19.4a were used, the corresponding side-on nematic elastomers being shown in Figure 19.4b.

As discussed in Section 19.2, in order to visualize at a macroscopic level the conformational changes occurring at a molecular level, all the macromolecules in

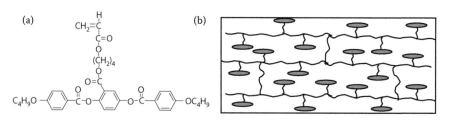

**FIGURE 19.4** (a) Chemical structure of a side-on LC monomer and (b) schematic representation of the targeted side-on LC elastomer.

the material have to be oriented parallel to each other in order to form a nematic LC
monodomain. The macromolecules must also be strongly associated (e.g., by covalent
cross linking) in order to prevent individual macromolecules from changing their
shapes and reorienting individually by sliding (a "memory" of the 3D macromolecular
organization is thus kept).

Different ways have been developed to prepare monodomain LC elastomers. The
first approach developed by Finkelmann et al. [19] uses a mechanical field. In a
three-stage process, a lightly cross-linked side-chain LC polysiloxane is firstly synthesized and then deformed with a constant load to obtain a uniform nematic director
orientation and a mechanically induced anisotropic conformation of the polymer
backbone. A second cross-linking reaction finally locks the out-of-equilibrium network anisotropy. The second approach employed by the group of Davis and Mitchell
[40] uses a magnetic field. It involves the cross-linking of a nematic polyacrylate
prealigned in a magnetic field. The third method uses the strategy of starting from LC
acrylate or methacrylate monomers, which can be completely aligned by using conventional techniques for LC alignment (such as electric, magnetic, or surface force).
The second step is a thermo- or photopolymerization, which ensures the formation of
polymer chains and cross-linking bridges at the same time [41–43]. Several works on
monodomain LC elastomers in the last decade have been done using the Finkelmann's
approach [44–47]. However, it is a complicated and a delicate process, which is well
suited only for the preparation of polysiloxane elastomers and which is not applicable
for micro-fabrication. Moreover, the level of anisotropy of the polymer chains reflects
not only the intrinsic anisotropy of the system but also the anisotropy induced by the
strain applied [6]. These disadvantages can be overcome by the two other approaches.
As we will see later, they are applicable for the preparation of micro- or nano-sized
artificial muscles. Also, monodomain LC elastomers prepared by using a magnetic
field or by photopolymerizing the aligned LC monomers, exhibit a level of chain
anisotropy, which reflects the intrinsic coupling between the mesogen units and the
polymer chains (equilibrium state).

We used the third approach [9] to prepare aligned side-on nematic elastomers.
The required alignment was obtained in the nematic phase of the monomers using
rubbed polymer alignment layers, deposited on the inner faces of a LC cell. The LC
cell was filled in the isotropic phase with a mixture composed of a nematic acrylate
monomer, a diacrylate cross-linking agent, and a photoinitiator. Upon slow cooling
to the nematic phase, the molecules aligned parallel to the surfaces along the rubbing

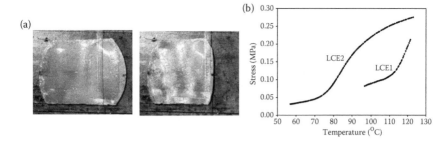

**FIGURE 19.5** Thermomechanical properties of the side-on LC elastomer. (a) Photos of a sample taken at $T=110°C$ (left) and at $T=125°C$ (right) (nematic–isotropic transition is around 120°C). (b) Iso-strain measurement (stress vs. temperature) at ~5% strain on heating the sample through the nematic to isotropic phase transition for LCE1 (see Figure 19.4) and LCE2 (Reprinted with permission from Thomsen III, D.L. et al. *Macromolecules*, 34, 5868–5875, 2001. Copyright 2001 American Chemical Society.)

direction, giving the desired uniform planar alignment. The sample was then irradiated with UV light in order to induce polymerization and cross-linking at the same time, "freezing" the orientation of the nematic phase in a nematic LC elastomer. After removal from the cell, the free-standing elastomeric material was obtained as a thin sheet (50–100 μm of thickness) with typical lateral dimensions of 1.5 cm × 3 cm.

When this sheet was heated close to its nematic to isotropic transition it started to contract along the nematic director axis (Figure 19.5). The typical contraction was around 35–45%. Preliminary mechanical characterizations were performed, which gave estimates of the generated force (around 210 kPa) and the speed of the thermo-mechanical effect (a few seconds for the contraction but a much longer time for the extension due to the low thermal conductivity of the elastomer) [9].

From this study, we showed that side-on nematic elastomers behave actually as real muscles with the stress and strain comparable to that of a skeletal muscle, the average values of the strain and stress for a skeletal muscle being 25% and 350 kPa.

At Naval Research Laboratory (Washington, DC), research continued on macro-scopic samples of these thermo-responsive side-on nematic LC elastomers. Contractile fibers were obtained [48]. Carbon-coated and blue-dye-doped LC elastomers were prepared in order to heat the sample rapidly by IR diode laser or 635 nm diode laser [49]. Mechanical properties were improved by stacking thin films of LC elastomers [50]. Tuning of the working temperature was investigated by using mixtures of side-on nematic LC monomers [51].

All these results are very promising, since we were able to prepare artificial muscles (thin films and fibers) with mechanical properties approaching those of biological muscle. However, all the systems described so far (including those by other groups) are "macroscopic," that is, their size is in the millimeter/centimeter range. For numerous potential applications involving surface responsive materials such as microfluidics or bio-inspired microrobots, micro or nanometer-sized "artificial muscles" are desirable. Our approach to make responsive nematic elastomers starts from nematic monomer mixtures, which are first aligned in a "pseudo monodomain," then photopolymerized and photocrosslinked to achieve a "freezing" of the molecular organization. This bottom up approach can easily been applied to the preparation of LC microactuators.

### 19.3.1.1.2  Liquid Crystalline Microactuators

Making use of a soft lithography technique called replica-molding [52], we have succeeded in creating micron-sized responsive pillars made of nematic LC elastomers [35] . The required soft poly(dimethylsiloxane) (PDMS)-made molds were prepared by standard photolithography techniques, which gave "soft molds" consisting of an array of holes with typical sizes of 20 $\mu$m in diameter and 100 $\mu$m in depth. Those soft molds exhibit two main properties, which are essential for the next step: their transparency to UV light and softness that facilitates mold "peeling."

The main difficulty in producing responsive pillars made of nematic LC elastomer was the creation of the nematic liquid single crystal elastomer [19]. If we wanted pillars to contract when the nematic elastomer underwent the nematic to isotropic phase transition, we must be able to align the nematic director, and thus the polymer backbone in a side-on nematic LC polymer, parallel to the long axis of the pillars (Figure 19.6). Due to our approach, such an organized structure can be obtained easily by the application, before the polymerization/cross linking step, of a suitably oriented magnetic field on the nematic mesophase by a small permanent magnet [13]. Practically, a small amount of a mixture comprised of a nematic side-on acrylate monomer, a cross-linking agent, and a photoinitiator is heated to the isotropic phase on a glass slide positioned atop a rare-earth permanent magnet of around 1 T (Figure 19.7). The soft mold is gently pressed down on the melted mixture, which fills the inner structure of the mold. The temperature is slowly decreased down to a temperature at which the mixture is in the nematic phase. During the cooling process, the magnetic field ensures the alignment of the nematic director parallel to the long axis of the pillars. The sample is then UV irradiated through the mold for a short period, and cooled to room temperature. The PDMS soft mold is peeled off, leaving a thin glassy polymer film covered by a regular array of pillars (Figure 19.8). Pillars, shaved off the surface with a razor blade, contract reversibly of around 30–40% when heated around the nematic to isotropic transition temperature of the elastomer (Figure 19.9).

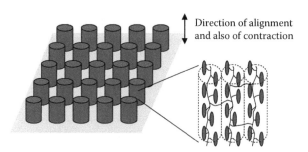

Direction of alignment and also of contraction

**FIGURE 19.6**  Schematic representation of an array of nematic elastomer-made pillars, showing the macromolecular organization in each pillar. (Reprinted with permission Buguin, A. et al. *J. Am. Chem. Soc.*, 128, 1088–1089, 2006. Copyright 2006 American Chemical Society.)

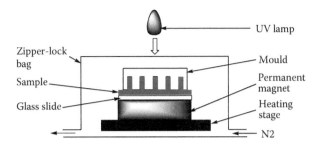

**FIGURE 19.7** Experimental setup used to prepare the responsive pillars. (Reprinted with permission from Buguin, A. et al., *J. Am. Chem. Soc.*, 128, 1088–1089, 2006. Copyright 2006 American Chemical Society.)

This experiment clearly demonstrates that we have successfully prepared for the first time an array of micron-sized nematic single crystal elastomeric pillars, which behave as micro-actuators. Moreover, the size reduction does improve thermal exchanges with the external medium and the time response of the microactuators becomes considerably shorter, in the order of a second.

By reducing the characteristic size of the patterns (involving more sophisticated microfabrication methods developed for biophysics applications [35]), we have obtained similar results with pillars of 2 μm in diameter and 7 μm in height. Those very small objects are much more difficult to manipulate individually than the previous ones.

Unfortunately, the accessible shapes for the microactuators prepared by a soft lithography approach are limited by the use of "a mold." Recently, the use of

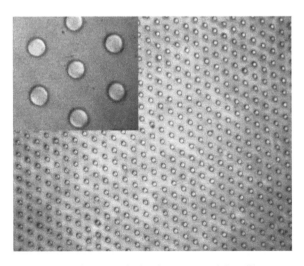

**FIGURE 19.8** Top view (under an optical microscope) of the pillar pattern obtained after peeling up the mold. (Inset) Zoom on the structure. (Distance between two adjacent pillars is around 30 μm.) (Reprinted with permission from Buguin, A. et al., *J. Am. Chem. Soc.*, 128, 1088–1089, 2006. Copyright 2006 American Chemical Society.)

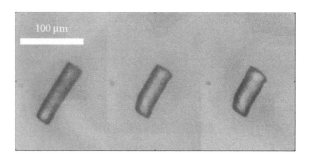

**FIGURE 19.9**  Contraction of an isolated pillar heated at 100°C, 110°C, and 120°C (from left to right). (Reprinted with permission from Buguin, A. et al., *J. Am. Chem. Soc.*, 128, 1088–1089, 2006. Copyright 2006 American Chemical Society.)

two-photon polymerization in the microfabrication of precise 2D and 3D microstructures made of polymers has been intensively explored [53]. One of the main limitations in the use of two-photon polymerization for 3D microfabrications remains the availability of efficient two-photon chromophores to induce two-photon polymerization. Using the same commercial photoinitiator we used before in UV photopolymerization, that is, Irgacure 369 [54], we succeeded in writing a simple pattern (e.g., a letter "E") in an oriented side-on nematic LC monomer by two-photon "IR" photopolymerization [55] (Figure 19.10). These preliminary results, which can be seen as a "proof of concept," open the road for the design of more "sophisticated" micro- and nano-actuators made of LC elastomers.

Although these results are very exciting, since we were able to prepare macro- and micro-actuators with mechanical properties approaching those of biological muscles, the main drawback of our approach is the stimulus used, that is, the temperature jump. It would be much more interesting from various viewpoints to use other stimulus for contraction in our side-on nematic LC elastomers. In the following paragraphs, we explore an approach to the creation of a photochemically driven LC elastomer muscle system.

### 19.3.1.2  Photo-Responsive Systems

As outlined previously, the motor for the contraction in side-on LC elastomers is the conformational change of the backbone at the nematic-to-isotropic transition. One can assume that any polymer chain shape change, induced by the disorganization or reorganization of the LC mesogens, triggered by any stimulus, can be used as a motor for contraction in LC elastomers. It is thus possible to produce elastomers responsive to other stimuli like light, mimicking the phenomenon that occurs at the thermal nematic-to-isotropic transition.

In the field of LCs, it has been known for years that in azo-containing molecules, a liquid crystalline to isotropic phase transition can be induced upon irradiation with light of suitable wavelength [56,57]. The light-induced phase transition is provoked by a *trans* to *cis* photoisomerization of the azo groups inducing a large change in the shape of the molecules, from a mesogenic rod-like shape for the *trans* isomer to a kinked nonmesogenic shape for the *cis* isomer (Figure 19.11).

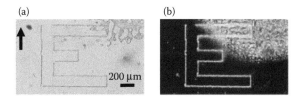

**FIGURE 19.10 (See color insert following page 304.)** Letter "E" written by two-photon photolithography in the nematic phase (63°C) of an aligned sample under confocal microscope (sample cell thickness: 10 μm; microscope objective: numerical aperture 0.45, magnification 50). Femtosecond laser (80 MHz at 780 nm) was used as excitation beam for two-photon polymerization (500 mW/cm$^2$). The observation was made between uncrossed polarizers (a) and crossed polarizers (b), at $T = 81.5°C$ where the polymerized part ("E" letter) was in the nematic state, while the unpolymerized part (dark background in (b)) passed already in the isotropic state. The width of the E-letter is 800 μm. The resolution of two-photon beam is 7 μm. The nematic director orientation is indicated by the arrow. (From Sungur, E. et al., *Opt. Express*, 15, 6784–6789, 2007. With permission from Optical Society of America.)

Starting from our thermally stimulated nematic side-on elastomer artificial muscle, our natural goal was to prepare a photochemically stimulated nematic azo side-on elastomer artificial muscle with a similar method as described above. In the first step, we designed an azobenzene-containing side-on monomer (Figure 19.12) and synthesized a homopolymer with this azo monomer by traditional radical polymerization [58]. The photochemical properties of the monomer and the polymer were studied. These studies showed that a nematic-to-isotropic transition was actually induced by

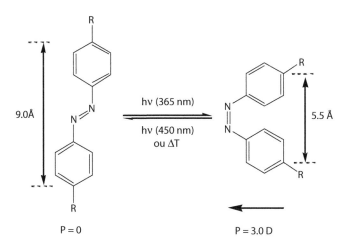

**FIGURE 19.11** Photochemical isomerization of azobenzene. The *trans* form is converted into the *cis* form by UV light irradiation, and the resultant *cis* isomer can return to the original *trans* form photochemically upon visible light irradiation or thermally in the dark. The *trans* and *cis* forms are very different in shape, in size, and in polarity.

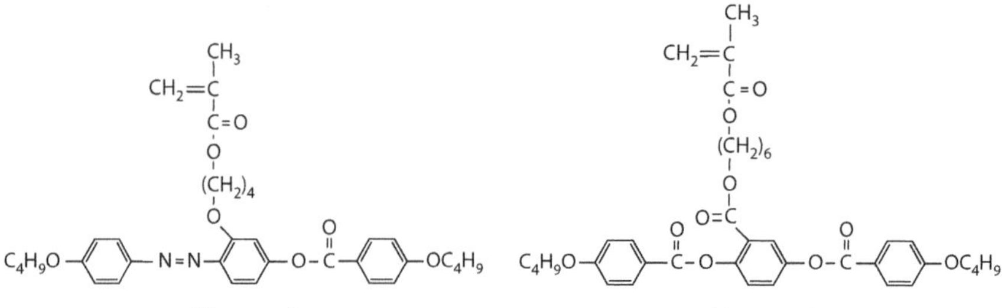

**FIGURE 19.12** Side-on LC monomers used in the photo-responsive systems.

UV irradiation in our side-on azo homopolymer, the isotropic-to-nematic back transition occurring spontaneously when the UV light was turned off. We considered then the preparation of aligned azo elastomers.

To our surprise, a careful literature search on the photochemical synthesis of pure azo LC elastomers gave no results. Obviously, there were technical difficulties to overcome in order to prepare such materials. We quickly realized that the problem was mainly related to the photopolymerization process we were planning to use in the preparation. Since the azo chromophore absorbs light strongly between 300 and 600 nm, the commercial UV initiators normally used were not applicable here. To be efficient, photoinitiators should be activated at wavelengths above 600 nm. We thus selected the following near infrared photoinitiator, 1,3,3,1',3',3'-hexamethyl-11-chloro-10,12-propylene-tricarbocyanine triphenylbutyl borate (CBC), first synthesized in Wang's group [59]. Red light ($\lambda > 600$ nm) was then used for the photopolymerization.

We prepared several oriented azo-containing methacrylate elastomers [22], from mixtures of monomers 1 and 2 (Figure 19.12), using 1,6-hexanediol dimethacrylate as cross linker. They were obtained as small sheets of yellow plastic with a thickness around 20 $\mu$m. The samples were held at a temperature of 70°C, well below their thermal N–I phase transition $T_{NI}$ ($T_{NI}$ is between 86°C and 90°C depending on the composition). When submitted to UV irradiation ($\lambda = 365$ nm), these films contracted quickly up to 12–18% (see Figure 19.13a for the contraction of E25% azo). Moreover, for all the elastomers the extent of contraction triggered by the UV irradiation was essentially the same as that found in the thermally induced N–I transition (see Figure 19.13b). As for the speed of the response, the E25% azo elastomer was the fastest with a characteristic time $\tau$ of 13 s under UV light of 100 mW/cm$^2$.

Complete reversal of the contraction took place after UV light was switched off because of the thermal *cis* to *trans* back reaction of the azobenzenes. This process was slow for all the studied elastomers (on the order of 30 min to one hour). Reverse *cis* to *trans* reaction induced by irradiation at 460 nm was attempted but gave a surprising result. A halogen lamp equipped with a hot mirror and a long-pass filter was used in order to obtain light with $\lambda = 450$–700 nm. The elastomer sheet E100% azo, contracted by exposure to UV light, was then illuminated with this visible light. The

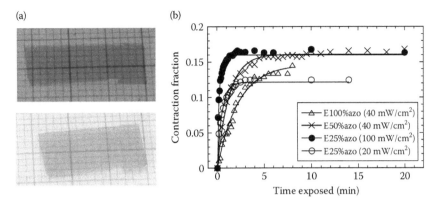

**FIGURE 19.13   (See color insert following page 304.)** (a) Elastomer E25% azo before UV irradiation (top) and under UV irradiation (bottom) (irradiation time: 130 s). (Background is a graduated paper.) (b) Contraction fraction ($F = (L_0 - L)/L_0$) of the azo elastomers as a function of time exposed ($t$) to UV light. The symbols are the experimental points. The lines are the fits with the function $F = F_{max}(1-e^{-t/\tau})$, where $F_{max}$ is the maximal contraction and $\tau$ the characteristic time of the contraction (time for reaching 63.2% of the maximal contraction). $\tau$ is 13, 31, 77, and 135 s for E25% azo (100 mW/cm$^2$), E25% azo (20 mW/cm$^2$), E50% azo (40 mW/cm$^2$), and E100% azo (40 mW/cm$^2$), respectively. (Li, M.-H. et al., *Adv. Mater.*, 15, 569–572, 2003. Copyright Wiley-VCH Verlag GmbH & Co. KGaA. Reproduced with permission.)

expected expansion was not observed and, more surprisingly, the spontaneous thermal expansion was even blocked by the exposure to visible light. We then directly illuminated the original elastomer sheet with visible light without previous UV irradiation. A small photo-contraction triggered by the visible light was observed. The reasons for this are not totally clear but might be related to the broadening and red-shifting of the absorption bands of the azo groups in liquid crystalline polymers that was reported previously [60]. Another explanation might be that, under visible light, the *trans–cis* isomerization through the n–π* mode is predominant compared to the *cis–trans* isomerization in our particular azo system.

Photo-mechanical effects using the photoisomerization of a chromophore as the active phenomena had already been observed in various systems [61–65]. Agolini and Agolino and Gay [61] reported a small stress increase in a film of semicrystalline poly-4,4′-diphenylazopyromellitimide at constant length under UV illumination. Smets and Blauwe [63] studied rubbery poly(ethyl acrylate) cross linked with a dimethacrylate containing two spiropyrane groups, and observed a 2–3% shrinkage upon irradiation of stretched samples at constant temperature. Eisenbach [64] used an azobenzene-containing cross linker to make a rubbery poly(ethyl acrylate), and obtained around 0.25% contraction on loaded film under UV illumination. In most of the experiments, azobenzene was used as chromophore. As shown in Figure 19.11, azobenzene and its derivatives can undergo photochemical *trans–cis* isomerization, and the *trans* form is thermodynamically more stable than *cis* form. From *trans* form to *cis* form, there are changes in shape, in size, and in polarity. In the first examples of photomechanical effects [64] the size change in the chromophore upon irradiation caused a change in the conformation of the adjacent chain segments, which was considered to be the main cause for the small contraction.

It is only recently that the azo chromophore was coupled with LC systems in the preparation of artificial muscles and large strains were obtained [21–23,66,67]. In our azo side-on LC elastomer muscles, we made use of the shape change of azobenzene occurring upon UV irradiation, which induced in turn the nematic-to-isotropic phase transition of the system. Finkelmann [21] and Hogan [66] used azobenzene-containing cross linkers and/or monomers to prepare monodomain end-on nematic siloxane elastomers by the mechanical alignment approach described above. In these cases, the *trans–cis* photoisomerization of the azo chromophores induced a lowering or suppression of the nematic order. The contraction of these elastomers along the nematic director reached 20% after hours of UV irradiation.

As stated previously, the preparation of monodomain azo LC elastomers is not a trivial problem. When starting from the azo LC monomers aligned by conventional techniques, the photopolymerization under UV light normally used is not compatible with the azo system because of its strong absorption in UV light. We synthesized for the first time aligned nematic azo side-on elastomers by photopolymerization using a near infrared photoinitiator. Recently, Ikeda's group prepared nematic end-on elastomers by thermopolymerization using a thermal initiator with a half-life of decomposition much longer than the time of alignment process [67]. They produced bending deformations with their system, using low UV light intensity ($1.5 \text{ mW/cm}^2$). It is believed that bending deformation is an effect of the limited penetration depth of the UV light, owing to the high molar extinction coefficient of the azo chromophores. In other words, volume contraction takes place only at the surface, induced by the photochemical changes in size and alignment order of the azobenzene moieties, the material working like a bimorph.

### 19.3.2 ARTIFICIAL MUSCLES BASED ON TRIBLOCK LC COPOLYMERS

As shown in the preceding section, a side-on polymer is a good candidate for the contraction segment. Therefore, the side-on nematic monomer shown in Figure 19.4a was chosen to construct the **N** domain in the triblock copolymer in order to prepare artificial muscles with a lamellar structure. Keeping in mind the three key characteristics of the triblock model: lamellar phase, monodomain, and cross linking (Figure 19.3), we first synthesized the triblock copolymers and studied their self-assembling properties in polydomain and monodomain samples [32,33]. At this stage, we used noncrosslinkable triblock copolymers in order to avoid the technical problems brought out by the possible cross linking in the course of physical investigation.

### 19.3.2.1 Non-Cross Linkable LC Triblock Copolymers

Atom transfer radical polymerization (ATRP) was chosen to prepare side-on **RNR** triblock copolymers because of its remarkable tolerance to functional groups present in the mesogens (Figure 19.14) [32,33]. Side-on liquid crystalline homopolymers **N** with unimodal molecular mass distributions of about 1.1 were successfully obtained, which exhibit crystal, nematic, and isotropic phases. Triblock copolymers with poly($n$-butyl acrylate) amorphous block were then prepared by ATRP using the homopolymers **N** as macroinitiators. They have rather well-defined structures with unimodal distributions around 1.3. These block copolymers, in which the weight ratios between LC and flexible blocks **N/R** are from 38/62 to 62/38, generate lamellar types of microphase segregation. The LC block in this microphase exhibits a nematic mesophase similar to that of the homopolymer. For the triblock copolymer, a supramolecular order–disorder transition is induced by the nematic–isotropic phase transition of the LC block and the nematic–isotropic transition temperature is depressed by 25°C relative to that of the homopolymer when **N/R** = 42/58. An important structural characteristic in these microphase segregated systems is that the lamella spacing decreases significantly from the glassy or crystal phase to the isotropic phase after a continuous shrinking over the all nematic range and a steep decrease at the nematic–isotropic transition [33]. This spacing variation with temperature will be the motor of the contraction for an artificial muscle made with this triblock copolymer.

Alignment of the triblock copolymers was achieved using a magnetic field (the sample was cooled slowly from isotropic phase to nematic phase of the **N** block under an electromagnetic field of 1.7 T) [33]. The long axes of rod-like mesogens were aligned parallel to the magnetic field because of the diamagnetism of the aromatic units [13,40,68]. Small angle x-ray scattering (SAXS) experiments on the aligned sample at room temperature showed that the supramolecular lamella were well oriented with layer normal parallel to the magnetic field. However, wide-angle x-ray scattering (WAXS) at room temperature showed the nematic structure in the LC sublayer was only partially aligned. This might be due to a rapid relaxation of the nematic orientation; therefore, an *in situ* cross linking in the **R** blocks would be necessary to lock-in the nematic orientation.

### 19.3.2.2 LC Triblock Elastomer as Artificial Muscle

In a following experiment, cross-linkable groups were introduced in the **R** blocks. The amorphous block was now a statistical copolymer, poly($n$-butyl acrylate)-*co*-poly(2-hydroxyethyl acrylate), which is potentially cross linkable (Figure 19.14). The molar masses of the homopolymer and of the block copolymer were: $M_n$=14,200 ($M_w/M_n$=1.08) for the homopolymer and $M_n = 23,600$ ($M_w/M_n = 1.3$) for the triblock copolymer. The weight ratio of nematic to amorphous block was about 60/40. The hydroxyl groups (–OH) in the amorphous block were finally transformed into polymerizable sites (acrylates) for the subsequent cross linking step. In the presence of a photoinitiator, this cross-linkable triblock copolymer was first aligned in a magnetic field (1.7 T) and then cross linked under UV light (365 nm). The aligned film sample (17 mm × 7 mm × 0.16 mm, the magnetic field is along the long side) is shown in Figure 19.15a. When heated from room temperature to 140°C the elastomer contracts by about 18% in total (Figure 19.15). The contraction is completely reversible when cooled down to room temperature. SAXS and WAXS experiments on the triblock elastomer at room temperature (Figure 19.16) indicated clearly an aligned nano-scaled lamellar structure and a rather well-aligned nematic sublayer.

Therefore, a muscle-like material with lamellar structure based on a nematic triblock copolymer was obtained for the first time [34] (Figure 19.17).

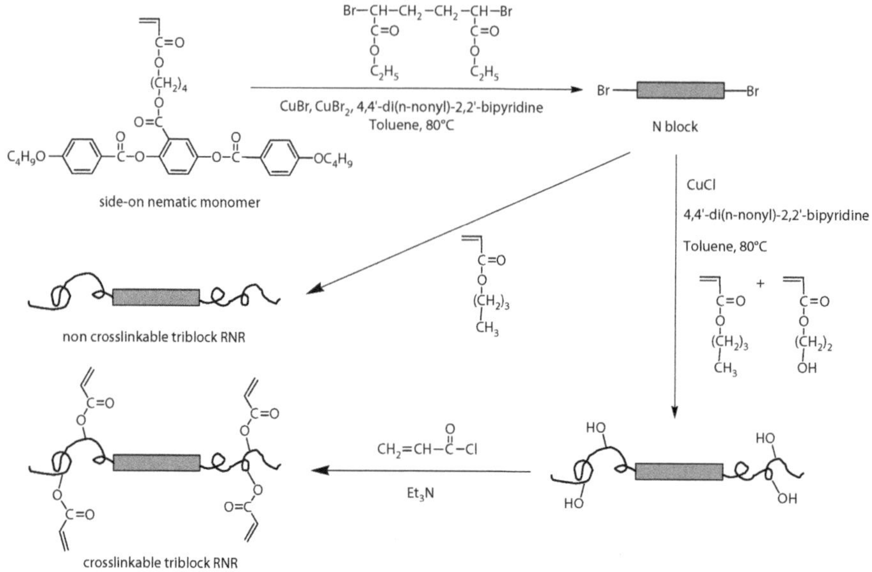

**FIGURE 19.14** Synthesis scheme of the noncross linkable and cross-linkable triblock copolymers.

In the above materials, the sample alignment was achieved by a strong magnetic field. The major disadvantage of this method is that the size of the sample is restricted by the distance between the poles of the electromagnet. Therefore, it was interesting to explore the possibility to align the side-on LC triblock copolymer by other techniques, such as shear field. Oscillatory shear was then applied to the noncrosslinkable triblock copolymer sample and SAXS experiments were simultaneously carried out [69]. It was shown that shearing the side-on nematic triblock copolymer at high frequency and strain amplitudes (i.e., $\omega = 100$ rad/s and $A = 10–70\%$) led to alignment of the lamellae with normal perpendicular to the shear velocity direction and to the velocity gradient direction (Figure 19.17). The rheological response was also investigated in detail. The decrease of lamella spacing upon the increase of temperature demonstrated by SAXS indicated that the triblock copolymer aligned by an oscillatory shear field could be used to prepare artificial muscles if cross-linkable groups were present. In this case, the sample size used in the shear flow experiment can be made larger than that used in the magnetic field experiment.

Work is in progress to prepare photoresponsive artificial muscles with lamellar structure, using azobenzene-containing side-on LC polymer as central block **N**.

## 19.4  CONCLUSIONS AND PERSPECTIVES

In this chapter, we presented our bottom-up strategy to make artificial muscles, or generally known as stimuli-responsive materials, using nematic side-on LC polymers as building blocks. A series of thermo-responsive or photo-responsive self-assembled

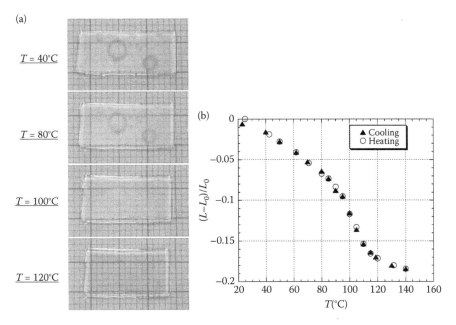

**FIGURE 19.15**  Reversible thermal contraction of the triblock LC elastomer. (a) Photos of a sample taken at different temperatures. The film floats on a thin layer of silicon oil and there is nearly no constraint during the contraction (The background is a graduated paper.). The circle patterns visible in the sample are superficial defects of orientation due to the bubble formation at the beginning of the film casting. These defects disappear and appear reversibly at the nematic-isotropic transition (near 110°C). (b). Contraction fraction $(L - L_0)/L_0$ as a function of temperature. (Li, M.-H. et al., *Adv. Mater.*, 16, 1922–1925, 2004. Copyright Wiley-VCH Verlag GmbH & Co. KGaA. Reproduced with permission.)

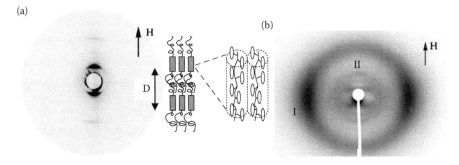

**FIGURE 19.16**  X-ray scattering patterns of the triblock LC elastomer aligned along the magnetic field **H**. (a) SAXS shows the aligned nano-scaled lamellar structure of the triblock elastomer, lamellar spacing being $D = 15.5$ nm. (b) WAXS shows the mesogen organization in the nematic sublayer. The crescent-like signals (I) along the equator give the nematic order parameter $S \approx 0.6$ and the average lateral distance between mesogens (0.44 nm). The inner signals (II) along the meridian with three visible orders give a distance of 2.6 nm, which corresponds to the mesogen length. (Li, M.-H. et al., *Adv. Mater.*, 16, 1922–1925, 2004. Copyright Wiley-VCH Verlag GmbH & Co. KGaA. Reproduced with permission.)

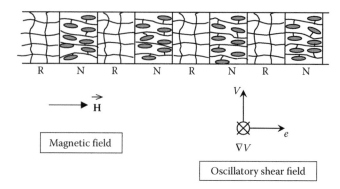

**FIGURE 19.17**   Side-on triblock elastomers can be aligned by magnetic field or by oscillatory shear field.

LC elastomers were prepared. The overall material response in these artificial muscles reflects the individual macromolecular response: the contraction/elongation of the material results from the individual macromolecular chain shape change, from stretched to spherical at the nematic to isotropic phase transition triggered by external stimuli. We have also demonstrated that this approach is particularly interesting for the development of micro- or nano-sized actuators. Using soft lithography and two-photon photochemistry we prepared for the first time micron-sized actuators with various shapes. We are now exploring new ways to prepare LC elastomer-made nano-sized actuators using "unconventional" approach such as amphiphilic block copolymers. We have recently discovered that amphiphilic block copolymers in which the hydrophobic block is a smectic LC polymer self-assemble in well-ordered nanofibers [70]. Building similar nanofibers from "stimulable" LC polymers is our next goal. All those approaches open the way to the realization of a broad range of new micro and nano active devices and active surfaces.

## ACKNOWLEDGMENTS

This chapter is dedicated to the memory of Professor P.-G. de Gennes, who introduced us to the field of "artificial muscles" in 1997, and who gave us so many enlightening advices over the years.

We would also like to thank the many coworkers whose names are mentioned in our references.

## REFERENCES

1. Bar-Cohen, Y. (ed.) *Electroactive Polymer Actuators as Artificial Muscles—Reality, Potential and Challenges*, 2nd edn. Bellingham, WA: SPIE Press, 2004.
2. Davis, F., Liquid-crystalline elastomers, *J. Mater. Chem.*, 3, 551–562, 1993.
3. Brand, H.R. and Finkelmann, H., Liquid crystalline elastomers. In *Handbook of Liquid Crystals* (eds. D. Demus, J. Goodby, G. W. Gray, H.-W. Spiess, and V. Vill), vol. 3, pp. 227–302. New York, NY: Wiley-VCH, 1998.

4. Mayer, S. and Zentel, R., Liquid crystalline polymers and elastomers, *Curr. Opin. Solid State Mater. Sci.*, 6, 545–551, 2002.

5. Warner, M. and Terentjev, M., *Liquid Crystal Elastomers*. Oxford Science Publications, Oxford University Press, New York, NY, 2007.

6. de Gennes, P.-G., Physique moléculaire—Réflexions sur un type de polymères némattiques, *C. R. Acad. Sci. Paris, Ser.* B, 281, 101–103, 1975.

7. de Gennes, P.-G., A semi-fast artificial muscle, *C. R. Acad. Sci. Paris, Ser.* II b, 324, 343–348, 1997.

8. Wermter, H. and Finkelmann, H., Liquid crystalline elastomers as artificial muscles, e-Polymers (http://www.e-polymers.org), 13, 1–13, 2001.

9. Thomsen III, D.L., Keller, P., Naciri, J., Pink, R., Joen, H., Shenoy, D., and Ratna, B. R., Liquid crystal elastomers with mechanical properties of a muscle, *Macromolecules*, 34, 5868–5875, 2001.

10. Lehmann, W., Skupin, H., Tolksdorf, G., Gebhard, E., Zentle, R., Krüger, P., Lösche, M., and Kremer, F., Giant lateral electrostriction in ferroelectric liquid-crystalline elastomers, *Nature*, 410, 447–450, 2001.

11. Köhler, R., Stannarius, R., Tolksdorf, C., and Zentel, R., Electroclinic effect in freestanding smectic elastomers, *Appl. Phys. A; Mater. Sci. Process*, 80, 381–388, 2005.

12. Spillman, C.M., Ratna, B.R., and Naciri, J., Anisotropic actuation in electroclinic liquid crystal elastomers, *Appl. Phys. Lett.*, 90, 021911, 2007.

13. de Gennes, P.-G. and Prost, J., *The Physics of Liquid Crystals*, 2nd edn., Oxford: Clarendon Press, 1993.

14. de Gennes, P.-G., Mechanical properties of nematic polymers. In *Polymer Liquid Crystals* (eds. A. Cifferi, W. R. Krigbaum and R. B. Meyer), pp. 115–131. New York: Academic Press, 1982.

15. Cotton, J.-P. and Hardouin, F., Chain conformation of liquid-crystalline polymers studied by small-angle neutron scattering, *Progr. Polymer Sci.*, 22, 795–828, 1997.

16. D'Allest, J.-F., Maissa, P., Ten Bosch, A., Sixou, P., Blumstein, A., Blumstein, R. B., Teixeira, J., and Noirez, L., Experimental evidence of chain extension at the transition temperature of a nematic polymer, *Phys. Rev. Lett.*, 61, 2562–2565, 1988.

17. Li, M.-H., Brûlet, A., Davidson, P., Keller, P., and Cotton, J.-P., Observation of hairpin defects in a nematic main chain polyester, *Phys. Rev. Lett.*, 70, 2297–2300, 1993.

18. Li, M.-H., Brûlet, A., Cotton, J.-P., Davidson, P., Strazielle, C., and Keller, P., Study of the chain conformation of thermotropic nematic main chain polyesters, *J. Phys. II France*, 4, 1843–1863, 1994.

19. Küpfer, J. and Finkelmann, H., Nematic liquid single crystal elastomers, *Makromol. Chem., Rapid Commun.*, 12, 717–726, 1991.

20. de Gennes, P.-G., Hébert, M., and Kant, R., Artificial muscles based on nematic gels, *Macromol. Symp.*, 113, 39–49, 1997.

21. Finkelmann, H., Nishikawa, E., Pereira, G. G., and Warner, M., A new opto-mechanical effect in solids, *Phys. Rev. Lett.*, 87, 015501-(1–4), 2001.

22. Li, M.-H., Keller, P., Li, B., Wang, X., and Brunet, M., Light-driven side-on nematic elastomer actuators, *Adv. Mater.*, 15, 569–572, 2003.

23. Yu, Y., Nakoto, M., and Ikeda, T., Directed bending of a polymer film by light, *Nature*, 425, 145, 2003.

24. Huang, C., Zhang, Q., and Jákli, A., Nematic anisotropic liquid crystal gels—self-assembled nanocomposites with high electromechanical response, *Adv. Funct. Mater.*, 13, 525–529, 2003.

25. Courty, S., Mine, J., Tajbakhsh, A.R., and Terentjev, E.M., Nematic elastomers with aligned carbon nanotubes: New electromechanical actuators, *Europhys. Lett.* 64, 654–660, 2003.

26. Urayama, K., Honda, S., and Takigawa, T., Electrooptical effects with anisotropic deformation in nematic gel, *Macromolecules*, 38, 3574–3576, 2005.

27. Yusuf, Y., Huh, J.-H., Cladis, P.E., Brand, H.R., Finkelmann, H., and Kai, S., Low voltage-driven electromechanical effects of swollen liquid crystal elastomers, *Phys. Rev. E*, 71, 061702(1–8), 2005.

28. Alberts, B., Johnson, A., Lewis, J., Raff, M., Roberts, K., and Walter, P., *Molecular Biology of the Cell*, 4th edn. Garland Science, New York, NY, 2002.

29. Salisbury, J.L. and Floyd, G.L., Calcium-induced contraction of the rhizoplast of a quadriflagellate green alga, *Sciences*, 202, 975–977, 1978.

30. Salisbury, J.L., Baron, A., Surek, B., and Melkonian, M., Striated flagellar roots: Isolation and partial characterization of a calcium-modulated contractile organelle, *J. Cell Biol.*, 99, 962–970, 1984.

31. Schiebel, E. and Bornens, M., In search of a function for centrins, *Trends Cell Biol.*, 5, 197–201, 1995.

32. Li, M.-H., Keller, P., Grelet, E., and Auroy, P., Liquid crystalline polymethacrylates by atom transfer radical polymerization at ambient temperature, *Macrom. Chem. Phys.*, 203, 619–626, 2002.

33. Li, M.-H., Keller, P., and Albouy, P.-A., Novel liquid crystalline block copolymers by ATRP and ROMP, *Macromolecules*, 36, 2284–2292, 2003.

34. Li, M.-H., Keller, P., Yang, J.-Y., and Albouy, P.-A., An artificial muscle with lamellar structure based on a nematic triblock copolymer, *Adv. Mater.*, 16, 1922–1925, 2004.

35. Buguin, A., Li, M.-H., Silberzan, P., Ladoux, B., and Keller, P., Micro-muscles: When artificial muscles made of nematic liquid crystal elastomers meet soft lithography, *J. Am. Chem. Soc.*, 128, 1088–1089, 2006.

36. Fontanille, M. and Gnanou, Y., *Chimie et physicochimie des polymères*, p. 198, Paris: Dunod 2002.

37. Leroux, N., Keller, P., Achard, M.F., Noirez, L., and Hardouin, F., Small-angle neutron scattering experiments on "side-on fixed" liquid crystal polyacrylates, *J. Phys. II*, 3, 1289–1296, 1993.

38. Leroux, N., Achard, M.F., Keller, P., and Hardouin, F., Consequences of the "jacketed" effect on mesomorphic and orientational properties of "side-on fixed" liquid crystalline polymers, *Liq. Cryst.*, 16, 1073–1079, 1994.

39. Lecommandoux, S., Achard, M.-F., Hardouin, F., Brûlet, A., and Cotton, J.-P., Are nematic side-on polymers totally extended? A SANS study, *Liq. Cryst.*, 22, 549–555, 1997.

40. Legged, C.H., Davis, F.J., and Mitchell, G.R., Memory effects in liquid elastomers, *J. Phys. II France*, 1, 1253–1261, 1991.

41. Strzelecki, L. and Liébert, L., Synthèse de nouveaux monomères mésomorphes. Polymérisation du para acryloyloxybenzylidene para carboxyaniline, Synthesis of new liquid crystalline monomers. Polymerization of *p*-acryloyloxybenzylidene *p*-carboxyaniline, *Bull. Soc. Chim. France*, 605–608, 1973.

42. Strzelecki, L., Liébert, L., and Keller P., Sur la synthèse et la polymérisation de monomères mésomorphes: Une série homologue de *p*-acryloyloxybenzylidène *p*-aminoalkylcinnamates, (On synthesis and polymerization of liquid crystalline monomers: an homologous series of *p*-acryloyloxybenzylidene *p*-amino (alkyl)cinnamates), *Bull. Soc. Chim. France*, 2750–2752, 1975.

43. Broer, D., Boven, J., and Mol, G.N., *In-situ* photopolymerization of oriented liquid-crystalline acrylates, 3 Oriented polymer networks from a mesogenic diacrylate, *Makromol. Chem.*, 190, 2255–2268, 1989.

44. Donnio, B., Wermter, H., and Finkelmann, H., A simple and versatile synthetic route for the preparation of main-chain liquid-crystalline elastomers, *Macromolecules*, 33, 7724–7729, 2000.

45. Hogan, P.M., Tajbakhsh, A.R., and Terentjev, E.M., UV manipulation of order and macroscopic shape in nematic elastomers, *Phys. Rev. E*, 65, 041720-(1–10), 2002.

46. Ahir, S.V., Squires, A.M., Tajbakhsh, A.R., and Terentjev, E.M., Infrared actuation in aligned polymer–nanotube composites, *Phys. Rev. B*, 73, 085420, 2006.

47. Clarke, S.M., Hotta, A., Tajbakhsh, A.R., and Terentjev, E.M., Effect of crosslinker geometry on thermal and mechanical properties of nematic elastomers, *Phys. Rev. E*, 64, 061702, 2001.

48. Naciri, J., Srinivasan, A., Joen, H., Nikolov, N., Keller, P., and Ratna, R.B., Nematic elastomer fiber actuator, *Macromolecules*, 36, 8499–8505, 2003.

49. Shenoy, D.K., Thomsen III, D.L., Srinivasan, A., Keller, P., and Ratna, B.R., Carbon coated liquid crystal elastomer film for artificial muscle applications, *Sensors Actuators*, 96, 184–188, 2002.

50. Spillman, C.A., Naciri, J., Martin, B.D., Farahat, W., Herr, H., and Ratna, R.B., Stacking nematic elastomers for artificial muscle applications, *Sensors Actuators A*, 133, 500–505, 2007.

51. Spillman, C.A., Naciri, J., Chen, H.S., Srinivasan, A., and Ratna, R.B., Tuning the physical properties of a nematic liquid crystal elastomer actuator, *Liq. Crystallogr.*, 33, 373–380, 2006.

52. Xia, Y. and Whitesides, G.M., Soft lithography, *Angew. Chem. Int. Ed.*, 37, 550–575, 1998.

53. Lee, K.S., Yang, D.Y., Park, S.H., and Kim, R.H., Recent developments in the use of two photon polymerization in precise 2D and 3D microfabrications, *Polym. Adv. Technol.*, 17, 72–82, 2006.

54. Schafer, K.J., Hales, J.M., Balu, M., Belfield, K.D., Van Stryland, E.W., and Hagan, D.J., Two-photon absorption cross-sections of common photoinitiators, *J. Photochem. Photobiol. A*, 162, 497–502, 2004. (Irgacure 369 has been used previously in two-photon polymerization of acrylates.)

55. Sungur, E., Li, M.H., Taupier, G., Boeglin, A., Romeo, M., Méry, S., Keller, P., and Darkenoo, K.D., External stimulus driven variable-step grating in a nematic elastomer, *Opt. Express*, 15, 6784–6789, 2007.

56. Leier, C., and Pelzl, G., Phase transitions of liquid crystalline modifications by photochemical isomerization, *J. Prakt. Chem.*, 321, 197–204, 1979.

57. Tsutsumi, O., Shiono, T., Ikeda, T., and Galli, G. Photochemical phase transition behaviour of nematic liquid crystals with azobenzene moieties as both mesogens and photosensitive chromophores. *J. Phys. Chem.* 101, 1332–1337, 1997.

58. Li, M.-H., Auroy, P., and Keller, P., An azobenzene-containing side-on liquid crystal polymer, *Liq. Cryst.*, 27, 1497–1502, 2000.

59. Zhang, S., Li, B., Tang, L., Wang, X., Liu, D., and Zhou, Q., Studies on the near infrared laser induced photopolymerization employing a cyanine dye-borate complex as photoinitiator, *Polymer*, 42, 7575–7582, 2001.

60. Creed, D., Photochemistry and photophysics of liquid crystalline polymers. *In Organic and Inorganic Photochemistry* (eds. V. Ramamurthy and K. S. Schange), pp. 129–194. Marcel Decker, New York, 1998.

61. Agolino, F. and Gay, F.P., Synthesis and properties of azoaromatic polymers, *Macromolecules*, 3, 349–351, 1970.

62. Van der Veen, G. and Prins, W., Photomechanical energy conversion in a polymer membrane, *Nat. Phys. Sci.*, 230, 70–72, 1971.

63. Smets, G. and de Blauwe, F., Chemical reactions in solid polymeric systems. Photomechanical phenomena, *Pure Appl. Chem.*, 39, 225–238, 1974.

64. Eisenbach, C.D., Isomerization of aromatic azo chromophores in poly(ethyl acrylate) networks and photomechanical effect, *Polymer*, 21, 1175–1179, 1980.

65. Blair, H. S., Pague, H. I., and Riordon, J. E., Photoresponsive effects in azo polymers, *Polymer*, 21, 1195–1198, 1980.

66. Hogan, P.M., Tajbakhsh, A.R., and Terentjev, E.M., UV manipulation of order and macroscopic shape in nematic elastomers, *Phys. Rev. E*, 65, 041720-(1–10), 2002.

67. Yu, Y., Nakoto, M., and Ikeda, T., Photoinduced bending and unbending behaviour of liquid crystalline gels and elastomers, *Pure Appl. Chem.*, 76, 1467–1477, 2004.

68. Hirschmann, H., Roberts, P.M.S., Davis, F.J., Guo, W., Hasson, C.D., and Mitchell, G. R., Liquid crystalline elastomers: Relationship between macroscopic behaviour and the level of backbone anisotropy, *Polymer*, 42, 7063–7071, 2001.

69. Castelletto, V., Parras, P., Hamley, I.W., Davidson, P., Yang, J., Keller, P., and Li, M.-H., A rheological and SAXS study of the lamellar order in a side-on liquid crystalline block copolymer, *Macromolecules*, 38, 10736–10742, 2005.

70. Piñol, R., Jia, L., Gubellini, F., Lévy, D., Albouy, P.A., Keller, P., Cao, A., and Li, M.-H., Self-assembly of PEG-b-liquid crystal polymer: The role of smectic order in the formation of nanofibers, *Macromolecules*, 40, 5625–5627, 2007.

# Index

## A

ac calorimetry, 441–442
  temperature dependence, 445, 446
Acrylate
  based monomers, 11
  cholesteric diacrylates, 28, 33, 334
  cholesteric film, 32
  diacrylate polymerization, 4, 5, 61
  hexagonal mesophases, 163
  LC diacrylate, 9, 15
  methyl spacers, 125
  moiety effect, 5
  monoacrylate, 4, 8, 11, 30, 39
  nematic monoacrylate, 334
  in photochemical patterning, 21
  polymerizable, 337, 579
  polymerization, 36
  polymerization limitations, 12
Actuator properties, 253
Adams and Warner (AW), 403
Affine layer deformation geometry, 507–508
AFM. *See* Atomic force microscopy (AFM)
AIBN. *See* Azoisobuthylnitrile (AIBN)
Alq3. *See* Tris(8-hydroxyquinoline)aluminum
    (Alq3)
Anisotropic
  beads and spring models, 452
  media, 95
Anisotropic cross-linked polymers, 157, 159
  film processing, 158
  orientation, 157–158
  photopolymerization, 158–159
Anisotropic phase separation, 207, 208–209.
    *See also* Phase separation
  microlens array fabrication, 212
  PSCOF structure, 208
Anisotropy, 470
  C$x$R family, 255
  high-elastic, 500
  length change, 258
  in mechanical properties, 254
  order network, 258
  order parameter, 257
  thermal expansion, 255, 257
Aromatic amines, 321
  hole transporting, 335
Artificial cilia, 279
  light-driven cilia, 280
  polymer cilia, 280
  polymer response, 282

processing steps, 278, 279
response to light, 281
strain response of printed cilia, 282
Artificial muscle, 565. *See also* Liquid crystalline
    elastomer (LCE); Nematic LC
    elastomers; Triblock copolymer
  azo chromophore in, 578
  LC microactuators, 572–574
  LC triblock elastomer, 579
  macroscopic LC actuators, 569–572
  materials, 565
  nematic side-on elastomer, 575
  striated, 568
  triblock LC copolymers, 578
Artificial muscle-like materials, 191. *See also*
    Cross-linked liquid crystalline
    polymers (CLCPs)
Astigmatism, 132
  in LC lens, 133
  and LC molecule alignment, 135
Atomic force microscopy (AFM), 186, 231
Atomistic models, 452
AW. *See* Adams and Warner (AW)
Azobenzene, 265
  as alignment layers, 156
  CLCP film, 197
  with LC properties, 185
  peak absorption, 265
  as photochromic molecules, 183, 184
  photodeformation of, 195
  photoisomerization, 154, 186, 265, 266,
    276, 575
  photon-mode photoresponse, 185
  shape deformations, 276
Azoisobuthylnitrile (AIBN), 474

## B

Bathocuproine (BCP), 323, 324
BCP. *See* Bathocuproine (BCP)
BCzVBI. *See* 4,4′-Bis(2-(9-ethylcarbazole-3-
    yl))ethylen-1-yl)biphenyl
    (BCzVBI)
Biaxial elastomer, 458. *See also* External field;
    Uniaxial elastomer
  constant-force MC simulations, 459, 460
  deformation-dependent quantities, 459
  heating/cooling runs, 460
  IB transition, 462
  N$_+$B transition, 461
  nematic part, 458

T - #0299 - 071024 - C8 - 234/156/28 - PB - 9780367383107 - Gloss Lamination